MACHINA CARNIS

MACHINA CARNIS

THE BIOCHEMISTRY OF
MUSCULAR CONTRACTION
IN ITS HISTORICAL
DEVELOPMENT

by

DOROTHY M. NEEDHAM
F.R.S.

*Fellow Emeritus of Lucy Cavendish
College, Cambridge*

CAMBRIDGE

AT THE UNIVERSITY PRESS

1971

CAMBRIDGE UNIVERSITY PRESS
Cambridge, New York, Melbourne, Madrid, Cape Town, Singapore, São Paulo, Delhi

Cambridge University Press
The Edinburgh Building, Cambridge CB2 8RU, UK

Published in the United States of America by Cambridge University Press, New York

www.cambridge.org
Information on this title: www.cambridge.org/9780521112673

First published 1971
This digitally printed version 2009

A catalogue record for this publication is available from the British Library

Library of Congress Catalogue Card Number: 75-142959

ISBN 978-0-521-07974-7 hardback
ISBN 978-0-521-11267-3 paperback

Fig. 85 is reproduced by courtesy of H. Fernández-Morán,
The University of Chicago
Fig. 45 is reproduced by permission of the American Heart
Association, Inc. and appeared in H. E. Huxley, 1961, The
Contractile Structure of Cardiac and Skeletal Muscle, *Circulation*, vol. 24, no. 2, part 2, The Myocardium – its biochemistry and biophysics

MEMORIAE
FREDERICI GOWLAND HOPKINS
PRAECEPTORIS OLIM ET AMICI
CONIVGI QVOQVE
IOSEPHO NEEDHAM
FAVTORI MONITORI ADIVTORI
PER ANNOS COMPLVRES
CONSTANTISSIMO
QVO SINE NIL PROFECISSET
HOC PIETATIS PIGNVS
D·M·N

Corporis membra in unam machinam ad usas vitales aptantur.

Asclepius, 27 (ca. 1st. cent. AD) in *Corpus Hermeticum*, II, p. 333

Iam vero quattuor elementorum diversitates contrariasque potentias nisi quaedam armonia coniungeret, qui fieri potest, ut in unum corpus ac machinam convenirent?

Boethius (d. AD 524), *De Institutione Musica*, I, 1

Partibus his constat humanae machina carnis,
  sed multipliciter quae numerentur habes:
portio prima caput, collus ac brachia, truncus,
  intera, sensus iners, femora, crura, pedes.

Eugenius of Toledo (d. AD 657), *Carmen* XLIII

CONTENTS

PREFACE

Looking back on the years spent in writing this book, I feel sometimes that it was done primarily for my own enjoyment. I wanted to visualise in a single perspective the path of man's knowledge about the function of muscles, progressing so slowly for so many centuries but then during the last seventy years reaching speedily towards the goal in a rush of great discoveries. Professor A. V. Hill, whose fundamental work constantly appears in these pages, tells me that he never took theories of contraction seriously; but perhaps he would not disagree with Dr William Croone who in the seventeenth century reckoned 'such speculations amongst the best entertainments of our mind'. Be this as it may, I hope that, in spite of its defects and omissions, this book will be useful to some of those for whom the fearful and wonderful phenomena of muscular movement retain all their fascination. We can respond to the words of Sir Thomas Browne (*Religio Medici* I. 13):

> Search while thou wilt, and let thy Reason go,
> To ransome Truth, even to th'Abyss below;
> Rally the scattered Causes; and that line
> Which Nature twists, be able to untwine...

DOROTHY M. NEEDHAM

The Master's Lodge
Gonville and Caius College
Cambridge
August 1970

ACKNOWLEDGMENTS

I have received most valuable help from the following who have read different chapters, and I record my gratitude to them.

Chapter 1: Dr G. H. Lloyd, Dr M. Teich, Dr L. G. Wilson; 2: Dr M. Teich, Dr L. G. Wilson; 3: Professor A. V. Hill, Professor Sir Rudolph Peters; 7: Professor H. H. Weber; 5: Dr J. R. Bendall; 8: Dr J. R. Bendall; 9: Professor A. V. Hill, Professor D. R. Wilkie; 10: Dr A. Weeds; 11: Professor Jean Hanson, Dr G. Offer; 12: Dr G. Offer, Professor S. V. Perry; 13: Professor A. F. Huxley; 14: Professor D. R. Wilkie; 15: Dr J. R. Bendall; 16: Dr M. Dixon; 17: Dr G. D. Greville; 18: Professor P. Randle, Dr R. M. Denton, Dr C. I. Pogson, Dr N. W. Wakid; 19: Professor S. V. Perry; 21: Dr R. J. Pennington; 22: Professor E. Baldwin, Professor J. C. Rüegg; 23: Professor Edith Bülbring, Professor G. Hamoir, Dr Claire Daemers-Lambert, Mrs C. F. Shoenberg; 24: Dr D. Kerridge, Dr B. M. Shaffer.

I also thank those who have so readily responded to my requests for advice on particular points: Dr D. F. Cheesman, Professor C. Crone, Dr C. L. Davey, Dr A. G. Debus, Professor M. Florkin, Dr I. M. Glynn, Dr P. Goodford, Professor E. Gutmann, Dr F. L. Holmes, Dr W. R. Keatinge, Dr D. R. Kominz, Dr H. W. Lissmann, Dr K. E. Machin, Dr A. J. Rowe, Dr K. Tipton, Professor Y. Tonomura, Dr D. C. Watts, Dr R. M. Young.

To Dr J. R. Bendall, Dr G. Offer, Mrs C. F. Shoenberg and Dr A. Weeds, I owe a particular debt for the benefit of numerous discussions with them. It was Dr Peter Dronke who most kindly unearthed the ancient manifestations of the machina concept which appear on page vi.

I must express my gratitude to the trustees of the Leverhulme Foundation for an award made in 1963 towards the expenses of preparing this book.

Warm thanks are also due to Miss Margaret Webb and Mrs Diana Brodie for their most expert secretarial assistance.

Finally, I should like to place on record the inspiration that I derived from working in the Cambridge Biochemical Laboratory from 1919 to 1963.

D.M.N.

ABBREVIATIONS

ADP	Adenosine-5-diphosphate	GTP	Guanosine-5-triphosphate
AMP	Adenosine-5-mono-phosphate	HMM	Heavy meromyosin
		IAA	Iodoacetic acid
ATP	Adenosine-5-triphosphate	IANH₂	Iodoacetamide
CTP	Cytidine-5-triphosphate	IDP	Inosine-5-diphosphate
CoA	Co-enzyme A	IMP	Inosine-5-monophosphate
Co-Q	Co-enzyme Q	ITP	Inosine-5-triphosphate
DEAE-cellulose	Diethylamino-ethyl-cellulose	LMM	Light meromyosin
		NAD	Nicotinamide-adenine dinucleotide (cozymase)
DNA	Deoxyribonucleic acid		
DSEM	Disulphide-exchanged myosin	NADP	Nicotinamide-adenine dinucleotide phosphate
EDTA	Ethylenediaminetetrace-tate	NEM	N-ethylmaleimide
		NTP	p-nitrothiophenol
EGTA	Ethyleneglycol-bis-β-amino-ethylether-N,N'-tetracetate	PCMB	p-chlormercuribenzoate
		Q, Q₂₇₅, CoQ	Co-enzyme Q
		RNA	Ribonucleic acid
ESF	EGTA sensitising factor	RPS	Relaxing protein system
FAD	Flavin-adenine dinucleo-tide	TNP	Trinitrophenol
		TPP	Thiamine-pyrophosphate
FDNB	1-fluoro-2,4-dinito-benzene	UDP	Uridine-5-diphosphate
		UDPG	Uridine diphosphate glucose
FeNH	Non-haem iron		
flpr	flavoprotein	UTP	Uridine-5-triphosphate
FMN	Flavin mononucleotide		

NOTE. All units of temperature in the text are °C

1

BRINGING MUSCLES INTO FOCUS; THE FIRST TWO MILLENNIA

ANTIQUITY AND THE HELLENISTIC AGE

In the development of thought about the bodies of men and animals there came a time when the age-old acceptance of undifferentiated body-substance, the biblical 'flesh of rams' or the meat on which Homeric heroes feasted, gave place to a realisation that it consisted of individual muscles. How early did this happen and when was the function of these muscles as instruments of movement realised? With these questions our story naturally begins.[1]

The Hippocratic collection of writings on medicine and its philosophy, by a number of writers of his school as well as perhaps by Hippocrates of Cos himself, was put together before the end of the third century B.C. and includes works of the two previous centuries, some indeed containing ideas from still earlier times.[2] There is thus no such thing as a single system of thought to be found in them; the different treatises of the Corpus, some sixty in number all told, represent several different, and even opposing, schools. Three of them have been attributed by some distinguished scholars to the great physician of Cos himself,[3] and eight more are considered to date from his time (460 to 380 B.C.).[4] The only certainly pre-Hippocratic one is the 'Sevens', a prognostic text which implies the humoral theory of disease and the doctrine of critical days.[5]

In these Greek writings the tendons (which were confused with nerves) were endowed with the power of causing movement. In fact the same word

---

[1] In what follows I have had the advantage of the advice of Dr G. H. Lloyd of King's College, Cambridge. We owe to Bastholm (1) an excellent history of muscle physiology which has also been of much assistance.

[2] Singer & Underwood (1).

[3] E.g. by W. H. S. Jones (1), 'Prognosis', 'Regimen in Acute Diseases' and 'Epidemics' I and III.

[4] 'Sacred Disease', 'Airs, Waters and Places', 'Diet', 'Head Wounds', 'Ancient Medicine', 'Nutriment', 'The Art' and 'Breaths'. The last three of these, though containing the earliest Greek mention of the pulse, were not from the Coan school itself.

[5] It may go back to the 6th cent. B.C. On the whole Corpus see conveniently Sarton (1) vol. 1, pp. 96 ff.; Castiglioni (1) pp. 151 ff. The substance of the famous 'Oath' may also be of the 6th cent. B.C.

neuron was used indiscriminately for both, just as *phlebes* was used in-differently for the veins and the arteries.[1] Thus: 'The bones give a body support, straightness and form; the nerves [tendons] give the power of bending, contraction and extension; the flesh and the skin bind the whole together and confer arrangement on it; the blood-vessels spread throughout the body, supply breath and flux and initiate movement.'[2] The last phrase of this sentence introduces the theory of *pneuma*, destined to have such influence in succeeding centuries. It was the main subject of one of the Hippocratic treatises already mentioned, the 'Breaths', certainly of the later fifth century B.C., but it is also important in another of the early works, the 'Sacred Disease' (epilepsy). There was certainly a connection here with the pre-Socratic philosophers, especially Diogenes of Apollonia (d. *ca.* 428 B.C.) who greatly emphasised the pre-eminence of the element Air in all Nature. He believed that the blood was everywhere accompanied by air (*pneuma*) in the vessels, and that sleep occurred when the air was driven down to the chest and abdomen.[3] Empedocles (d. *ca.* 430 B.C.) also associated blood and air very closely in his theory of respiration,[4] but Diogenes was probably more indebted to Anaximenes (d. *ca.* 494 B.C.), who had seen in air the source of all the other elements and the substrate of all change.[5] *Pneuma* is also prominent in the 'Nature of Man',[6] another Hippocratic treatise, probably written by Polybus of Cos, about the beginning of the fourth century B.C., reputedly Hippocrates' son-in-law and successor as head of the Coan school.[7]

Polybus' book is the main source for the other basic biological idea of the Hippocratic writers, the humoral theory. This no doubt originated in a sense from the concept of Anaximander (fl. 560 B.C.) and Empedocles that all matter was composed of four 'roots' (elements) – fire and water; earth and air.[8] These are pairs of opposites, and Empedocles added two 'original causes', love (*philia*)and hate (*neikos*), or forces of attraction and repulsion, to explain their combination and splitting apart.[9] But the development of medical theories was complicated. While the 'Nature of Man' certainly expounds four humours, blood, yellow bile, black bile and phlegm, there are

[1] It is interesting that in ancient Chinese writings the word *chin* bore just the same ambiguity as *neuron*. For an account of the Chinese equivalent to the Hippocratic Corpus see Needham & Lu (1).
[2] 'On the Nature of the Bones', tr. Littré (1) IX, p. 183; eng. auct.
[3] Freeman (1) pp. 279 ff., (2) pp. 87 ff.
[4] Freeman (1) p. 195. It is interesting that Empedocles visualised a blood–air interface advancing and retreating within vessels and pores, closely similar to what goes on (as we know today) in the tracheal respiration of insects. It was in connection with this that Empedocles used his famous demonstration of the wine-pipette.
[5] Freeman (1) pp. 65 ff.   [6] Filliozat (1) p. 189.
[7] Sarton (1) vol. 1, p. 120.   [8] Freeman (1) pp. 56, 181.
[9] Freeman (1) pp. 172 ff., (2) pp. 51 ff.; Leicester (1).

traces of a two-humour system (bile and phlegm only) in earlier books such as the 'Airs, Waters and Places' and the 'Sacred Disease'.[1] Moreover the physicians were sometimes very critical of the adoption of any ideas from the philosophers. The whole polemic of the 'Ancient Medicine' is directed against this, and Empedocles is even mentioned by name in it.[2] That did not prevent the doctors from borrowing from natural philosophy of course, and the 'Nature of Man' seems rather like a deliberate attempt to synthesise the idea of four humours with that of the four elements.

Similar notions originated in the much earlier Vedic writings of India from the fourteenth century B.C. onwards, as Filliozat has shown.[3] The physiological ideas contained in them were elaborated and systematised in the Ayurvedic Corpus, especially the *Suśruta-samhita*[4] and the *Caraka-samhita*.[5] Here we find the living body considered as composed of the 'elements' air (*vāyu*), fire (*tejas*), water (*ap*) and earth (*pṛthivī*).[6] Even in the earliest texts breath or *prāṇa* (with a role closely comparable to that of *pneuma*) was divided into five or seven varieties, distinguished, when they came to be defined, by the parts of the body they served.[7] One of these, *vyāna*, ran through all the limbs and explained their movement.[8]

The later Hippocratic writers pictured all the parts of the body as composed of four humours.[9] These were: blood, hot and wet (corresponding to air); yellow bile or *chole*, hot and dry (corresponding to fire); black bile or *melanchole*,[10] cold and dry (corresponding to earth); and phlegm or saliva, cold and wet (corresponding to water). These humours were afterwards considered to be characteristic of the liver, gall-bladder, spleen and lungs respectively.[11] Health depended on the right balance (*krasis*) between these four. Of course the thought of the Hippocratic schools was not in reality as clear-cut as this. There are many different humoral theories in the Corpus.[12] That in the 'Nature of Man' is closest to the scheme just outlined, and unlike some other authors who derived three of the humours from blood,

[1] As also in the 'Affections' and 'Diseases I', books of the Cnidian school.
[2] Para. 20, tr. Adams (1) vol. 1, p. 175.   [3] Filliozat (1) pp. 46 ff.
[4] Datable between the 2nd cent. B.C. and the 2nd cent. A.D., in its present form by the 7th cent. A.D.
[5] Datable between the 1st cent. B.C. and the 3rd cent. A.D., in its present form by the 8th cent. A.D.
[6] Filliozat (1) pp. 20 ff.
[7] Filliozat (1) pp. 141 ff. Completing the circuit of the Old World *pneuma* and *prāṇa* had a close equivalent in Chinese culture, *ch'i*, of immense significance in all ancient and medieval East Asian biology and medicine (see J. Needham (1) vol. 2).
[8] Filliozat (1) p. 23.   [9] Cf. Leicester (1); Jevons (1).
[10] According to Jevons, this idea was probably derived from the observation of the lower part of a blood clot.
[11] Cf. Singer (1); Singer & Underwood (1).
[12] This shows itself even in secondary and propaedeutic sources. Singer (1) p. 8 has fire associated with blood and air with yellow bile; Singer & Underwood (1) p. 46, reverse this.

Polybus considered all four independent and 'congenital'. Although the association with organs is late, it does occur in the Cnidian 'Diseases IV', but in this case the four humours include water instead of black bile, and the associations are all different, blood connected with the heart (not the liver), yellow bile with the liver, water with the spleen and phlegm with the head rather than the lungs. In the thought of Aristotle (384 to 322 B.C.) all substances were made of primary matter and on this matter different forms could be reversibly impressed.[1] The fundamental properties or 'qualities' were hotness, coldness, moisture and dryness; and by combining these in pairs the four elements (fire, air, earth and water) were obtained.[2] The relation between the qualities, the elements (or roots) and the humours is illustrated in fig. 1.

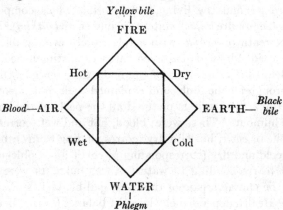

Fig. 1. The relationship, according to Aristotle, between the Qualities, the Elements and the Humours. (Singer & Underwood (1).)

The roles attributed to the *pneuma* and the humours in one presentation (the 'Sacred Disease') can be seen in the two quotations following.

For when a person draws in breath by the mouth and nostrils, the breath (*pneuma*) goes first to the brain, then the greater part of it to the internal cavity, and part to the lungs and part to the blood-vessels, and from them it is distributed to the other parts of the body along the blood-vessels; and whatever passes to the stomach cools it and does nothing more. But the air which goes into the lungs and the blood-vessels is of use (to the body) by entering the brain and its ventricles, and thus it imparts sensibility and motion to all the members; so that when the blood-vessels are excluded from the air by the phlegm and do not receive it, the man loses his speech and intellect, and the hands become powerless and are contracted, the blood stopping and not being diffused as was its wont...[3]

[1] Hence the philosophical sanction for all Hellenistic and mediaeval alchemy.
[2] *De Partibus Animalium*, 646a 14 ff., tr. Ogle (1).
[3] *De Morbo Sacro*, tr. Littré (1) VI, p. 373; eng. Adams (1) vol. 2, p. 850, mod. auct.

Again:

It is the brain which is the messenger to the understanding. For when man draws the breath (*pneuma*) into himself, it passes first to the brain, and thus the air is distributed to the rest of the body, leaving in the brain its acme, and whatever has sense and understanding. For if it passed first to the body and last to the brain, then having lost in the flesh and veins the judgment, it would be hot, and not at all pure, but mixed with the humidity from the fleshy parts and the blood, so as to be no longer pure.[1]

The idea of a metabolic activity of the flesh is well pictured in this quotation: 'The flesh draws upon both the stomach and the environment; it is clear that the whole body breathes in and breathes out. The little blood-vessels warmed by being overcharged with blood raise up the hot (or burnt) material and excrete it immediately: as yellow bile if the fatty element predominates; as black bile if blood predominates.'[2]

The protection afforded by the flesh against extremes of heat and cold was regarded as one of its main functions, since changes in temperature were believed to have serious effects on the balance of the humours. In the Hippocratic treatise 'On Diseases' we find:

When these humours (bile and phlegm) are set in movement and moistened, the individual, whether drunk or not, is seized with shivers; the side, which naturally is the part of the body most deprived of flesh, and which, far from having anything inside which supports it, is adjacent to a cavity, the side, we say, particularly feels the cold...the flesh which is at the side and the venules draw themselves together and contract, and what there is of bile and of phlegm in the flesh itself or in the venules of the flesh is, largely or totally, secreted inwards towards the warmth, because towards the outside the flesh is compact. These (the bile and phlegm)...cause intense pain, become warmed up, and by the heat attract to themselves bile and phlegm out of the veins and the neighbouring flesh.[3]

It is perhaps curious that the 'contraction of the flesh' due to cold, here described, was not thought of in the context of muscular motion. Muscle twitching and convulsions were also commented upon, but put down to movement in the blood vessels, transmitted to the muscles.

For Aristotle, the flesh was characterised by its divisibility in any direction, unlike the tendons and blood-vessels, but particularly by being the most important organ of the sense of touch. 'An animal [he says] is by our definition something that has sensibility, and chief of all the primary sensibility, which is that of Touch; and it is the flesh, or analogous substance, which is the organ of this sense.'[4] He regarded the sinews as responsible for motion, as we read:

[1] *De Morbo Sacro*, tr. Adams (1) vol. 2, p. 856.
[2] *Epidemiorum*, tr. Littré (1) v, p. 323, eng. auct.
[3] *De Morbis*, tr. Littré (1) vi, p. 193, eng. auct.
[4] *De Partibus Animalium*, 653b 23 ff., tr. W. Ogle (1).

The movements of animals may be compared with those of automatic puppets which are set going on the occasion of a tiny movement; the levers are released, and strike the twisted strings against one another...Animals have parts of a similar kind, their organs, the sinewy tendons to wit and the bones; the bones are like the wooden levers in the automaton, and the iron; the tendons are like the strings, for when these are tightened or released movement begins.[1]

Aristotle gave the 'more honourable part' to the heart rather than to the brain,[2] and communication with the rest of the body he regarded as due to the vessels full of blood containing *pneuma*.

Now experience shows us that animals do both possess connatural[3] spirit (*pneuma*) and derive power from it...And this spirit appears to stand to the soul-centre or original in a relation analogous to that between the point in a joint which moves, and (that which is) unmoved. Now since this centre is for some animals in the heart, in the rest in a part analogous with the heart, we further see the reason for the connatural spirit being situate where it actually is found...We see that it is well disposed to excite movement and to exert power; and the functions of movement are thrusting and pulling. Accordingly the organ of movement must be capable of expanding and contracting; and this is precisely the characteristic of spirit. It contracts and expands naturally and so is able to pull and to thrust from one and the same cause, exhibiting gravity compared with the fiery element, and levity compared with the opposites of fire.[4]

It is in the work of Herophilus of the Alexandrian school (early third century B.C.) that we find, as Bastholm rightly says, the first hint of the responsibility of the muscles for movement, and the first attempt to distinguish between the various cord-like organs which had all been classed together under the name of *neura*.[5] This work is known to us only through the writings of Galen, Rufus of Ephesus and one or two other later writers The following extracts illustrate these points.

Herophilus assigns the motive power of the body to the nerves [or sinews], the arteries and the muscles.[6]

We shall consider again whether twitching is something that affects only the muscles as Herophilus thought or whether it affects the skin and arteries too...[7]

If one believes Herophilus there are nerves of voluntary movement which arise from the brain and the dorsal marrow [medulla], others which are inserted some from one bone to another [ligaments], some from one muscle to another [aponeuroses], and finally others which attach (them to) the joints [tendons].[8]

---

[1] *De Motu Animalium*, 701 b 1 ff., tr. A. S. L. Farquharson (1). Some idea of the automatic puppet theatres of the Hellenistic age can be gained from Beck (1).

[2] Because it was more central to the body (*De Partibus Animalium*, 665 b 19 ff.).

[3] Innate or congenital.

[4] *De Motu Animalium*, tr. A. S. L. Farquharson (1) 703 a 9 ff.

[5] See the study of Dobson (1).

[6] Galen, *De Historia Philosophica* (Pseudo-Galen), tr. Kühn (1) XIX, p. 318; eng. Wright.

[7] Galen, *De Tremore, Palpitatione, Convulsione et Rigore*, tr. Kühn (1) VII, p. 594; eng. Wright.

[8] Rufus of Ephesus, *Opera*, tr. Daremberg & Ruelle (1) p. 185, eng. auct.

In another place Galen says:

And here I blame Praxagoras[1] and Herophilus, the former for calling tremor an affection of the arteries, and the latter for trying to show that it originates in the nervous system. Praxagoras was far from the truth, and Herophilus was mistaken in referring the affection of the faculty to the instruments. For he knew that the nervous system, not the arterial, is subordinate to voluntary motions; but since it is not the body of the nerve itself which causes their motion, this being but an instrument, and the moving cause the power which passes through the nerve, here I blame him for not distinguishing power and instrument...Now in the case of dead bodies, neither muscles nor nerves are subject to any such affections as Herophilus and Praxagoras suppose, but all their motion ceases when the soul departs, muscles and nerves being but its instruments; so it is not the property of either muscle or nerve to produce motion, but only of soul.[2]

But since we know from Galen himself that Herophilus 'placed the dominant principle of the "soul" in the ventricles of the brain'[3] he was not really open to much criticism; all the less so indeed if he was really trying to say that there may be some conditions of uncontrollable tremor due to faults in the conducting nerve-channels rather than the voluntary activity of the 'soul'. Herophilus distinguished between arteries and veins (noting the great difference in thickness of their walls), and realised that the arteries contained blood. He investigated the pulse, and put it down to contraction of the arteries (resulting from a stimulus from the heart) followed by elastic return.[4]

Erasistratus, a younger contemporary of Herophilus, went further and distinctly recognised the muscles as organs of contraction.[5] He made further elaboration of the *pneuma* theory, postulating two kinds both coming from the air. This air, passing through the lungs and pulmonary veins, was drawn into the left ventricle of the heart during diastole; there a particular *pneuma* (the *pneuma zootikē* or *spiritus vitalis*), was formed from it. This vital spirit was pumped during systole to other parts of the body by way of the arteries; Erasistratus considered that the arteries contained no blood, the latter travelling only in the veins. When the *spiritus vitalis* reached the brain it was changed into a second kind, the *pneuma psychikē* or *spiritus animalis*.[6] 'Erasistratus says that the animal spirit comes from the head, the vital from the heart.'[7] The *spiritus animalis* was thus distributed through the (hollow tubular) nerves to the muscles, and here for the first time we find a theory of the mechanism of contraction, one which was to

---

[1] Praxagoras of Cos, fl. 340 to 320 B.C., physician and anatomist (cf. Sarton (1) vol. 1, p. 146).
[2] Galen, *De Tremore*..., tr. Kühn (1) VII, p. 605, eng. Dobson (1).
[3] Galen, tr. Kühn (1) XIX, p. 315; cf. Dobson (2).
[4] Dobson (1).   [5] Dobson (2).
[6] 'He wrote accurately', said Galen, 'about the brain's four ventricles' (Kühn (1) V, p. 602).
[7] Galen, *De Hippocratis et Platonis, Placitis* II, tr. Kühn (1) V, p. 281, eng. Wright.

have influence centuries later.[1] 'Erasistratus says that the muscles, if they are filled with *pneuma*, increase in breadth but diminish in length, and for this reason are contracted.'[2]

Erasistratus is credited with the discovery of the bicuspid and tricuspid valves and seems to have considered the heart as a unidirectional pump.[3]

The heart, when it is dilated, is filled by the inrush into a vacuum; but the arteries when they are filled, are dilated; and they are filled by the *pneuma* sent from the heart. Both must occur at the same time, the dilation and filling, but he thinks that the one is the cause of the other, in the heart the dilation, and in the arteries the filling, as is observable elsewhere. A blacksmith's bellows are filled because they are dilated; sacks, wineskins and so on are dilated because they are filled.[4]

This distinction between active dilatation and passive expansion (with the implicit corollaries of positive contraction and mere elasticity respectively) was one of considerable insight. Some uncertainty remains whether Erasistratus was the first discoverer of the cardiac valves, for the Hippocratic book 'On the Heart' displays some knowledge of them; this however is now considered on linguistic grounds to be post-Aristotelian and therefore very little, if at all, anterior to the time of Erasistratus.[5] In any case he glimpsed the function of the valves though he thought that they prevented the regurgitation of *pneuma* rather than blood.

Rufus of Ephesus (early second century B.C.) was outstanding for his isolation of muscles by dissection. Erasistratus had regarded the muscles, like other organs, as built up by an aggregation of fine particles of blood around the *triplokia* (or *vasa triplicia*) – a basal fibre structure of nerves (or sinews), veins and arteries.[6] Rufus now recognised muscle as a tissue built on a particular pattern with a specific function – that of voluntary movement.

The muscle is a firm and dense body, not simple but resulting from an interlacing of nerves, veins and arteries, not deprived of sensibility; it is the organ of voluntary movement.[7]

The flesh is the solidified part which, in the viscera, is found between the vessels; it is at the same time a sort of tissue and a kind of packing between the network of the vessels so that there should be no spaces between them; then there is the flesh of the muscles, fibrous and resistant; and finally the coagulum which forms in wounds and is found in the cavities of the bones.[8]

[1] E.g. in the ideas of Descartes and Borelli, considered on pp. 14 & 23 below.
[2] Galen, *De Locis Affectis*, tr. Kühn (1) VIII, p. 429, eng. Wright. Cf. IV, p. 707.
[3] Wilson (1); Dobson (1); cf. Galen (Kühn (1) V, pp. 166, 206, 548 ff.).
[4] Galen, *De Differentia Pulsum*, tr. Kühn (1) VIII, p. 703, eng. Wright; cf. V, p. 562.
[5] Abel (1).   [6] Galen, tr. Kühn (1) II, p. 96, III, p. 538, XIV, p. 697.
[7] Rufus of Ephesus (attrib.), 'On the Anatomy of the Parts of the Body', in *Opera*, tr. Daremberg & Ruelle (1) p. 184, eng. auct.
[8] Rufus of Ephesus, 'On the Names of the Parts of the Body' in *Opera*, tr. Daremberg & Ruelle (1) p. 164, eng. auct.

We come now to Galen (129 to 201 A.D.), the writer of the greatest works of Western antiquity on animal anatomy and physiology. For Galen true muscles were to be defined as organs of voluntary movement, and the heart, uterus, oesophagus, etc. were classed as muscle-like. He made it clear that a muscle has only two possibilities – contraction and relaxation; the latter he regarded as a purely passive movement brought about by the contraction of the antagonist. '...The natural activity of the muscles consists in contracting and withdrawing upon themselves, and lengthening and relaxation takes place when the antagonist muscles pull and draw towards themselves.'[1] The possibility for an organ like the tongue to move in several directions depended on the presence of several different muscles.[2] Galen also performed experiments to show the effect upon movement of cutting off the muscle from communication with the spinal cord.[3] Although he was clear that the muscle mass had the power of movement he believed that tendon also took an active part.[4]

Galen's detailed dissections of muscles went far beyond anything that had gone before, and 'myology' was placed by him on a permanently scientific basis. Even though his dissections were made mostly on animals rather than on the human body his descriptions read strangely like those in a modern anatomical handbook. This may be illustrated by quoting a little of his account of the extrinsic muscles of the tongue.[5]

Should you wish to dissect all the tongue muscles separately in the body of a dead animal, as I am about to describe for you, then you must, I say, commence by reflecting the skin over the neck and the lower portion of the mandible. Next remove the muscle which is called the 'muscular carpet' [M. platysma myoides]... When you have reflected it you will see the peculiar muscle of the mandible, which is the one that is tendinous in its middle portion [M. digastricus], and simultaneously with it there will appear firstly the muscle of the tongue that is called the 'transversely directed' one [M. mylohyoidens], whether you like to call it one muscle with two parts, or else two muscles associated closely and united with one another...Then pass on to the 'oblique' muscle [M. hyoglossus with M. chondroglossus] which in apes has its source and origin on the lower rib [greater cornu] of the bone which resembles the letter Λ [lambda] of the Greek script [the hyoid bone], one on either side of the neck...

And so on at length. No other contemporary civilisation carried anatomy to the height attained by the indefatigable Pergamene physician. After all, the

---

[1] 'On Muscular Movements' in 'Oeuvres anatomiques, physiologiques et medicales', tr. Daremberg, II, p. 334, eng. auct.
[2] Ibid. II, ch. IV and v.   [3] Ibid. II, p. 323.   [4] Ibid. II, p. 327.
[5] In ch. 7 of book 10 of his work 'On Anatomical Procedures' (Duckworth, Lyons & Towers (1) pp. 56 ff.). Only the first eight and a half books of this survived in Greek and were in recent years retranslated by Singer (2); the remaining six and a half came down only through the Arabic, whence they were put into German by Max Simon and English by W. L. H. Duckworth, hence the above-mentioned publication. It is of interest that John Caius printed a revised Greek text of the first portion at Basel in 1544.

morphological description of muscles was an indispensable preliminary to the analysis of the contractile function of muscle. In later times it would be necessary, for example, to distinguish between red and white muscles, smooth and striated muscles, etc., and biochemists would want to select equal and opposite anatomical pairs of muscles.

A fundamental change which Galen made in the physiological scheme of Erasistratus was his demonstration that the arteries contain not air (*pneuma*) but blood. There has been much debate on Galen's conception of the *pneuma* theory; one interpretation may be described thus.[1] Air taken in by the lungs goes to the left ventricle where it plays its part both in maintaining the innate heat (essential for life as Galen emphasised) and in providing 'refreshing cooling' lest the heart should become overheated. Most of the air returns to the lungs and is expired with some waste substances. Galen regarded the *pneuma psychikē* or *spiritus animalis* as an exhalation of the blood produced under the influence of the innate heat. He believed that waves of dilation passed through the walls of the arteries causing the sucking in of blood from the veins, and of air both from the heart and from the exterior through the pores of the skin. The gently maintained heat in the arteries produced in the brain blood rich in *spiritus animalis*; the latter passed to the muscles along the nerves.

These Galenic conceptions were destined to hold the field in Western Europe with no alternative and little criticism for the next 1300 years. Further research on minor Latin writers, and especially on Byzantine and Arabic contributions (to say nothing of cultures further east), may well discover some interesting developments, but by and large the influence of Galen in the field of muscle contraction as in other realms of physiology and medicine reigned unchallenged until the Renaissance.

THE RENAISSANCE AND THE SEVENTEENTH CENTURY

In what has so far been considered we may discern, running through the whole, four threads of enquiry. First, the identification of the functional motile tissue – whether muscle itself or tendon. Secondly, the function of the flesh, apart from the problem of movement – its protective action, its sensitivity to touch and its metabolism (which on primitive views was bound up with the humoral theory). Thirdly the important matter of the inciting influence reaching the effective motor organ (whether muscle or tendon) and the channel (whether the blood-vessels or the nerves) by which the influence travelled; this was the province of the *pneuma* theory. Fourthly the problem of the morphology of the muscle tissue gradually descried, in so far as this could be studied through fine dissection without

---

[1] That of Wilson (1); see also Bastholm (1) pp. 87 ff.

the aid of the simplest microscope. Just raising its head, as we see in the work of Erasistratus, was the ambition to explain the mechanism by which the muscle shortened.

The ideas of Aristotle concerning the four qualities and elements, combined with the earlier Hippocratic concept of humours, were used by physicians throughout the middle ages; before their gradual abandonment in the seventeenth century, however, some new concepts had been added to them. It is worth while to mention these since, as will be seen, they clearly influenced the theories of contraction elaborated by seventeenth-century workers.

Aristotle taught that under the influence of the sun two vapours were emitted from the earth, one *anathumiasis* moist, the other hot, dry and smoky.[1] If these vapours were imprisoned in the earth, the former gave rise to the metals, while the latter gave rise to the *fossiles* or refractory minerals.[2] In later times alchemists identified the moist vapour with mercury, the dry with sulphur. Paracelsus (1493 to 1541) considered all matter to be ultimately composed of the four Aristotelian elements, but immediately of the *tria prima*, salt, sulphur and mercury. These identifications were not to be taken literally but as abstractions of qualities: the salt embodied the principle of incombustibility and non-volatility; the mercury the principle of fusibility and volatility; the sulphur the principle in virtue of which substances take fire.[3] Important for us is the paper by Debus (1) in which is described Paracelsus' concept of a 'Sal nitric food'. In the *Liber azoth*, besides the idea of a vital fire or sulphur, there is the theory of a nitrous salt or saltpetre, having amongst other functions that of an aerial nutriment for the muscles. In the century succeeding the death of Paracelsus many workers were concerned with the Sal nitric from an alchemical and often mystical point of view. In the sixteen seventies we shall come again upon the nitro-aërial particles, this time in Mayow's interpretation of his experiments on muscle.

The influence of these ideas on the writings of Thomas Willis and of John Mayow on muscle contraction will be discernible later. An interesting passage in Willis' work 'Of Feavers' (1684) reads as follows:

The mass of the Blood, by the opinion of the Antients, was thought to consist of four humours, to wit, Blood, Phlegm, Choler and Melancholy...This opinion, though it flourished from the time of Galen, in the Schools of Physicians, yet in our Age, in which the Circular motion of the Blood, and other affections of it were made known, before not understood, it began to be a little suspected...because these sort of humours do not constitute the blood, but what are so called (except the Blood) are only the recrements of the blood, which ought to be continually

[1] *Meteorologica*, 341 b 6 ff., tr. H. D. P. Lee (1) p. 29.   [2] Eichholz (1).
[3] These ideas can be studied in detail in the books of Pagel (1), Debus (2), Sherwood Taylor (1), Holmyard (1) and Leicester (2).

separated from it: For in truth the Blood is an only humour; not one thing about
the viscera, and another in the habit of the body...and wheresoever it is carried
through all the parts of the body it is still the same...But as these humours
commonly so called, are made out of the other Principles, viz. Choler out of Salt
and Sulphur, with an admixture of Spirit and Water; and Melancholy out of the
same with an addition of Earth; and as the blood is immediately forged out of
these kind of principles, and is wont to be resolved sensibly into the same, I
thought best...to bring into use these celebrated Principles of the Chymists, for
the unfolding of the Nature of the Blood and its affections.[1]

In a later variant 'acid' and 'alkali' were taken as two fundamental and
opposing principles of matter, 'a duality replacing the Aristotelian quartet.'[2]
Such theories were held by Franciscus Sylvius (working at Leiden from 1658
to 1672), an influential figure in the application of chemistry to biological
processes. He made a study of the nature of salts, realising that they resulted
from a union of acid and base. Since potassium carbonate was the commonest
alkali available, acid–alkali neutralisation, effervescence and fermentation
became amalgamated to virtually a single concept in Sylvius' theories. It is
clear, as we shall see, that these ideas influenced Borelli in the picture he
made of the mechanism of muscle contraction in the sixteen seventies. The
ebullient acid–alkali concept was refuted both by Robert Boyle[3] and by
Boerhaave in the early eighteenth century, though as Jevons has pointed
out, something of it remained in the latter's ideas of acescence and alkes-
cence, characteristic of plants and animals respectively. From the middle of
the seventeenth century atomistic chemistry took over more and more, as
witness the words of Boyle (1) concluding the 'Sceptical Chymist' (1679): 'For
by that (the former part of my discourse) I hope, you are satisfied, that the
Arguments, brought by Chymists to prove that all Bodies consist of either
Three Principles, or Five, are far from being so strong as those that I have
employed to prove, that there is not any certain and Determinate number
of such Principles or Elements to be met with Universally in all mixt
Bodies.'

In returning to the muscle story we may mention first Vesalius, whose
great work *De Humani Corporis Fabrica* was published in 1543. His ideas of
the fine structure of muscle are shown in the following quotation.

But if nature had simply divided the nerve and tendon in this way, and had not
stuffed the spaces in between, the interstices of the fibres, with a soft substance
that provided for the fibres' body some kind of structure and a secure foundation,
the muscles could not preserve these fibres unbroken and unharmed for even the
shortest period of time. Now that foundation and body is simple flesh covered
with fibres, which is put into the fibres in exactly the same way as the experts in
cheese-making put milk into baskets and other vessels when they are curdling it.
So imagine that the fibres that flow from the nerve and tendon correspond to the

[1] T. Willis, 'Of Feavers', tr. S. Pordage (1).   [2] Jevons (1).   [3] Boas (1) p. 154.

rushes, the blood to the milk itself and the flesh to the cheese. For as cheese is made from milk, so is flesh from blood. Yet your picture will be nearer the truth in proportion as you imagine more and closer packed rushes or channels thrust through the cheese and not only stretched along the sides of the cheese.[1]

Vesalius abandoned the idea of flesh as the organ of sensation, deciding from his own observations that this faculty is seated in the skin. As Michael Foster (1) showed in his famous lectures, Vesalius saw clearly that contractile power resides in the actual muscle substance:

Muscle therefore, which is the instrument of voluntary movement..., is composed of the substance of the ligament or tendon divided into a great number of fibres and of flesh containing and embracing these fibres. It also receives branches of arteries, veins and nerves, and by reason of the presence of the nerves is never destitute of animal spirits so long as the animal is sound and well. Now I do not regard this flesh as merely a foundation or basis, as it were a bed or support by which the fibres and the above-mentioned divisions of the nerve are held together. Nor do I with Plato and Aristotle (who did not at all understand the nature of muscle) attribute to the flesh so slight a duty as to serve, after the fashion of fat or grease or some sort of clothing, the purpose of lessening the effects of heat in summer and of cold in winter. On the contrary I am persuaded that the flesh of muscles, which is different from everything else in the whole body, is the chief agent, by aid of which (the nerves, the messengers of the animal spirits not being wanting) the muscle becomes thicker, shortens and gathers itself together, and so draws to itself and moves the part to which it is attached, and by help of which it again relaxes and extends, and so lets go again the part which it had so drawn.[2]

Continuing the work of Vesalius, Fallopius (1523 to 1562) was keenly interested in fibrous tissue in connection with movement. Bastholm remarks that it was he who first enunciated the functional significance of the fibre: 'These...things are to be observed, as I have said, in every part that moves itself, and they cannot exist unless the part itself has fibres, and indeed internally fleshy fibres. For...motion requires a fibrous nature in the actual body that is moved; since (experience shows) whatever moves itself does so by contraction or extension.'[3]

Fabricius ab Aquapendente (1537 to 1619), pupil of Fallopius and teacher of William Harvey, carried out fine dissections of muscles, and he also emphasised the difference between their fibrous nature and that of other organs. Although Fabricius does write of the contractile function of the flesh he yet feels obliged, on account of its softness and weakness, to lay the main duty on the tendons:

So the twofold nature of the tendon was called for by the twofold use that was proposed, so that it was constructed in one part unitary, alone and compact, in

---

[1] Edition of 1543, p. 220, tr. Wright. The analogy with cheese and cheese-making played a considerable part in ancient and mediaeval biological speculation (see J. Needham (2), passim), as Vesalius was no doubt well aware.
[2] Foster (1) pp. 69–70.   [3] *Opera omnia*, vol. I, p. 94, tr. Wright.

the other split into fibres and divided. For where it appears unitary and compact and cartilaginous, there it is stronger and employs this opportune strength in lifting and moving bones; for this reason the tendon almost alone has always been recognised as the mover of bones. But where it is seen to be split into fibres and divided, there it is softer and more suitable for the bringing about of contraction; this happens more in the belly and the middle and even the beginning of the muscle, where contraction is required to a greater degree.[1]

It is interesting to notice that William Harvey in his notes[2] and lectures[3] on muscle returns to the view that the contractile material is flesh and not tendon or sinew. He remained entirely under the influence of the Aristotelian *pneuma* as the causation of movement.

This is not the place to discuss the whole upsurge of science in the seventeenth century. The new chemistry, typified by Boyle and Mayow, as well as the old, influenced the thought of those concerned with muscle function in this period; the science of mechanics as founded by Galileo and the mathematics of Descartes and Newton sent out their inspiration; in physiology Harvey's demonstration of the circulation of the blood had profound effects, disposing in the end (though not at once) of the ancient ideas of the distribution of nutrient substances and spirits by ebb and flow in the veins and arteries.[4]

Descartes was a mathematician and philosopher – no physiologist; but he applied himself to the explanation of the working of the whole body as a machine. 'We must admit' wrote Foster (1) 'that he did succeed in showing it was possible to apply to the interpretation not only of the physical but also of the psychical phenomena of the animal body, the same method which was making such astounding progress when applied to the phenomena of the material world.' Descartes still used the concept of animal spirits, but he treated them as a very subtle fluid amenable to physical laws; he also postulated that the rational soul was something added to and independent of bodily mechanisms, including that of the brain. His ideas (1646 to 1647) with regard to muscular movement can be illustrated as follows:

Now according as these spirits enter thus into the concavities of the brain, they pass thence into the pores of the substance, and from these pores into the nerves; where according as they enter or even only as they tend to enter more or less into the one rather than into the others, they have the power to change the shape of

[1] *Opera omnia anatomica et physiologica*, p. 404, tr. Wright.
[2] *De Motu Locali Animalium*. Ed. and tr. Whitteridge (1).
[3] *The Anatomical Lectures of William Harvey*. Ed. and tr. Whittridge (2).
[4] Galen had admitted such a tidal flow in part, but expressed himself obscurely on the motions of the blood, hence some misrepresentation of his beliefs by modern historians of science. Fleming (1, 2) has attempted to rectify the matter. Galen's assumptions of pores in the intraventricular septum of the heart, and of a partial ineffectiveness of the mitral valve, are not in dispute however.

the muscles in which these nerves are inserted, and by this means to make all the limbs move. Just as you can have seen in the grottoes and foundations in the gardens of our kings, that the same single force by which the water moves, coming out from its source, is enough to move diverse machines and even to make them play instruments or to pronounce words according to the different disposition of the tubes which conduct it.[1]

And truly one can well compare the nerves of the machine which I describe with the tubes of the machines of these fountains, its muscles and its tendons with the other diverse engines and springs which serve to move them, and its animal spirits with the water which puts them in motion, of which the heart is the source and the concavities of the brain the reservoir. Further, the respiration and other such actions, which are natural and usual to it, and which depend on the course of the spirits, are like the movements of a clock or mill, which the ordinary course of the water can render continuous.

For the single cause of all the movement of the limbs is that some muscles shorten and that some muscles lengthen... ; the single cause that makes a muscle shorten rather than its antagonist is that there comes a certain amount (it may be little) more spirit from the brain to it than to the other. Not that the spirits that come immediately from the brain suffice alone to move the muscles, but they determine the other spirits which are already in these two muscles to come out very promptly from one of them and to pass into the other; by means of which the one which they leave becomes longer and limper; and the one which they enter, being promptly inflated by them, shortens and pulls on the limb to which it is attached. This is easy to imagine provided one realises that only very little of the animal spirits comes continually from the brain towards each muscle. There is always a quantity of other spirits enclosed in the same muscle which move about there very quickly, sometimes swirling round only in the place where they are (when they do not find passages open for their exit) and sometimes flowing into the antagonist muscle, according as there are little openings in each of the muscles, by which the spirits can flow from one to the other, and which are so disposed that, when the spirits which come from the brain towards one of them have even little more force than those which go towards the other, they open all the entrances by which the spirits of the other muscle can pass into this one, and close at the same time all those by which the spirits of this muscle can pass into the other. Thus all the spirits contained before in these two muscles are assembled in one of them very quickly and thus inflate and shorten it, while the other lengthens and relaxes.[2]

In the intellectual climate which had thus been created, much observation, experiment and cogitation were expended during the latter half of the century, by several contemporaries, on the mechanism of muscle contraction. We may next consider therefore the work of Croone, Borelli, Willis, Stensen and Mayow.

---

[1] Once again we see the influence of mechanical simulacra of life on physiological thinking (cf. p. 6 above). This occurred in other civilisations also (see J. Needham (1) II, p. 53 ff.).
   Chapuis (1) describes the 'hydraulic gardens' of the 17th cent. and their automatic puppet-plays.
[2] 'Oeuvres', ed. Cousin, p. 347, eng. auct.

William Croone's first work on muscular movement, *De ratione motus musculorum*, was published anonymously in 1664. From experiments on nerve section Croone deduced that something (for which he still used the word spirit) necessary for contraction of the muscle, passed along the nerve from the brain. In his mind the word had a new connotation, reminiscent of the interpretation used by Descartes: 'As often therefore as I say the Animal Spirits, I mean the most subtle, active and highly volatile liquor of the nerves, in the same way as we speak of spirit of wine or salt...those spirits in the nerve called Animal, are nothing else than a rectified and enriched juice of this kind.'[1] He pictured the muscle as containing countless tendinous fibres, which were loosely arranged in the belly of the muscle (where more abundant flesh occupied the spaces), but joined together at each end to form a rope-like structure. The following paragraphs describe his point of view.

What, I ask, is more likely than that the force is carried with this liquor or rather, that this liquor itself, or animal spirits of the fibres, is struck out from the branchlets of the nerves by some impulse? If this be so it will also be highly probable that from the mixture of this liquor or spirit with the spirits of the blood there occurs continuously a great agitation of all the spirituous particles which are present in the vital juice of the whole muscle, as when spirit of wine is mixed with the spirit of human blood. For I have stated above that all parts of living things are swollen with a certain vivifying and spirituous liquor;...and no one is such a novice in chemistry as not to know how great a commotion and agitation of the particles is accustomed to occur from different liquors mixed with each other, as may be discerned in the example just mentioned, and also in common water with oil of vitriol or butter of antimony dissolved with spirit of nitre and in an almost infinite number of other cases of this kind.

Moreover, since it has been shown already that all the motions of the spirits in the parts of a muscle are contained in certain spaces, and in addition, that these spaces are broader and more definite where the belly of the muscle is...I say that the agitation of the spirits which is made by the different liquors within the heaving membranes of the muscle necessarily impels them with a great effort through straight lines towards A and C (the ends of the muscle). And since, in the intervals or spaces, they always strike against the smaller ends, they turn back on themselves and are accumulated in greater quantity around the middle or belly of the muscle...and hence the muscle begins to swell.[2]

It is interesting that Croone carried out an experiment from which he drew support for this hypothesis. He showed that when a weight of 7 pounds was hung at the bottom of an empty bladder, and water was poured into it, the weight was raised.

Just about this time the Danish anatomist Nicholas Stensen was actively working on muscle structure and movement. In 1664 he published *De musculis et glandulis observationum specimen* and three years later *Ele-*

---

[1] W. Croone (1), tr. Wilson (2).   [2] Ibid.

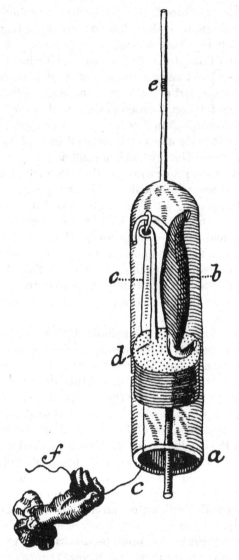

Fig. 2. Swammerdam's experiment (see text).

mentorum myologiae specimen seu musculi descriptio geometrica.[1] He described how the heart and other muscles contain no parenchyma but are
composed of motor fibres which may be dissected into most minute fibrils.
The fibres run from one tendon to the other and the middle parts are of a

[1] Ed. Maar (1).

different nature, soft and broad, making up the fleshy part of the muscle.[1]
So striking are his descriptions that some commentators (e.g. Foster (1) and
Wilson (2)) have felt that they strongly suggest he had made microscopic
observations. Walter Charleton (1) a few years earlier had said 'The flesh is
of such a nature as to yield easily to the *pneuma* coming in,'[2] and Stensen too
was clear that the fleshy part of the fibres alone, and not the tendinous part,
was concerned in contraction. Stensen's assertion that the heart is a muscle
with the same structure and function as other muscles was a very important
contribution, since it undermined the ancient idea of the heart as the
source of vital heat; even Harvey had laid emphasis on 'innate heat' in the
blood of the heart as the prime cause of distension in diastole.[3]

Up to this time it was tacitly assumed that the contracted muscle, hard
to the touch and appearing swollen, had increased in volume. Charleton[4]
however, in 1658, had expressed the view that its breadth increased as its
length diminished and so the volume was unchanged. Stensen (2), in his
book of 1667, expounded a complicated geometrical theory which would
allow of contraction without change in volume. This depended on the
representation of the fleshy part of the muscle as forming a parallelepiped
with oblique angles, and of the tendons at each end as forming rectangular
prisms.

Wilson suggests that Stenson emphasised this constancy in volume
knowing of the experiments of his friend Jan Swammerdam, made about
1663 but published only some sixty years later. Swammerdam[5] placed a
frog's muscle, still attached to its nerve, in a tube drawn out at one end to a
capillary containing a drop of water (fig. 2). The nerve was attached to a
wire which passed out through a fine hole in the stopper of the tube. Movement of the wire caused irritation of the nerve and contraction of the muscle,
but the bubble in the capillary scarcely moved – indeed it fell rather than
rose. Then Francis Glisson, in his work *De ventriculo* published in 1677,
described[6] another experiment to prove that there was no increase in
volume on contraction. In this case a strong muscular man inserted his
arm into a large tube filled with water and carefully sealed except for a small
glass side-tube; contraction of the muscles even led to a slight fall of water
in this tube. This experiment has often been attributed to Glisson himself,
but he made no claim to it. As Hierons & Meyer (1) have pointed out, it was
described in the Register of the Royal Society as carried out by Dr Jonathan
Goddard in 1669 (see fig. 3), though it was not published until 1756 in
Birch's *History of the Royal Society*.[7]

---

[1] Foster (1) p. 71.   [2] Tr. Wright.
[3] E.g. the 'Second Essay to Riolan', tr. Franklin (1).
[4] See Portal (1) III, p. 86.   [5] Swammerdam (1) p. 127.   [6] Glisson (1) p. 167.
[7] Fig. 3 shows the diagram which appears in Goddard's original paper – a slightly different version from that in the Royal Society's Register Book.

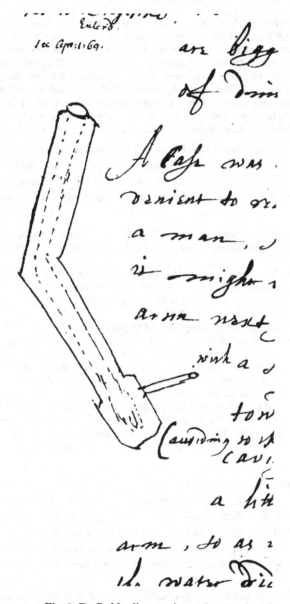

Fig. 3. Dr Goddard's experiment (see text).

In his later work Stenson expresses caution about the muscle fluids:

Concerning the fluids of muscle our knowledge is so uncertain as to be non-existent...Many people talk of the animal spirits, the more subtle part of the blood, the juice of the nerves – but these are mere words expressing nothing. Those who would continue further would introduce salty and sulphureous parts, or those analogous to spirit of wine; which may be true perhaps, but are neither certain nor sufficiently distinct. Just as the substance of this fluid is unknown to us, similarly its movements are uncertain; seeing that it has not yet been established definitely, either by arguments or experiments, whence it comes, by what means it proceeds or whither, once departing, it is received.[1]

In his later publication (1675) Croone (2) was much influenced by Stensen. He had now elaborated a theory which would permit the operation of the mechanism already proposed, but would not involve increase in volume of the muscle as a whole. With reference to his previous publication he writes:

An Objection rose, That we did not see any such conspicuous swelling in the belly of a muscle; to answer which, and explicate the hypothesis more clearly, I made this farther addition.

I supposed each carnous fibre as AE to consist of an infinite number of very small globules, or little bladders, which for explication sake I here express as so many little triangles, e.g. four in this fibre, ALB, BMC, CLD, DOE, all opening into one another at the points A, B, C, D, E.

I did not only suppose, but endeavoured to prove, that from the artery of each particular muscle, the nourishing Juice of the muscle was thrown out and extra-vasated to run at large among the Carnous Fibres, and insinuating itself by the constant pulse of the heart was driven on, and after mixing with another liquor it meets with between the Fibres in the Muscles, came to be strain'd through the coat of each globule or little bladder into the cavity of it: And likewise, that from each ramification of the Nerve within the Muscle, that second sort of Matter much more fluid and active than the former is extravasated, and these mixed together as I said, enter into each little Bladder, and by these constant agitations, ebulli-tion or effervescence...keeps these Globules or small vesicles always distended.

Croone then went on to calculate that the distension of four little bladders would be as effective in raising the weight from F to E as distension of one great one, AKE, and that 'if instead of four little bladders in a fibre we substitute 4000 that fibre shall raise the Weight in the space EF with 1/4000 part of the swelling; and so, of 1/400000 etc.' (See fig. 4a.)

Croone was also influenced by Leeuwenhoek's account of the appearance of muscle fibres under his microscope, as he emphasises in the concluding paragraph of this paper (see fig. 4b).

It is true that in 1674 Leeuwenhoek[2] wrote of the smallest muscle ele-ments as composed of 'conjoyned globules'. But in 1682 (the first to

[1] Stensen (2) tr. Wilson (2).
[2] Leeuwenhoek (1) I, p. 111. Letter to Mr N. Oldenburgh.

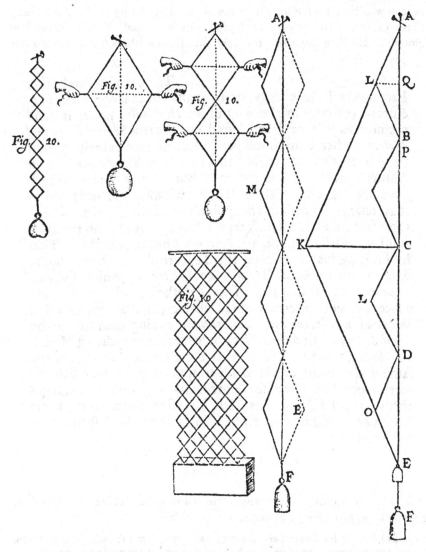

Fig. 4a. Diagram illustrating the lectures of William Croone, after *Philosophical Collections* (Royal Society) ed. Robert Hooke, 1680. This figure follows the words of Croone: 'as if we shall suppose four little bladders only for Example's sake each of them being distended from Q to L, shall all together raise the weight F to E, as well as one great swelling C to K'.

describe the transverse striations of the skeletal muscle fibre) he said:[1] 'The case now is that I often imagined that I could distinctly see that the fleshy fibres of which the greater part of a muscle consists were composed of globules.' He then described the very thin muscle fibres (he estimated one

Thefe and feveral other Particulars I did endeavour to make out at large in thofe Lectures; yet only in the way of an *Hypothefis*, not as if I did prefume to beleive I had found out the true Secret of *Animal Motion*, when I am almoft perfuaded, no Man ever did or will be able to explicate either this or any other *Phenomenon* in Nature's true way and method: But becaufe I reckon fuch Speculations among the beft Entertainments of our Mind, I may chance to collect and publifh fometime or other what I then faid, and have lately try'd about it.   And I am the more willing, Firft, becaufe Mr. *Lewenhook* has fince told us, That he finds by his Microfcope the Texture of a carnous Fibre to be of innumerable fmall Veficles or Globules, which gives an appearance of reality to the faid Hypothefis, and them, becaufe a fheet or two, and two or three fchemes, of that long expected Work of *Borelli*, *de Motu Animalium*, having been fent to the *Royal Society*, I find there fome Schemes for explicating Mufcular Motion, the very fame with thofe I make ufe of, and the fame Experiment of the Bladder applied in another Scheme to this occafion likewife: How he has explain'd or manag'd them, as yet I know not, but his ufing them, has made me have a better opinion of this Thought than elfe I fhould ever have had;

Fig. 4b. Part of a page from Croone's paper in Hooke's
Philosophical Collections, 1680.

million to the square inch) in which 'rings and wrinkles' became apparent, giving the appearance of globules. He went on:

Since I observed this I have been able to make out why our fingers, arms and legs, nay our entire body cannot lie stretched out straight, when reposing, but must be slightly bent...I also conceived the reasons of the movements, the stretching and the shortening of our muscles, that is to say, that when a muscle is stretched the fibres that constitute the muscles show no wrinkles or rings, but when a muscle is not stretched, but contracted and thick, each flesh-fibre is full of rings and wrinkles.

[1] Leeuwenhoek (1) III, p. 385 ff. Letter to the Royal Society of London, addressed to Mr R. Hooke.

Fig. 5[1] pl. gives one of his drawings of a fibre from fish muscle, together with his description of it. From his observations he calculated that such a fibre might contain some 3000 filaments; 'who can tell', he wrote, 'whether each of these filaments may not be enclosed in its proper membrane and contain within it an incredible number of still smaller filaments?'

The great physiological work of Alfonso Borelli, De motu animalium, was published posthumously in 1680–1. There is no doubt that Borelli knew of Croone's earlier work,[2] and Wilson considers that Borelli, like Croone, but not perhaps knowing of Croone's later work, had set out to find a way in which inflation might bring about contraction without increase in volume of the muscle; 'in muscles, by diligent inspection, the volume does not seem to be increased, for the length of the muscle is instead contracted and shortened, and indeed the breadth and thickness do not seem to be increased but retain the same measurement'.[3] Borelli's picture of muscle structure is very similar to that elaborated by Croone, the fibres in the muscle consisting of chains of rhombs of inextensible material, collapsed in the resting muscle but capable of inflation so that they become 'shortened, hardened and swollen'. Borelli also pictures the nerve fibre as tense with spirits or succus nerveus, coming from the brain; thus any disturbance could cause exudation of a few drops into the fleshy mass of the muscle, and lead to a violent chemical reaction between the nerve juice and the blood in the muscle. He describes the latter as saline, the former as abounding in alkaline salts. 'Salty fluids and acid liquors mixed with saline liquors of another kind; namely fixed or alkaline, excite boiling and effervescence.' 'Thus they can excite heat and boiling nearly instantaneously in the fibres or spongy tubules of the muscles, or their interstices, whence inflation, hardness and contraction of the muscles follow.'[4]

Still another contemporary of Croone contributing to the contraction problem was Thomas Willis (1621 to 1675). He was an outstanding anatomist of that time, especially concerned with the brain and nervous system and first adumbrated his theory in De Cerebri Anatome in 1664. The following quotation from De Motu Musculorum (published in 1670) shows his point of view to be rather similar to that of Croone in De ratione motus musculorum:[5]

Therefore as to muscular motion in general, we shall conclude after this manner... that the animal Spirits being brought from the head by the passage of nerves to every muscle and (as it is very likely) received from the membranaceous fibrils, are carried by their passage into the tendinous fibres, and there they are plentifully laid up in fit Store-houses; which Spirits, as they are naturally nimble and

[1] Leeuwenhoek (2) II, pt. 3, p. 113.   [2] Wilson (2).
[3] Vol. II, p. 27, tr. Wilson (2).   [4] Vol. II, p. 64, tr. Wilson (2).
[5] For an interesting appraisal of the mutual influence of workers concerned with muscle see Hierons & Meyer (1).

elastick, wherever they may, and are permitted, expanding themselves, leap into the fleshy fibres; then the force being finished, presently sinking down, they slide back into the Tendons, and so vicissively. But whilst the same animal spirits, at the instinct given for the performing of motion, do leap out of the tendinous fibres into the fleshy, they meet there with active particles of another nature supplied from the blood, and presently they grow mutually hot; so that by the strife and agitation of both the fleshy fibres, for that they are lax and porous, are stuffed up and driven into wrinkling, from all which being at once wrinkled and shrivelled up, the contraction of the whole muscle proceeds; the contraction being finished, the sincere or clear spirits, which reside or are assuaged, go back for the most part into the tendinous fibres, the other particles being left within the flesh; the loss or wasting of these the blood supplies, as the nerves do those.

If it be demanded, of what nature, to wit whether spiritous, saline, as may be believed, or of any other disposition, the animal spirits, derived from the Brain into the Muscles, may be; and then whether the other Latex, immediately carried to them from the blood, is sulphureous or nitrous. Concerning these, because it appears not to the sense, we shall pronounce nothing rashly or positively.[1]

The mention by Willis of 'Latex sulphurous or nitrous' brings us to the key work of John Mayow in 1674, on the substance in the air responsible for both combustion and respiration – his nitro-aërial particles. In 1660 Robert Boyle[2] was doing experiments which demonstrated the necessity of air for combustion. Robert Hooke[3] continued such experiments and, besides confirming the general principle that combustible substances would only burn in the presence of air, showed that certain of them, e.g. charcoal and sulphur, would burn if heated with nitre in absence of air. He concluded, in accord with the commonly accepted view of that time, that inherent in air is a substance resembling (or even identical with) saltpetre. Mayow now applied these ideas to the living organism and showed that when a small animal was confined in air over water, its respiration was accompanied by a diminution in volume of the air. He was also interested in muscle and wrote 'It is probable that this aerial salt is altogether necessary for every movement of the muscle.'

The nitro-aerial particles, he stated, seemed 'in a high degree to fit the character of animal spirits'. He pictured them as transmitted by means of respiration from the air to the mass of the blood and thence to the brain; here the nitro-aerial spirit was separated from the blood and stored, to be despatched when the need arose via the nerves to the muscles. For contraction, interaction of these particles with another kind was necessary – particles 'of a saline–sulphureous quality' carried to the muscle in the blood:

I think, namely, that sulphureous and saline particles brought to the highest volatility in the mass of the blood by its continuous fermentation...and most

---

[1] Willis (2), tr. Pordage (1).   [2] McKie (1).
[3] Hooke (1); and Birch (1) for the description of Hooke's experiments before the Royal Society.

intimately joined together, are separated from the blood by the action of the muscular parenchyma and stored up in the motor parts for setting up their contraction. For we may note that no small loss of fat takes place in the more violent exercises, and that it almost wholly disappears in long-continued hard work; while yet, on the other hand, animals indulging in ease and free from hard work become very obese, and fat is deposited on their muscles in quite sufficient abundance. Whence we may gather that the sulphureous particles of the blood, of which the fat is formed, have some share in the production of muscular contraction.[1]

The effervescent reaction taking place, with generation of heat, between the two sorts of particle was of a very special nature; he wrote:

So that for effecting the contraction of the muscles there is required an excitement of the elastic particles, of a kind that can be accomplished instantaneously and without any sort of coagulation. And indeed I do not know if there be in the nature of things any other such fermentation but the singular case of the effervescence of nitro-aerial and saline-sulphureous particles, which mutually, as their nature is, excite themselves to a rapid motion. We must therefore conclude that it is from that that muscular contraction proceeds.[2]

He also pictured that supply of nitro-aerial particles was essential for the proper fluidity, fermentation and motion of the blood. (A difficulty arises here since it is not made clear what prevented interaction in the blood of the two kinds of particle.)

From examination of muscle boiled for some time, Mayow concluded that parallel, fleshy fibres were inserted obliquely into the tendons and that a great number of fibrils ran transversely to the fibres, joining them together. Contraction was ascribed to these fibrils, the fleshy fibres acting as a filter to remove the saline-sulphureous particles from the blood. The contraction itself he believed was due to contortion of the fibrils 'caused by nitro-aerial particles set in motion and even pretty intensely warmed in the motor parts'. He gave the analogy of a 'very fine music string' held above a lighted candle; this contraction also was put down to 'nitro-aerial particles bursting out of the flame of the lamp'.[3]

Mayow died in 1679 at the age of thirty-six. His work was fundamental to the ideas concerning utilisation in contraction of that part of the air which we now know as oxygen; of great interest too is his observation of the actual using-up of a body constituent in the process.[4] But his contribution was ignored by his contemporaries and forgotten for more than a hundred years, only being noticed again after the rise and fall of the phlogiston theory and the recognition of oxygen.[5]

[1] Alembic Club Reprint No. 17, ch. 3.   [2] Alembic Club Reprint No. 17, ch. 4.
[3] Alembic Club Reprints No. 17, chs. 2 and 6.
[4] For an account of the development of concepts of respiration in the 17th cent. see Wilson (3).   [5] Cf. Bastholm (1) p. 209.

During the 150 years intervening between the work of Vesalius and of Mayow we can see the gradual realisation of the primary role of the muscles in contraction and of the importance of their fibrous structure for this function. The idea of the bringing of animal spirits to the muscles by way of the nerves came to be very generally held, though the nature of these spirits and that of the change caused by their arrival in the muscle allowed of diverse speculations. During this period Francis Glisson (1597 to 1677) introduced the conception of intrinsic irritability in tissues, and this property was intensively studied with particular regard to muscle by Albrecht von Haller in the middle of the next century. About that time too interest in electrophysiology was to arise. To these themes we shall return later on.[1]

After the late years of the seventeenth century little change took place for a long time in the ideas of the mechanism of muscle contraction, though we should mention the variants of the main idea, such as that inspired by Newton, in which the *succus nerveus* was replaced by a nerve 'aether' transmitting oscillations.[1]

The century and a half succeeding the work of Boyle in 1660 was a time of great chemical advances. Workers on muscle during the late eighteenth century and the nineteenth century were much occupied with the application of this chemical knowledge to the living organism: oxidation reactions, analyses of tissues, the fate of nutrients, catalytic processes and so on. During these times the *succus nerveus* came to be forgotten, and new types of chemical theorisation (more sophisticated but just as speculative as their predecessors) held the field. Theories of contraction from about 1850 also owed much to improvements in microscopy.[2]

[1] Ch. 13.   [2] Ch. 7.

2

MUSCLE METABOLISM AFTER THE CHEMICAL REVOLUTION; LACTIC ACID TAKES THE STAGE

The greater part of the eighteenth century brought no fundamental contributions to the elucidation of contractility in living organisms. A deeper understanding of the chemistry of inorganic and organic matter was really pre-requisite for this, and great strides now began to be made in these directions.

THE CHEMICAL BACKGROUND[1]

The quantitative study of gases (which had begun with Robert Boyle in 1660) was continued vigorously during this next century, and interpreted in terms of the phlogiston theory of Stahl, enunciated in 1697. This theory explained combustion as due to the presence in combustible material of a principle of inflammability (sometimes credited with negative weight) termed phlogiston; material supporting combustion did so in virtue of its power to absorb phlogiston, and this principle was lost during combustion. The work of Black on fixed air (carbon dioxide) in 1755; of Cavendish between 1766 and 1784 on fixed air, inflammable air (hydrogen) phlogisticated air (nitrogen) and dephlogisticated air (oxygen); and of Priestley on dephlogisticated air may be specially mentioned. In 1774 Priestley prepared purified dephlogisticated air (to which Lavoisier a little later gave the name oxygen) by heating red oxide of mercury; he showed that this gas was better than common air for supporting combustion and life.

Lavoisier had also intensively studied calcination, and had shown that in this process tin for example gained in weight, while the air in which it was contained lost equally in weight and also diminished in volume. He therefore replaced the proposition

$$Calx = metal - phlogiston$$

by

$$Calx = metal + air.$$

He recognised Priestley's new gas as that part of the air active in his experiments, and continued his work to produce the oxygen theory which

[1] Cf. Partington (2) vol. III; Lowry (1); Lieben (1).

[27]

threw great light on all previous studies of combustion. He included oxygen, nitrogen and hydrogen in his list of some thirty elements in 1789.

The way was now open for the study of the gaseous exchange of the living organism, which Lavoisier himself inaugurated. In 1780 and 1784 he published two memoirs with Laplace,[1] in which the gas exchange of guinea-pigs kept over mercury was measured – the carbonic acid by increase in weight of alkali, the oxygen by diminution in volume of the bell-jar contents. Heat output was also measured by means of the melting of ice, and the conclusion was drawn that the change of pure air into fixed air was responsible for the greater part of animal heat.

Advance in the knowledge of chemical combination also depended on the greater attention paid to quantitative methods. The concepts of atoms, equivalents, molecules, chemical formulae, constant proportions in chemical combination, chemical affinity and the laws of mass action, are associated with the names (amongst many others) of Berthollet, Richter, Dalton, Berzelius, in the period about 1785 to 1813.[2]

THE FOUNDATIONS OF MUSCLE BIOCHEMISTRY

RELATION OF THE NUTRIENTS TO THE MATERIAL OF THE ANIMAL ORGANISM. The years between 1780 and 1840 saw the gradual growth of knowledge of the elementary composition of animal and plant substances, this growth depending on the gradual improvement of analytical methods. By 1786 Berthollet (1) had concluded that animal material contains nitrogen: he showed that on heating or on putrefaction a volatile alkali was given off and experimentation convinced him that this substance (ammonia) contained inflammable air and phlogisticated air. A little later the presence of nitrogen in plant material, but in smaller amount, was noted.[3] In 1789 de Fourcroy (2) described his separation of the animal substances fibrin, albumin and gelatin, finding characteristic proportions of nitrogen in each. The idea of 'animalisation' grew up – a process in the animal body consisting in the removal from a vegetable nutrient of part of its carbon and hydrogen elements.[4] Cuvier (1)[5] in 1805 wrote of his conception of the origin of muscle fibres in the following words:

The last of these filaments, or the most delicate fibres that we can see, do not appear to be hollow; one cannot see that they contain a cavity; and it seems that

---

[1] Lavoisier & Laplace (1).   [2] Leicester (2).   [3] E.g. de Fourcroy (1).
[4] E.g. de Fourcroy, *Philosophie chimique*, Paris, cited by Holmes (1) who gives a detailed account of the importance of elementary analysis in the origins of physiological chemistry.   [5] See pp. 90–2.

one can regard them as the most simple associations of the essential molecules of the fleshly substance. In fact, they are formed, one could even say crystallise out before one's eyes, when the blood coagulates; for when a muscle has been disembarrassed, by boiling and maceration, of the blood, the other humours, and in general of all the substances foreign to its fibre, that it could contain, it presents a tissue filamentous, white, insoluble, even in boiling water, and resembling, by all its chemical properties, the substance that remains in the clot of blood, after one has removed the coloured part by washing. This matter has, above all, by the abundance of nitrogen which enters into its composition, a character of animality perhaps more marked than other animal substances. The elements of the fibrous substance appear then so closely brought together in the blood, that a little repose suffices for their coagulation; and the muscles are without doubt, in the living state, the only organs capable of separating this matter from the mass of the blood, and of appropriating it to themselves.

In fact, the respiration removing above all hydrogen and carbon from the blood, it augments the proportion of nitrogen; and as one knows that it is the respiration which supports muscular irritability, it is natural to think that it does this by augmenting the quantity of the substance in which alone this irritability resides.

It was not until 1816 that Magendie (1) demonstrated that nitrogenous constituents in the diet are essential for life; before this time it was often suggested that nitrogen was absorbed by the animal from the air. The more accurate methods of elementary analysis from about 1814 onwards enabled Prout (1), in his work between 1815 and 1830, to distinguish three classes of nutrient – the three great staminal principles – saccharine, oleagenous and albuminous. Representatives of these were present in all animals and plants; thus it seemed that only small changes were needed to transform each type of food into the related compound in the animal tissues or fluids. Then in 1837 Mulder (1) began the work which showed the very close similarity in proportions of carbon, hydrogen, oxygen and nitrogen to be found in albumin, fibrin and casein of animal origin, as well as in the albumin of plants. He suggested the presence of a common 'protein' radical, and that changes from one albuminous substance to another only involved small changes in the amount of other elements (e.g. sulphur and phosphorus) attached to the protein.

In the early eighteen twenties Despretz (1) and Dulong (1), using better technique than Lavoisier's, made simultaneous measurements of the gas exchange and the heat production in order to attempt a quantitative comparison of the oxygen uptake and CO_2 production, the heat of combustion of the carbon and the heat produced by the animal. Despretz concluded that respiration is the principal cause of the heat production, but Dulong was prepared to consider some other source of heat to explain the discrepancy (amounting to 10 to 20%, after assuming a value for water production in respiration) which both observed.

Liebig (3) in his book[1] of 1842 expressed his certainty that heat production in the animal body was due solely to oxidation of the carbon and hydrogen of the food. He wrote: 'It signifies nothing what intermediate forms food may assume, what changes it may undergo in the body, the last change is uniformly the conversion of its carbon into carbonic acid, and of its hydrogen into water; the unassimilated nitrogen of the food, along with the unburned or unoxidised carbon, is expelled in the urine or in the solid excrements'.[2] His ideas concerning the origins of tissues, including muscle, are shown in the following statement: 'The second quality of the blood, namely, the property which it possesses of becoming part of an organised tissue, and its consequent adaptation to promote the formation and growth of organs, as well as to the reproduction or supply of waste in the tissues, is owing, chiefly, to the presence of dissolved fibrine and albumin'.[3] This conception was general in Liebig's day, and it was supported by chemical analyses from many sources which showed the great similarity in composition of the fibrin and serum albumin to that of a number of animal tissues.[4]

One may reflect how many centuries had had to pass before the ideas of Erasistratus on the origin of muscle from blood particles could take on a chemical form. Nonetheless, these later ideas were mistaken also, of course. Yet the time was now almost ripe for the discovery in the early twentieth century[5] of protein breakdown and synthesis in the animal body.

CATALYSIS AND BIOLOGICAL PROCESSES. Before turning to consider the intensive studies in the nineteenth century on muscle metabolism during rest and work, we may briefly mention a few of the names associated with the contemporary progress in knowledge of catalytic activity. Mitscherlich (1, 2) studied several examples of 'contact action' with metallic substances or with acids, e.g. the breakdown of hydrogen peroxide in the presence of manganese, or of alcohol to carbon dioxide and water in the presence of sulphuric acid. He drew attention to an analogy between yeast fermentation (which he recognised as due to living cells) and this contact action; and Berzelius (1) in 1836 made the important suggestion that in organic as well as inorganic nature catalytic substances are active. He instanced the ferment in malt converting starch to dextrin and sugar. These effects had been noticed much earlier. Thus Dubrunfaut (1) in 1830 reported that malt

---

[1] Holmes (2) in his Introduction to the 1964 reprint of this book, has given a clear and graphic account of Liebig's experimental work and particularly of his theorisation; the latter went far beyond experimental bases and was intended to raise questions suitable for experimental attack. In this it was successful, and its influence was wide, as is shown by the vigorous controversies of that time.

[2] *Animal Chemistry*, p. 21.   [3] Ibid. p. 172.

[4] Ibid. Notes 7 and 27. It is surprising that no mention is made of analyses of skeletal muscle, though the 'middle membrane of arteries' was used.

[5] Lieben (1).

extract worked as well as malt itself in bringing about saccharification of starch. Payen & Persoz (1) in 1833 gave the name diastase to the ferment which they prepared by alcoholic precipitation from malt extract. Berzelius (1) wrote 'Catalytic strength seems to consist essentially in the fact that bodies through their mere presence may awaken affinities slumbering at that temperature, so that as a result the latter arrange the elements in a neighbouring body in other relations, through which a great electrochemical neutralisation is caused...Here it suffices to have shown the existence of catalytic force by a sufficient number of examples. If we turn with this idea to the chemical processes in living nature, then we receive a quite new illumination here.' In 1836 also Schwann (1) studied an extract of gastric mucous membrane, showing its ability, in the presence of acid, to dissolve coagulated protein or muscle tissue, and to bring about some ill-defined changes in the properties of these materials. He established that acid alone did not suffice, and gave the name pepsin to the 'digestion principle', which he regarded as a specific substance, working in very small quantities through 'catalytic or contact action'.

Justus von Liebig had his own idea of catalytic activity, suggesting a mechanism whereby the catalyst transmitted atomic movements to the substrate. In 1839 writing on the cause of fermentation and putrefaction (which he regarded as purely chemical processes going on in organisms which had ceased to live) he said: 'This cause is the capability which a body undergoing breakdown or combination (that is to say, engaged in chemical activity) possesses, to call forth in another body mixed with it, the same activity, or to make it capable of undergoing the same change which it itself undergoes.'[1] But he argued that the same cause active in fermentation and putrefaction must have a most important role in life processes of the animal organism; as we shall see, he used these ideas in his explanation of muscle metabolism.

In 1836 to 1838 both Cagniard-Latour (1) and Schwann (2) clearly described the yeast cell and its budding, and expressed their conviction that fermentation was a process consisting of the action of the living cell on chemical substances.[2] This conception was of course strongly supported later by the brilliant work of Pasteur (1857) whose experiments with many different micro-organisms led him to regard 'all true fermentations as accompanied by the formation, development and increase of living cells'. The instance of simple hydrolytic activity in cell-free extracts Pasteur disregarded.[3] A similar attitude was adopted by many biologists – but not

[1] J. von Liebig (1).
[2] The first suggestion that the formation of alcohol was due to the living activity of yeast was apparently made by C. P. F. Erxleben in 'Ueber Guete und Staerke des Biers', Prague, 1818, p. 69. (Personal communication from Dr M. Teich.)
[3] Lieben (1) p. 236.

indeed by Schwann. In a paper of 1870 Liebig (2) admits that yeast consists of plant cells, but protests that in presenting us with 'an act of life' Pasteur puts in place of an explanation a fact which itself needs explanation. In 1877 Traube (1) described ferments according to a concept he had held since 1858, as '...chemical compounds similar to proteins which, although till now not prepared pure, doubtless like all other bodies possess definite chemical structure, and through expression of definite chemical affinities call forth alterations in other bodies'. He repudiated Liebig's idea of ferments as actively disintegrating bodies, able to communicate their motion to passive bodies; and considered insufficient Pasteur's hypothesis of fermentations as activities of vital forces in lower organisms.

In 1878 Kühne (5) introduced the name enzyme for soluble unorganised ferments. In 1897, after the death of both Liebig and Pasteur, Eduard Buchner (1) was able, in carefully controlled experiments, to show typical yeast fermentation using cell-free extract of yeast.

These enzymic studies bring up the pressing problem confronting nine-teenth-century biochemists – the choice between the vitalist and the mechanist approach to the understanding of the basis of life. The term 'vital force' seems to have been used first by F. C. Medicus in 1774;[1] more than twenty years later J. C. Reil (1) in the first volume of his *Archiv der Physiologie*, devoted a long paper to its consideration. He writes:

Vital force means the relation of several individual phenomena to a special kind of matter which we meet only in living nature amongst plants and animals. The most general characteristic of this peculiar form of matter is a characteristic kind of crystallisation. In other respects we can give no genetic definition of this force, so long as chemistry has not made known to us more exactly the elements of organic matter and their properties.

The addition of a foreign material to an animal body and the resulting transforma-tion of the added material constitutes a characteristic (animal) crystallisation of the animal material. The animal substance bursts forth into vessels, nerves, skin, muscle fibres and so on, just as cooking salt does into cubical crystals.

Teich (1) in an interesting lecture discussing this question, concludes that Reil firmly believed that the phenomena of life would in course of time be explicable in terms of chemistry. As he also points out, the attitude of Berzelius (2) was not so clear. The following quotation from the Lehrbuch der Chemie (1827)[2] illustrates this: 'This *something* which we call vital force lies quite outside inorganic elements and does not belong to one of their original properties such as gravity, impermeability, electrical polarity, etc.; but what it is, how it comes into being and how it passes away we do not comprehend.' Liebig also wrote often of the 'vital force' in a living animal appearing as a cause of growth, resistance to external agencies and as a cause of motion. It seems clear that he regarded this vital force as subject to laws and accessible to study in the same way as force in the form of mechanical or electrical phenomena.[1] He sums up his conclusions in the following words:

So is it with the vital force, and with the phenomena exhibited by living bodies. The cause of these phenomena is not chemical force; it is not electricity, nor magnetism; it is a force which has certain properties in common with all causes of motion and change in form and structure in material substances. It is a peculiar force, because it exhibits manifestations which are found in no other known force.

Schwann believed that the laws of physics and chemistry could be applied to the investigation of living processes[2]. He was deeply religious and in his *Mikroskopische Untersuchungen*[3] he wrote in 1839: 'The adaptation to a purpose which is characteristic of organised bodies differs only in degree from what is apparent also in the inorganic part of nature; and the explanation that organised bodies are developed, like all the phenomena of inorganic nature, by the operation of blind laws framed with the matter, cannot be rejected as impossible. Reason certainly requires some ground for such adaptation, but for her it is sufficient to assume that matter with the powers inherent in it owes its existence to a rational Being. Once established and preserved in their integrity, these powers may, in accordance with their immutable laws of blind necessity, very well produce combinations, which manifest, even in a high degree, individual adaptation to a purpose.'

EARLY WORK ON MUSCLE METABOLISM, INCLUDING THE
INOGEN CONCEPTION

In 1789, Séguin and Lavoisier (1) showed by experiments on man (Séguin himself) that the oxygen consumption is increased by work. It is clear from the following quotation from this memoir that Lavoisier subscribed to the view of that time that oxidation took place in the blood:

Now as vital air cannot be converted into carbonic acid except by addition of carbon: as it cannot be converted into water without the addition of hydrogen; as this double combination cannot operate unless the vital air loses a part of its specific heat, it results that the effect of respiration is to extract from the blood a portion of carbon and hydrogen, and to deposit in their place a portion of its specific heat, which during the circulation distributes itself with the blood in all parts of the animal economy and keeps this temperature nearly constant.

Liebig (3) in his textbook of 1842 relates the colour changes of blood to its content of an iron compound which combines with and releases oxygen. He goes on to say, 'In the animal organism two processes of oxidation are

[1] *Animal Chemistry*, pp. 195–232.
[2] See Florkin (1, 2); Teich (1).   [3] Schwann (3) p. 188.

going on; one in the lungs, the other in the capillaries. By means of the former...the constant temperature of the lungs is kept up; while the heat of the rest of the body is supplied by the latter.' He does, however, picture a small portion of the oxygen as used for repair of tissue which has produced mechanical force, i.e. of muscle. 'As long as the vital force of these parts is not conducted away and applied to other purposes, the oxygen of the arterial blood has not the slightest effect on the substance of the organised parts; and in all cases only so much oxygen is taken up as corresponds to the conducting power, and, consequently, to the mechanical effects produced.'[1]

It was only with the study of muscle deprived of its circulation that conclusions could be drawn about the part played by the muscle itself in oxidation and other aspects of metabolism. Such experiments were first done by Helmholtz (1) in 1847 measuring, by means of thermocouples, heat production with excised frog muscle; he found this to rise upon stimulation for 2 to 3 minutes, and was satisfied that contraction heat was really produced in the muscle and was not due to increased blood flow to the muscle. With regard also to the site of oxidation Georg v. Liebig (1) in 1850 used muscles from frogs which, while living, had been injected with distilled water to remove blood. He found carbon dioxide production and oxygen consumption to be similar in these muscles to that in muscles from frogs with normal circulation; he concluded that carbon dioxide was formed from part of the oxygen absorbed as a living activity of the muscle outside the body. Oertmann (1) later (1877) used living frogs in which the blood had been replaced by 0.75 % NaCl, and which lived for 1 to 2 days. Great care was taken to provide oxygen by saturating the salt solution and keeping the frogs in an oxygen atmosphere. Again the changes in carbon dioxide and oxygen were similar to those in normal frogs, and Oertmann was 'obliged to think that the overwhelmingly great part, if not the whole, of the substances normally undergoing oxidation are bound to the tissues'.

It is interesting that as early as 1861 Traube (2) was emphasising that oxygen passed in solution through the capillary wall and united with the muscle fibre in loose combination, which was in such a state that the oxygen was readily given up to other substances in the muscle fluid. He expressly says that this process is to be regarded as of the same type as that in the ferment processes he had earlier discussed He writes further: 'The forces of movement, however, which oxygen develops in muscles, nerves, spinal cord and brain owe it only to the specific construction and chemical make-up of the apparatus in which the oxidation processes go on, that they appear not in the form of heat but in their special form, as a hitherto unexplained vital activity.'

We come now to the experiments which had such great influence on the

[1] Ibid. p. 223.

thought of those concerned with muscle in the later nineteenth century – the experiments showing contraction in absence of oxygen. Georg v. Liebig had already, in the work quoted, shown that excised muscle can maintain irritability for some hours in an atmosphere of nitrogen or hydrogen. In 1856 Matteucci (1) tried to remove carbonic acid from freshly prepared muscle *in vacuo*, and examined the subsequent carbonic acid production in hydrogen. He found that even during rest a little carbonic acid was given off and this amount was increased on contraction He took this to demonstrate that the oxygen which gives rise to the carbonic acid was not the oxygen of the air, and found it necessary to assume that the oxygen was in the muscle in a combined state. Hermann (2) in 1867 described many experiments in which he showed that though the gas pumped off from resting muscle contained no oxygen, muscles kept *in vacuo* gave off carbonic acid and the amount was increased on tetanus. By subjecting control muscles and tetanised muscles to evacuation at 50° and treatment with phosphoric acid, he believed he had shown that carbonic acid was newly formed on contraction without oxygen. Hermann also believed that carbonic acid formation was stopped by heating the muscle to 70°; this represented, in his view, a 'fixation' of the complex living material – an idea which played a large part in his theories and in those of Pflüger.

In 1875 Pflüger (1) found that frogs could support several hours in oxygen-free nitrogen, apparently functioning normally and moving on stimulation. His collaborator Stintzing (1) a few years later demonstrated that rabbit muscle heated to 100° gives off 100 volumes of carbonic acid/ 100 g muscle. He found more carbonic acid given off upon treatment of the muscle by both acidification and heating than upon acidification alone. The extra evolution of carbonic acid on heating was believed to be due to the breakdown of the complex living material. This result of course did not tally with that of Hermann, but both were used as evidence for the same conception. We shall come later to the critical appraisal of such experiments.

Before describing in more detail what came to be known as the theory of the 'inogen molecule', we may conveniently consider the development of ideas on the association of protein breakdown with muscle contraction. Thus in 1842 Liebig (3) wrote:

The sum of the mechanical force produced in a given time is equal to the sum of force necessary, during the same time, to produce the voluntary and involuntary motions...The amount of azotised food necessary to restore the equilibrium between waste and supply is directly proportional to the amount of tissues metamorphosed. The amount of living matter, which in the body loses the condition of life, is, at equal temperatures, directly proportional to the mechanical effects produced in a given time. The amount of tissue metamorphosed in a given time may be measured by the quantity of nitrogen in the urine.[1]

[1] *Animal Chemistry*, pp. 244–5.

The connection of oxidation with the process visualised here is shown in the following passage:

The act of waste of matter. . .occurs in consequence of the absorption of oxygen into the living parts. This absorption of oxygen occurs only when the resistance which the vital force of living parts opposes to the chemical action of the oxygen is weaker than that chemical reaction; and this weaker resistance is determined by the abstraction of heat, or by the expenditure in mechanical motions of the available force of the living parts.[1]

This view, that protein breakdown was necessarily concerned in contraction, was generally held in the first half of the nineteenth century; after 1860 experimental results began to accumulate which were difficult to reconcile with it. Traube, in the paper already cited (2), expressed his belief that muscular work was not dependent on protein breakdown; proteins were necessary for provision of ferments, and for repair, which went on even at rest. The work of Fick & Wislicenus (1) in 1865 made a turning-point. Climbing the Faulhorn (1956 m), they performed each 150000 m kg of work on a nitrogen-free diet. But this external work was, they realised, only part of the work performed by the body; and allowing for the 'static work' involved and for waste heat, they assessed the total energy expenditure as nearer 300000 m kg. From the nitrogen excreted in the urine could be calculated the protein used and thence the energy available on its combustion (allowance being made for the urea excreted) – only some 70000 m kg.[2] They concluded that the protein utilised was far from adequate to supply the necessary energy.

Pettenkofer & Voit also (1), in experiments in the following year on man, found that muscular work was without appreciable effect on the nitrogen excretion. However, their deductions were different from those of Fick & Wislicenus. They wrote: 'We think that through oxygen uptake and steady breakdown of protein, tension forces are built up, which also during rest are gradually consumed, and which at will can be changed into mechanical work. During the latter, the oxygen is incited to combine with carbon-containing substances (fat) not belonging to the muscle, which then burn with production of the same heat as outside the body'. And they used the analogy of a millstream always running from which the miller can choose how much water he shall divert to pass over his wheel.

Liebig (4) also was entirely unconvinced, making the suggestion in 1870 (based on experiments of Parkes (1)) that the excess nitrogen excretion connected with work might be considerably delayed. But Voit (1) about this

---

[1] *Animal Chemistry*, p. 243.
[2] These calculations are given by Nasse (1) p. 329. With the data available to them at the time, Fick & Wislicenus calculated about 105000 m kg as possibly provided by the protein combustion. See Lieben (1) p. 137.

time had shown that on a high protein diet there was considerable oxidation of protein, and pointed out how rapidly the extra nitrogen of a heavy protein meal after fasting is excreted. He could find no evidence that extra nitrogen excretion occurred during rest days after work.[1]

In the conception of a large, complex, living molecule responsible for tissue activities, of which muscle contraction was the outstanding example, Liebig and Pflüger were both influenced by two main *idées fixes*: that oxygen must be built into the molecule since contemporary oxygen supply was unnecessary for contraction; and that protein breakdown must accompany contraction. In the suggestions advanced by Hermann, and much later (1893) by Fick (1), the complex oxygen-containing molecule is pictured, but its protein need not break down.

Hermann (2)[2] wrote in 1867:

We see clearly that not oxygen uptake, not true oxidation, is the basis of muscle work, but a process which has nothing to do with oxygen uptake.

Breakdown of a nitrogen-containing body is indeed the basis of muscle action, but of the breakdown products, amongst which are included carbonic acid, fixed acid and a jelly-like protein body (myosin), at least the last (perhaps also the fixed acid) is concerned in the rebuilding of the original body, and only with regard to the carbonic acid is it certain that it leaves the muscle and the organism.

He points out the similarity of his views to those of Traube. The myosin to which he refers is the substance first described by Kühne (1) in 1859, separating out in the form of a gel in muscle press-juice,[3] and named by him a few years later (2). The term 'inogen' was first used by Hermann, and his interpretation of it is found in his *Physiologie des Menschen*:

The simplest expression for the chemical processes during onset of rigor and the active state is therefore probably the following: The muscle contains at any moment a store of a complicated N-containing substance, dissolved in the muscle contents and plasma (which one can designate for sake of brevity the energy-generating or 'inogen' substance) which is capable of splitting with development of energy; the products of the splitting are, amongst others: CO_2, sarcolactic acid,

[1] It is interesting that later, better-controlled and more extended experiments have shown many instances where a small increase in N excretion accompanies and follows hard work (sometimes over several days). Thus Cathcart & Burnett (1) found an increase of about 15 % in the total N over 10 days after 4 days' work. There is no question that the greater part of the energy is supplied by combustion of non-nitrogenous material, and some of the small amount of extra N excreted may originate from non-protein constituents of the muscle such as purine bases and creatine. Cathcart (1) emphasises that since the muscle machine is composed of protein it is surprising that more signs of wear and tear are not demonstrable after great activity. He concludes that anabolic processes must play an important part in masking the catabolism, and points to the established fact that steady work can lead to hypertrophy of muscle.

[2] See pp. 67, 91, 92, 100. For further discussion see Hermann (3); and the chapter by Nasse (1) in Hermann's *Handbuch* (4).

[3] See ch. 7, p. 132 below.

perhaps glycerophosphate and a gelatinous protein body separating out and later contracting firmly.[1]

Liebig (4), in 1870, described his conception of energy provision for muscular movement in the following way:

All parts of the animal body arise from inner alterations of protein...in which oxygen has a causal part and one can assume that, if these products of the protein are sources of energy, the movement which they produce depends, not on their combustion and the transformation of heat into movement, but on the tensile force (*Spannkraft*) (which was piled up in them during their formation) becoming free on their breakdown.

Liebig was of the opinion that substances of much higher *Spannkraft* than myosin were formed from protein with absorption of heat on interaction with oxygen; and that when such compounds were split the absorbed energy was converted into its mechanical equivalent. The nitrogenous substances of high *Spannkraft* he pictured as resembling creatine. The heat production accompanying contraction might be due to oxidation of N-free substances.

Pflüger (1) in 1875 thought of the living molecule as containing cyan compounds formed from food protein; at death these returned to stable amide formation. He described cyanic acid, $HOCN$ (on account of its high heat of combustion) as 'a half-living molecule'. Summarising his hypothesis of energy production he wrote:[2]

The life process is the intramolecular heat of the most highly unstable protein molecule dissociating with formation of carbonic acid, water and compounds resembling amides; this protein molecule is formed in the cell substance which continually regenerates it, and it grows by polymerisation.

As the intramolecular swinging changes the attractions atoms come into relation with one another which otherwise did not work on one another, so one understands the sudden appearance of strong pulling forces as these atoms attract each other. If such attracting parts lie in an ordered series and if the attraction arises at the same moment in the whole series, so forces can be generated in this way significant for the muscle twitch.[3]

Pflüger considered that the living protein, in the neighbourhood of its carbohydrate radicals, did not differ from ordinary protein; oxidation of the living protein was chiefly concerned with these carbohydrate radicals, and its regeneration depended on metabolism of fat and carbohydrate.

Fick's 1893 version of a possible explanation of muscle activity shows the influence of T. W. Engelmann's (1) microscopic observations on muscles.

---

[1] Hermann (1) p. 231 of the 5th edition, 1874. (I have not seen earlier editions.)
[2] Pflüger (1) p. 343. Also Pflüger (2).
[3] Pflüger (1) p. 329.

In the muscle fibre are small separate discs of crystalline texture in orderly arrangement above one another. We will picture now on the under side of each disc a carbon atom, on the upper side an oxygen atom, in such position that the intrinsic chemical affinity between the oxygen atom of the one disc and the carbon atom of the one above it still cannot work...Now through the stimulus the molecule in the upper disc to which the carbon atom belongs, is so shifted that it is vertically over the oxygen atom of the next disc, and so comes so near to it that the strong chemical affinity...is active. As however the carbon and oxygen atoms are chemically bound to other atoms of their discs, and the discs are also mechanically rather stiff bodies, so the attractive force will not, as would be the case with free atoms, cause a violent oscillation round the common centre of gravity, in other words produce heat, but the carbon atom will draw the upper disc, the oxygen atom the lower disc, together, so that they approach each other, in so doing pressing out the fluid lying between them.

One must understand that the atoms hanging next to the C- and O-atoms detach themselves from the chemical structure of the discs, and so a molecule becomes free in the surrounding fluid. At the same time one must suppose that more C- and O-atoms come out from the inside of the discs to move to the places where those first considered have been. The discs will then obviously, under the influence of the elastic forces of the fibre itself and of constantly stretching external forces, again be drawn away from one another – the fibre is again lengthened.[1]

EARLY APPLICATION OF THE FIRST AND SECOND LAWS OF THERMODYNAMICS TO THE ORGANISM

The Law of the Conservation of Energy was enunciated by J. R. Mayer (1) in 1842, and Liebig in his book of the same date (3) assumes that in living organisms, as in nature in general, the various forms of power are inter-convertible without loss.[2] He considered (though it was by no means proven at that time) that there was no other source of heat in the animal body than the chemical action between the elements of the food and oxygen. He wrote: 'The contraction of muscles produces heat; but the force necessary for the contraction has manifested itself through the organs of motion in which it has been excited by chemical changes. The ultimate cause of the heat produced is therefore to be found in the chemical changes.'[3] He compared this dependence of energy provision in the animal on chemical break down with the quantitative dependence of electrical force produced upon the amount of acid reacting with zinc in a voltaic cell.[4] Mayer (2) wrote in rather different vein in 1845:

Whilst the fibres bend, and the muscle, without suffering an alteration in volume, shortens, work is produced to greater or smaller degree; at the same time in the

[1] Fick (1).
[2] It is uncertain whether during the preparation of this book, Liebig knew of the work of Mayer. See Partington (2) p. 314.
[3] *Animal Chemistry*, p. 33. [4] Ibid. pp. 215–18.

capillaries of the muscles an oxidation process takes place to which a heat production corresponds; of this heat with the action of the muscle a part becomes 'latent' or expended, and this consumption is proportional to the work performance. . . The muscle, to speak in familiar terminology, uses heat in *status nascens* in performing work.

He thus took the muscle as a heat engine.

In 1847 Helmholtz (2) published a thorough study of the conservation of energy in its various forms, ending with a brief consideration of this question in living organisms. He felt satisfied that the same law applied here, the indications being that the heat of combustion and transformations of the food material provided the same quantity of heat as that given off by the animal. Some twenty years later Rubner (1) with the advantages of improved calorimetric and chemical estimation methods, and with greater knowledge of the products excreted, was able to establish that the assumption of Liebig and of Helmholtz was indeed valid for the resting animal.[1] In 1880 Danilewsky (1) produced evidence concerning the active animal. He compared the heat production during contractions in which the load was removed from the muscle before relaxation with the heat production under the usual conditions in which the load fell back during relaxation, doing on the muscle an equivalent amount of work which was changed into heat. Calculation showed an average value of 535 μcal/g mm – not unsatisfactory agreement with Joule's value of 425 for the mechanical equivalent of heat.

During the latter part of the century the muscle was often regarded as a heat engine, and this point of view was put forward as late as 1895 by Engelmann (2) in his Croonian Lecture.[2] Attempting to make this theory conform with the Second Law (Clausius,[3] 1850) according to which the fraction of heat available as free energy in a reversible system is proportional to the temperature fall in the system, Engelmann did not hesitate to assume temperatures of about 140° at very localised sites in the muscle. Fick (1, 2, 3, 4) had been actively combating such views for twenty years before that time. He had demonstrated (2) that the efficiency of human and horse muscle is at least 20 %. In 1893 he showed (1) that the postulation of very hot loci could not solve the thermodynamic problem, besides being unacceptable physiologically. Without such postulation the muscle could not be more than 1 % efficient (2). By the turn of the century, the conception of the muscle as depending directly on chemical forces was generally accepted.

[1] See also the discussions of Fick (2, 3).
[2] Ch. 7, p. 141 below. [3] Clausius (1).

DISCOVERY OF LACTIC ACID AND GLYCOGEN IN MUSCLE

Hermann suggested, as we have seen, that the 'fixed acid' formed both on tetanus and rigor was lactic acid, a compound which had already had quite a long history in muscle. Acid formation was first observed in the muscles of hunted stags by Berzelius. Writing in 1847,[1] he refers to this work as published in Swedish in 1807 and describes the method whereby he demonstrated at that time that the acid was identical with that from milk. C. G. Lehmann (1)[2] in 1850 wrote in his textbook 'It is now forty-two years since Berzelius recognized the existence of free lactic acid in *muscular fluid*... Berzelius thought that he had convinced himself that the amount of free lactic acid in a muscle is proportional to the extent to which it has been previously exercised.' Wislicenus distinguished it from fermentation lactic acid by its optical rotation in 1884. Dr L. G. Wilson has pointed out to me that a very early observation of lactic acid in muscle, with discussion of its origin, was made by Claude Bernard (1) in 1855.[3] He found no sugar in calf foetuses, but if the muscles or lungs were left in water at room temperature, the liquid became very acid owing to lactic acid production. If the macerated tissue was kept at a low temperature, or if 30 % alcohol was added, then lactic acid formation ceased and sugar could be detected. He connected this sugar formation and breakdown with development, comparing the changes with those going on in a germinating seed. These observations may have played a part in his discovery of glycogen in muscle a few years later. Du Bois-Reymond in 1859 recorded that fresh resting muscle had a neutral or slightly alkaline reaction, which changed to an acid reaction on activity or death.[4] Ranke (1) in 1865 found by titration of extracts, that muscle removed from the circulation reaches in time a constant acid maximum; muscles tetanised in the living animal gave a lower maximum, from which he deduced that during activity the acid-forming substance was used. Injection of lactic acid into frogs (curarised to avoid the stimulating effect on nerve) brought on fatigue. With regard to the origin of this lactic acid, other views than those of Hermann may be found, and glycogen as its source was mooted by Nasse (2) in 1869. Glycogen in muscle was first described in 1859 by Claude Bernard (2), particularly in young and growing animals, but was later shown to be normally present in adult muscle. Nasse found that glycogen (estimated as copper-reducing substance after ferment treatment) was less in the muscle after rigor; he assumed that it would also be reduced by activity. When he compared carbohydrate loss with lactic acid content he found (making in the glycogen values corrections indicated by later work using acid hydrolysis instead of ferment action)[5] the former

---

[1] Berzelius (3) p. 586. [2] Vol. I, p. 98. [3] Bernard (1) p. 249.
[4] See Lieben (1) p. 200. [5] Nasse (3).

to be greater. He concluded that it was likely that the lactic acid was derived from the glycogen and pointed out (1) that such a part played by glycogen must modify Hermann's views. In his *Leçons sur le Diabète*, published in 1877, Claude Bernard[1] wrote that in the normal state 'sugar is never formed in the muscles; all our experiments could not be more decisive on this subject; the glycogen of muscle undergoes incessantly a lactic fermentation, and this in the living animal as in the corpse, is the only transformation of muscle glycogen.'

Gleiss (1) in 1887 was interested in the different behaviour of the muscles of toad and frog, and of the slow red muscles of some mammals as compared with fast white muscles. He sought an explanation in following lactic acid production on stimulation.[2]

The work of Gad (1) in 1893 followed a different train of thought. He had observed that the height of a muscle twitch showed a minimum at 19°, and suggested that two chemical processes were involved, differently affected by temperature. Very tentatively he proposed breakdown of sugar to lactic acid as the first reaction, oxidation of lactic acid as the second. The fibre was pictured as made up of transverse discs, kept separate in the resting muscle, but able by some change induced by the intermediate product (e.g. a change in surface tension) to draw together and thus cause contraction. Gad expressed doubt about the suitability of lactic acid for this role, since so little of the energy of carbohydrate breakdown seemed to be released at this stage. Relaxation would be due to the removal of the intermediate by oxidation.

We must turn now to the work of Fletcher which led directly to the great developments of the present century.

[1] Bernard (2) p. 428.   [2] Ch. 19.

3

THE RELATIONSHIP BETWEEN
MECHANICAL EVENTS, HEAT PRODUCTION
AND METABOLISM; STUDIES BETWEEN
1840 AND 1930

FROM THE LIQUIDATION OF INOGEN TO THE FIRST
BALANCING OF THE THERMOCHEMICAL BOOKS

In 1898 Fletcher, coming with an open mind to a subject in danger of being stifled with theorisation, published his first paper on survival respiration of excised muscle.[1] This work may be regarded as the real beginning of quantitative muscle biochemistry. The new, rapid and comparatively micromethod of carbon dioxide estimation that he used made it possible to study the gas evolution over far shorter periods than formerly, and so to avoid the complications of putrefactive changes.[2]

Fletcher first observed the behaviour of frog muscle kept in air or nitrogen. There was an initial fall in carbon dioxide output, followed by a small steady evolution during some hours; then an important acceleration accompanied by shortening – the onset of rigor. From the shape of the time curve, Fletcher explained the early fall as due to outward diffusion of carbon dioxide already present in the muscle; the plateau as due to evolution of the gas displaced from carbonates by slow production of acid. The survival carbon dioxide production of resting muscle in oxygen was some four times that in nitrogen.[3] Stimulation to contraction in nitrogen had little effect unless the muscle was pushed to fatigue; but in oxygen there was always increased output roughly proportional to the number and degree of contractions. Oxygen delayed loss of irritability and onset of *rigor mortis*, prevented fatigue and could bring about recovery. Thus one support of the inogen theory – that the events of contraction were independent of contemporary oxygen supply – was removed. In 1902 Fletcher (3) summed up the situation in the following words:

[1] Fletcher (1).
[2] This method was adapted from that of the botanist F. F. Blackman, to whom he expressed great indebtedness.
[3] Fletcher (2), Fletcher (3).

...the chemical processes of muscular activity, like the survival processes of resting muscle, do not reach their natural end in the production of CO_2 without an adequate oxygen supply. With or without oxygen available at the moment, the excised muscle gives rise slowly if at rest, rapidly during activity, to bodies believed to be precursors of CO_2, and known to be poisonous, whose action is marked by the onset of fatigue and hastening of *rigor mortis*. It is reasonable to suppose that the beneficial action of oxygen, in delaying both fatigue and *rigor mortis*, is a sign of the greater completeness with which in its presence the metabolic products are expressed in the form of liberated CO_2.

His experiments with Hopkins on frog muscle and his own work on rabbit muscle convinced him by 1913 that anaerobically the yield of lactic acid was high enough to account by simple displacement for the carbon dioxide formed.[1]

The other support for the inogen theory was found, as we have seen, in the effects of heat on carbon dioxide production. Hermann saw in his failure to find such production in muscle heated to 70° a proof for the fixation of the labile molecule postulated; Pflüger and Stintzing on the other hand, finding at higher temperatures more carbon dioxide evolved than could be displaced by acid treatment, considered this as evidence for the destruction by heat of the labile molecule. Fletcher pointed out that the techniques used in the early experiments of this kind were not really such as to allow definite conclusions to be drawn from them.

In his first paper, Fletcher had found that heating to 70° or 100° was associated with carbon dioxide evolution; the effect at any temperature over 40° included all the effects which might have been produced at lower temperatures. In a later paper[2] (containing much interesting discussion), the following values were found for carbon dioxide production/100 g of muscle:

Heat rigor at 40°	35 to 40 ml
Further heating to 100°	35 to 40 ml
Scalding (which goes on with less acid formation than in rigor)	60 to 70 ml

Excised muscle after survival in oxygen exhibited the same discharges as fresh muscle; after survival in nitrogen or after rigor brought on by chloroform, the discharge at 40° might be abolished, but the discharge at 100° was unaffected. They concluded that the carbon dioxide discharged at 40° was pre-existent in the muscle and expelled from combination by the acid formed. The carbon dioxide expelled at 100° was also pre-existent, but held by muscle colloids (perhaps in combination with amino-acid groups in proteins); its expulsion depended on dissociation by heat of such combinations, or on protein coagulation.

[1] Fletcher & Hopkins (1); Fletcher (4).
[2] Fletcher & Brown (1).

In 1907 the classical paper of Fletcher & Hopkins (1) appeared, on lactic acid in amphibian muscle. By improved estimation methods but fundamentally by care in treatment of the muscle during extraction, really reproducible values were for the first time obtained for the lactic acid content in different states of the muscle. As Fletcher & Hopkins said, '...it is notorious that, quite apart from the question of the oxidative removal of lactic acid – which has not previously we think been examined – there is hardly any important fact concerning the lactic acid formation in muscle which, advanced by one observer, has not been contradicted by some other.' Earlier workers had attempted to aid quantitative extraction by chopping the muscle, and to avoid rapid survival changes by previously treating it with boiling water or with alcohol. Fletcher & Hopkins showed how all these procedures called forth immediate acid production in the irritable tissue. As is well known, the method they found most suitable was to cool the muscle to 0°, then to crush it rapidly in ice-cold alcohol and grind with sand. They showed that uninjured frog muscle contained at most 0.02 % of lactic acid; on death rigor, however attained, there was a maximum of about 0.4 %; and on stimulation to complete fatigue, a maximum of about 0.2 %. Lactic acid accumulated slowly in excised muscle in nitrogen, but not demonstrably in oxygen. Moreover, if fatigued muscles were allowed to recover by being kept in oxygen, their lactic acid disappeared. In order to throw light on the metabolic processes involved, they carried out the experiment depicted in fig. 6. From this it is clear that repeated periods of stimulation and recovery did not affect the heat rigor maximum, which was as high after nine periods of stimulation as when elicited from control resting muscles. Fletcher & Hopkins were inclined to interpret this result as indicating that the store of unknown lactic acid precursor available for immediate use could be renewed from stable reserve material during recovery. Alternatively they suggested that the lactic acid was not oxidised but reconstituted. They pointed out that 'lactic acid is a substance still possessed of high potential energy, and it is not easy to see...how the energy of a substance, oxidised after it has appeared as a product of the *spaltung* processes associated with contraction, can contribute to the sources of contractile activity'.

About 1912 A. V. Hill began his important work on heat production in muscle. Heat output fell quickly after excision to a constant value in nitrogen and the presence of oxygen greatly increased it.[1] With contraction in oxygen the heat production was not only greater but much prolonged.[2] This was in accord with the observation of Verzar (1) (measuring the changes in oxygen saturation of blood leaving gastrocnemii of anaesthetised cats) that the oxygen consumption of active muscle took place largely after

[1] Hill (1).   [2] Hill (2).

the contraction was over.[1] Peters (1) found the heat production in rigor to be 1.7 cal/g muscle (425 cal/g of lactic acid), in fatigue to be 0.9 cal/g – values in the same ratio as the lactic acid production of rigor and fatigue. Hill (1) in 1912 calculated that if the lactic acid appearing were produced from the same weight of glucose, then from the difference in the heats of combustion only 170 cal would be expected per g of lactic acid. From these facts he suggested that a substance of more 'free energy' than glucose must

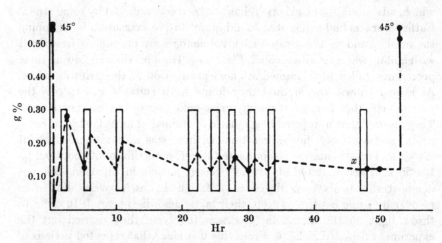

Fig. 6. The relation of the heat-rigor lactic acid 'maximum' to the survival history of muscle.

Four estimations of lactic acid due to heat rigor are shown, two at the beginning in the case of resting muscles, two at the 53rd hour, in the case of inexcitable muscles, which had gone through nine periods of severe stimulation alternated with periods of rest in an oxygen atmosphere. The enclosed areas represent time periods of stimulation by strong interrupted shocks. x = loss of excitability. Temp. 15 °. Continuous lines show course of acid loss as actually determined. Dotted line shows presumed course of acid loss and gain during other alternate periods. (Fletcher & Hopkins (1).)

be the precursor and that during oxidative recovery energy was used for building up a fresh supply of this substance. He did indeed find[2] that the heat of the recovery phase, though difficult to measure accurately, was similar to that of the contraction phase – only about 450 cal/g of lactic acid removed, while the heat of oxidation of 1 g of lactic acid is no less than 3700 cal.

Meanwhile Parnas & Wagner (1) had thrown some light on the lactic acid precursor by finding glycogen disappearance in fatigue and rigor equivalent

---

[1] D. K. Hill (1) many years later, using a differential oil volumeter, at 0° showed accurately that the O_2 intake of stimulated frog sartorius occurs entirely after activity.
[2] Hill (3).

to the lactic acid formed. They missed however the glycogen synthesis (later found by Meyerhof) which takes place during recovery. Parnas (1) also measured the oxygen uptake during recovery and believed (though here he was mistaken) that this was enough to account for all the lactic acid disappearing. He, like Hill, found the heat production to be much less than that expected.

Fletcher & Hopkins (2) in their Croonian Lecture of 1915 summed up and interpreted the position as it then appeared. Their previous idea of renewal of lactic acid precursor during recovery seemed now unnecessary since the work of Kondo (1) (showing inhibition by acid of lactic acid formation in muscle press-juice) suggested that its formation in fatigue and rigor was regulated by increase in hydrogen ion concentration rather than by lack of precursor. They saw clearly that the main reservoir of energy in the muscle was carbohydrate, and they were unwilling to postulate its conversion before use into a substance of higher chemical potential. Instead they suggested that:

in a system of colloidal fibrils, or of longitudinal surfaces, into relation with which H-ions of lactic acid lie ready to be brought, we have a potential of energy which may be discharged as work, with or without heat, on the development of a new state of tension in the fibrils, whether tension due to imbibition or to added surface tension along the longitudinal surfaces. The observed heat production of anaerobic contraction may be in part due to the exothermic molecular change which yields the free acid from its precursor, and in part due to the resultant change in colloidal surfaces or substances upon the delivery to them of the acid ions. Upon recovery by oxidative removal of the lactic acid, the energy of combustion is discharged in part as heat and in part returned to the muscle in the restoration of the initial potential. In this restoration will be involved the separation of the acid ions from the colloidal fibrils, by which the condition will be given for the return of the fibrils to their former tension. The contractile act may call, not only upon the chemical energy liberated when sugar becomes lactic acid, but also, and perhaps to a greater degree, upon the energy derived from the oxidation of the lactic acid, residing in the physico-chemical system of the muscle, which was produced during the previous contraction.

It had seemed before this time that the use of carbohydrate in muscle contraction might be a very wasteful process. It is true that considerably more energy appeared in the contraction phase than could come from the breakdown of glycogen to lactic acid (170 cal/g of lactic acid); but even the amount evolved (450 cal/g of lactic acid) represented only about 12 % of the total energy arising from oxidation of the equivalent amount of glycogen. Although it was established that not all the heat of combustion of the lactic acid disappearing during oxidative recovery was given out, it was not clear exactly how much was stored or in what form. The epoch-making researches of Meyerhof, beginning during the First World War, quickly changed the picture in this respect, as we shall see.

Meyerhof (1) first showed that during oxidative recovery from fatigue the respiratory quotient of the frog gastrocnemius was 1.02. From this he concluded that the oxygen taken up must be used for oxidation of the lactic acid formed, or of carbohydrate, and could not be stored in oxygen-rich compounds; thus these results supported those of Fletcher & Brown against the revival of the inogen theory in any form. He went on to find that after the piling up of lactic acid during rest or contraction with exclusion of oxygen, the increased oxygen consumption on recovery was equivalent to only one-third or one-quarter of the lactic acid disappearing.[1] He confirmed Parnas & Wagner in finding glycogen loss during fatigue equivalent to the lactic acid formed; but during recovery he found that the glycogen content increased by the difference between the total lactic acid disappearing and the amount calculated as burnt by the oxygen used.[2] Moreover, he found that the heat given out in the recovery period was clearly less than corresponded to the heat of combustion of the lactic acid oxidised: it was in fact only about half of that to be expected, and the absorbed heat tallied with the heat given out in the anerobic contraction period.

Using round figures for the calories produced or removed per g of lactic acid, Hopkins (2) described the new situation thus in his Herter Lectures of 1921:

Heat of anaerobic contraction	Heat of oxidative recovery	Total heat of activity in oxygen	Heat of combustion of material burnt
400	500	900	900

This is a balance sheet which any auditor would pass as satisfactory.

Meyerhof[3] studied many aspects of this balance sheet. In anaerobic rest, the glycogen loss was equal to the lactic acid formed; in aerobic rest to the oxygen uptake, no lactic acid accumulating. He investigated the relation between tension or work and lactic acid formation under a variety of anaerobic conditions. In given circumstances, provided fatigue did not go too far, there was good parallelism.[4] During recovery in oxygen the lactic acid content diminished; the increase in oxygen uptake above the resting metabolism corresponded to only one-third to one-quarter of that to be

---

[1] Meyerhof (2).

[2] Meyerhof (3). It is perhaps interesting to recall that in 1921 Foster & Moyle (1), at the request of Professor Hopkins, re-investigated a few of these problems, in particular some cases where the results of Meyerhof and Parnas were at variance. In this work, instead of the new micro method for lactic acid used by Meyerhof for which one frog gastrocnemius sufficed, they used the macro method of which Hopkins had so much experience. This depended on the final weighing of isolated zinc lactate, and necessitated the use of some five to twenty limb pairs for each estimation. The findings of Meyerhof were confirmed, much to Professor Hopkins' satisfaction.

[3] The methods he used may be found described together in the last chapter of his book (Meyerhof (4)).

[4] Meyerhof (5, 6).

expected for complete oxidation of the lactic acid disappearing. He used the term oxidation quotient,

$$\frac{\text{Moles of lactic acid disappearing}}{\text{Moles of lactic acid oxidised}},$$

to express the relation between lactic acid disappearing and oxygen uptake. Its value was usually 3–4. In this recovery period resynthesis of glycogen went on, in earlier experiments[1] equivalent to about one-half of the lactic acid disappearing. Meyerhof & Lohmann (13) later, in experiments in which lactate was added to the circumambient Ringer solution, found much higher syntheses – four to seven times the lactic acid equivalent of the oxygen uptake. When samples of chopped muscle were compared, one fully oxygenated and the other oxygen-free, much more lactic acid accumulated in the latter, indeed nearly four times as much as would correspond to the oxygen uptake of the former. This was taken as an indication of carbohydrate synthesis from lactic acid in chopped muscle, although analysis of the oxygenated sample showed only small diminution in carbohydrate loss. Meyerhof explained this by resynthesis only to some intermediate stage.[2]

It is only with the knowledge (to be gained during the next two decades) of the role of adenosinetriphosphate in glycogen metabolism and of the phenomena of oxidative phosphorylation,[3] that explanation of the mechanism of these results on glycogen resynthesis can be attempted. We shall return to this in chapter 18.

Hill and Peters in their early work had taken the value 450 for the calorific quotient of lactic acid (the calories produced in muscle per g of lactic acid formed anaerobically).[4] Using later methods, both for lactic acid estimation and for measurement of heat production, Meyerhof found consistently a lower value – e.g. 370 cal on electrical stimulation at temperatures between 14° and 22°.[5] Later work also revised the early figure for the oxidation quotient to about 4.7.[6] Hill (4) in 1928, using series of twitches, compared total heat in oxygen with total heat in nitrogen, and found the ratio to be 2.07. He took the calorific quotient of lactic acid as 385, from which the total oxidative heat would be 797 cal. The oxidation of 0.9 g of glycogen (the amount yielding 1 g of lactic acid) he took as producing 3815 cal, and the ratio of these two values, 4.8, is the oxidation quotient – a result in good agreement with Meyerhof's.

Puzzled as the earlier workers had been by the great discrepancy between the calorific quotient and the heat to be expected from the breakdown of glycogen to lactic acid, Meyerhof re-investigated the heat of combustion of glycogen[7] and also made the first direct determination of the heat of com-

[1] Meyerhof (3).   [2] Meyerhof (7).   [3] Chs. 6 and 17.   [4] Hill (3).
[5] Meyerhof (8). Also Meyerhof, McCullagh & Schulz (1).
[6] Meyerhof & Schulz (1).   [7] Meyerhof & Meier (1).

bustion of lactic acid[1] – the value till then available had been obtained by calculation from measurements on the ethyl ester. He found 3790 cal for 0.9 g of highly purified glycogen, and 3602 for 1 g of lactic acid, using the crystalline zinc salt. The difference thus is 188 cal and 182 cal per g of lactic acid, almost half the heat production, remained to be accounted for.

Meyerhof (8, 9) in 1924 suggested that the main part of this unexplained heat might be due to reaction of the lactic acid with the muscle proteins. It was known (by calculation from the van't Hoff equation) that, on the alkaline side of its isoelectric point, an amino acid such as glycine had a very high negative heat of dissociation, – 13000 cal per mole. The values for the isoelectric points of myosin and myogen, and for the pH of fatigued muscle were investigated about this time. H. Weber (1) for the isoelectric points gave 6.3 for myogen and 5.1 for myosin, though these values were uncertain owing to the important effects of traces of salt. Meyerhof & Lohmann (1) found for frog muscle containing 0.28 % of lactic acid a pH value of 6.6, and Furasawa & Kerridge (1) for cat muscle containing 0.3 % of lactic acid a value of 6.26. Thus it seemed that even in advanced fatigue the muscle proteins would remain on the alkaline side of their isoelectric point. Meyerhof now studied the heat produced when lactic acid was added to various amino-acid and protein solutions, including solutions of muscle proteins. After the necessary corrections, the heat of de-ionisation was found, +12000 cal per equivalent or 140 cal per g of lactic acid. Neutralisation of the lactic acid by bicarbonate or phosphate led to much less heat production. Hill (5) found 27 cal per g of lactic acid for the former, and Meyerhof 19 cal per g for the latter, using disodium phosphate.

In order to examine this question in the muscle itself, Meyerhof studied the heat production when frog muscle was placed in a solution containing valerianic acid. This acid was chosen because it has a dissociation constant similar to that of lactic acid, passes readily into the muscle and provokes little if any lactic acid formation in the muscle, indeed quickly inhibits it. The amount entering the muscle was found from the fall in concentration in the external solution. Since valerianic acid is not very soluble, a suspension had to be used; when corrections had been made for heats of solution, dilution and dissociation, as well as for the heat of production of any lactic acid formed, a figure was obtained of 9900 cal per mole of valerianic acid entering the muscle – or per mole of lactic acid, since equivalent amounts of the two acids would give about the same H ion concentration. This corresponds to 110 cal per g of lactic acid. Meyerhof considered that this low value as compared with the value for isolated protein (140 cal) was due to neutralisation of nearly one-third of the acid by bicarbonate and phosphate. In the experiments of Meyerhof & Lohmann (1) alcohol-extract-

[1] Meyerhof (8).

able ash was estimated in resting and fatigued muscle, and conclusions were drawn about the part played by phosphate in lactic acid neutralisation. However, these results were difficult to interpret, as Meyerhof (4) later pointed out, since no account could be taken at that time of phosphocreatine breakdown.

The figure of 110 cal per g of acid was confirmed by Stella (1) using carbon dioxide at one atmosphere pressure as the acid to penetrate the muscle. The thin sartorius muscle was used and diffusion was complete in a few minutes, instead of in the six hours necessary in the experiments with valerianic acid and 30 g of muscle.[1]

Of the heat set free on production in the muscle of 1 g of lactic acid (370 cal), there thus remains about 70 cal unaccounted for, since the chemical reaction glycogen – lactic acid yields 188 cal and neutralisation of the acid 110 cal. (see p. 94).

Like Fletcher and Hopkins, Meyerhof regarded lactic acid as part of the machinery of the muscle as well as part of the fuel – as Hopkins[2] said 'at one stage...concerned with the liberation rather than with the supply of energy'. Meyerhof's conception[3] had more precision than theirs, since he distinguished clearly between the postulated events of relaxation and recovery. Contraction was called forth by the release, upon glycogen breakdown, of a certain concentration of lactic acid on the *Verkürzungsorten* ('shortening places') of the fibrils; the energy provided came partly from the physical change resulting and partly from the chemical reaction. Relaxation entailed the removal of the lactic acid from the 'shortening places' to the *Ermüdungsorten* ('fatigue places'), by a process providing at least as much free energy as was developed by the physical process at the 'shortening places', and thus the fibrils would be restored to their original condition. This relaxation process he explained as involving neutralisation by and adsorption on the sarcoplasmic proteins. During oxidative recovery, energy was retained in the muscle to make possible removal of the lactic acid from the 'fatigue places', and resynthesis of the major part of it to glycogen.

MECHANICAL EVENTS AND HEAT PRODUCTION; THE PROBLEM
OF THE NATURE OF THE ACTIVE MUSCLE

Improvements in the methods for estimation of lactic acid, carbohydrates and oxygen were continually being made, but no chemical method could compare with the delicacy and rapidity of the measurement of heat produc-

---

[1] Hill & Woledge (1) in 1962 showed that Stella's value probably needed correction for an erroneous calibration number, so that it should be given as about 85 cal per g of lactic acid.　　　　　　　　　　　　[2] Hopkins (2) p. 19.
[3] Meyerhof (5) pp. 271 ff. For his development of this theory see p. 143, ch. 7.

tion. By 1920 Hill & Hartree (1), with improved thermopile and galvano-meter, were able to record photographically temperature changes in the muscle of less than 10^{-6} °. The response was relatively rapid and control curves could be made by electrical heating of dead muscle. By a numerical method of analysing the photographic records, the time relations of the heat production could be investigated, even in the rapid processes immediately after excitation. At this time, Hill and Hartree were already concerned with the distribution in time of the heat production resulting

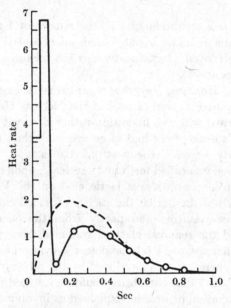

Fig. 7. Single isometric twitch of frog sartorius at 0 °. Full line: results of heat analyses; broken line, tension curve. (After Hartree (1).)

from a single twitch or short tetanus in oxygen or anaerobically. By 1928 Hill (4) could give accurately the relation of the prolonged recovery heat in oxygen to the initial heat; and by 1933 Hartree (1) reported a quantitative analysis of the initial heat in a single twitch (fig. 7). Such measurements of heat production, whilst telling us nothing about the actual chemical reactions going on, are most valuable to the biochemist, in indicating at what stage such reactions are to be expected. If some idea of the nature of these reactions is already entertained, then light can be thrown on their extent and possibly on the interrelationships of different suspected reactions.

It should be noticed that in the experiments to be discussed (unless it is expressly stated otherwise), the heat measured represents the whole of the energy set free. With an isometric contraction, potential energy (measured as tension × muscle length) is converted entirely into heat on relaxation. With an isotonic contraction, when the muscle relaxes still carrying the load, work is done on the muscle; this work is equivalent to that of lifting the load, and heat is produced about equivalent to the mechanical energy disappearing.

It had been shown as early as 1914 by Weizäcker (1), using single twitches of the frog semi-membranosus, that the magnitude of the initial heat was independent of oxygen supply; he concluded that it was due entirely to non-oxidative processes. Hill & Hartree (1) in 1920 confirmed this in showing that the time relations of the initial heat production at 0° were also the same during the first few seconds, whether the conditions were aerobic or anaerobic.[1] With the frog sartorius they were able to break up the heat production caused by a 2 second tetanus into four phases: (i) the initial rapid production, diminishing gradually; (ii) a small heat production maintained as long as stimulation lasted; (iii) a rather sudden evolution during the later stages of relaxation; (iv) in oxygen a large but slow production for some minutes after contraction was over. The same picture, apart from the maintenance of heat (ii), was found after a single twitch.

It is necessary now briefly to turn to the influence of studies investigating the purely mechanical aspects of contraction and the relationships between these and heat production. First we may mention here a pioneer experiment in physiology, made in 1835. Schwann found with fresh frog gastrocnemii that the force produced on stimulation was greatest when the muscle was at normal body length, and diminished at lengths less than this. This experiment caused a great sensation amongst physiologists of the time and was long known in Germany as the *fundamentale Versuch* of Schwann. Müller[2] pointed out that the muscle could be considered as obeying the law governing elastic bodies, and du Bois-Reymond is quoted[3] as saying that this was the first time that a vital force had been examined in the same way as a physical force and that the laws of its action were mathematically expressed. Müller emphasised that this result ruled out any theory of muscular contraction depending on attraction of any known kind between particles, since such an attraction would increase in strength as the particles approached each other.

E. Weber (1) was concerned in the eighteen forties with experiments on the elasticity and extensibility of resting and stimulated muscle. These led him to formulate the conception of the excited muscle as a new elastic

---

[1] See also the results of D. K. Hill (2, 3), 20 years later, using more exact procedure.
[2] This work was not published independently by Schwann, but is described by Müller (1) in his *Handbuch*; vol. ii, pp. 59–62. For a different interpretation see ch. 11
[3] Florkin (1) p. 40.

body – that is with elastic properties different from those of the resting muscle – and new natural length. He wrote:[1]

Every vital influence in the muscle fibre should first be considered as an alteration in its elasticity, just as every effect of heat on a steel spring is interpreted as an alteration of the elasticity of the spring; to this alteration, as soon as it has been determined, all disturbances of the equilibrium or movements can be attributed without further consideration of its origin. The basis of all observed phenomena in a muscle fibre would then lie entirely in its elasticity; this elasticity of the muscle fibre would not however be considered to be unalterable or merely alterable, as with inorganic bodies, by physical influences, such as temperature; but also to be alterable under the influence of the animal life-force.

This comparison, enunciated in very general terms, of the active muscle to a stretched spring had long-lasting influence, as we shall see, which can be traced in many of the models of muscle contraction entertained during the next hundred years.

Questions concerning the quantitative relations of the energy mobilised in the two types of contraction, isometric (in which the stimulated muscle is held at the initial length) and isotonic (in which it is allowed to shorten), and the effects of work performance on muscle metabolism were greatly debated in the latter half of the nineteenth century. Heidenhain (1) found in 1864 that during contraction of a loaded muscle the total heat production (including that evolved when the load fell back during relaxation doing work on the muscle) increased with increase in load. He wrote:[2]

If the muscle stimulated to maximum twitches is loaded with increasing weights, the work performed by the muscle as well as the heat developed by it increases up to a certain limit of the load...As work performance and heat are the two forms in which the living forces of the active muscle make their appearance, one can give this law the following general form: the total sum of the *Spannkraften* which ...are converted into living forces is not constant but variable with the load on the muscle; it increases up to a certain limit with increasing load.

He did however find maximum heat production usually in isometric contraction, and this observation led him to conclude[3] that heat production cannot be due to friction between moving muscle particles. To test whether the size of metabolism rose and fell like the sum of the living forces, Heidenhain[4] crushed muscle in boiled tissue extract containing litmus – according to du Bois-Reymond, with whom Heidenhain had worked, muscles kept in water spontaneously generated lactic acid, but not if they were kept in *Kochsaft*. The degree of acidity causing alterations in colour was measured by titrating the litmus solution with oxalic acid. He compared muscles

[1] E. Weber (1) pp. 104–5.
[2] Quoted in O. Frank (1) p. 443. I have not been able to see Heidenhain's book.
[3] See Meissner (1) p. 429.   [4] See Meissner (1) p. 433.

stimulated in the same way but differently loaded, and found more strongly acid reaction with greater load up to a point – beyond this both living force and acidity fell off.

Fick (7), in experiments performed in 1871, had been inclined to support the concept of Weber and to discount the objections raised in Heidenhain's work against the over-riding importance of elasticity changes on contraction. But in his book (2) some ten years later he expressed a different attitude.[1] He described Heidenhain's results as astonishing and most important, 'throwing a quite new light on the inner nature of the muscle fibre.' Fick (5) asked himself the question whether upon isometric stimulation a full store of potential energy was produced ready for mechanical performance under the right conditions. To test this point he held the muscle after stimulation till maximum tension was developed and then released it to shorten under load. More heat was developed in the isotonic than in the corresponding isometric contractions. He wrote that he believed one could say 'with a simple twitch, isometric in the beginning, there is not in this first stage provision of a store of mechanical potential energy, convertible in the second stage into mechanical performance without further chemical processes.' Further evidence for this point of view was found in experiments done in 1892 (6) in which stimulated muscles were stretched. Less heat was evolved when work was done on the muscle than when work was done by the muscle. He wrote: 'This appears to me to show that it is just the act of the performance of work, i.e., the shortening under tension, which demands in the muscle the chief consumption of chemical energy.'

A different point of view was supported in 1902 by Blix (1) who always found the greater heat production with isometric contraction. Thus he concluded that the length of the muscle is of prime importance and wrote 'Everything which hinders the chemical process called forth by the stimulus, diminishes the heat formation in the muscle following on the stimulus. As causes of hindrance may be named: low temperature, lack of food store, restriction of the surfaces which with work become chemically active.'

In 1913 A. V. Hill, using his greatly superior techniques, began to re-investigate this question. He remarked how variable were the results on which Fick based his claims and showed how gravely open to criticism was the work of the earlier experimenters; e.g. in their apparatus the muscle after shortening could never have come back to the original position on the thermopile. He overcame many difficulties which they had not recognised. Working mainly with semi-membranosus and gastrocnemius muscles of the frog, he found less heat given off when the muscle shortened before or during tension development; but the heat output was unaffected by the shortening

[1] P. 179.

when this took place only after full tension development. He interpreted these results as meaning that 'under certain conditions the initial process of contraction consists largely if not entirely of the liberation of free potential energy manifested as tension energy in the excited muscle; and that this potential energy can be used indifferently for the accomplishment of work or the production of heat'.[1] The stimulated muscle was here again regarded as a new elastic body and this idea predominated for the next ten years. The results of Hill & Hartree (1) in 1920 seemed to fit in with this picture. They measured the maximum work (by means of an inertia system) and the total heat production from the end of the stimulus. These two values increased in a similar manner as the duration of the stimulus increased, both becoming constant at a certain limiting duration; they concluded that both were derived from the same source, the potential energy liberated on excitation. Hill regarded the prospective mechanical or potential energy as stored in an unstable chemical compound, which after activity could be rebuilt in the presence of oxygen.[2] Hill and his collaborators[3] about 1922 introduced the idea of viscosity or visco-elasticity as playing a dominant part in regulating the speed of contraction, so that the muscle was sometimes likened to a stretched spring operating in a viscous medium.

Then Fenn (1, 2) working in Hill's laboratory in 1923, obtained results in surprising agreement with the ideas of Fick. Using frog sartorius muscles and an inertia lever allowing work to be measured separately, he found that when the muscle lifted a weight through increasing heights, or increasing weights through the same height, the increase in heat production was roughly proportional to the work done. Whenever a muscle shortening on stimulation did work in lifting a load, an extra amount of energy was mobilised which was not found on isometric contraction (see fig. 8).

These results led Fenn (2) to a different analogy for performance of work by the muscle – that of the raising of a weight by means of a chain and windlass, in which process energy is mobilised as needed. He wrote: 'Energy liberated by contraction of a single muscle fibre for a given stimulation is not dependent solely upon the initial mechanical and physiological conditions of the muscle, but can be modified by the nature of the load which the muscle discovers it must lift.' (1). Fenn got similar results to those of Fick on stretching the muscle during the contraction periods – a decrease in the energy liberated. In an experiment on the gastrocnemius his results were in agreement with those of Hill – increased heat production when the muscle was prevented from shortening for lengthening periods of time after stimulation. He suggested that the apparent differences in behaviour might be connected with the different anatomical arrangement of their fibres. This suggestion was borne out by the observations of Martin (1) in 1928, com-.

---

[1] Hill (7); Hill (8).   [2] Hill (7); cf. p. 46 above.   [3] E.g. Hill (34).

paring gastrocnemius, semi-membranosus and tibialis anticus with the sartorius. The last is the only one made up of parallel straight fibres; in the three others it seems likely that the relatively great amount of the 'isometric' heat compared with the isotonic heat might be due in fact to the shortening of individual fibres, although the muscles were held firmly at both ends.

Hartree and Hill in the later nineteen twenties devoted much effort to gaining further light on the Fenn effect.[1] Hill (10) studied the tension/heat

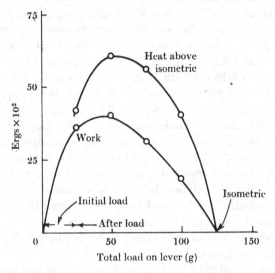

Fig. 8. Variations of work and heat in units of 100 ergs in isotonic contractions against increasing loads. The heat curve represents only the heat in excess of the isometric heat, which was 193×10^2 ergs. (Fenn (1).)

ratio in the sartorius muscle in isometric tetanic contractions. The variation in this ratio with length, he concluded, must depend on the nature of the mechanism for converting chemical into mechanical energy. He suggested an explanation for the Fenn effect. The active muscle, as an elastic body, contains a store of free energy; in relaxation this is dissipated as heat if not previously used to do work. But a third possibility could exist – that in a muscle held in a tetanic contraction in a state of active tension some of this mechanical potential energy might be used in beginning to carry out the recovery process without appearing as work or heat. He wrote: 'The pre-

[1] Hill (33) has referred to his realisation in his Nobel Lecture of 1923 (just before the publication of Fenn's first paper) that the concept of the stimulated muscle as simply a new elastic body must be wrong, since more energy was needed to do more work. Nevertheless, he says, he continued for some time to think in terms of the visco-elastic theory.

sence of such a mechanism would explain why extra energy is liberated, in shortening under load, in proportion to the work done. The energy used to do external work is not available to assist recovery during relaxation, and the total breakdown is greater.'

Some further work in Hill's laboratory seemed to support the conception of the muscle mechanism as an elastic system operating in a viscous medium. Gasser & Hill (1) studied the sartorius in maximum tetanic contraction opposed by the reaction of an inertia lever, the speed of shortening being varied by changing the equivalent mass on the lever. They found that the work done decreased as the rate of shortening rose, although in fact the relation was not linear as might have been expected from a purely viscous effect. They imagined the active muscle as an elastic network containing viscous fluid. Then in 1927 Levin & Wyman (1), impressed by the fact that tension never rises or falls instantaneously with stretch or release, and by the non-linearity of the relation between speed and shortening, suggested the presence of an element with free elasticity undamped by viscosity, in series with the main damped elastic component. But Fenn & Marsh (1) in 1935 used after-loading conditions such that the undamped elastic component had already completed its changes in length and tension. In this way effects due to the damped contractile component alone could be studied, and it was found that the relation between speed and force was still not linear but exponential. They believed that this exponential effect was in some way concerned with the development of extra energy for work, and that the loss of tension on increased speed of shortening was due chiefly to delay in chemical reactions and not to friction.[1]

In 1929 Hartree & Hill (2) with tetanic stimulation of the frog sartorius found like Fenn more energy liberated when work was done, the excess energy being equal to or greater than the work. They were not able however to find at this time the Fenn effect with twitches. In 1930 Hill (9) made the further important demonstration that, whether for twitch or tetanus, the heat production depends both on the amount of work done and on the length of the fibres before and during contraction. The nature of the observed effect therefore must be influenced by the size of the load, since this affects not only the work done, but also the degree of shortening and therefore the length of the fibres. In tetanic contraction of the sartorius the range of loads in which isometric heat is greater than isotonic may be so small as to be difficult to detect. With large loads Hill was able to show with single twitches also that the isotonic heat might be 85 % greater than the isometric. He was now prepared to say that it was untrue that in a twitch the energy set free depended only on the conditions at the moment of stimulation; release at a much later stage might affect the energy provision

[1] Ch. 9, p. 175.

and the isotonic heat could vary almost from one-half to double the iso-metric. The conflicting results of earlier workers can thus be put down partly to the different anatomical arrangement in the different muscles used, and partly to variation in the range of the initial loads.

It is interesting to find that Hill's results were in good agreement with chemical observations made just at the same time. E. Fischer (1), using a differential volumeter, showed for sartorius muscle that isotonic twitches with large loads had a greater oxygen consumption than isometric twitches; while isotonic twitches with small loads needed less oxygen than did isometric twitches. The results of Meyerhof (5) and of Rothschild (1) on lactic acid formation in isotonic and isometric conditions can also be interpreted on the same lines.

As A. V. Hill has said[1] it was not possible to go farther with the analysis of these complicated relationships until a method was available for following accurately the whole time course of heat production during actual shortening. This waited upon the invention about 1938 of the 'protected thermopile' and the advances then made possible will be described later.[2]

THE DISCOVERY OF THE DELAYED ANAEROBIC HEAT AND ITS NEGATIVE PHASE

Hartree & Hill (1) in early work had noticed a considerable delayed anaerobic heat production after relaxation, occupying about ten minutes; this was much smaller than the delayed heat in oxygen, but might be equal to some 50 % of the initial heat.[3] During the next ten years there were many efforts to elucidate this, and as methods always improved (better thermopiles, avoidance of over-stimulation, etc.) the delayed anaerobic heat became progressively less and less, but was still substantiated. Furusawa & Hartree (1) in 1926 estimated it to equal about 12 % of the initial heat; Blaschko (1) had a little earlier given a figure of about 10 % of the initial heat for the delayed anaerobic heat after a tetanus, and it was found in rather larger amount by Cattell & Hartree (1) after a series of twitches. Its interpretation, in terms of simultaneous chemical reactions must be deferred until after discussion of carbohydrate and phosphocreatine metabolism during anaerobic recovery.[4]

In 1932 Hartree (2) also observed at 0° in frog muscle after a short anaerobic tetanus a *negative* delayed heat production up to about 8 % of the initial heat. After 1 second stimulus this was complete in about 30 seconds; it was not found after long stimulation or at higher temperatures and then was probably masked by the greater and more rapid positive recovery heat. Some attempted explanations will be described later.[4]

[1] (11) Pp. 64 and 149.   [2] Ch. 9.   [3] See also Hartree & Hill (3).   [4] Ch. 5.

With regard to the mechanical efficiency of muscle, early measurements of Hartree & Hill (3 a) in which work done and total energy expenditure were measured for the anaerobic process, gave the low value of 26 % for this initial process, or about 13 % taking into account the recovery process. In later work, in which certain sources of error could be avoided and when conditions for optimally efficient working had been clarified, Hill (35) found for the whole cycle with frog sartorius maximum efficiency of 20 % – not much less than for man. This applied to both tetanus and twitch. Meyerhof (6) had earlier compared the work performed anaerobically with the lactic acid formed. Using a calorific quotient of 400 cal per g, he found maximum efficiency of about 45 %.

4

THE INFLUENCE OF BREWING SCIENCE
ON THE STUDY OF MUSCLE GLYCOLYSIS;
ADENYLIC ACID AND THE AMMONIA
CONTROVERSY

FERMENTATION AND GLYCOLYSIS

As early as 1912 Embden,[1] encouraged by the knowledge of cell-free fermentation in yeast juice, became interested in the possibility of obtaining from muscle a soluble system capable of forming lactic acid from carbohydrate. He used an adaptation of the technique of Buchner (1). Dog muscle, obtained with cooling precautions from the animal under narcosis, was minced and frozen; it was ground with sand, mixed with kieselguhr to obtain a slightly damp mass, and squeezed in a Buchner press. When the resulting press-juice was incubated at 40° for 2 hours, lactic acid formation took place. No increase in formation was found on addition of glycogen, glucose, inositol or alanine. This was very surprising, in view of the well-known production of lactic acid on perfusion of a glycogen-rich liver; or of a glycogen-poor liver, if the perfusing blood contained glucose.[2] Embden suggested the name 'lactacidogen' for the unknown precursor.

Embden recalled that early work had often connected muscular exercise with increased formation of free phosphate. Thus G. J. Engelmann (1) in 1871 had first recorded increased phosphoric acid excretion in the urine after very strenuous work; this observation was confirmed by some later workers of the nineteenth century, but the effect could not be found by others.[3] Mindful also no doubt of the already known importance of phosphate in fermentation in yeast press-juice,[4] Embden and his collaborators proceeded to look for changes in inorganic phosphate concentration which might accompany lactic acid formation in muscle press-juice.[5] They did in

---

[1] Embden, Kalberlah & Engel (1).
[2] Embden himself had contributed to such studies. See von Noorden & Embden (1).
[3] For literature on this subject, see the paper by Embden & Grafe (1) who themselves in 1921 found increased phosphate in the urine after work in man.
[4] Wróblewski (1); Harden & Young (2).
[5] Embden, Griesbach & Schmitz (1) and numerous later papers in the same volume.

many cases find an increase which was sometimes, but by no means always, about equivalent in amount to the lactic acid formed.

Efforts were next made to isolate the lactacidogen. Muscle press-juice was found to contain a phosphate compound with reducing properties, and by 1921 Embden & Laquer (1, 2) had prepared from dog muscle an osazone compound identical with the phenylosazone made from yeast hexosediphosphate. They were thus for a time satisfied that lactacidogen was hexosediphosphate. A little later it was found that the amount of ester in muscle *brei* could be greatly increased by addition of citrate or fluoride, particuarly the latter.[1] Then in 1924 Embden & Zimmermann (1) prepared hexosediphosphate as the crystalline brucine salt (identical with the fermentation salt) from muscle press-juice to which glycogen and fluoride had been added. They realised that their earlier identification of the hexose ester from muscle with the fermentation hexosediphosphate had not been conclusive since, when the osazone is made from the diphosphate, one of the phosphate groups is split off.[2] When this question of the nature of the lactacidogen present in muscle itself was re-investigated in 1927, Embden & Zimmermann (2) obtained the crystalline brucine salt of a hexosemonophosphate different from either of the known hexosemonophosphates (see p. 67). They found no hexosediphosphate.

On the assumption that the lactic acid and inorganic phosphate formed in muscle under various conditions were derived from the same compound, Embden and his collaborators in the early nineteen twenties carried out many series of experiments to determine the amount of lactacidogen present in various types of muscle treated in various ways.[3] This they did by measuring the phosphate set free when the muscle *brei* was heated in bicarbonate solution for some hours at 40 or 55°. Although cases were found of muscles with lowered 'lactacidogen content' after activity, e.g. in dogs after strychnine poisoning or in the white muscle (in contradistinction to the red muscle) of rabbits after strenuous work, such fall was often lacking. Embden remained unshaken in his belief that a phosphorylated intermediate in carbohydrate metabolism yielded both lactic acid and free phosphate; when the lactic acid formed greatly exceeded the phosphate, this must, he thought, be due to a rapid 'assimilatory process' involving re-esterification of the phosphate with carbohydrate.[4]

The same theme is found in the paper of Laquer (1) who, using isolated hind-limbs of frogs, saw no increase in inorganic phosphate after stimulation to exhaustion. Embden & Lawaczeck (1) did in 1922 find a temporary increase in free phosphate during contraction, but their experiments are

---

[1] Embden & E. Lehnartz (1).
[2] W. J. Young (1).　　　　　　　　　　　　[3] Embden, Schmitz & Meincke (1).
[4] Embden, Schmitz & Meincke (1); Cohn (1).

difficult to interpret since a curious method of obtaining the contracted muscle and its control was used. One muscle (A) in the maximally contracted state was plunged into liquid air; the corresponding muscle (B) from the opposite limb was stimulated to fatigue, allowed to recover for an interval of time, and then frozen in the same way as (A). The comparison was thus between fully contracted muscle, and muscle fatigued enough not to respond by contraction to the cold stimulus of the liquid air. (They had observed that a fresh unfatigued muscle put into liquid air contracted strongly and froze in this state.)

Later, in 1928, after the recognition of hexosemonophosphate as the ester contained in the muscle, Embden & Jost (1) turned to a method of lactacidogen estimation based on separation of this ester by precipitation in ammoniacal alcohol in presence of magnesium, and its estimation by reducing power. It had become clear by this time that pyrophosphate breakdown (see p. 84) would contribute greatly to the inorganic phosphate release in the conditions of their earlier method. Using the same procedure with liquid air stimulation as just described, they found again a momentary reversible decrease in lactacidogen upon contraction. Later still, after the experiments of the Eggletons and of Lohmann (to be described, see pp. 80, 84) as well as after those of Wilhelmi (1) in their own laboratory, they did agree that during long stimulation there is an increase in lactacidogen above the initial contraction value. This they put down to an over-compensating synthetic process.[1]

Meyerhof (7) now emphasised the importance of phosphate in lactic acid formation. Kondo (1) for press-juice and Laquer (1) for muscle *brei* had shown that in alkaline medium (such as 1–2% bicarbonate) much more lactic acid was formed than in a neutral medium. Meyerhof had also found that presence of phosphate greatly increased the lactic acid maximum with chopped muscle, and that no other buffer acted in the same way.

In 1926 Meyerhof (10) began his fruitful work on glycolysis in a soluble system prepared simply by crushing the cooled frog or rabbit muscle at −1° under isotonic KCl solution or distilled water. The cell-free extract obtained on filtering would produce lactic acid from a number of carbohydrates[2] – best from added starch or glycogen[3] – continuing for many hours at a rate about the same as that for the same amount of chopped muscle. The much lower activity with the fermentable sugars was reminiscent of earlier results of Laquer (2, 3)[4] using muscle *brei* under various conditions. On dialysis of the extract the glycolytic activity was lost, but could be restored on addi-

---

[1] Embden & Jost (2).   [2] Meyerhof (11).
[3] The failure of Embden to find lactic acid formation from glycogen was probably due to the greater thermo-lability of the system with glycogen than with hexosediphosphate (see Meyerhof (4) p. 144).   [4] Also Laquer & Meyer (1).

tion of a boiled extract from yeast or muscle. With hexosediphosphate as substrate, lactic acid and phosphate were formed from it in equimolecular proportions but the reaction rate fell off long before breakdown was complete. Fig. 9 shows the effect of adding starch at this stage – a very rapid formation of lactic acid accompanied by esterification of the phosphate

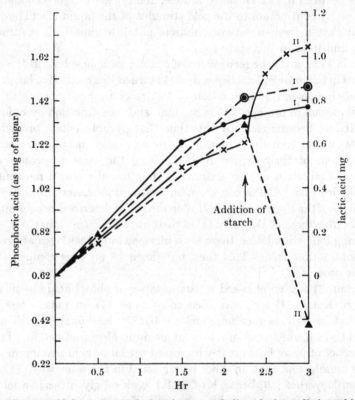

Fig. 9. Lactic acid formation and phosphate metabolism during splitting of hexosediphosphate in muscle extract, and during subsequent starch addition. Curve I: ●—●, lactic acid formation from hexosediphosphate; ⊙- - - - ⊙, phosphoric acid formation from hexose diphosphate. Curve II: ×—× and ▲- - - ▲, lactic acid and phosphate formed from hexosediphosphate during 2 hr 15 min; starch then added. (Meyerhof (11).)

present. Meyerhof also investigated the effect of adding fluoride to extracts containing glycogen; this led to great accumulation of hexosediphosphate, as had been observed earlier by Embden & E. Lehnartz (1) in press-juice. The latter had considered that the fluoride ions had a favourable effect on the synthesis of lactacidogen; Meyerhof now showed that the piling up of the ester was due rather to inhibition of glycolysis, including inhibition of the ester breakdown.

Meyerhof's results were nevertheless in agreement with Embden's conception of a phosphorylated carbohydrate as a stage in the breakdown of carbohydrate to lactic acid. However, on the evidence of such experiments as that shown in fig. 9, Meyerhof suggested that hexosediphosphate was not on the normal direct pathway, but a stabilisation product of a nascent ester which underwent glycolysis much more rapidly.

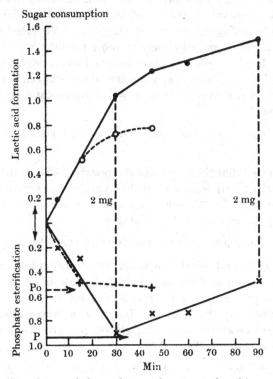

Fig. 10. Metabolism of 2 mg of glucose by muscle extract + hexokinase.

    (i) ●—●, lactic acid formation   ⎫
            ×—×, phosphate esterification ⎬ in presence of added phosphate
The vertical dotted lines correspond to utilisation of the whole of the added glucose.

    (ii) ○ - - - ○, lactic acid formation   ⎫
            + - - - +, phosphate esterification ⎬ without phosphate addition

P_0 - - - > shows the phosphate content of experiment (i); P→ shows the phosphate content of experiment (ii). (Meyerhof (12).)

Such a point of view Meyerhof found to be reinforced by his experiments on glycolysis of the hexoses.[1] He found a great acceleration of the slow glycolysis of these compounds when a preparation from yeast was added (see fig. 10). This activator was prepared by precipitation with 50%

                    [1] Meyerhof (12).

alcohol at $0°$ of an aqueous extract of toluene-plasmolysed yeast. It was heat-labile and sensitive to acid or alkali treatment. Meyerhof considered that it was probably the enzyme responsible for bringing about the conversion of the sugar into the active form. He later named it 'hexokinase'.[1] In the presence of the hexokinase rapid glycolysis, running parallel with phosphate esterification, took place (stage I) followed after a time by a period (stage II) in which glycolysis went on much more slowly and was accompanied by equivalent release of free phosphate. Addition of free phosphate greatly increased the rate in stage II. As Meyerhof pointed out, these two stages corresponded closely to those familiar in yeast fermentation of glucose. Here the effect of phosphate had long been known.[2] Harden & Young[3] had found in stage I a close correlation between carbon dioxide formation and phosphate esterification, which they expressed by the well-known equation

$$2C_6H_{12}O_6 + 2PO_4HR_2 \longrightarrow 2CO_2 + 2C_2H_5OH + 2H_2O + C_6H_{10}O_4(PO_4R_2)_2 \ (R = Na \ or \ K).$$
$$(1)$$

They had suggested that the slow stage II (appearing when either phosphate or sugar was exhausted) depended on the hydrolysis of hexosediphosphate to provide reactant for further breakdown according to equation (1). Equation (2) thus expressed the slow stage:

$$C_6H_{10}O_4(PO_4R_2)_2 + 2H_2O \longrightarrow C_6H_{12}O_6 + 2PO_4HR_2.$$
$$(2)$$

Harden & Young (4) had also shown in 1906 the necessity in yeast fermentation for a water-soluble, heat resistant, dialysable coferment. This was later named co-zymase by von Euler & Myrbäck (4).

After the discovery by Robison (1) that a hexosemonophosphate (Robison ester) often also occurs during glucose fermentation an alternative form of equation (1) was proposed:[4]

$$3C_6H_{12}O_6 + 2PO_4HR_2 \longrightarrow 2C_6H_{11}O_3PO_4R_2 + 2H_2O + 2CO_2 + 2C_2H_5OH.$$
$$(3)$$

The happenings in muscle extract after addition of a hexosemonophosphate, Robison ester, Embden ester or Neuberg ester (the latter made by splitting off one phosphate group from the fermentation hexosediphosphate) were found by Meyerhof & Lohmann (2) to be broadly similar to those after addition of glucose, with the difference that there was in the rapid stage a correlation of the formation of two molecules of lactic acid with one molecule of phosphate esterified. No hexokinase was needed.

[1] Meyerhof (4). A thorough account of his views on glycolysis at this time is given in this book.
[2] E.g. Wróblewski (1); Harden & Young (1).
[3] Harden & Young (2); Harden (1).
[4] Harden (1) pp. 69, 139; Harden & Henley (1).

Meyerhof[1] described the events of glycolysis or fermentation in a rather different way from that of Harden & Young. According to his formulation the equations ran:

$$2C_6H_{12}O_6 + 2HR_2PO_4 \longrightarrow 2C_6H_{11}O_5(R_2PO_4)^* + 2H_2O \longrightarrow$$
$$2C_3H_6O_3 + 6_6H_{10}O_4(R_2PO_4)_2 + 2H_2O \qquad (4)$$

where * signifies active form. But Harden was not convinced. As late as 1930 he expressed (2) his belief that 'a coupled reaction of some kind occurs, as the result of which the introduction of phosphate groups into certain sugar molecules – either into the same molecule or one each into two different molecules – induces the decomposition of another one'. Harden noted that Meyerhof & Suranyi (1) themselves had found that the phosphorylation of carbohydrate in muscle extract was accompanied by a small evolution of heat, and he suggested that this liberation of energy might have some significance for the coupled reaction. Meyerhof[2] on the other hand considered that 'the whole behaviour of the natural monoesters indicates that a compound corresponding to them is formed as an intermediate in sugar breakdown in muscle extract and yeast juice, and that the breakdown of this ester is the rate-determining factor during the "phosphate" period'. Meyerhof did not believe that the active ester in question was any of the known monophosphates, but pictured that the Embden ester in muscle and the Robison ester in yeast were formed as stabilisation products of a monoester *in statu nascendi*; hexosediphosphate would be formed according to the above equation when the phosphate of one active monoester was transferred to a second, only the first molecule undergoing glycolysis. It was noteworthy, he said, that the muscle extract maintained its power to break down hexosediphosphate in conditions which had led to loss of such power with starch or glycogen – after long storage in ice; after short warming at 37° and after dialysis.[3] These facts suggested that the hexosediphosphate was a (stabilised) intermediate, since it could not be breaking down via the free carbohydrate.[4]

The idea of 'stabilisation stages', he emphasised, had a purely physiological significance, and implied no connection with the stability of the esters concerned to chemical treatment such as acid hydrolysis.[5]

The preparation and properties of hexosediphosphate (fructose-1,6-diphosphate) and of the Embden, Robison and Neuberg esters were described by Meyerhof & Lohmann (2), Lohmann (1), Harden (1) and Meyerhof (4). The Embden and Robison esters are very similar and each consists of about 80 % aldosephosphate with about 20 % ketosephosphate. Neuberg ester is

---

[1] Meyerhof & Lohmann (2); Meyerhof (4).
[2] Ibid.
[3] Meyerhof (11).
[4] Meyerhof (12), (4) pp. 145–7.
[5] Meyerhof (4) p. 159.

mainly fructosephosphate, with a very small aldosephosphate component. Lohmann (20) showed that with the Embden ester an equilibrium is concerned, the same proportion of the aldo and keto forms being very rapidly reached, in presence of muscle extract, from either the keto-ester or aldo-ester.

Meyerhof[1] in 1927 brought evidence that put a new interpretation on the slow stage II of hexosediphosphate breakdown. Harden & Young (3) had shown a very marked accelerating effect of low concentrations of arsenate (about 10^{-3}M) on the rate of fermentation of sugar in the absence of phosphate; and also on the rate of fermentation of hexosediphosphate. They concluded that the likely explanation in both cases was an increased rate of hydrolysis of hexosediphosphate, providing reactants for equation (1). Meyerhof however (and his results were confirmed by Macfarlane (1) in Harden's laboratory) found that arsenate had no effect on the rate of hydrolysis of hexosediphosphate by yeast preparations freed from co-enzyme; it thus appeared that the hexosediphosphate itself underwent an accelerated fermentation or glycolysis.[2]

Another theory of fermentation mechanism influential at that time was put forward by von Euler and his collaborators in Sweden.[3] Their results had led them to the conclusions (i) that co-zymase was concerned in activating oxido-reduction processes; (ii) that co-zymase was not necessary for the initial phosphorylation of carbohydrate. Nilsson wrote:

The conversion of hexosemonophosphate follows by means of an oxido-reduction activated by co-zymase, two molecules of hexosemonophosphate entering at the same time into the reaction, whereby results a splitting of the 6-carbon chains of both ester molecules into two 3-carbon chains. The two triosephosphate residues unite to give hexosediphosphate. Through the oxidoreduction a redistribution of energy takes place, so that on the one hand an energy-poor stable compound (hexosediphosphate) is formed, on the other hand there is formation of an energy-rich phosphate free residue, ready to break down.

No factual evidence for such a 'dismutation' seems ever to have been brought forward, and the real contribution of this school is rather to be found in connection with the later stages of carbohydrate breakdown.[4]

From all the work of this period on yeast fermentation and muscle glycolysis under various conditions it became clear that the relationship of alcohol and carbon dioxide or of lactic acid formed to phosphate esterified could show considerable variations from that demanded by the original Harden & Young equations. Formation of hexosemonophosphate as well as diphosphate of course affected these relations, and the percentage of the

---

[1] Meyerhof (12); Meyerhof & Lohmann (2).   [2] Cf. p. 121, ch. 6.
[3] Von Euler & Nilsson (1); von Euler & Myrbäck (1, 2); Nilsson (1).
[4] Ch. 6.

monophosphate could be very variable. Meyerhof (10) pointed out that with glycogen or starch as substrate, the phosphate esterified might be very much more than equivalent to the lactate formed. The case of the living yeast cell is interesting. It shows no response to phosphate but its fermentation rate is high, about double that of the maximum phosphate-stimulated rate with yeast-juice. Harden[1] suggested that this must depend on a balance of enzymes such that the supply of free phosphate is maintained at the optimal. As we have seen, normal muscle was found by Embden & Zimmermann (2) to contain hexosemonophosphate but no hexosediphosphate, and this was abundantly confirmed in later work. Another striking feature is the variation under different conditions in requirement for co-enzyme. All these relations could only be understood in the light of later discoveries – the cyclic process concerned, the easy reversibility of several of the reactions, the different enzymes phosphorylating hexose and polysaccharide, and the requirement for two distinct co-enzymes.[2]

DELAYED LACTIC ACID FORMATION. Before leaving the present discussion of carbohydrate metabolism it will be convenient to mention the very interesting observation of Embden and his colleagues, first made in 1924,[3] of the continued formation of lactic acid when relaxation was over. After a 10 sec anaerobic tetanus, the lactic acid content of the muscles might go on increasing for 5–30 sec. The results were variable, but the delayed increase might be as much as 40 % of the amount present immediately after cessation of the contraction. These results were sharply criticised by Meyerhof & Lohmann (3), and by Suranyi (1) in Meyerhof's laboratory, on the grounds that the strong direct stimulation applied to the muscles had brought about a pathological condition. In a later paper, taking account of these criticisms, Embden[4] used a series of single twitches of thin frog semi-membranosus, indirectly stimulated; the one muscle contracted in nitrogen, the other corresponding muscle from the same animal, in oxygen, under otherwise comparable conditions. If the moles of lactic acid oxidised were calculated from the oxygen uptake of the aerobic muscle and the moles of lactic acid formed from the lactic acid present in the anaerobic muscle, the oxidation quotients derived were very high, varying between 7 and 10 instead of the usual 5 (see page 49). Embden explained this as caused by increase in the numerator, due to delayed extra lactic acid formation *between* the contractions in the anaerobic muscles. Meyerhof[5] again criticised these experiments, this time on the grounds that oxygen supply was inadequate, and therefore the denominator was too low. Embden & E. Lehnartz (2) replied by

[1] Harden (1) p. 183.   [2] Ch. 6.
[3] Embden (2); Embden, Hirsch-Kauffmann, Lehnartz & Deuticke (1).
[4] Embden, Lehnartz & Hentschel (1).   [5] Meyerhof & Schulz (1).

obtaining similar large quotients in experiments with sartorius muscles scarcely 1 mm. in thickness. In 1931, E. Lehnartz (1) returned to the more direct method of demonstrating the point, and confirmed that after tetani with indirect stimulation in nitrogen and strictly comparable conditions for the corresponding control muscle from the same animal, there was 20–30 % more lactic acid formed during the five minutes after relaxation. By this time Meyerhof (13), too, had been able to demonstrate the delayed production. Thus he found with 5 sec tetanus at 3° as much as two-thirds of the lactic acid formed after relaxation; with a 20 sec tetanus about one-third. With one 10 sec tetanus at 18° about half the lactic acid was delayed. In explanation of his earlier failures, Meyerhof says that in the former experiments the tetanic tension developed and the lactic acid formation were scarcely half those obtained in the present work, this being due either to sub-maximal stimulation or to some difference in preparation of the muscle. In these conditions he supposed the delayed formation to be much smaller and within experimental error.

We shall consider a little later[1] the significance of the delayed lactic acid for energy provision.

THE FINDING OF ADENYLIC ACID; AMMONIA IN CONTRACTION AND RECOVERY

We must turn now to another phosphate-containing compound of muscle which received much attention in the nineteen twenties. Already in 1914 Embden & Laquer (1) had observed in crude preparations of lactacidogen the presence of a compound with a soluble barium salt, containing phosphorus, nitrogen and pentose. Adenine was identified as a product of its breakdown. In 1927 Embden & Zimmermann (3) obtained from rabbit-muscle press-juice a crystalline preparation which, from its analysis and melting point, they took to be identical with adenylic acid already known from hydrolysis of yeast nucleic acid (designated at that time 'yeast adenylic acid') – adenine-ribose-phosphate.[2] Inosinic acid, the corresponding hypoxanthine nucleotide, had long been known to appear in muscle extracts, having been prepared crystalline from this source by Liebig (5) in 1847; its constitution was described by later workers.[3]

Embden[4] therefore now (in 1927) suggested that the inosinic acid of aqueous muscle extracts might arise by deamination of the adenylic acid of fresh muscle. He and his collaborators did indeed find that frog or rabbit

---

[1] Ch. 6.
[2] Levene (1).
[3] Levene & Jacobs (1), in 1911.
[4] Embden (3); Embden, Riebling & Selter (1); Embden & Wassermeyer (1).

muscle *brei* on standing produced ammonia, this production coming to an end in three or four hours. It was not affected by adding urea, but increased greatly when adenylic acid was added. On stimulation of isolated frog muscle there was increase in ammonia content, the greater the more the fatigue. Active frogs contained more ammonia in their muscles than quiescent ones, and the high ammonia content fell when the frogs were induced to rest by being put in the dark. Embden[1] believed that this disappearance of ammonia could also take place in isolated muscles. He and his colleagues found this 'reversibility' not always easy to demonstrate. One argument used was the claim that more ammonia was found at the end of a series of twitches in rapid succession than at the end of an exactly similar series at greater intervals, the assumption being made that during the pauses between twitches reamination could occur.

Embden & Wassermeyer (2) obtained some evidence supporting the view that adenylic acid was the source of the ammonia. They knew that adenylic acid and inosinic acid were precipitated from deproteinised extract by treatment with copper sulphate and chalk. By comparing the ammonia present in muscle brei after 4 hours incubation at 40° with the nitrogen content of the copper–chalk precipitate obtained from the unincubated brei plus the initial ammonia content, they concluded that the ammonia had come from a source containing just five times as much nitrogen as could be split off as ammonia...thus that the ammonia came from the amino group of the adenylic acid. G. Schmidt (1) in Embden's laboratory separated from muscle extract two enzyme solutions, one causing the deamination of adenylic acid only, the other of adenosine only amongst the substrates tried. An interesting outcome of this work was the realisation of the chemical difference between the muscle and yeast adenylic acids, since the latter was not deaminated by the muscle enzyme. Embden & Schmidt (1, 2) a little later found several other divergences, including differences in optical

---

[1] Embden, Carstensen & Schumacher (1); M. Lehnartz (1).

behaviour, in ease of splitting off phosphate on acid hydrolysis; and in degree of furfural formation by the Hoffmann method.[1]

Very shortly after Embden, Parnas (who had for some years been interested in the ammonia content of blood)[2] independently turned his attention, as the result of a chance observation, to ammonia formation in muscle. A few months after Embden's first publication,[3] Parnas & Mozołowski (1) described experiments in which they showed for many different types of vertebrate muscle that trauma (for instance heat or caffein rigor, grinding with sand and water) rapidly led to a trauma maximum of ammonia formation. This ammonia formation was inhibited at high pH values so that low initial values were very conveniently obtained when the muscle was ground with sand and saturated borate solution (pH 9.3) at 0°. Stimulation to fatigue also led to accumulation of ammonia, though to a smaller extent than did trauma. Intact excised muscle or muscle *brei* showed a slow post-mortal ammonia formation. Parnas, Mozołowski & Lewiński (2) found a striking constancy of the ammonia formation per unit of tension in anaerobic and aerobic conditions. So also did Nachmansohn (1), in anaerobic experiments. However it appeared later[4] that this held only under a given set of physiological conditions and marked variations were shown when these were changed (see p. 74 below). Parnas and his collaborators had found difficulty in determining the source of ammonia in circulating blood; they now suggested that it might originate in the working muscle, and found that exercise of the forearm in man led to increased ammonia content (by some 200–300 %) in the blood of the cubital vein.

In 1929 Parnas (2) very reasonably remarked that in considering the origin of so ubiquitous a substance as ammonia, which might be derived from so many sources, the idea that it was formed in muscle solely from adenylic acid required stringent proof. He also pointed out that if Embden were right in this assumption, it must mean that by far the greater part of the purine content of the muscle was in the form of nucleotide and little as nucleic acid, a novel idea at that time. Parnas now worked out methods for estimation in small amounts of muscle of the total amino- and hydroxy-purines, as well as of aminopurine- and hydroxypurinenucleotides and nucleosides, free adenine and hypoxanthine, and the nucleic acid adenine and hypoxanthine. Table 1 shows the various fractions in two samples of the same muscle, one before, the other after, traumatic ammonia formation.

---

[1] Muscle adenylic acid is adenosine-5-phosphate (see ch. 6); 'yeast adenylic acid', prepared as in the experiments of Embden and Schmidt by alkaline hydrolysis of yeast nucleic acid, is a mixture of adenosine-2-phosphate and adenosine-3-phosphate (see Brown, Fasman, Magrath & Todd (1)). Yeast of course does contain the same *free* adenylic acid as muscle.

[2] See refs given in Parnas, Mozołowski & Lewiński (1).   [3] Embden (3).

[4] Parnas (7); Parnas & Lewiński (1).

Since Fraction c_1 contained nucleic acid and nucleoprotein N as well as purine N, total guanine was assessed separately. This was found to be about 0.5 mg% for the frog. The total extractable adenine (which included guanine) amounted in winter frogs on the average to 26 mg %. If now we assume that all the guanine is contained in nucleic acid, and is accompanied there by an equivalent amount of adenine, we see that the ratio of nucleotide aminopurine to nucleic acid aminopurine is about 25:1. From table 1 it is also clear that the adenine nucleotide fraction falls by an amount (24 mg %) corresponding to five times the well-known traumatic ammonia maximum (5 mg %). Table 2 shows the change in total adenine content as a result of stimulation, and the relation of the ammonia release to this change. These Polish experiments left no doubt that the traumatic and fatigue ammonia had its origin in the adenylic acid; further that the quantitatively important nucleotide in resting muscle was not inosinic acid as generally supposed.

TABLE 1. *Adenine and hypoxanthine derivatives in frog muscle*

P	a_1	a_2	b_1	b_2	c_1	c_2	a_1+c_1	a_2+c_2	$a_1+b_1+c_1$	$a_2+b_2+c_2$
II 37.9	28.2	2.6	0.2	0.2	5.2	0.45	33.4	3.1	33.6	3.3
I 30.2	4.5	20.4	0.9	0.6	3.2	0.5	7.8	20.9	8.6	21.5

Adenine and hypoxanthine derivatives in frog muscle; series II, before traumatic ammonia formation; series I, after. a_1, nucleotide adenine; b_1, nucleoside + free adenine; c_1, adenine of nucleic acid and nucleoprotein. a_2, nucleotide hypoxanthine; b_2, nucleoside and free hypoxanthine; c_2, hypoxanthine of nucleic acid and nucleoprotein. P, total purine N content. All expressed as mg N % (Parnas (2)).

TABLE 2. *Behaviour of the purine bases on muscle activity*

	I		II	
	Fresh	Fatigued	Fresh	Fatigued
NH_3-N content, mg %	0.3	1.3	0.8	2.6
Adenine-N content, mg %	36.3	29.4	44.5	33.8
Hypoxanthine-N content, mg %	4.0	7.7	3.6	12.1
Increase in NH_3-N, mg %		1		1.8
1/5 decrease in adenine-N mg %		1.3		2.14
1/4 increase in hypoxanthine-N mg %		0.95		2.37

After (Parnas (2)).

Parnas could however find no evidence for Embden's belief that rapid disappearance of ammonia formed on fatigue took place – i.e. that reamination occurred and that the reversibility was important in contraction.[1]

[1] Embden, Carstensen & Schumacher (1).

Nachmansohn (1) had also failed to find anaerobic removal of ammonia. Parnas in 1930 made a vehement attack on the Embden school[1], making use of recently introduced statistical methods to demonstrate the wide scatter in their published results for resting ammonia content, and the likelihood that the ammonia disappearance after work was too small to be of meaning. Parnas (2, 7) however had found that when frog muscle performed isometric twitches aerobically for many hours without fatigue, the adenylic acid deamination could lag far behind the free ammonia formation. He interpreted this as meaning that oxidative deamination of other substances (perhaps amino acids) could bring about resynthesis of the adenylic acid from inosinic acid. Embden & M. Lehnartz (1) were able to confirm the high ammonia production in the circumstances described; but they found this little increased by work and were inclined to put it down to autolysis. They asked by what mechanism ammonia produced by oxidative deamination should reaminate inosinic acid when ammonia produced from adenylic acid could not?

In work which came a little later (1931) (see ch. 6) it became clear that adenylic acid is present in resting muscle as adenosinetriphosphate. This compound is not deaminated by the muscle enzyme,[2] and in the light of this it seems probable that the oxidative conditions were to a large extent preventing breakdown of the adenine nucleotide by making possible rapid rephosphorylation (see ch. 17). At this time[3] only rephosphorylation by anaerobic processes (from phosphocreatine and phosphopyruvate) was considered (see ch. 6) and all the phenomena of oxidative rephosphorylation remained to be discovered.

Conclusive demonstration that the inosinic acid formed on contraction can be reaminated in the muscle had to await the use of isotopes. In 1947 Kalckar & Rittenberg (1) administered ^{15}N ammonium citrate to rats and showed incorporation of the isotope into the ATP isolated from the muscle. In 1955 an enzyme system forming adenylosuccinate was studied by Carter & Cohen (1, 2) in a protein fraction from yeast autolysate. They suggested that it might be concerned in introducing the amino group into purine

$$\text{COOH.CH.CH}_2\text{.COOH}$$

Fumaric acid + AMP $\rightleftharpoons$ [structure] $+ H_2O$ (2)

ribose-5-phosphate

---

[1] Parnas, Lewiński, Jaworska & Umschweif (1).
[2] See work in 1933 and 1934 of Mozołowski & Sobczuk (1) and Parnas, Ostern & Mann (1).
[3] E.g. Parnas & Lewiński (1).

nucleotides and found (3) adenylosuccinate to be formed in pigeon-liver homogenate from inosinic acid in presence of an energy source and aspartic acid; while Abrams & Bentley (1) showed that aspartic acid was specific for this synthesis in a soluble enzyme extract from bone marrow. Lieberman (1) about the same time, using a purified enzyme preparation from *Escherichia coli*, found that guanosine triphosphate was the specific energy source required.

$$\text{Aspartic acid} + \text{inosinic acid} + \text{GTP} \longrightarrow \text{adenylosuccinate} + \text{GDP} + \text{H}_3\text{PO}_4 \quad (2)$$

Later Newton & Perry (1, 2) showed some similar effects with rabbits. When muscle *brei*, or extract of a powder made from the soluble fraction of a muscle homogenate, was incubated with inosinic acid in the presence of ^{15}N aspartic acid, formation of labelled ADP and ATP took place. With the extracts, practically all the isotope found in the adenine nucleotide was located in the 6-amino group. From such extracts a compound was isolated which was highly labelled and the properties of which (U-V absorption spectrum; atoms per cent excess of ^{15}N) suggested that it was adenylosuccinate, formed as an intermediate in the reaction. No amination took place with ammonium chloride as the source of nitrogen. The aminating activity of the muscle appeared to be much less than its deaminating activity; it was difficult to demonstrate the net formation of adenine nucleotide unless the deaminase activity was reduced (e.g. by the acetone drying of the muscle powder before extraction). Davey (1, 2) has also studied the reamination system in extracts of acetone-dried powder of beef and rabbit muscle. He observed the marked activating effect of small amounts of GTP, and from his experiments concluded that GDP bound to the synthetase could be phosphorylated by ATP and then act as energy source for the synthesis. Reamination of inosinic acid would thus be mediated by reaction (2) followed by the reverse of reaction (1).

With regard to the fate of the AMP released by the adenylosuccinase in reaction (1), the participation of the enzyme myokinase (or adenylate kinase)[1] catalysing the reaction $\text{ATP} + \text{AMP} \rightleftharpoons 2\ \text{ADP}$ must be considered. It has to be remembered (i) that the K_m for myokinase[2] is 2.6×10^{-4}, while for the deaminase[3] it is 14×10^{-4}; (ii) that the specific activities seem to be in favour of the myokinase – a turnover number of 25000 moles/min mole of enzyme at 25° to be compared with the value of 18300 at 30° for the deaminase. The relative concentrations of the enzymes are not known, but the values given above suggest that myokinase could phosphorylate the major part of the AMP formed in the resting muscle, thus protecting it from the deaminase; whilst the free energy of the ATP dephosphorylation

---

[1] Colowick & Kalckar (1); Kalckar (1); see ch. 8.
[2] Noda (1).   [3] Currie & Webster (1).

is needed for contraction any AMP formed would fall a victim to the deaminase.[1]

The demonstration by Embden in 1924 that lactic acid formation might take place in considerable measure *after* contraction was over made the first rift in the lactic acid theory of contraction, according to which the acid production seemed satisfactorily to explain both energy provision and contractile mechanism. Embden rejected the lactic acid theory and swung to the other extreme in postulating a change to the alkaline side as the primary event eliciting contraction. In papers between 1925 and 1930[2] for example he instanced ammonia formation from adenylic acid and dephosphorylation of hexosemonophosphate[3] as two initial reactions leading to rise in pH; by 1928 the breakdown of phosphocreatine could be included.[4] He pictured exothermic colloidal processes as providing the direct source of energy, and suggested that the sudden rise in pH could affect the colloids concerned. As evidence for such colloidal changes on activity he instanced the diminution in the power of phosphate ester synthesis, in presence of fluoride ions, in muscle *brei* from fatigued muscle as compared with formation in *brei* from resting muscle;[5] and the decrease in solubility of the muscle proteins after strenuous activity.[6] Lactic acid formation (together with resynthesis of the adenylic acid and lactacidogen) would restore the initial pH. Energy supplied by chemical changes, notably lactic acid formation, after the contraction, had the function of recharging, so to speak, the physical accumulator.

As we shall see, the idea that a pH change, either to acid or alkaline side, was of direct importance for contraction soon had to be abandoned. The significance of the deamination of adenylic acid is still obscure. It may be that the main function of the deaminase is avoidance of the deleterious pharmacological effects of accumulating AMP.[7] Another function has also been suggested by Dydynska & Wilkie (1).[8]

In the later years of this decade the discovery of phosphocreatine and of adenylpyrophosphate in muscle laid the foundation of our present knowledge of the complex chains and cycles of chemical reactions which supply energy to the muscle machine.

[1] I am indebted to Dr C. L. Davey for discussion of this question.
[2] Embden (3, 4); Embden & Jost (4).
[3] For the alkaline change on this dephosphorylation see Meyerhof & Lohmann (4).
[4] See ch. 5.   [5] Embden & Jost (4).
[6] Deuticke (2, 3).   [7] Ch. 19.   [8] Ch. 14.

5

THE DISCOVERY OF PHOSPHAGEN AND
ADENOSINETRIPHOSPHATE; CONTRACTION
WITHOUT LACTIC ACID

THE DISCOVERY OF PHOSPHAGEN AND EARLY IDEAS OF ITS FUNCTION

In 1927 P. Eggleton & G. P. Eggleton (1), and independently Fiske & Subbarow (2), reported the existence in muscle extracts of a phosphorus compound, very labile especially in acid solution; the figures for inorganic P content of muscle found by earlier workers, using methods involving acid treatment of the extracts, were therefore open to grave doubt.

Eggleton & Eggleton, using the Briggs method (1), in which the colour due to reduced phosphomolybdate is allowed to develop during 30 min in acid solution, found that the increase in colour during this time with inorganic phosphate solutions was only some 5%; but with extracts from resting frog's muscle the increase was several 100%. They proposed the name 'phosphagen' for the labile substance. The value for the true inorganic P of resting muscle, found by extrapolation back to zero time when the rate of colour development was followed, amounted to about 25 mg/100 g muscle; the phosphagen P content to about 60 mg/100 g. Estimations made in neutral or slightly alkaline solution (as in the Bell–Doisy (1) method or by precipitation with magnesia mixture) gave results approximating to the extrapolated values of the Briggs method. In rapidly induced fatigue the true inorganic phosphate increased at the expense of phosphagen P, though not all the phosphate of the disappearing phosphagen was found as inorganic P. In aerobic recovery,[1] phosphagen quickly reappeared at the expense of inorganic P, during a time when little lactic acid removal had yet taken place. In rigor the phosphagen disappeared entirely, and more inorganic P was formed than its equivalent.

Fiske & Subbarow (1) had worked out in 1925 a modified form of the method of Briggs, needing only 4 min for full colour development with inorganic P solutions. When they applied this method to muscle extracts[2] they found that colour development was not complete for about 30 min;

---

[1] P. Eggleton & G. P. Eggleton (2).   [2] Fiske & Subbarow (2).

[77]

from this behaviour they deduced the presence either of a labile P compound or of an inhibitor of the colour formation with inorganic P (many of which do occur). The fact that greater amounts of extract gave no greater delay led them to consider the former explanation more probable. They found in resting cat muscle 60–75 mg of labile P/100 g, and only 20–25 mg of true inorganic P. Like the Eggletons, they observed the loss of labile P on stimulation, and also its recovery to some extent during a period of rest after stimulation. Most important: in the labile material partly purified by

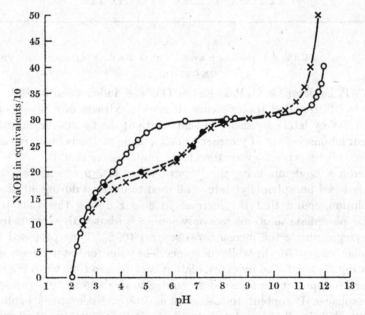

Fig. 11. Electrotitration curve of phosphocreatine. ○—○, phosphocreatine; ×—×, the same after hydrolysis; ●----●, mixture in the same concentrations of creatine and phosphate. (Meyerhof & Lohmann (8).)

treatment with various precipitants, the labile P was always accompanied by an equivalent amount of creatine. This purification was carried further in later work,[1] and by 1929 Fiske & Subbarow were satisfied that the whole of the labile phosphate occurred in the muscle as phosphocreatine, which could be isolated in 70 % yield as the crystalline calcium salt. They showed with this material that the secondary dissociation constant of phosphocreatine was much greater than that of orthophosphate; consequently a marked change to the alkaline side takes place on hydrolysis of the compound. This is shown in the electrotitration curve of Meyerhof & Lohmann (fig. 11). In this fact Fiske & Subbarow saw an important function of

[1] Fiske & Subbarow (3, 4).

phosphocreatine breakdown – that of neutralising lactic acid formed on contraction and thus mitigating fatigue.

This substance creatine[1] had been the subject of much investigation and speculation ever since its preparation in crystalline form from meat extract was described by Chévreul (1) in 1835. Liebig (5) in 1847 showed that it could be obtained from several kinds of muscle, though not from other organs he tested. There was much difference of opinion during ensuing years as to whether or not the creatine content of muscle increased on activity[2] – this question was naturally of importance in the controversy that we discussed in chapter 2 concerning the dependence of muscular work on protein catabolism.[3] Not long before the discovery of the phosphorylated form in which creatine exists in resting muscle, Schlossmann (1) and Tiegs (3) showed that diffusible creatine increased in the muscle during contraction. About this time also Riesser (1) was impressed by the distribution of creatine – present in greater amount in rapidly contracting white muscles than in slow red muscles, this distinction also holding when summer frogs were compared with winter frogs, or warm blooded animals with the more sluggish poikilotherms. Tiegs pictured the creatine in resting muscle as possessing a ring formation which only on excitation gave rise to the free amino group. This basic substance could then neutralise lactic acid, thus bringing about relaxation.[4] It reverted to the cyclic form on oxidative recovery, but since it was diffusible some could escape from the muscle. There was indeed by 1927 already much in the literature pointing to an important function for creatine in contraction.

Eggleton & Eggleton (2) had been inclined to regard phosphagen as a carbohydrate ester, giving rise to lactic acid as well as phosphate (though not to be confused with Embden's lactacidogen); and they suggested that the phosphagen formed in oxidative recovery was derived from glycogen. In a later paper (3) they agreed with Fiske & Subbarow that the labile compound extracted was a compound of creatine. Though free and combined creatine could not be very accurately distinguished, they found a much higher proportion of the creatine combined in resting than in fatigued or rigor muscle. They considered however that the compound extracted might be a breakdown product of the phosphagen contained in the muscle itself.[5] This view they based partly on the inability of phosphagen to diffuse

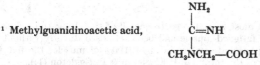

[1] Methylguanidinoacetic acid,

[2] A. Hunter (1) p. 173. [3] Lieben (1) p. 215.
[4] Compare e.g. the theory of Meyerhof described in ch. 3.
[5] G. P. Eggleton & P. Eggleton (1).

out of living muscle (though, as they said, this might be due to selective impermeability) and partly on certain considerations concerned with heat production which we shall presently discuss.

Eggleton & Eggleton[1,2] found phosphagen in all the types of vertebrate striated muscle they tested, but little in smooth muscle or heart, and none in invertebrate muscle. They worked out a useful method[1] for separating the various phosphate fractions in muscle by means of the different solubilities of the barium salts. Some typical results by this method are given in table 3 for resting and fatigued muscle. The 'soluble ester' increasing in amount on fatigue was broken down on incubation with bicarbonate buffer, and was considered to be identical with Embden's lactacidogen, hexosemonophosphate. They deprecated the use of the term lactacidogen, since in their experience this substance increased during activity while lactic acid is being formed. It was formed at the expense of the phosphagen P, only part of the latter appearing as inorganic P.

TABLE 3. *The effect of rapidly induced fatigue on skeletal muscles of the frog*

	Total phosphate	Ortho-phosphate	Phos-phagen	Pyro-phosphate	'Soluble' esters	'Insoluble' esters
1. Resting	142½	24	40	20	5	11
Fatigued	142½	38	18	23½	12	8½
Change	.	+14	−22	+3½	+7	−2½
2. Resting	105	20	39	21½	5½	14
Fatigued	105½	37½	16½	24	10	12
Change	.	+17½	−22½	+2½	+4½	−2
3. Resting	121½	19½	40	24	4½	12
Fatigued	120½	40½	5	24	15	15½
Change	.	+21	−35	0	+10½	+3½
4. Resting	134	18½	49	20½	2	10
Fatigued	125½	53½	4½	18½	11	12½
Change	.	+35	−44½	−2	+9	+2½
5. Resting	138½	20½	47½	21	1	10
Fatigued	138½	34½	26	21	7½	11
Change	.	+14	−21½	0	+6½	+1
6. Resting	112½	15	42½	22	9	11½
Fatigued	111½	38	12	21½	15½	12
Change	.	+23	−30½	−½	+6½	+½
Mean change	.	+21	−29½	+½	+7½	+½

The muscle pairs were in most cases the gastrocnemii of the Hungarian (giant) species of *Rana temporaria*. The fatigued muscle had been tetanised directly without load for 1–2 minutes. The total phosphate values in mg P/100 g of muscle; the five fractions expressed as percentages of the total. (G. P. Eggleton & P. Eggleton (1)).

[1] G. P. Eggleton & P. Eggleton (1).   [2] P. Eggleton & G. P. Eggleton (3).

Meyerhof and his collaborators quickly entered the phosphagen field with the important discovery in 1927[1] that the hydrolysis of phosphocreatine is accompanied by heat output – about $12\,500$ cal/g mole H_3PO_4 split off.

Considerable difficulties now confronted biochemists and physiologists in accepting at their face value the results of the metabolism of phosphocreatine *in vivo* during contraction. Meyerhof & Lohmann (5) were at first ready to see in the heat output during phosphocreatine breakdown the source of the still unexplained fraction of the caloric quotient of lactic acid.[2] It was true that, in agreement with Eggleton & Eggleton (2) they found the ratio

$$\frac{\text{mg phosphagen } H_3PO_4 \text{ disappearing}}{\text{mg lactic acid formed}}$$

diminished during a contraction series, being about 1.5 at the beginning, 0.75 for a medium degree of fatigue (about 0.2 % lactic acid) and then diminishing still further. The phosphocreatine breakdown for medium fatigue gave on calculation an increase in heat production of 120–130 cal/g of lactic acid formed and thus they concluded that the caloric quotient for lactic acid was completely explained. This contribution to the caloric quotient would not of course be constant throughout contraction to fatigue; they commented: 'It was earlier shown that the caloric quotient falls with piling up of lactic acid; that it increases with very short stimulation remains to be shown.'

Nachmansohn (1) in Meyerhof's laboratory examined in detail the change in K_z (the isometric time coefficient) for phosphocreatine breakdown during tetanus:

$$\frac{g \text{ tension} \times cm \text{ muscle length} \times sec}{\text{mg } H_3PO_4 \text{ split off}}.$$

It rose from 24 after two 5 sec tetani to 100 after six 5 sec tetani. All the evidence at the time[3] showed constancy for the corresponding K_z value for lactic acid formation over a wide range of tension production. Moreover, Nachmansohn (1, 2), as well as Gorodissky (1) independently in Embden's laboratory, showed about this time that anaerobic phosphagen re-synthesis to the extent of about 30 % can take place after relaxation. In Nachmansohn's experiments the muscles were stimulated and frozen in liquid air at the moment of relaxation and at intervals of a few seconds thereafter (see fig. 12). In parallel experiments Nachmansohn found no post-relaxation lactic acid production. Thus this endothermic re-synthesis could apparently go on without any accompanying energy-yielding reaction, while Hartree & Hill (3) just at this time reported that they could find no evidence of heat absorption after activity. Nachmansohn therefore postulated that it was

[1] Meyerhof & Lohmann (5); see also Meyerhof & Lohmann (6); Meyerhof & Suranyi (1).
[2] Ch. 3, p. 49.   [3] Nachmansohn quotes e.g. Suranyi (1).

only a re-stabilisation – in other words that the phosophocreatine re-synthesised had never really broken down.[1] Meyerhof & Lohmann (7) also reported experiments in which resynthesis of phosphocreatine was observed in extracts at alkaline reaction without lactic acid formation or any other known reaction to provide energy. They suggested that phosphocreatine goes first with little energy release to an intermediate which chemically could not be distinguished from the end products, but which could be converted to phosphocreatine again without energy expenditure.[2]

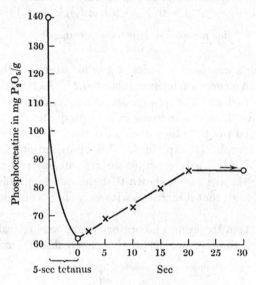

Fig. 12. Breakdown and anaerobic resynthesis of phosphocreatine with a 5 sec tetanus. (Nachmansohn (1).)

Hill (12) in a paper published almost simultaneously with those just mentioned, advocated the same point of view. He first showed that, according to a range of papers from the Meyerhof school, K_m (the isometric coefficient) for lactic acid

$$\frac{g \text{ tension} \times cm \text{ muscle length}}{mg \text{ lactic acid formed}},$$

[1] It is interesting that Gorodissky, while mentioning the current concept of de-stabilisation, rather inclines to the view that 'an intravital process of very considerable endothermy' takes place. She does not mention the much-disputed claim of the Embden school that post-relaxation lactic acid formation can take place, though its proof (at any rate after tetani) by Lehnartz (1) followed soon after. She apparently did not know of the experiments of Hartree and Hill.

[2] It is difficult to see how the increased synthesis which they also obtained on adding free creatine to the extract could be explained in this way.

could be regarded as constant during a series of twitches. He then demonstrated stringently that the isometric heat coefficient, Tl/H^1 changed very little (only some 5 %) as the result of previous activity. Hill concluded that in a single twitch the processes leading to the development of tension and heat were the same as in a long series of twitches. It was of course difficult to fit early exothermic hydrolysis of phosphagen, falling off later in the series, into this frame; such a changing breakdown, accompanied by constant lactic acid formation, could not have failed to show itself with Hill's methods by a rise of Tl/H with advancing fatigue. As Hill said, 'It seems necessary to assume either that purified phosphagen has very different thermal properties from the substance existing in living muscle, or that the breakdown does not really occur; perhaps the breakdown should be regarded rather as an "unstabilisation" of some kind, allowing the phosphagen to be broken down by the chemical treatment necessary for its estimation, to which it is normally resistant'. These flirtations with vitalism proved before long to be unnecessary, as we shall shortly see (p. 94).

Another series of observations influenced Meyerhof in his readiness to discount the participation of phosphagen breakdown in the energetics of contraction – the evidence which seemed to suggest that degree of phosphagen 'breakdown' was concerned with the rate of the excitation processes in the muscle. Thus Nachmansohn (1, 2) had shown that with curarised muscle the K_z value was already high (indicating restricted phosphagen breakdown) even with a single 2 sec tetanus, and changed little with further tetani. He went on (3) to consider phosphagen breakdown in relation to the 'chronaxie' concept of Lapicque (1), which provided a quantitative measure for the rate of the excitation process in excitable organs. The chronaxie gives the minimum duration of a constant current of definite strength necessary to call forth response, the current intensity used being double that which on infinite duration would bring a response. This threshold intensity was termed the 'rheobase'. The chronaxie is then the time in sigma needed for excitation by means of a current twice the rheobase. Nachmansohn found that for a number of different states the increase in chronaxie was correlated with decrease in the phosphagen breakdown associated with a given tension production. Experimental series in which this diminished phosphocreatine breakdown was accompanied by increased excitation time included the effects of curarisation, changes in temperature, influence of fatigue, influence of nerve degeneration. Veratrin treatment, which decreases chronaxie, was found to be accompanied by increased phosphagen breakdown, while strychnine leaves both unaltered. Ammonium salts, such as trimethyloctylammonium iodide, had effects similar to those of curare but

$$1 \quad \frac{\text{g tension} \times \text{cm length}}{\text{g cm initial heat}} .$$

more marked. Consideration of such facts, which they interpreted to mean that phosphagen breakdown conditioned the size of the chronaxie, led Meyerhof & Nachmansohn (1) in 1928 to consider that phosphagen had only indirect participation in contraction; on this and other grounds they withdrew the earlier expressed opinion on its energetic role.[1]

THE DISCOVERY OF ADENYLPYROPHOSPHATE

In 1928 Lohmann (2, 3) had described the estimation of pyrophosphate in muscle and the isolation of the inorganic compound. He was engaged in a study of the various phosphate esters and had worked out a method whereby they were characterised and estimated (in presence of each other) by means of their rate of hydrolysis in N HCl at 100°.[2] When this method was applied to protein-free muscle extracts, it was found that the release of inorganic phosphate was extremely rapid in the first few minutes and fell off sharply thereafter, with a flattening of the hydrolysis curve after about 7 min. The hydrolysis constant in the steep part of the curve was about 250×10^{-3}, to be compared with 23×10^{-3} for the first phosphate group of hexosediphosphate, 3.5×10^{-3} for its second, and 0.2×10^{-3} for the Embden ester. If the estimations were carried out with muscle *brei* which had stood for 1–2 h at 40°, this rapid phosphate release was not seen. The responsible substance was first obtained by Lohmann as inorganic pyrophosphate; it was isolated by means of barium precipitation at neutral reaction and purified through the lead and copper salts.[3] Lohmann pointed out that in the method up to that time used by Embden[4] for estimation of lactacidogen, 75 % of the phosphate released was derived not from hexosemonophosphate but from pyrophosphate.

The wide distribution of pyrophosphate in many kinds of plant and animal cells was emphasised by Lohmann (4); in muscle the content remained constant (about 30 mg pyrophosphate P/100 g of muscle) under normal physiological conditions, including moderate stimulation (5). With long stimulation or with rigor there was a fall.

Lohmann (5) observed that muscle extracts, especially those prepared at −1 to −3°, showed rather special behaviour when incubated in dilute bicarbonate solution at 20°. There was an induction period of about 10 min before the steady rate of hydrolysis supervened; this induction period was not observed at 38°. It was deduced that this phenomenon depended on some special state of the substrate, since inorganic pyrophosphate (or the pyrophosphate isolated from muscle) added to the extract broke down from the beginning at a steady rate. A little later in 1929 Lohmann (6, 7) found

[1] Meyerhof & Lohmann (5).   [2] Lohmann (1).
[3] Lohmann (3).   [4] Ch. 4, p. 62.

that, provided the barium precipitation was carried out in a faintly acid solution, the pyrophosphate was obtained in combination with adenylic acid. The purified compound yielded on acid hydrolysis (10–15 min in N HCl at 100°) 1 mole of adenine, 1 mole of pentosephosphoric acid, and 2 moles of inorganic phosphate; on neutral hydrolysis of the barium salt, adenylic acid and pyrophosphoric acid were obtained. The isolation of adenylic acid, previously announced by Embden & Zimmermann (3),[1] had been due to their treatment of the deproteinised muscle extracts with $Ca(OH)_2$. In 1929 Fiske & Subbarow (5) also, independently, prepared a compound from muscle which accounted for most of the muscle purine and contained the phosphate which Lohmann had called pyrophosphate. The purified silver salt had the formula $C_{10}H_{13}O_{13}N_5P_3Ag_3$, and besides purine and carbohydrate contained 3 moles of inorganic phosphate, two of which were readily removed on acid hydrolysis.

Adenylpyrophosphate, or adenosinetriphosphate (ATP) as it is now usually called, has since then emerged as a substance of primary and fundamental importance for the energetics of living function, not only of muscle but of all other tissues and cells. The long story of the development of our knowledge of its operations can only be told in later chapters, but its discovery is mentioned here in order to complete with this important detail the picture held of the make-up and metabolism of resting and contracting muscle at the end of the year 1929 – just before the major advance which led to the discovery of contraction without lactic acid in iodoacetate-poisoned muscle and the role of phosphocreatine. The story of this was as follows.

CONTRACTION WITHOUT LACTIC ACID

In 1929 in Copenhagen Lundsgaard (1) began his study of phosphocreatine metabolism in muscle poisoned with iodoacetic acid. In a personal communication Professor Lundsgaard has explained that in the late twenties he was working on the specific dynamic action of amino acids; of these, glycine gave the strongest effect per unit weight on metabolism. At that time the formula of thyroxine was given as iodine-substituted tryptophan, and he decided to try what happened with iodine-substituted glycine. The very striking effects he obtained on muscle behaviour led him to concentrate attention on this tissue.

When iodoacetic acid, neutralised with sodium carbonate, is given by intravenous injection to a rabbit in a dose of about 50 mg/kg, a definite train of symptoms follows. There is a latent period of 5–20 min, during which the animal behaves normally; then usually it falls suddenly on one side, making active, struggling movements, and the whole musculature goes

[1] Ch. 4.

immediately into strong rigor. Respiration becomes impossible and the animal dies; after death it remains as stiff as a piece of wood. With the frog, the course of events is similar but the rigor comes on more gradually.

Lundsgaard noticed that this rigor seemed to be brought on by activity of the muscles, for with frogs after urethane or curare narcotisation, iodoacetate injection caused no such change in the muscles. If the nerve to one hind limb was cut before the injection, the rigidity spread through all the body with the exception of this limb; but after a short series of contractions caused by electrical stimulation of the severed nerve, this limb too became stiff.

Lundsgaard had had a predecessor whose very similar results were unknown to him when he began this work. Pohl (1) in 1887 was engaged in a study of the dependence of physiological action on chemical constitution. With a very small dose of bromacetic acid he found that an animal would go into complete rigor; this did not happen with curarised animals. From this observation he expected to find that the bromacetate acted on the nerve endings, as an antagonist to curare. But after section of the nerve there was no rigor, although the nerve endings as well as the muscle were exposed to the poison. He suggested as explanation a change in excitability of the muscle fibres so that the muscle responded to a normal stimulus in an abnormally intense manner by contracture; and that the increased irritability did not express itself if the stimulus was lacking. He observed that acidification of the muscle could not be demonstrated with litmus. Curiously enough, he found little effect with iodoacetate, but Lundsgaard found the two substances to act in a very similar manner. Further, as Lundsgaard pointed out, Schwartz & Oschmann (1, 2) in 1925 had observed that bromacetate rigor was unaccompanied by increase in either free phosphate or lactic acid. They did not investigate further than this and did not realise that the power to form lactic acid was completely suspended.

Lundsgaard also was quickly struck by the failure of the muscle in rigor to become acid, even when kept for some hours at 40°.[1] He found that poisoned muscles in rigor showed no lactic acid formation; nor was lactic acid formed when a denervated limb was stimulated till rigor came on. Lundsgaard clinched this by a well-planned series of experiments on pairs of frogs in which, the day before, the plexus lumbalis on both sides had been severed. Iodoacetate injection was made into the dorsal sac of one frog. As soon as the fore-limbs became stiff, the gastrocnemius of one side was removed and frozen in liquid air; the gastrocnemius of the other side,

---

[1] This contraction without lactic acid formation was welcomed by Embden (3) as support for the importance of alkaline reaction in causing contraction. With bromacetate-poisoned muscle he and Norpath (1) in 1931 showed very marked ammonia formation during activity (ch. 4).

in situ, with a load of 40 g was stimulated with 2 shocks/sec. 100–150 contractions could be elicited before the muscle went into rigor; it was then removed and frozen. The control frog without iodoacetate was treated in exactly the same way, the number of stimuli corresponding to that needed for the poisoned muscle. The results are shown in fig. 13 (pl.) and table 4.

TABLE 4

	Normal			Poisoned with IAA		
Rest	17.9	27.1	40.0	14.9	14.0	25.0
Work	77.1	74.5	100.0	15.4	10.0	16.9

Three experiments of the kind shown in fig. 13; results of lactic acid estimations in mg %. (Lundsgaard (1).)

TABLE 5

		Lactic acid	Inorganic P	Phosphocreatine
Normal	Rest	25	21	61
muscle	Work	84	29	46
IAA-poisoned	Rest	16	29	57
muscle	Work	15	28	0

(Lundsgaard (1)) Results in mg of lactic acid or P%.

It might be mentioned here (though fuller discussion will come in chapter 15) that Claude Bernard had noticed alkaline rigor in animals deprived of muscle glycogen by starvation.[1] Hoet & Marks (1) also had studied rigor in rabbits after prolonged thyroid treatment and after death from insulin convulsions. In such rigor muscle there was no glycogen, very low lactic acid content and the pH was 7.0–7.2.

Lundsgaard's next step was to investigate the phosphate metabolism. The phosphocreatine breakdown was far greater in the poisoned muscles – the content had in fact fallen to zero, while in the normal muscle the fall was only some 25 % (table 5). The inorganic P did not increase in the poisoned muscle, and further experiments in which the new reduction method of Embden & Jost (1) was used to estimate hexosemonophosphate showed a great increase of ester in the poisoned muscle. No change was at this time found in the pyrophosphate. Taking heat formation in the reactions as a guide to free energy available, Lundsgaard calculated from the lactic acid production of the normal muscle an energy requirement of

[1] Bernard (1) p. 429.

20 000 cal/100 kg of muscle. It was not possible in these preliminary experiments to measure exactly the work performance, but from the records it could be judged that the poisoned muscles had performed about the same amount of work as the controls, probably a little more. The only energy-yielding reactions available according to the analyses were phosphocreatine breakdown and hexosemonophosphate formation. The phosphocreatine breakdown could yield about the required 20 000 cal, the carbohydrate phosphorylation an additional 9000 cal.

Lundsgaard concluded that the recognition of phosphocreatine breakdown as an energy-yielding process in these experiments could not be avoided. He pointed out that further evidence for effective hydrolysis of the phosphocreatine was given by the appearance of its phosphate in a new compound. Since Eggleton & Eggleton[1] as well as Davenport & Sacks (1) had reported for normal muscle on contraction the appearance of some part of the phosphocreatine P as hexosemonophosphate P, this would point to the true breakdown of at least some of the phosphocreatine there also. Lundsgaard went on to suggest that in normal muscle contraction phosphocreatine breakdown directly yielded energy, while lactic acid formation provided energy for its continual resynthesis; lactic acid formation would get under way when a certain concentration of free phosphate (due to phosphocreatine breakdown) had accumulated. The level of this breakdown found at the end of contraction would then depend on the limit at which lactic acid formation began to be called upon. He thought that the restricting action of such drugs as tetramethylammonium chloride on phosphocreatine breakdown could be explained by postulating in their presence the onset of lactic acid formation at a lower phosphate concentration, so that phosphocreatine resynthesis would begin sooner.[2] Lundsgaard advocated return to the view that the heat produced in phosphocreatine breakdown could contribute to the calorific quotient. With regard to the anaerobic phosphocreatine resynthesis, he suggested that the delayed lactic acid formation found on over-stimulation[3] might be an indication that a smaller formation more difficult to detect might occur after normal stimulation. If the resynthetic process went on with 100 % efficiency the 'revival of the concept' of delayed anaerobic heat would not be necessary.[4]

In 1930 Lundsgaard (2) went to work in Meyerhof's laboratory in Heidelberg. There he was able to make much more definite the connection he had wished to draw between phosphagen breakdown and tension produc-

---

[1] P. Eggleton & G. P. Eggleton (3); G. P. Eggleton & P. Eggleton (1).
[2] Later work on the complicated equilibria intervening in phosphagen resynthesis shows that this simple explanation cannot be correct; nevertheless the experiments and interpretations placed upon them are of interest in following the trains of thought at that time.
[3] Ch. 4.   [4] P. 94 below.

tion (fig. 14). In isolated muscles under strictly anaerobic conditions after iodoacetate poisoning, the phosphagen isometric coefficient

$$\left(K_{m(P)} = \frac{\text{length in cm} \times \text{tension in g}}{\text{g } H_3PO_4 \text{ split off}}\right)$$

remained almost constant, in contrast to its behaviour for normal muscle, with only very slight increase. The same was true for the isometric time coefficient $K_{z(P)}$ after tetanus. Again, iodoacetate-poisoned muscles after treatment with curare or trimethyloctylammonium iodide as well as IAA gave normal $K_{m(P)}$ and $K_{z(P)}$ values; these results supported Lundsgaard's earlier idea that the apparently small phosphagen breakdown in presence

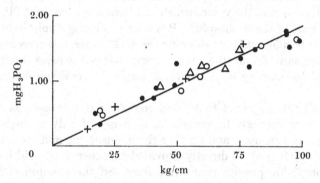

Fig. 14. Relation between tension and phosphocreatine splitting.
● *Esculenta* poisoned with IAA.
+ *Esculenta* poisoned with IAA + curare.
○ *Esculenta* poisoned with IAA + trimethylacetylammonium iodide.
△ *Temporaria* poisoned with IAA. (Lundsgaard (2).)

of these drugs was due to a readier resynthesis. E. Fischer (2) in 1928 had shown that H/Tl has the same value for normal muscles as for those poisoned with tetramethylammonium chloride. In the meantime Meyerhof & Nachmansohn (2), without knowing of these experiments of Lundsgaard, had indeed found that anaerobic phosphagen resynthesis in curarised muscles was absolutely as great as in normal muscles, and so relatively much greater.

Lundsgaard (3) had found in experiments with IAA-poisoned yeast that oxidative metabolism could continue when anaerobic metabolism was impossible; and also that the oxygen uptake of live frogs continued little changed after IAA injection. He now found that, with carefully controlled degree of poisoning, the sartorius of one side provided with oxygen would perform much more work than the corresponding muscle in nitrogen. The former could show undiminished tension production at the time when the latter had passed into rigor. Moreover, the oxygenated muscle had a phosphagen content many times that of the anaerobic one. Thus, even in IAA-poisoned muscle oxidative resynthesis of phosphagen can occur.

Lundsgaard (2) also investigated by Lohmann's hydrolysis method the esters formed on phosphagen breakdown in poisoned muscle, finding values which gave roughly 25% of hexosemonophosphate and 75% of hexosediphosphate. At this time also Lipmann (1) showed that poisoning of frogs by means of fluoride injection gave a similar result to that obtained with IAA – a limited number of contractions followed by rigor, these events being accompanied by phosphocreatine and ATP breakdown and hexose esterification. The presence of this hexosediphosphate was unexpected and necessitated revision of the values which had been accepted for ATP, for an appreciable part of the diphosphate P is released during 7 min hydrolysis, and this must be allowed for in assessing the ATP. When this correction was made, Lundsgaard found that both work and subsequent rigor were accompanied by varying degrees (which might be considerable) of ATP splitting.

Lundsgaard (2) suggested that in so complicated a process as muscle contraction it is simplest to assume that energy of only a single quite definite origin can be used, and that for the moment one must reckon that phosphagen splitting alone directly provided this energy. He did however make the prophetic proviso that there remained the possibility that in normal muscle both processes, phosphagen splitting and lactic acid formation, placed their energy at the disposal of a third unknown process.

After the publication of Lundsgaard's first work, Lipmann & Meyerhof (1) undertook a new study of the pH changes in muscle during anaerobic rest and contraction, and oxidative recovery. They saw that previous results on muscle extract or *brei* had not taken into account the breakdown of phosphagen now known to take place when such preparations were made and (from the electrotitration curves of Meyerhof & Lohmann (8)) to involve a shift to the alkaline side in media at pH about 6.3–7.4; Lipmann & Meyerhof used a new method depending on the measurement of CO_2 exchange with intact frog muscles (sartorius) contained in Warburg manometers under various conditions. If acid equivalents are formed in the muscle, or basic equivalents disappear, the CO_2 pressure will increase, and the change can be measured; with formation of basic equivalents, CO_2 will be taken up from the airspace. The muscles were suspended in nitrogen or oxygen, containing a known percentage of CO_2, or in Ringer solution under these atmospheres, the solution containing bicarbonate (0.02–0.03 M) and phosphate (0.01 M at the desired pH). The muscles could be stimulated *in situ*, and the effect of anaerobic stimulation at different pH values upon the CO_2 exchange is shown in fig 15. It can be seen (i) that when the medium

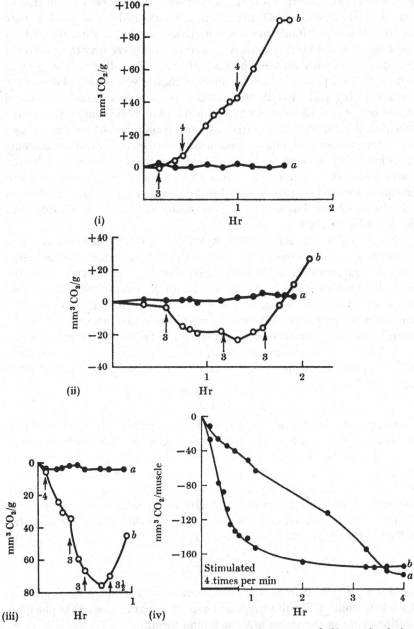

Fig. 15. Course of the alterations in CO_2 pressure (a) on rest and (b) on stimulation. (i) With 0.85 % CO_2 in N. Muscle in Ringer solution with 0.03 M bicarbonate and 0.01 M phosphate, pH 8.4. Stimulated for 3 to 4 min at each arrow. (ii) With 5 % CO_2 in N. Muscle in Ringer solution with 0.03 M bicarbonate, 0.01 M phosphate, pH 7.4. Stimulated for 3 min at each arrow. (iii) Muscle in 15 % CO_2 in N without liquid, pH 6.7. Stimulated 3 to 4 min at each arrow. (iv) Sartorii from IAA-poisoned frog in 34.5 % CO_2 in N, without solution. Stimulation was for 45 min. 4 times a min. (Lipmann & Meyerhof (1).)

was definitely alkaline (pH 8.4) the only effect observed was an acid change
in the muscle, shown by CO_2 evolution; (ii) with medium at pH 7.4 there
was with the first stimulation series a change to the alkaline side, which
after two more stimulation series was reversed and gave way to an increase
in acidity; (iii) with medium of acid pH (6.7) the alkaline change was
marked. In IAA-poisoned muscle, the only change was to the alkaline side,
whether during rest or with stimulation. With normal muscle, chemical
estimations of phosphocreatine breakdown and of lactic acid and ammonia
formation showed that these three reactions could account for the changes
in acid-base equilibrium found manometrically (the change due to ammonia
formation being only about 10 % of that due to phosphocreatine breakdown). The manometric change when IAA-poisoned muscle was used was
accounted for by phosphocreatine breakdown, pyrophosphate breakdown
and ester synthesis. The acid change on aerobic restitution of phosphocreatine could be clearly seen.

Two years later Meyerhof, Möhle & Schulz (1) described experiments in
which the tension production as well as the CO_2 exchange was measured
with a frog sartorius inside the manometer. Since the muscles used were too
small to allow for the chemical estimations, the K_m values for lactic acid and
phosphocreatine were determined on other sartorii stimulated in a parallel
manner in the same gas mixtures as used in the manometers. Of special
interest was the initial period (comprising 150 twitches in 30 min) during
which the curve for gas exchange fell sharply owing to alkali formation.
Here the phosphagen breakdown was most important, and the isometric
coefficient for CO_2

$$\frac{\text{g tension} \times \text{cm length}}{\text{ml } CO_2}$$

was compared with the K_m value for phosphocreatine. It was found that
the observed values always showed a rather greater alkali production than
expected from the phosphagen breakdown. There had been some indications
of this in the earlier paper. The same was seen with IAA-poisoned muscle,
but there the excess alkali formation was less. On the other hand, in the
later part of the curve dominated by lactic acid formation, good agreement
was found for calculated and observed values for the ratio of the isometric
coefficient for CO_2 to that for lactic acid:

$$\frac{K_{m(CO_2)}}{K_{m(L)}}.$$

The discrepancy with regard to alkali formation has not been explained;
Meyerhof, Möhle & Schultz suggested that it might indicate that phosphocreatine exists in the tissue in some bound form.

Lipmann & Meyerhof concluded, in agreement with Lundsgaard, that
such manometric experiments left no doubt of the reality of phosphocrea-

tine hydrolysis during contraction. They discussed the significance of these results as argument against any theory calling in a change in pH at the 'Verkürzungsorten', whether to the acid side (as had been held by Meyerhof himself) or to the alkaline side as postulated by Embden. They also pointed out how a fall in pH causes on the one hand increased rate of phosphocreatine breakdown and increased alkaline change per unit of breakdown; on the other hand hinders lactic acid formation. Thus these effects of any pH change operate together towards maintenance of neutral pH.

Meyerhof & Lohmann in 1928 had felt that, if the behaviour of phosphagen had significance for contraction in vertebrate striated muscle, then its counterpart should be found in invertebrate muscle. Recalling Kutscher's dictum (1) that in invertebrate muscle the place of creatine is taken by arginine, Meyerhof & Lohmann (6) looked for and found phosphoarginine in crustacean muscle.[1] This compound was labile, but the conditions for its breakdown were rather different from those for phosphocreatine breakdown. Thus while the latter is steadily increased by increasing acid concentration, and is catalysed by molybdenum ions, phosphoarginine breakdown shows its maximum rate of hydrolysis at 0.01 N acid, and the hydrolysis is inhibited by Mo ions. It was thus missed by the method used by the Eggletons. This crustacean phosphagen was found to behave physiologically in an analogous manner to vertebrate phosphagen. Lundsgaard (4) found with IAA-poisoned *Maia squinado* (the spider-crab) that phosphoarginine behaved exactly as phosphocreatine in frog muscle.

In a paper in 1931 Lundsgaard (5) discussed the energetics of normal muscle contraction. The outstanding contradictions were (a) the constancy of T/H, lactic acid formation remaining fairly constant per unit of tension and phosphagen breakdown markedly falling off (Nachmansohn (1) for example, had found in a single 2 sec tetanus enough phosphocreatine breakdown to supply all the necessary energy); (b) the failure to find lactic acid formation during anaerobic phosphagen resynthesis.

Lundsgaard now re-investigated, with certain improvements in technique such as estimation of lactic acid and phosphocreatine on the same muscle, the question of the tension coefficients for lactic acid and phosphocreatine in a series of single twitches and for a short series of tetani. With single shocks he found, like Nachmansohn (2), a rising value for

$$\frac{\text{kg tension} \times \text{cm muscle length}}{\text{g } H_3PO_4 \text{ split off}}$$

with increasing tension performance; but he found also a fall in the corresponding lactic acid coefficient which had escaped the earlier workers. The value for T/H, obtained from the measured tension and the heat calculated

---

[1] Also Meyerhof & Lohmann (7, 8).

from the chemical changes was constant, and in good agreement with the value obtained in Hill's myothermic measurements (table 6). With tetani, Lundsgaard found less change in the lactic acid coefficient with increasing number of stimulation periods; but he also found much less change in the phosphocreatine coefficient than Nachmansohn had found, the breakdown with a single 2 sec tetanus being less than Nachmansohn had reported. Careful investigations of this discrepancy led him to conclude that Nachmansohn's high value might be due to his having used estimations in sartorius muscles for resting values, while using gastrocnemi for stimulation. Again T/H was constant. Meyerhof & Schulz (2) independently found the reciprocal change in $K_{m(L)}$ and $K_{m(P)}$ values, and concluded that the heat production could be satisfactorily explained throughout contraction to fatigue on the basis of 270 cal/g lactic acid formation plus de-ionization and 120 cal/g phosphagen H_3PO_4 split off.

Lundsgaard also investigated the question of delayed lactic acid production, and found quite definitely that this occurred. Thus after a 5 sec tetanus more than half the lactic acid formation was postrelaxation. He clearly discusses the difficulties involved in this demonstration. The figures obtained for lactic acid formed and phosphocreatine synthesised averaged to show formation of four moles of phosphocreatine per mole of glycogen breaking down. He calculated[1] from the heats of reaction accepted at that time that the synthesis was carried out with an efficiency of about 100 %, and the small delayed anaerobic heat he could explain by the neutralisation of acid formed.[2]

As we have already seen, both Lehnartz (1) and Meyerhof (13) just at this time also showed convincingly the delayed anaerobic lactic acid formation after tetanic contraction. Neither Meyerhof nor Lundsgaard could accept the evidence of Lehnartz (1) for delayed formation after a series of twitches, and it seems that lactic acid formation is rapid enough to be completed in the interval between the most rapid stimuli which can be used as distinct shocks.

The first study of the time course of heat production in a very short contraction (0.5 sec tetanus) of muscle poisoned with halogen-substituted acetate (in this case bromacetate) was that of E. Fischer (5) in 1930 who found practically no difference comparing results before and after poisoning of the same muscle. He also showed (5, 6) that the ratio Tl/H where H is the total initial heat, was not affected by poisoning with bromacetate, and similar observations with IAA were made independently by Meyerhof, Lundsgaard & Blaschko (1).

---

[1] In these calculations Lundsgaard ((5) p. 325) considers the allowance to be made for the diminution in the protein de-ionisation contribution, owing to neutralisation of a portion of the lactic acid by base equivalents formed on phosphocreatine breakdown.
[2] Chs. 3 and 6.

TABLE 6. *Lactic acid formation and phosphocreatine breakdown after direct stimulation with 16 and 68 single shocks/min*

Date		Muscle weight (g)	Muscle length (cm)	Tension performance (g cm/g)	Twitches /min	Lactic acid (mg/g)	Inorg. P (mg P_2O_5 /g)	Phosphocreatine (mg P_2O_5 /g)	Lactic acid formation (mg/g)	Phosphocreatine breakdown (mg H_3PO_4 /g)	K_{mL}	K_{mP}	T'/H
10 January	a	·	·	·	·	0.16	0.69	1.69	·	·	·	·	·
	b	2.125	4.05	52	16	0.36	1.10	1.27	0.20	0.58	260	90	7.4
10 January	a	·	·	·	·	0.10	0.85	1.45	·	·	·	·	·
	b	1.680	3.8	57	16	0.34	1.26	1.05	0.24	0.55	237	108	7.5
12 January	a	·	·	·	·	0.15	1.00	1.31	·	·	·	·	·
	b	1.857	3.9	140	16	0.96	1.80	0.50	0.81	1.12	173	125	6.7
12 January	a	·	·	·	·	0.15	0.86	1.39	·	·	·	·	·
	b	1.704	3.7	126	68	0.95	1.53	0.69	0.80	0.97	158	115	6.4
14 January	a	·	·	·	·	0.08	0.68	1.75	·	·	·	·	·
	b	2.410	4.4	162	16	0.99	1.86	0.58	0.91	1.62	178	100	6.3
14 January	a	·	·	·	·	0.11	0.70	1.69	·	·	·	·	·
	b	2.150	4.3	156	68	1.00	1.58	0.80	0.89	1.23	176	128	6.8
14 January	a	·	·	·	·	0.10	0.70	1.64	·	·	·	·	·
	b	1.520	3.7	69	16	0.35	1.29	1.08	0.25	0.77	284	90	7.6
14 January	a	·	·	·	·	0.10	0.73	1.50	·	·	·	·	·
	b	2.070	4.0	75	68	0.54	1.01	1.19	0.44	0.43	170	175	7.3

a, resting muscle; b, stimulated. (Lundsgaard (5).)

Closer examination by these last workers (2) in 1931 showed that the constancy of the isometric heat coefficient held only for the beginning of a contraction series with poisoned muscle; when contractions were followed over a longer time there was excess heat formation and the coefficient fell from about 5.5 to about 3.8 in Hill's units (p. 83). The behaviour in this respect was thus quite different from that of normal muscle. While in a short stimulation the heat produced was not much greater than that to be expected from the phosphagen breakdown, the average value for a long series was much higher and during the end period rose to about three times this expected value. Fischer (6) had shown that the onset of bromacetate rigor was attended by considerable heat output. Meyerhof, Lundsgaard & Blaschko considered that ATP breakdown[1] (already noticed by Lundsgaard in the poisoned muscle and now further studied) was to some extent responsible for this extra heat, though unknown reactions must be playing a part and becoming more important as phosphagen breakdown lessened. They were of opinion that so far as the actual contraction heat was concerned, probably nearly all was accounted for by phosphagen breakdown.

THE NEGATIVE DELAYED HEAT

Something should be said here of the negative delayed heat, first observed by Hartree (2) in 1932.[2] Bugnard (1) got indications of it after single twitches in nitrogen and D. K. Hill (3) in 1940 reported it after tetani at 0° in normal and poisoned muscles. Meyerhof & Schulz (3) had suggested in 1935 that it was a reflection of the endothermic resynthesis of phosphocreatine at the expense of phosphopyruvate, but its persistence in IAA-poisoned muscle ruled out this explanation.

This curious phenomenon has been confirmed much more recently by A. V. Hill (36), who observed it at 0° and at room temperature, in oxygen and in nitrogen, in normal and poisoned muscles. He suggested that it may be the result of still unknown chemical reactions accompanying contraction. The evidence for these comes from the work of Hill & Kupalov (1) and of Meyerhof[3] in 1930. Hill & Kupalov (measuring vapour pressure changes) found the increase in osmotic pressure of fluids of a muscle on stimulation was some 20% greater than could be accounted for by changes to be expected (from values in the literature) in lactate, inorganic phosphate, phosphocreatine and other phosphate compounds. Similarly Meyerhof, following the depression of freezing point of the muscle during fatigue and rigor found this about 30% greater than was explicable by the chemical

---

[1] For the measurement of heat output on ATP breakdown see p. 98.
[2] Ch. 3.
[3] Meyerhof (26); Meyerhof & Grollman (1).

reactions he measured. Meyerhof & Grollman found that lactate added to muscle gave just the expected depression of freezing point. Another possible cause of the negative heat considered by Hill lies in the rise in pH observed by Distèche (1) to continue for as much as half a minute after contraction ceased. Such a rise in pH, with consequent increased ionisation of weak acids including proteins, would lead to absorption of heat. Hydrolysis of phosphocreatine in the muscle does of course bring about increase in alkalinity; but there is no evidence for post-relaxation phosphocreatine breakdown in normal muscle so the long drawn-out nature of the effect seen here would suggest some diffusion barrier allowing the hydrolysis only gradually to affect the general pH. Against this explanation is the fact that the appearance of the negative heat is hastened by rise in temperature.

The lactic acid era in theories of direct energy provision for contraction lasted more than twenty years; the phosphagen era hardly five or six. Already in 1934 the first indications of the primary role of ATP appeared, and evidence for this accumulated convincingly from studies *in vitro*. However it was only after some thirty years more that definite proof for this in the intact muscle was obtained.

6

ADENOSINETRIPHOSPHATE AS FUEL
AND AS PHOSPHATE-CARRIER

THE CO-ENZYME FUNCTION OF ATP

In 1931 Meyerhof, Lohmann & Meyer (1) published their first observations on the co-enzyme function of ATP in the glycolytic system of muscle. Muscle *kochsaft* was prepared from extract which had been allowed to autolyse for 1 h at 37° before being boiled; such a *kochsaft* was incapable of restoring activity to a muscle extract which had lost its glycolytic power after many hours dialysis. Replacement of only a small part of this autolysed *kochsaft* with *kochsaft* prepared from fresh extract led to considerable lactic acid formation, though this small quantity had little effect alone. This suggested that the co-ferment system was made up of an autolysable part and a non-autolysable part. Earlier experiments of K. Meyer (1) had suggested that easily hydrolysable phosphate esters with insoluble barium salts were concerned in the co-enzyme activity, and the autolysable, easily hydrolysed component was now identified as ATP. Lohmann (8) showed a little later than the stable component was Mg. With purified ATP he confirmed Meyerhof's earlier finding with *kochsaft* that more was needed by the system the less the substrate was phosphorylated. The same was true for Mg.

The important implications of these observations became clearer with the experiments of Meyerhof & Lohmann in 1932 (9). It was already known that phosphocreatine synthesis could take place in muscle extracts in circumstances such that lactic acid formation could not supply more than a fraction of the necessary energy.[1] It was clearly necessary to look for some other energy-yielding reaction. With this in mind, Meyerhof & Lohmann determined the heat of hydrolysis of ATP and the deamination of adenylic acid. They found for the hydrolysis 170 cal released/g of H_3PO_4 split off, or 25000 cal/g molecule; for the deamination 8000 cal/g molecule. Since the heat of hydrolysis of phosphocreatine was known to be about 12000 cal/g molecule, it is evident that its synthesis (2 moles from 1 mole of ATP) might go on with little heat exchange. Experiments in which phosphocreatine synthesis was measured in glycogen-poor extracts to which ATP was added,

---

[1] Meyerhof & Lohmann (7); E. Lehnartz (2).

showed in general good agreement between the heat directly found and that calculated from the chemical changes (including a small lactic acid formation). Further, a synthesis of ATP from adenylic acid and free phosphate could be shown in actively glycolysing extracts, and the dependence of the phosphocreatine formation in such extracts on the presence of ATP was clearly seen.

The resumption of glycolysis or carbohydrate oxidation observed on adding ATP to inactivated muscle or yeast extracts was widely interpreted as indicating a transfer of phosphate from the ATP to glycogen, hexose or hexosemonophosphate substrate. This is seen in the papers of Lohmann (19), Meyerhof & Lohmann (14), Parnas, Ostern & Mann (2) and Euler & Adler (1) though separate investigation of these initial stages was not possible at that time.

As late as 1935 Meyerhof clung to the early idea (see chapter 4) of the known esters as stabilisation products of active intermediates.[1] Thus he and Kiessling[2] now envisaged the possibility that the hexosediphosphate formed on interaction with ATP was not the normal ester, but retained some of the 'Spaltungsenergie' of the ATP. Similarly the triosephosphate[3] formed was regarded by them not as a known form but as a breakdown product of the primary labile hexose ester. As we have seen, Meyerhof[4] considered hexokinase as an activator capable of converting hexoses into an active form. It was only in this year that he showed (16) that it was in fact an enzyme bringing about phosphorylation of the glucose by means of ATP, in the absence of inorganic phosphate and of any possibility of formation of inorganic phosphate from the ATP.[5] By 1937 he seems to have abandoned this concept of active intermediates and explained the accumulation of hexosediphosphate as due simply to lack of phosphate acceptor in the chain of reactions occurring. About the same time Parnas[6] spoke out strongly against the introduction into biochemical processes of 'stabilised products' and 'hypothetical labile forms'. He wrote 'We shall avoid describing as "stabilisation products" intermediate products well attested by fact but unwelcome for our conclusions; as if they were substances which could arise by spontaneous conversion of labile intermediates by a process outside the course of factual chemistry. I see in the whole field of biochemistry no single product of which one could correctly assume this.'

Meyerhof had been criticised in 1931 by Hahn (1) for paying too little attention to free energy; but Meyerhof and Lohmann were well aware that in the functioning of such systems as we are considering, free energy and

---

[1] Also Nilsson (1) for a presentation of this point of view.
[2] Meyerhof & Kiessling (3).   [3] See p. 107 below.   [4] Meyerhof (15) p. 65 above.
[5] In 1935 also Euler & Adler (1) with yeast preparations showed the presence of an enzyme bringing about phosphorylation of hexose by ATP.
[6] Parnas (5) p. 68 above.

not the heat of the reactions was concerned.[1] However methods at that time were not available for measurement of the free energies, and it was assumed that the difference between the heat of reaction and the free energy available under favourable conditions was not great. Hahn had also criticised Meyerhof's use of the phrase 'energetic coupling', a term unknown to thermodynamics; chemical coupling had been stringently defined by Ostwald (2) as expressible in each case by a single, stoichiometric equation. Hahn remarked that many endothermic reactions needing free energy acquire this from the heat of the surroundings. If such a reaction is accompanied by an exothermic reaction, the temperature fall will be masked, but there is no coupling, since no special exothermic reaction is needed. Hahn seems to suggest that, as far as was known, such an arrangement could operate in oxidative resynthesis of glycogen from lactic acid. Meyerhof (14)[2] now explained that 'energetic coupling' was an intentionally vague expression used only when the nature of the postulated intermediate steps was unknown. One certainly could not speak of chemical coupling in the case of the oxidative synthesis of carbohydrate, since no definite stoichiometric formulation of events could be given. Meyerhof's apparent expectation that such a formulation might eventually be possible has been fully justified by later work.

In the paper with Lohmann[3] there is an interesting passage, which throws light on their ideas of chemical groups as carriers in the coupled anaerobic reactions:

the present experiments lay the foundation of the thesis that the endothermic synthesis of phosphocreatine can take place through a coupling of this process with the exothermic and spontaneous breakdown of ATP, whilst the resynthesis of ATP out of adenylic acid and inorganic phosphate is made possible through the energy of lactic acid formation. One may also assume here a coupling of the synthesis with the metabolism of the intermediate hexose esters and so see in the phosphate groups contained in all these compounds, the unique carriers of the chemical coupling process. This would at once make understandable how the ATP acts as co-enzyme of the lactic acid formation, in that the taking up of phosphate by the hexose stands in connection with the giving up of phosphate on the part of the ATP; or the giving up of phosphate on the side of the ester is concerned with the rebuilding of the pyrophosphate groups on the adenylic acid. This conception would also make possible a special meaning for the fact that for the anaerobic splitting of hexoses intermediate esterification with phosphoric acid is above all necessary.

At the same time it is interesting to notice that perhaps Hahn's suggestion that Meyerhof was setting up a biological (as opposed to a thermodynamic conception) was not entirely wide of the mark. For Meyerhof & Lohmann

---

[1] Also Meyerhof (14).   [2] Also Meyerhof (15) p. 41; von Muralt (3).
[3] Meyerhof & Lohmann (9) p. 460.

were tempted, in cases where the energy provision from pyrophosphate breakdown would not suffice for the phosphocreatine synthesis to provide the deficit by means of the energy afforded on deamination of free adenylic acid. It is not easy to see how any chemical coupling could underlie this particular energy transfer.

Between 1932 and 1935 Lohmann (11, 12) worked out the chemical constitution of ATP and proposed the formula which is still accepted today. He based this on the following considerations (besides those already mentioned concerning the products of acid and alkaline hydrolysis)[1].

(a) Alkaline titration showed that ATP contained three primary acidic groupings and one secondary, adenosinediophosphate two primary and one secondary. On hydrolysis to adenylic acid and free phosphate the expected degree of acidification was found (see fig. 16). (b) Inosinicpyrophosphate and diphosphate could be prepared from ATP and ADP by treatment with nitrous acid; this showed that, contrary to the opinion of Barrenscheen & Filz (1) the pyrophosphate grouping was not attached to the amino group. (c) The formation of a deep blue complex copper salt of ATP in alkaline solution, and a positive Boiseken reaction indicating complex reaction with boric acid. These reactions had been used by Klimek & Parnas (1) to demonstrate the presence in adenylic acid of two neighbouring and stereo-chemically comparable hydroxyl groups – clearly their presence in ATP would preclude the attachment of the pyrophosphate grouping to carbon atom 3 of the pentose, an arrangement which had been suggested by Satoh (1). Lohmann (12) had realised that in a molecule containing so many hydroxyl groups, these tests were not conclusive. In 1945, however, Lythgoe & Todd (1) confirmed the presence of the α-glycol grouping by titration with sodium metaperiodate, and considered Lohmann's structure to be established.

[1] See p. 84.

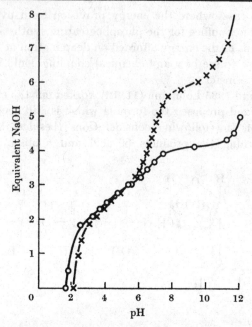

Fig. 16. Electrotitration curve of ATP, ○—○; and after 3 hr hydrolysis at 100°, ×—×. The free ATP possesses four titratable acid valencies; after hydrolysis, six. (Lohmann (11).)

PHOSPHOCREATINE BREAKDOWN AS A RECOVERY REACTION

This chapter will be almost entirely devoted to studies of the enzymic reactions concerned with phosphate transfer, including the further elucidation of the co-enzyme function of ATP. The purpose in mind here is to give as clear a picture as possible of the state of knowledge in 1939, when the work of Engelhardt & Lyubimova first brought ATP breakdown into direct relation with the structural proteins. But we must briefly here return to the muscle itself to consider another important and closely reasoned paper by Lundsgaard (6) in 1934; this gave the first proof that phosphocreatine breakdown, like lactic acid formation, had the status of a recovery process.

This work began with Lundsgaard's re-investigation in 1934 of the possibility of anaerobic phosphagen resynthesis immediately after relaxation in IAA-poisoned muscles. He had already in careful experiments failed to find this.[1] Now, influenced by Meyerhof & Lohmann's recent demonstration that phosphocreatine synthesis could go on in muscle extracts at the expense of ATP breakdown, he tried again. In experiments at 22° again he found no phosphagen synthesis, but rather a slight continuous fall after

[1] Lundsgaard (2).

relaxation. Thinking that any injury to the poisoned muscles which might be occasioned by stimulation would be less at low temperatures, he tried similar experiments at 2°. He then found very clearly a delayed phosphagen *breakdown* (table 7); with a 25 sec tetanus, about 30% occurred after relaxation. Since then a large part of the breakdown was not in the contraction period, it was to be deduced that the whole was concerned in some unknown energy-requiring restitution process. Lundsgaard noticed that poisoned muscles which had become fatigued on stimulation, after some minutes of anaerobic rest showed renewed power of tension production.

TABLE 7. *Phosphocreatine and pyrophosphate content of symmetrical muscles after 25 sec tetanus at 2° (in mg P_2O_5/g of muscle)*

A. Directly frozen		B. Frozen 3 min after relaxation	
Phosphocreatine	Pyrophosphate	Phosphocreatine	Pyrophosphate
1.25	0.58	0.59	0.51
1.31	0.71	1.09	0.78
1.14	0.56	0.93	0.68
0.97	.	0.79	.
0.96	.	0.72	.
1.05	.	0.84	.
1.27	0.60	0.98	0.61
1.10	0.56	0.74	0.64
1.35	0.60	0.96	0.57
1.43	0.66	1.07	0.57
Av. 1.18	0.61	0.87	0.62

Delayed breakdown of phosphocreatine averages 0.31. (Lundsgaard (6).)

This also he took to mean that even in IAA-poisoned muscle synthesis of the unknown energy source was possible. Further, experiments with single twitches showed that towards the end of a series more energy was used than could be supplied by the phosphagen breakdown, which by then was becoming smaller.[1] These time relations too suggested that the unknown energy-yielding reaction was closer than phosphocreatine breakdown to the contraction mechanism, only making itself obvious as the phosphagen breakdown fell off. Lundsgaard considered the possibility of ATP breakdown as this initial reaction. He felt obliged to dismiss this idea for two reasons. First, the evidence from Meyerhof & Lohmann's experiments on extracts showed the opposite effect to that needed here – they of course had observed the use of ATP to build up phosphocreatine; and secondly, in conditions where it seemed that a 'special' energy source besides phosphagen breakdown must be postulated, he found pyrophosphate breakdown was

[1] Compare p. 96.

very variable and might or might not occur. He considered the ATP break-down as particularly concerned with the last stages of contraction and the onset of rigor. It is interesting to notice in connection with this that Mozołowski, Mann & Lutwak (1), in Parnas' laboratory, found in 1931 using IAA-poisoned muscle that by far the greater part of the ammonia production did not accompany the contraction, but occurred rather suddenly when almost all the phosphocreatine had broken down (see fig. 17). It was not known at that time that the muscle deaminase scarcely touches ATP,[1] but it seems now that the outburst of ammonia formation was connected with the appearance of free adenylic acid in the muscle when the phosphorylating activity of phosphocreatine was no longer possible.[2]

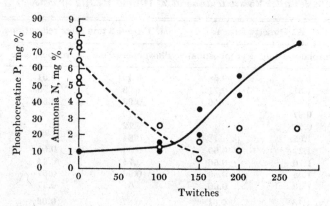

Fig. 17. Ammonia formation and phosphocreatine disappearance in IAA-poisoned muscles. ●—●, ammonia-N content. ○ - - - - ○, phosphocreatine P content. (Mozołowski, Mann & Lutwak (1).)

The turning-point in the realisation of the function of ATP breakdown as the energy-yielding reaction closest, of all those known, to the muscle machine, came with the work of Lohmann (9) in this same year (1934). He showed conclusively that dialysed muscle extract contained no enzyme capable of bringing about the hydrolysis of phosphocreatine. The splitting off of phosphate from this compound was the result of two enzymic activities, which he formulated thus:

$$ATP = adenylic\ acid + 2H_3PO_4\ (+25000\ cal) \tag{1}$$
$$Adenylic\ acid + 2\ phosphocreatine = 2\ creatine + ATP \tag{2}$$

Summing these reactions: 2 phosphocreatine = 2 creatine + $2H_3PO_4$ (+24000 cal).

This participation of ATP was clear from the following observations. (*a*) Phosphocreatine was dephosphorylated by dialysed or inactivated

---

[1] See p. 74.
[2] For the connection of ATP disappearance with onset of rigor see ch. 15.

extracts only if a catalytic amount of adenylic acid or ATP was added. (b) The esterification reaction could be demonstrated in extracts in which the ATPase had been inactivated by more prolonged dialysis, and many days standing at 0°. (c) The ATPase activity alone was found in alkaline extracts (pH 9) where the esterification reaction was hindered.

Lohmann deduced that phosphocreatine breakdown in muscle contraction must be preceded by ATP breakdown, since no adenylic acid (according to the methods of detection of that time) was contained in resting muscle.[1] The values known for the heats of hydrolysis showed that the resynthesis of ATP, like that of phosphocreatine during lactic acid formation, must go on with almost 100 % efficiency.

The esterification shown in equation (2) above is of course the reverse of the reaction already studied in alkaline muscle extracts by Meyerhof & Lohmann (9), when phosphocreatine was synthesised at the expense of ATP; but it did not necessarily follow that it was a single reversible reaction. It was now shown by H. Lehmann (1) that a freely reversible reaction was indeed involved, and with extracts long-dialysed and aged to get rid of side reactions, he studied the reaction expressed by the following equation:

$$\text{ATP} + 2 \text{ creatine} = 2 \text{ phosphocreatine} + \text{adenylic acid.}$$

He found that the equilibrium depended markedly on pH (being shifted to the right at more alkaline reactions) and could be reversibly altered by changing the pH. Changes in concentration of individual components altered the equilibrium in the way to be expected. However, calculation of the equilibrium constant on the basis of a trimolecular reaction gave unsatisfactory results, and Lehmann turned to the possibility that the ADP discovered in crab muscle about this time by Lohmann (10) was playing a part.

Lohmann had shown that dialysed extracts of washed crab muscle split off only one phosphate group of ATP, with formation of adenosinediphosphate. Formation of adenylic acid took place only after the addition of Mg.[2] Dephosphorylation was thus in two stages and he found the same to be true for vertebrate muscle. Soon after Lehmann (2) found satisfactory bimolecular equilibrium constants for the reversible reactions of ATP with arginine and creatine, in crab and frog muscle extracts respectively, to give ADP and the phosphagen. The esterification of creatine by ADP in rabbit muscle extract to give adenylic acid and phosphocreatine seemed probable,

---

[1] This argument loses its force when we remember that ADP rather than adenylic acid is the compound involved in equations (1) and (2) (see p. 106), since by more recent methods ADP can be found in resting muscle. Nevertheless, Lohmann's deduction was correct.

[2] As we shall see, activation of the adenosinetriphosphatase also requires Ca or Mg and Ca ions; presumably enough was left in the dialysed extract for the needs of this enzyme.

he remarked, but could not be established owing to interfering reactions. It seems, however, from later work that only ATP, not ADP, can normally act as phosphate donor, and only ADP, not adenylic acid, as phosphate acceptor in phosphate-transferring reactions.[1] The success of the early experiments using adenylic acid, including those discussed in this chapter, must have been due to the presence of small amounts of ADP or ATP, and of the enzyme myokinase which catalysing the reaction

$$2ADP \rightleftharpoons ATP + AMP.$$

THE FORMATION OF 3-CARBON COMPOUNDS IN GLYCOLYSIS

So far we have paid no attention to the intermediate stages in glycolysis between hexose esters and lactic acid. This question of formation of 3-carbon compounds had quite early been studied in yeast fermentation, and in the late twenties began to be seriously examined in muscle.

In 1913 Neuberg & Kerb (1) found that pyruvic acid was fermented with production of alcohol and CO_2 by a large number of different strains of yeast and different sorts of yeast preparations. This led them to regard pyruvic acid as a likely intermediate in carbohydrate breakdown, perhaps formed by dismutation of methyl glyoxal. Neuberg, in experiments with digested muscle *brei* and muscle extract (1, 2), observed that methyl glyoxal was readily converted into lactic acid, and similar observations were made by Dakin & Dudley (1). Neuberg & Kerb went on to ask, if pyruvic acid is the oxidation product of methyl glyoxal, what is the corresponding reduced product? Glycerol suggested itself and in 1914 Neuberg & Kerb (2) offered for fermentation the scheme which follows. It is significant in embodying two ideas concerning the mechanism of carbohydrate breakdown which have stood the test of time – that of the oxido-reduction stages, and that of cycles, seen here in the reaction of the acetaldehyde formed with the earlier constituent of the chain, the methyl glyoxal. Later Neuberg & Reinfurth (1) and Neuberg & Kobel (1) were able to ascertain with fresh yeast in presence of sodium sulphite as trapping agent for the pyruvate, and also with toluene-treated dried yeast preparations without trapping agent, that equivalent amounts of pyruvate and glycerol accumulated.

$C_6H_{12}O_6 \longrightarrow 2CH_3CO.CHO + 2H_2O$
(sugar)   (methyl glyoxal)

2 methyl glyoxal + $2H_2O \longrightarrow CH_2OH.CHOH.CH_2OH + CH_3CO.COOH$
                              (glycerol)   (pyruvic acid)

Pyruvic acid $\longrightarrow CH_3CHO + CO_2$
                  (acetaldehyde)

Methyl glyoxal + acetaldehyde + $H_2O \longrightarrow$ pyruvic acid + C_2H_5OH
                                                            (alcohol)

---

[1] E.g. Colowick & Kalckar (1); Bücher (1); Boyer, Lardy & Phillips (1).

The counterpart for glycolysis to this scheme would be:

$$2CH_3CO.CHO + 2H_2O \longrightarrow 2CH_3CHOH.COOH$$
(lactic acid)

Then in 1929 Utewski (1) showed that pyruvate added to muscle *brei* disappeared during autolysis and that more lactic acid was formed than in its absence; and Vogt (1) in Neuberg's laboratory found that methyl glyoxal accumulated in muscle juice after hexosediphosphate addition, provided the juice was prepared from muscle digested to remove coenzymes. Hahn[1] (1) also, in a series of papers, dealt with the formation of pyruvate in muscle *brei* from many added substrates, including hexosediphosphate, hexosemonophosphate and lactate. This was particularly marked with washed *brei* to which a hydrogen acceptor, such as methylene blue, was added, but also happened in unwashed *brei* without added acceptor. In 1931 Case & Cook (1), in Hopkins' laboratory, began a study of intermediates in carbohydrate breakdown in muscle. Using sodium sulphite as trapping agent, they found pyruvate formation in muscle *brei* incubated in phosphate buffer under aerobic or anaerobic conditions; the amount was increased by addition of glycogen but not by addition of lactate. Some formation could be shown in the absence of sulphite. Particularly interesting was their finding with frog muscle that the pyruvate content was increased by stimulation. For some years the view was rather generally accepted that methyl glyoxal was probably a stage in normal muscle glycolysis.[2]

In 1933 Embden, Deuticke & Kraft (1) began work which led to fundamental changes in conceptions of fermentation and glycolysis. They were investigating the difficulty hydrolysable esters (believed to be hexose esters)[3] which accumulated in muscle in presence of fluoride. On purification they got a beautifully crystalline salt which turned out to be the secondary barium salt of phosphoglyceric acid. Thus, though fluoride hindered dephosphorylation, it did not prevent splitting of the hexose. Phosphoglyceric acid added to fresh muscle *brei* went very readily to pyruvic and phosphoric acids. Embden now suggested that hexosediphosphate yielded a triosephosphate which underwent a Cannizaro dismutation to give phosphoglyceric acid and glycerolphosphate. He and his collaborators found (*a*) that either of these substances alone added to muscle *brei* gave little lactic acid; (*b*) but that when added together they caused greatly increased lactic acid formation; (*c*) that little phosphate was liberated from glycerolphosphate alone, but its addition with phosphoglyceric acid greatly increased free phosphate production; (*d*) glyceric acid was not formed as an intermediate in pyruvate

---

[1] Hahn, Fischbach & Haarmann (1); Hahn (2).
[2] E.g. Case (1).
[3] Lipmann & Lohmann (1).

formation from phosphoglyceric acid. From these observations they proposed the following scheme in which lactate formation is accompanied by triosephosphate formation, the latter entering again into the cycle.

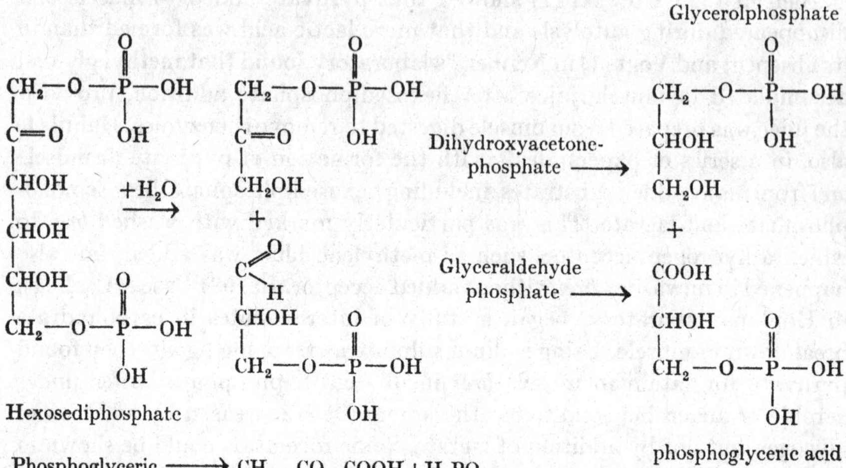

$$
\begin{array}{lll}
 & & \text{Glycerolphosphate} \\
\text{CH}_2\text{—O—P—OH} & \text{CH}_2\text{—O—P—OH} & \text{CH}_2\text{—O—P—OH} \\
\text{C=O} \quad \text{OH} & \text{C=O} \quad \text{OH} \quad \text{Dihydroxyacetone-} & \text{CHOH} \quad \text{OH} \\
\text{CHOH} \quad +\text{H}_2\text{O} & \text{CH}_2\text{OH} \qquad \text{phosphate} \longrightarrow & \text{CH}_2\text{OH} \\
\text{CHOH} \quad \longrightarrow & + & + \\
\text{CHOH} \quad \text{O} & \text{C} \qquad \text{Glyceraldehyde} & \text{COOH} \\
\text{CH}_2\text{—O—P—OH} & \text{CHOH} \qquad \text{phosphate} \longrightarrow & \text{CHOH} \quad \text{O} \\
\qquad \text{OH} & \text{CH}_2\text{—O—P—OH} & \text{CH}_2\text{—O—P—OH} \\
\text{Hexosediphosphate} & \qquad \text{OH} & \qquad \text{OH}
\end{array}
$$

Phosphoglyceric acid $\longrightarrow$ $\text{CH}_3\text{—CO—COOH} + \text{H}_3\text{PO}_4$

phosphoglyceric acid

Pyruvate + glycerolphosphate $\longrightarrow$ lactate + glyceraldehydephosphate

Shortly before this time, Meyerhof had become interested in pyruvate metabolism in muscle, partly because some of the results we have been considering seemed to be in contradiction to his own work on the equivalence of glycogen disappearance and lactic acid formation. Meyerhof & McEachern (1) confirmed Case & Cook's finding that pyruvate is present in normal muscle. They found however that without trapping agent the pyruvate amounted to only about 4 % of the lactate, with sodium sulphite to about 8 %. With very finely divided muscle these percentages were increased to 10 % and 35 % respectively; the amount of lactate formed was reduced by the amount of the pyruvate increase. When glycogen was added, the pyruvate increase was in proportion to the lactate formed; with carbohydrate-free muscle extracts it was clear that the pyruvate was formed from added glycogen, but not from added lactate. When they came to consider the reduced product they observed that the phosphate set free from hexosediphosphate was roughly equivalent only to the lactate plus pyruvate; the reduced product must therefore be still in the phosphorylated form. They suggested glycerolphosphate, and on applying Lohmann's method for study of hydrolysis constants, they found a difficultly hydrolysable ester formed, which was identified as α-glycerophosphate. It was at this point that the paper of Embden, Deuticke & Kraft appeared,[1] and Meyerhof & McEachern made an addendum to their paper, welcoming the complementary nature of the two pieces of work.

To return for a moment to methyl glyoxal; Neuberg & Kobel (1) had found that its enzymic change to lactic acid needed a co-enzyme; and Lohmann (14) had shown this to be glutathione.[2] The co-enzyme system (ATP plus Mg) of glycogen breakdown was ineffective with added methyl glyoxal. Such evidence against this compound as an intermediate was not of course entirely convincing, since it was postulated that the test substance (the synthetic form) might be different from a reactive form produced *in vivo*. Meyerhof & McEachern now remarked that their results and those of Embden, Deuticke & Kraft left no room for methyl glyoxal as a normal intermediate.

We turn now to the effect of fluoride. Lohmann (13) had first observed in muscle extract treated with fluoride the conversion of added hexosediphosphate (or of hexosemonophosphate or glycogen with uptake of phosphate) into much more difficultly hydrolysable ester. He made no claim that it was a single compound, but simply that it had a composition very like that of hexosediphosphate. With extracts without fluoride both hexosemono- and diphosphates could be of course detected during glycolysis; in certain circumstances, notably on very high dilution of the extract, again very similar difficultly hydrolysable ester appeared. Meyerhof & Kiessling (1), following up some preliminary observations of Meyerhof & McEachern (1), found the following effects on adding pyruvate plus glycerolphosphate to muscle extract: In the absence of fluoride, twice as much lactic acid appeared as pyruvic acid disappeared; in the presence of fluoride the pyruvate decrease was equivalent to the lactic acid increase. Moreover it could be shown that fluoride prevented the dephosphorylation of phosphoglyceric acid. Meyerhof & Kiessling interpreted these results in accordance with the scheme of Embden, Deuticke & Kraft. Iodoacetic acid they found to prevent the interaction of the pyruvate and glycerolphosphate. A little later, in 1934, Lohmann & Meyerhof (1) showed that in the dephosphorylation of phosphoglyceric acid with pyruvate formation an intermediate compound was formed. In absence of co-enzyme (for which they used ATP plus Mg) a phosphorylated acid much more easily hydrolysable than phosphoglyceric acid accumulated; this was purified as the silver barium salt and shown by its properties, particularly its reaction with iodine, to be

---

[1] Embden died suddenly in July 1933. This was his last paper to appear during his lifetime. Detailed descriptions of the work (including demonstration of the formation of glycerolphosphate in fluoride-poisoned muscle) are to be found in the following posthumous papers: Embden, Deuticke & Kraft (2); Embden & Deuticke (1, 2); Embden & Ickes (1); Embden & Jost (3); Lehnartz, E. (3).

[2] γ-L-glutamyl-L-cysteinyl-glycine.

a phosphorylated pyruvic acid. This new acid was in temperature-dependent equilibrium with phosphoglyceric acid (70% of the latter at 20°). Its formation from phosphoglyceric acid was prevented by concentrations of fluoride as low as 0.01–0.005 M, but much higher concentrations were necessary to prevent its dephosphorylation.

PHOSPHOPYRUVATE AND PHOSPHATE TRANSFER

Parnas & his collaborators, in their work on ammonia formation in muscle in 1934, had been struck by the fact that when fresh muscle was ground with phosphate solution, the traumatic ammonia formation seen upon grinding with water was avoided.[1] They concluded that conditions favouring lactic acid formation would allow conversion of adenylic acid (the source of ammonia) into ATP and its maintenance in the phosphorylated form for some time. As we have seen,[2] ATP is not deaminated by the muscle enzyme. This interpretation was borne out by the fact that poisoning of the *brei* with either fluoride or iodoacetate prevented the delay in ammonia formation. They decided to try what products of carbohydrate breakdown could, in the presence of these poisons, ward off ammonia formation. They found that in presence of fluoride either phosphoglyceric acid or pyruvic acid was effective; in presence of iodoacetate, only phosphoglyceric acid. Remembering the experiments of Mozołowski, Mann & Lutwak on ammonia formation in IAA-poisoned muscle entering rigor, they concluded that phosphate from a phosphorylated substance between phosphoglyceric acid and pyruvate (phosphopyruvate) was transferring phosphate to creatine, and that the phosphate then passed, by Lohmann's reaction, to adenylic acid. Synthesis of phosphocreatine was indeed found.[3] The effectiveness of pyruvate in fluoride-poisoned *brei* (provided that phosphate was present) led to their suggestion (3) (unfounded as we shall see) that phosphopyruvate could be synthesised from pyruvate and free phosphate. In the light of the experiments described below[4] it seems that suppression of ammonia formation depended on the ATP formed in the process of coupled oxidoreduction and phosphorylation. As for the efficacy of phosphoglycerate in presence of fluoride, this was a puzzle, since, as we have seen, fluoride inhibits both formation of phosphopyruvate and its dephosphorylation. Mann (2) solved this by showing that the inhibition depended on the relative concentrations of phosphoglycerate and fluoride. Conditions could be found where, even with 0.018 M fluoride, pyruvate was formed.

Experiments with muscle *brei* were of course very difficult to interpret, and in their later work the Parnas school used dialysed muscle extracts.

---

[1] Parnas, Ostern & Mann (1).   [2] Ch. 4.
[3] Parnas, Ostern & Mann (2).   [4] See p. 119.

Almost simultaneously, in 1935, Ostern, Baranowski & Reis (1, 2) and Needham & van Heyningen (1, 2) (in Hopkins' laboratory) showed that phosphoglyceric acid was not dephosphorylated in absence of adenylic acid and that its phosphate was transferred (presumably via phosphopyruvate) to adenylic acid, then from the ATP formed to creatine. This point was also realised in Meyerhof's laboratory at almost the same time.[1] Needham & van Heyningen clearly stated the view that the co-enzyme function of the adenylic compounds in glycolysis lay in their ability to act as both phosphate acceptor and phosphate donor. They emphasised that this co-enzyme had been found unnecessary at stages involving no phosphate transfer:

Hexosediphosphate $\longrightarrow$ 2 triosephosphate.[2]
Phosphoglyceric acid $\longrightarrow$ phosphopyruvic acid.[3]
Pyruvic acid + glycerolphosphate $\longrightarrow$ lactic acid + phosphoglyceric acid[4]

Parnas & Baranowski (1) also spoke of 'this circulation of phosphate' and the donor-acceptor aspect of the participation of the adenylic compounds.

The formula of phosphopyruvic acid had been left vague, between the two possibilities

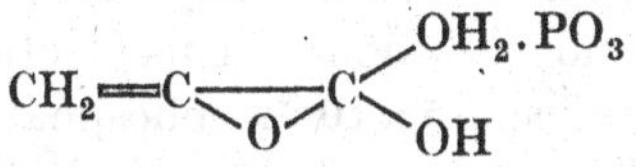

$$CH_2\!\!=\!\!C\underset{O}{\overset{OH_2.PO_3}{<}}C\!\!<\!\!^{OH_2.PO_3}_{OH} \quad \text{and} \quad CH_2\!\!=\!\!C(OH_2.PO_3)\!\!-\!\!COOH$$

the fact that only a very small optical activity was observed (perhaps due to a trace of impurity) being in favour of the latter, phosphoenolpyruvate. This question prompted an investigation into the constitution of the phosphoglyceric acid from which it came, since the formation of this as substituted in position 3 was an assumption depending on its origin via a trisephosphate from fructose-1,6-diphosphate. Meyerhof & Kiessling (2) in 1935 found that the naturally occurring phosphoglyceric acid, whether in yeast fermentation or in muscle glycolysis, was an equilibrium mixture of the 3- and 2-acids. By fermentation of the racemic 3- or 2-acid, the optical activity of the biologically inactive form was found $-$ +15.0 for the former and $-$23.6 for the latter. Since the optical activities of the biologically active acids would be equal and of opposite sign, the amount of each in an equilibrium mixture could be calculated. Purified ($-$) 3-phosphoglyceric acid was obtained by the action of muscle extract, in presence of NaF, on the racemic 2-acid, thanks to the much greater insolubility of its barium salt. Purified (+) 2-acid could be obtained from the ($-$) 3-acid by treating

[1] Meyerhof & Lehmann (1); Lehmann (1
[2] Meyerhof & Lohmann (11).
[3] Lohmann & Meyerhof (1).
[4] Meyerhof & Kiessling (1).

again with muscle extract and fluoride, followed by separation of the barium salts. The equilibrium constant

$$K = \frac{\text{[3-phosphoglyceric acid]}}{\text{[2-phosphoglyceric acid]}}$$

varied between 7.3 and 2.3 with temperature between 60° and 0°.[1] It seems then that a chain of events is concerned in the formation of pyruvate from phosphoglyceric acid:

$$\text{3-phosphoglyceric acid} \xrightarrow{\text{phosphoglyceromutase}} \text{2-phosphoglyceric acid} \xrightarrow{\text{enolase}} \text{phosphopyruvic acid}$$
$$\xrightarrow{\text{adenylic acid}} \text{pyruvic acid} + \text{ATP.}$$

The intermediation of 2-phosphoglyceric acid of course supports the phosphoenol formula for phosphopyruvic acid.

Warburg & Christian (1) in 1941 isolated and crystallised enolase. They observed (as had indeed been remarked by Lohmann & Meyerhof) that, in absence of phosphate, fluoride inhibition was greatly weakened. A kinetic study in presence of Mg phosphate and fluoride convinced them that enolase was being inhibited by very low concentrations of Mg fluorophosphate. It was only twenty years later that this suggestion was actually tested by Peters, Shorthouse & Murray (1), who found no inhibition at concentrations below 5×10^{-2}M, and then only a slight effect. 5×10^{-3}M phosphate + 5×10^{-3}M fluoride inhibited completely. The mechanism of these effects remains mysterious.

Since it had been shown that in presence of phosphopyruvate breakdown, synthesis of ATP could go on from adenylic acid, Meyerhof & Schulz (3) asked what part of the energy of glycolysis might be available at this point. They showed that on hydrolysis of phosphoglyceric acid or phosphopyruvic acid, only 8300 cal is liberated/mole. The formation by this means of phosphocreatine from creatine via adenylic acid would therefore be an endothermic reaction, and this was found. The reaction proceeded with absorption of 2700 cal/mole of phosphate transferred.

ALDOLASE

Meyerhof & his collaborators in 1934 and 1935 were also investigating another reversible reaction in glycolysis – the production of triosephosphate from fructosediphosphate. As the product of this reaction Meyerhof & Lohmann (10) at first identified only dihydroxyacetonephosphate, finding equilibrium to be established in dialysed extracts from either side. The equilibrium constant

$$K = \frac{2 \text{ triosephosphate}}{\text{hexosediphosphate}}$$

[1] For further work into the mechanism of this phosphoglyceric mutase see p. 116 below.

was strongly temperature-dependent, increasing nearly 100-fold between 0° and 70°: thus, according to the equation of van't Hoff's isochore, an endothermic reaction was indicated. This question was studied further by Meyerhof & Lohmann (11, 12) and by Meyerhof & Schulz (4) who found a value of −14000 cal/g mole hexosediphosphate.

A little later, Meyerhof (15) found that the kinetics of the reaction did not correspond to a bimolecular synthesis, but the rate-determining reaction behaved in both directions in a monomolecular fashion. His first suggestion was that this intermediate reaction might be the conversion of the Harden & Young ester into a labile form. It was at this point that Meyerhof & Kiessling (3) published their paper on the 'main path' of lactic acid formation. Here they showed that it is not necessary to postulate continual dismutation of triosephosphate to give glycerolphosphate and phosphoglyceric acid. The oxidoreduction between pyruvate and triose-phosphate itself was found to be as rapid in dialysed muscle extract as normal glycolysis and far more rapid than the oxidoreduction between pyruvate and glycerolphosphate.[1] Moreover the former reaction could go on in systems (such as extract of muscle acetone powder, as had been remarked by Aubel & Simon (1)) which could show active glycolysis but could not carry out the latter reaction.

Meyerhof & Kiessling (4) had been interesting in glyceraldehydephosphate partly because it appeared in the original scheme of Embden, Deuticke & Kraft, partly because Smythe & Gerischer (1) had shown with the synthetic preparation of Fischer & Baer (1), that the dextrorotatory component was fermented to lactic acid, and they had themselves shown[2] that in presence of fluoride, phosphoglyceric acid can arise from it. They now found that in dialysed muscle extract, added synthetic glyceraldehydephosphate went readily (to the extent of one-half) to dihydroxyacetonephosphate. Further work showed[3] that the latter in dialysed muscle or yeast extract could undergo an aldol condensation with a variety of aldehydes. The conclusion was inescapable that the glyceraldehydephosphate was concerned in this stage of glycolysis. Meyerhof (19) in 1938 clinched the matter by capturing the triosephosphate formed during aldolase activity by means of hydrazine and showing that it consisted of equal parts of glyceraldehydephosphate and dihydroxyacetonephosphate. The hexosediphosphate breakdown should thus be formulated as in the formula overleaf.

It was at first thought that the isomerase reaction (shown to have its equilibrium far to the side of dihydroxyacetonephosphate) could be the monomolecular limiting reaction, but Meyerhof (27) has more recently

---

[1] This opinion was also reached by Parnas, Sobczuk & Mejbaum (1) as a result of experiments by the method of suppression of NH_3 formation.
[2] Meyerhof & Kiessling (1).   [3] Meyerhof, Lohmann & Schuster (1, 2).

pointed out that in the meantime it has been found for many enzymes that, because of competitive affinity of reactant for enzyme, a bimolecular type reaction may have a monomolecular action rate.

$$
\begin{array}{lll}
& & \text{dihydroxyacetonephosphate} \\
\mathrm{CH_2O.H_2PO_3} & & \mathrm{CH_2OH_2PO_3} \\
| & & | \\
\mathrm{CO} & & \mathrm{CO} \\
| & & | \\
\mathrm{HOCH} & \text{(aldolase)} & \mathrm{CH_2OH} \quad \uparrow\downarrow \text{ isomerase} \\
| & \rightleftharpoons & + \\
\mathrm{HCOH} & & \mathrm{CHO} \\
| & & | \\
\mathrm{HCOH} & & \mathrm{HCOH} \\
| & & | \\
\mathrm{CH_2O.H_2PO_3} & & \mathrm{CH_2OH_2PO_3} \\
\text{Hexosediphosphate} & & \text{glyceraldehydephosphate}
\end{array}
$$

ESTERIFICATION OF INORGANIC PHOSPHATE

PHOSPHORYLATION OF GLYCOGEN. The mechanism of uptake of inorganic phosphate in muscle itself and in muscle extract raised interesting questions. We have seen for the latter[1] that with glycogen or hexosemonophosphate as substrate, the esterification of free phosphate can be very marked. Up to rather late in the decade, opinions as to the reactions involved in the esterification seem to have been uncertain; as we shall see, phosphate is taken up at two distinct stages of glycolysis, and the unravelling of the mechanisms proved complicated. Meyerhof until 1932 for example was inclined to the view that the initial phosphorylation of glucose to give the mono- or diester went on by direct transfer of phosphate from ATP to an active hexose (formed as the result of hexokinase activity); he emphasised however that this reaction had never been stringently tested for. Formation of a similar active hexose seems to have been pictured as the first stage in glycogen breakdown.[2] The Parnas school visualised the formation of fructosediphosphate by direct transfer from ATP to glycogen.[3] But when the carbohydrate esterification was at expense of inorganic phosphate, this co-enzyme mechanism left unexplained the mechanism of phosphate uptake.

Clarification began in 1935 when Parnas & Baranowski (1) made the observation that in extracts thoroughly dialysed at room temperature, if Mg was added, glycogen could be phosphorylated to a difficultly-hydrolysable ester by inorganic phosphate in absence of ATP. Ostern & Guthke (1) confirmed the suggestion that the ester formed was hexosemonophosphate – the Embden ester. They also showed that the further phosphorylation of this ester to hexosediphosphate did only take place at the expense of ATP.

[1] Cf. p. 64 above.   [2] Lohmann (16) p. 919   [3] Parnas (4).

Lundsgaard (7), in investigating the glycosuric effect of the glucoside phlorizin, had found with fluoride-poisoned muscle *brei* that the former drug, in a concentration of M/50, completely prevented esterification of glycogen. Ostern, Guthke & Tersakovec (1) now showed with autolysed and dialysed muscle extract, that M/200 phlorizin had no effect on the formation of hexosediphosphate from hexosemonophosphate and ATP, while it inhibited to 75 % the phosphorylation of glycogen by inorganic phosphate.

As regards intact muscle, C. F. Cori & G. T. Cori (1) found in 1936 that soaking of frog muscle in solutions containing adrenaline and iodoacetate led to accumulation of hexosemonophosphate accompanied by decrease in inorganic phosphate and only slight drop in phosphocreatine. On treatment with iodoacetate alone, hexosediphosphate accumulated; there was decrease in phosphocreatine but no liberation of inorganic phosphate. They further observed,[1] using minced frog muscle (washed with water so that it had lost the power to form lactic acid, and contained only traces of acid-soluble organic phosphate), that in presence of isotonic phosphate buffer the hexosemonophosphate content rose. This increase was greatly enhanced by addition of small amounts of ATP or adenylic acid, which they regarded at that time as transferring inorganic phosphate to the carbohydrate.[2] The ester first formed was not glucose-6-phosphate; it was readily hydrolysable in acid, and its lack of reducing activity indicated an aldose-1-phosphate, presumably glucose-1-phosphate. C. F. Cori, Colowick & G. T. Cori (1) prepared the synthetic compound, and found properties to agree with those of the natural ester; added to dialysed frog-muscle extract, both went rapidly to glucose-6-phosphate. Parnas (5) pointed out that this reaction of inorganic phosphate with glycogen could be formulated as in the figure.

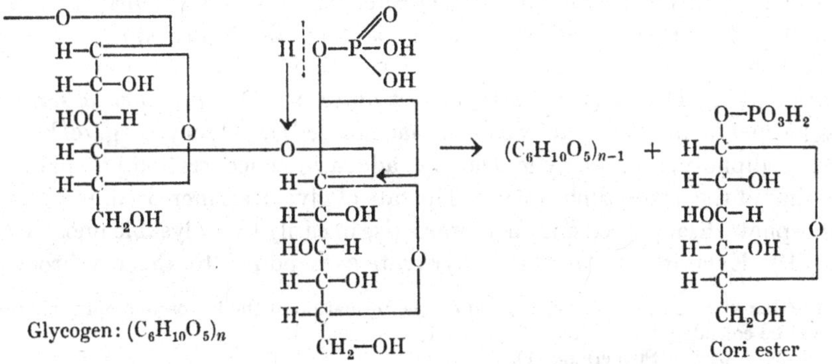

Glycogen: $(C_6H_{10}O_5)_n$ $\rightarrow$ $(C_6H_{10}O_5)_{n-1}$ +   Cori ester

---

[1] G. T. Cori & C. F. Cori (1).
[2] But see ch. 18.

The mechanism did not involve water, as in the hydrolysis of ATP or phosphocreatine but might be termed a 'phosphorylysis'.[1]

A few years later Sutherland, Colowick & Cori (1)[2] found that the conversion of glucose-1-phosphate to glucose-6-phosphate by the enzyme phosphoglucomutase was not an irreversible reaction; equilibrium was reached with about 6% of the glucose-1-phosphate still present. In 1948 Leloir and his collaborators[3] noticed in yeast fermentation that this enzyme needed the presence of a thermostable factor. They obtained evidence[4] that this co-enzyme was glucose-1,6-diphosphate and suggested a mechanism whereby phosphate was transferred from the 1-position in the diphosphate to the 6-position in the 1-ester. A molecule of glucose-6-phosphate was thus formed at the expense of a molecule of the diphosphate, but another molecule of the latter was generated to continue the reaction. Cori and his collaborations[5] had been puzzled by the lack of reaction of highly purified glucose-1-phosphate with phosphoglucomutase while it readily reacted with phosphorylase; they saw an explanation in this co-enzyme requirement and quickly brought confirmation for the muscle enzyme. They synthesised the glucosediphosphate and by use of glucose-1-phosphate labelled with ^{14}C and ^{32}P brought evidence for the mechanism suggested by Leloir and his collaborators. It is interesting that Sutherland, Posternak & Cori (1) went on to show that the phosphoglyceric mutase also needed a co-enzyme – 2,3-diphosphoglyceric acid – and that the mechanism of phosphate transfer was exactly analogous to that operating in the phosphoglucomutase reaction.

COUPLED OXIDOREDUCTION AND ESTERIFICATION. We come now to a second and more complicated method of phosphate esterification, that connected with the oxidoreduction stage. One of the first to mention observations of this sort was Nilsson (1) referring, without comment, to the uptake of inorganic phosphate accompanying the oxidoreduction when pyruvate was added to a glucose fermentation held up by some specific poison. Then Dische (1, 2) in two short notes in 1934 and 1936, described experiments in which to washed, haemolysed erythrocytes glycolysing hexosediphosphate, adenylic acid was added in concentration comparable to that of the hexosediphosphate. The rate of glycolysis increased, inorganic phosphate disappeared and there was a rise in easily hydrolysable phosphate (ATP). Experiments in which pyruvate was added to the erythrocyte

---

[1] Further work on phosphorylase, and on the formation of the hexosemonophosphates, will be considered in ch. 18.
[2] Also Colowick & Sutherland (1).
[3] Caputto, Leloir, Trucco, Cardini & Paladini (1).
[4] Cardini, Paladini, Caputto, Leloir & Trucco (1).
[5] Sutherland, Cohn, Posternak & Cori (1).

suspension containing glucose and fluoride showed oxidoreduction (formation of equivalent amounts of lactic and phosphoglyceric acids), formation of the difficultly hydrolysable ester being at the expense of inorganic phosphate. Dische assumed that ATP was acting as intermediate.[1] This phosphate uptake was inhibited by iodoacetate. Meyerhof & Kiessling (5) also observed for yeast fermentation of glucose poisoned by fluoride that addition of phosphopyruvate started a series of reactions expressed thus:

2 mole phosphopyruvate + 2 mole hexose + 2 mole H_3PO_4 ⟶
2 mole CO_2 + 2 mole alcohol + 1 mole hexosediphosphate + 2 mole phosphoglyceric acid.

They interpreted this to mean an oxidoreduction (between pyruvate and triosephosphate formed) which was accompanied by esterification of glucose by means of free phosphate. Without attempting to suggest a mechanism, they argued that free energy was probably available at the oxidoreduction stage, and remarked on the light which such phosphate uptake threw on the Harden–Young equation.[2] Similar results with muscle extract were found a little later.[3] Schäffner & Berl (1) also reported such observations with yeast fermentation. In both cases the phosphate uptake was prevented by iodoacetate.

It is necessary now to say something of the history of co-zymase since the discovery by Harden & Young of a dialysable activator essential in fermentation. Euler and his colleagues in researches over many years studied this co-enzyme and as we have seen,[4] they came as early as 1926 to the view that it was concerned with oxidoreduction. Their purification studies showed that it consisted to a very large extent of material with an adenine-nucleotide structure,[5] although it was clearly different from muscle adenylic acid. Myrbäck (1) in 1934 obtained muscle adenylic acid from it by treatment with 0.01 N alkali. Occasional confusion between the action of the two co-enzymes can probably be explained by (a) presence of free adenylic acid as impurity in co-zymase preparations; (b) the necessity under certain conditions to add a phosphate acceptor at the oxidoreduction stage, as we shall see. In 1931 Warburg & Christian (2) discovered in red blood cells a third co-enzyme, and a great step forward was taken in 1935 when they showed[6] that this was a dinucleotide containing adenine and a pyridine base (nicotinic acid amide), and three phosphate groups. Its activity in removing hydrogen from a substrate (glucose-6-phosphate with formation of phosphohexonic acid) was shown to depend on the reduction of the pyridine ring. Euler and his collaborators[7] as well as Warburg & Christian (3) then investigated co-zymase itself, and found that it also was a similar

---

[1] Also Dische (3) for experiments on intact washed erythrocytes.
[2] See p. 66.   [3] Meyerhof & Kiessling (3).
[4] Ch. 4.   [5] Euler & Myrbäck (3).
[6] Warburg, Christian & Griese (1).   [7] Von Euler, Albers & Schlenk (1, 2).

dinucleotide, but contained only two phosphate groups. Warburg & Christian worked out a 'Gärtest' to show the mechanism of co-zymase action. They made two protein fractions, A and B, from yeast juice; these used together could bring about the following reactions (ATP and co-zymase both being present):

1 mole hexosemonophosphate + 2 mole acetaldehyde + 2 mole $H_2O \longrightarrow$
$\qquad$ 1 mole pyruvate + 1 mole phosphoglyceric acid + 2 mole alcohol
accompanied by

1 mole hexosemonophosphate + 1 mole $H_3PO_4 \longrightarrow$
$\qquad\qquad\qquad$ 1 mole hexosediphosphate + 1 mole H_2O.

The combined fractions could not in their opinion bring about dephosphorylation of phosphoglyceric acid. If the acetaldehyde were omitted, reduced co-zymase was formed and could be estimated by the now familiar absorption at 340 mμ. Addition of acetaldehyde then led to oxidation of the reduced co-enzyme. Warburg believed that the hexosemonophosphate was itself oxidised by the co-enzyme, with splitting into pyruvate and phosphoglycerate.

Meyerhof, Kiessling & Schulz (1) re-investigated Warburg's 'Gärtest', since production of pyruvate in this way (not via phosphopyruvate) was contrary to their experience. They found that protein B *did* contain enolase; its action on phosphoglyceric acid however could only be demonstrated in presence of a phosphate acceptor – and this would be the reason it was missed by Warburg & Christian. The reason for dephosphorylation of only about half the phosphoglyceric acid would also be deficiency of acceptor, since the substrate hexosemonophosphate was already partly esterified, and had to take up the phosphate esterified in connection with the oxidoreduction as well as that from the phosphopyruvate. Meyerhof, Kiessling & Schulz, as the result of a long series of experiments under many different conditions of the 'Gärtest' concluded that rapid hexosemonophosphate breakdown involved hexosediphosphate thus:

(1) 1 mole hexosediphosphate ($\longrightarrow$ 2 mole triosephosphate)
$\qquad$ + 1 mole hexosemonophosphate + 1 mole H_3PO_4 + 2 mole acetaldehyde $\longrightarrow$
2 mole phosphoglyceric acid + 1 mole hexosemonophosphate + 2 mole alcohol

followed by:

(2) 1 mole hexosemonophosphate + 1 mole phosphoglyceric acid ($\longrightarrow$ phospho-
$\qquad\qquad\qquad\qquad\qquad\qquad\qquad\qquad\qquad$ pyruvate) $\longrightarrow$
1 mole pyruvate + 1 mole hexosediphosphate.

The phosphoglyceric acid formed in equation 1 was considered to arise from the phosphorylation of inorganic phosphate accompanying the oxidoreduction; and transfer of phosphate to take place via phosphopyruvate and adenylic compounds. Warburg's protein-A was shown to contain the enzymes needed for this phosphate transfer; all the other necessary enzymes were contained in his protein-B.

Meyerhof, Ohlmeyer & Möhle (1) turned their attention to co-zymase function in muscle extract. They showed that it was essential for the oxidoreduction; this had not been recognised earlier because only with very long dialysis could the last traces be removed. In a preliminary paper Meyerhof (17) had examined in detail the coupled phosphorylation in muscle extract. He showed that the phosphate uptake depended on the presence of the adenylic system, and calculated that about half the energy of carbohydrate breakdown must be available at the oxidoreduction stage, so that synthesis of ATP was perfectly feasible. Presence of a phosphate acceptor was necessary for rapid reaction, and for this acceptor Meyerhof used creatine. Needham & Pillai (1)[1] very shortly after and independently showed this coupled reaction using adenylic acid in acceptor, not catalytic, amounts. The overall reaction may be expressed thus:

2 triosephosphate + 2 pyruvate + $2H_3PO_4$ + AMP ⟶
$$2 \text{ phosphoglyceric acid} + 2 \text{ lactate} + \text{ATP.}$$

They found that this method of phosphate uptake, unlike the phosphorylation of glycogen, was not affected by phlorizin. Both Meyerhof and these workers showed the necessity for co-zymase in the coupled reaction, and also that it was at this oxidoreduction stage that IAA exerted its inhibitory action. They emphasised that knowledge of the coupled oxidoreduction–phosphorylation greatly lessened the gap between results *in vitro* and *in vivo*. In the breakdown of one molecule of glycogen to give two molecules of lactic acid, four molecules of ADP are phosphorylated to ATP. One of these is needed to replace the ATP used in phosphorylation of hexosemonophosphate, leaving three for phosphorylation of creatine. The figures given by Lundsgaard (5) for the anaerobic recovery period for lactic acid formed and phosphocreatine synthesised average to show a resynthesis of four moles of phosphocreatine/mole of glycogen.[2] Meyerhof (20), writing in 1944, accepted this relationship for the living muscle, but found it puzzling. The question does not seem to have been investigated further.

A more detailed consideration of the energy relations at the oxidoreduction stage may be found in the papers of Meyerhof, Schulz & Schuster (1), Meyerhof, Ohlmeyer & Möhle (1, 2) and Meyerhof (18). It was not possible to measure the heat production of the single reaction

Triosephosphate + pyruvate ⟶ phosphoglycerate + lactate

because of the presence of isomerase in the extract. But from knowledge of the total heat production of the formation of lactic and phosphoric acids

---

[1] Also Needham & Pillai (2).   [2] Ch. 5.

from hexosediphosphate, and of the heat produced at the other intervening stages, it was calculated that the heat production of the oxidoreduction would be about 8000 cal/mole lactic acid formed. This would be increased by neutralisation of the acid formed to about 16000 cal. Thus the actual formation of ATP at this stage would be an endothermic reaction (as is its formation at the expense of phosphopyruvic acid) but the cooling effect would be masked by the heat due to neutralisation. Meyerhof, Schulz & Schuster had obtained results under certain conditions which suggested that phosphate uptake could accompany re-oxidation of reduced co-zymase. Later work,[1] however, showed this interpretation to be erroneous, and Needham & Lu (1) directly testing the point, found no esterification during the reaction between reduced co-enzyme and pyruvate.

Already the earlier work of Meyerhof, Schulz & Schuster (1) had suggested, by showing the necessity for phosphate acceptor in order that the coupled reaction might run, that reversible reactions were involved. Green, Needham & Dewan (1) then found, with dialysed extract of muscle acetone powder, that the reverse reaction

$$\text{phosphoglyceric acid} + \text{lactate} \longrightarrow \text{triosephosphate} + \text{pyruvate}$$

could take place in the presence of 0.07 M cyanide to trap the products. The reaction was accompanied by an unexplained release of inorganic phosphate, which seemed to be dependent on the reversed oxidoreduction, since like the latter, it disappeared in absence of co-zymase or presence of iodoacetate. Shortly afterwards Meyerhof, Ohlmeyer & Möhle (2) showed that the reaction ran:

$$\text{phosphoglyceric acid} + \text{lactic acid} + \text{ATP} \longrightarrow \text{triosephosphate} \\ + \text{pyruvate} + H_3PO_4 + \text{ADP}$$

– in fact, that the whole coupled oxidoreduction–phosphorylation was reversible, even in absence of trapping agent. The inorganic phosphate noticed by Green, Needham & Dewan was no doubt to be explained by the presence of traces of adenylic compounds in their co-zymase.

Adler & Günther (1) in von Euler's laboratory, studied further the reaction between glyceraldehydephosphate (for which they used hexosediphosphate as source) and co-zymase. They found the reduction of the co-enzyme to come to a standstill before it was completed, as though at an equilibrium point; but this equilibrium was not influenced by adding the supposed end product, phosphoglyceric acid. In presence of arsenate, however (which both Meyerhof, Kiessling & Schulz (1) and Needham & Pillai (1, 2) had shown to uncouple phosphorylation from the oxidoreduction reaction) the reduction of co-enzyme went quickly to completion. Adler &

---

[1] Meyerhof, Ohlmeyer & Möhle (2).

Günther postulated the formation of an unknown intermediate, and light was soon thrown on this by the work of Warburg & Christian (4) in 1939, and of Negelein & Brömel (1) in Warburg's laboratory.

Warburg & Christian prepared pure crystalline glyceraldehydephosphate dehydrogenase, and were thus for the first time able to ascertain that glyceraldehydephosphate was indeed the substrate in the coupled oxido-reduction–phosphorylation; while Negelein & Brömel (1) isolated the product and found it to be diphosphoglyceric acid. This acid, which differed markedly in its properties from the 2,3-diphosphoric acid already known, was extremely readily hydrolysed and was shown by its ultra-violet absorption spectrum to be an acid anhydride. It was thus the 1,3-acid, and Warburg & Christian proposed the following mechanism:

glyceraldehydephosphate + H_3PO_4 ⟶ glyceraldehyde diphosphate
glyceraldehydediphosphate + pyridinenucleotide ⟶ 1,3-diphosphoglyceric
                              acid + dihydropyridinenucleotide.

It was left open whether the phosphorylation of the aldehyde group was spontaneous or enzymic.[1] Bücher (1) in Warburg's laboratory, also prepared a specific protein which brought about the transfer of phosphate from diphosphoglyceric acid to ADP. Warburg explained the action of arsenate as due to the replacement by arsenate of the phosphate attached to the carboxyl group, with formation of a very unstable arsenophosphoglyceric acid, which broke down at once spontaneously to give monophosphoglyceric acid. This presentation of Warburg's held the field for the next twelve years. Meyerhof & Junowicz-Kocholaty (1) however were inclined to reject this proposal since in a very careful kinetic study they could find no evidence for the participation of a diphosphate of glyceraldehyde in the equilibria set up by aldolase and isomerase.[2] Moreover they could find no sign of formation of a second asymmetric C atom. About 1951 quite new ideas began to emerge, and it now seems clear that the glyceraldehyde-phosphate combines with the enzyme and formation of an acyl derivative of the enzyme follows. The acid is then removed by phosphorolysis, to give the 1,3-diphospho acid.[3]

FORMATION OF FREE PHOSPHATE

The power of dialysed, non-glycolysing muscle extract to dephosphorylate ATP, hexosediphosphate (to hexosemonophosphate) and inorganic pyro-phosphate was early shown by Lohmann (15),[4] and in the study of reactions

---

[1] As Warburg wrote (Warburg & Christian (4)) '...kann es nicht anders sein als dass zunächtst Phosphat und Fischer ester (reversibel) unter Bildung von 1,3-Diphospho-glycerinaldehyd reagieren, ob von selbst oder an dem Protein, bleibe dahingestellt'.
[2] Also Meyerhof & Oesper (1).   [3] Ch. 17.   [4] Also Pillai (1).

where phosphate transfer was needed the importance of ATPase in providing continuous supply of acceptor was recognised. In intact muscle, contraction was always associated with formation of free creatine and free phosphate, the latter being rather less than the former.[1] Although attention during this decade was mainly centred on phosphate transfer in muscle metabolism, the work of Lohmann and of Lundsgaard clearly suggested that the first chemical reaction yielding energy might be ATP dephosphorylation. D. Needham (1) remarked in 1938 that 'the problem of transferring energy from the ATP to the myosin micellae is still unsolved'. This problem will be considered in the chapters following.

CONCLUSIONS CONCERNING METABOLIC CHANGES IN CONTRACTION AND RECOVERY

We may conclude with two schemes prepared by Needham (1) in 1938 summing up work from all the sources we have been considering up to that time; one illustrates the contraction process, the other the changes involved in anaerobic recovery.

THE ENERGY-RICH PHOSPHATE BOND

The year 1941 is memorable for the appearance of two reviews, by Lipmann (2) and Kalckar (4), in which the fundamental concepts of physical and organic chemistry and of thermodynamics were applied to consideration of the accumulated knowledge of intermediate reactions in glycolysis and fermentation.

Lipmann, setting out from the well-established fact of the utilisation of free energy from phosphocreatine hydrolysis for muscle contraction and from the known possibilities of phosphocreatine resynthesis in muscle, introduced the term 'energy-rich phosphate bond' ($\sim$ph). An alternative mode of expression which he suggested is that of 'high group potential'; by utilisation of the energy of certain metabolic processes, certain such groups can be prepared and then transferred into desired situations.

As we have seen, around 1930 Meyerhof & Lohmann (5, 9) and Meyerhof & Schulz (3) had concerned themselves with measurements of the heat production associated with phosphocreatine and ATP hydrolysis and with various intermediate reactions in glycolysis, assuming that $-\Delta H$ might be a guide to $-\Delta F$, the free energy liberated. In this way they gained some knowledge of energetic coupling.

Lipmann now divided phosphate esters into two classes: (1) those in which the phosphate was linked to an alcohol OH group, and for which the

[1] Ch. 5.

Contraction scheme

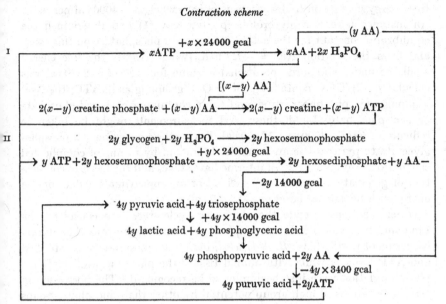

I

$$\rightarrow xATP \xrightarrow[\;+x\times 24\,000\text{ gcal}\;]{} xAA + 2x\ H_3PO_4$$

$(y\ AA)$

$[(x-y)\ AA]$

$2(x-y)$ creatine phosphate $+(x-y)\,AA \longrightarrow 2(x-y)$ creatine $+(x-y)\ ATP$

II

$2y$ glycogen $+2y\ H_3PO_4 \longrightarrow 2y$ hexosemonophosphate

$\rightarrow y\ ATP + 2y$ hexosemonophosphate $\xrightarrow[\;+y\times 24\,000\text{ gcal}\;]{} 2y$ hexosediphosphate $+y\ AA-$

$-2y\ 14\,000$ gcal

$\rightarrow 4y$ pyruvic acid $+4y$ triosephosphate

$\downarrow\ +4y\times 14\,000$ gcal

$4y$ lactic acid $+4y$ phosphoglyceric acid

$\downarrow$

$4y$ phosphopyruvic acid $+2y\ AA \leftarrow$

$\downarrow -4y\times 3400$ gcal

$4y$ puruvic acid $+2y ATP$

Reactants disappearing: $2(x-y)$ creatine phosphate; $2y$ glycogen.
Reactants appearing: $2(x-y)\ H_3PO_4$; $4y$ lactic acid.

Anaerobic recovery scheme

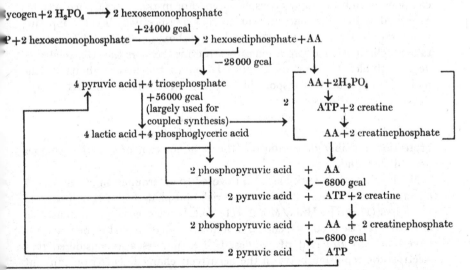

lycogen $+2\ H_3PO_4 \longrightarrow 2$ hexosemonophosphate

$+24\,000$ gcal

P $+2$ hexosemonophosphate $\xrightarrow{\hspace{2cm}} 2$ hexosediphosphate $+AA$

$-28\,000$ gcal

4 pyruvic acid $+4$ triosephosphate

$+56\,000$ gcal
(largely used for
coupled synthesis)

4 lactic acid $+4$ phosphoglyceric acid

$AA + 2H_3PO_4$

$\downarrow$

ATP $+2$ creatine

$\downarrow$

$AA + 2$ creatinephosphate

2 phosphopyruvic acid $+$ AA

$\downarrow -6800$ gcal

2 pyruvic acid $+$ ATP $+2$ creatine

2 phosphopyruvic acid $+$ AA $+$ 2 creatinephosphate

$\downarrow -6800$ gcal

2 pyruvic acid $+$ ATP

Reactants disappearing: glycogen; creatine; phosphate.
Reactants appearing: lactic acid (4 mole); creatine phosphate (6 mole).

free energy change on hydrolysis was only -2000 to -4000 cal per mole. An instance here was glycerolphosphate; Kay (1) had determined the equilibrium constant for the action of intestinal phosphatase on this ester, and from the relation $\Delta F^0 = -RT\ln K$ (where ΔF^0 is the free energy available under standard conditions) a value for ΔF^0 of -2280 cal was calculated. (2) Those containing the P–O–P grouping as in ATP; the N–P grouping as in phosphocreatine; and those containing carboxyl-phosphate or enol-phosphate. In all these, ΔH measurements might be taken to indicate a ΔF^0 of -8000 to $-12\,000$ cal. Taking the conversion of 2-phospho-glycerate to pyruvate as an example, Lipmann, by means of complicated calculations (involving ΔF^0 of intermediate stages and the heats of combustion of glycerate and pyruvate) arrived at an approximate value for the energy-rich phosphate bond of $-11\,250$ cal.

Other reactions outside the glycolytic cycle were also considered by Lipmann and Kalckar. For example it was already known[1] that during oxidation of pyruvate by *B. delbrückii*, inorganic phosphate was esterified and ATP formed; while with animal tissues the phenomena of oxidative phosphorylation were just beginning to be recognised.[2] The possession in certain situations of high group potential by other than phosphate groups (e.g. by the methyl group of choline and methionine) was also suggested by Lipmann.

Kalckar discussed the chemical basis for the high group potential in terms of differences in resonance stabilisation between the energy-rich compounds and their hydrolysis products. For instance, resonance systems exhibited by the phosphate and carboxyl ions respectively become opposing when the two groups are combined as carboxylphosphate. This simultaneous elimination of two resonating structures renders the ester relatively very unstable. This instability in relation to the high stability of the hydrolytic products is responsible for the large amount of free energy liberated on hydrolysis.

Lipmann (7) has commented on the importance of the protection of the potentially labile bond in the phosphoryl linkage against hydrolysis – protection possibly depending on the accumulation of negative charges around the bond.

Much subsequent work has been devoted to determination of the factors affecting the ΔF^0 value for the energy-rich phosphate bond. It was stressed by Oesper (1) and by Hill & Morales (1) that, besides resonance stabilisation of the hydrolytic products, energy of ionization plays a part if the reactants have markedly different pK values. Hill & Morales also considered that electrostatic repulsion between the negatively charged centres on adjacent P atoms in the molecule might make an important contribution to the

[1] Lipmann (3).   [2] Ch. 17.

energy content. Burton (1) found that the affinity of Mg^{2+} for ATP is greater than for ADP, and that consequently ΔF^0 for ATP hydrolysis in absence of Mg^{2+} is more negative (by about 1600 cal) than in presence of Mg^{2+}. It is thus clear that the ΔF^0 of hydrolysis of such a bond is not dependent on localised bond energy, but upon the difference in free energies between products and reactant. Nevertheless the use of the term energy-rich phosphate bond continues as a very convenient form of shorthand.[1]

As the result of the application of this newer knowledge together with the use of more refined methods, the value of ΔF^0 has been revised, and more recent results agree on -7000 to -9000 cal at physiological temperatures and pH. A few examples may be given. Thus Levintow & Meister (1), using an enzyme from peas, calculated -8000 cal for the ΔF^0 of ATP hydrolysis in the reaction[2]

$$\text{Glutamate} + \text{ATP} \rightleftharpoons \text{glutamine} + \text{ADP} + \text{H}_3\text{PO}_4.$$

Burton (2) from calculations of the ΔF^0 of hydrolysis of acetylcoenzyme A gave the figure of -8250 cal. The overall reaction concerned here[3] is

$$\text{ATP} + \text{coenzyme A} + \text{acetate} \rightleftharpoons \text{ADP} + \text{H}_3\text{PO}_4 + \text{acetylcoenzyme A}.$$

Robbins & Boyer (1) obtained a value of -7600 by combining the results of measurements of the hexokinase and glucose-6-phosphate equilibria

$$\text{ATP} + \text{glucose} \rightleftharpoons \text{glucose-6-phosphate} + \text{ADP}$$
$$\text{glucose-6-phosphate} + \text{H}_2\text{O} \rightleftharpoons \text{glucose} + \text{H}_3\text{PO}_4.$$

It is to be noticed also that the value for ΔH of ATP hydrolysis accepted at present is much lower than that (12500) given by Meyerhof & Lohmann (9). At that time they had to use muscle extracts as the ATPase; they subjected these to partial inactivation in order to ensure breakdown of preformed ATP and to minimise other reactions such as lactic acid formation from residual glycogen. The calorimetric measurements of Podolsky & Morales (1) more than twenty years later were carried out with purified myosin ATPase and careful allowance was made for the heat associated with neutralisation by the buffer of the H^+ liberated. The value for ΔH of the hydrolysis was only -4700.

In order to assess the free energy released under conditions *in vivo* the ΔF^0 values given above for standard conditions must be corrected by a term involving the actual concentrations of the reactants and products, since

$$\Delta F = \Delta F^0 - RT \ln \frac{(\text{ADP})(\text{H}_3\text{PO}_4)}{(\text{ATP})}.$$

[1] For a good account of 'phosphoric acid anhydrides and other energy-rich compounds' see the review of 1960 by Huennekens & Whiteley (1).
[2] Also Benzinger, Hems, Burton & Kitzinger (1).   [3] Ch. 16.

It is difficult to guess how much effect this will have, since the concentrations of these substances will vary with the metabolic state; also they may be compartmentalised or may be subject to extensive binding. Burton (3) discussing this question gives a value of -11000 to -12000 cal based on the concentrations in the whole tissue but emphasises that this is only a very rough guide to the proportion of metabolic energy in actual fact transferred to ATP or made available on its hydrolysis in the living cell.

Wald (1) has emphasised the biological selection (which has become apparent) of S and P for group and energy transfer reactions. He raised the question of the basis of the selection of these two elements and has made a very interesting attempt to find an answer in terms of their electron orbitals and the nature of the bonds which they can form.

7

EARLY STUDIES OF MUSCLE STRUCTURE
AND THEORIES OF CONTRACTION,
1870 TO 1939

EARLY ADVANCES IN THE MICROSCOPY OF MUSCLE

In the immediately preceding chapters attention has mainly been concentrated on the processes of energy provision for contraction; speculations concerned with the nature of the muscle machine itself were vigorously canvassed in the seventeenth century as we have seen, but it is necessary now to consider in detail the intensive work on this question which began early in the nineteenth century.

This work followed two main lines: first, the microscopic examination of muscle sections and fibres by ordinary and by polarised light; secondly, since the muscle structure must consist mainly of protein, the biochemical examination of the extracted proteins. Early in the twentieth century, the methods of X-ray diffraction were called upon, and about the middle of the century the phase contrast, interference and electron microscopes began to play their part. This chapter then, leading up to 1939 when Engelhardt & Lyubimova (1) made their pregnant discovery of the adenosinetriphosphatase activity of the structural protein myosin, will be concerned with the microscopic structure of the muscle machine and the nature of its protein composition.

After the first microsopic examination of muscle by Leeuwenhoek[1] in 1674 and his discovery of the cross-striations in 1682, there was little progress for more than 100 years. This is to be correlated with the fact that during the eighteenth century, though much experimentation went on, no optical but only mechanical improvements were made in the microscopes generally available. Early in the nineteenth century however, achromatic objectives became readily obtainable, and by the middle of the century the regular use of immersion objectives began. By 1866 Abbe and Zeiss were able to produce achromatic objectives corrected for chromatic aberration at three wavelengths as well as for spherical aberration. Improvements in methods of preparation of the biological material took place, as well as in

[1] Ch. 1.

[127]

methods of illumination; by 1850 it was possible to use an electric light
source. Such advances went far to overcome earlier criticisms justly made
of microscopy that everyone could see what he wanted to see.[1] Nevertheless
histologists remained greatly dissatisfied with the results obtainable on
muscle. This is illustrated for example in 1853 by the differences in interpretation put forward by Kölliker (1) in his *Manual* and by the translator-
editors[2] in their lengthy footnote. As late as 1909 Hürthle (1) was lamenting
the 'labyrinth of contradictions' which in his opinion was the outcome of
attempts to study contraction phenomena mainly in fixed, not living,
muscle.[3] Although it has to be remembered further that most observers
remark on the great variability of the picture they saw, nevertheless the
results which are given below are typical of those which became generally
recognised, and many of which have indeed found a remarkable degree of
confirmation in the electron microscope.

In 1840 Bowman (1) published a very thorough examination of the
'minute structure and movements of voluntary muscle'. He describes the
muscle as made up of primitive fasciculae (fibres) composed of fibrillae and
showing along their length regularly alternating light and dark patches (see
fig. 18 pl.). The fibrillae, he says, had the appearance of strings of beads –
the beads constituting the light bands, the intervals between them the dark
bands. This description agrees, as he remarks, with that given by Schwann[4] at
about the same time. Bowman argues cogently against an earlier view
(which had indeed begun with Leeuwenhoek)[5] that the light and dark lines
were due to the presence of transverse or spiral threads wound round the
fibres; he as well as Schwann demonstrated that the finest fibrils that could
be seen on splaying out of the finest bundle showed just the same striation.
Like Schwann he regarded the striations of the fibres as due to the lateral
parallelism of the 'beads' of contiguous fibrillae. He observed the breaking
up of the fibres under various treatments, longitudinally into fibrillae and
transversely into discs (see fig. 19 pl.). By the use of cross-sections he also
countered the view that the fibres were hollow tubes. Schwann had doubted
that the zigzag flexures often seen in fibres and considered very important
by some earlier observers were the only cause of contraction. Bowman was
emphatic that the flexures were to be regarded as the natural position into
which the fibres were thrown if, on elongation after contraction, they were
not at once stretched by antagonistic muscles; they were to be regarded
therefore as associated with relaxation not contraction. He observed under
the microscope a frog fibre electrically stimulated till it would no longer

[1] For much information on which this paragraph is based see Rooseboom (1), tr. van Loo.
[2] See vol. II, p. 239.
[3] Barer (1) in 1948 however considered the evidence from fixed and stained sections
perhaps more reliable.
[4] Müller (1) vol. II, p. 33.   [5] Bowman (1) p. 461.

react to the stimulus; of the slow spontaneous movement which followed he wrote 'In that form of contraction which takes place as the last act of vitality the transverse striae, that is the discs of the fasciculus, approach each other, become thinner and expand in circumference, in other words, the contractility of muscle is independent of any inflexion of its fasciculae and resides in the individual segments of which these are composed'. Bowman also gave the name sarcolemma to the delicate structureless sheath surrounding the fibre, though he attributed its discovery to Schwann. Fig. 20 pl. shows a muscle fibre (from the staghorn beetle) in different degrees of contraction.

In 1857, Brücke (1) was a pioneer in the application of polarised light to microscopic studies. He noticed that of the two substances in the muscle fibre giving the striated appearance only one, that with the higher refractive index, was doubly refracting with positive sign; thus the bands showing dark in ordinary light showed bright when viewed in polarised light with crossed nicols. By the action of acid or alkali, or on heating, muscles lost their double refraction. Brücke concluded that this property was due to presence of small solid bodies, longitudinally aligned, of higher refractive index than the isotropic medium in which they were imbedded. He proposed the name *Disdiaklasten* for these.

In 1868 Krause (1) and Hensen[1] described the dark line, the *Querlinie* in the middle of the light band. Krause mentions that it had been seen as early as 1848 by Quekett. In Krause's conception the elementary parts of the fibre were the *Muskelkästchen* or little caskets, each consisting of a solid membrane, the Q line, followed by one-half of the light band followed by the dark band, then one-half of the next light band, after which another Q line was reached. According to his observations, on contraction the light bands became smaller, while the dark bands and Q lines approached each other. He assumed that the light bands were of fluid nature; the 'muscle prisms' of neighbouring dark bands, like small magnets, drew together, displacing the fluid and moving the Q lines; the prisms, each swimming suspended in the fluid in its casket, remained unaltered in form during the contraction. Krause (2) later described the 'prisms' as rigid muscle rods.

Hensen is also credited by Engelmann (5) with being the first to describe the light region (now known as the H zone) in the centre of the dark band. However Engelmann points out that it must have been seen by Bowman, since it appears in one of his figures.[2] It is interesting that this figure (of muscle from the staghorn beetle) seems also clearly to show the Q line. The figure is only referred to in the text as 'another unusual appearance' (see fig. 21 pl.).

In a long series of papers between 1873 and 1895 T. W. Engelmann (1–6)

[1] Engelmann (5) p. 47.   [2] Bowman (1) plate xv, fig. 20.

made many observations on fixed muscle fibres, resting and contracted, and developed his ideas on contraction.[1] He stated that the shortening forces were situated exclusively in the anisotropic bands (3), which he found to increase progressively in volume while the volume of the isotropic bands decreased. As regards actual shortening, this he saw mainly in the isotropic bands, the anisotropic undergoing little change in width (4). We shall come later to his interpretation of these observations in his inbibition theory which had great influence.

Much careful work was done on both fixed and fresh material by Rollett.[2] In the former the greater decrease in length of the I band than of the A band[3] during the early stages of a contraction wave can be clearly seen. However Hürthle (1), in 1909, making a photographic study of fresh and frozen-dried fibres, concluded that on contraction shortening took place in the A bands; with fixed material he could confirm the greater shortening of the I band, but took this for an artifact. A. F. Huxley (1) has pointed out the extremely low ratio of the I to A band width in Hürthle's so-called resting fibres – only $1:8$ as compared with the more usual ratio given by other workers of $1:2$; and suggests that these fibres were already contracted to the point where further shortening must involve the A band.

Another study on contraction in which the shortening of the A band (with lengthening of the I band) was recorded was that of Buchthal, Knappeis & Lindhard in 1936; here single frog fibres were stimulated with a microelectrode, and instantaneous microphotographs were taken. A. F. Huxley (1) has discussed the optical difficulties inevitable in attempts to determine the limits of the A and I bands with the ordinary microscope, whether ordinary or polarised light is used, in an object so thick as a muscle fibre. Certain delimitation in the resting and contracted states had to await further development in microscopy when, as we shall see, the shortening of the I band in normal contraction was established. A striking phenomenon (only to be explained by the electron microscope) observed by many early workers was that of striation reversal on contraction and formation of contraction bands. Speidel (1) in 1939 gave a good account of the steps in full contraction leading to the appearance of these bands in living fibres. Engelmann (6) had much earlier noted that they showed no double refraction. In this condition the neighbourhood of the Z line appeared as the densest part of the fibril. Jordan (1) in 1933 held that the deeper staining reaction and dark appearance of the A bands in ordinary light was not due to the presence of the anisotropic substance, but to the presence of other

[1] Ch. 2.   [2] E.g. Rollett (1, 2, 3).
[3] With regard to terminology, it should be noted that the anisotropic and isotropic bands came to be known as the A and I bands respectively; while the term Q line was replaced by Z line. The length of myofibril bounded by two Z lines is known as the sarcomere.

substances (possibly salts) which moved on contraction from the A band to the Z line and back again on relaxation.

Hürthle had also doubted the real existence of the Z line. He rejected the common view that it was a membrane binding together the fibrils into fibres. This led him to wonder at the regular arrangement of the fibrillae to give the striae of the fibres; he could only suggest that 'the hypothetical force or process which segments the fibrils in the same manner governs the whole fibre arrangement'. It is interesting that in 1934 W. J. Schmidt (1) still felt obliged to postulate binding of the fibrils by means of the Z line and the M line (the dark line in the centre of the H zone) saying 'Without cross binding the regular arrangement of the corresponding member of all fibrils at the same level (in spite of changes during contraction and relaxation) would be incomprehensible'. However, Barer (1) in 1948 and A. F. Huxley (1) in 1957 have brought together the evidence against the existence of a membrane across the fibre in the position of the Z line. Barer remarked that the problem of the alignment remained a mystery; he suggested three possibilities – surface forces, long-range colloidal forces or hydrogen bonds. Huxley considered that the Z line might well act in each fibril to keep the actin filaments in register; he pointed out that material to be seen crossing the whole fibre at the level of the Z line of each fibril in many muscles is part of the sarcoplasmic reticulum – concerned, as we shall see, not with support but with conduction of excitation. It might, however, contribute to maintenance of the general organisation of the fibril arrangement.[1]

In 1923 Stübel made a closer study of the phenomena of double refraction in muscle, basing his work on the theories of Wiener (1) according to which double refraction may be shown by a system containing elongated parallel particles imbedded in a medium of different refractive index from their own, even if the particles themselves are of isotropic material. This rod double refraction is higher the greater the difference between the refractive indices of the rods and the medium. Stübel (1) now found, working with single frog-muscle fibres fixed in alcohol or 10 % formalin, that the double refraction of the fibre fell to a minimum with changing media as the refractive index of the medium was raised, or lowered, to a value of about 1.45. He concluded that positive rod double refraction was operating, and that, since the double refraction did not fall to zero, the rods themselves had intrinsic double refraction.

As early as 1881 von Ebner (1) had recorded a fall in double refraction during isotonic contraction, and he wrote again in 1916 (2) upholding this observation against criticism from Bernstein who, believing that the isotropic bands shortened, maintained that double refraction had nothing

[1] For later work see ch. 11. It is now known that the Z and M lines do indeed contain structures linking respectively the actin filaments and the myosin filaments within the fibril.

to do with contraction. Von Ebner's observations are rather difficult to interpret because of the changes in shape of the muscle during isotonic contraction, but in 1932 von Muralt (1) found this decrease with isometric contraction also.

Before considering the more important of the numerous theories of muscle contraction elaborated during this seventy years of groping after the fine structure of muscle, we must describe the development of our knowledge of the muscle proteins.

PARTS OF THE MUSCLE MACHINE; FIRST STUDIES ON THE MUSCLE PROTEINS

Kühne published his first observations (1) in 1859 and his final method of preparation of 'myosin' is described in his book (2) of 1864. Frog muscles were freed from blood by injection of 1 % salt solution, then removed and frozen. After 3 hours the mass was cut up and pounded to a snow; on thawing it gave a syrupy liquid which was filtered through a linen cloth. A drop of this 'muscle plasma', falling into water at 0° gave a white, opaque precipitate which would redissolve in salt solution, more readily at a higher salt concentrations, for example 10 % NaCl, and was reprecipitated by great excess of salt. In 0.1 % HCl a precipitate was also formed at first but soon re-dissolved; this precipitate was insoluble in salt solution. The liquid could be kept many days at 0° but on being brought into a warm room it quickly coagulated. Kühne had no doubt that the coagulum he described, to which he gave the name myosin, was 'identical with the substance which separates out in muscle undergoing death rigor'. He found that myosin could be extracted from muscle in rigor provided that 10 % NaCl (nearly 2M) was used.

Kühne also recognised the presence of albumins in muscle serum, the liquid separating from the myosin clot, and in the water filtered off after the precipitation of myosin by dilution. The serum readily coagulated if treated carefully with acid and then left for an hour at 25–30°. He considered that the phenomena of death rigor were enhanced by this serum coagulation consequent upon acid formation in the muscle after death. He believed that presence of blood hastened the clotting of muscle plasma, fibrin ferment playing a part here.

Halliburton (1) in 1887 applied methods similar to those of Kühne to mammalian muscle with similar results. He extended Kühne's observations on temperature coagulation of the various protein fractions and attempted by this means to separate the different muscle proteins. He distinguished two constituents of the myosin clot – one precipitated by heat at 47° the other at 56° from a solution of the clot in 5 % MgSO$_4$ solution. The former was also precipitated at lower salt concentrations than the latter. These he called paramyosinogen and myosinogen respectively.

Halliburton considered that he had evidence for participation in the gelation process of a 'myosin-ferment' analogous to the fibrin-ferment in blood clotting. He prepared it in a similar way: the muscle in rigor was chopped and the pieces were kept under absolute alcohol for some months. They were then dried over sulphuric acid and powdered. An aqueous extract of this powder, when added to muscle plasma or to a solution of myosin, greatly accelerated the coagulation. Finck (2) has very recently drawn attention to the astonishing fact (hitherto unnoticed) that Halliburton had actually made a preparation of the second myofibrillar protein, actin, and with it obtained from his myosin the much less soluble complex, actomyosin. As we shall see in chapter 8, for further clarification of the nature of these proteins and their interaction muscle biochemists had to wait more than fifty years. Halliburton could not confirm any effect of fibrin-ferment on the rate of muscle–plasma clotting.

In 1895 von Fürth (1) attempted to estimate paramyosinogen, myosinogen and albumin in muscle plasma by differential heat coagulation. He found that myosinogen was not precipitated by dialysis against water, and was therefore not a globulin as Halliburton had thought. Since it could thus be no precursor of Kühne's globulin, myosin, he proposed for it the name myogen, and this term remained in use for many years for the soluble fraction from muscle, which is now known to contain many different proteins. For Halliburton's paramyosinogen, he preferred to use the old name myosin.

In a long series of experiments (2) comparing the effects of various toxic agents on the precipitation of myogen and myosin solutions with their effect in causing rigor in living animals, he could find no proof for attribution of rigor to protein coagulation.

There was much controversy in the late nineteenth and early twentieth centuries concerning the nature of the fibrillar protein. Kühne had regarded it as present in the form of a sol which could be pressed out of the muscle. Van Gehuchten (1) in 1886, using gold-impregnated sections, described the sarcoplasm as organised and structured, forming a network of mathematical regularity to which he attributed contractility.[1] The fibrils he saw as amorphous and half-fluid. As late as 1925 Wöhlisch (1) put forward a theory[2] depending on the colloid-osmotic properties of a protein solution contained within the fibrils.

On the other hand many workers, influenced in a number of different

---

[1] For consideration of the nature and function of the sarcoplasmic reticulum and particles responsible for this appearance see chs. 13, 16 and 17.
[2] See p. 144 below.

ways, believed in the solid nature of the fibrils. Kölliker (2) in his textbook (first published in 1850) which went through many editions, was a strong upholder of the real existence of the fibrils and fibril bundles in living muscle. He also described a granular substance between the fibrils, binding them together, to which he gave the name sarcoplasm. Retzius (1)[1] in 1880 wrote 'In considering the manner of contraction I go all the way with Kölliker when he asserts "that everything we know about the processes urges us to the assumption that on contraction of all contractile elementary parts alterations of preformed molecules (Brücke's Disdiaklasten, Engelmann's Inotagmen) play the chief role through alterations in shape or arrangement, which assumption would not rule out a part played by chemical processes."' Engelmann[2] (3) in 1873 laid stress on the importance of solid, regularly arranged particles in the anisotropic band, and described them as 'without doubt the chief source of myosin according to microchemical reactions'. Rollett (2) was emphatic that the regular arrangement apparent in the fibres (both in the resting and excited states) was only explicable if solid material was contained in both anisotropic and isotropic parts; he could not picture the solid Disdiaklasten carrying out their function swimming in a fluid isotropic environment. Biedermann, who published in 1927 an excellent review (3) of the histochemistry of striated muscle covering the period from 1875, was also of this opinion. Hürthle (2) in 1907 described experiments in which pieces of muscle were subjected to slow centrifugation; he could see on subsequent microscopic examination no heaping up of the fibril contents such as would be expected if they were of fluid or only semi-solid nature. When a fresh fibre was damaged with a sharp instrument so that only half the sarcolemma was cut through, nothing came out of this opening and the undamaged half of the fibre remained normal in appearance.

We may turn now to studies aimed at elucidating the state of the myosin contained in the living muscle. T. H. Huxley (1) in 1880 recorded that in living or fixed muscle the ability to polarise light shown by the interseptal zones (A bands) was lost on treatment with salt solution or dilute acid. Since it was known that these reagents could dissolve 'the peculiar constituent of muscle, myosin' he concluded that the substance of the interseptal zones was mainly myosin. Schipiloff & Danilewsky (1) about this time also formed the opinion that the double refraction of the muscle discs depended on the crystalline state of the myosin. They found myosin solutions to be isotropic, but carefully dried drops to be doubly refracting. Little double refraction was found in the tissue after thorough extraction

---

[1] This paper, besides describing the author's work on the fine structure of several different types of muscle, gives an excellent historical account of such observations over the previous fifty years.   [2] Engelmann (3) p. 174.

with ammonium chloride solution. Biedermann (3) describes many researches, including early work of his own, showing how little effect upon the microscopic appearance was caused by prolonged treatment of muscle fragments with water or physiological salt solution; the fibrillar arrangement and cross-striation were still well seen; moreover the double refraction was preserved. If however the frog sartorius muscle, carefully held at resting length, was treated with 10 % ammonium chloride or with dilute formic acid solution containing ammonium or potassium chloride, there was diminution or complete disappearance of the birefringence of the fibrils. Botazzi & Quagliariello (1) described in 1912 the innumerable particles seen in muscle press-juice by means of the Zsigmondy ultramicroscope. They regarded these as pre-existing granules of myosin, explaining both spontaneous coagulation of press-juice and death rigor as due to aggregation of these particles.

In the nineteen twenties the time was ripe for a more detailed biochemical study of the properties of the muscle proteins. In 1925 H. H. Weber (2) pointed out the indefinite nature of von Fürth's evidence that myosin was a protein *sui generis* and not a denaturation product of myogen. He brought evidence for the former point of view in finding the isoelectric point of myosin to be at pH 5.2, while that of myogen under the same conditions was at pH 6.3. In this work he prepared the myosin in solution by extracting the residue (at pH about 7.3) after removal of the press-juice; or as a precipitate by diluting the press-juice at slightly acid pH. Myogen was prepared by dialysing the press-juice at pH about 6.0 and removing the precipitate.

In 1929 Bate Smith (1) became interested in the views of Kühne and Halliburton on *rigor mortis*, since they implied the existence of a liquid plasma in living muscle. As the result of his attempts to repeat their experiments under the conditions described by them, he came to the conclusion that plasma capable of gelation could only be expressed from the cold or frozen muscle if a small amount of added salt was present. In his own experiments, the addition to the muscle snow of 1 % of its weight in NaCl was enough to ensure the obtaining of a plasma which readily clotted on standing. He suggested that the success of the early workers depended on contamination of their muscle with small amounts of salt from their freezing mixtures. It thus seems that the squeezing out of a plasma capable of gelation depends on conditions leading to solution of a weak gel already present in the muscle; a change in the nature of this gel to a more solid and less soluble form would represent the phenomenon of rigor mortis.[1] Indeed he found during the course of rigor development a falling off in the ability to form clottable plasma. Bate Smith concluded that the most likely cause of the gelation of the plasma obtained from melting muscle snow, was

[1] Bate Smith (2)

dilution (by the melting ice crystals) of the solution of muscle proteins in the more concentrated solution of salts derived both from the muscle itself and from the accidentally added NaCl. He found no connection between clotting and fall in pH, indeed good clotting was obtained in plasma at pH 7 from muscle rendered glycogen-poor by insulin injection – just as rigor had been observed by Hoet & Marks (1) without acidity increase in animals

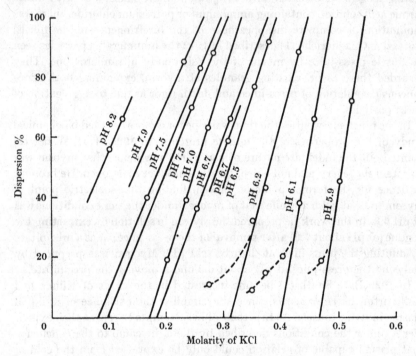

Fig. 22. Dispersion of myosin in KCl solutions, pH 5.9 to 8.2.
(Bate Smith (2).)

dying in insulin convulsions.[1] Bate Smith (2) went on to a precise study of the solubility of myosin. The myosin was purified by the method of Edsall (1) and from the effects of salt concentration, pH and total protein concentration it was concluded that in living rabbit muscle 90% of the total myosin must be in the gel form (see fig. 22).

Meanwhile Edsall[2] (1) had purified 'muscle globulin' by repeated resolution and reprecipitation and studied its properties, including its acid- and base-binding capacity, its solubility under various conditions and its very high viscosity. The anomalous nature of this viscosity (i.e. its dependence

---

[1] Ch. 15, p. 369   [2] Also Salter (1).

on shear rate) was described by von Muralt & Edsall (1). In their work also (1, 2) a very important property of myosin solutions came to light – their double refraction of flow, measured in the annular space between two concentric cylinders, one of which rotated. This effect was ascribed primarily to orientation of an isotropic particle, and secondarily to photoelastic effects. The measurements of the angle of isocline indicated that the responsible muscle–globulin particles were of uniform size and shape. Von Muralt & Edsall concluded that myosin was the source of the double refraction of the A bands, and that the myosin particle was the rod-shaped element to be inferred from Stübel's data.

Enough was now known of the properties of myosin to allow a quantitative estimation of this and other protein fractions in muscle. Meyer & Weber (1) in their careful separation found about 40 % of the total protein as myosin both in red and white rabbit muscle. They also showed the presence of a new muscle protein, which formed 17 % of the total and which they called globulin X; this could be separated from myosin by its precipitation only below 0.005 M KCl, instead of at 0.04 M, and was distinguished also by having a different isoelectric point and no double refraction of flow. It was considered to be situated in the sarcoplasm, whether as gel or sol.

Bate Smith (3) found considerably higher myosin content; he confirmed the existence of globulin X, and also found small amounts of an albumin with properties distinct from those of the myogen fraction.[1]

	Bate Smith (%)	Meyer & Weber (%)
Myosin	63	40
Globulin X	9	20
Myogen and myoalbumin	10	20
Stroma	17.5	20

An entirely new approach was the making of artificial filaments of myosin. In 1935 Weber (3) and Noll & Weber (1) studied the double refraction of myosin threads produced by squirting myosin solution through a fine orifice into a large volume of water. These delicate threads contained only about 1 % of protein and were feebly birefringent. On stretching and drying they became strongly birefringent.[2] They were then soaked in 0.1 M phosphate, pH 7.4 and allowed to take up water until they contained about 20 % of protein, equivalent to the content of muscle fibre. Such threads showed both rod and intrinsic double refraction, and calculated on the same

[1] Later work on the distribution of the muscle proteins will be considered in ch. 9.
[2] Such threads had already been used by Boehm & Weber (1) in their X-ray diffraction studies (see p. 139).

protein basis the double refraction of the natural fibre and the thread was in the relation of 0.4 to 1.0. Noll & Weber believed they had evidence for uniform protein content in the A and I bands in the natural fibre; and since the I bands were slightly longer than the A bands, the relation of the optical constants was in harmony with the view that the double refraction of the fibre was due to ordered arrangement of myosin in the A band. It should be noticed in passing that in 1934 Schmidt (1) had shown that the I band also has birefringence, though only about 10 % of that of the A band.

APPLICATION OF THE X-RAY DIFFRACTION METHOD TO ELUCIDATION OF MUSCLE STRUCTURE

Hertzog & Jancke (1) in 1926 were pioneers in this field. Using quickly-dried, stretched sartorius muscle of the frog, they found a fibre diffraction pattern with strong repeat spots at about 10 Å. Improvement on such photographs only came some years later, after experience had been gained from work on other fibrous materials. Astbury[1] has described how the elucidation of the structure of cellulose led on to the results of Meyer & Mark (1) on silk fibroin, the first protein to give a good X-ray photograph. Basing their interpretation on the work of Emil Fischer, they concluded that the silk fibres were composed of polypeptide chains (in the case of this protein consisting mainly of glycine and alanine) lying parallel to the fibre axis and associating together to form long, thin, pseudo-crystalline micells. The repeat unit along the chain they found to be 7 Å – the length of a glycylala-nine residue. The repeat units in the two directions at right angles (that is, the distance of separation of the chains) were 4.6 and 5.2 Å. It was at the time disappointing that X-ray photographs from other fibrous proteins, such as hair, collagen and muscle, did not allow of this interpretation; but in 1930 Astbury & Woods (1) discovered the reversible intramolecular transformation which occurs on stretching hair keratin, the stretched form giving the picture associated with extended polypeptide chains. Astbury & Woods suggested that the unstretched keratin must be in equilibrium in some regularly folded configuration (the α-form); this could be opened up on stretching to about twice the original length, to give the β-form. The dimensions given by Astbury (1) for β-keratin were 3.38 Å in the direction of the fibre length; 4.65 for the 'backbone spacing' and 9.8 in the direction at right angles, the side-chain spacing. This last dimension was identified by Astbury & Woods (2) as depending on the side-chains in their study of the disruptive action of steam on stretched keratin. The unstretched form was characterised by a marked repeat of 5.15 Å along the fibre axis and two chief side spacings at 9.8 and 27 Å.

[1] E.g. Astbury (1).

In 1931 Boehm (1) returned to the attempt to investigate muscle by this method. Working with dried muscles, he observed like Hertzog & Jancke only one repeat period, that of 10 Å; in isotonic contraction the pattern vanished, but on isometric contraction there was a slight sharpening. Boehm & Weber (1) in 1932 then got X-ray diagrams of artificial myosin threads stretched and dried; of dried sartorius preparations contracted by about 23 %; and of sartorius dried under tension. From the similarity of these pictures they concluded that the Röntgen diagram of muscle was entirely due to ordered myosin. In 1939 H. Weber (13) cited observations made with Boehm in which myosin threads, swollen to 20 % protein by soaking in physiological salt solution, gave an X-ray diagram indistinguishable from that obtained with a resting or slightly stretched fresh sartorius muscle.

Astbury (1) in 1934 had drawn attention to the striking resemblance of X-ray photographs of muscle to those of unstretched hair – suggesting that the main chains of the myosin molecule were in a folded configuration. A little later Astbury & Dickinson (1) succeeded in showing the $\alpha\beta$ transformation with air-dried myosin films stretched after moistening. Very soon afterwards they were able to show (2) the $\alpha\beta$ transformation by stretching muscle itself – using frog sartorius and the foot muscle of *Mytilus edulis*. In their work on the action of steam on keratin, Astbury & Woods (2) had shown that in the labile state keratin was characterised by greatly enhanced power of contraction – as much as 40 % below its normal length. They called this phenomenon supercontraction, and Astbury & Dickinson (3) now found that unstretched myosin film exposed to steam contracted by about 20 %.

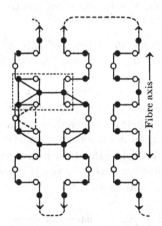

The close packing of side-chains in the α-fold of keratin and myosin. —— represents the direction of the main-chain; ●, represents a side-chain pointing *up* from the plane of the diagram; ○, represents a side-chain pointing *down* from the plane of the diagram.

They therefore suggested that the normal contraction of muscle may be compared with the supercontraction of keratin.

Astbury's earlier theory of the structure of the α-form involved hexagonal rings;[1] but later when in 1941 it became apparent that such a structure could not accommodate the side-chains, intramolecular folds, as shown overleaf, were proposed.[2]

THEORIES OF MUSCLE CONTRACTION, 1870 TO 1939

As we have seen, it was rather generally accepted that the fibrils of the muscle cell made up its contractile element, but there were other points of view. Kühne (3) in his Croonian Lecture of 1888 argued persuasively for the contractile role of the protoplasmic network, the sarcoglia, interpenetrating the striped mass which he called the rhabdia, and to which he attributed only elastic properties. He says 'We have almost in our own person lived to see the old anticipation of a single kingdom of living things become gradually an established truth through the discovery of the cell'; and after speaking of the movement of protoplasm, for example in free-living protozoa, he continues:

I could wish that this view might be accepted as an hypothesis. As far as I can see it does not contradict experience, for it only puts back the muscle nearer to protoplasm and to all that is contractile, and so far coincides with experience that we find muscles in the same measure less elastic and more sluggish in protoplasmic movement the richer they are in sarcoglia, as in the case of the red muscles, nucleated and rich in glia, which contract more slowly but with greater power than the white muscles poorer in glia which are quick and spring-like, and also the sluggish embryo muscles, in which glia predominates because as yet but little protoplasm has been converted into rhabdia...

Botazzi (1, 2)[3] also pressed the claims of sarcoplasmic contraction, in addition to fibrillar contraction, in striated and smooth muscle.

We have already, in chapter 2, while considering early ideas of energy provision for contraction, described the pictures formed by Liebig and by Fick in explaining muscle activity; and also the simile introduced by Hill in the nineteen twenties, whereby the contracting muscle was likened to a stretched spring operating in a viscous medium. Of the many theories of contraction put forward in the late nineteenth and early twentieth centuries based on changes in the fibrillar material, we can discuss only some of those which had most influence in their time.[4]

[1] Astbury (2).
[2] Astbury & Bell (1). See also p. 182. For later, now universally accepted, evidence for the helical nature of the chains see ch. 10.   [3] Also ch. 23.
[4] For a comprehensive study in 1933 of contraction hypotheses see the review by von Muralt (2).

IMBIBITION THEORIES. As we have seen, Engelmann (3) believed that he had evidence for a decrease in volume of the isotropic band and an increase in volume of the anisotropic. He accounted for these changes by postulating a passage of water from the isotropic to the anisotropic bands, followed by imbibition of this water by the long, narrow, cylindrical or prismatic elements arranged parallel to the axis of the fibre. These changed towards a spherical form and in so doing shortened. He thus regarded the more pronounced shortening of the I band as merely passive. Engelmann was convinced that contractility in general was linked to the presence of doubly refracting, uniaxial particles with their optical axes in the direction of shortening; to test this he examined for double refraction many other systems such as cilia and spermatozoa (1). Further, Engelmann[1] regarded the muscle as a heat engine; in pursuit of this idea, he studied the swelling and shortening effects when a violin string was placed in pure water, and the enhancement of these effects when water was replaced by 0.25% lactic acid. Thus release of lactic acid in the muscle could initiate the turgescence. He considered however that provision of energy in this way would be quite inadequate, and held to the heat-engine conception even though he was obliged to postulate that certain infinitesimal parts must reach on contraction temperatures as high as 140°. Again using the violin string model, he demonstrated the rise of turgescence and the shortening due to heating. 'We may conclude,' he said, 'that chemical contraction by turgescence of the inotagmata [Brücke's *disdiaklasten*] is most likely a constant concomitant of the thermal contraction of living muscle, but that compared with the latter, in a single contraction, at least of striated fibres, the former is of little or no consequence as regards the shortening effect.' Engelmann thus placed the burden of energy provision for the alteration in shape of the imbibing molecular complex on unknown chemical reactions causing the localised high temperatures.

It is interesting that the imbibition theory, shorn of its thermodynamic aspects, was welcomed by a number of workers.[2] In 1912 Pauli (1) substituted increasing electrostatic forces for rise of temperature as the reason for the imbibition; he supposed that the number of ionized groups increased because the contractile protein changed from a more or less isoelectric state to a highly cationic state, by reason of the hydrogen ions produced on lactic acid formation. There was a variety of views as to the source of the imbibition fluid. For example, McDougall (1) in 1897 examined the fibrils (particularly of insect flight muscles) and described the longitudinal series of sarcomeres of which they were made up, and the flattened doubly bulging appearance of each contracted sarcomere. He explained this by an increase in pressure inside the sarcomere, the longitudinal walls and the transverse membrane

---

[1] Engelmann (2) the Croonian lecture of the Royal Society, 1895.
[2] Von Fürth (3) pp. 539 ff.; Biedermann (3) p. 430.

(the Z line) bounding each being inextensible, while a septum across the centre of each (the M line) was extensible. As an explanation he favoured increase in volume of the sarcomere contents due to passage of fluid from the sarcoplasm. Meigs (1) ten years later warmly supported this hypothesis. McDougall suggested that the osmotic current was caused by breaking up of the inogen molecule into a number of smaller parts. On the other hand, von Fürth in 1919 considered that the water shift took place entirely within the A band – 'that the ultramicroscopic elements, the myosin granules of Botazzi, became capable of swelling at the cost of surrounding fluid under the influence of newly-formed lactic acid.' These myosin granules he pictured as contained within the doubly-refracting *Stäbchen* or rodlets and he connected the postulated changes in these with the change in double refraction long known to take place during isotonic contraction. He calculated that the degree of shortening to be expected from such a system and the energy available in it were adequate for the muscle's needs; he welcomed as of great meaning for the theory of muscle contraction the work of J. R. Katz (1) who showed that the energy of imbibition processes in certain colloid systems *in vitro* might be to a great extent available for external work.

As regards relaxation, on Engelmann's view cessation of the heat-yielding reactions would allow lengthening of the muscle fibres simply by the elastic powers of parts passively moved during their shortening. On views such as that of von Fürth, removal of lactic acid and water from the imbibing protein would be necessary, and for this process energy would be required. Von Fürth was prepared to believe that the greatest energy requirement fell at this stage of restitution of the initial mechanical potential.

H. H. Weber (4) in 1927 criticised the imbibition theory, emphasising the importance of the fact[1] that lactic acid formation is self-hindered, the enzyme system being inhibited at a pH value of about 6.0. This made very unlikely the suggestion of von Fürth (4) that explosive formation of lactic acid should shift the H^+ concentration at the localised shortening sites, with production of protein cations. Further it was pointed out[2] that the idea that great release of free energy accompanied hydration of the muscle proteins emanated from experiments on the effects of hydration on water-poor gels of hydrophilic colloids. The conditions in the cell are quite different, but the assumption was still made that there also changes in hydration forces of several atmospheres took place, produced by a change in the number of ionised and therefore highly hydrated groups. Experiments concerning the dependence of hydration of proteins on pH and ionisation demonstrated that such dependence did not exist. These results were soon explained by the evidence provided by Weber (14) that isoelectric protein exists as zwitter ions and not as uncharged molecules.

[1] Kondo (1); Laquer (1).   [2] Weber & Nachmansohn (1).

THE SURFACE TENSION THEORY. The suggestion that contraction depended on changes in surface tension had as we have seen been vaguely put forward by Gad in 1893 and later as a possibility by Fletcher & Hopkins. A precise formulation was attempted by Bernstein (1, 2, 3) early in the present century.[1] He pictured muscle shortening and its energy provision as depending on increase of surface tension of minute ellipsoid forms (connected together by elastic threads) making up the fibrils. These ellipsoids would decrease in surface, becoming spheres, with release of free energy. He attributed (1) initiation of this change to lactic acid formation, and in 1915 considered (3) that if one took the surfaces in question as those of the smallest elements, then the energy available should be enough. Hill (13) was at first inclined towards this theory, but later calculation[2] convinced him that the amount of lactic acid formed during contraction was enough to cover, with a monomolecular layer, only 2 % of the surface area calculated by Bernstein. The latter had found support for his theory in the fact that the maximum response in a muscle twitch is greater at lower temperatures than at high, since surface tension changes have a negative heat coefficient.[3] Hill however pointed out that this was not a valid argument, since the total energy produced during a twitch also varied in the same way, and for a given chemical change the tension energy set free was independent of temperature. Also for tetanic contractions the tension development and heat liberated were both greater, not less, at higher temperatures.

Garner (1) in 1925 produced a variant of the surface tension theory in the idea that shortening of the anistropic segments resulted from the deposition on them of a monomolecular solid film of fatty acid produced (owing to lactic acid formation) from esters of fatty acids, lipids, etc. Such long chain acids might be expected to cover a much greater surface than lactic acid.

A THEORY INVOLVING DEHYDRATION OF PROTEIN. In 1924 Meyerhof (28) was on the other hand inclined to regard removal of water from protein as the essential process in contraction. If one considers the 'shortening protein' at the resting pH of the muscle as present, partly at any rate, in the form of an ionised salt, then with formation of lactic acid the following reaction would occur:

$$H^+L^- + B^+P^- \longrightarrow B^+L^- + HP$$

where B $\equiv$ basic group; P $\equiv$ protein; L $\equiv$ lactate.

Since the un-ionised form of the protein is the form showing the least swelling, Meyerhof suggested that its formation in this way would involve a drawing together of the structural protein, that is a contraction. The

---

[1] For an interesting discussion (1912) of the early literature in Macallum (1).
[2] Hill (14).   [3] Bernstein (3).

alkali–lactate would then distribute itself in the muscle substance and take no further part in the mechanical process. The rest of the muscle protein (which should have an isoelectric point of about 5.0) would react with the shortening protein, giving up to the latter part of its own alkali. In this process its low isoelectric point would not be reached and it would suffer no change of form, but the shortening protein would ionise anew and relax. Weber (1, 2) dealt also with this *Entquellung* theory, pointing out that it demanded (a) an isoelectric point above pH 6.0 for the shortening protein, since in fatigue the pH of the muscle does not fall below about 6.3; (b) that since he found a value of pH 6.3 for the isoelectric point of the myogen and a value of 5.1 for the myosin this would make the myogen the candidate for shortening protein and the myosin for fatigue protein. But (apart from considerations of muscle structure) there would be grave objections to taking myogen as the shortening protein because, while myosin does flocculate reversibly on de-ionisation as required of the shortening protein, myogen either remains in solution or undergoes an irreversible denaturation.

THE COLLOIDOSMOTIC THEORY. Wöhlisch (1) in 1925 criticised the theory of Meyerhof as well as the acid imbibition theory on the grounds that these supposed effects of acid were not compatible with known properties of the muscle proteins, and that lactic acid formation was largely post-contractile.[1] He went on to propound a theory of his own, picturing the fibril as composed of a system of invisible ultrafibrils with elastic and semipermeable walls, their natural length being less than that of the resting muscle. These contained a protein solution with higher osmotic pressure than the surrounding medium, and thus in the resting muscle their elastic membranes were kept stretched. If now on stimulation a substance was set free (perhaps a breakdown product of lactacidogen) causing decrease of the colloidosmotic pressure in the ultrafibrils (or increase in that of the bathing fluid) the ultrafibrils must contract. The situation would be reversed on removal of the stimulating substance.

MEYER'S THEORY INVOLVING CHANGES OF CHARGE. In 1929 K. H. Meyer (1) followed up the important work on silk fibroin by a theory of the contraction mechanism of fibrous proteins including muscle.[2] He pictured these fibres as made up of protein chains which might be about 400 Å long, though only about 4 Å wide. Regarding the isoelectric proteins as zwitter ions and postulating in addition an alternating arrangement of acid and basic groups, he says of these chains 'With the presence of ionisable groups

---

[1] Ch. 4.
[2] This theory may be regarded as an extension of the views on inner salt formation in proteins put forward by Meyerhof (28).

...ionisation will lead to straightening, de-ionisation to crumpling. With amphoteric electrolytes contraction will occur at the isoelectric point, stretching on departure from this point.' He calculated the amount of lactic acid formed during the raising of 10 kg/cm² of muscle through 10 Å. The value he found, 10^{12} molecules, should equal the number of valency chains in action/cm². From this calculation he concluded that only 5% of the total muscle protein need be involved, and therefore that the theory had no impossible consequences. Meyer further calculated that each free amino acid group in the stretched chain was likely to be on the average 17 Å away from the neighbouring carboxyl groups, while in the folded chain these groups could approach as near as about 2 Å to one another; the theory thus permitted contraction down to 15% of the initial length to be compared with the maximum shortening of intact muscle to about 30%. Weber (5) objected to this conclusion on the grounds that at the resting pH the protein chains, at any rate of myogen, would not be fully stretched by repulsion between negatively-charged groups, since at this pH (7.4) two-thirds to three-quarters of the positively-charged groups remain ionized. He could find likely only some 5–30% shortening. Meyer (2) replied – with the too familiar type of argument – that since only 5–10% of the muscle protein was involved, it remained possible that just these contractile chains might at pH 7.4 have discharged their basic groups. Meyerhof[1] in 1930, already prepared for the part which phosphocreatine breakdown might play in contraction, saw that Meyer's hypothesis would have to be strongly modified if this breakdown must be regarded as occurring between the mechanical change and lactic acid formation.

We must turn to the discovery of the complex nature of 'myosin' (an association of the two proteins we now know as myosin and actin); to the properties of these two proteins and to the special interactions of actomyosin and ATP – studies which occupied the attention of muscle biochemists during the decade from 1940 to 1950.

[1] Meyerhof (4) pp. 296 and 304.

8

INTERACTION OF ACTOMYOSIN AND ATP

THE MACHINE AS ENZYME AND THE FUEL AS SUBSTRATE

The discovery by Engelhardt & Lyubimova[1] in 1939 of the ATPase activity of myosin opened a new era in muscle biochemistry. Lundsgaard (8) had suggested that breakdown of ATP might be associated with restoration of the contractile substance, and D. M. Needham (1) that possibly ATP had some special spatial relationship to the myosin micellae.[2] But the idea of the enzymic activity of the muscle machinery itself was an entirely new one, and the Russian workers fully realised its implications. They endeavoured to free the myosin from enzymic activity by repeated washing and re-precipitation but instead the activity rose to a fairly constant level. The purified myosin split off only one phosphate group, yielding ADP. They remarked on the great heat-lability of the ATPase, its activity being lost in 10 min at 37° and compared this with the low coagulation temperature of myosin known since the time of Kühne;[3] they also noticed the similar sensitivity to acids of myosin as protein and as ATPase. These similarities served to increase the probability of the identity of the ATPase and myosin, but Engelhardt & Lyubimova considered that no final decision could be taken.

In 1941 Engelhardt, Lyubimova & Meitina (1) were the first to test the effect of ATP on myosin threads. These, though containing only about 2 % of protein, showed a certain amount of tensile strength; they were immersed in fluid and connected with the lever of a torsion balance so that when tension (about 200 mg) was applied the extensibility could be measured. They found that addition of 5×10^{-3}M ATP caused considerable increase, 50–100 %, in the extensibility. The explanation of these pioneer experiments will emerge later.[4]

Confirmation of the enzymic results of Engelhardt & Lyubimova soon came from the results of Szent-Györgyi & Banga (1), D. M. Needham (3), Bailey (1) and Banga (1). Their findings also stimulated much work on the complementary aspect of the phenomenon – the effect of ATP upon the properties of myosin.

---

[1] Engelhardt & Lyubimova (1); Lyubimova & Engelhardt (1).
[2] Also Needham (2).   [3] Ch. 7; Bate Smith (3).
[4] See p. 161 below; for good accounts in English of the work of the Moscow group see Engelhardt (1, 2).

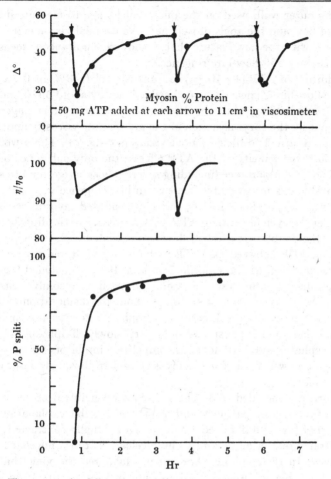

Fig. 23. Three successive falls and recoveries of flow-birefringence of a myosin sol treated three times with adenosinetriphosphate, with two successive falls and recoveries of relative viscosity (expressed in percentage of initial value) and estimations of inorganic phosphate liberated by ATPase activity of the myosin during the first cycle. (Dainty *et al.* (1).)

The work of J. Needham and his collaborators began to appear in 1941.[1] They studied four variables in myosin solutions under different conditions: the intensity of flow birefringence; the angle of isocline; the relative viscosity; and the extent of the anomalous viscosity. They used an elaborate technique of high precision involving (*a*) a small cell with central fixed pillar

---

[1] Needham, Shen, Needham & Lawrence (1); Needham, Kleinzeller, Miall, Dainty, Needham & Lawrence (1); Dainty, Kleinzeller, Lawrence, Miall, Needham, Needham & Shen (1).

and rotating outer wall, used on the microscope stage to measure double refraction of flow and the angle of isocline; (b) a co-axial viscometer of the Couette type, with the new feature that it allowed simultaneous measurements of viscosity and flow-birefringence.[1]

Upon addition of ATP $(4 \times 10^{-3}\text{M})$ to a myosin sol (1–3% in 0.5M KCl or LiCl) the flow-birefringence was reduced at once by about 50% and the viscosity by about 14%, while the flow anomaly remained unchanged. As ATP was split by the enzymic action of the myosin, the viscosity and flow-birefringence returned to their original values (see fig. 23). These workers remarked on the specificity of the ATP effects: the only substance out of many of biological importance (including several phosphate esters) to show to a comparable extent reversible decrease in birefringence was ITP. This compound was dephosphorylated by twice-precipitated myosin, as shown by Kleinzeller (1), even faster than ATP, with formation of the diphosphate; its effect on the double refraction of flow also persisted with thrice precipitated myosin. ADP behaved like ATP with once-precipitated myosin, but its effect was not seen with the purified protein. It was concluded that the once-precipitated myosin was contaminated with the soluble enzyme myokinase, which, as we have seen, brings about the dismutation of ADP to ATP and adenylic acid.[2] Inorganic pyrophosphate was not split by myosin, and inorganic triphosphate only very slowly. Triphosphate (but not pyrophosphate) and ADP acted as competitive inhibitors for the effect of ATP on the flow-birefringence of myosin; and triphosphate also on its ATPase activity.[3]

These workers concluded that ATP caused a reversible decrease in the axial ratio of the myosin particles, and considered possible explanations for this effect: that the myosin molecules themselves actually contracted; that a sliding parallel retraction took place in micellae; that intermicellar forces played a part. In their opinion the evidence favoured the view that the change in birefringence was a manifestation of enzyme–substrate combination.

PARTS OF THE MUSCLE MACHINE: RECOGNITION OF THE PROTEINS MYOSIN AND ACTIN

While the above work was in progress, an intensive study of myosin began in 1941 in the laboratory of A. Szent-Györgyi in Szeged, Hungary. Szent-Györgyi[4] had been much impressed by the difference in properties of an extract made with Weber solution[5] $(0.6\text{M KCl} / 0.01\text{M Na}_2\text{CO}_3 / 0.04\text{M}$

---

[1] Lawrence, Needham & Shen (1).   [2] Chs. 4 and 6; also p. 152 below.
[3] For the inhibitory effect of ADP on myosin ATPase (which was not examined by Needham et al.) see Kalckar (5).
[4] Banga & Szent-Györgyi (1).   [5] Meyer & Weber (1).

NaHCO$_3$) according to whether the muscle was left at 0° only 20 min for extraction, or overnight in contact with the extractant. In the first case the supernatant after centrifuging was a thin fluid which became slightly more viscous on storage at 0°; Szent-Györgyi suspected an effect here of the disappearance of ATP.[1] On the other hand, the tissue left extracting overnight yielded a semi-solid gel, from which the tissue particles could not be centrifuged down, except after dilution with salt solution. On addition of ATP (2.5×10^{-4}M) the gel liquefied, assuming the appearance of the 20 min extract. Banga & Szent-Györgyi studied in the simple Ostwald viscosimeter the viscosity with and without ATP of the two types of extract, comparing samples with the same content of myosin. The 20 min extract (myosin A) showed low viscosity and no effect on ATP; the 24 h extract (myosin B) had high viscosity reduced by ATP addition to the same low value as that of the myosin A solution. It may be recalled that about this time Schramm & Weber (1) for the first time examined myosin solutions in the ultracentrifuge. Purified Edsall–Weber myosin had two components: L-myosin, sedimentation constant $S_{20} = 6$; and S-myosin, $S_{20} = 20$, rising to about 36 on standing. Both components showed strong double refraction of flow. Schramm & Weber raised the question whether two myosins were concerned, or whether the heavier one was an aggregated form of the lighter.

The next step was Szent-Györgyi's study (1) of the effect of ATP on myosin threads and suspensions. If a myosin B thread was suspended in an aqueous filtered extract of muscle it contracted within 30 sec to less than half its length, becoming at the same time proportionately thinner. If the aqueous extract was stored overnight, it showed none of this stimulatory action unless ATP was added to it; ATP dissolved in water was ineffective unless certain ions, e.g. K$^+$ and Mg^{2+}, were also present. Gerendás (1) showed that in this isodiametric shortening and decrease in volume of the thread much of its water (up to 98 %) was lost, so that a thread containing initially 1.5 % of protein might after contraction contain 50 %. Such experiments as these, suggesting an imitation of muscle contraction by means of isolated biochemical substances, made a deep impression on Szent-Györgyi. He also worked with suspensions of myosin B at low salt concentrations, about 0.1 M KCl. Such suspensions were fairly stable and settled only slowly; if ATP (2.5×10^{-4}M) was added, the precipitate immediately became roughly granular and settled quickly; Mg^{2+} enhanced the reaction. This striking effect Szent-Györgyi called superprecipitation.

[1] It should be explained that owing to war conditions even the earlier work of Needham *et al.* was not known to Szent-Györgyi until the middle of 1942, and then only indirectly. The whole work of the Cambridge group was finished by the end of 1942, but the main part could not be published until 1944. The Cambridge group knew nothing of the work of the Szeged group until 1945, when J. Needham and A. Szent-Györgyi met in Moscow at the 225th anniversary celebrations of the U.S.S.R. Academy of Science.

Then followed the elucidation by Straub (1), also at Szeged, of the difference between myosins A and B, and the preparation of a new muscle protein which he called actin. It was shown that the viscous myosin B resulted from the mixing of actin with myosin A, the viscosity and double refraction of flow running parallel with the actin content of the solution of actomyosin (as we may now call it) up to a certain maximum. Straub called this process 'activation'. The effect of ATP on myosin B or on artificially prepared actomyosins was to be explained by the splitting of the complex to give the two constituent proteins. Myosin B was much less soluble in salt solutions than myosin A.[1]

Straub proceeded by first removing myosin A by means of short extraction with Weber's solution. The washed residue was then treated with acetone and after removal of the acetone allowed to dry at room temperature. From this dried residue the actin could be extracted by distilled water. Straub also found that if the muscle residue after the extraction of myosin A was very thoroughly ground with sand, activation of added ATP-free myosin A solution resulted – in other words the actin was now extractable. The only other way he found of liberating the actin was by the acetone treatment described above, and then only if the muscle structure had been previously loosened up by the treatment with the alkaline Weber's solution.

Straub (2) went on to show that actin could exist in two forms; each could react with myosin, but only one could give the highly viscous actomyosin of myosin B. He called these two forms active and inactive actin. The combination of inactive actin with myosin[2] was deduced from the fact that when active actin was added to a mixture of myosin and inactive actin, no rise of viscosity followed. He showed that inactive actin could be transformed into active actin by addition of small amounts of salt. The process could be to some extent reversed by dialysis under certain conditions. Inactive actin solution had low viscocity, while active actin solutions showed high and anomalous viscosity and double refraction of flow; it was concluded that the inactive form was a globular protein, the active form on the other hand being fibrous and consisting of rod-shaped particles. Szent-Györgyi (10) introduced the terms G- and F-actin respectively.[3] Erdös (2) showed that the presence of actin as well as myosin in protein threads was necessary for contraction, the contractibility increasing with the actin content within a certain range.

---

[1] A. Szent-Györgyi (4). Since according to the Arrhenius relationship, the logarithm of the relative viscosity is proportional to the concentration, Portzehl, Schramm & Weber (1) in 1950 introduced the use of the term 'ATP sensitivity', putting

$$\frac{\log \eta_{rel} - \log \eta_{rel\ \text{ATP}}}{\log \eta_{rel\ \text{ATP}}} \times 100$$

to express the changes observed in different myosin solutions on ATP addition.   [2] Ch. 12.

[3] The work of the Szeged school was summarised by A. Szent-Györgyi (2) in a supplement of *Acta Physiol. Scand.*

THE ATPASE ACTIVITY OF MYOSIN AND ACTOMYOSIN

Much earlier work had of course been done on the dephosphorylation of ATP by muscle *brei* or muscle extracts.[1] In one case as we have seen, that of crab muscle, it had been observed by Lohmann (10) that the washed residue, as well as the dialysed extract, could attack ATP, splitting off only one phosphate group and giving ADP. This activity of the muscle residue could now be interpreted as depending, in large part at any rate, on its myosin content.

Engelhardt & Lyubimova did not consider the question of activation by metal ions. The later work of D. M. Needham (3) showed that the good activity of the myosin in their experiments was due to their having used the calcium salt of ATP as substrate. Bailey (1) made a very careful comparison of ionic effects and at pH 9.1 found Ca^{2+} and Mn^{2+} to activate efficiently while Mg^{2+} and Ba^{2+} had little or no effect.[2] Banga (1) comparing myosin A and B suspended in 0.01 M KCl found that, while the latter is activated by either Ca or Mg ions, the former is activated by Ca^{2+} but not by Mg^{2+}. Mommaerts & Seraydarian (1) stated that purified myosin was always inhibited by Mg^{2+}, but with increasing actin present activation was obtained if the concentration of other salts was low, e.g. 0.01 M KCl. Spicer & Bowen (1) mentioned experiments in which synthetic actomyosin was activated by Mg^{2+} in the absence of Ca^{2+}, provided the KCl concentration was less than 0.03 M. Kuschinsky & Turba (1) recorded that when actomyosin was in the gel form, the ATPase activity was increased by Mg^{2+}; when in the sol form by Ca^{2+}. Using homogenised glycerol-extracted muscle in 0.1 M KCl, Sarkar, Szent-Györgyi & Varga (1) found 5×10^{-4} M $MgCl_2$ gave maximal ATPase activation; in the same conditions 5×10^{-3} M $CaCl_2$ activated only 42 %, and the addition of Ca^{2+} in the presence of Mg^{2+} had no enhancing effect.

---

[1] Ch. 6.

[2] It is difficult to decide whether early workers on 'myosin' were dealing with myosin A or (more probably) with myosin B containing variable, often rather low, amounts of actin. Thus in the work of Engelhardt & Lyubimova (2) failure of their myosin solution to show any effect of ATP on viscosity indicates that they were using myosin A.

The same thing is suggested for the work of D. M. Needham and of Bailey, in the light of the lack of Mg^{2+} activation at the low ionic strength they used. This may have been due to the filtration of the turbid extracts through a thick layer of paper pulp, and short time of extraction. On the other hand, J. Needham and his collaborators, and von Muralt & Edsall (p. 137), must have used solutions containing a considerable amount of actomyosin. This may be explained in the former case by the LiCl extraction used, since Bate Smith (4) found LiCl solutions to be amongst the most effective extractants for muscle globulin (probably because of their softening effect on collagen); in the latter case Edsall (1) emphasised fineness of grinding and the careful maintenance of the pH well above 7.0 during the one to two hours of extraction – a rather longer period than that usually given by earlier workers. The myosin of Greenstein & Edsall (1) probably contained a high proportion of actomyosin, since in purification it was reprecipitated at an ionic strength of 0.15 or higher.

In 1952 Hasselbach (2) went systematically into the effects of changes in Mg and Ca ion concentration (at pH 7.0 and ATP concentration about 10^{-3}M) on the ATPase activity of myosin and actomyosin (figs. 24 and 25). Actomyosin gel was activated by Mg^{2+} up to a concentration of 3×10^{-4}M; beyond this concentration activity fell off. Myosin and actomyosin sols were inhibited by Mg^{2+}; they were more strongly activated by Ca^{2+} than was the actomyosin gel. Raising the concentration of ATP to above 10^{-3}M in presence of 10^{-4}M Mg^{2+} led to marked falling off in actomyosin ATPase activity. Activity also fell off markedly as the ionic strength was raised above about 0.1 M.

Taken all together, these results show that myosin associated with actin requires different conditions for enzymic activity from those needed in the absence of actin. The fact that this influence of actin is only to be seen in presence of low KCl concentrations, when the protein is in the gel form, is to be explained by the fact that the presence of higher concentrations of other salts reinforces the influence of ATP in removing actin from the myosin. The results of Hasselbach make precise the conditions in which Mg inhibition is to be expected.

Lyubimova & Engelhardt (1) and Bailey (1) had found the pH optimum for ATPase activity of their myosin preparations in the neighbourhood of 9.0. Mommaerts & Green (1) working with highly purified myosin later recorded a sharp optimum at pH 6.5 and a trough of activity at about pH 7.0; they considered that the peak about pH 9 did not indicate a real pH optimum there, the falling off in activity above this pH being probably due to irreversible inactivation. Banga (1) found a single optimum at pH 6.5 with myosin B, with zero activity at pH 9. Kielley & Meyerhof (1) in 1948 similarly observed with myosin Mg^{2+}-activated, in presence of added actin, a single pH optimum – in their case placed at about 7.5 – and very low activity at pH 9. With Ca^{2+}-activation on the other hand the activity fell only slightly on the alkaline side of the pH 7.5 optimum, then rose again to high values in the more alkaline region.

As regards the substrate specificity of myosin, we have already mentioned the results of J. Needham and his collaborators. D. M. Needham (1) found it had no phosphatase activity with certain monophosphates, and Bailey (1) that it was inactive with a number of organic pyrophosphates. Later work[1] has shown that it attacks all the nucleoside triphosphates so far tested, and it is best regarded as a nucleosidetriphosphatase. We may pause to consider here the properties of the enzyme myokinase, which as we have seen was responsible for the removal of the second phosphate group from ATP when crude myosin preparations were used. It is interesting in its unusual resistance to heat and to acids. Thus 10 min heating at 100° in

---

[1] Blum (2); Kielley, Kalckar & Bradley (1).

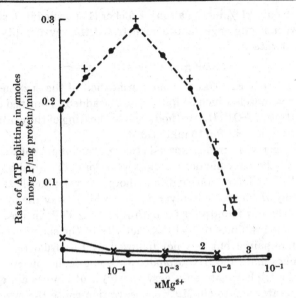

Fig. 24. Influence of Mg^{2+} on ATPase activity of actomyosin (curve 1) and L-myosin (curve 2) at I = 0.07 to 0.1; and on actomyosin sol, I = 0.6 to 0.9 (curve 3). Temp. 20°; pH 7.0; ATP conc. 0.9×10^{-3}M. (Hasselbach (2).)

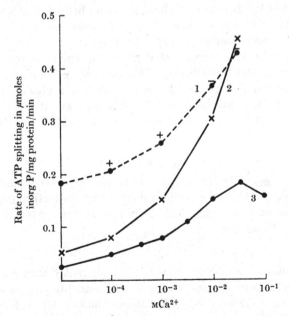

Fig. 25. Influence of Ca^{2+}; (otherwise as fig. 24, except that ATP conc. 1.4×10^{-3}). (Hasselbach (2).)

0.1 N HCl caused only 21 % inactivation.[1] Kalckar (1) purified it, mainly by fractionation with strong acids, and demonstrated the reversibility of the reaction which it catalyses:

$$2\ ADP \rightleftharpoons ATP + AMP.$$

Laki (1) independently carried out some purification of the enzyme in the aqueous extract of muscle which he found must be added to purified myosin to activate splitting of ADP. His method included heating of the centrifuged extract in 0.1 N acetic acid for 15 min at 36°.

In 1947 Mommaerts (4) made a special study of the divorce of the initial visocity fall from the enzymic dephosphorylation of ATP. Straub (3) had already shown that under certain conditions inorganic pyrophosphate caused the viscosity drop which was only reversed by washing away the pyrophosphate. Mommaerts now found that for both ATP and ITP, in presence of 0.001 M Mg^{2+} (which inhibits the ATPase activity of the sol) the viscosity decrease was unimpaired but was not followed by the usual rise. Again, inorganic pyrophosphate and triphosphate, while not dephosphorylated (or in the case of the triphosphate only very slowly) could in certain circumstances of temperature and high Mg^{2+} concentration cause the unreversed viscosity fall. Thus combination with the pyrophosphate group seems to be a condition for the dissociation. Mommaerts' experiments suggested that Ca^{2+}, though speeding up the enzymic action and recovery, could prevent or lessen the decrease in viscosity, which on the other hand was promoted by Mg^{2+}.[2] These effects of ions were confirmed and made more precise by the work of E. H. Bárány, Edman & Palis (1) using an electronic recording viscosimeter, with which viscosity changes within fractions of a second could be measured. Also Hasselbach (2) in 1952 observed that actomyosin, treated with sodium metaphosphate to remove traces of alkaline-earth metals from the protein, lost both its ATPase activity and its dissociability by ATP. Addition of Mg^{2+} restored both properties, addition of Ca^{2+} only the former.

MYOSIN AS AN ENZYME

Lyubimova & Engelhardt (1) used the device of expressing the unit of activity of the enzyme in terms of the volume of a hypothetical gas (in μl) equivalent to the amount of inorganic P (in μg) set free per mg of myosin per hour at 37°. Thus;

$$Q_P = P \times \frac{22.4}{31}.$$

They found values of 500–850, similar to those of Bailey with carbonate/bicarbonate buffer; with glycine buffer, which probably co-ordinates inhibitory heavy metals, Bailey found Q_P values of 3000–6000. Values of this

---

[1] Colowick & Kalckar (1).   [2] Also Acs, Biro & Straub (1).

order, as Bailey pointed out, though low, are not exceptionally so, and bear comparison with the specific activities of a number of other enzymes.

The conception of the myosin of muscle as an enzyme in its own right did not pass unchallenged. One of the few claims made to have separated the enzyme activity from myosin itself was that of Polis & Meyerhof (1) in 1947. They treated myosin solutions with a low concentration of lanthanum acetate. Nearly all the protein was precipitated, and after centrifuging and washing this was subjected to elution with 0.5 M KCl/0.03 M bicarbonate/ 0.001 M ATP/0.01 M KCN. The specific ATPase activity of the eluate was 2–4 times as high as that of the original myosin. The Q_P was 5000–6000, thus about the same as Bailey's best values, but values of 10 000–15 000 were also found. It is interesting that even as late as 1950 H. H. Weber (10), although regarding the effects of ATP on the colloidal properties of myosin as basic for the contraction process, kept an open mind with regard to the possibility that its enzymic effects might all be due to other adsorbed proteins. Dubuisson (3) at this time considered the ATPase to be probably only coupled with myosin. With regard to the results of Polis & Meyerhof it seems possible that (a) some removal of impurities was going on; (b) that there may have been some separation of myosin fractions in slightly different states of aggregation or configuration. Also it is possible that the non-enzymic catalysis of ATP dephosphorylation by lanthanum ions, later observed by Bamann, Fischler & Trapmann (1) and by Selwyn (1) may have been playing a part. It should be noticed that the highest value claimed was not higher than that found by Mommaerts & Green (1) for purified, freshly-prepared myosin. Tsao (1) also concluded in 1953 that, contrary to earlier indications, butanol treatment of myosin gave no evidence of detachment of small molecules of ATPase from myosin. In 1944 Ziff & Moore (1) had found that electrophoresis of Greenstein–Edsall myosin even after 60 h showed only one main component containing at least 90 % of the ATPase activity. Later work has shown that myosin, freed from all traces of actomyosin, can be separated into fractions of different ATPase activity by chromatography on diethylaminoethylcellulose columns. Thus Perry (1) in 1960 found the later-eluted fractions to be higher in ATPase activity, and considered this was at any rate partly attributable to the removal of impurities, amongst them adenylic deaminase and ribonucleoprotein. It is true to say that in the intensive work of recent years upon many aspects of the properties of myosin no evidence casting doubt on its ATPase function has been brought forward.

THE ACTION OF ATP ON ACTOMYOSIN SOLS

The change in viscosity on the addition of ATP to a solution of actomyosin had been interpreted by the Szent-Györgyi school as due to dissociation of the protein; this interpretation was in harmony with the increase in viscosity obtained on mixing actin and myosin; early arguments brought forward in its favour were observations on light-scattering and on the imitation of the ATP effect by urea and other disaggregating substances.[1] However, on this important point better evidence was needed and several workers tackled the problem with, for a time, discordant results.

Jordan & Oster in 1948 (1) concluded from light-scattering experiments by the method of Debye that there was no dissociation but possibly an increase in coiling of the actomyosin particles after ATP addition. Mommaerts (3) in 1951 came to the opposite conclusion, finding with solutions of purified myosin, to which different amounts of actin had been added, that the viscosity after ATP addition was equal to the sum of the viscosities of the two constituents. Further, in studies with the ultracentrifuge, he showed purified myosin to have sedimentation constant $S_{20w} = 6.0$. When actin was added, the actomyosin formed sedimented very rapidly; on addition of ATP, the actomyosin disappeared and the original free myosin was restored. Rather similar observations were made by Portzehl, Schramm & Weber (1) and by Johnson & Landolt (1) with the ultracentrifuge method. It is to be noted, however, that no F-actin peak was reported in any of these cases.

A few years later Blum & Morales (1) described light-scattering studies made with the extrapolation method of Zimm, which gives particle weight independently of particle shape. They observed that particles in myosin B solutions (5 or 24 h of extraction, three times precipitated at 0.06 M KCl) showed with constant molecular weight a reversible shape change on addition of ATP – an extension followed by contraction when the ATP was removed. These results were challenged by Gergely (1) using the same Zimm method but more precisely prepared protein solutions. He examined both reconstituted actomyosin (made from myosin and actin each highly purified) and natural actomyosin prepared by extraction of rabbit muscle with Weber–Edsall solution and purified by repeated precipitation at $\mu = 0.3$ to remove myosin.[2] Special precautions were taken to eliminate inert aggregates. The results were consistent with a dissociation of actomyosin caused by ATP. Gergely was able to explain how the presence of

---

[1] Mommaerts (1, 2). Mommaerts emphasised that the ATP effect was different from that of the other agents, in that the physical properties of the protein obtained approximated to those of myosin. For the effects of the disaggregating agents see ch. 10.

[2] See p. 194.

large ATP-insensitive aggregates in an actomyosin solution might lead by the use of the Zimm method to results giving apparent elongation of the particles on ATP addition.

In 1959 Morales and his collaborators,[1] making use now also of the ultracentrifuge, somewhat modified their attitude. By sedimentation analysis they showed that 5 h myosin B consisted of 35 % of heavy components and 65 % of free myosin. The heavy components could be divided into two classes: ATP or inorganic pyrophosphate depolymerised the lighter class, while the sedimentation behaviour of the heavier class was unaltered. The latter class showed inflation by Zimm's method at constant weight-average molecular weight. On the other hand 24 h myosin showed no free myosin peak till after ATP addition.

Somewhat earlier A. Weber (1) had obtained good evidence for the dissociation. She used actomyosin freed from myosin, centrifuged for 3 h in the presence of ATP and Mg^{2+} at $100\,000g$. Myosin in the top part of the supernatant was identified by its sedimentation constant, the characteristics of its ATPase activity, its reaction with actin and its lack of viscosity change on ATP addition. The amount of myosin obtained showed that most of the actomyosin must have disappeared. The pellets were washed and extracted with 10^{-4}M ATP; very little protein dissolved, but the solution did give characteristic reactions of actin with myosin. Another good piece of evidence was the finding by Martonosi[2] that salting out of actomyosin solutions with ammonium sulphate gave only one peak of precipitation; in the presence of inorganic pyrophosphate two peaks were obtained, at ammonium sulphate concentrations corresponding to those needed for precipitation of free myosin and actin.

Blum & Morales (1) had regarded the protein particle as behaving as a unit, so that for the purposes of ATP interaction it did not matter whether it consisted of myosin or actomyosin. However, in the later work of this group[3] it was found with 24 h myosin B centrifuged in the presence of ATP that the supernatant became progressively less capable of 'repolymerisation' after removal of the ATP. They concluded that 'some "cement", possibly actin' was being removed in the centrifuging.

Thus by 1959 there was general consensus of opinion about the correctness of the early view of dissociation. Various reasons were suggested for the lack of an actin peak in the ultracentrifuge – for example that it was due to the very rapid sedimentation of the F-actin during acceleration or to the very polydisperse nature of the protein. The smallness of the amount of F-actin extractable from the pellet was attributed by A. Weber to denaturation. The formation under certain conditions of G-actomyosin was

[1] Gellert, von Hippel, Schachman & Morales (1); von Hippel, Gellert & Morales (1).
[2] Mentioned in Gergely (2).   [3] Von Hippel, Gellert & Morales (1).

described in 1964 by Johnson & Rowe (4); it is difficult to distinguish in the ultracentrifuge from myosin, and its presence might help to explain the fate of the actin. In 1960 Holtzer, Wang & Noelken (1) found a way out of these difficulties, by centrifuging myosin B solutions in media such as 0.6 M KI or KCNS, known to give complete depolymerisation of F-actin to G-actin. In such media they did find two peaks – the main one corresponding to free myosin ($S = 5.9$); this was followed by a clearly visible small peak (with $S_{20w} = 3.5$) corresponding to pure G-actin.

Two other matters which have been studied in actomyosin sols may be briefly mentioned here; they will be important when we come again to discuss mechanisms of contraction.

First we may consider the minimal amount of ATP needed in relation to the myosin to give maximal effects. Earlier workers[1] gave amounts as high as 1 mole of ATP per 2300 g of myosin. This was a gross over-estimate, no doubt due to the rapid loss of the added ATP by enzymic activity. The viscometric experiments of Mommaerts (4) in 1947, done in presence of Mg^{2+} to inhibit the myosin ATPase, gave a value of about 1 mole of ATP per 360000 g of myosin. More satisfactory results, because obtained with uninhibited enzyme by means of the very rapid luciferin–luciferase method, were recorded in 1960 by Nanninga & Mommaerts (1). The ratio here was 1 mole of ATP to 420000 g of myosin. This is important since, as we shall see,[2] a one-to-one molar ratio will appear.

Secondly, consideration by Mommaerts (5) of the dissociation curve of the myosin/ATP system, obtained by observing in the presence of Mg^{2+} the percentage of maximum effect given by increasing ATP additions, led him to the suggestion that the ATP binding goes on with liberation of free energy, the value of $RT \ln K$ being about -10000 cal. From kinetic considerations also, based mainly on light-scattering experiments, Ouellet, Laidler & Morales (1) have considered the binding to be a reaction providing free energy. Morales, Botts, Blum & Hill (1) gave a value of -6600 for $\Delta F°$ of binding; Nanninga & Mommaerts (2) making use of the firefly method calculated later a value of -11000 cal.

Thus the interaction of ATP and actomyosin in sols may be very briefly summed up as involving (1) dissociation of the actomyosin upon combination of the myosin with ATP, this combination taking place with release of free energy; (2) the enzymic breakdown of the ATP; (3) the recombination of myosin and actin when the ATP concentration is sufficiently reduced.

[1] Dainty *et al.* (1); Johnson & Landolt (1).
[2] Ch. 10.

INTERACTION OF ACTOMYOSIN GELS AND ATP

In the study of this system much has been done with gel suspensions at low ionic strength, showing superprecipitation (which can be taken as a model of contraction), and with those actomyosin threads already described. More useful (since measurements of tension can be made with them) have been glycerinated fibres as prepared by Szent-Györgyi (3). Rabbit psoas muscle *in situ* was divided into a number of fibre bundles about 1 mm in diameter. These were tied to a thin stick, cut out at resting length and placed in 50 % glycerol for 24 h at 0°; they were then removed from the stick, spent one more day at 0°, and afterwards were stored at −20°. Before use, the bundles were transferred to 20 % glycerol to mitigate their stiffness and were then divided into strips 0.2–0.5 mm in diameter and washed in salt solution. By this treatment water is removed very gradually and uniformly throughout the bundle; most of the crystalloids are washed out and about 50 % of the sarcoplasmic proteins. A sort of actomyosin skeleton of the muscle fibres remains. The semi-permeable membrane of the cells is destroyed, so that the fibres no longer react to electrical stimulation, but ATP penetrates and brings about contraction.

Soon afterwards H. H. Weber and his colleagues made use of *single* fibres extracted with glycerol and water;[1] they also prepared oriented actomyosin threads of controlled protein content.[2] It was clear that for any quantitative study of the interaction of ATP and actomyosin fibres the thickness of the fibre was crucial, since the concentration of ATP within the fibre depends on the balance between the rate of diffusion of the ATP inwards and the rate of its hydrolysis by the actomyosin. In the steady state the concentration of ATP on the outside (denoted by C) is related to the concentration at the centre (denoted by I) according to the Meyerhof–Schulz formula:[3]

$$C = Ar^2/4D + I$$

where A is the rate of splitting, r the radius of the fibre and D the diffusion constant of ATP within the fibre. A. Weber (2) first made calculations of the maximum thickness permissible for the fibre model assuming that D was about the same within the fibre as in aqueous solution. But Hasselbach (1) later, by determining the limiting thickness of glycerol–water-extracted muscle slices allowing maximum hydrolysis at a given ATP concentration, was able to calculate the diffusion constant within the fibre and found it to be only one-hundredth of that observed on free diffusion. It must also be

[1] A. Weber (2).
[2] For the method of accomplishing this see Portzehl (1). The threads were soaked in glycerol–water mixture, then allowed to dry gradually at 0° and finally stretched overnight on the dilatometer. Before use the thread, still on the dilatometer, was immersed in the experimental medium and the glycerol quickly diffused out.
[3] Meyerhof & Schulz (1).

remembered that the rate of ATP hydrolysis decreases at very low ATP concentrations. According to the calculations of H. H. Weber & Portzehl (1) the diameter for glycerinated skeletal fibre preparations in a physiological concentration of ATP, about 5×10^{-3}M, should ideally not exceed $30\,\mu$. With a fibre bundle $500\,\mu$ in diameter the concentration of ATP in the centre would be zero. In such a case, as A. Weber pointed out, the fibres of a large central core remain stiff, playing no part in contraction and even hindering it. Tension measurements have been successfully carried out with single glycerinated fibres $50\text{--}60\,\mu$ in diameter, for example by Briggs & Portzehl (1); here it was calculated that the ATP-free core would be 20–50 % of the cross-section; but it is much less likely than in a thicker fibre to be in a condition of rigor since it has an ample supply of ADP, a good plasticiser in the presence of enough Mg^{2+}. For such preparations from smooth muscle or for actomyosin threads a greater diameter was permitted $(60\,\mu)$ owing in the first case to the low ATPase activity, in the second case to the loose structure. The Weber school showed that glycerinated fibres can develop about the same maximal tension as the muscle from which they were made; glycerinated threads produce about one-tenth of this amount of tension. Many characteristic features of muscle contraction are reproduced in these fibres and threads.

Studies of the effects of ATP have also been made with comminuted glycerinated fibres or isolated fresh myofibrils some $2\text{--}3\,\mu$ in diameter.[1] Here there is no diffusion problem and enzymatic activity under different conditions can be accurately studied. Correlation of ATPase activity with rate or degree of shortening observed under the microscope is sometimes usefully made, but it is impossible to assess how far these mechanical effects can be taken as an indication of the degree of power to produce tension.

As we have already seen, the first recorded effect of ATP on actomyosin gels was that observed by Engelhardt, Lyubimova & Meitina with loaded actomyosin threads – an increase in extensibility. Soon afterwards Szent-Györgyi described the opposite effect – the rapid contraction of unloaded threads placed in ATP solution. This paradox was resolved by Buchthal and his collaborators[2] in 1947. They confirmed the observations of Szent-

---

[1] E.g. Schick & Hass (1) and Perry (3). In the former case trypsin, in the latter case collagenase, was used to help in break-up of the cells.

[2] Buchthal, Deutsch, Knappeis & Munch-Petersen (1). It should be noticed that these workers did not interpret their results as supporting the views held by Szent-Györgyi of the myosin thread plus ATP as a model of contraction. This was because (a) of discrepancies they found between contractility and ATPase activity – discrepancies no doubt due to limitations in experimental procedures at the time; (b) of differences between the behaviour on ATP addition of the elasticity modulus with changing length in the thread as compared with the muscle itself. This is hardly surprising – it is to be expected that cohesion of the reprecipitated protein molecules will be slight compared with that in the original muscle structure.

Györgyi on rapid isodiametric shortening and shrinkage of myosin B threads placed in ATP solution; threads which had been dried and then soaked in salt solution before use on the other hand shortened with increase in diameter and no volume change.[1] Now they found that these same threads with a load of 5–180 mg lengthened on ATP treatment. They considered that the release of linkages played a part in both cases.

While of course the same degree of dissociation cannot take place in a gel as in a sol, it is likely that the same bonds are affected by binding of ATP to the myosin. Szent-Györgyi (3) has emphasised this dual role of ATP on the one hand causing contraction, on the other showing a plasticising effect. Fibre bundles contracted after treatment with ATP were hard and opaque, but on renewal of the ATP (3×10^{-3}M) they became momentarily soft, flexible and extensible, before hardening again as the fresh ATP was used up. Bozler (1) in 1951 was the first to get a full cycle of contraction and relaxation: a glycerinated fibre bundle was used at high ATP concentration (0.02M) and the contraction was brief. A concentration of 0.003M ATP caused only contraction. Bozler made it clear that on washing out of ATP after the contraction, tension in the fibre bundle was not maintained but gradually dropped. He put this down to slow breaking of bonds (not present in the unstimulated fibres) within the contractile elements.

The effect on precipitation and solubility of the joint action of KCl and ATP had been studied by Szent-Györgyi (4) using a solution which contained mainly myosin and comparatively little actomyosin. This question has been re-investigated by Spicer (1) using reconstituted actomyosin made from purified myosin and one-third of its weight of actin. The strength of the gel was assessed by pouring and the contractibility of the precipitate by inspection; the results could be expressed in a semi-quantitative manner. In the presence of 0.5×10^{-3}M ATP, with increasing KCl concentration there was increasing contractility of the gel formed up to about 0.17M. Between 0.24 and 0.32M there was no superprecipitation but gel formation, which was reversible. Above 0.32M the sol remained nearly clear. With 0.15M KCl the pH optimum for superprecipitation was 6.7; at pH 8.6 gel was formed and at 9.1 the sol remained clear. Thus a twofold effect of the nucleotide could be detected – immediate clearing followed by gelation or superprecipitation, according to the salt concentration.

Just as Mg^{2+} promotes and Ca^{2+} antagonises the fall of viscosity of actomyosin solutions with ATP, so, as A. Szent-Györgyi (2, 4) early showed, Mg^{2+} favours gel formation and superprecipitation, while Ca^{2+} has the opposite effect. Spicer confirmed this, finding for instance that Mg^{2+} extends the range of KCl concentration over which gel formation takes place, while

---

[1] See also the rather similar experience of Gerendás (1) with actomyosin threads partly denatured by treatment with 10^{-3}M ZnSO$_4$.

Ca^{2+} restricts it; and much less ATP was needed to cause gelation or superprecipitation when Mg^{2+} was present, much more when Ca^{2+} was present, than in the absence of these ions. Further discussion will be needed of such complex effects. We shall see[1] that Mg-activated ATPase can support contraction while this is not possible with Ca-activated ATPase; further that there are two centres on the myosin, one concerned with actin binding, the other with Mg-activated ATPase activity which is only possible in presence of actin.[2] It is uncertain whether ATP^{4-} or $Mg^- ATP^{2-}$ is the substrate for the MgATPase; but Mg ions are necessary for the activation of this centre and seem also (in their plasticising effects) to play a part in the dissociation of actin;[3] Ca ions antagonise this plasticising effect. Then we must make preliminary mention of the fact, realised only in 1959, that for the MgATPase activity a trace of Ca^{2+} (about 10^{-6}M) is also necessary.[4]

THE EFFECT OF SH REAGENTS ON THE PROPERTIES OF MYOSIN SOLS AND SUSPENSIONS

The presence of SH groups in myosin seems to have been noticed first by Arnold (1) in 1910 when he found that the protein prepared according to Halliburton[5] gave a positive although not very strong nitroprusside reaction. Hopkins in the early nineteen twenties (3) was intensely interested in the SH groups of well-washed minced muscle protein, finding them to undergo repeated oxidation and reduction in the presence of glutathione. In 1936 Mirsky (1) first showed that the number of active SH groups, giving the nitroprusside reaction, on the proteins of minced muscle or of any of its protein fractions (including myosin) was small compared with the number found after denaturation. Todrick & Walker (1) soon afterwards found the free SH groups in native myosin to be 0.27 %; Greenstein & Edsall (1) gave a rather higher value (0.42 %) for the stage at which porphyrindin treatment abolishes the nitroprusside reaction, and showed that on denaturation by treatment with concentrated guanidine hydrochloride the active SH content rose to 1.15 %. Neurath and his collaborators[6] considering in 1944 the question of denaturation in connection with SH groups, concluded that in the native protein also these groups exist in their normal form, but may not all be detected by specific reagents because of their protection by the steric configuration. Upon denaturation, all become freely accessible.

In 1944 a number of different workers studied the inhibition of myosin ATPase by sulphydryl reagents – mild oxidising agents such as H_2O_2, or mercaptide-forming substances such as p-chloromercuribenzoate. The in-

---

[1] See p. 166.   [2] Ch. 12.
[3] See pp. 154, 291.   [4] Ch. 13.   [5] P. 132.
[6] Neurath, Greenstein, Putnam & Erickson (1).

hibitory effects could be reversed by treatment with cysteine or reduced glutathione.[1] Singer & Barron (1) found that abolition of all freely-reacting SH groups caused inhibition of only about 12 % of the ATPase activity; abolition of 86 % of the SH groups led to total enzymic inhibition. The important paper of Bailey & Perry (1) appeared in 1947. Like Singer & Barron they found ATPase activity of myosin to be inhibited by oxidation of, or substitution in, the SH groups. Their results however were quantitatively different in that they found the major part of the inhibition connected with the disappearance of the more reactive SH groups (assessed by

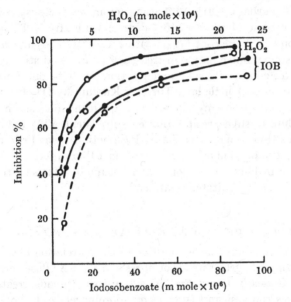

Fig. 26. Inhibition of actomyosin formation and ATPase activity as a function of oxidant concentration. -●—●—●-, actomyosin formation; -○--○--○-, ATPase activity; IOB = iodosobenzoate. (Bailey & Perry (1).)

a carefully controlled iodine method) and concluded that only a small residuum of inhibition depended on the removal of the less reactive ones. Bailey & Perry also found inhibition coming on gradually by alkylation with iodoacetamide. It appears that the SH groups of myosin are peculiarly resistant to alkylation and this would explain the earlier negative results of D. M. Needham (3) and of Singer & Barron (1) when looking for inhibition with iodoacetic acid or iodoacetamide. Bailey & Perry tested, by means of viscosimetric measurements, the ability of myosin after treatment with SH reagents to react with actin. Using H_2O_2, iodosobenzoate or chlormercuri-

[1] Mehl (1); Ziff (1); Singer & Barron (1).

benzoate, they found this ability to fall off in parallel with the fall in enzymic activity (fig. 26), and deduced that certain centres, of which SH groups form part, were concerned in reaction with either ATP or actin, the affinity of the former for the enzyme being great enough to allow complete displacement of the latter. Singer & Barron had remarked that the effect of sulphydryl reagents (known to be attacking myosin) upon the ATPase activity argued for the identity of the myosin and the ATPase; this argument was strongly reinforced by Bailey & Perry's finding of a stoichiometric relation in inhibition of actomyosin formation (which must be a property of the myosin) and of the ATPase activity.

Kuschinsky & Turba (1, 2, 3) in 1951, using actomyosin gel suspensions and observing the volume of the centrifuged precipitate after ATP addition, found inhibition of volume contraction with 2×10^{-4}M Salyrgan,[1] reversible with 10^{-3}M cysteine. Since Salyrgan-treated F-actin could still react with myosin, this interference with superprecipitation was to be put down to loss of myosin SH groups. A little later Turba & Kuschinsky (1) observed that short treatment of an actomyosin sol with oxarsan[2] in absence of ATP resulted in failure of superprecipitation on subsequent ATP addition and dilution; but in these conditions the ATPase activity was little affected. This was the first indication of the important fact that different myosin SH groups are concerned in combination with actin and in ATPase activity and can be distinguished by suitable treatment.[3]

CONTRACTION AND ATP BREAKDOWN *IN VITRO*

Much work has been done on the question of the correlation between contraction of actomyosin gels, threads or fibres and their ATPase activity. As early as 1944 Godeaux found that in actomyosin threads treated with certain war gases and vesicants (such as monobromacetate) loss of contractile power ran parallel with diminution in coloration with nitroprusside; Buchthal and his collaborators[4] similarly observed loss of the power to contract caused by treatment with SH-combining reagents. Korey (1) making use of glycerinated fibre bundles 500–1000μ in diameter and measuring their contraction kymographically, found that soaking the fibres in iodosobenzoate, Mapharsen or mercuric chloride (in concentration between 10^{-3} and 10^{-4}M) led to disappearance of the contraction. The inhibition with the first two compounds was reversed by 10^{-2}M cysteine. The ATPase of the fibres was originally 25–30μg of P liberated/mg 15 min; this fell to about 4μg when contraction was inhibited and rose with cysteine to about 7 μg.

---

[1] Complex mercury compound of salicyl–allylamide–Na–acetate.
[2] *m*-Amino-p-hydroxphenyl arsenite.   [3] Ch. 12.
[4] Buchthal, Deutsch, Knappeis & Munch-Petersen (1).

In 1952 Portzehl (2) made a study of the effects of Salyrgan on tension production in single glycerinated fibres $50\,\mu$ in diameter. The fibre was caused to contract with ATP and, after using up its supply, remained stiff and short. If Salyrgan (about 10^{-3}M) was now added no effect was seen, but ATP (about 10^{-4}M) addition in the presence of the Salyrgan caused relaxation not contraction. Only after the addition of cysteine in large excess was contraction again possible (fig. 27). Weber and his colleagues[1] brought forward much further evidence for the close correlation between tension production in glycerinated fibres and their ability to dephosphorylate ATP.

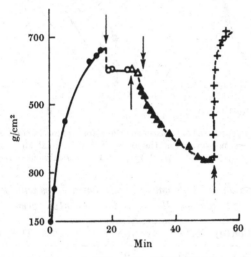

Fig. 27. Behaviour of a glycerinated fibre with ATP and Salyrgan. Contraction caused by 1.7×10^{-3}M ATP. At ↓ ATP washed out. At ↑ 6.6×10^{-4}M Salyrgan added. At ↓ 1.7×10^{-3}M ATP added. At ⤊ addition of 6.7×10^{-2}M cysteine. (Portzehl (2).)

A. Weber & H. H. Weber (1) showed in 1951 that tension production in glycerinated fibres was dependent on ATP concentration which had an optimum value varying with temperature. Heinz & Holton (1), using glycerinated fibres 70–80 μ in diameter then found that the fall in ATPase activity and in tension production began together at the same ATP concentration: 1–2×10^{-2}M at 20°; 5×10^{-3}M at 0°. Allowance was made for diffusion conditions. Ulbrecht & Ulbrecht (1, 2) have studied this question also using glycerinated fibres (40–100 μ in diameter) from smooth muscle. From the yellow adductor muscle of *Anodonta* it was possible to prepare fibres thin enough to allow complete penetration of ATP and which yet would not tear under maximal isometric tension. With these the maximal power produced at 20° was seven times as great as at 0°, and the same ratio

[1] A. Weber & H. H. Weber (1); Heinz & Holton (1); Ulbrecht & Ulbrecht (1).

held for ATP splitting measured in suspensions of myofibrils. They suggested that the temperature coefficient of the mechanical power was determined by the temperature coefficient of the rate of ATP hydrolysis. The significance of these results has also been discussed by Weber & Portzehl (1).

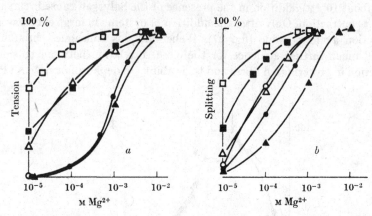

Fig. 28. The effect of magnesium concentration on tension production and nucleoside triphosphate hydrolysis, in percentage of the behaviour at optimal Mg^{2+} concentration. □, ATP; ■, acetyl-ATP; ○, UTP; ●, ITP; ▲, GTP. (Hasselbach (3).)

TABLE 8. *Tension development in glycerinated fibres with different nucleoside triphosphates; the dependence on Mg-concentration*

Mg-concentration (M)	ATP	Acetyl-ATP	CTP	UTP	ITP	GTP
$\sim 10^{-11}$*	10	0	.	.	.	.
$\lesssim 10^{-5}$†	15	.	.	.	.	.
$\gtrsim 10^{-5}$	60	30	18	0	0	0
10^{-4}	90	55	60	.	5–10	<3
10^{-3}	100	80	80	.	30	15
5×10^{-3}	100	87	90	75	45	.
10×10^{-3}	100	90	95	.	50	30

* In presence of 10^{-2}M EDTA.
† After previous treatment with EDTA. The tension development by ATP under optimal conditions is put equal to 100. $T = 21°$; ionic strength $= 0.1$, nucleoside triphosphate conc $= 5 \times 10^{-3}$ M. (Hasselbach (3).)

These experiments were all carried out with Mg (0.003M) in the medium; in 1952 Hasselbach (2) mentioned the very significant observation made by H. H. Weber that, in the absence of Mg ions, Ca ions did not mediate tension production by glycerinated fibres though mediating ATPase activity. A few years later Hasselbach (3, 4) made a very thorough study of the dephosphorylation of various nucleoside triphosphates by actomyosin gels, and of the effect of these triphosphates on tension production by and plasticisation of glycerinated fibres and fibre bundles (fig. 28). The importance of Mg ions for the two latter effects was very clear. Table 8 shows the predominant place of ATP in tension production, especially at low Mg concentrations. For ATPase activity of actomyosin at low ionic strength the triphosphates could be arranged in the same order but for Mg-activated ATPase activity of myosin or actomyosin at high ionic strength the order was different – GTP > ITP > UTP > CTP $\gtrsim$ ATP. The importance of Mg^{2+} and the inefficacy of Ca^{2+} for contraction were also found by Ashley, Avasimavicius & Hass (1) using glycerinated myofibrils and measuring shortening under the microscope. After being depleted of Mg^{2+} by dialysis, the fibrils did not contract, although they did still split ATP; addition of Ca-ATP increased splitting but did not elicit contraction; addition of Mg-ATP caused instantaneous contraction. Bendall (1) also, studying ATPase activity of glycerinated myofibrils, noticed instant shortening on addition of Mg^{2+} to the ATP-containing medium, but no such effect with Ca^{2+}.

Many experiments on the application of ATP to filaments and fibres have on the other hand been interpreted as due rather to the binding of ATP than to its dephosphorylation. Thus Korey (1) was inclined to interpret his observation on Mg^{2+}-activated contraction of fibre bundles as due to increased ATP adsorption in presence of Mg^{2+}, citing Szent-Györgyi (5).[1] Bowen (1, 2) found a number of conditions in which myosin B threads ($200–300\mu$ in diameter) showed lack of correspondence between rate of shortening and ATPase activity. Thus addition of 10^{-2}M $MgCl_2$ made shortening more intense while 10^{-2}M $CaCl_2$ inhibited it, yet this concentration of Mg^{2+} inhibited ATPase activity, and Ca^{2+} increased it three- to fourfold. Again Bowen & Kerwin (1) found with glycerinated fibre bundles ($300–500\mu$ in diameter) that increasing concentration of KCl up to 0.3 M retarded dephosphorylation but accelerated shortening. All these facts were regarded as favouring the importance of ATP binding for contraction.[2] Relying on Blum's revaluation (1) of the problems of ATP penetration, Bowen considered that diffusion did not play any significant part in these experiments. In connection with these findings several points must be remembered. For example the conflicting results of Bowen might be explained by the presence of high and variable amounts of myosin in his myosin B threads – playing no part in contraction but giving ATPase results at variance with those of the actomyosin. Again, Hasselbach (2) *did* find activation of the ATPase of myosin B threads with 10^{-2}M Mg; and with regard to the accelerating effect of 0.3 M KCl on contraction, Perry (2) has pointed out that this could be due to a dehydrating effect on the glycerinated fibres, changing their mechanical properties, quite apart from any effect on

[1] See p. 50 (2nd ed.).   [2] Also Bowen (3).

their ATPase activity: also the ATPase was measured only after homogenisation and might not coincide with that of the fibres themselves.

In attempts to determine a quantitative correlation between contraction and ATP breakdown it has to be remembered that in some methods of testing for this (for example free shortening of isolated filaments) very little energy supply is necessary for maximum effect; while in others (loaded fibres exerting maximum tension) the response measured requires maximum energy provision. Also, as we have seen, binding of ATP as well as its breakdown can produce free energy; instances of theories of contraction involving the availability of the energy of binding will be considered in chapter 9.

It seems then that the action of ATP on actomyosin gels takes place in several stages: first the binding of ATP to myosin leading to the loosening of certain linkages; then ATP dephosphorylation causing a fall in ATP concentration at certain sites. Evidence for the importance for contraction of a low level of ATP comes from the work of Marsh (1, 2), who used muscle homogenates made up of fibre bundles, and followed the water loss and shrinkage of the fibre fragments by measuring their volume after centrifugation. He found a sharp diminution in volume when the ATP concentration within the fibres (which was very different from the concentration in the medium) had fallen to about 0.05 mg of labile ATP-P per ml of fibres. The third stage can now follow – the linkage formation between actin and myosin involved in the contraction. We shall see how fundamental to our ideas about contraction, not only in muscle models but in the living muscle itself, is the process of continually repeated making and breaking of these links.

9

SOME THEORIES OF CONTRACTION
MECHANISM, 1939 TO 1956

BACKGROUND TO THE THEORIES

MUSCLE STRUCTURE. Before introducing some of these conceptions entertained during the ten or fifteen years after the discovery of the interaction of myosin, actin and ATP, we may consider the re-orientation of ideas concerning interpretation of visible muscle structure. This closer look was necessitated by the discovery of actin, the more exact knowledge of the relative quantities of the muscle proteins and the early observations by means of the electron microscope. As we have seen,[1] 'myosin' had been allotted by Noll & Weber (1) in 1935 to the A band, and the double refraction of the fibre had been explained as due to the rod and intrinsic double refraction of this protein. Weber (6) in 1956 remarked that this would mean that the I band must consist of other proteins – perhaps including globulin X and stroma. The assumption however was frequently made that it consisted of disordered myosin. Some observers recorded that the I band rather than the A band shortened on contraction, but the general opinion seems to have been that the material of the A band was that primarily concerned in the mechanism of movement, the changes in the I band being passive.

It is striking to see how many of the observations of classical histology were confirmed by the electron microscope – for example, the A and I bands, the H zone, and the Z and M lines could all be distinguished.

We may take first the new ideas stimulated by the electron microscope study made in 1946 by Hall, Jakus & Schmitt (1) using thin myofibrils after disintegration of formalin-fixed muscle. Perhaps the most striking finding was that the filaments seen (varying between 50 and 250 Å in width) extended continuously in relatively straight lines through both the A and the I bands, remaining straight also in contracted muscle; they appeared to pass through several successive sarcomeres. The authors concluded that the filaments themselves must shorten, and that on contraction there was no gross spiralling or folding on the scale of dimensions visible in the electron microscope. Further, since the filaments, presumably myosin, were seen in both bands, it seemed impossible any longer to ascribe the very low

---

[1] Ch. 7.

double refraction of the *I* bands (only 10% of that of the *A* bands) to the presence of myosin in an unoriented state as compared with an orderly array of myosin micelles in the *A* bands. They therefore suggested that material of negative double refraction was also present in the *I* band, compensating for most of its positive double refraction. Thirdly, the high density of the *A* bands in both the electron microscope and the light microscope was attributed, as in earlier work, to the presence there of the '*A* substance', possibly characterised by relatively high salt concentration. Indeed, Macallum (1) had reported a localisation of potassium in the *A* band, and potassium has a high affinity for phosphotungstic acid (used in the electron staining). However, Draper & Hodge (1) a little later showed that formalin-fixed fibrils contain very little potassium. The *A* substance was regarded as closely associated with the filaments, since no interfibrillary material could be observed. On contraction the phenomenon of contraction-band formation (already familiar with the light microscope) was seen, the *A* substance now being concentrated around the *Z* line.[1] At the same time these electron microscope observations dispelled the idea that the *Z* line was composed of collagen. A further important observation was the beaded appearance of the filaments, with a fine cross-striation, at intervals of roughly 400 Å. This axial periodicity, apparent also in X-ray diffraction diagrams,[2] has had various interpretations and its meaning is not yet entirely clear.[3]

A number of suggestions were made as to the possible nature of a constituent having negative double refraction. Dempsey, Wislocki & Singer (1) ascribed the role to lipid substances, finding difficultly-extractable phosphatides particularly in the *I* bands. In 1947 Matoltsy & Gerendas,[4] in Szent-Györgyi's laboratory, described the disappearance of positive double refraction in the fibrils, and the appearance of negative double refraction in the *I* bands of muscle pieces fixed after removal of the myosin and then the actin by successive extractions with KCl and KI solutions. Microscopic structure and all double refraction was lost on further extraction with Weber solution containing 30% urea. This final extract had maximum absorption at 2650 Å, and the 'N-material' it contained was taken to be a nucleoprotein; it was further claimed that the solution showed strong double refraction of flow.

In neither of these cases could the existence be upheld of a compensatory negatively doubly refracting constituent in the *I* band. Thus Perry (10) found the ribonucleic acid P content of isolated myofibrils to be only 0.05%, and the ether–ethanol soluble material only 5%. These amounts seem far too small to be of significance in this connection. Moreover Dubuisson &

---

[1] Pp. 130–1.   [2] See e.g. p. 226.   [3] Ch. 11.
[4] Matoltsy & Gerendás (1, 2); Gerendás, Szarvas & Matoltsy (1); Gerendás & Matoltsy (1).

Fabry-Hamoir (1), after repeating the extraction procedure of Matoltsy & Gerendás, were unable to find double refraction of flow, positive or negative, in the Weber-urea solution; the same negative result was the experience of Hasselbach & Schneider (1). It seems likely that the nucleoprotein found by Matoltsy & Gerendás came from nuclei or cytoplasmic granules.

Draper & Hodge (1) in an important paper in 1949 confirmed many of the observations of Hall, Jakus & Schmitt. They drew attention to the correspondence between the axial periodicity (250–450 Å) of the protein filaments and that seen in muscle after micro-incineration,[1] to be attributed to localisation of salts – in particular of calcium and magnesium; earlier work of Scott & Packer (1) had shown the retention of these salts in micro-incinerated muscle sections, and the evaporation of potassium and sodium from the tissue under the same conditions. As Scott (1) and Scott & Packer (1) had found, the ash was densest in the A band. Draper & Hodge considered that there was a rough correlation between axial period in the protein filaments and the state of contraction, estimated by sarcomere length. They suggested that the ultimate contractile unit was the portion of each filament contained in one axial period, i.e., between two fine cross-striations. They saw the fibrils as collapsible tubes made up of the filaments or protofibrils, and believed the A substance to be arranged on the inner surfaces of these tubes. They found the electron microscope evidence to be in harmony with the classical picture of migration of A substance to the Z line with reversal of striation during contraction.

In their electron-microscope study of myofibrils in 1950, Rozsa, A. Szent-Györgyi & Wyckoff (2) concluded that the whole appearance of the filaments was compatible with the hypothesis that they consisted of actin threads, covered by some second substance. The threads had the diameter they had seen earlier (1) in actin filaments, and in clear spaces free from overlying material a periodicity of 250–300 Å, similar to that seen in the actin threads, could sometimes be made out. They suggested that the overlying substance might consist of exceedingly fine threads of myosin in lengthwise association with the thicker threads of F-actin. Their shadow-casting technique showed that the outer ends of the A bands were much denser than the rest of the fibril; this they put down partly to the presence of salts, partly to an unspecified 'A band substance'. Ashley and his collaborators[2] in 1951, working with glycerinated myofibrils, noted that the A substance could be removed by extraction with salt solutions capable of dissolving myosin, and suggested that the A substance was myosin.

As H. H. Weber (6) has emphasised, the volume of the A band in most vertebrates would be about 45 % of the fibre volume. If the whole of the myosin plus part of the actin were concentrated there, the protein content

---

[1] Draper & Hodge (2, 3).   [2] Ashley, Porter, Philpott & Hass (1).

of the A band would have to be very considerably higher than in the rest of the muscle. If the filaments visible in the electron microscope consisted of actin, then only a small part of the fibrillar protein (some 20 %) could exist in a visible form. Weber also inclined to the view that the myosin, in a structure capable of conferring double refraction but not of resolution in the electron microscope, was situated in the A bands. The alternative possibility of the presence of a substance with negative double refraction in the I bands had been by this time (1956) discredited. Weber doubted the identity of the A substance with myosin, since movement of this protein by diffusion would be too slow, and its movement by electric cataphoresis would mean postulation of potential difference of 100 V/cm between the M and Z lines.

In 1942 Caspersson & Thorell (1), using a specially constructed quartz microscope, investigated the ultra-violet absorption spectra of muscle fibres at a wavelength of 2600 Å, the region of maximal absorption by the adenine nucleotides. With insect fibres, having specially long sarcomeres, a banded appearance was readily seen and when these photomicrographs were compared with photomicrographs taken in polarised light, it transpired that the strongly absorbing striae corresponded to the I bands. D. K. Hill (6) in 1964 confirmed this localisation, using autoradiography with toad striated muscle, after injection of tritium-labelled adenine nucleotide. It appeared that about 50 % of the adenine nucleotides were concentrated in this way. Caspersson & Thorell discussed the possible significance of this localisation of ATP, in connection with its energy-providing function when in contact with myosin, which they took as occupying the A bands. In particular, it appeared that in fatigued muscle the sharp differentiation was partly smudged out, as though some of the adenylic acid had migrated to the A band (see fig. 29 pl.).

HEAT PRODUCTION AND MECHANICAL CHANGES. Those working on muscle contraction at this period were seriously concerned to decide whether energy provision from ATP breakdown was necessary in the contraction or relaxation phase. Hill (15) had shown as early as 1938 that after an isotonic tetanus the work done on the load during shortening appeared during relaxation as heat, equivalent to the work done. When the load was removed before relaxation, little heat appeared in the relaxation phase. The same results were obtained (6) on isotonic twitches ten years later with more refined techniques, and it was established that the relaxation heat appeared simultaneously with disappearance of mechanical energy, and that there was no measurable heat production in the relaxation phase when the muscle was without load (fig. 30). Although these results were compatible with the view of relaxation free from the necessity of metabolic energy provision, they could not be taken as proof. Thus Needham in 1950 (4) emphasised the possible analogy with the anaerobic recovery period,[1] a period in which much chemical reaction is proceeding with little heat wastage. However, at this time A. V. Hill (6, 16, 37) gave evidence from mechanical experiments that there was no detectable active lengthening of the muscle during relaxation. These results were on frog sartorius with straight parallel fibres and Hill emphasised that, with muscles of complex anatomical structure, elastic constraint might force the relaxed fibres to lengthen. Bendall (9) much later described the high degree of spontaneous relaxation shown e.g. by the

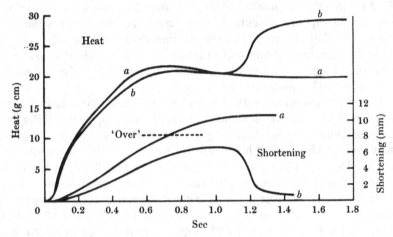

Fig. 30. To show 'relaxation heat'. Records of the heat production and shortening in consecutive twitches of a pair of toad's semi-membranosus at 0°. (a) 20.7 g on the ergometer lifted over (and thus applying no further load to the muscle). (b) 25.7 g on ergometer not lifted over; the work was dissipated as heat in the muscle as it relaxed. (Hill (6).)

sterno-mandibularis, which contains 10 % of its dry weight as collagen. It must be remembered also that in 1940 Ramsey & Street (1), with single muscle fibres suspended in Ringer solution between horizontal supports, observed a quick return to resting length after stimulation provided the shortening was not greater than 30 % of the resting length. In 1964 Podolsky (1)[2] described fibres in which part of the sarcolemma had been dissected away; these were suspended in oil and contraction was caused by Ca injection. Relaxation followed and Podolsky considered the presence in both fibres and fibrils of 'elastic systems acting in parallel with a force generator'.[3]

It is important here to consider further the great advances made by A. V. Hill and his collaborators in measurement of heat production and

[1] See p. 94.   [2] Also Podolsky & Constantin (1).
[3] See also Hanson (4); Parsons & Porter (1).

mechanical change in muscle under different conditions. Clearly any definitive picture of the chemical events and changes in fine structure underlying contraction must take into account both the magnitude of the heat production and its timing in relation to the mechanical response.

As we have seen,[1] A. V. Hill in 1922 regarded visco-elasticity as playing an important part in muscle contraction, and he and his collaborators recorded quantitative observations in favour of this conception. However, as he has described (38), the work of the ensuing years brought doubts of any simple relationship between speed of shortening and viscosity. For one thing, resting muscle showed little resistance to stretch, so that it was necessary to suppose an entirely new elastic system in the stimulated muscle. Also, further experiment showed that the relationship $P_0 = P + kv$ (where P_0 = maximum tension in the spring at zero load; P = load with spring shortening at velocity v, k = constant) did not hold for contracting muscle and further postulates became necessary.

Then in 1938, working with new and refined techniques (including the protected thermopile which avoided errors due to temperature differences along the muscle when shortening occurred) Hill (15) made important discoveries which changed his views. He now found that shortening was accompanied by heat production proportional to the change in length, its amount being independent of the load and of the speed of shortening.[2] During an isotonic contraction work is done; the rate of total energy liberation (work + shortening heat) was found to be a linear function of the load, decreasing as the load increased. The rate of total energy liberation may then be written:

$$(P+a)v = b(P_0 - P) \tag{1}$$

where v = velocity of shortening in cm/sec, P = actual load, P_0 = maximum load (isometric contraction), a = shortening heat/cm, and b = increase in total energy rate per g decrease in load.

Both a and b were constant for a given temperature and the relationships of equation (1) were verified experimentally.

Equation 1 may be written:

$$(P+a)(v+b) = b(P_0+a) = \text{constant.} \tag{2}$$

an equation (usually known as the 'characteristic equation') relating speed and load, which could be investigated by purely mechanical methods. The values for v plotted against P fell on a hyperbolic curve showing the falling off of v as P increased[3]; this curve was the same as that calculated from

---

[1] Ch. 3.

[2] It is interesting to remember that Fenn (2) had sometimes noticed greater extra heat liberation than that corresponding to the work done. Hill (11) has remarked that with the instruments at his disposal he could not have separated (better than he did) the extra energy due to shortening from the extra energy liberated as work.

[3] Compare the similar results of Fenn & Marsh (1) in 1935 (ch. 3).

equation (2), when the values for a and b, previously found in the thermal measurements, were used. These relationships led Hill to the conclusion that active muscle could no longer be regarded as analogous to a stretched spring operating in a viscous medium. Its slower shortening under greater force, expressed in equation (2), was not due to viscosity, but to regulation of the reactions providing energy, as had been adumbrated by Fenn.[1]

Strong evidence against the viscous elastic theory, and against any idea that the shortening heat was the result of degraded work, came with Hill's demonstration in the same paper that when a stimulated muscle was slowly stretched, the work done in stretching did not increase the amount of heat produced in the muscle – again a finding indicated in the work of Fenn. The work done on the muscle was not accounted for either as extra heat or as mechanical potential energy of increased tension, and this observation seemed to indicate the slowing of reactions normally associated with activity. In his memorable book 'Trails and Trials in Physiology'[2] Hill has paid tribute to Fenn's intuitive grasp of the problems involved; this led him, in spite of the uncertainty of the results available with the crude apparatus of the time, to draw out the right conclusions. Verification of these conclusions had to wait on the slow process of technical improvements.

Hill pictured a mechanism whereby the control exercised by the tension in the muscle on the rate of energy release at any moment could be operated.

Imagine that the chemical transformations associated with the state of activity in muscle occur by combination at, or by the catalytic effect of, or perhaps by passage through, certain points in the molecular machinery, the number of which is determined by the tension existing in the muscle at the moment. We can imagine that when the force in the muscle is high[3] the affinities of more of these points are being satisfied by the attractions they exert on one another, and that fewer of them are available to take part in chemical transformation. When the tension is low[4] the affinities of less of these points are being satisfied by mutual attraction, and more of them are exposed to chemical reaction. The rate at which chemical transformations would occur, and therefore at which energy would be liberated, would be directly proportionate to the number of exposed affinities or catalytic groups, and so would be a linear function of the force exerted by the muscle, increasing as the force diminished.[5]

This formulation supposes that the links are *either* concerned in holding the tension *or* in producing energy. D. M. Needham (4) in 1950 suggested one rather different, in which energy was produced by chemical interaction between groups situated along the protein chain.

[1] See Fenn (2) and p. 56 above.   [2] Hill (11).
[3] As in isometric contraction.   [4] As in isotonic contraction.
[5] For consideration of these ideas in terms of the sliding mechanism see H. E. Huxley (9) p. 450.

For shortening a given distance, the 'shortening heat' is the same, whether the shortening is slow or fast. But the rate of shortening depends on the load, being slower the greater the load; thus at slower rates of shortening between two given lengths more work is done and more energy must be produced, since the heat remains the same. If this energy production is the result of the interaction of the same groups at different rates of shortening we must suppose that at the slower rates, repeated interaction takes place.[1]

In this paper of 1939 Hill also suggested that in isometric contraction the actively contracting parts of the muscle do actually shorten at the expense of stretching of passive undamped elastic parts in series with them; this series elastic component, important as a buffer when muscle passes abruptly from the resting to the active state, had been postulated by Levin & Wyman (1) and was studied by Hill (17) in 1950 and later for example by Jewell & Wilkie (1) and Abbott & Mommaerts (1).

Hill (18, 19, 20) went on to study the heat production in a single isotonic twitch under small load. This heat was made up of two parts – the activation heat and the shortening heat. The activation heat started off at maximum rate about 2 milliseconds after the stimulus in frog's muscle at 20°, well before the shortening. It fell off gradually to a constant value while the muscle continued to shorten, then decreased to zero. In the slower tortoise muscle contracting isometrically, it began at 60 milliseconds at 0°, and was followed only at 90 milliseconds by the contraction. The heat (considered up to the end of shortening and not including relaxation) consisted of $A + ax$, where x = the shortening in cm, a = the heat of shortening/cm, and A = the activation heat. A was about one-half to one-third of the total heat in contractions with small load. The heat produced during a maintained contraction, the maintenance heat, was studied by Abbott (1). It is the summed effect of the heats of activation of the successive responses. Hill described the activation heat as 'presumably the thermal sign of some chemical process by which the muscle is put, and maintained for an interval, in a state of activity, whatever that means, of readiness to shorten, to exert a force, or to do mechanical work'. He emphasised that these results were evidence against the view of contraction as an endothermic process, and that the clear preceding of the mechanical response by heat output at maximum rate was not consonant with the conception of the mechanical energy as derived from a reservoir, where it remained latent until released by stimulation.[2]

The latency relaxation, a transient and very small lengthening or decrease in tension of the muscle following very rapidly upon the stimulus, was first

---

[1] Compare H. H. Weber (9) ch. 11 below.
[2] See ch. 3; the present-day explanation of the activation heat in terms of release of Ca ions, is discussed in ch. 14.

observed by Rauh (1) in 1922, and was much studied later by Sandow (1). In 1949 D. K. Hill (4) found that the latency relaxation, which in frog's muscle at 18° began about 2 milliseconds after the stimulus, and reversed to contraction a few milliseconds later, was accompanied by an increase in transparency of the muscle, beginning at exactly the same time. During the next few years A. V. Hill (21, 22) found that the earliest manifestation of the mechanical response, beginning at about the same moment as the activation heat, was an increase in rigidity of the muscle, which became less extensible[1] (not more so, as had been assumed by Sandow). A. V. Hill considered the early transparency change as the physical sign of altered molecular pattern in the muscle proteins; the heat of activation as the sign of chemical changes by means of which the mechanical and physical changes were brought about. The shortening heat, he surmised, was perhaps set free in reactivating a system the activity of which had been dissipated by shortening. With regard to the latency relaxation, he considered (22) that the weight of evidence was in favour of its origin in a parallel elastic structure – possibly the sarcolemma.[2]

More recently further work has been done on the various phases of heat production in contraction. Some corrections in the earlier results concerning the shortening heat have been made and some new interpretations suggested. A description of these will find its place in chapter 14.

It is now time to consider the theories of the nineteen forties and early nineteen fifties. It must be admitted that in a short chapter it is possible to deal only with a selection of the many which flourished in this period. These have perhaps been chosen on a rather subjective basis, being the ones which mainly kept in mind the basic importance of the ATP–actomyosin relationship and so seem to the writer to be in the main-stream of advance.

THE THEORIES

THEORIES INVOLVING RELAXATION–ENERGISATION VIA ATP. Engelhardt (2) took the view that ATP splitting went on during relaxation and proposed in 1946 a tentative scheme of successive events: on stimulation ATP, which had been held in combination with other proteins, was transferred to myosin; molecular or micellar changes, responsible for contraction, followed in the myosin; ATP was dephosphorylated and the ADP (having a lower affinity for myosin) dissociated from the myosin which returned to its original form. Engelhardt (1) was influenced in taking this position by the observations described earlier[3] in which he and his collaborators had found increase in extensibility on application of ATP to acto-

---

[1] This very early decrease in extensibility had been recorded by Gasser & Hill (1) in 1924.
[2] For later interpretation see ch. 13. [3] See p. 146.

myosin threads, this change apparently depending on ATPase activity, in that both effects were abolished by treatment with 10^{-5}M Ag^+.

These experiments also affected the viewpoint of other investigators, e.g. Shen Shih-Chang (1) who agreed with Engelhardt's interpretation, and Sandow (2) who introduced some new considerations. Sandow was particularly interested in the precontractile latency relaxation. He assembled evidence from effects of pH and temperature upon the magnitude of the latency relaxation on the one hand and the ATPase activity of myosin on the other; from these observations he was led to suggest that Engelhardt's extensibility effect was an expression of latency relaxation,[1] that combination of myosin and ATP was necessary for the latency relaxation, that ATP breakdown began at once at this stage and that the energy so released began to activate the myosin directly for contraction. This scheme is thus, as he said, a form of contraction–energisation.

In 1946 Varga (1), working in Szent-Györgyi's laboratory at Szeged, put forward a theory of contraction based on the idea that each actomyosin particle could exist either in the relaxed form or in the completely contracted form; that the degree of contraction of the whole muscle depended on the relative number of contracted particles; and that the change between the two forms could be treated as a reversible chemical reaction, with equilibrium point dependent on temperature, heat being absorbed on shortening of the particles.

Relaxed actomyosin $\rightleftharpoons$ contracted actomyosin.

The equilibrium constant K of this reaction was found by measuring the percentage contraction (in the presence of ATP using in the earlier experiments psoas strips after freezing and thawing,[2] in later work glycerinated fibre bundles) at different temperatures. At $16°$ there was maximum contraction, taken as meaning the presence of 100% of particles in the contracted form; at $0°$ there was no contraction, indicating 100% of relaxed particles. Other values of K were found for intermediate temperatures, and when $\ln K$ was plotted against $1/T$ a straight line was obtained – as was to be expected if K represented an equilibrium constant.[3]

These experiments and their interpretation were extended by Szent-Györgyi (3, 6, 7) and his collaborators.[4] The total maximal work which a muscle could perform at different temperatures was measured; this work

---

[1] But see ch. 13, p. 336

[2] In the freezing-thawing process semi-permeable membranes are destroyed and the muscle's own ATP becomes freely available to cause contraction; in some of these experiments the thawed muscles were washed and ATP was added to give the contraction.

[3] This relationship would, of course, also be found if K represented a rate-constant.

[4] Hajdu & O'Sullivan (1); Hajdu (1, 2); Gergely (3); Gergely & Laki (1); Varga (2).

was taken as equal to ΔF, the free energy change of the reaction, and after various assumptions was equated with ΔF^0, the standard free energy change, the mole of myosin reacting being taken (for various reasons listed by Szent-Györgyi)[1] as 70000. ΔF^0 obtained in this way was found to agree approximately with ΔF^0 calculated from $\Delta F^0 = RT \ln K$ (based on the shortening experiments). The experiments of Varga (2)[2] were taken to demonstrate the reversibility of the reaction. A glycerinated psoas fibre, connected to a lever, was immersed in 3×10^{-3}M ATP in Ringer solution at $0°$; replacement by similar fluid at $15°$ led to immediate increase in tension, which fell again at once when the muscle was returned to the low temperature. Szent-Györgyi regarded contraction as dependent on depolarisation of charges in the actomyosin complex in presence of ATP, accompanied by decrease in hydration; this new complex was unstable and went over into an energy-poorer shortened form.

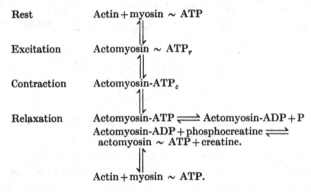

However, cogent arguments against the involvement of an equilibrium arose out of the work of Weber and his colleagues. These we shall discuss a little later.

We have already mentioned[3] the theory of Meyer (1) in which the changes in length of protein chains depended on changes in pH causing changes in charge on the molecules. In 1948 Riseman & Kirkwood (1) suggested an alternative charging mechanism – the phosphorylation by ATP of OH groups in hydroxyamino acids in the protein chains. At pH 7.0 a unit negative charge would thus be imparted to each of these sites, and electrostatic repulsion would keep the myosin chains stretched. Thus as a result of degradation of the high-energy phosphate bond in ATP, free energy would be stored in the myosin molecules in the resting state in the form of negative configurational energy of extension. Dephosphorylation of the myosin molecules would remove the negative charge and release free energy as

[1] A. Szent-Györgyi (11) p. 65.   [2] Also Varga (3).
[3] Ch. 7, p. 144.

mechanical work on contraction of the structure.[1] They calculated that phosphorylation sites at 100 Å would be necessary, and that the myosin content of serine and threonine was adequate for this. This suggestion eliminated the need for an environmental pH change, and linked up with the new ideas of energy provision.

It is interesting that at the end of his life K. H. Meyer (3, 4) again took up the question of contractility in animal tissues. He called attention to the tension–stretch curves recorded by Hill (23) for resting and tetanised muscle, the former resembling the curve for rubbery substances, the latter showing very marked differences. He drew the conclusion that tension production by stretching a muscle at rest included a component resembling that in rubber in which the retracting force is due to thermal action;[2] but in the active muscle electrostatic forces (operating on the filament structure as indicated by results with X-ray diffraction and double refraction) played the decisive part. Meyer pictured the protein threads as incompletely oriented and held stretched in the resting muscle by repulsive charges.

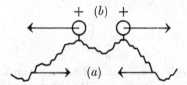

The attracting force a (involving randomisation of particles) was, as in rubbery solids, increased by temperature rise; the force b (involving orderly folding) being practically independent of the temperature. This would explain the observation made by Wöhlisch (2) in 1941 that resting fibres shortened when the temperature was raised; but in active muscle the force b (now attracting owing to change in sign) became far more important than force a. Meyer was willing to consider phosphorylation, according to Riseman & Kirkwood, rather than pH change as responsible for the change of charge on the filaments. He emphasised (4) that, since muscle in tetanic

---

[1] A similar theory (but without specification of the site of phosphorylation) was sketched by Needham *et al.* (1) and by Kalckar (2) in 1941. D. M. Needham in 1950, continuing her discussion (see p. 176 above) of the possible interaction in shortening groups along the protein chains, and of repeated interaction at slower rates of shortening said 'When speculations are made as to the timing of ATP breakdown, it is usually supposed that this is confined either to the contraction phase or to the relaxation phase (in the latter case its energy being used to restore energy-rich protein linkages). If we suppose that, when work is done, before a pair of groups can react together a second time, they must have been put back into their original state by means of free energy provided by reaction with ATP, we see that ATP breakdown could begin within the contraction phase, even though it were associated with restoration of the chains.'

[2] P. 184 below.

contraction was itself continuously producing heat, the temperature coefficient of the force generated in the stimulated muscle could not be a valid basis for determining the origin of this force (whether due to changes in entropy or in interatomic attraction).

Pryor (1) brought criticisms against Meyer's original theory, as also against that of Riseman & Kirkwood; in particular there seemed no reason why the electrostatic forces invoked should not cause an equal expansion (or contraction) in all directions, since neighbouring chains would be expected to influence one another. Bailey (3) later found some difficulty in contemplating such theories as that of Meyer, since any modification of electrostatic forces must work in relation to the very great total charge, and to the H-bonding of the Pauling–Corey α-model. (See p. 196.)

THEORIES INVOLVING CHANGES IN INTERNAL ENERGY OF AN UNSPECIFIED NATURE. Astbury (3), writing in 1945, concentrated attention entirely on the myosin partner, explicitly suggesting 'that it is not really critical at this stage whether what has hitherto been called myosin is a genuine individual or not.' He was very sceptical about the use of actomyosin threads, with their synaeresis on ATP addition, as a model for muscle and its contraction, and other workers expressed the same doubts. As we see it now, the loss of water in such threads is simply a squeezing out on contraction, but Szent-Györgyi regarded the loss of hydrate water (though not of intermicellar water) as an essential part of the mechanism.

In studying the effects of contraction upon the X-ray diffraction pattern of frog sartorius Astbury was obliged to make use of iodoacetate contracture to obtain the contracted state, since with the apparatus available at that time, it was not possible to maintain the fresh unpoisoned muscle in contraction long enough for X-ray photography. No deep-seated changes were observed in the diffraction pattern of the contracted muscle. In maintaining his thesis that in myosin, as in other members of the keratin–myosin–fibrinogen group of fibrous proteins, contraction depended on folding of protein chains, he relied on two arguments. First, that the change observed in angular dispersion was so small that changes in length of the muscle could not be put down to simple disorientation of a micellar or fibrillar structure. Secondly, that the regular X-ray diffraction pattern given by resting muscle could arise from a relatively small proportion of better-ordered crystalline aggregates, in series with less well-ordered components, and that it was just the latter which would respond more readily to give a change in molecular form. We have seen that Astbury & Dickinson (3) in 1940 put forward the hypothesis that normal contraction was due to the same change in folding of protein chains as took place in supercontraction of keratin. In his Croonian Lecture (3) of 1945 he suggested that in contraction the square folds of the α-form were prolonged sideways to give a β-pattern at right angles to the long axis. A frog sartorius contracted in water at 60° did give this cross β-pattern, which was also described by Rudall (1) in X-ray diffraction studies of epidermis, fibrinogen and myosin films contracted by heat. Much later, in 1958, Pautard (1) reported that the synaeresis of actomyosin (in the form of thin transparent films) upon addition of ATP was accompanied by indications of a transformation of part of the protein into the cross β-configuration. By 1950 Astbury (5) was ready to recognise the importance of actin. He pictured that, in the plastic resting muscle, the myosin chains and actin corpuscles were relatively free. On stimulation side-to-side union took place between myosin chains in the supercontracted form and linear arrays of the actin corpuscles (i.e. F-actin). This combination brought about the change ' to a state almost like that of a crystalline solid '.

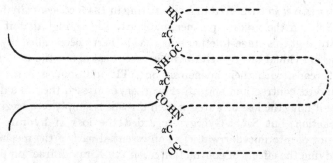

Astbury was firmly of the opinion that the driving force for the folding of the chains in muscle contraction would turn out to be due to changes in the internal energy and not to changes in entropy. This conviction was based on the highly specific configuration of the proteins, and especially on the high proportion of polar groups in myosin which made it very unlikely that the chains would fall into random configurations as they do in rubber. Edsall (2) was of this opinion and it received support from experiments by H. J. Woods (1) on the thermodynamic behaviour of myosin strips, oriented by stretching and dried at the stretched length. According to the Wiegand–Snyder equations (originally worked out for rubber)

$$K = \left(\frac{dU}{dL}\right)_T - T\left(\frac{dS}{dL}\right)_T = K_u + K_s \tag{1}$$

$$\left(\frac{dS}{dL}\right)_T = \left(-\frac{dK}{dT}\right)_L \tag{2}$$

where K = the restoring force, tensile strength or load acting on a fibre extended to length L; T = temperature; U = internal potential energy; S = entropy of the fibre. By equation (1) the restoring force is analysed into two components, K_u depending on the interatomic inner energy, K_s depending on entropy. Equation (2) permits the measurement of the entropy contribution by measuring the temperature coefficient of the restoring force provided (as noted by Meyer) that the fibre material is *not* itself generating heat. With the oriented myosin strips Woods found the entropy contribution was very small.

THEORIES DEPENDING ON ENTROPY INCREASE FOR THE ENERGY OF CONTRACTION. In an interesting short paper in 1950 Weber & Weber (2) took the view that the glycerol-extracted muscle fibre contracting in ATP solution could act as an entropy machine. Setting out from the great temperature-dependence of the tension-production they made use of the Wiegand–Snyder equation to calculate that the thermokinetic component of the tension might be 10–50 times as great as the total tension; thus the thermokinetic tension was used for the greater part, they suggested, in overcoming affinity forces which kept the fibre in the stretched form. Since the simplest way in which the kinetic energy of molecular movement could be converted into mechanical work would be by a curling up of the molecular chains, they suggested that the main if not the only role of ATP (known to plasticise and to dissociate actomyosin) was reversibly to decrease the cohesion forces in the stretched structure. They also envisaged (as alternative explanation of the temperature effects) the possibility of an endothermic reaction between protein and ATP, leading to a well-ordered contraction – perhaps on the lines of Astbury's folding theory. In their paper of 1951, however, as we shall see,[1] they put forward a quite different point of view, one for which evidence was built up in the ensuing years, largely in Weber's own laboratory.

An unusual theory about the same time was that of Pryor (1, 2) based on an energy cycle in a rubber-like substance, with ATP as plasticiser altering the elastic properties in the same way as heat alters those of rubber. Tendons immersed in mercury–potassium iodide solution contract, and if they are held at constant length develop tension. It was suggested that the solution acted as a plasticiser, weakening attractive forces and allowing protein molecules to curl up under the influence of thermal agitation. A model was built in the form of a wheel with 24 spokes made of tendon; the wheel was suspended in a narrow trough containing 2M mercury-potassium iodide solution to the level of the axle, above that, very dilute solution. The lower spokes, immersed in the strong solution, contracted and stretched the upper spokes bringing the wheel of out of balance and causing it to turn. This brought the contracted spokes into the dilute layer in which they gave up reagent and were extended in their turn by the contraction of

[1] P. 185 below.

the lower spokes. The energy would be derived from the increase in entropy of the solution on dilution, which went on with large absorption of heat. In muscle, actomyosin would be the working substance and ATP the plasticiser would be removed not by dilution but by chemical decomposition. The change in concentration of ATP would be very small, compared with the changes in the model, but the effect might be enhanced by repeating cycles, successive segments of the same protein molecule extending one another, as one spoke in the model extended another.

This ingenious idea was refuted by A. V. Hill (24) in 1953, when he showed that, while resting muscle had analogies with rubber (over an important range of length and tension) giving out heat when the tension was raised and absorbing heat when the tension was lowered, active muscle was different. When the tension of a muscle in isometric contraction was rapidly lowered, there was an immediate and proportional rise in temperature. 'Active muscle', A. V. Hill wrote, 'behaves instantaneously as though it possesses only very short-range elasticity and a high elastic modulus. The process of physical shortening is in no way related to the elasticity of the materials of which it is constructed any more than the properties of a motor are related to the elasticity of the materials of which it is constructed. In muscle the nature of the motor still remains unexplained.'

In the conception of Morales & Botts (1, 2), first appearing in 1952, the length of the myosin filaments in the resting state was again taken as a compromise between the contracting effect of intramolecular Brownian movement and the extending tendency of a net charge repulsion. They considered (1) that the Riseman & Kirkwood theory would not account for the marked and immediate effect of ATP addition in causing shortening of actomyosin threads *in vitro*, and new points of detail were now suggested. The net charge on the protein was taken as positive in the resting muscle, owing to the well-known capacity of myosin to adsorb certain divalent cations;[1] when upon stimulation the anion of ATP was adsorbed on the myosin, discharging it, the filaments could contract gaining configurational entropy; dephosphorylation of the ATP would give ADP, not so strongly adsorbed, the negative charge would be removed and extension would become possible again. Morales, Botts, Blum & Hill (1) suggested that the free energy release associated with ATP adsorption meant that a considerable fraction of the free energy loss of the system took place before the hydrolysis step. In a series of papers Morales and his collaborators subjected this model to different types of experimental test. Thus Botts & Morales (1) using extruded threads of Szent-Györgyi's myosin B obtained length–tension and temperature–tension curves which convinced them that positive tension was entirely due to the entropy contribution. On the other

[1] P. 195.

hand, with threads which had been dried and then rewetted, they obtained results in agreement with those of Woods with actomyosin film. This they put down to 'vulcanisation' of the actomyosin by hydrogen bonding during drying. A study of the light-scattering of myosin B solutions in 0.6M KCl, 0.001M $CaCl_2$, pH 7.0 (conditions in which the myosin was negatively charged) led to the conclusion (as we have seen)[1] that the effect of added ATP on the particles was to cause not a dissociation, with change in molecular weight, but an extension. This was interpreted as due to the increased negative charge. Like Astbury, Morales[2] treated myosin B as an individual protein, and he was influenced in this by Laki's views (3, 4) that myosin itself was built up of sub-units comprising actin, tropomyosin and an unidentified protein. The support adduced from a study of myosin-catalysed phosphorolysis of homologues of ATP will be discussed below.[3]

The demonstration by the Morales group (after Gergely's criticisms of their results) that in the myosin B solution which they used only a part of the protein (less than 25%, made up of the heaviest particles) showed extension, left Morales (1) still emphasising the importance of the ion–polyelectrolyte effect. In this review in 1959 he agreed that it was not permissible to put the onus of the ion–polyelectrolyte effects he needed for contraction upon the adsorption of ATP to the myosin, the myosin not reacting with a large enough amount of ATP to cause the diffuse negative charge invoked. But he said: 'Nothing...has weakened my intuitive conviction that the tension-generating device in excited muscle will prove to be a mechanically continuous structure...nor am I ready to relinquish my faith in Coulombic interactions as the most rapidly generated and the most long-ranged of the forces that the transducer could employ.'

THEORIES DEPENDING ON CONTRACTION–ENERGISATION VIA MYOSIN-ATP COMBINATION OR ATP BREAKDOWN. In their paper of 1951 (1) Weber & Weber recorded their conviction that a number of experimental results obtained did not fit with the conception of the temperature dependence of the tension as a thermokinetic equilibrium. For example, according to the Wiegand–Snyder equation applied to changes in tension in a rubber-like body, the increase in restoring force with temperature, $1/K \cdot \Delta K/\Delta T$, should be 1/273 for 1° rise in temperature. But the increase found for stimulated muscle was very much greater. This fact might have been explained by the plasticising action of ATP effectually providing an increased number of molecules, but the evidence was against this; for example it would then be expected that with decreasing ATP con-

---

[1] P. 156.
[2] Morales, Botts, Blum & Hill (1).
[3] P. 186.

centration the region of most marked tension increase with temperature would be shifted to higher temperatures. This was not the case.

Weber & Weber then turned to examine the possibility that free energy from hydrolysis of ATP was the cause of the contraction, and that the temperature dependence of the tension production was due to the temperature effect on the velocity of the chemical reaction. No thermodynamic equilibrium would be involved, but a steady state with a level of activity depending on temperature. This point of view fitted much better with the facts. We have already discussed the results obtained in Weber's laboratory at that time, showing the similar optimal ATP concentration for ATPase activity and for tension production at various temperatures and the effect of Salyrgan in preventing both ATPase activity and contraction. Several other cases have also been cited which showed the dependence of tension on the degree of activity of Mg-activated ATPase.[1] Weber & Portzehl (1) have clearly set out this evidence in a review of 1954.

In a review (finished in 1953 but not published till 1956) Weber (6) referred to the work of Ströbel (1) who had shown that when the total double refraction fell during contraction, the decrease concerned the intrinsic double refraction, the rod double refraction remaining unchanged. Taking this into consideration with the other details of fine structure discussed earlier, he was of opinion that contraction depended on re-arrangement of smaller particles, not recognisable in the electron microscope, within the filaments.

Support for the theory of Morales & Botts[2] had been found in the work of Blum (2) who compared the behaviour of a number of nucleoside triphosphates with myosin B in 0.6 M KCl. He measured the rate of dephosphorylation and the degree of binding of substrate to enzyme – the latter both by phosphate analysis and by the light-scattering method. He concluded that, in the presence of either Mg or Ca, the constant K (the equilibrium constant of the reaction: substrate plus enzyme $\rightleftharpoons$ enzyme–substrate complex) decreased for the nucleoside triphosphates in the order

$$ATP > UTP > ITP > TPP,$$

while the maximum velocity of the ATPase, in present of Ca or Mg gave the order ITP > UTP > ATP > TPP. Friess & Morales (1) also studied TPP under similar conditions; they found that, though TPP is a quite ineffective contracting agent, its rate of dephosphorylation was as much as 20 % of that of ATP; but the extent of binding of TPP was estimated as only about 7 % of that of ATP. From these results it might be assumed that the outstanding position of ATP

---

[1] Pp. 164 ff.

[2] For description of the kinetic analysis used see Ouellet, Laidler & Morales (1).

in connection with contraction was concerned rather with its binding reaction than with its ATPase activity. But it is clear from figures given earlier[1] that the nucleoside triphosphate order in Blum's ATPase experiments agrees with that found in Hasselbach's (4) on myosin sol and on actomyosin at high ionic strength, and is quite different from the order found with actomyosin gel at low ionic strength. Only the latter however can be regarded as relevant to contraction. Weber & Portzehl (1) and Weber (6) moreover contested this idea of the dependence of contraction on ATP binding, on the grounds that it could give no explanation of the inhibitory effects of low temperature and of over-optimal ATP concentration.

An interesting table prepared by Tonomura, Kubo & Imamura (1) in 1965 lists some 25 analogues of ATP, with (1) their effect in causing contraction of isolated myofibrils, in 0.05 M KCl, 3 mM $MgCl_2$; (2) their rate of hydrolysis in presence of myosin B at low ionic strength, Mg-activated.[2] The correlation was on the whole good, though one instance is given (that of dimethyl ATP) when contraction was caused, but little dephosphorylation was observed.

THEORIES INVOLVING CHANGES IN ACTIN. It is interesting to notice in connection with the tendency to neglect consideration of actin in contraction theories that a number of claims have been made (for example that of Kafiani & Engelhardt (1)) that myosin threads made from pure myosin films could contract in the absence of actin. But in 1957 this claim was withdrawn,[3] the contractile effect having probably depended on pH changes. In 1959 Zaalishvili & Mikadze (1) also found that while threads made from compressed actomyosin film would contract down to 50 % of their length on addition of ATP, there was no shortening effect with myosin threads similarly made.

We come now, however, to a theory in which actin was given an important role. Mommaerts & Seraydarian (1) in 1947 calculated that myosin ATPase could not be considered to act rapidly enough to supply the energy needed by the muscle if this energy was all channelled through the ATPase. They focused attention on myosin ATPase, which they found of course to be activated by Ca but inhibited by Mg, to the extent of 90 % even with a Mg/Ca ratio as low as 0.2. They considered a ratio of about 4 to be likely in the muscle. The required activity they calculated to be some 70 times as high as that found *in vitro*. When they turned to myosin B they found

---

[1] Ch. 8.
[2] Also Ikehara, Ohtsuka, Kitagawa & Tonomura (1).
[3] Engelhardt (3); Kafiani & Poglazov (1).

activation by Mg, but only at ionic strength of about 0.01 – much lower than that of muscle. Mommaerts (9, 14) calculated further that in 1 g of muscle, containing 3 % of actin, 5×10^{-7} moles of ATP would be dephosphorylated on polymerisation of the actin,[1] and this was just about the amount needed to provide energy for a single twitch. He therefore proposed that resting muscle contained myosin and G-actin in combination; in the primary event of contraction the actin reacted with a stoichiometric amount of ATP, dephosphorylated this during polymerisation and formed F-actomyosin, which contracted. Relaxation then involved rephosphorylation of ADP by a phosphate donor. Although polymerisation of actin and its reversal are slow processes *in vitro*, Parrish & Mommaerts (2) found certain conditions (for example after freeze-drying of the actin solutions) in which both processes were extremely rapid, and this encouraged them to postulate conditions *in vivo* in which the cycle could be rapid enough to provide energy for a twitch, or a series of partial cycles to provide energy for a tetanus.

In 1960 Martonosi, Gouvea & Gergely (2) sought for evidence for such a process but without success. They injected ^{32}P-labelled inorganic phosphate *in vivo* into rabibts, rats and pigeons, and at intervals after the injection studied the specific activity of the ADP bound to actomyosin, presumably to the actin. It may be expected that the γ- and β-phosphate groups of the intracellular ATP would become labelled, but even after several hours the specific activity of this bound ADP was only a fraction of that in the ADP in a trichloracetic extract of the whole muscle. Moreover, exercise or electrical stimulation of the muscles did not affect the time taken for incorporation. Such negative evidence is not conclusive since it is possible, for instance, that the ADP of the F-actin might be rephosphorylated by or exchanged with the ATP or ADP of some localised pool. We shall find again, in the consideration of the sliding-filament model of contraction, interesting postulations of the $F \rightleftharpoons G$ transformation or of cyclical interruption and reformation of the actin filaments.[2] However the evidence of other workers for the ATPase activity of actomyosin, under conditions resembling those within the cell, does not agree with Mommaerts & Seraydarian's estimate. Perry (3) in 1951 questioned the grounds of their arguments, since he found that isolated myofibrils had Mg-ATPase activity in 0.1 M KCl solution, while after solution of the fibrils in M KCl the Mg^{2+} effect of activation in 0.1 M KCl was changed to one of inhibition. According to Hasselbach (5) the Mg^{2+} concentration in the muscle fluid is not greater than 8×10^{-3} while the Ca^{2+} concentration is extremely low; Bendall (2) has given $I = 0.154$ for mammalian muscle, based on much earlier work of Dubuisson (12). As we have seen in the work of Hasselbach cited in

[1] See p. 213.   [2] Chs. 12 and 14.

chapter 8, good actomyosin ATPase activity was found at $I =$ about 0.1 and 10^{-2} or 10^{-3}M Mg^{2+}.[1] Bendall (1) using suspensions of glycerinated rabbit myofibrils found that at 37° in the presence of 4×10^{-3}M Mg^{2+}, raising the ionic strength from 0.1 to 0.15 had no inhibitory effect on the ATPase. Mommaerts (10) himself has recently remarked that later findings on polymerisation of G-actin-ADP under certain conditions, or even of nucleotide-free G-actin,[2] show that the stoichiometric relations on which his early theory was based only represent a limiting case.

The idea of a sliding-filament mechanism was first suggested by H. E. Huxley (4) in 1953; we shall see that during the following years of this decade, when evidence for the sliding filaments was massively accumulating, the theories based on molecular folding in the direction of the long axis (often associated with the necessity for energy provision in the relaxation phase) died very hard. A number of the ideas used in this folding mechanism will be met again in a different context when the actual details of the sliding filament process are considered. But we first must consider the remarkable properties of three proteins concerned in the contraction–relaxation process – actin, myosin, and tropomyosin.

[1] Also Hasselbach (4).   [2] Ch. 10.

10

ON MYOSIN, ACTIN AND TROPOMYOSIN

SOLUBILITY AND EXTRACTABILITY OF THE STRUCTURAL PROTEINS

In chapter 8 we were concerned with the growth of knowledge of actomyosin–ATP interactions from 1939 to about 1953. In chapter 9 we considered the theories of contraction (with their interesting variations) which resulted from the realisation of the importance of the actomyosin–ATP relationship. Here again the period chosen terminated about 1953, because at this time the idea of the sliding-filament mechanism began to emerge; as evidence has accumulated, this has gradually replaced almost all other postulated mechanisms. The story of this will occupy chapter 11. In the present chapter I want to discuss the properties of the individual structural proteins, to which another was added in 1946 by the discovery of tropomyosin by Bailey (2).

We have already discussed the results of earlier workers[1] who estimated 'myosin', myogen, globulin X[2] and stroma protein in muscle. After the discovery of actin and actomyosin, Balenović & Straub (1) were the first to try to estimate, albeit in an indirect way, the amount of actin present.

As the basis of this method they used the formation of actomyosin when actin was added to excess of myosin, the actomyosin being assessed by the fall in viscosity on addition of ATP. This decrease in specific viscosity as a function of the specific viscosity in presence of ATP, Straub (1) termed the 'activity' of unknown 'myosin' solutions, and he took the activity of myosin B solutions prepared from muscle in a standard way as 100%. To estimate the actin content of muscle, the residue after removal of myosin A was extracted for 24 h with excess of myosin A solution and the activity of this extract was then measured. Balenović & Straub were struck by the observation that such extracts never had activity greater than about 100% (like that of myosin B), but by adding actin preparations to myosin A, activities up to 170% were obtained. They concluded that this anomaly depended on a competition for bond formation with the actin, between the extracting myosin on the one hand and the proteins of the muscle residue

---

[1] P. 137.
[2] Globulin X appears to consist of material from the sarcoplasmic reticulum (personal communication from Dr J. R. Bendall).

on the other. They estimated thus that the muscle contained 25–30 mg of actin/g wet wt or 12–15 % of the muscle protein; they considered that there was enough actin fully to activate, under conditions *in vivo*, the myosin present in the same amount of muscle.

Some years later, in 1951, Hasselbach & Schneider (1) undertook direct estimation of free myosin and free actin, based on quantitative preparation. To allow complete dissociation of the two proteins from one another, the extracting fluid ($\mu = 0.6$, pH 6.3) contained 0.9×10^{-2}M potassium pyrophosphate. With this solution, three to four extractions of muscle (passed through an ordinary mincer, then a Latapie mincer) led to complete removal of the L-myosin and the globular proteins. This extract (after removal of the pyrophosphate by dialysis) showed no fall in viscosity with ATP and was thus actin-free. If now the muscle *brei* was further disintegrated by 2–4 min in the Waring blender, a further 13 % of the total protein could be extracted at any ionic strength between 0 and 0.6, at pH greater than 6.0. This protein had no ATP sensitivity, but addition of myosin led to development of high sensitivity. From this and other properties Hasselbach & Schneider concluded that the material was F-actin. They took the view that the difficulty in extracting actin was not one of solubility, but of the permeability of the fibre structure to the long actin filaments. Their values in percentage of the total protein were:

Globular protein	L-myosin	Actin	Stroma protein
28	38	13–15	19–21

51–53

They suggested that the lower value for 'myosin' found by Meyer & Weber (1)[1] was due to breaking up of actin by the highly alkaline reaction they used, and its passage into the supernatant, to be estimated with the globular proteins. Jakus & Hall (1) had indeed found in 1947 with the electron microscope that actin at pH about 7.0 showed numerous filaments, but even at pH 7.5 these had mainly disappeared and given place to a relatively homogeneous granular background. Tsao[2] later found that the whole of the tropomyosin[3] was included in the actin fraction which should therefore be reduced to about 10 %.

The laboratory of Dubuisson in Liège about 1945 took up the study of the muscle proteins by means of the Tiselius–Longsworth electrophoretic method. Experiments were first carried out on extracts made at low ionic strength ($I = 0.15$, pH 6.5) which would contain myogen; some dozen constituents (or groups of constituents) could be distinguished.[4] Identification of only one of these was attempted at that time[5] – that of band h

---

[1] P. 137. [2] Quoted in Bailey (3). [3] P. 227 below.
[4] Dubuisson & Jacob (1); Jacob (1, 2, 3); also Bosch (1). [5] Jacob (3).

with the myoalbumin of Bate Smith (3); later work on various muscle enzymes has permitted their allocation to the different parts of the electrophoretic pattern.

Bailey (1) in 1942 had found the electrophoretic diagram of 'myosin' to contain only one peak; Ziff & Moore (1) a little later, as we have seen, using Greenstein–Edsall myosin found it to give electrophoretically one main component which contained more than 90% of the ATPase activity. In 1946 Dubuisson began his investigation of the proteins soluble only at higher ionic strength. Diagrams of these did indeed present a great contrast to those of the more soluble proteins. Whereas the latter showed more and more components as the conditions of pH or ionic strength below 0.15 were varied, Dubuisson (1) constantly found only two or three components in myosin prepared according to Greenstein & Edsall (1), with mobilities of 2.7, 2.5 and 2.1×10^{-5} cm/V sec. These he called respectively α-, β-, and γ-myosins. It was noticeable that the turbidity of the solutions followed component α. He then found (2) that this component in 0.25 M KCl could be precipitated at 23–27% saturation with ammonium sulphate, while component β was precipitated at 35–39% saturation. By repeated precipitation each was obtained homogeneous. By comparison of the properties of α- and β-myosins with those of Szent-Györgyi's actomyosin and myosin, it became clear that these proteins corresponded.[1]

Dubuisson also considered the effect of fatigue on the extractability of his myosins. Many years before, about 1930, the solubility of the muscle proteins in a variety of extractants had been studied in Embden's laboratory.[2] Deuticke (2) found, on very severe stimulation, decrease of as much as 30% in the amount of protein extracted with 0.09 M potassium phosphate, pH 7.2, containing 0.03 M KI. Upon recovery in oxygen, normal solubility was regained. The correlation between degree of change in solubility and amount of tension production led him to suggest that the cause of the former was a change, at the moment of contraction, in the colloidal state of protein (perhaps only one particular protein) specially concerned in contraction. Kamp (1) in H. H. Weber's laboratory, using buffered KCl solution as extractant, showed in 1940 that the Deuticke effect did indeed concern only 'myosin'. He found that the solubility decrease accompanied iodoacetate contraction, and concluded that the colloidal change in the myosin in normal contracting muscle preceded the glycolytic process, which indeed he suggested might up to a point bring about its reversal. When now Dubuisson tested the effect of fatiguing the muscle to exhaustion upon whole muscle extracts (made with KCl solutions at $I = 0.35$), the electrophoretic pattern showed the water-soluble proteins to be unaffected, but much less α-myosin

---

[1] Dubuisson (3).
[2] Deuticke (1, 2); Hensay (1) cf. p. 76.

and β-myosin were extracted.[1] Dubuisson (6) also studied the behaviour of β-myosin on short extraction of resting muscle at 0.35μ, and found it more readily extracted the more finely the muscle was divided. However, no matter how finely divided the fatigued muscle, the extraction of α-myosin was somewhat reduced and that of β-myosin greatly reduced, compared with the resting muscle. Dubuisson suggested that these findings implied for the β-myosin 'its localisation in structures difficult of access', where it existed in the form of a complex, the dissociation of which could be brought about by salt solutions according to their ionic strength. Fatigue affected this dissociability, but not the structural accessibility.

In these extracts of muscle during contraction or fatigue a new protein was found,[2] which Dubuisson called contractin, and which later turned out to be identical with γ-myosin,[3] already observed in much smaller amount in Greenstein–Edsall myosin from normal muscle. There was some evidence[4] that the decreased extractability of α and β, and the marked appearance of γ-myosin depended to some extent on the degree of shortening of the muscle at the time of extraction, whether the shortening was caused by electrical stimulation, or e.g. by IAA contracture. That β-myosin had not disappeared from the contracted muscle was shown by the fact that extraction with certain salt solutions (pyrophosphate, potassium iodide) at ionic strength equivalent to the KCl solution ordinarily used, brought out the normal amount.[5]

It thus became clear that several factors were involved in the quantitative extraction of actin and myosin from muscle, whether as the individual proteins or as the actomyosin complex. 1. Comminution is important in breaking up the long fibres and the structures surrounding them – supporting connective tissue and sarcoplasmic reticulum;[6] probably also in breaking up the myosin and actin filaments themselves, particularly the latter. Bailey (3)[7] has drawn a parallel between fibrinogen and F-actin, both soluble at low ionic strength, but often difficult to get into solution without, for example, blender treatment. 2. It is not surprising that, to be effective, extracting solutions must be used of higher ionic strength than those needed to dissolve the isolated proteins *in vitro*. A steep concentration gradient will thus be created into the broken fibre particles, and the withdrawal of the actomyosin and myosin will be facilitated. 3. Agents which promote actomyosin dissociation and F-actin depolymerisation, such as pyrophosphate and potassium iodide, will assist in extraction, particularly in fatigued or rigor muscle, where the ATP concentration will be low.

[1] Dubuisson (4, 5); Crepax, Jacob & Seldeslachts (1).
[2] Dubuisson (4).   [3] Dubuisson (7).
[4] Dubuisson (3); Crepax, Jacob & Seldeslachts (1).
[5] Dubuisson (7).   [6] See ch. 13.
[7] See Bailey, p. 1010. Weber & Portzehl (2) also discuss the question of extractability.

Dubuisson (3) reported that salt solutions to which ATP had been added (in amount equal to that in resting muscle) showed no improved capacity to extract α- and β-myosins from a pulp made from stimulated muscle. He took this to mean that the binding forces holding the β-myosin in fatigued muscle must be qualitatively different from those acting in resting muscle. However it seems inevitable that the added ATP would be rapidly broken down by the very active ATPases of the muscle pulp, and so prevented from exerting any effect.

MYOSIN

A word may be said about purification. In 1949 Portzehl, Schramm & Weber (1) prepared L-myosin[1] homogeneous in the ultracentrifuge. Myosin A preparations were made, according to Szent-Györgyi, redissolved and reprecipitated at $\mu = 0.3$. By this means the actomyosin, which has a higher salting-in threshold than free myosin, was precipitated and could be removed. The L-myosin was precipitated at $\mu = 0.04$. The purified β-myosin of Dubuisson, equivalent to Weber's L-myosin, may, as we have seen, be prepared by ammonium sulphate precipitation; but for general purposes Weber's method is preferable. Tsao (1) found myosin purified by ammonium sulphate precipitation to have a Q_P value only about 25% of the initial value. This was probably due to the presence of traces of inhibitory heavy metals in the ammonium sulphate. The purification of myosin has been much improved in recent years by the use of ion exchange chromatography. For example Harris & Suelter (1) in 1967, using cellulose phosphate, were the first to separate the AMP deaminase activity of myosin from its other activities.

PROPERTIES OF MYOSIN

Amino acid composition. Amino acid analysis was carried out by Bailey (2) in 1948 and by Kominz, Hough, Symonds & Laki (1) using the Moore & Stein method in 1954. The myosin as prepared by Bailey contained very little actomyosin, and the latter protein was specially removed from the preparation of Kominz *et al.* by superprecipitation with ATP at $I = 0.12$. The results are assembled by Bailey (3) and serious discrepancies between the two sets are found for only two or three amino acids. Myosin is rich in charged groups – 34% of the total amino acid residues, to be compared with a median value of 25% for some 20 proteins.[2] From the amino acid analyses and the titration curve of Mihalyi (1) Bailey (3) has calculated that at physiological pH there will be 154 anionic groups and 134 kationic groups per 10^5 g of protein.

Bailey (5) went on to assay the terminal amino acids by means of the

[1] P. 149.   [2] Bailey (4).

fluorodinitrobenzene method of Sanger. Only traces were found and the conclusion was drawn that myosin consisted of cyclic polypeptide sub-units. Later, however, the finding by Locker (1) of C-terminal isoleucine disproved this hypothesis. In 1964 Offer (1)[1] was able to explain the absence of N-terminal groups by demonstrating the presence of terminal N-acetyl in a digest of myosin made with pronase (a proteolytic enzyme of low specificity). The yields of peptide showed that at least two N-terminal sequences N-acetyl-Ser-Ser-Asp-Ala-Asp must occur per myosin molecule. Kielley (1) in 1965 gave a preliminary report of the presence of one N-terminal histidine per 200000 g of myosin.[2] Gaetjens, Cheung and Bárány (1) however confirmed the negative result of Bailey.

Fifteen SH groups are contained in 200000 g of myosin. The importance of certain of these in interaction with actin and with ATP will be described in chapter 12.

Isoelectric point and ion-binding power. Szent-Györgyi[3] early stressed the marked ion-binding tendency of myosin, and much work has been done since on this question.

Erdös & Snellman (1) gave the electrophoretic mobilities for myosin A at different pH values and different KCl concentrations. No change in mobility was found on altering the latter, so that these ions were either not bound or bound in equal amounts. On the other hand, the charge on the protein was greatly changed in presence of Ca^{2+} or Mg^{2+}. This agrees with the results of Ghosh & Mihalyi (1) who investigated the effect of addition of various salts on the pH of myosin solutions. Little change was caused by KCl, but divalent metals and certain anions had marked effects. Mihalyi (1) also found little effect of KCl on the titration curve of myosin. Later Lewis & Saroff (1) measured directly, at 5° by means of permselective-membrane electrodes, the K^+ and Na^+ taken up by myosins A and B from solution. They found both ions were bound at 5°, the latter more strongly than the former; on warming to room temperature the binding properties of the protein were lost. Nanninga (1) measured Mg^{2+} and Ca^{2+} uptake by the meromyosins (products of controlled trypsin treatment of myosin)[4] by the dialysis equilibrium method. Both ions were strongly bound, as had been predicted by Ghosh & Mihalyi for myosin; Nanninga considered that his results on the meromyosins argued similar behaviour for myosin, and described these proteins as amongst those with strongest affinity for these two ions.

Most workers[5] agree that the isoelectric point of myosin is at pH 5.4–5.7.

---

[1] Also Offer (2).   [2] Also Kielley, Kimura & Cooke (1).
[3] E.g. A. Szent-Györgyi (10).   [4] Pp. 199 ff.
[5] Hollwede & Weber (1); Erdös & Snellman (1); Mihalyi (1).

The claim of Sarkar (1) to have shown very great changes in isoelectric point on KCl addition has been criticised by Hamoir (1). Marked effects are found in presence of Mg^{2+} or Ca^{2+}: Nanninga gives a value of 8.7 for myosin in 10^{-2}M $MgCl_2$, and Erdös & Snellman suggest from their observations a value above 9 in 0.1 M $CaCl_2$.[1]

Configuration of the peptide chain. In 1950 Bragg, Kendrew & Perutz (1) published the results of a survey of a large number of possible models for α-keratin, crystalline haemoglobin and myoglobin; these models conformed to accepted bond lengths and angles and had the folds held by hydrogen bonds. When comparison was made of the vector of the models with two-dimensional Patterson projections based on X-ray data, configurations of the type suggested by Astbury & Bell (1)[2] for α-keratin were in closest agreement. However, the three-dimensional projection of haemoglobin gave anomalous results and the authors stressed that the conclusions were tentative. About the same time, Pauling & Corey,[3] who had made very precise studies[4] of interatomic distances and bond angles found in amino acids and peptides, turned their attention to the protein chain, and made the new suggestion of a hydrogen-bonded spiral configuration. They based this on (*a*) their use of dimensions more accurate than those employed by any previous workers; (*b*) the planar nature of each residue because of the resonance of the double bond between the C=O and the C=N positions.

$$\begin{array}{c} H \searrow \\ C \nearrow \end{array} N \text{---} C \begin{array}{c} \nearrow C \\ \searrow O \end{array}$$

Abandoning the assumption (made by Bragg, Kendrew & Perutz) that each turn of the spiral must contain an integral number of residues, they found that a helix with 3.7 residues per turn, each hydrogen-bonded to the third residue from it in each direction in the chain, best represented the structure of α-myosin and similar fibrous proteins.

In 1951 Perutz (1) drew attention to a hitherto unobserved reflection in proteins of the keratin–myosin–fibrinogen group; this reflection was from planes perpendicular to the fibre axis at a spacing of 1.5 Å, corresponding to the repeat of the amino acid residues along the chain. This spacing excluded all models except that of Pauling, Corey & Branson, but was in perfect accord with the 3.7 residue helix. The 1.5 Å reflection was also found by H. E. Huxley & Perutz (1) in photographs of dried sartorius muscle.

---

[1] For the importance of Mg and Ca ions in contraction see chs. 8 and 13.
[2] Ch. 7, p. 139.   [3] Pauling & Corey (1, 2); Pauling, Corey & Branson (1).
[4] For a review of this work see Corey & Pauling (1).

The difficulty remained that Pauling & Corey's α-helix failed to explain the 5.15 Å arc on the X-ray photographs of α-keratin as a true meridional reflection. The α-helix of synthetic polypeptides gave strong reflections on the 5.4 layer line, but displaced from the meridian. Crick (1) then pointed out that by tilting the helix this difficulty could be removed; thus the α-helix twisted into a super helix or coiled coil with pitch of about 18° would give the observed meridional reflection at 5.1 Å. Crick further suggested that without this deformation of the α-helix, packing of the side-chains of two adjacent helices would be difficult. He pictured the side-chains of the helix as knobs on the surface of a cylinder alternating with holes, i.e. spaces into which the knobs of an adjacent helix could fit. With the close arrangement of polypeptide chains in proteins and the smaller less flexible side-chains (as compared to poly-γ-methyl-L-glutamate, for example) packing of straight helices would be difficult; but Crick showed that by this deformation into coiled coils the knobs could be made to interlock systematically.

Pauling & Corey (3) had independently decided that an α-helix for a polypeptide chain involving repeated sequences of different kinds of amino acid residues would not be expected to have a straight axis. A protein might consist of helices wound round each other to form three- or seven-stranded cables, and Pauling & Corey showed how this arrangement could answer for the 5.15 meridional arc. Other evidence also in favour of such structures was brought together, and as we shall see, the coiled coil structure for the myosin molecule came to be accepted. In 1961 Simmons and collaborators[1] (1) by measurements of anomalous rotatory dispersion (Cotton effect) found the myosin molecule to be 63 % helical. This value was also found by Mommaerts (15) using measurements of the ultra-violet circular dichroism.

The molecular weight. The first tentative determination of the molecular weight of 'myosin' (actin-contaminated) by H. H. Weber & Stöver (1) in 1933, using osmotic pressure measurements, gave very high and indefinite values – 0.6–1.2×10^6. In 1952 Weber & Portzehl (2) calculated a value of 850000 for L-myosin from the sedimentation coefficient $S_{20}^{\circ} = 7.1$ (measured at about 20°)[2] and the diffusion constant (measured at 0°)[3] and this was the best available for some years.

About 1953 the dependence of the apparent sedimentation coefficient of myosin on temperature in the ultracentrifuge was noticed;[4] Laki & Carroll (1) then found that myosin solutions left at room temperature for 2 h

---

[1] Simmons, Cohen, A. G. Szent-Györgyi, Wetlaufer & Blout (1).
[2] Portzehl, Schramm & Weber (1).   [3] Portzehl (1).
[4] Parrish & Mommaerts (1); Blum & Morales (1).

underwent marked changes, the sedimentation coefficient S_{20w}, rising from 6.15 (run in the ultracentrifuge at 5°) to 8.2 (run at 29°); at this stage the change showed some reversibility but longer incubation produced further and apparently irreversible changes. H. Holtzer (1) also found this spontaneous change with increase in sedimentation coefficient. Light-scattering observations showed that the molecular weight increased without change in radius of gyration and he concluded that side-to-side aggregation occurred. The same changes went on even at 4° but more slowly. Further work by Lowey & A. Holtzer (1) showed that a stepwise process was concerned, in which as many as eight molecules might join side to side. The aggregates were not affected by ATP or by high ionic strength. A. Holtzer & Lowey (2) in 1959, using preparations very rapidly made and working in the cold, obtained myosin giving a single peak in the ultracentrifuge, $S_{20w} = 6.4$. By the method of Zimm, light-scattering experiments gave a molecular weight of 493000 and a radius of gyration 477 Å. In an earlier paper (1) they had shown that light-scattering, viscosity and sedimentation data indicated that the molecule was a rod 1620 Å long and 26 Å thick. A single helical chain of molecular weight 500000 should have dimensions about 6250 Å and 10–15 Å; suggestions were made for the possible arrangement of the helices.

Johnson & Rowe (1, 2) a little later studied many properties of myosin at 20–25° and at 4°; changes in light-scattering and sedimentation results confirmed H. Holtzer's suggestion of side-to-side dimerisation. From a very careful study (3) of the intrinsic viscosity, a value of 2.5 ± 0.03 dl/g was found. With this value, together with the sedimentation and diffusion coefficients, and the partial specific volume, the properties of the hydrodynamically equivalent ellipsoid were calculated. A value for the molecular weight of 540000 was obtained and an axial ratio of 34.5. Johnson & Rowe pointed out the agreement between this value from sedimentation diffusion data and those depending on light-scattering, while the discrepancy between such values and those obtained by the Archibald sedimentation equilibrium method became increasingly obvious.[1]

This difficulty seems to have been resolved by Mueller (1) who showed in 1964 with the latter method that the mean apparent molecular weights must be related to the prevailing meniscus concentration and not to the bulk concentration. When this was done, a molecular weight of 524000, in good agreement with that from the other methods, was obtained.

More recent molecular weight determinations have been made by the sedimentation method. Tonomura, Appel & Morales (1) in 1966 used low

---

[1] See Mommaerts & Aldrich (1), Holzer & Lowey (1), both of whom obtained values about 420000 by the Archibald method; with the same method Kielley & Harrington (1) found 619000.

speed equilibrium, analysing the data by the midpoint method of van Holde & Baldwin, and obtained values of 480 000. In this procedure presence of low n-mers of myosin would increase the molecular weight. Similar values have however been reported by Gersham[1] in 1967, using the high speed equilibrium technique of Yphantis (1). This method separates low n-mers from monomeric myosin and gives a minimum value for the molecular weight; it also separated low molecular weight impurities and sub-units of myosin. The high values (near 600 000) obtained by Woods, Himmelfarb & Harrington (1) in 1963 were for myosin purified by ammonium sulphate fractionation. These higher values have been attributed by some workers to aggregation of the myosin at high salt concentration. The majority of recent determinations favour a value of about 500 000, but there is still uncertainty about the true molecular weight of the myosin monomer.

EFFECTS OF DISSOCIATING AND PROTEOLYTIC AGENTS ON THE MYOSIN MOLECULE: EARLIER WORK. It was natural that so large a molecule should present a challenge to biochemists, curious to know how some 5000 amino acids might be put together in it. The first attempts to break up the molecule were by the use of urea to disrupt secondary linkages. Weber & Stöver (1) reported briefly that by means of treatment with 45 % urea the molecular weight fell to 100 000, measured by osmotic pressure; and Edsall, Greenstein & Mehl (1) in 1939 found that myosin solutions treated with urea or guanidine hydrochloride decreased in double refraction of flow and viscosity. They explained this by the breaking up of very long molecules, though they noted that the breakdown products retained globulin properties. Snellman & Erdös (1) also examined the degrading effect of urea applied over long periods; a very poly-disperse product of low viscosity was obtained. In 1953 Tsao (1) carried out an investigation of the effects of urea, guanidine hydrochloride, and dilute acid and alkali on purified myosin. Depolymerisation with 6.7 M urea at pH 6.5 was very slow and reached a steady state only after some months. Two subunits, isolated by ammonium sulphate fractionation, were found. One, constituting 92 % of the whole, had an average particle weight of 165 000 (determined by osmotic pressure and fluorescence polarisation methods); the other, of 16 000.

A more rewarding way at the time of attacking the myosin molecule was by use of carefully controlled trypsin treatment. In 1950 Perry (4) and Gergely (4) independently studied these effects.[2] Perry had been using the Schick & Hass (1) method for fibril preparation and found that on storage, traces of trypsin remaining caused marked changes in the myofibrillar proteins. The turbid myofibril suspension became clear; the ATPase

---

[1] See Dreizen, Gersham, Trotta & Stracher (1).
[2] Also Perry (3) and Gergely (5).

activity was maintained and even enhanced, but on addition of ATP to an extract in M KCl the usual decrease in viscosity was not seen. Similar results were found with a dilute myosin gel treated at 0° for a few hours with crystalline trypsin, 0.0045 gm/ml. Gergely made similar observations, noting also a fall in specific viscosity of the myosin solution; on fractionation by means of pH change, about 90 % of the ATPase activity was found in the supernatant, with 2–3-fold increase in specific activity.

In 1953 Mihalyi and A. G. Szent-Györgyi examined the trypsin effects in detail. They showed[1] by ultracentrifugal analysis that very rapid changes went on – with 0.05 mg of trypsin/ml at 23° the myosin was broken down completely in about 10 min to give a new sedimentation diagram with two peaks, one component being faster, the other slower, than the original myosin. Actomyosin was also rapidly attacked; here also the slow component was seen, but it appeared that the actin (only very slowly digested) remained in combination with the fast component. At the end of this rapid phase of digestion of myosin, only about 3 % of the total nitrogen was not precipitable with trichloracetic acid. The ATPase activity was unchanged and, contrary to the original view of both Perry and Gergely, ability to react with actin remained. The increase in specific viscosity on stepwise addition of actin was less than with native myosin, and the drop with ATP was less; but the amount of actin needed to saturate the myosin was the same in both cases. Addition of actin to the digest allowed the separation in the ultracentrifuge of a supernatant containing the light fraction making up about 42 % of the original protein. The pellet was dissolved, the actin was removed by centrifuging in presence of ATP and Mg^{2+}, and the fast fraction (about 60 % of the whole) was recovered in this supernatant. Practically all the ATPase activity and actin combining power were in this fraction, and the specific activity was correspondingly increased. A. G. Szent-Györgyi (2) isolated the two sub-units by precipitation with ammonium sulphate. He termed them light meromyosin (LMM) and heavy meromyosin (HMM), finding the molecular weights by sedimentation and diffusion to be 96000 and 232000 respectively. The LMM formed paracrystals; it had the same solubility properties as myosin itself in KCl and ammonium sulphate solutions; HMM was water-soluble (see fig. 31 and tables 9 and 10).

Later work showed that myosin can also be fragmented in a very similar manner by other enzymes – chymotrypsin was used by Gergely, Gouvea & Karibian (1); subtilisin by Middlebrook (1) and snake venom by Kominz.[2] The implications of these findings for myosin structure will be considered later.

The mechanism of the rapid phase of enzymic breakdown of myosin has

[1] Mihalyi & Szent-Györgyi (1, 2); Mihalyi (2).
[2] Kominz, personal communication to Laki (2).

been the subject of much discussion. Laki (2) interpreted the effects of trypsin, chymotrypsin and snake venom as due to true proteolysis, quoting the finding of C-terminal lysine on the LMM formed in the first case,[1] of C-terminal phenylalanine in the second case[2] and C-terminal alanine in the third case.[3] These are the C-terminal acids to be expected according to the

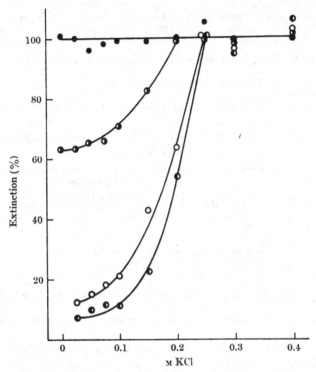

Fig. 31. Solubility curves in KCl solutions. ○, myosin; ◑, L-meromyosin; ◓, digested meromyosin; ●, H-meromyosin. Extinction at 280 mμ, expressed in percentage of maximal value. (A. G. Szent-Györgyi (2).)

specificity of the enzymes concerned. Mihalyi & Harrington (1) followed the trypsin digestion by means of hydrogen ion and non-protein N production; and changes in viscosity, sedimentation properties and optical rotation. They drew from these data a picture of the myosin molecule as consisting of a small number of parallel chains, some segments of which were more uncoiled than the rest. They suggested that the trypsin attacked a more uncoiled region much more rapidly than the rest of the molecule, and that the

---

[1] Gladner & Folk (1).
[2] Gergely, Kohler, Ritschard & Varga. Meeting of the Bioph. Soc. 1958, p. 46 (mentioned by Laki (2)).   [3] Kominz, personal communication to Laki (2).

TABLE 9. *Sedimentation constants of fractions of trypsin-digested myosin*

Material	Protein concentration (mg. per ml)	S_{20}* Slow fraction	Fast fraction
Digested myosin	5.14	2.62	5.37
Digested myosin + actin	5.14†	2.70	ca. 80.0
Isolated slow fraction	2.04	2.67	
Isolated fast fraction	3.66		5.70

* S_{20} expressed in S (Svedberg units) and calculated to 0.6 M KCl and 0.0067 M neutral phosphate buffer.
† The figure gives only the concentration of the digested myosin.
(Mihalyi & Szent-Györgyi (2).)

TABLE 10. *Comparison of ATPase activity of myosin, digested myosin, and isolated fractions*

Material	Phosphate liberated per mg protein in 5 min at 35° (γ)
Myosin	83.7
Digested myosin	80.7
Fast fraction + actin*	130.0
Isolated fast fraction	131.6
Isolated slow fraction	11.5

* The protein introduced with actin was omitted in the calculations. (Mihalyi & Szent-Györgyi (2).)

other enzymes also split rapidly through this same region, attacking each its specific links there. Analysis of their data for hydrogen ion production led them to the view that during the initial period of tryptic digestion two first-order reactions of very different velocities were proceeding. The faster reaction (breaking of the random regions) was essentially complete before a significant fraction of the slower had taken place. They calculated that 64 peptide bonds were actually cleaved in the rapid reaction and 240 in the slow reaction per molecule of myosin; out of a total of 516 susceptible bonds[1] this represented 59%, and a very trypsin-resistant core seemed to exist. Determination of α-amino groups formed agreed approximately with the number of bonds broken according to the results with the pH stat. They further suggested therefore that certain bonds in the coiled chains were also enzymically split in the slow reaction even in the initial period, and they cited the experiments of Szent-Györgyi & Borbiro (1), on urea attack on

[1] Kominz, Hough, Symonds & Laki (1).

LMM, in support of this view. Szent-Györgyi & Borbiro had found in 1956 that LMM was much more susceptible to urea depolymerisation than myosin itself. After a few hours in 5 M urea at room temperature the LMM was broken down into somewhat heterogeneous units of molecular weight about 5000, which they called protomyosins. No change in carboxyl group concentration could be detected during this depolymerisation, which was by them not regarded as due to peptide-bond breakage.

But Mihalyi & Harrington took the protomyosin formation as the result of enzymic breaks, only becoming apparent on the urea treatment, in the LMM chains, and considered that secondary forces between the chains could be strong enough to hold the original configuration, even after breakage of a number of peptide bonds.

However, Middlebrook (1) did not agree with this conception. With the Akabori technique[1] for determining C-terminal lysine, he could find only 0.1 equivalent per mole of trypsin-produced LMM; in the protomyosins he found similar N- and C-terminal residues, no matter which enzyme had produced their parent LMM. He later (2) studied myosin after acetylation with acetic anhydride, by which means the ε-amino groups of lysine were acetylated. Acetyl–lysine peptide bonds are not hydrolysed by trypsin; nevertheless the acetylmyosin after the rapid phase of digestion with trypsin gave two peaks in the ultracentrifuge similar to those of LMM and HMM. Middlebrook suggested that the proteases first modified the myosin by breaking secondary bonds, and that only the later slow degradation phase was due to specific hydrolysis. This explanation ignores the fact that acetyl myosin can still be cleaved by trypsin at the arginine residues. Offer (3) has pointed out that Middlebrook's thesis rests mainly on his data for C-terminal residues, since the amino acids he found as N-terminal in the protomyosins are so prevalent in myosin that little can be deduced from the fact that they were found in approximately the same yields after each type of digestion.

As we have seen, Szent-Györgyi (2) gave the molecular weights of LMM and HMM as 96 000 and 232 000. Using the then current molecular weight for myosin, he suggested that the myosin molecule contained four molecules of LMM and two of HMM. Lowey & Holtzer (2) in 1959 found higher values for the molecular weights – 126 000 and 324 000; and with the smaller molecular weight of myosin favoured by determinations at that time (420 000–490 000) a one to one ratio seemed to fit the case. There was also a difference from the earlier results in the distribution of the meromyosins as calculated from area measurements on the sedimentation diagrams. After corrections for radial dilution and for the Johnston–Ogston effect, they

[1] This involves digesting the protein with anhydrous hydrazine and leads to the liberation of only C-terminal residues as free amino acids.

obtained 25 % LMM and 75 % HMM, in fair agreement with the suggested molar ratio. In 1964 Mueller (1) using the Archibald method as previously mentioned found molecular weights of 151000 and 350000 which fitted well for a one to one ratio in the myosin molecule of 524000 by the same method.

We must turn now to consider the light which more recent studies of the fragmentation of myosin have thrown upon its molecular structure.

FRAGMENTATION OF THE MOLECULE: LATER WORK. In 1960 A. G. Szent-Györgyi, Cohen & Philpott (1) described the fractionation of LMM by alcohol precipitation and the preparation of LMM Fraction 1. This protein had properties very similar to those of LMM and the same molecular weight. From optical-rotation studies it appeared to be a fully-coiled α-helix; from viscosity data the length of the molecule was assessed at 750 Å, and since the calculated length of the fully-extended molecule would be about 1500 Å a two-chain structure seemed probable. Lowey & Cohen (1) could recover as LMM Fr. 1 about 70 % of the original LMM, and regarded the latter as consisting mainly of LMM Fr. 1. The X-ray diffraction diagram and rotatory dispersion data indicated a very highly α-helical coiled-coil structure for LMM Fr. 1, and thus suggested a rigid rod-like model, of which the length (from light-scattering data) would be 770 Å and the unhydrated diameter (from viscosity data) about 20 Å. Thus a two-stranded molecule consisting of two helices coiled to form a larger helix was envisaged. On the other hand HMM had been found by Cohen & Szent-Györgyi (1) to be only 50 % helical; the correlated data from light-scattering and viscosity were best interpreted in a model of non-uniform mass distribution in a molecule 740 Å long and about 30 Å wide. From the length of the myosin molecule it must be supposed that the two constituent molecules are attached to each other linearly; from the HMM moiety the globular part or parts could stick out and constitute the connecting bridges to actin[1] – a fundamental part of the sliding filament mechanism, as we shall see.

By 1961 actual observations made in the electron microscope by Rice (1) with the shadow-casting technique had shown myosin molecular particles to be some 1100 Å long and 10–30 Å wide. They often had a thickened portion at one or both ends. Later observations[2] by Rice (2) and H. E. Huxley (1) gave the dimensions more definitely[3] as length most frequently 1600 Å, the diameter of the rod being 15–20 Å; the length of the globular region (only at one end in single molecules) was 150–250 Å, its width about 40 Å. LMM particles appeared as rods, many about 900 Å long; with HMM, globular particles were seen, similar in size to those in the intact myosin, each with a rod-shaped tail attached. The length of the latter was variable,

---

[1] Also Cohen (1). Cf. p. 245 below.   [2] Also Zobel & Carlson (1).
[3] It must be remembered that these dimensions are of dried particles.

but about 750 Å (see figs 32–4 pl.). These data suggest a myosin molecule made up of one molecule each of LMM and HMM joined end to end, and fit well with the facts we have been considering.[1] In 1964 Lowey (1) set out to examine the nature of the tail region, bearing in mind the stability to acid shown by LMM, and the likelihood that the globular region of the HMM would be more susceptible to acid than its fibrous part. She found that after titration of a solution of HMM to below pH 4.0 it showed in the ultracentrifuge two peaks with Svedberg units 3.8S and 10S. The former value resembles that of LMM, and the substance was found to have high helix content – 73 %. Moreover its amino acid composition showed a strong resemblance to that of LMM, particularly in the low proline and the high basic and acidic groups as compared with HMM.[2]

The significance of γ-myosin remains uncertain. In 1958 Azzone (1) isolated from the dystrophic muscle of E-deficient rabbits a protein having properties of γ-myosin as described by Dubuisson; further it had high viscosity and double refraction of flow but no ATPase activity and no ATP sensitivity. Its increase in dystrophy was much more marked than in contracture; they conjectured that it was a denaturation product of actin or myosin. Later Kay & Pabst (1), also working with E-deficient rabbits, determined the molecular weight (49000) and showed that the molecule was 85 % helical. The molecular weight and axial ratio were compatible with a double-helix model. If γ-myosin is not a new protein but a degradation product, then LMM seems the most likely source.

In 1960 Kielley & Harrington (1) returned to the attack on myosin by means of the disaggregating agents, choosing 5 M guanidine hydrochloride. After treatment at room temperature the molecular weight had fallen from 619000 to 206000, a figure which invites comparison with Tsao's 160000 unit. Further work[3] with 12 M urea at 40° showed that the breakdown product was monodisperse on starch gel electrophoresis. These molecular weights suggested that all parts of the molecule contained three individual polypeptide chains. The molecular weight values accepted at that time for LMM and HMM (126000 and 324000) together with their own value for the molecular weight of myosin (619000) seemed to favour the presence of two LMM and one HMM in the myosin molecule. Using also the known results of light-scattering, optical rotation and X-ray diffraction measurements, Keilley & Harrington suggested a model. In this a three-stranded helix (1650 Å long) consisted, for 1040 Å of its length, of two LMM molecules linearly joined, with a small amorphous segment between them. HMM, also three-stranded, was joined also linearly and through another small

---

[1] The arrangement of the myosin molecules in the myosin filament, and the interaction of myosin and HMM molecules with actin filaments are considered in ch. 11.

[2] Also pp. 209–10 below.   [3] Small, Harrington & Kielly (1).

amorphous segment, but was folded back upon itself, adding only 610 Å to the length. The amorphous segments provided for the known effects of enzymic breakdown.

In 1963 however Woods, Himmelfarb & Harrington (1) pointed out that the proportion of HMM (70%) calculated from measurements of the Schlieren peaks in the ultracentrifuge was too high for this representation. Moreover, no sign of a peak for HMM–LMM, which might be expected, was ever seen. They therefore pictured only one molecule each of LMM and HMM held together by the short amorphous region. Young, Himmelfarb & Harrington (2) the next year gave molecular weights of 162 000 and 362 000 respectively for LMM and HMM.

In the work of this group only one size of breakdown product with disaggregating agents was recognised; but the finding by Tsao of a much smaller unit in small quantity has been substantiated by several workers. Kominz and his collaborators[1] in 1959 found such material when treating myosin at pH 10.0. They remarked on its very high proline and phenylalanine content and its similarity in this respect to the small sub-unit obtained on urea treatment. Dreizen, Hartshorne & Stracher (1) in 1966 showed that myosin is polydisperse in 5 M guanidine hydrochloride and obtained two components whose molecular weights were determined in the mixture as 200 000 and 46 000 by high-speed equilibrium. The low molecular weight components had escaped detection by earlier workers using guanidine hydrochloride because of the rapid diffusion and the low protein concentrations used. High concentrations of myosin are needed to show the asymmetry of the peaks in sedimentation velocity experiments and double sector cells must be used because of the curved base lines obtained in high salt concentration. Again the low molecular weight material cannot be detected in molecular weight determinations at low equilibrium speeds because of the non-ideality of the high molecular weight species. The molecular weight of the low molecular weight species has been re-determined in the absence of the high molecular weight species, and values close to 21 000 were obtained.[2] Then Gersham, Dreizen & Stracher (1), using alkali treatment, obtained three light sub-units electrophoretically distinguishable and all of average molecular weight about 20 000 – thus in close agreement with the values with guanidine hydrochloride. The major breakdown product was fibrous, of molecular weight 418 000; it dissociated on dialysis against 5 M guanidine to give a molecular weight of 215 000. They suggested that a reasonable model for myosin (which would be consistent with Mueller's molecular weight) would contain an axial core of two fibrous units forming a double helix and extending into the head region, the latter containing also the three globular sub-units.

[1] Kominz, Carroll, Smith & Mitchell (1).   [2] Locker & Hagyard (1).

Soon afterwards Oppenheimer and his colleagues[1] carried out the succinylation of myosin, thus introducing a high negative-charge density which disrupted the molecule. Two fractions were obtained, a light one and a main one having a molecular weight not far from that of myosin itself. They noted the high proline content of the former, which would preclude its arising from the helical part of the molecule.[2] Bárány & Oppenheimer (1) indeed found that with succinylation of the meromyosins LMM had unchanged molecular weight while the product with low molecular weight arose from HMM. The experience of Locker & Hagyard (1), who used acetylation or carboxymethylation at neutral pH to cause depolymerisation, was similar. The main structure was practically intact but there was dissociation of about 15 % as small sub-units. Polyacrylamide disc electrophoresis gave three components in similar amounts, molecular weights 17 000, 19 000 and 20 000; two of these were found in acetylated HMM but not in the acetylated LMM. Such experiments show that the light chains are not hydrolytic fragments of the heavy chains.

It is to be noticed that C-terminal isoleucine occurs in HMM[3] and in the light component isolated after alkali treatment.[4] The similarities of the amino-acid composition of the alkali-[4] and urea-[5]produced light components to that of HMM[6] also reinforces the argument from the stoichiometric relations that the light components are not impurities but real constituents of HMM.

Meanwhile much further work was done on the effects of proteolysis. In 1960 Mueller & Perry (1)[7] showed by chromatography on DEAE cellulose that standard HMM preparations contained about 85 % of a main component together with a minor heterogeneous fraction, more slowly sedimenting. Both fractions had ATPase activity. Similarly Kakol, Gruda & Rzysko (1) found two fractions making up 65 and 35 % respectively of the initial protein, the latter having about twice the specific ATPase activity of the former. Mueller & Perry (3) then carried further the study of trypsin effects on HMM, using higher concentrations of the enzyme. Whereas trypsin-treated myosin gives rise to two components appearing together in the course of digestion, trypsin-treated HMM was converted to a single major component sedimenting more slowly. This they called subfragment 1; in the later slower phase of digestion it gave rise to a more heterogeneous fraction, subfragment 2. Subfragment 1 had both ATPase activity and actin-combining power. Young, Himmelfarb & Harrington (1) in 1965

[1] Oppenheimer, Bárány, Hamoir & Fenton (1).
[2] For the relation between proline content and α-helix content see Szent-Györgyi & Cohen (1).
[3] Sarno, Tarendash & Stracher (1).   [4] Kominz, Carroll, Smith & Mitchell (1).
[5] Middlebrook & Szent-Györgyi, cited in A. G. Szent-Györgyi (4).
[6] Lowey & Cohen (1).   [7] Also Mueller & Perry (2).

confirmed the formation of subfragment 1; they found it to make up about 80% of the HMM molecule and to have a molecular weight of 117000–121000. Its enzymic activity was about one-third of that of HMM on a molar basis. They pictured it as arising from the dissociation of one HMM molecule into three similar sub-units, each possessing an active site and each derived from one of the three polypeptide chains previously postulated by Kielley & Harrington.

These findings however did not agree with those of Mueller (2) and Jones & Perry (1) who found higher molecular weight values for subfragment 1 – 112000–170000 and 129000 respectively; more important, in both cases the ATPase (on a nitrogen basis) was markedly increased over that of the parent molecule – by 100% as found by Jones & Perry. Szent-Györgyi (2) in 1952 had noted that little of the cysteine of myosin occured in LMM; Jones & Perry now found that subfragment 1 was richer than HMM in cysteine. They considered it likely that one subfragment 1 molecule was obtained from one molecule of HMM, representing the intact thickened portion of the original molecule. They recalled the evidence for binding of only one ATP or pyrophosphate molecule/molecule of myosin.[1] Mueller was inclined to interpret his results in a similar way. Tokuyama, Kubo & Tonomura (1) were also of the opinion that their tryptic subfragment (M.W. 180000, of ATPase activity equal to that of myosin or HMM on a molar basis) carried the only ATPase active centre of the myosin. It had already been found[2] that two lysine residues in myosin (situated in different amino-acid sequences) reacted with 2,4,6-trinitrobenzenesulphonate, depression of ATPase activity resulting. It now appeared that both these specific lysine residues occurred in HMM, but only one of them was found in sub-fragment 1. Similarly Yagi & Yazawa (1) isolated a subfragment of molecular weight 147000 – 40% of that of HMM – having 80% of the total ATPase activity. Nihei & Kay (1)[3] have recently prepared by means of papain digestion (which splits myosin directly to subfragment 1) a particle of molecule weight 110000 having three times the specific Ca-ATPase activity of myosin. They suggest that the myosin molecule could contain one or two, but not three, of such units.

An important paper appeared in 1967 by Slayter & Lowey (1), who examined myosin preparations in the electron microscope by the rotary shadowing technique. The myosin heads then appeared in the majority of the molecules as bipartite, each lobe being about half the size of the heads as ordinarily seen. The two heads could assume a variety of positions, and it was suggested that they were attached by a single strand to the more rigid rod portion. With a papain preparation of HMM subfragment 1, the particles

---

[1] Ch. 8.   [2] Ch. 12.
[3] Also Kominz, Mitchell, Nihei & Kay.

consisted of single lobes resembling those just described. This clearly is very substantial evidence in favour of a two-chain structure (see fig 35 pl.).

Lowey, Slayter, Weeds & Baker (1), like Nihei & Kay, found their highly purified subfragment 1 (molecular weight 115000) to have high specific ATPase activity – three times that of HMM tested as the Mg-activated enzyme in presence of actin. They point out that, on a molar basis, HMM and subfragment 1 have similar actin-activated ATPase activity and that this implies that only one globule in the HMM head may be active at a time. The measurements by Nanninga & Mommaerts (1) of the minimum ATP concentration necessary for maximum viscosity change in actomyosin solution indicated only one active site for myosin. The same interpretation can be placed on the finding by Young (1) that one subfragment 1 particle or one HMM molecule combined with only one F-actin monomer. The method used here was treatment of the myosin derivative with increasing amounts of very pure actin, followed by centrifugation and examination of the supernatant. It would thus seem that, in the normal condition of the active enzyme, binding of ATP to one site can block the binding site on the other chain, though both are potentially enzymic. On the other hand Luck & Lowey (1), using highly purified myosin inhibited by the presence of Mg-ions and 1.5 M NaCl, found, by measuring ADP disappearance from the medium, that about 1.5 moles were bound per mole of myosin. Again Schliselfeld & Bárány (1) using similar conditions, observed that 1.7 moles of ATP (^{32}P labelled at the γ-phosphate) were taken up per mole of myosin. Thus it seems that the inactive enzyme can bind nucleotide at two sites.

An interesting observation was made in 1967 by Lowey, Goldstein, Cohen & Luck (1), in their work on the proteolysis of myosin and the meromyosins by a water-insoluble, negatively-charged derivative of trypsin. It seems that meromyosins of somewhat different lengths may appear, depending on the digestion conditions. They surmise that the usual picture of a random non-helical part of the molecule susceptible to attack by several enzymes might be erroneous. Possibly the different enzymes attack specific linkages in the helix situated several 100 Å apart. This would mean dispensing with the idea of the sensitive region as a sort of hinge; instead the helical coiled coil, extending throughout the rodlike portion, would act as an elastic element with uniform power of bending. The work of Balint and his collaborators would fit in with this idea.[1] Fragmentation of L-meromyosin by prolonged trypsin treatment gave three kinds of distinct fragments of molecular weights 112000, 84000 and 56000 (lengths 750000, 580000 and 410000 Å). This led to the suggestion of a periodic structure for the entire length of the protein.

Lowey, Goldstein, Cohen & Luck (1) have also isolated HMM subfragment 2

[1] Bálint, Szilagyi, Fekete, Blazso & Biro (1).

by treatment of HMM with the water-insoluble trypsin; they showed it to have molecular weight about 60000 and to be the α-helical structural portion of HMM, i.e. the tail. Subfragment 2 is soluble at neutral pH at all ionic strengths. It seems possible (as we shall see)[1] that the difference from LMM indicated by this characteristic may play an important part in allowing interaction of the globules of subfragment 1 with actin.

Consideration of the molecular weights of subfragments 1 and 2, 115000–120000 and 60000 respectively, shows that they sum to good agreement with the molecular weight of HMM (330000–340000) if two molecules of the former and one molecule of the latter are assumed and about 30000 is allowed for peptide formation during the HMM breakdown.

Lowey, Slayter, Weeds & Baker (1) also described the preparation (by means of papain digestion at low ionic strength) of 'myosin rods' i.e. whole myosin molecules deprived only of the two terminal globules of HMM subfragment 1. Measurements in the electron microscope of the length of the rods gave 1360 Å, while the molecular weight by sedimentation equilibrium was 220000. Account being taken of a mean amino-acid residue weight of 115, and a 1.5 Å rise/residue for the α-helix, the length of a single α-helical chain was calculated to be 2900 Å. Thus again there was evidence for the two-chain molecule. The absence of proline in these rods reinforces the concept of a uniform elastic element rather than the presence of a mechanical hinge.

Further information about the number of strands making up the myosin molecule and the arrangement of the amino acids in them can be gleaned from a study of the number and nature of the peptides formed on exhaustive tryptic proteolysis. Trypsin specifically attacks peptide bonds in which the carbonyl is contributed by L-arginine or L-lysine. For a molecular weight of 524000 and the known number of susceptible arginine and lysine bonds, formation of some 670 different peptides would be possible. In the work of Perry & Landon (1) only 120 were found. Allowance has to be made for about 10 % of free arginine and lysine appearing, indicating regions where these two amino acids are adjacent, but even so the number of unique peptides is definitely less than would be expected from three or even two dissimilar chains. It seems that certain sequences must repeat along the chains. With HMM and LMM the situation is the same – only 120 peptides out of a possible 390 were found for the former, 75 out of 200 for the latter. The finding for LMM (when corrected for free arginine and lysine) would approximate to the requirements of two identical chains. It is striking that Perry & Landon found that the peptides of LMM were present also in HMM, the latter containing others as well.

In 1961 Kielley & Barnett (1) labelled myosin SH groups with ^{14}C N-ethylmalemide, and prepared autoradiographs of the peptides formed on

[1] Ch. 11, p. 306.

exhaustive tryptic digestion. About fifteen labelled peptides were found; since forty-four half-cystine residues are present in 530000 g of myosin, this result was taken as indicating the presence of three similar polypeptide chains in the molecule. Weeds & Hartley (1)[1] however found evidence for at least twenty-two unique SH sequences, a result supporting a structure of two identical sub-units. A similar inference can be drawn from Weeds' study (1) of the number of unique thiol sequences in LMM Fr. 1.

The significance of the short chains is still problematic and it remains possible that some part may be tightly bound impurity. There is however a considerable amount of evidence favouring the existence of these light chains as part of subfragment 1. Thus Lowey, Slayter, Weeds & Baker (1) make preliminary mention of experiments on highly purified material in which each subfragment 1 yielded on alkaline treatment at least one light chain. The experiments of Trotta, Dreizen & Stracher (1) concerning specific SH groups[2] point in the same direction.

The findings of Offer (1, 2) showed that myosin must contain at least two chains with the terminus N-acetyl-serine-serine-aspartic-alanine-aspartic. Sarno, Tarendash & Stracher (1) have recently found that the yield of C-terminal isoleucine reported by Locker (1.7 moles/mole of myosin) can be increased to 4 moles, all located in the HMM. Offer & Starr (1) have now shown that the N-acetyl termini also are in HMM, but none in the short chains. This must mean that the isoleucine C-termini are contained in the short chains. Thus there could be four short chains or perhaps more probably two short chains each terminating in Ile-Ile. The C-terminal of the long chains and the N-terminal of the short chains are at present unknown.

Weeds (1, 2) has made a detailed study of the light chains. By fractionation of the light alkaline component on DEAE-cellulose, followed by electrophoresis on cellulose acetate, three components were found; two of these had much amino-acid sequence in common, including identical thiol sequences. Peptides from ^{35}S-cystine-exchanged myosin showed 2–3 equivalents of this light-component thiol group/mole of myosin.

Myosin in 0.5 M KCl was then treated by Weeds (3) with 5,5-dithiobis-(2 nitrobenzoic acid) (DTNB). On dilution to 0.05 M KCl the myosin was precipitated, the light chains remaining in the supernatant. After re-solution and exhaustive dialysis against buffer containing dithiothreitol to remove the blocking agent, the regenerated myosin contained all the original ATPase activity. There was evidence that the DTNB light-chain material extracted by this method, and inessential for enzymic activity, consisted mainly of a single component.

The next step was incubation of the regenerated myosin with 2 M LiCl at pH 11.0; the recovered myosin now had no ATPase activity. The alkali

---

[1] Also Weeds & Hartley (2).　　　　　　　　[2] Ch. 12, p. 288.

light-chain material obtained in this way was different from the DTNB light chains, as was shown by the different amino acid sequences; on gel electrophoresis it gave two major bands, each containing the same thiol sequence. Thus it seems that the DTNB light chains (amounting to about half the low molecular weight material) may be simply contaminants. The two different alkali light chains found, necessary for enzymic activity, probably arise from two isoenzymes of myosin, present in the mixed fast and slow muscles used (see fig. 39 pl.).

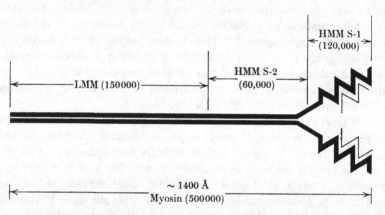

Fig. 37. Schematic representation of the myosin molecule. The figures shown in brackets are molecular weights. (Lowey, Slayter, Weeds & Baker (1).)

We see then that electron microscope observations and the results of physical measurements favour a 2-chain structure for myosin; the chemical evidence is not decisive, but here again the 2-chain structure seems strongly indicated. However this question may be resolved, it is evident that myosin is an extraordinary protein, combining as it does globular and fibrous properties in the one molecule. The fibrous part provides strength and rigidity in the LMM and flexibility in the HMM subfragment 2. The globular part is the seat both of the enzymic reaction providing energy by ATP breakdown and of the specific groups whereby interaction with actin is achieved. Fig. 37 sums up knowledge of the structure of the myosin molecule, but does not show the double helix of the fibrous portion.

ACTIN

This section will be devoted to consideration of the chemical and physical behaviour of actin molecules and polymers, with only brief mention of the location of actin, in the fibrous form, in the muscle cell. The actin filaments and their function will be described in chapters 11 and 12.

As we have seen, it was already known by 1945 (a) that actin existed in two forms, one non-viscous which, by addition of inorganic ions up to about 0.1 M, could be converted into a very viscous form; (b) that this fibrous form interacted with myosin to give the highly viscous actomyosin. By 1948 Straub and his collaborators[1] had made a number of further observations which turned out to be of much significance.

With regard to the action of ions, they found that there seemed no difference between the effects of Na^+ and K^+; in presence of 0.1 M Na^+ or K^+, Mg^{2+} (0.01 M) had a marked accelerating effect on polymerisation, while Ca^{2+} (0.003 M) had some retarding effect. Even after prolonged dialysis or repeated washing of the isoelectric precipitate with acetate buffer at pH 5.0, actin preparations contained calcium; the magnesium content was small and variable.

Another important point was the susceptibility of actin to oxidation; a number of oxidising agents – methylene blue, ferricyanide, cystine – decreased the stability of G-actin, and addition of dilute permanganate prevented polymerisation.

These workers also found that G-actin on dialysis against water, or on repeated washing of the isoelectric precipitate, lost its power to polymerise and to form actomyosin. This inactivation could be prevented by adding to each washing a dilute solution of boiled actin or boiled muscle juice. The presence in the actin of a prothetic group was suggested. The next year Pettko & Straub (1) showed that ATP could always be detected in actin solutions, and in 1950 Straub & Feuer (1) found that the ATP was a bound functional group of the G-actin and that its removal by prolonged dialysis led to loss of polymerisability. Evidence for its bound state was given by its relative inaccessibility to enzymes, e.g. potato apyrase. The amounts of ATP concerned were very small for estimation by the methods usual at that time, and Straub & Feuer turned to a highly specific pharmacological method, depending on the action of ATP or ADP on quinine-treated frog's heart, and also the method (specific for ATP only) depending on viscosity changes in actomyosin solutions. They found its amount to be about 2 g molecules/70 000 g protein. When G-actin was converted to F-actin, ATP concentration decreased and inorganic phosphate was formed. They showed, by myokinase treatment, that ADP was also formed and believed it was bound to the actin though this was not demonstrated. Reversible depolymerisation could take place if the F-actin was dialysed against ATP solution in the absence of other salts, bound ATP reappearing in the process. They introduced the practice, still in use, of adding small amounts of ascorbic acid to prevent oxidative denaturation during these procedures. Straub & Feuer also suggested the possibility that the bound ADP of F-actin might be rephosphorylated during depolymerisation in the absence of added

[1] Feuer, Molnár, Pettkó & Straub (1).

ATP, the necessary energy being provided by the structural change in the protein; but this idea has not stood the test of time.[1]

Guba (1) made a special study of the effects on depolymerisation of salt concentration higher than 0.1 M. With 0.5 M KCl or KBr, in 30 min some fall in viscosity and double refraction of flow was seen. With 0.5 M KI depolymerisation was instantaneous. Straub (1) had already noticed this exceptional behaviour of iodides.

In 1951 Feuer & Wollemann (1) revised the figure for bound ATP of G-actin, placing it now as about 1 molecule/molecule of the protein. They confirmed that the ADP formed during polymerisation could not readily be attacked by apyrase or myokinase, and concluded that it was bound. It could be removed by treatment with large amounts of apyrase or by long dialysis against salt solution. Such ADP-free actin was still polymerised and could combine with ADP. About the same time Laki, Bowen & Clark (1) independently reached the conclusion that ATP was concerned in actin polymerisation; this arose from their observation that polymerisation could be prevented by incubation with glucose and hexokinase, and that polymerisability could be partly restored by adding ATP.

PROPERTIES OF GLOBULAR AND FIBROUS ACTIN. The detailed study of the two forms has presented great difficulties. The thixotropic nature of F-actin made difficult any accurate measurements in the Tiselius apparatus or ultracentrifuge. Active G-actin without polymerisation cannot be studied conveniently because, without addition of salt, charge effects cannot be avoided, and even a very small degree of polymerisation seriously interferes with the interpretation of measurements of size and shape. Dubuisson (8) obtained an electrophoretic mobility of -6.3×10^{-5} cm/V sec for F-actin at pH 7.4, and of -4.7×10^{-5} for G-actin irreversibly depolymerised by NaI. Johnson & Landolt (1) found Straub actin to be very disperse in the ultracentrifuge, but the two main components had sedimentation coefficients (in terms of Svedberg units) of 5–7S, and 45–60S.

Feuer, Molnár, Pettkó & Straub (1) had given for G-actin a molecular weight of 70000 based on the tryptophan content (now known to be greatly in error) but they also had evidence for a value of this order from passage of the protein through membranes of graded porosity. In 1951 both Mommaerts (6) and A. G. Szent-Györgyi (1) showed that Straub actin was indeed only some 50 % pure. By repeated ultracentrifugation and re-solution of the F-actin pellet Mommaerts obtained, after finally depolymerising by dialysis against 10^{-4} M ATP, a G-actin monodisperse in the ultracentrifuge and which polymerised completely on salt addition. Light-scattering measurements (7) gave 57000 for the molecular weight.

[1] A. G. Szent-Györgyi (3); Martonosi, Gouvea & Gergely (1).

Other procedures in purification which have proved useful include that of A. G. Szent-Györgyi, depending on extraction with KI and fractionation with alcohol; that of Tsao & Bailey (1) in which extraction of other material (including tropomyosin B) from the dried muscle residue was diminished by extracting with 30 % acetone instead of water; and that of Bárány, Biro, Molnár & Straub (1) in which use was made of the precipitation of F-actin by means of 0.025 M $MgCl_2$ and which freed the actin from a number of contaminating enzymes. Drabikowski & Gergely (1) later showed that tropomyosin contamination could be greatly reduced by extraction of the dried residue at 0° instead of at room temperature. It seems that this had been the practice in Mommaerts' laboratory, but since the temperature was not specifically mentioned other laboratories had performed the extraction at room temperature. To obtain highly purified material, combinations of these methods have been used, and recently starch gel electrophoresis[1] and gel filtration[2] have given good results.

In 1960 Kay (1) made a careful study of G-actin (purified by the method of Tsao & Bailey) in water (containing 10^{-4} M ATP) and in 0.5 M KI solution. The molecular weight was found by light-scattering (66000) and by sedi-mentation-viscosity (61000). Kay's value for the intrinsic viscosity was 0.14 dl/g in 0.5 M KI, considerably lower than the value given by Tsao (2). Grant and his colleagues[3] pointed out in 1964 that values even as low as 0.1 are much higher than would be expected for a globular protein; indeed Bailey (3) had been inclined to reject the idea that G-actin was really globular. The value found by them, 0.036 dl/g, was obtained on G–ADP actin,[4] only slowly polymerisable at 0°; it is comparable with that of serum albumin. From negative-contrast electron microscopy the axial ratio of individual G-actin molecules has been given by Rowe (1) as about 3.[5] The estimate of the molecular weight by Rees & Young (1) in 1967 gave a lower value – 46000. This was obtained by the method of molecular exclusion chromatography giving a high degree of particle homogeneity. Adelstein, Godfrey & Keilley (1) in 1963 had found a molecular weight of 28000 in 5 M guanidine HCl, a result suggesting that the G-actin molecule was composed of two polypeptide chains. Rees & Young however found the molecular weight remained unchanged in concentrated guanidine HCl, and Adelstein et al. now agreed that the molecular weight in these conditions was near 50000.[6] Evidence also for the low molecular weight comes from the results of Johnson, Harris & Perry (1) who have recently detected 3-methylhist-idine in actin from mammalian, fish and bird skeletal muscle. Calculation

---

[1] Carsten & Mommaerts (1); Kraus, van Eijk & Westenbrink (1).
[2] Adelstein, Godfrey & Kielley (1).
[3] Grant, Cohen, Clark & Hayashi (1).
[4] P. 220 below.   [5] P. 231 below.   [6] Quoted by Rees & Young.

from the content of this amino acid gives a minimum molecular weight of 47000. Values of over 60000 may indicate contamination with a partial oligomeric form, and this could markedly increase the viscosity.

Kominz, Hough, Symonds & Laki (1) have given the amino-acid composition of actin; the high proline content was in conformity with the low helix content.[1] Carsten (1), using the preparation method of Carsten & Mommaerts (1) for actin of high purity free from tropomyosin, found her results in good agreement with those of the earlier workers when correction was made in the latter for the estimated 10 % tropomyosin contamination. Carsten found by reaction of G-actin with iodoacetate in 8 M urea that the SH content was seven groups in a molecular weight of 61–62000; this was in agreement with values found a little earlier by Tonomura & Yoshimura (1) and by Katz & Mommaerts (1). Carsten found no S–S groups but Kraus, van Eijk & Westenbrink (1) found that their carefully purified G-actin, homogeneous on starch-gel electrophoresis after reduction with β-mercaptoethanol, before reduction gave four bands. They suggested the existence of a small number of disulphide bridges which might cause the formation of molecules of various types, differing in areas of chain linked. In a later paper Carsten (2) confirmed, by means of the interchange reaction with N,N'-bis-2,4-dinitrophenyl-L-cystine, the presence of seven SH groups. After tryptic digestion of alkylated actin labelled with iodoacetate-1-^{14}C, peptide maps gave five radioactive peptides with some radioactivity remaining at the origin. Again no cystine could be found, even after disappearance of some of the SH groups during aggregation.

Carsten & Katz (1) also made a comparative study (by means of sedimentation coefficients, electrophoretic patterns, amino-acid composition and peptide maps) of actins prepared from a wide range of animals. The actins from mammals and birds were very similar, while those from frog, fish and *Pecten* (especially the last) showed differences from the mammalian.

Locker (1), by means of the carboxypeptidase and hydrazine methods, found a single phenylalanine C-terminal group. This was confirmed by Laki & Standeart (1) and by Kraus and his collaborators (1). Attempts to find N-terminal amino acids at first gave negative results; but in 1956 Alving & Laki (1) and Gaetjens & Bárány (1) independently were successful in demonstrating N-acetyl-Asp-Glu in pronase digests. One mole of acetyl/60000 g of actin was found, accounting for most of the acetic acid appearing on hydrolysis of the actin.

We may turn now to the study of F-actin. In 1963 Johnson, Napper & Rowe (1) using F-actin made by the Straub method and purified according to Mommaerts, found in sedimentation studies three components – a gel-forming component, a component F_1, of sedimentation coefficient 40S; and

[1] P. 222 below.

a third component F_2, of sedimentation coefficient 98S. The diameter of F_1 was calculated to be about 62 Å, and it was suggested that these particles were filaments such as those seen about the same time in the electron microscope by Hanson & Lowy,[1] diameter about 80 Å. There was evidence that F_2 was a side-to-side dimer. Yoshimura, Matsumiya & Tonomura (1) also at this time made a thorough examination of the course of polymerisation of G- to F-actin, finding a number of intermediate stages.

In 1965 Maruyama, Hama & Ishikawa (1) published the account of a very interesting investigation of F-actin treated in different ways. They made F-actin from myofibrils by extraction with 0.6 M KI (after previous extraction of myosin) and dialysis against 0.1 M KCl in presence of ATP. This actin had a sedimentation coefficient of 40S (thus apparently corresponding to the F_1 of Johnson *et al.*) and showed no component of 100S. Flow birefringence studies showed the particle length was much less than that of Straub actin. Upon repeated purification the actin showed change in properties towards those of Straub actin. The same feature of short particle length had been observed by Noda & Maruyama (1) with actin prepared from actomyosin by ultracentrifugation in presence of ATP.

Maruyama (1) suspected the presence of a factor inhibiting network formation in solution. He succeeded in making from actin supernatants a protein preparation which, added to solutions of Straub actin, diminished network formation as shown (*a*) by the slowness of return of flow birefringence and viscosity after submission of the solution to a velocity gradient; (*b*) by the much more marked effect of sonication in diminishing double refraction of flow and viscosity; (*c*) by the effect of sonic vibrations in changing the sedimentation coefficient from 100 S to 40 S – a change which did not happen in absence of the factor. Thus the 'actin-factor' was considered to inhibit the formation of gel-forming actin and of F_2-actin. The absence of the factor (now known as β-actinin)[2] in F-actin made by Straub's method could not be explained, but the suggestion was made that, in presence of some other material, the factor was bound in some insoluble form as a result of the acetone treatment. Maruyama remarked on the interesting fact that the particle length, about 1μ, obtained in presence of the factor, was about the length of the I filaments in the sarcomeres, and speculated on the possible role of the factor during myogenesis.[3]

It is interesting that preliminary reports of factors apparently concerned in controlling the size of actin particles have been made from other laboratories. Thus Graham (1) and Graham & Wilson (1) found that if certain precautions of rapidity of preparation and fineness of mincing were taken,

---

[1] P. 226 below.   [2] Ebashi & Maruyama (1).

[3] For a discussion of another factor, α actinin (also reported to have some effects on actin) see ch. 12.

actin in the F-form instead of the usual G-form could be directly extracted by means of water from the acetone-dried muscle residue. This actin, which they called Fn-actin, was shown to sediment between 22 and 27 S. Purified Fn-actins were unstable in water, breaking down to G-actin (subsequently polymerisable). This behaviour suggested that the crude extracts contained a stabilising factor. It was then shown that the diffusate of crude supernatants contained a factor which might influence the complex equilibrium between polymers of actin in solution. Also Pápai, Székessy-Hermann & Szöke (1) in Hungary briefly reported in 1964 that actin prepared from dystrophic rabbits by the Straub method was only partly polymerisable, and much of the protein remained in the supernatant on purification by Mommaerts' method. This supernatant could inhibit the polymerisation of G-actin from normal animals, but had no influence on the viscosity of F-actin. Further investigation showed that a factor with these effects could also be prepared from normal muscle; it appeared to be a protein, extractable only at high ionic strength.

THE PART PLAYED BY NUCLEOSIDE PHOSPHATES, SH GROUPS AND DIVALENT METAL IONS IN THE BEHAVIOUR OF ACTIN. A. G. Szent-Györgyi (3) in 1951 confirmed the dephosphorylation of ATP on polymerisation; Mommaerts (8) soon after confirmed the 1:1 molar ratio for nucleotide and G-actin,[1] and also the finding that ADP could be almost completely removed from F-actin (by 20 days' dialysis) without depolymerisation. He identified by chromatographic analysis the F-actin–ADP, and showed that here also one mole was present per mole of actin monomer.

The work of Perry (11) at this time is of interest in bringing the actin question into relation with the situation *in vivo*. Perry estimated in washed myofibrils the amount of ATP, ADP and AMP. ADP was present in much greater amount than the other two compounds and he calculated that if all of it were associated with the actin of the fibrils, it would be present there in a roughly 1 to 1 molar ratio. A similar result was later obtained in a more elaborate study by Seraydarian, Mommaerts & Wallner, (1). Hasselbach (5) also found the ADP of muscle almost exclusively bound to actin, and that the ADP content of actin in the fibrils was as great as in isolated F-actin.

In 1960 Martonosi, Gouvea & Gergely (1), using ^{14}C ATP, found that the ATP on G-actin was completely exchangeable within a few minutes; after polymerisation the radioactivity was found in the F-actin. During depolymerisation of the F-actin, there was no evidence for re-phosphorylation

---

[1] Also Ulbrecht, Grubhofer, Jaisle & Walter (1) for estimation of nucleotide content by paper chromatography in highly purified actin, and the finding of 1 mole/62000 g protein.

of the ADP, either by exchange with added ATP or from inorganic phosphate. They pictured the reactions as follows:

$$\text{G-actin} + \text{ATP} \rightleftharpoons \text{G-actin-ATP}$$

$$\text{G-actin-ATP} \xrightarrow{\text{salt}} \text{F-actin-ADP} + \text{H}_3\text{PO}_4$$

$$\text{F-actin-ADP} \xrightarrow{\text{salt removed}} \text{G-actin ADP}$$

$$\text{G-actin-ADP} + \text{*ATP} \rightleftharpoons \text{G-actin*ATP} + \text{ADP}.$$

Asakura (1) also came to the same conclusion – a dissociation equilibrium between G-actin and ATP – as the result of his experiments in which removal of excess free nucleotide from G-actin (by mild treatment with ion-exchange resin) led to gradual loss of activity of the actin.

In 1957 Oosawa and his collaborators began an intensive study of the nature of actin polymerisation, following its course under conditions of very low inorganic ion concentration, so that the rate could be slowed and varied. They propounded a theory[1] which involved (a) formation, by means of non-directional, short-range attractions, of aggregates of charged macromolecules; (b) change of these under suitable conditions into the fibrous state, owing to long-range electric (Debye–Huckel) repulsion between them. According to this theory, the G $\longrightarrow$ F transformation would be a reversible fibrous condensation phenomenon, similar to gas–liquid condensation. Evidence for this point of view included the following observations, obtained by the study of actin in low Mg^{2+} concentrations (0.2–3 mM), by means of light-scattering, viscosity, osmotic pressure and flow birefringence measurements: (a) the existence of a critical concentration of G-actin (its level depending on the Mg^{2+} concentration) below which no polymerisation occurred, and above which all G-actin in excess of the level was converted to F-actin: (b) the co-operative nature of the process was emphasised by the fact that the rate of polymerisation was proportional to the third or fourth power of the G-actin concentration: (c) addition of a small amount of F-actin led to 'explosive' increase in rate of polymerisation, presumably because it acted as a nucleus. The light-scattering and viscosity measurements indicated at low actin concentrations the presence of an intermediate stage in F-actin formation; as the Mg^{2+} concentration was increased the actin filaments became longer, and at high Mg^{2+} concentrations side-to-side aggregation took place. The rate of change, as well as the critical concentration depended on the Mg^{2+} concentration.[2] The Japanese workers pictured the various states of the actin molecule as in equilibrium with one another, so that a steady state was established. If this were so, and if the change

[1] Oosawa (1, 2).
[2] Asakura, Hotta, Imai, Ooi & Oosawa (1); Oosawa, Asakura, Hotta, Imai & Ooi (1); Kasai, Asakura & Oosawa (1).

from G- to F-actin were associated with ATP dephosphorylation, then continuous liberation of inorganic phosphate would be expected, and this they found.[1] In 1961 Oosawa & Kasai (1), from thermodynamic and kinetic analysis, suggested that F-actin polymerisation was mainly helical, although simple linear polymers did often occur in certain conditions.[2]

Gergely, Gouvea & Martonosi (1) criticised some aspects of this work. They were unable to confirm the critical actin concentration at any given Mg^{2+} concentration, finding (by means of ultracentrifugation) the degree of polymerisation to depend solely on the Mg^{2+} concentration and to be independent of the actin concentration. They also queried the equilibrium idea, since they found that ^{14}C–ATP was incorporated to a far less extent than was equivalent to the inorganic phosphate set free. However, they admitted that this could result if only a limited number of bound ADP groups on the outside of the F-actin particles were able to take part in the continuous exchange. These would quickly become labelled and so not contribute further to the increase in radioactivity although continuing to take part in the liberation of phosphate. This suggestion of Gergely *et al.*, led Asakura (1) to attempt the loosening of the rigid F-actin structure by means of sonic vibration. In this he was successful. The ADP of the F-actin was no longer tightly bound (or inaccessible) but could be substituted by ATP when this was added.

About the same time Grubhofer & H. H. Weber (1) published an important paper in which they described the results of rapid depolymerisation of F-actin in salt-free solution by means of a few seconds treatment in a Teflon homogeniser. The resulting G-actin contained bound ADP; it could polymerise completely on salt addition, although more slowly than G-actin–ATP, without splitting off of phosphate. The affinity of the G-actin for ADP was low and it was soon lost spontaneously or very rapidly on treatment with an exchange resin. The nucleotide-free actin would not polymerise; if ATP was added quickly it regained the power, but with delay it became irreversibly denatured. Hayashi & Rosenbluth (1, 2) using Grubhofer & Weber's method of G-actin–ADP preparation, found its polymerisability slow at 0° but at room temperature quite similar to that of G-actin–ATP. It should be noticed that Drabikowski, Maruyama, Kuehl & Gergely (1) found the viscosity of this rapidly depolymerised actin was higher than that of G-actin and also that the sedimentation diagram contained a faster sharp boundary. It was now clear that, contrary to the general opinion since the experiments of Straub and his collaborators, splitting of ATP was not necessary for polymerisation. Two cases showing this had indeed already been described: A. G. Szent-Györgyi (3) had found that actin

---

[1] Asakura, Hotta, Imai & Ooi (1); Asakura & Oosawa (1).
[2] Also Oosawa, Asakura & Ooi (1).

depolymerised in 0.6 M KI could repolymerise simply on dilution provided ATP *or* ADP was added before depolymerisation; and Prágay (1) had described a protein fraction obtained from dried muscle residue by very short extraction which polymerised normally yet split no ATP and on depolymerisation gave a G-actin without ATP content.

The possibility had often been discussed that the G-actin monomers were held together in F-actin by the ADP prosthetic group. Such suggestions are found for example, in the work of Tsao & Bailey (1), Bárány, Spiró, Köteles & Nagy (1), and Asakura & Oosawa (1). Grubhofer & Weber now considered that the only argument for a covalent bond between the monomers had disappeared with the finding that a reaction involving ATP dephosphorylation was dispensable in the polymerisation process; the reversible dependence on the ionic conditions argued for polymerisation by means of electrostatic forces, and they pictured the ADP as held by steric interference, the monomers lying too thickly to allow its escape. This point of view was further justified when, in 1964, Bárány, Koshland, Springhorn, Finkelman & Therattil-Anthony (1), polymerising actin in presence of $H_2^{18}O$, found that the nucleophilic attack was on the terminal P atom of the ATP. When ADP was liberated from F-actin–ADP by trichloracetic acid in presence of $H_2^{18}O$, no labelling was found in the ADP. Covalent reaction of the ADP with actin seemed to be thus precluded as the cause of the tightness of binding. As early as 1952 Mommaerts (16) had argued for the electrostatic nature of the intermolecular reaction.

In 1962 Kasai, Asakura and Oosawa (2) were ready to agree that, although a sufficient amount of ATP is necessary for the maintenance of G–F equilibrium in actin solution (with concomitant ATP dephosphorylation) this dephosphorylation was not necessarily coupled to bond formation between monomers. In 1963 Kasai & Oosawa (1) obtained by prolonged dialysis preparations of F-actin with low nucleotide content – sometimes as little as 10 % of the normal amount, which still had normal optical and rheological properties. This form of F-actin could only be obtained if the protein contained bound Ca^{2+} not Mg^{2+}. It was concluded that the ADP group in normal F-actin did not participate in bonding of monomers, but existed confined in a sort of cage made of the polymerised structure, the bonds between G-actins acting as shutters to this cage. Thus depolymerisation and sonic vibration could open shutters, while it would seem that ATP (which accelerates polymerisation) was a rate-regulating factor for bond formation, the fast closing of the shutters being accompanied by ATP splitting. Kasai, Nakano & Oosawa (1) have since found that polymerisable G-actin free from both nucleotides and divalent metal, could be obtained by EDTA–Dowex^{-1} treatment in 50 % sucrose solution under carefully controlled conditions, although in water nucleotide-free monomers immediately lost polymerisability. The function of the nucleotide was considered to be (a) stabilisation of the G-actin structure; (b) regulation in some way, possibly coupled with dephosphorylation, of the rate of bond formation between the monomers, without itself being included in the bond.

Evidence has accumulated that certain other nucleotides may in some conditions replace adenine nucleotides in reactions with actin.[1] For example Iyengar & H. H. Weber found that, although the affinity of other nucleotides for G-actin was less than that of ATP (in the order ATP > ITP > UTP and ADP > CTP, GTP and ITP), yet with any of these nucleotides at a concentration of 10^{-4}M, 2×10^{-5}M G-actin solution (myokinase-free) was completely polymerised within 10 min. Inorganic phosphate was split off from each of the triphosphates.

In 1962 Nagy & Jencks (1) made a study of the optical rotatory dispersion of G-actin. Their purest preparation (assessed by amount of inorganic phosphate liberation on polymerisation to be 98% pure) gave rotatory values from which an α-helix content of 29% was calculated; removal of ATP by means of EDTA lowered the rotatory values, but still not to the level characteristic of a completely random coil; this latter change (reversible if the treatment was not prolonged) only happened with high concentrations of urea or guanidine HCl. A structural change resulting from ATP removal was also made evident by the much greater subsequent susceptibility of the protein to attack by trypsin or chymotrypsin. Nagy & Jencks summed up the situation in this way:

$$\text{G-actin-ATP} \underset{\text{rapid}}{\rightleftharpoons} \text{ATP} + \text{G-actin}$$
$$\downarrow \text{rapid in absence of ATP}$$
$$\text{G''-actin} \underset{\text{urea}}{\rightleftharpoons} \text{G'-actin}$$

They pictured the ATP in G-actin as perhaps holding the protein together by bridging through helices or folds of protein, in the sort of way suggested by Shifrin & Kaplan (1) for co-enzymes.

Later, in 1965, Higashi & Oosawa (1) studied the ultraviolet absorption spectra of G- and F-actin. The difference spectra at 270–310 mμ suggested that tyrosine and tryptophan residues which in G-actin were exposed to the solvent, in F-actin formation became folded inside the polymer. The difference spectra below 240 mμ indicated conformational change in the polypeptide backbone – possibly an increase in α-helix.[2] These intramolecular changes ran closely parallel to the intermolecular changes measured by increase in flow birefringence during polymerisation. When nucleotide

---

[1] Bárány, Spiró, Köteles & Nagy (1); A. G. Szent-Györgyi (3); Martonosi & Gouvea (1); Bárány & Finkelman (2); Martonosi (1).

[2] Oosawa, Asakura, Higashi, Kasai, Kobayashi, Nakano, Ohnishi & Taniguchi (1).

(whether ADP or ATP) was bound to the G-actin, there was a similar but smaller spectral shift, indicating perhaps an intramolecular structure of the G-actin intermediate between that of the nucleotide-free monomers and that of the monomers in the F-actin molecule. Nagy (1) has also found that there was no difference in optical rotatory dispersion between G-actin–ADP and G-actin–ATP.

So far we have concentrated attention mainly on the part played by adenine nucleotide in actin transformations; but much work has also been done on the role of divalent metals and of SH groups. Two generalisations emerge: first that the binding of ATP and divalent metal and the intactness of the SH group were often found to be interdependent;[1] secondly, that heterogeneity of the SH groups with regard to their role could be detected.

The first investigation of SH groups in actin was that of Bailey & Perry (1) in 1947. On treatment of actin with dilute iodine the nitroprusside reaction disappeared when 0.8–0.9 % apparent cysteine had been oxidised; actomyosin formation with this actin was inhibited only 5 %. Similarly, a concentration of H_2O_2 which gave 75 % inhibition of actomyosin formation when applied to myosin, gave only 10% when applied to actin. The cysteine content found in this work corresponded to about 4 moles/mole of G-actin, and was the value also found in the analysis of Kominz, Hough, Symonds & Laki (1), but later work as we have seen above put it at 6–7 moles/mole of actin.

The first studies of the importance of the SH groups in polymerisation were made by Kuschinsky & Turba between 1950 and 1952.[2] They found inhibition of polymerisation by treatment with Salyrgan or Cu-glycine, reversible by cysteine. Salyrgan-treated F- or G-actin could combine with myosin.[3] Tsao & Bailey (1) observed that only two of the actin SH groups could react with N-ethylmaleimide (NEM) and that this treatment left polymerisation unaffected; they confirmed the finding of Kuschinsky & Turba that treatment with a heavy metal reagent (in their case chlormercuribenzoate) abolished the power to polymerise. A little later Bárány, Spiró, Köteles & Nagy (1, 2) confirmed the heterogeneous nature of the SH groups indicated in the work of Tsao & Bailey, finding that in F-actin there were two less SH groups per 57000 of actin capable of reacting with Salyrgan. This could be due to steric hindrance, but might well indicate reaction during the polymerisation of two of the SH groups. It was also found that the presence of 6×10^{-3}M ATP (especially if Mg^{2+} were also present) could protect F-actin from depolymerisation by various agents – Salyrgan, urea, etc. The possibility was discussed that the actin monomers were held

---

[1] For a recent study in this field see Kitagawa, Drabkowski & Gergely (1).
[2] Kuschinsky & Turba (1, 2): Turba & Kuschinsky (1).
[3] For the part played by SH groups of both actin and myosin in actin/myosin interaction see chs. 8 and 12.

together by H-bonds between SH and NH_2 groups, the latter possibly on the protein itself, possibly on the nucleotide.

It became apparent that the loss of polymerisability caused by treatment with mercurials was often accompanied by loss of the ability of G-actin to bind ATP. This was found by Bárány, Nagy, Finkelman & Chrambach (1); by Martonosi & Gouvea (2); and by Strohman & Samorodin (1). Tonomura & Yoshimura (1) in a study of effects of p-chlormercuribenzoate on actin found that two SH groups could react instantaneously with PCMB, and also react with NEM, without effect on polymerisation. The slower reaction with PCMB of the remaining SH groups led to inhibition of G-actin polymerisation, or to depolymerisation of F-actin, as well as removal of the ATP and much of the Ca^{2+} from the G-actin. Drabikowski and his collaborators[1] further found that there were conditions in which polymerisability could be lost, while ATP binding was not disturbed to the same extent. They therefore considered that three classes of SH groups should be distinguished – those rapidly reacting; those taking part in the slower reaction involving both inactivation of G-actin and loss of ATP; and a third group reacting slowly but more directly involved in ATP binding than in polymerisation. About the same time Katz & Mommaerts (1) made a very thorough investigation of the number and behaviour of actin SH groups, coming to conclusions compatible with those of Drabikowski *et al.* They found 6 groups/mole of 60000 and the number did not increase in concentrated urea solution. Titration with NEM gave only two groups in G-actin or in the F-actin monomer; but on amperometric titration with silver ions six were found in G-actin, only four in F-actin. Blocking of four SH groups with Ag in G-actin left polymerisation intact, but blocking of five prevented polymerisation. Katz & Mommaerts classified the SH groups as follows: (1) two reacting with NEM and very rapidly with p-chlormercuribenzoate or Ag^+; (b) two which can react with Ag^+ or Hg^{2+} without loss of polymerisability; (c) two slowly reacting groups, the integrity of which was necessary for the G $\longrightarrow$ F transformation.

Other researches concerned particularly with the role of bound Ca^{2+} must be mentioned. Hasselbach (5) in 1957 had confirmed the early observation of the binding of Ca^{2+} to actin. In fibrils 60 % of the Ca^{2+} was bound to actin, 35 % to myosin; 50 % of the Mg^{2+} was bound to actin and 50 % to myosin. In 1961 Martonosi & Gouvea (1) made the interesting observation that treatment of G-actin with EDTA led to inhibition of polymerisation accompanied by loss of the bound ATP. Strohman & Samorodin (1) and Tonomura & Yoshimura (2) also independently showed the same thing; the latter workers demonstrated that the bound Ca^{2+} was actually lost during the EDTA treatment. Maruyama & Gergely (1) confirmed this, and added

---

[1] Drabikowski, Kuehl & Gergely (1); Drabikowski & Gergely (2).

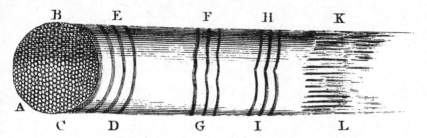

Fig. 5 (see p. 23). One of Leeuwenhoek's drawings. He described it as follows: 'These fibres being divested of their membranes I saw in them the same turnings, foldings and wrinkles, as I have before noted...; sometimes they appeared as BECD; at another time as at FG and also as at HI. Upon cutting these fibres transversely, I very plainly saw the ends of those multitudes of filaments of which the inside of a fishy fibre is composed.' (Leeuwenhoek (2).)

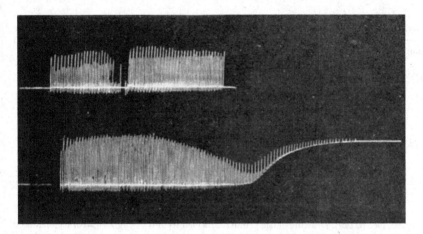

Fig. 13 (see p. 87). The upper curve is for contractions of normal muscle (the break being due to a short technical hitch interfering with the lever); the lower curve is for IAA-poisoned muscle. (Lundsgaard (1).)

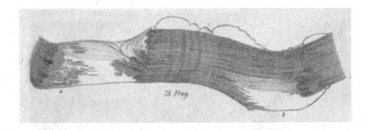

Fig. 18. (see p. 128). Drawing of frog muscle 'primitive fasciculus' contracting in water, the two ends being fixed. Magnified 300 × . (Bowman (1).)

(facing page 224)

Fig. 19. (see p. 128). Drawing of fractured primitive fasciculus from the Boa; showing the transverse striae and the fibrillae presenting a series of light and dark points. Magnified 300 ×. (Bowman (1).)

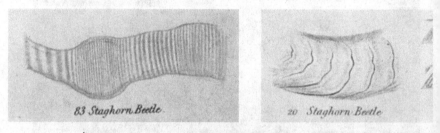

Fig. 20. (see p. 129). Drawing of the primitive fasciculus of the Staghorn Beetle contracted in different degrees. 'The thickening of the fasciculus everywhere corresponds with the amount of the approximation of the striae.' Magnified 300 ×. (Bowman (1).)

Fig. 21 (see p. 129). Drawing of 'unusual appearance of the striae in a fasciculus of the Staghorn Beetle, observed immediately after death'. Magnified 300 ×. (Bowman (1).)

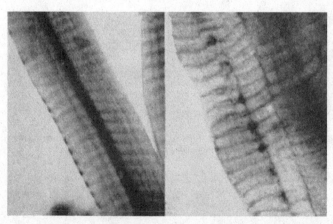

Fig. 29 (see p. 172). The ultraviolet absorption spectra (wavelength 2570 Å) of living muscle fibres from *Drosophila*. Left: strongly working fibre; right: feebly working fibre. (Caspersson & Thorell (1).)

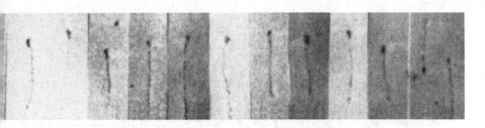

Fig. 32 (see p. 205). Myosin molecules from shadowed preparations.
Magnification 100000 ×. (H. E. Huxley (1).)

Fig. 33 (see p. 205). A preparation of L-meromyosin, as seen by the shadow-casting technique. Magnification 30880 ×. (H. E. Huxley (1).)

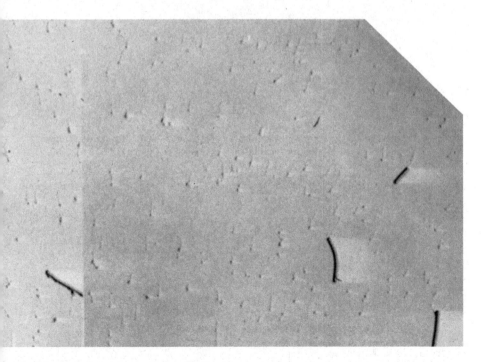

Fig. 34 (see p. 205). A preparation of H-meromyosin, as seen by the shadow-casting technique. Magnification 38600 ×. (H. E. Huxley (1).)

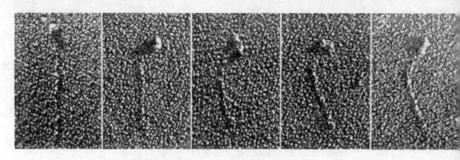

Fig. 35 (see p. 209). Rotary shadowed molecules of myosin. Magnification 175000 ×. (Slayter & Lowey (1).)

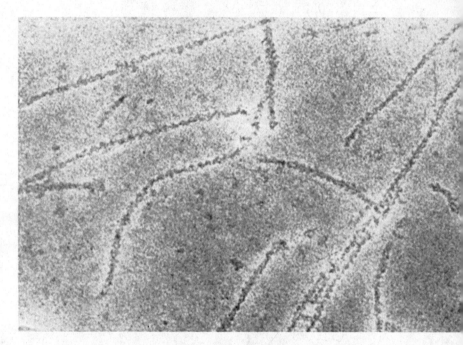

Fig. 36 (see p. 226). Actin filaments negatively stained with uranyl acetate. Magnification 425000 ×. (Hanson & Lowy (1).)

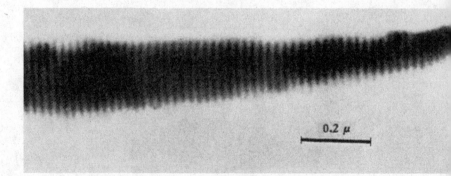

0.2 μ

Fig. 38 (see p. 231). Tropomyosin paracrystal, stained with phosphotungstic acid. Protein prepared from pig stomach muscle. (Tsao, Kung, Peng, Chang & Tsou (1).)

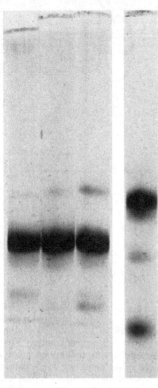

(1) (2) (3)

(a)   (b)

Fig. 39 (see p. 212). (a) Polyacrylamide gel electrophoresis of carboxymethylated light chains removed from myosin by DTNB treatment. (1) In the absence of urea; (2) in the presence of 1 M urea; (3) in 2 M urea. (b) Gel electrophoresis of carboxymethylated light chains removed from DTNB-treated myosin by alkali. (Weeds (3).)

Fig. 40 (see p. 239). Myofibrils photographed in phase contrast. Magnification 4000×. 1–4: the same four sarcomeres of one fibril photographed during contraction induced by ATP from rest length down to 50 % rest length, when contraction bands have formed. 5 and 6: stretched fibril (115 % rest length) before (5) and after (6) extraction of myosin. 7–9: fibrils after extraction of myosin. 7: rest length; 8: 90 % rest length; 9: 75 % rest length. (Huxley & Hanson (1).)

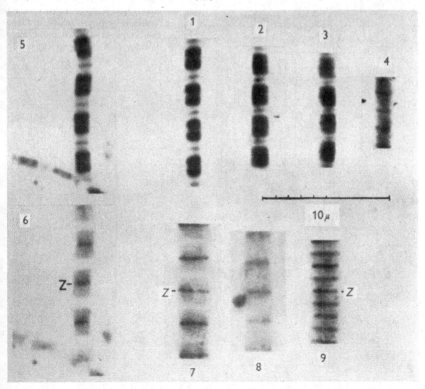

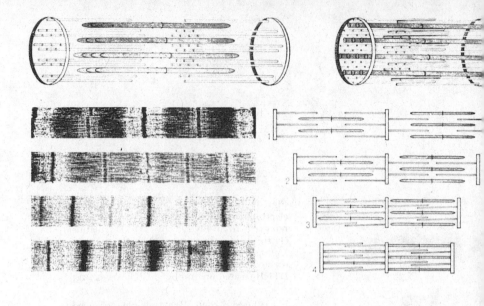

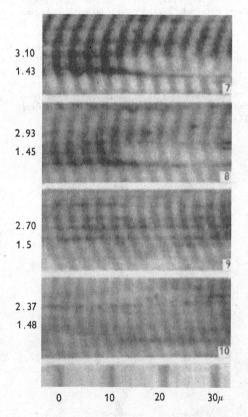

Fig. 41 (see p. 240). Schematic illustrations of fibril in longitudinal section, at consecutive stages of contraction. First the H zone closes (1,2), then a new dense zone develops in the centre of the A band (3,4) as thin filaments from each end of the sarcomere overlap. (H. E. Huxley (14).)

Fig. 42 (see p. 240). Muscle fibre during short isotonic contraction. Positive compensation (A bands dark). In each case the upper figures refer to the sarcomere length in μ, the lower to the A band length. (A. F. Huxley & Niedergerke (2).)

3.10
1.43

2.93
1.45

2.70
1.5

2.37
1.48

0   10   20   30μ

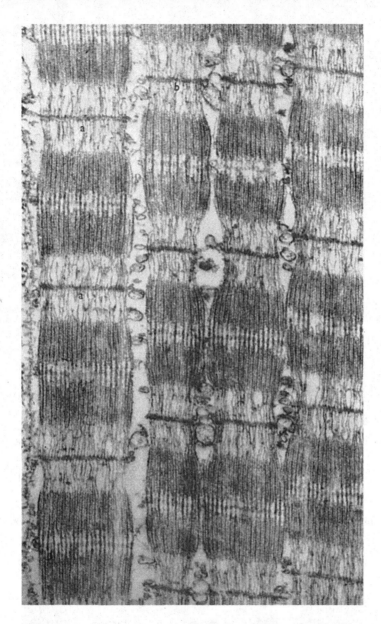

Fig. 43 (see p. 244). Low power view in the electron microscope of a thin section through striated muscle fibres. Examples of sections parallel to the 10$\bar{1}$0 lattice planes are marked (*a*), and of sections parallel to the 11$\bar{2}$0 lattice planes (*b*). The former show simple alternation of primary and secondary filaments; the latter show two secondary filaments between each pair of primary filaments. Magnification 45 000 ×. (H. E. Huxley (5).)

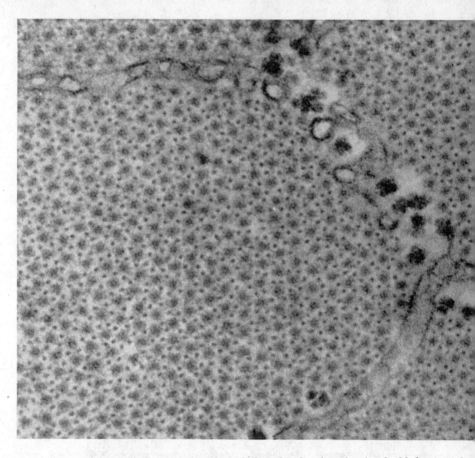

Fig. 44 (see p. 244). Cross-sections through the *A* band region, showing double hexagonal array of primary and secondary filaments, and the cross-bridges between them. Magnification, 150 000 ×. (H. E. Huxley (5).)

Fig. 45 (see p. 245). Heavily contracted muscle, showing double overlap of the thin filaments. Magnification 150 000 ×. (H. E. Huxley (15).)

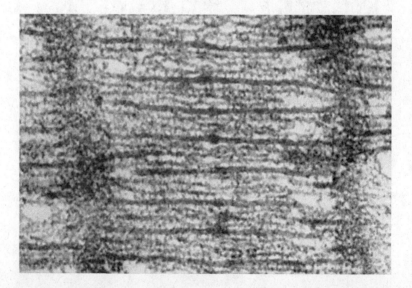

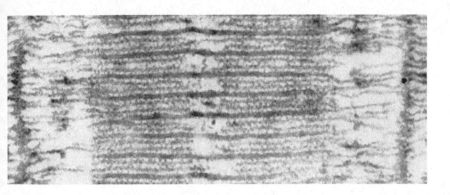

Fig. 46 (see p. 244). Section through a sarcomere in the 1120 direction showing thin filaments passing from the Z line into the A band. Magnification 100000 ×. (H. E. Huxley (5).)

Fig. 49 (see p. 249). I segment treated with HMM. The thin filaments now show the characteristic polarised structure, and all filaments on the same side of the Z line point in the same direction. Those on the other side of the Z line point in the opposite direction. Magnification 87200 ×. (H. E. Huxley (1).)

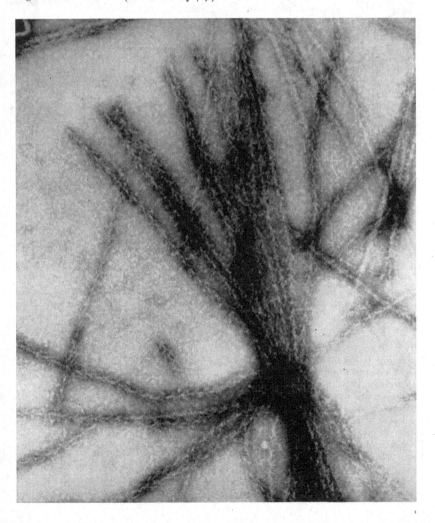

Fig. 50 (see p. 249). Synthetic myosin filaments showing characteristic pattern of bare central shaft and projections all along the rest of the filament. Magnification 145000 ×. (H. E. Huxley (1).)

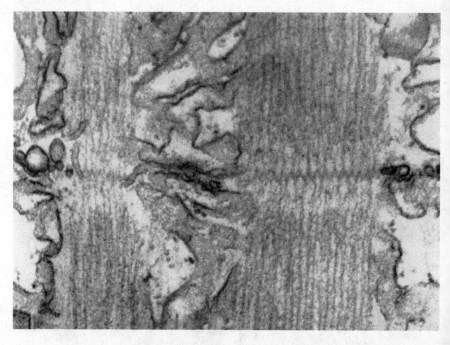

Fig. 51(a) (see p. 250). Longitudinal section of fibril in the region of the Z line. A clear zigzag is apparently formed by the branching of each actin filament into two dense lines which form an angle with it. Magnification 54000 ×. (Franzini-Armstrong & Porter (2).)

(b) (see p. 250). A drawing interpreting the structure of the Z disc. Viewed in the direction of the single arrow, the thickenings of the membrane are disposed to form a network; viewed in the direction of the double arrow, the actin filaments appear to pass straight through the Z disc. (Franzini-Armstrong & Porter (2).)

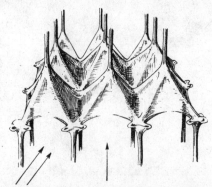

Fig. 52(a) (see p. 250). M bridges and M filaments in cross-section. Six cross-cut M
filaments marked by arrows. Magnification 250000 ×. (Knappeis & Carlsen (2).)

(b) (see p. 250). Diagrams to show arrangement of A filaments (A), M filaments
(Mf) and M bridges (Mb); in the lower one the actin filaments are also shown as white
circles. (Knappeis & Carlsen (2).)

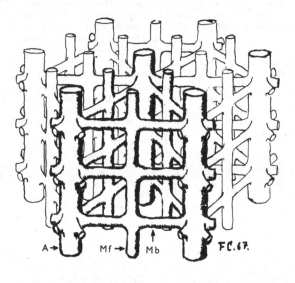

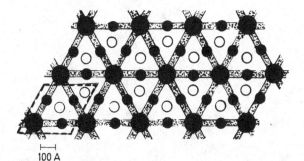

100 A

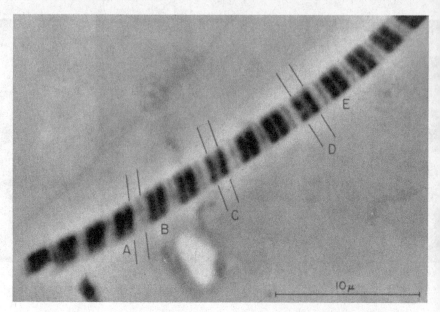

Fig. 57 (see p. 261). Myofibril which has been irradiated with 270 mμ ultraviolet light for 30 sec in the indicated areas and subsequently contracted with ATP. The two full *A* band irradiations (C and D) which prevented contraction permit comparison of contracted (E) and non-contracted sarcomeres. The sarcomere in which *I* filaments were released (see text) at the *Z* line shows a normal *A* band with shortened *H* zone (B), but the half *I* band, where the filaments were irradiated, has not shortened. The adjacent sarcomere, receiving full half *I* band irradiation and partial *A* band irradiation owing to the microbeam width (A), shows an asymmetrically positioned, shortened *H* zone. (Stephens (2).)

Fig. 73 (see p. 332). Electron micrograph of muscle from swim-bladder of *Opsanus*, cut longitudinally. The triads consist of a slender intermediate tube, about 30 mμ in diameter, flanked by two larger lateral channels about 100 mμ in diameter. These run across the myofibril in planes perpendicular to the page. The longitudinally oriented branches of the two triads in the same sarcomere are continuous in the region of the *H* band, but often appear to be discontinuous at the level of the *Z* band (arrows). Magnification 60 000 ×. (Fawcett & Revel (1).)

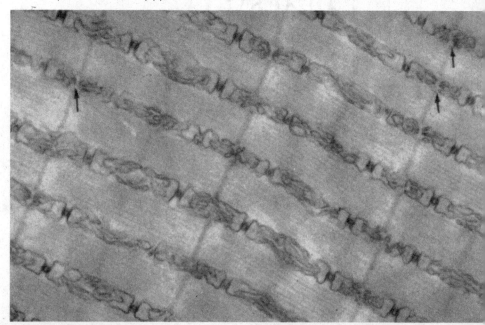

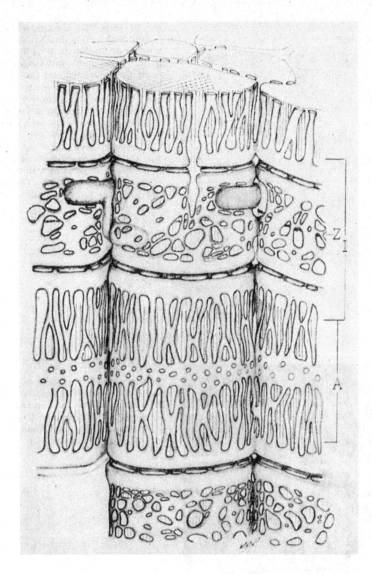

Fig. 72 (see p. 332). A schematic summary and interpretation of observations on the sarcoplasmic reticulum of rat sartorius muscle. A complete sarcomere with associated reticulum is pictured as part of the central fibril. Sarcosomes, shown in less than the normal number, occupy a well-defined position relative to the Z line. They are pictured with their long axes oriented circumferentially with respect to the fibril and covered or not by the close reticulum of the *I* band region. Magnification *ca.* 32000 × . (Porter & Palade (1).)

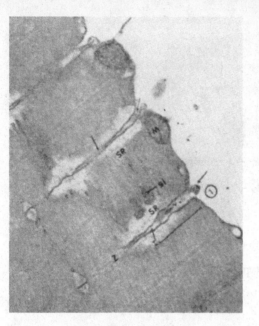

Fig. 74 (see p. 333). Electron micrograph of longitudinal section through the segment of muscles of *Molliensis* fixed in glutaraldehyde. The section is just tangential at one point to a myofibril and thus provides a face view of the sarcoplasmic reticulum (*s.r.*). The sarcolemma borders, at the right side of the photograph, the periphery of the fibre, and can be followed (arrow) as it penetrates at the *Z* line level (*z.*) between the larger terminal sacs of the *s.r.* to form the *T* (*t.*) system. Strands of cytoplasm remnants of connections with the adjacent fibre, penetrate within the *T* system. The line indicates 1 μ. (Franzini-Armstrong & Porter (2).)

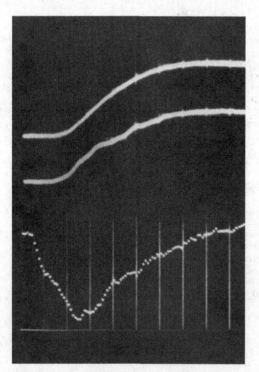

Fig. 75 (see p. 336). The time relations between tension development (upper trace), light scattering (middle trace) and the kinetics of the Ca-murexide complex (measured by absorption at 470 mμ). Time scale: 25 msec/division. Toad sartorius 12°. (Jöbsis & O'Connor (1).)

Fig. 85(a) (see p. 391). Negatively stained heart mitochondria. Flattened mitochondria membranes showing regular arrangement of repeating particulate components 62000 × l.

(b) (see p. 391). Profile view of cristae with arrays of elementary particles (EP) in enlarged segment of 85(a), 420000 ×. (Fernández-Morán, Oda, Blair & Green (1).)

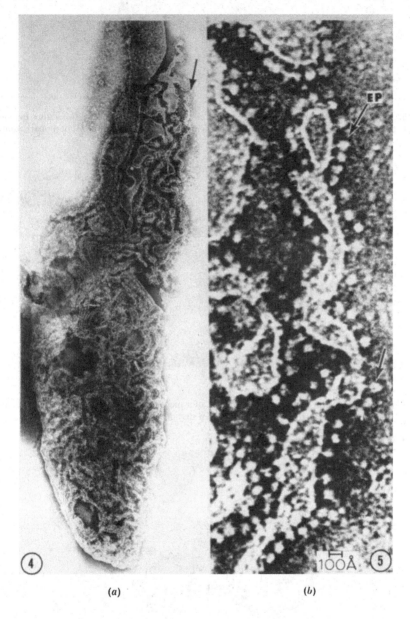

(a)   *(b)*

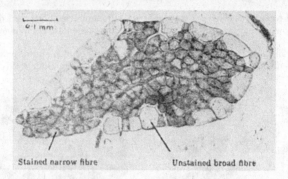

Stained narrow fibre   Unstained broad fibre

Fig. 90 (see p. 455). (*a*) Photomicrograph of a transverse section of a fasciculus in the pectoralis major muscle of the pigeon, stained with sudan black B for fat demonstration.

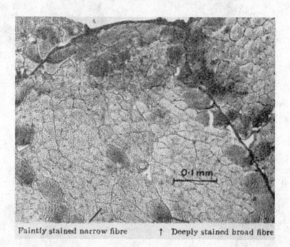

Faintly stained narrow fibre   ↑ Deeply stained broad fibre

(*b*) Photomicrograph of a transverse section of a piece of pectoralis muscle of the pigeon, stained with Best carmine for glycogen demonstration. (George & Naik (1).)

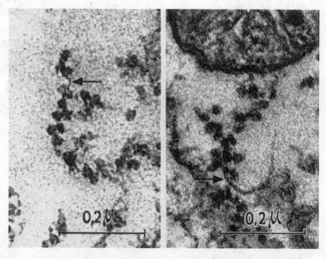

Fig. 95 (see p. 477). Polysome. A thick filament 100 Å in diameter is attached to the extremity of an aggregate of ribosomes. (See arrows.) Magnification 126 000 × . (Henson-Stiennon (1).)

Myosin, actin and tropomyosin 225

that the loss of polymerisation was much slower than the Ca^{2+} removal. Bárány and his collaborators[1] in 1962 studied the stability of the Ca^{2+} binding. After dialysis of G-actin against excess of Ca^{2+}, followed by repeated treatment with Dowex 50, one g atom of bound Ca^{2+} was found per 62000 g of actin. This was exchangeable against labelled Ca^{2+}, but only partly replaceable by dialysis against Mg^{2+}. With F-actin the bound Ca^{2+}, like the bound nucleotide, was much more difficult to remove than with G-actin; during sonic vibration, however, in presence of ATP it could be removed by EDTA. The actin had now lost its property of reversible depolymerisation, but in 0.1 M KCl it maintained its viscosity and its power to react with myosin; it still contained about one mole of nucleotide per mole of actin monomer. Significant also for questions of the stability of the actin molecule were the observations of Asakura (1) that, with G-actin depleted of ATP by resin treatment, lifetime was doubled by adding 0.3 mM Ca^{2+}; and of Strohman & Samorodin (1) that the exchange of bound ATP was much slowed by addition of 5×10^{-5} M Ca^{2+}.

Many suggestions were of course made that the ATP could be bound directly through SH groups or through the bound Ca^{2+} or in a two-point binding by both; but the possibility was also generally recognised that reaction of the SH groups with added reagents might so change the structure of the molecule as to decrease its power to bind ATP and divalent metals. Both calcium and nucleotides seem essential for the integrity of the actin structure.

PARTICIPATION OF OTHER GROUPS IN ACTIN POLMERISATION. By specific modification of a small number of histidine and tyrosine residues Martonosi & Gouvea (2) were able to affect the polymerisation of actin. Thus illumination of actin at slightly alkaline pH in presence of methylene blue (conditions which favour the oxidation of histidine) led quickly with G-actin to loss of power to polymerise, and with F-actin to depolymerisation. The oxygen uptake corresponded to oxidation of two groups per actin molecule. The actin retained its power to combine with myosin (as shown by production of protein insoluble at 0.06 M KCl) and to activate the Mg^{2+} ATPase of myosin. Reaction of one tyrosin group per molecule with diazotised sulphanilamide caused complete loss of capacity to polymerise and to react with myosin.

On the other hand Tonomura, Tokura & Sekiya (1) found treatment of G-actin with trinitrobenzenesulphonate did not stop polymerisation, but did prevent actomyosin formation. Combination of one mole of the reagent with one mole of protein took place rapidly, and evidence from the

---

[1] Bárány & Chrambach (1); Bárány & Finkelman (1); Bárány, Finkelman & Therattil-Anthony (1).

change in absorption spectrum suggested that the ϵ–NH_2 group of lysine had reacted.

X-RAY DIFFRACTION AND ELECTRON-MICROSCOPE STUDIES ON ACTIN. In 1945 Bear (1), after a wide comparative study of dried muscle by means of X-ray diffraction, reported the occurrence of fibrils of two distinct structural patterns, which he called Type I and Type II. Type I, which has since been identified as that of paramyosin,[1] was restricted in its occurrence but all muscles examined contained Type II. This gave a series of moderate- and small-angle reflections on or near the meridian, indicating an axial period of either 350 or 420 Å; it was concluded that this pattern was given by the 'myosin' of that time.[2] However, Astbury (4) in 1947 showed that these reflections were given by isolated films of F-actin, which did not give the well-known α-pattern. Thus the full muscle-diagram was composed of the wide-angle α-pattern of myosin proper, together with the moderate-angle pattern of actin. These results were confirmed by Cohen & Hanson (1).

Electron-microscope examination of actin, began, as we have seen,[3] with the work of Jakus & Hall; Astbury, Perry, Reed & Spark (1) also about the same time photographed the change from globular to fibrillar formation. In neither of these cases was any structure visible in the fibrils formed. An advance was made when in 1949 Rozsa, Szent-Györgyi & Wyckoff (1) found that the single actin filaments produced on gradual polymerisation of very dilute actin solutions seemed to be associations of ellipsoidal rodlets, joined obliquely end to end; these were much too large (300 Å by 100 Å) to be individual G-actin molecules. By the middle of the nineteen fifties it was known that the muscle fibril contains two types of filament and that the thinner of these contains actin.[4] In 1963 Hanson & Lowy (1, 2) obtained electron micrographs of the thin filaments isolated from various types of muscle (see fig. 36, pl.). By negative staining it was clearly seen that each filament consisted of two helically wound strands composed of similar subunits, approximately spherical. The crossover points of the two strands were spaced at 340 Å along the filament, the number of subunits per turn of each strand being very close to thirteen. From its size and probable density (1.3) the weight of each sub-unit was calculated to be 68400: so each sub-unit was identified as a G-actin molecule. Exactly similar appearances were given by filaments prepared from isolated F-actin. The apparent sub-units of Rozsa et al. were probably portions of the filament between two crossover points. By negative-contrast electron microscopy Rowe (1) has recently got pictures of individual molecules of G-actin,

[1] P. 234.   [2] E.g. Hall, Jakus & Schmitt (1).
[3] P. 191.   [4] Ch. 11.

showing the length to be 105 Å, the width 27–37 Å, thus the axial ratio about three.

The electron-microscope results of Hanson & Lowy should be correlated with the descriptions of large axial spacings earlier found, with the X-ray diffraction method, by Selby & Bear (1) and by Worthington (1, 2) for actin filaments *in situ* in various types of muscle. These workers considered a possible helical structure; the former, from their values for the length of the unit cell of either 350 Å or 406 Å, were uncertain between thirteen and fifteen residues per turn; the latter preferred fifteen per turn.

Very recently Huxley & Brown (1) with greatly improved apparatus and very well oriented frog-sartorius muscle concluded, from their low-angle X-ray diffraction pattern, that there was compelling evidence for the value of 2×360–2×370 Å for the pitch of the double helix. This excludes the possibility of thirteen, fourteen or fifteen sub-units per turn, and necessitates the acceptance of a non-integral double helix. It is relevant that Hanson (1) has shown that the value obtained in the electron microscope for the axial period of actin filaments (both natural and synthetic) differs with the fixative and electron stain used. Thus while the value given earlier by Hanson & Lowy was confirmed by the method they used, higher values (about 380 Å) were found with specimens prepared in uranyl acetate.

Huxley & Brown also described a 380 Å meridional reflection, assigned to the thin filaments but clearly not due to actin. They discuss its possible relation to the additional protein factors known to be associated with actin – particularly tropomyosin and troponin.[1]

We see then that some twenty-five years of endeavour have given a good idea of the nature of the actin filament, and of the varying degrees of polymer formation of which this protein is capable. The involvement of adenine nucleotides and divalent cations in these structural changes is clear, as well as that of SH groups in the polymerisation and in the vital matter of interaction with myosin. It seems however that, though reversible changes in actin structure have from time to time been suggested as taking part in the contraction process, there is no striking evidence for such changes. It may be that the importance of the facts accumulated lies rather in defining the conditions under which the actin filaments can maintain the integrity necessary for the playing of their role.

TROPOMYOSIN

DISCOVERY, PREPARATION AND PURIFICATION. In 1946 Bailey (6)[2] isolated from skeletal and cardiac muscle a fibrous protein which he justly described as unusual. It was prepared by washing away the sarcoplasm of

---

[1] Ch. 12.   [2] Also Bailey (2).

finely minced muscle and dehydrating the residue by means of ethanol and ether. The extract obtained by soaking the residue in M KCl at pH 7.0 contained this new protein, which he called 'tropomyosin', and other proteins which could be precipitated at pH 4.5. When this precipitate was dispersed in water at pH 7.0 the tropomyosin passed into solution, but much of the other protein was denatured and insoluble. Such of it as dissolved could be removed by precipitation at 40 % saturation with $(NH_4)_2SO_4$, while the tropomyosin only came down between 40 % and 70 % saturation. It was purified by repeating the isoelectric precipitation and salting-out process; the yield was about 2.6 % of the total muscle protein. This manner of preparation indicated that the tropomyosin was contained in the fibrillar and not the sarcoplasmic regions of the muscle. This was confirmed by Perry (6) when in 1953 he isolated tropomyosin from washed myofibrils by prolonged extraction with 0.075 M borate at pH 7.1; Perry & Corsi (1) found that 10–12 % of the myofibrillar protein was tropomyosin.

Although tropomyosin is water-soluble, its extraction from rabbit muscle by means of water or dilute salt solution was only possible if the muscle was in a highly divided state. Tsao (quoted by Bailey (3))[1] thus found that in Hasselbach & Schneider's method for actin extraction the tropomyosin also appeared in the extract. Hamoir (2, 3) who was engaged in 1951 in an ultracentrifugal study of fish myosin, observed that two proteins with the same solubility properties as tropomyosin could be extracted; this he did by means of salt solution at ionic strength 0.45, pH 5.0, without previous treatment with organic solvents. One of these proteins (with sedimentation coefficient of 2.5) seemed to be identical with Bailey's tropomyosin; the other, extracted in larger amount, had a higher sedimentation coefficient – 3.6–5.1. Acting on a suggestion from Bailey, Hamoir found the latter to be an association of tropomyosin with pentose nucleic acid. Nucleotropomyosin had similar solubility to tropomyosin; the percentage of nucleic acid varied from one preparation to another between 15 and 20 %, and on standing at pH 3.5 partial dissociation took place, leaving a fraction richer in nucleic acid. By Hamoir's method the yield of tropomyosin was only 0.1 % of the wet weight; but on treatment of the muscle residue by Bailey's method, a further 0.3 % (practically free from nucleic acid) was extracted, and the yield was thus raised to a value comparable to that from rabbit muscle (0.47 %).

The question arose whether the nucleotropomyosin existed as such in the muscle, being dissociated by the treatment with organic solvents and strong salt solution; or whether it was formed as an artefact during the extraction. Hamoir inclined to the former view. However Perry & Zydowo (1, 2) isolated chromatographically a ribonucleoprotein in small amount

[1] Bailey (3) p. 1011.

from well-washed myofibrils. The base analyses of the RNA were similar to those given by Hamoir (3) for the RNA component of fish nucleotropomyosin, but in their nucleoprotein little or no tropomyosin was present, as judged by viscous behaviour.[1] They were of the opinion that nucleotropomyosin should be regarded as a complex of tropomyosin with a ribonucleoprotein, rather than with ribonucleic acid alone. Sheng & Tsao (1) by extracting finely minced muscle with M NaCl (containing 0.01 M sodium pyrophosphate) obtained extracts from which nucleotropomyosin could be isolated by slight modification of Bailey's isoelectric and $(NH_4)_2SO_4$ precipitations. Sheng, Tsao & Peng (1) found that the two components of such nucleotropomyosins, prepared from various types of muscle, migrated independently, whether in the Tiselius apparatus or on paper electrophoresis; they took this to mean that the nucleic acid and tropomyosin did not exist in the tissue in the combined state. Later work on the function and location of tropomyosin supports this conclusion.

It is interesting to notice that, though in the preparation of tropomyosin the method of dealing with the muscle somewhat resembles that worked out by Straub for actin preparation, tropomyosin was discovered before the isolation of actin was reported in England.[2] The tropomyosin prepared by this method was homogeneous by electrophoretic and sedimentation behaviour in contrast to actin which, as we have seen, proved very difficult to purify and especially to free from tropomyosin. The resistance of tropomyosin to denaturation, in the sense that it is not rendered insoluble by treatment with organic solvents, heat or acid, is one of its unusual properties, very helpful in purification. Another is its ready crystallisability, though this is inadequate as a purification procedure. When a concentrated solution (2–3 %) at pH 7.0 was dialysed against 1.6 % $(NH_4)_2SO_4/0.01$ M acetate buffer at pH 5.4, so that the pH gradually fell, Bailey obtained large flat square or hexagonal crystals from rabbit skeletal tropomyosin or pig heart tropomyosin. Sheng & Tsao, working with nucleotropomyosin from various animals and types of muscle, obtained in different cases the platelike crystals or, for example from uterus muscle, clusters of fine needles. Both types of crystal have also been obtained by Kominz, Saad, Gladner & Laki (1) with tropomyosin preparations from various sources.

THE SIZE, SHAPE AND METHOD OF POLYMERISATION OF THE TROPOMYOSIN MOLECULE. Tropomyosin is soluble in water at pH above 6.5 or below 4.5, giving extremely viscous solutions. Bailey (2, 6) found that this viscosity fell rapidly on adding salt, being a function of the ionic strength. The particle weight was shown by osmotic pressure measurements to vary from 135000 in 0.06 M phosphate to 53000 in 6 M urea.[3] An

---

[1] See below.   [2] Bailey (3) p. 995.   [3] Tsao, Bailey & Adair (1).

X-ray diffraction study by Astbury, Reed & Spark (1) gave a typical α-diagram. From the intramolecular configuration of an α-type fibre and the minimum particle weight, Tsao, Bailey & Adair (1) calculated that the molecule would have dimensions of length 385 Å, mean width 14.5 Å, axial ratio 26, if the molecule contained two parallel chains. Approximate values calculated from viscosity results, in the same media as those used for osmotic pressure measurements, gave about 25 for the axial ratio.

Bailey (2) had carried out an amino-acid analysis of the pure protein. It was characterised, like myosin, by a high percentage of polar side-chains – 65% for tropomyosin and 57% for myosin. In terms of charges of both positive and negative type, the value was higher than for any other protein analysed. Two interesting points arise here. In the first place, the building up of long fibrils can readily be explained in terms of electrostatic interaction between the charged molecules. The light-scattering data of Kay & Bailey (1), who used solutions at pH 2.0 and pH 12.0 to ensure depolymerisation, confirmed the monomer weight of 53000, and the constancy of the ratio molecular weight; length of particle indicated that, at least up to the hexamer stage, aggregation was end to end. Secondly, the comparison with myosin suggested to Bailey that here we had a prototype of myosin – hence the name; he pictured the possiblity that it might function as a building stone in the elaboration of myosin.

The amino acid analyses of Bailey (by individual determinations), of Kominz, Hough, Symonds & Laki (1) by the Moore & Stein method, and of Jen & Tsao (1) by Sanger's method agree well together for rabbit tropomyosin. All the tropomyosins examined were characterised by absence of tryptophan and very low proline content.

The search for terminal groups makes a long story. Bailey (5) had found no N-terminals and this suggested to him a cyclic structure. But in 1954 Locker (1), using the Akabori and carboxypeptidase methods, reported a single C-terminal group, isoleucine. In 1957 Kominz, Saad, Gladner & Laki (1), using carboxypeptidase, examined the tropomyosin C-terminal groups in a number of different types of mammalian muscle. Their results indicated two C-terminals per molecule – isoleucine and serine in the protein from striated muscle, asparagine and leucine in the uterus-muscle protein. Soon afterwards Jen, Wen & Niu (1) confirmed Locker in finding a single isoleucine C-terminal group in a number of tropomyosins from different sources. They considered that these results disproved the claim that serine was the C-terminal residue of a second peptide chain. This interpretation was apparently accepted by the American workers, as Saad & Kominz (1) in a later paper refer only to a single C-terminal group.

Failure to find N-terminal amino acids was also the experience of Jen & Tsao (1) and of Chibnall & Spahr (1), the latter using methoxycarbonyl chloride as condensing agent, in the hope that this smaller molecule would meet with less steric hindrance than fluorodinitrobenzene. However in 1961, with the idea of disrupting internal H-bonds which might interfere with the activity of the N-terminal groups, Saad & Kominz (1) treated tropomyosin with fluorodinitrobenzene in presence of 6 M urea at 50°. They found one glutamic acid (or glutamine) N-terminal residue per mole. But Jen, Hsü & Tsao (1) could not confirm this. Then in 1966 Alving, Moczar & Laki (1) found on pronase digestion evidence for an N-acetylated peptide chain. Reverting to the previous findings in their laboratory of two C-terminal groups, they now again suggested two chains.

In 1957 Cohen & Szent-Györgyi (1) applied Moffat's theory of the optical rotatory dispersion of helical macromolecules to rabbit tropomyosin. The values found were characteristic of a fully coiled, right-handed α-helix;[1] treatment with 8 M urea caused disappearance of the helical structure. Chang & Tsao (1) found a rather lower helical content – 77 % – which in presence of the H-bond forming agent, chloroethanol, was increased. They confirmed the unfolding of the molecule on urea treatment, and added that the urea effects were largely reversible. Thus the helical content and the polymerisability were restored when the urea was removed, but the re-covered tropomyosin would not crystallise – as had been noted much earlier by Tsao, Bailey & Adair. The preliminary experiments of Cohen & Holmes (1), with frog sartorius and rabbit psoas, gave X-ray diffraction diagrams having coiled-coil near-equatorial intensity distribution; this could be due to light meromyosin and tropomyosin. Also Rowe (1) in 1964 made a preliminary study of monomeric tropomyosin in the electron microscope, using the mica-replication method with platinum shadowing. The particles seen were approximately 400 Å long, and of diameter fitting that expected for a twofold α-helical model. The hydrodynamic and light-scattering studies of Holtzer, Clark & Lowey (1) in 1965 gave the tropomyosin molecule as a rod 490 Å long by 20 Å in diameter and of weight 74000. These values fitted well with the hypothesis of a double-stranded, α-helical coiled coil.

Electron micrographs by Tsao and his collaborators[2] of needle-shaped paracrystals of tropomyosin from various sources showed a main repeat period of 400 Å (see fig. 38 pl.); with gizzard tropomyosin (which had the highest polymerisability and a molecular weight of 150000) an 800 Å period was also seen. They took these periods to be functions of the molecular length. Cohen & Longley (1) had similar results with gizzard tropomyosin paracrystals.[3] Miller (1) in an X-ray examination of dried tropomyosin

[1] Also Simmons, Cohen, Szent-Györgyi, Wetlaufer & Blout (1).
[2] Tsao, Kung, Peng, Chang & Tsou (1); Peng, Kung, Hsiung & Tsao (1).
[3] P. 236.

films found a smaller repeat unit – 228 Å; this probably indicates over-lapping of molecules under these conditions.

Jen & Tsao (1) found some variation in arginine content, which was for example 19.4 % of the total nitrogen in duck-gizzard tropomyosin and only 10.9 % in pig-heart tropomyosin. They suspected some correlation between arginine content and polymerisability. Tan & Tsao (1) indeed showed a little later that by using arginase to detach the amidinyl groups of the arginyl side-chains they could reduce the polymerisability; when 35 % of the amidinyl groups were lost, polymerisation was impossible. On the other hand, 75 % of the ε-amino groups of lysine could be acetylated without impairing the power to polymerise.

The nature of the sulphur-containing groups in tropomyosin has been the subject of much discussion. In the early analyses of Bailey (2) three cystine/2 residues were found; Kominz, Hough, Symonds & Laki (1) found no SH groups, and this led to the suggestion that one or two S–S bridges might be involved in the molecular structure. However Locker (1) on oxidising with performic acid found no fall in molecular weight such as might be expected if this were the case. Then in 1957 Kominz, Saad, Gladner & Laki (1) reported that all the tropomyosins they examined contained two free SH groups. In this work they used highly purified $(NH_4)_2SO_4$ and they suggested that the SH had previously been missed on account of its oxidation in the presence of traces of heavy metal contaminants of this salt. In 1959 A. G. Szent-Györgyi, Benesch & Benesch (1), calculating from the difference between the SH value and the cysteic acid value obtained on performic acid oxidation, reported two SH groups and one S–S linkage.

The possibility of the linking of two chains in the molecule by a disulphide bridge was raised again by the interesting results of Woods (1). He determined the molecular weight by equilibrium sedimentation in 8 M urea in presence of 0.1 M β-mercaptoethanol, obtaining the value of 34000 – about half the molecular weight of the original protein.[1] In urea solution alone the molecular weight did not fall to this value. Woods at the time suggested the covalent linking of two chains but later experiments (2), determining molecular weights under a variety of conditions, led him to a different interpretation: that the disulphide groups were probably not present in native tropomyosin, though they might appear in some cases during preparational procedure.[2] The effect of concentrated urea he took to be the separation of the polypeptide chains, which might then (in absence of SH-masking reagents) unite by means of S–S linkages. This linking-up appeared to be prevented by EDTA which would stop the metal-catalysed oxidation

[1] Compare Cohen & Longley (1).
[2] Also Mueller (1) for a similar suggestion. He prepared tropomyosin in presence of dithiothreitol and found all the cystine/2 residues of the protein in the form of SH.

of SH to S–S. A little earlier Drabikowski & Nowak (1) had described experiments pointing in the same direction. The protein after very rapid preparation contained 4 SH groups/mole, but two of these were very liable to oxidation, perhaps owing to particular steric conditions. In concentrated urea solution (where as we have seen the protein becomes unfolded) all the SH groups appeared; in 0.6 M KCl solution only three reacted; in water (where the protein is polymerised) only two. These effects also might be due to steric hindrance, but they suggested that the difference between the SH content in KCl solution and in water might mean that one SH group was involved in polymerisation. Tan, Sun & Lin (1) in 1958 had discussed the possible connection between SH groups and polymerisability, noting that performic acid oxidation of these groups or their blocking by metal ions or organic mercurials retarded polymerisation. The Polish workers however found that interaction with certain sulphydryl reagents led to increased viscosity of tropomyosin solutions. Further work to clarify the role of the SH groups is thus still needed.

Tsao, Bailey & Adair, as we have seen, found the particle weight of tropomyosin to be higher in salt solution (this was true even at molar salt concentration) than the monomer weight in urea solution. Kay & Bailey (1), by the light-scattering method and using much more dilute protein solution down to 0.04 %, found the state of aggregation to depend on the ionic strength of the medium and there was no dependence of the molecular dissymmetry on the protein concentration. It was pointed out by Asai (1) and by Ooi, Mihashi & Kobayashi (1) that such observations as these made it difficult to consider the polymerisation as a reversible physical equilibrium. This difficulty they found could be overcome by working with sufficiently dilute protein solutions. Thus Asai, using measurement of electric birefringence, was able to work at concentrations as low as 0.01 %, and found then that even under salt-free conditions, the tropomyosin was completely depolymerised to monomers at infinite dilution. Ooi et al. also, using the light-scattering method, were able to make observations on very dilute protein solutions, and found like Asai a marked dependence of molecular weight on protein concentration below 0.05 %; on extrapolation to zero concentration the monomer weight was 55000 even in 0.03 M KCl. Theoretical analysis of their results led both Asai and Ooi et al. to the conclusion that, besides the main linear aggregation there was some side-to-side overlapping. Asai calculated that the particle length at infinite dilution tended to a value of 400 ± 50 Å, independent of the solvent conditions, in good agreement with the estimate of Tsao, Bailey & Adair. In view of the highly charged nature of the tropomyosin molecule, interest had often been expressed in its dipole moment; Asai was now able to assess this and found a permanent dipole moment of 390 Debye units at pH 7.0 and infinite dilution.

THE DISCOVERY OF INVERTEBRATE TROPOMYOSIN OR PARA-
MYOSIN. It is necessary at this point to mention the discovery by Bailey (7)
(just ten years after his discovery of tropomyosin) of a new form of this
protein. We have already mentioned the Type I fibrils, giving a characteristic X-ray diffraction pattern, first observed by Bear. Such fibrils from
clam adductors were also studied in the electron microscope by Hall, Jakus
& Schmitt (1), who had named the responsible protein paramyosin (2).
Bailey set out to characterise paramyosin chemically. It was known to be
specially abundant in the slow smooth parts of lamellibranch adductor
muscles, and Bailey extracted it (at ionic strength 0.5) in good yield, 30 %
of the muscle protein, from this muscle in *Pinna nobilis*. The solution was
very viscous, and needle-shaped crystals were formed on lowering the
ionic strength to 0.3. Amino acid analysis showed its composition to be
similar to that of vertebrate tropomyosin – it contained no tryptophan or
proline, but lysine had been partly replaced by arginine and glutamic acid
partly by aspartic acid. Its high viscosity and double refraction of flow
showed it to be a very asymmetric molecule, and it gave a well-defined
α-X-ray diffraction pattern. Soon afterwards Kominz, Saad & Laki (1)
isolated the two forms of tropomyosin from the smooth adductor of the
clam. They were separated by $(NH_4)_2SO_4$ fractionation, the new invertebrate tropomyosin coming down at 28–40 % saturation, the well-known
form at 45–65 % saturation. Kominz *et al.* suggested the designations
tropomyosin A for the former and tropomyosin B for the latter; but we
shall rather retain the terms paramyosin and tropomyosin. Since paramyosin has not been found in vertebrate skeletal muscle, with which we are
now primarily concerned, we shall defer its further consideration to a later
chapter.

THE ROLE OF TROPOMYOSIN. Bailey (6) had considered whether tropomyosin, in view of its similarities in composition and properties, might be
either an artefactual breakdown product or a natural building-stone of
myosin. He showed that the former could not be the case by subjecting
myosin to the treatment involved in tropomyosin preparation from muscle
and finding no tropomyosin formed. Attempts to investigate the building-
stone hypothesis have met with little success; such evidence as has been
adduced in its favour comes mainly from considerations of amino-acid
composition of myosin, actin and tropomyosin, and of fragments of myosin
and meromyosin prepared in different ways.[1] Experiments *in vivo* have
given negative results. Velick (1) injected labelled phenylalanine into rabbits,
and afterwards isolated from the muscles a number of proteins, including
myosin, tropomyosin and actin. The degree of labelling of the myosin and

[1] E.g. Laki (3, 4); Kominz, Carroll, Smith & Mitchell (1).

tropomyosin gave no indication of a simple precursor relationship. Needham & Williams (1) also found equally rapid incorporation of labelled amino acids into uterine tropomyosin and actomyosin after injection of the acids into rabbits in late pregnancy, when active synthesis of actomyosin was known to be going on in the uterus. Again D. S. Robinson (1), comparing the tropomyosin content of chicken embryo and adult muscle, found no greater concentration in the former, such as might have been expected on the building-stone hypothesis. As we shall see, a structural role for tropomyosin has emerged.

Evidence has accumulated in recent years for some sort of association between tropomyosin and actin. Perry & Corsi (1) had noticed that during long treatment of isolated myofibrils with solutions of low ionic strength, actin and tropomyosin were extracted in rather similar proportions under a variety of conditions, e.g. at different pH values. In 1962 Martonosi (2) showed the presence of three components in the sedimentation of partially polymerised actin. One of these, on isolation, proved to contain about 40 % of tropomyosin. Martonosi pointed out various indications of some form of interaction between the two proteins: the fact that no tropomyosin boundary was visible in the sedimentation diagram under these conditions; an increase in viscosity when tropomyosin was added to F-actin; in contrast the lack of any apparent interaction when tropomyosin was added to denatured actin. About the same time Laki, Maruyama & Kominz (1) also obtained evidence for interaction between the two proteins. This took place only under certain conditions of salt concentration – above the threshold required for F-actin formation; tropomyosin was then incorporated into F-actin gel, up to a limit of 25 % – about one mole of tropomyosin to three of actin. Maruyama (2) confirmed these findings in a flow-birefringence study. He found no evidence for any binding with G-actin. Since the tropomyosin was rapidly bound to F-actin, it seemed likely that there was no co-polymer formation, but that the tropomyosin particles became attached on the outer surface of the F-actin. We have already seen[1] that workers with purified actin had had to reckon with tropomyosin contamination. From these later results it seemed that this contamination was perhaps not merely a nuisance, but might have significance in muscle function – Maruyama pointed out that association of the two proteins was to be expected under physiological conditions.

We shall consider in chapter 12 the evidence for the location of tropomyosin in the muscle fibre. From the appearance in the electron microscope of the lattice of the Z line and of fragments of tropomyosin crystals it has been suggested that part of the Z line might consist of tropomyosin. There is also some evidence that tropomyosin in filamentous form extends into the

[1] P. 215.

I filaments. In this connection the observations of Cohen & Longley (1) on the effects of divalent ions on the manner of aggregation of tropomyosin are significant. With rabbit tropomyosin, above a critical concentration of Ca of Mg ions (0.01 M) and at pH 7.0–9.0, fibres were formed. In the absence of these divalent ions and at pH about 5.4, crystal nets appeared. They suggest that in the developing myofibril a gradient of these cations could be responsible for the aggregation of the protein in the crystalline form in the *Z* line and in the fibrous form (with 400 Å period) in the *I* band.[1]

Finally we should mention the recent realisation of the part played by tropomyosin in the 'troponin complex', which is essential in the regulation by Ca ions of contraction and relaxation. This also will be discussed in chapter 12.

---

[1] Cf. Hanson (1) and p. 297.

11

THE SLIDING MECHANISM

DISCOVERY OF THE SLIDING MECHANISM

The first hints of the sliding-filament mechanism of contraction were given by the low-angle X-ray diffraction patterns obtained by H. E. Huxley with living and glycerol-extracted muscle. As we have seen, high-angle X-ray diffraction patterns had not differentiated between resting and contracted muscle; on several occasions the need for the low-angle technique had been emphasised, but before 1953 this had been used only on dried material, as in the work of Astbury (quoted by MacArthur (1)) in 1943 and of Bear in 1945. H. E. Huxley (2) began in 1951 by showing with dried muscle that stretching by 60 % had no effect on the repeat pattern found by Bear (of which the largest spacing was about 58 Å); he surmised that if the structure producing this pattern existed in living muscle and if this structure were affected by stretching, then the effect could not be on the 58 Å units themselves but must consist in the pulling apart of the long molecules containing these units. Equipped with newly designed apparatus Huxley (3) went on to study the effect of stretch in living rabbit and frog muscle. Here the axial pattern at very low angles showed reflections corresponding to orders of a 420 Å repeat pattern, of which previous workers had seen only the reflections at 50–60 Å. It seemed likely at that time that this long axial spacing must arise from the actin,[1] since it was reminiscent of that seen (about 300 Å) by Rozsa, Szent-Györgyi & Wyckoff (1) in electron micrographs of actin filaments, while myosin in the electron microscope showed no axial spacings. Astbury (4) also had found X-ray reflections corresponding to the orders of a 54 Å repeat in fibrous actin but not in myosin. On the other hand, the axial periodicity (250–450 Å) in the electron micrographs by Draper & Hodge (1) of toad myofibrils appeared very clearly in both A and I bands. Huxley now found that stretching by 40 % did not change the axial pattern. The transverse X-ray examination of living muscle showed long molecules in hexagonal array parallel to the fibre axis and about 450 Å apart. In absence of ATP, as in *rigor mortis*, IAA contracture or in glycerol-extracted

---

[1] It was only much later that the distinction between the actin period of *ca.* 370 Å and the myosin period of 429 Å was clearly established; Elliott & Worthington (1) and Worthington (1) in 1959 recognised that the period in question here was probably due to myosin.

muscle, the second line of the equatorial X-ray diagram (the 225 Å reflection) became very strong. This indicated a region of very high electron density at a specific place between the long molecules, which remained 450 Å apart. When ATP was present, the electron density in between the long molecules was relatively uniform. Huxley concluded (14) that a double array of filaments was present, some consisting of myosin, some of actin; and that the changes in relative intensity of reflections showed the fixation, in absence of ATP, of significant amounts of material at specific sites suggesting the formation of links between the actin and myosin.

In an electron-microscope study in 1953 Huxley (4) used the new thin-sectioning technique of Hodge, Huxley & Spiro (1). Now for the first time two sets of filaments were seen in both cross and longitudinal sections. Cross-sections through the A band showed large filaments, 110 Å in diameter, in hexagonal array, 200–300 Å apart, with smaller filaments (about 50 Å in diameter) lying fairly symmetrically, each in between three of the primary filaments. Sometimes 'bridges' appeared to extend between a primary and a secondary filament. In the H zone there were only large filaments, about 140 Å in diameter; the I band was disoriented in structure but sometimes thin filaments could be seen there. Longitudinal sections gave corresponding results; it was not possible at this time to trace the thin filaments of the I band into the A band, but sometimes thin filaments could be seen lying between the thick ones.

The single set of filaments seen in earlier longitudinal studies, e.g. of Hall, Jakus & Schmitt (1), using fragmented material dried down on the grid were probably composite: in the I band probably consisting of thin filaments adhering to one another, in the A band of thin filaments adhering to one of the thick. Such composite filaments would appear continuous through the sarcomeres. Huxley now suggested explicitly that extensibility depended on the sliding past each other of the two sets of filaments, the thin filaments being pulled out of the A bands; he also raised the possibility that contraction might depend on an analogous process.

In considering the extensibility of muscle, Huxley referred to the study in Weber's laboratory of the plasticising effect of ATP on glycerol-extracted muscle in conditions where the ATP concentration could be maintained. Thus Portzehl (2) had shown that the modulus of elasticity of glycerinated fibres fell to 40 % or less of its value on treatment with 3–6×10^{-3}M ATP. This led Huxley to the suggestion that the essential condition for extensibility was that the actin and myosin filaments should not be linked together, and that this state of affairs required the presence of ATP.

Hanson & Huxley (1) had already found evidence that the thick filaments in the A band consisted of myosin while the actin was contained in the thin filaments. This evidence depended on differential extraction of the muscle proteins (fig. 40, 7–9 pl.). After treatment for 20 min with Hassel-bach & Schneider solution or Guba–Straub solution for myosin extraction, examination of the myofibrils in the phase-contrast or electron microscope showed that the *A* band had largely disappeared, except for two fine lines one at each end of the original *H* zone. In the electron microscope it appeared that thick filaments still remained in the *H* zone. The whole fibril was only very weakly birefringent. Further treatment of the residue, after breaking up in a blender, with 0.6 M KCl at pH 6.0 for 18 h (as used by Hasselbach & Schneider for actin extraction) led to disappearance of all organised structure. Hasselbach (8) had also at just the same time carried out the extraction of myosin without actin from myofibrils, and shown in the electron microscope the disappearance of the material of the *A* band. The *I* filaments appeared unchanged, and showed the approximately 400 Å periodicity of Huxley. Hasselbach also noted that in maximum contraction the contraction bands were extracted while the actin filaments remained. Viscosity measurements on the protein extracts showed the presence of myosin, but only some 5 % of actin.

We have already seen how much difference of opinion there was (owing to the limitations of the light microscope used with whole muscle fibres) on the question of the shortening of the *A* or the *I* band on contraction. The suggestion of Huxley would now involve shortening of the *I* band and disappearance of the *H* zone. To test this suggestion (with avoidance of optical artefacts) Huxley & Hanson (1) used, with the phase-contrast microscope, isolated myofibrils only 2 μ in diameter made by blending glycerol-extracted rabbit psoas. (Hanson (2) had already shown how suitable was this material for study of changes in striation on ATP con-traction.) In resting fibrils the *I* band measured about 0.8 μ and the *A* band about 1.5 μ. On contraction by means of ATP addition (fig. 40, 1–4, pl.), the *I* band shortened until it completely disappeared, the *A* band remaining constant in length, but changing in density. The *H* zone, in resting muscle low in density, became at first indistinguishable from the rest of the *A* band and was then replaced (at about 80 % of rest length) by a narrow region denser than the rest of the band. By the time the *I* bands had disappeared (at about 65 % of rest length, normally the limit of physiological shortening) contraction bands had formed at the lines of contact between adjacent *A* bands. With shortening down to about 30 % of rest length, the dense zone in the middle of the sarcomere split into two parts which then merged with the incoming contraction bands. Preparations from stretched muscle (fig. 40, 5, 6 pl.) usually contained fibrils with long *I* bands and long *H* zones; no significant change in length of the *A* bands could be found, and no change in axial period of the thin filaments.

Electron microscopy on thin sections of stretched and contracted muscle confirmed the behaviour of the bands. The thick filaments remained straight even after contraction band formation which seemed to result from the folding up of their ends. As a driving force the formation of actin–myosin links was suggested, when ATP (which had displaced actin from myosin) was enzymically split; the possibility of repetitive conformational changes in the myosin filaments (without overall changes in length) was also visualised as somehow causing movement of the *A* filaments. Fig. 41 (pl.) gives a schematic picture of these events.

At just this same time (1953) Huxley & Niedergerke (1) were studying the striation changes in living muscle fibres during contraction, using an improved interference microscope which gave a reliable image of the fibre and its striations. They also found that stretch and isotonic contraction (effected by electrical stimulation) caused changes in width of the *I* band, the *A* band remaining constant. Isometric contraction led to no change in either band (fig. 42 pl.). They recalled the postulation of Krause (1, 2) that the anisotropic material of the *A* bands was in the form of submicroscopic rods of definite length, and the recent identification of this material as myosin. Further, the existence of filaments in the *I* band, extending into the *A* band, had been reported by electron microscopists; these considerations suggested to them the hypothesis that during contraction actin filaments are drawn into the *A* bands between the myosin rodlets, with generation of force at each of a series of points in the overlap region. In these terms they discussed the known effect of stretching in causing a fall in the amount of isometric tension obtainable;[1] and the relation to be expected between narrowness of striation and high speed of contraction (as in arthropod muscle).[2]

In the model pictured at this time by Huxley & Hanson[3] the actin filaments extended from each *Z* line to the edge of the *H* zone, where it was postulated, each actin filament was attached to an elastic component (not actually seen) the *S* filament.[4] The latter provided continuity between the actin filaments associated with one *Z* line and those associated with the next. When fibrils were stretched in presence of a high plasticising ATP concentration the *I* bands increased in length and so also did the *H* zone; stretching of the fibrils during contraction (presence of low ATP concentration) on the other hand, did not lead to change in length of the *H* zone but there was a decrease in density of the *Z* line. These observations were interpreted as meaning that in the first type of stretch the fibrils were in a semi-plastic state resembling that in relaxed muscle; while in the second type of stretch only the series elastic component of A. V. Hill could be extended; it was suggested that the neighbourhood of the *Z* line was the site of this component.

Some very interesting experiments of H. E. Huxley's were described by

---

[1] P. 245 below.   [2] Ch. 22, p. 534.   [3] Also Hanson & Huxley (2).
[4] For further discussion of the *S* filament see below p. 251, and ch. 12.

Hanson & Huxley (2) in which the 'backbones' left in fibrils after the extraction of myosin (and believed to consist of actin) were irrigated by solutions containing myosin. The fibrils became denser, except in the gap corresponding to the H zone; these 'reconstituted ghosts' could actually contract when the medium was changed to salt solution containing ATP, the Z lines being drawn together.

A GENERALISED PICTURE OF ITS MODE OF ACTION

In 1955 Hanson & Huxley (2) used this new knowledge of myofibrillar structure together with the biochemical facts available on actomyosin/ATP interaction and the quantities of these proteins present in the muscle, to put together suggestions for the sequence of the events of contraction in, for example, rabbit muscle.

First, calculations were made of the number of molecules likely to be present in each filament of actin and of myosin – 450 in the former case, 230 in the latter. These calculations were based on the biochemical estimates of the muscle content of myosin and actin;[1] the molecular weights;[2] and the filament lengths, 15000 Å for myosin, 10000 Å for actin in one half-sarcomere. These lengths were measured in the phase contrast not the electron microscope, because the technique of preparation for the electron microscope entails some changes in dimensions.[3] Since six actin filaments surround each myosin filament, and each actin filament is surrounded by three myosin filaments, it was suggested that the myosin filaments must have six molecules in parallel longitudinally, working out to a repeat of 392 Å, $(15000 \times 6/230)$, while the actin filaments might have three molecules in parallel, giving a repeat of 132 Å $(10000 \times 3/450)$. It was further suggested that links between the two proteins occurred with the same periodicity as that of the myosin, and that this would account for the bridging appearance between filaments with a 400 Å repeat seen by Hodge, Huxley & Spiro (1). Intervening sites on the actin molecules would be unlinked, but could join up if the filament slid along 132 or 264 Å. A *branch* of the myosin molecule was pictured as contracting to pull the actin filament along 132 Å, unhooking and re-extending to link with the next molecule 132 Å further along the actin filament; all this without overall change in the length of the myosin filament. Another possibility would be that of progress of the actin in short steps; this might be effected by slight increase in length of the actin when bound to the myosin, decrease in length when the link at one end was broken, and re-attachment further on.

---

[1] Hasselbach & Schneider (1).   [2] Weber & Portzehl (2); Tsao (2).
[3] Huxley & Hanson (1).

The energy relations were also considered. Thus the maximum tension per unit area which a muscle can exert at about rest length is about 4 kg/cm^2; from the X-ray information on filament arrangement it could be calculated that this means 3.3×10^{-5} dynes tension in each actin filament. If the energy for work done during shortening was derived from ATP splitting (taking 10 000 cal set free per mole) then hydrolysis of one molecule of ATP would provide for 2 Å shortening. Clearly several ATP molecules would be needed for the size of step we have been considering in the first model suggested above; if each link were made as the result of splitting of one ATP molecule, then at least some 60 links would be active in each step of shortening. Huxley (5) suggested later that the minimum step distance might be 54 Å – on the assumption that this axial distance between nodes in the helical structure of actin filaments[1] represented actin monomers. Since the A band filament contains approximately 230 myosin molecules, the splitting of one ATP per myosin molecule would allow, at maximum tension, a relative movement of approximately 460 Å, the bridges being made and broken and successive alighting points on the actin for any one projection being 460 Å apart.

FIRST REACTIONS TO THE THEORY

The belief in a single type of myofilament (cf. pp. 169 ff. above) was very persistent, as was shown by three electron-microscope studies in 1956. Hodge (1) considered the myofilaments to be of actin only and agreed that myosin was localised in the A band, being responsible for the double refraction on account of the longitudinal orientation of its particles. He described, especially in insect flight muscle, dense filamentous cross-bridges uniting each filament to its six nearest neighbours, regularly spaced along the filaments with a periodicity of 250–400 Å. He suggested that the presence of these neutralised the double refraction due to the filaments in the I band. He supposed that in contraction the actin filaments shortened by interaction with myosin and were drawn into the A band; the axial spacing appeared to decrease with shortening of the sarcomeres.

Sjöstrand & Andersson (1) also saw only continuous filaments, but considered that these were much thicker in the A than in the I bands, and that the thickness as well as the distance between connecting bridges in the A bands varied with the degree of shortening. They interpreted the appearance in the A band at different degrees of shortening as due to a coiling of smaller 'filamentous' sub-units, the pitch of the helix varying with the degree of shortening.

Spiro (1) identified the filaments largely with actin, the myosin again

[1] See the X-ray diffraction data of Selby & Bear (1).

being thought to be present in the form of a dispersion of macromolecules between the actin filaments. With splitting of ATP, he suggested, actin and myosin could unite by deposition of myosin, so that the thin filaments might become thicker and were gradually drawn further and further into the A band.

Scepticism about the nature of the cross-striation was also apparent in some biochemical studies. Thus Szent-Györgyi, Mazia & Szent-Györgyi (1) found with washed fibrils from glycerol-extracted psoas muscle that the use of salt solutions designed for myosin extraction (Hasselbach–Schneider or Guba–Straub solutions) gave an extract in which some 30 % of the protein was not myosin. Phase contrast microscopy showed that the A band had disappeared and it was suggested that the unknown protein, not precipitated on dialysis against 0.04 M KCl, was the 'A band substance' responsible for the density and double refraction of the A band. De Villafranca (1) studied this material but without being able to find that it corresponded to any known protein.

TABLE 11

Protein fraction	As % of total protein in fresh muscle	As % of total protein in glycerol-ex-tracted muscle	As % of total protein in washed fibrils
Glycerol-extracted	Up to 9	.	.
Rest of soluble	.	28	.
Total soluble	34	.	.
Myosin + X-protein	41	45	62
Myosin ⎱ by dilution	34	37	51
X-protein ⎰ by dilution	7	8	11
Myosin ⎱ by dialysis	29	32	44
X-protein ⎰ by dialysis	12	13	18
Residue	25	27	38

(Hanson & Huxley (3).)

In 1957 Huxley & Hanson (2) used the interference microscope to determine *in situ* the relative amounts of protein in the bands of glycerol-extracted myofibrils. A modified Hasselbach–Schneider solution selectively extracted the material of the A band, amounting to about 55 % of the total fibrillar protein. There was also a small change in the I bands, indicating extraction at the same time of about 10 % of other fibrillar proteins. Parallel experiments by Hanson & Huxley (3) on large-scale extraction of protein from washed glycerinated myofibrils (table 11) showed that about 62 % of the protein was extracted. Of this, 83 % was precipitated on dilution of the extract to ionic strength 0.04, leaving only some 17 % of unknown protein. Thus about 51 % of the total protein of the myofibrils seemed to be myosin – in good agreement with the results of interference microscopy; and the combined experiments located at least 80 % (probably the whole) of the myosin in the A band.

The nature of the 'extra protein' has been carefully studied by Perry and his collaborators,[1] using fresh washed myofibrils. Perry & Corsi (1) found that extraction according to Szent-Györgyi, Mazia & Szent-Györgyi (1) followed by dialysis to ionic strength of 0.06, left protein in solution only amounting to 3–7 % of the total extracted. Perry & Zydowo (1) found this soluble protein to consist of a mixture which could be resolved by chromatography on diaminoethylcellulose into four main fractions – some sarcoplasmic protein, as shown by traces of aldolase activity; tropomyosin up to 22 %; an unknown globulin; and a ribonucleic fraction. Corsi (1) using a similar method of myofibril extraction followed by dialysis found about 7 % of the extracted protein to be soluble at low ionic strength; of this the main component, according to electrophoretic, viscosity and salting-out observations, was tropomyosin. Perry & Corsi (1) also extracted myofibrils with buffer solutions of low ionic strength (e.g. 0.078 M borate, pH 7.1); this extract contained tropomyosin, an inactive form of actin and an unidentified protein together making up about 40 % of the total myofibrillar protein. The actin extracted was electrophoretically identical with F-actin inactivated by dialysis against the borate buffer.

From the results of extraction with a variety of salt solutions followed by examination of the fibrils in the phase contrast microscope, Corsi & Perry (1) concluded that myosin was absent from the I band and Z line, and that actin and tropomyosin were present in both A and I bands.

ELECTRON-MICROSCOPE EVIDENCE FOR THE DOUBLE ARRAY OF FILAMENTS AND FOR THE CROSS-BRIDGES

In 1957 Huxley (5) was able to bring definitive evidence for the double array of filaments in striated muscle (fig. 43 pl.). From the array seen in cross-section (fig. 44, pl.), two thin filaments would be expected in longitudinal sections between each pair of thick filaments. But, as he showed, these can only be clearly visible if the sections are not more than 150 Å thick and if the longitudinal sections are cut parallel to certain planes of the filament lattice, conditions not fulfilled by earlier workers. Moreover, in many instances, thin filaments could now be clearly seen to pass continuously from the Z line into the A band, terminating at the edge of the H zone (fig. 46 pl.). Besides showing very beautifully the longitudinal arrangement of the two sets of filaments, electromicrographs in this paper show the arrangement of bridges linking the actin and myosin. These

[1] Also Perry (6).

bridges were seen always to have the same relation to the filaments – always to be at right angles – whether the muscle was stretched or contracted. This must imply that during the sliding the bridges remain attached for only a short distance, becoming detached and attached again during the change in length. Huxley commented on the tendency of the I filaments, not very closely packed together, to depart slightly from linearity. He was inclined to explain the low birefringence of the I bands as due to their containing much less oriented material and this not so perfectly aligned; moreover, presence of a higher concentration of soluble proteins might occur here in the larger space between the filaments, with the effect of increased refractive index of the surrounding solution and decrease in form birefringence.

A very important suggestion also made in this paper was that the bridges consist of the heavy meromyosin parts of the filament myosin; this was based of course on A. G. Szent-Györgyi's finding that the H-meromyosin fragments contain all the ATPase activity and actin-combining power of the myosin molecule. Calculation at that time seemed to show that the number of bridges per gram of muscle corresponded roughly with the number of myosin molecules.[1] Fig. 45 (pl.) from a later paper (11) shows greatly contracted muscle, with a double overlap of thin filaments between the thick ones.

A point of cardinal importance for the sliding-filament mechanism is the constancy of the length of the A and I filaments. This constancy was questioned, for example by Carlson, Knappeis & Buchthal (1). A very careful study of the matter was therefore made by Page & Huxley (1) in 1963, using resting and excited frog muscle under different conditions. They came to the conclusion that any difference in length of the filaments could be entirely accounted for by preparational procedures; the different conclusion of Carlson et al. could be explained by their assumption that if the preparational procedures did not affect the sarcomere length, then the filament lengths also would be unaffected by them.

H. E. Huxley (7) has described the disoriented, crumpled thick filaments, responsible for the contraction bands when they become forced against the Z lines.[2]

RELATION OF TENSION TO SARCOMERE LENGTH

A. F. Huxley & Niedergerke (1) remarked in 1954 that if the myosin filaments were $1.5\,\mu$ long, and the actin filaments $2\,\mu$ long ($1\,\mu$ on each side of the Z line), then isometric tension should fall in a linear manner as the fibre was stretched over the range of sarcomere lengths from 2 to $3.5\,\mu$; this would follow from the hypothesis that tension is proportional to the number

---

[1] But see p. 254 below.　　　　　　[2] See discussion of this paper, p. 28.

of molecular sites in action between the two types of filament. As they pointed out, Ramsey & Street (1) had observed in 1940 with frog fibres that the maximum tension decreased on stretching approximately in proportion to the degree of stretch beyond resting length, and fell off also at shorter than resting lengths. Much earlier A. V. Hill (10) had found tension production to be at a maximum at about resting length of frog sartorius, falling off on either side.[1]

In 1961 A. F. Huxley & Peachey (1) working with isolated fibres under stretch, found that when the fibre was stimulated electrically to isotonic tetani, cine-microphotography showed that there was no shortening at points in the fibre where striation spacing was greater than about 3.5μ. Parallel electron-microscope studies showed that at this sarcomere length there was no overlap of the two sets of filaments.

In 1964 Podolsky (1) raised the question whether the normal type of membrane activation would remain adequate when the sarcomeres exceeded a certain length. He therefore tried a more direct method of activation,[2] in which single fibres were stripped of their sarcolemma and 0.1 mM Ca^{2+} was applied in a micro-pipette with tip 1μ in diameter. The ends of the fibre were gripped by jewellers' foreceps mounted on mechanical microscope stages, and the sarcomere lengths could be varied by relative movement of the mechanical stages. In good agreement with the results of Huxley & Peachey he found that the sarcomere length above which contraction would not take place was 3.7μ. Thus the change in behaviour at this critical length is not due to an effect on the activation process but to an effect on the contraction mechanism, and is well explained by the necessity for overlap.

At this time also Gordon, A. F. Huxley & Julian (1)[3] using single frog fibres succeeded in relating tetanic tension production to sarcomere length and thus to the degree of overlap of the filaments. The apparatus was designed to allow tension measurements to be made on the middle region of the fibre where the sarcomere spacing was uniform. Fig. 47 shows the results. The filament lengths used for finding the degree of overlap were taken from the work of Page & H. E. Huxley (1) and Huxley (1). It can be seen that for sarcomere lengths above 2μ the tension is directly proportional to the number of bridges overlapped by thin filaments. The change in tension production at about 2μ may be caused by collision of thin filaments in the centre of the sarcomere, or the beginning of competition of actin filaments for bridges on the same myosin filament, or to both causes. Also, owing to the polarised structure of the filaments,[4] actin filaments intruding

---

[1] It is interesting to remember the 'fundamental experiment' of Schwann in 1835, when he showed that the force produced by a muscle diminishes with its shortening below resting length (see ch. 3).

[2] Ch. 13.   [3] Also Gordon, Huxley & Julian (2, 3); Edman (1).   [4] Ch. 12.

into the other half of the sarcomere might find interaction impossible for steric reasons. The change about 1.6μ is probably caused by contact between thick filaments and the Z line.

Interesting confirmation of these ideas concerning the sliding mechanism came in the work of Ward, Edwards & Benson (1) on the relation between sarcomere length and Mg-activated ATPase activity. They used glycerinated fibre bundles, less than 200 μ in diameter, which had been stretched to the

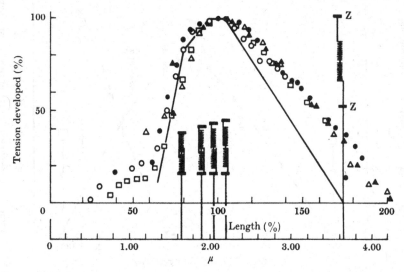

Fig. 47. 'The full line summarises our measurements of tetanic tension, related to the lower horizontal scale. It is superimposed on the results of Ramsey & Street ((1) fig. 5), taking their 100 % as equivalent to the middle of our plateau, or 2.10μ. There is good agreement over much of the range; the higher tensions found by Ramsey & Street at great lengths are attributable to the shortening of the ends, where the striations are narrower, and those at short lengths to development of the delta state in contractions of long duration.' (Gordon, Huxley & Julian (1).) The diagrams indicate the degree of overlap (see text).

desired length before glycerination. The length of overlap of the actin and myosin filaments was measured in sections in the electron microscope, and sarcomere length in unfixed bundles by phase contrast microscopy. The ATPase activity decreased as the sarcomere length increased beyond resting length, and the results thus indicated that this enzymic activity, necessary for tension production, depended on the opportunity for interaction between the myosin and actin. Similar results were obtained a little later by Y. Hayashi & Tonomura (1) who also included some points showing the falling off of the enzymic activity at fibre lengths less than the resting value. Experimental results on intact muscle were also in harmony. Infante,

Klaupiks & Davies (1) found with frog rectus abdominis that the maximum phosphocreatine breakdown on isometric contraction was at resting length, and fell off, like the tension, on either side of this length. Similarly they found (2) with frog sartorius muscle, poisoned with fluorodinitrobenzene (to inhibit the creatine phosphokinase)[1] that ATP breakdown and tension production showed this same correlation with muscle length (fig. 48).

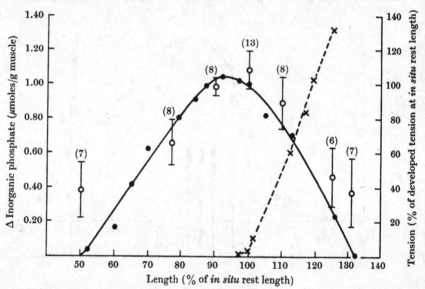

Fig. 48. Relation between formation of inorganic phosphate and muscle length in frog sartorii performing single isometric contractions at 0°. In some experiments FDNB was used. ○, change in inorganic P ± s.e.; ●, developed tension; ×, passive tension in grams (numerically equal to units on right ordinate). The numbers in parentheses are the number of muscle pairs used. (Infante, Klaupiks & Davies (2).)

NATURAL AND SYNTHETIC MYOSIN FILAMENTS AND THEIR INTERACTION WITH ACTIN

H. E. Huxley (1) in 1963 developed a technique for disintegrating muscle in a 'relaxing' medium[2] (containing ATP and EDTA) designed to prevent formation of cross-links between actin and myosin. It was thus possible to obtain preparations of individual thick and thin filaments which were then examined in the electron microscope with the negative staining technique.[3]

The thin filaments corresponded closely with those described by Hanson

[1] Ch. 14.   [2] Ch. 13.
[3] See also Huxley (6); Jakus & Hall (1) had described rather similar filamentous aggregates as early as 1947, but had not realised their significance.

& Lowy (1) from striated and smooth muscles and with filaments in preparations of purified actin. Most interesting results were obtained by examining the course of interaction between these filaments and H-meromyosin. The H-meromyosin molecules arranged themselves with a periodicity of about 366 Å along the actin filament, all oriented like arrow heads in one direction. This was interpreted as probably meaning that the HMM molecules had combined with the actin (one to each of the G-actin units) forming a double helix of HMM molecules wound around the original actin filament. A similar appearance with approximately similar periodicity was seen in filaments deposited from a solution of natural actomyosin in 0.6 M KCl. It is interesting to notice that this suggested combining ratio of 1 HMM or myosin molecule per G-actin is very different from the equilibrium combining ratio *in vitro* (3.7:1 by weight for myosin:actin, or a molar ratio of myosin:G-actin of 1:2–3).[1] The arrangement of the arrow heads showed a structural polarity of the actin filaments; this was emphasised by the fact that when such filaments were observed still attached to both sides of an isolated Z line, the arrows pointed in opposite directions on the two sides – always away from the line (fig. 49 pl.).

The thick filaments were about 110 Å in diameter, and generally 1.5–1.6 μ long, with tapering ends. On their surface they bore a number of irregular projections which were present right out to the tips, but absent from a central region about 0.17 μ long. In the middle of this bare zone the filaments were slightly thicker. Thus these isolated filaments had exactly the same appearance as the A band filaments seen in sections.

Synthetic myosin filaments were prepared by lowering the ionic strength of a myosin solution to about 0.15, when rod-shaped particles readily became visible with negative staining. These particles were of varying dimensions, sometimes being as large as natural A band filaments; they must thus have represented aggregates of molecules. They showed the irregular projections and the central bare patch, the latter forming a greater proportion of the whole length in shorter filaments. Huxley suggested that the myosin molecules also showed polarity, arranging themselves on either side of the filament centre with the globular portion pointing away from the centre and the tails overlapping (fig. 50 pl.). The bare patch would be the part where tails only, overlapping in anti-parallel, were present; the tapering ends would be caused by decreasing number of molecules. Huxley suggested that the directionality of movement, which results in pulling of the myosin filaments towards the Z line, may be locally determined by the orientation of the actin or the myosin, or both. We shall return to these questions.

[1] Ch. 12, p. 285.

THE Z LINE AND THE M LINE; THE S FILAMENTS

The structure of the Z line itself was specially studied by Knappeis & Carlsen (1) and by Franzini-Armstrong & Porter (1) in 1964. It was found that the I filaments ended on each side of the Z line in rod-like projections composed of very thin filaments; the I filaments were thus not continuous through the Z line, but one filament on one side lay between two filaments on the other side (figs 51 a and b, pl.). Cross-sections through the Z region showed that the I and Z filaments formed a tetragonal pattern which could be interpreted to mean that each I filament on one side faced the centre of the space between four I filaments on the other side, interconnection being made by four Z filaments.[1] Huxley (8) a little later pointed out the remarkable resemblance between this approximately square lattice of the Z line and the structure seen in the electron microscope in fragments of tropomyosin crystals after negative staining. Thus some part of the Z line might well consist of tropomyosin; other material seemed to be present also, in the spaces between the fine filaments. If these fine filaments leaving the actin filaments do consist of tropomyosin, this would fit in with the various other indications[2] of the participation of tropomyosin in the structure of the I filaments.[3]

The region of the M line is interesting. Huxley (7)[4] in 1965 showed with glutaraldehyde-fixed material that the narrow region of higher density is due to cross-connections between the thick filaments – probably three sets, 200 Å apart – occurring only in this region of the A band.[5] In 1968 Knappeis & Carlsen (2) obtained electron micrographs (figs. 52 a and b pl.) showing a rather elaborate structure for the M line. Three to five arrays of transverse bridges connected each filament with its six neighbours; there were also filaments parallel to the A filaments linking each set of M bridges together. This arrangement would not only preserve the position of the A filaments but would keep the sliding I filaments each to its own place in the M region where they are not guided by A filament projections.[6]

Several workers have noticed that antisera prepared against structural muscle proteins contained an antibody specific for the M line. Masaki, Takaiti & Ebashi (1) have now prepared (by ammonium sulphate fractionation of a modified Hasselbach–Schneider extract) a protein (the ' M substance') which was electrophoretically homogeneous and gave a single precipitation line when tested immunoelectrophoretically against material

---

[1] Also Reedy (1).   [2] Pp. 235 and 297.
[3] For further work on the Z line proteins see ch. 12.
[4] See p. 28 of the Symposium at which it was given, for the discussion of this paper.
[5] Also Pepe (2).
[6] Dobie (1) (cited by Knappeis & Carlsen) as early as 1849 remarked on a structure in the middle of the A band differing in light transmission from the other parts.

containing M line antibody. Antibody prepared against the M substance and labelled with fluorescent dye stained only the M line. This protein exists in two forms – with sedimentation coefficients of about 8 and 12 S. Electron-microscope study showed that addition of the substance, particularly the heavy form, to myosin in 0.1 M KCl caused increased rate of lateral association of the myosin aggregates; this effect was exerted on the LMM. It seems likely that this property reflects its physiological role. On either side of the M line is a region of lower density than the rest of the H zone; the probable cause of this is the absence of cross-bridges. Sonnenblick, Spiro & Cottrell (1) had reported in 1963 that, with the papillary muscle of the cat heart, there was no increase in the H zone with stretching up to 45 % above resting length. Huxley (8, 7) re-investigating this question, found that the H zones behaved normally, and suggested that Sonnenblick *et al.* had mistaken for the H zone the lower-density area (now known as the pseudo-H-zone) just described, which is of constant length.

The possible presence of fine elastic 'S' filaments connecting the ends of the actin filaments through the H zone was suggested by Hanson & Huxley (2). Carlsen, Knappeis & Buchthal (1) using muscle stretched beyond the point of any overlap of the A and I band filaments, saw in the electron microscope fine filaments crossing the 'gap'. Such filaments were also reported by Sjöstrand (1) who described them as only 30 Å or less in diameter and apparently continuous with the A band filaments. Huxley[1] later expressed doubt of their real existence, owing to the finding of the double overlap of the I band filaments and also to the failure of his attempts to see them in the electron microscope.[2] A number of workers have however recently described their finding of very thin filaments, remaining after extraction of myosin (or of both actin and myosin), running through the whole sarcomere and connecting the Z lines. Such work has been largely on insect muscle[3] but Guba (1) reports them in rabbit psoas and claims to have prepared a new protein, fibrillin, from them. Garamvölgyi (3) who saw such filaments in bee muscle, is inclined to believe that they form the basic framework of the myosin filaments and this also is the opinion of Guba. Hoyle (1) describes electron-microscope evidence for T filaments (or very thin filaments) in a number of muscles from different types of animal, including rabbit and frog sartorius. All these workers consider that some elastic properties of muscle can only be explained by taking account of such filaments; much of the evidence for their existence is based on experiments on the elasticity and

[1] Huxley (7) p. 28 of the discussion; Huxley (10).
[2] A. F. Huxley (personal communication) has pointed out that these filaments might well exist unseen, if they consisted of material resembling the rubber-like, transparent, optically isotropic protein resilin described by Weis-Fogh (1) in insect cuticle.
[3] Ch. 22.

other mechanical properties of muscles or muscle fibres.[1] This is a very interesting but obviously very difficult field, and much more work is needed, especially for convincing electron-microscope demonstration.

THE MYOFIBRILLAR STRUCTURE IN REST AND CONTRACTION

We have seen that the early low-angle X-ray diffraction work of H. E. Huxley showed a 420 Å period at that time attributed to actin. Elliott & Worthington (1) and Worthington (1) in 1959 recognised that this was probably associated with myosin. In frog sartorius, rabbit psoas and blowfly muscle (examined in both the wet and the dry state) they had evidence of an axial period of 435 Å; this period did not fit the actin pattern. The characteristics of the reflections concerned showed that they emanated from thicker filaments, and these reflections remained when the actin reflections disappeared on formaldehyde treatment. Worthington also reported (2) in insect muscle a non-meridional layer-line near 400 Å, which from its characteristics seemed to arise from projections from the filaments – he believed the actin filaments. Elliott (1) however in 1964 with living frog sartorius showed that the period of this layer line was actually 432 Å and coincided with that of the meridional component of the 432 Å repeat due to myosin. The projections therefore were associated with myosin not actin.

The very accurate measurements recorded by Huxley & Brown (1) in 1967 for the low-angle X-ray diagram of vertebrate striated muscle have thrown much light on the structure of the myosin filament in the living muscle.[2] The marked off-meridional layer-line reflections with a repeat of 429 ± 0.6 Å and the strong-meridional layer-line reflections with a repeat of 143 ± 0.29 indicate an arrangement of projections in an approximately helical manner with a helical pitch of 429 Å and an axial repeat of 143 Å. Since cross-bridges can be seen in the electron microscope at approximately 400 Å intervals, and since each myosin filament is surrounded by six actin filaments, an arrangement of cross-bridges in pairs (in a 6/2 helix) as shown in fig. 53 would seem to fit these data. It is interesting to notice that in order that this structure may correspond with the detailed observations, it must be assumed that the sub-units have their centre of gravity at a radius of approximately 105 Å; since the radius of the myosin-filament backbone is only about 80 Å, some structure jutting out from the thick filaments must dominate the layer-line pattern. Such a structure would of course be the cross-bridges.

Pepe (2, 3) using his results from antibody staining in fluorescent and electron microscopy, has recently described a model to illustrate the packing

---

[1] E.g. Foulks & Perry (1); Belágyi & Garamvölgyi (1); Garamvölgyi & Belági (1, 2).
[2] For an excellent review of X-ray diffraction studies of muscle see Hanson (3).

of the myosin molecules in the filament. This is based (1) on the availability
of the antigenic sites from which, as we shall see,[1] deductions can be made
as to the closeness of packing; (2) on the restraining effect of the M line
connections on the orientation of the myosin filaments; (3) on the triangular
profiles of the thick filaments seen in cross-sections through the pseudo-H-
zone.

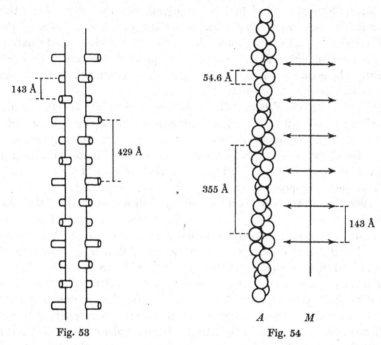

Fig. 53   Fig. 54

Fig. 53. Diagram showing the arrangement of cross-bridges on 6/2 helix. Helical repeat is
429 Å, but true meridional repeat (i.e. periodicity of the variation in density of structure
projected on to long axis of filaments) is 143 Å. (H. E. Huxley & Brown (1).)

Fig. 54. Diagram illustrating relative longitudinal positions of cross-bridges and actin
monomers in myosin and actin filaments. Note very close approximation to repeat of
longitudinal alignment after ~ 710 Å intervals, if actin sub-unit repeat is 54.6 Å along
either chain, if actin chains are helices of pitch 2 × 355 Å, and if cross-bridge repeat is
143 Å. (Huxley & Brown (1).)

Fig. 54 illustrates a structure with the actin helix containing 13 sub-units
per turn. But Huxley & Brown have now shown that this helix is non-
integral;[2] they calculate that for resting muscle only two near-matches are
likely on a given actin and myosin filament in each half A band. This may
have a functional significance in reducing the actin-activated ATPase
activity to its very low resting value.

[1] Ch. 12.   [2] See p. 227.

We have seen (page 245 above) that in 1957 Huxley (5) suggested that the number of bridges might be taken as equal to the number of myosin molecules per gram of muscle. It later became clear that there was a large discrepancy here and this is still unresolved. Recently Hanson and Offer[1] have made the following calculation, based on the best data available. From the number of projections seen by X-ray diffraction the content of myosin in the myofibril should be 60 mg/ml, but the mass of the thick filament obtained from extraction studies suggests a value of 90 mg/ml. This would mean 1.5 myosin molecules to the cross-bridge, or alternatively the presence of extra protein (myosin or protein sufficiently similar to myosin to be estimated as such after extraction) within the thick filament, not participating in bridge formation.

Huxley & Brown also describe a strong meridional reflection with a periodicity of 442 ± 2 Å, the orientation and nature of which show that it must arise from the myofibrils. Since a transverse periodicity of this size is seen in the H zone in electron micrographs of stretched myofibrils, its origin would seem to be in the thick filaments. The presence there of an extra, still unknown, component besides myosin is thus indicated.

In 1963 Elliott, Lowy & Worthington (1) made a close study of the small-angle equatorial X-ray reflections, using living frog and mammalian muscle. The nature of the reflections showed that they were caused by two types of filament of different diameters; it was concluded that they arose from the A band, and that the changes in their intensity with sarcomere length, when the muscle was stretched, depended on the degree of overlap of the A and I filaments. This provided the first evidence from *living* muscle that the amount of overlap between filaments varies with sarcomere length. This work confirmed the presence in living muscle of the filament lattice deduced by Huxley from his early work on rigor muscle, and showed that the change in the relative intensities of the reflections seen there was not confined to onset of rigor, being also in the living muscle the result of conditions favouring regularity of arrangement of the thin filaments.[2] Huxley & Brown, following rigor changes in low-angle axial X-ray reflections, and Huxley (11) studying changes in low-angle equatorial X-ray reflections, have concluded that the results show a moving out of the centres of mass of the cross-bridges and their attachment to the actin filaments.

Elliott, Lowy & Worthington also confirmed Huxley's early finding (14) of the constant volume of the filament lattice during stretch. Further work by Elliott, Lowy & Millman (1) (in which improved apparatus allowed observations on contracting muscle also) showed that this constancy of volume was maintained from the point at which the actin filaments abutted on each other at the centre of the A band to the point at which the

---

[1] Personal communication.   [2] Pp. 237–8 above.

actin filaments leave the A band. It was found for resting frog muscle that, over lengths where contraction was possible, the centre-to-centre distance between the actin and myosin filaments decreased with stretch, varying between 190 and 260 Å. Then, using contracting toad muscle, they found similar change in filament spacing with sarcomere length; but in the contracting muscle the centre-to-centre distance was constantly some 6–12 Å less than in resting muscle. Thus during change of sarcomere length there is no adjustment to maintain a constant distance between the A and the I filaments. This implies the involvement of long-range forces, or the possibility of variation in the length of the bridges between the two types of filament.[1]

A little earlier Millman & Elliott (1) had shown for molluscan muscle that, during stimulation for twelve hours, there was no change in the X-ray pattern of actin or paramyosin – thus no change in arrangement of molecules within the filaments such as would be involved in a change in their length. The possibility of small cyclic changes was not eliminated. A similar conclusion was drawn by Elliott, Lowy & Millman (1), in the paper just mentioned, for toad sartorius. This muscle was stimulated about once a minute for three days, a shutter arrangement in the path of the X-ray beam operating so that the shutter was open only when the tension was above a certain value. No change greater than 1 % was seen. Just at the same time also Huxley, Brown & Holmes (1), with newly introduced apparatus and working on resting and actively contracting frog sartorius with total exposure times of less than 40 min, brought evidence for the constancy (within 1 %) of the axial spacings on contraction, and thus for the constancy of length of the filaments. They also emphasised that any changes in configuration or periodicity must be restricted to a short part of the filament and must change in position with time.

The important change which *was* observed, was the large decrease during contraction in the intensities of the off-meridional parts of the myosin layer lines – indicating a movement of the bridges. This change in intensity was confirmed by Elliott, Lowy & Millman in an addendum to the paper just mentioned.[2]

Reedy, Holmes & Tregear (1) also have recently observed a remarkable difference in the arrangement of the cross-bridges in the resting flight muscle of the tropical water-bug *Lethocerus maximus*, as compared with the muscle in rigor. Sections of the muscle in the relaxed state showed in the electron microscope the cross-bridges usually perpendicular to the filaments; while in sections of muscle in rigor, the bridges were slanted at an angle, and in symmetrical pairs joined the myosin to the actin. The chevron pattern thus produced resembled the arrowhead arrangement mentioned earlier in

[1] Pp. 256, 306 below.   [2] Also Elliott, Lowy & Millman (2).

Huxley's work on combination of actin filaments with H-meromyosin or myosin molecules. Here again a polarisation effect was obvious, all the chevrons on the actin filaments pointing away from the Z line.

A further detailed discussion of the interpretation of the changes on contraction in the low-angle X-ray diagram of frog sartorius has been given by Huxley & Brown (1). Not only is there the large decrease in regularity of helical arrangement of the cross-bridges, but also a smaller decrease in intensity of the 143 Å meridional reflection, indicating some longitudinal movement or tilting of the bridges. Thus marked azimuthal and possibly radial movement of the cross-bridges takes place; these movements are unsynchronised as is shown by the absence of new reflections, such as those appearing during rigor. There is also evidence that, though a high proportion of the cross-bridges is affected by the disorder, at any given moment only a proportion of them is attached to actin filaments.

Huxley & Brown consider the important question of the ability of the cross-bridges to extend over the distance between the A and I filaments – some 190 Å at rest length of the muscle, but increasing with decrease in sarcomere length, the force per cross-bridge remaining constant.[1] The known specific effects of actin on myosin ATPase activity indicate that close contact of the cross-bridges with the proper sites on the actin is essential. Taking into account also the changes in X-ray pattern seen with onset of rigor, Huxley & Brown tentatively suggest that these requirements could be met if the myosin molecule contained a flexible portion (as a little earlier suggested by Pepe (3)[2], fig. 55) so that the distance of projection of the globular head of the cross-bridge could vary. This flexible portion was pictured as situated at the junction of LMM and HMM, the cross-bridge moving as a whole from this point. But it is here, as Huxley (11, 13) has emphasised, that the importance emerges of the special nature of the HMM tail. As we have seen,[3] this HMM subfragment 2 is water-soluble, in contrast to LMM. If now this linear portion (some 400 Å long) is not bonded all along its length to the surface of the thick filament and moreover is of a flexible nature, then it could move as required to bring the subfragment 1 globules into contact with the actin; further, over a large range of filament spacing, the globules could then approach the actin sites always at the same orientation, a matter which may be essential for their proper interaction. Huxley (13) has discussed possible ways in which the two subfragment 1 heads, acting independently or in conjunction, could develop the sliding force.

We have already considered the evidence[4] indicating that *in vitro* only one HMM subfragment globule is active at a time in ATP hydrolysis and actomyosin formation. This suggests that *in vivo* only one subfragment 1 head can at any time make the necessary contact with actin.

[1] Gordon, Huxley & Julian (3).   [2] Pp. 258 ff. below.   [3] Ch. 10.   [4] Ch. 9.

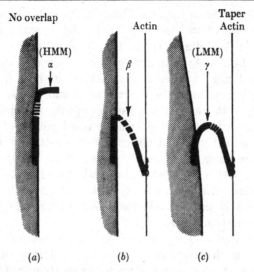

Fig. 55. Representation of actin–myosin interaction at different places along the thick filament. (a) A myosin molecule in the thick filament where no overlap occurs with the actin filaments. The HMM sites (α) are available for antibody staining. (b) and (c) Myosin molecules in the thick filament where overlap occurs with the thin filaments. No HMM sites are available for antibody staining. (b) A myosin molecule in the precisely packed region of the thick filament. No LMM sites (γ) are available for antibody staining, but trypsin-sensitive sites (β) are. (c) A myosin molecule near the tapered end of the thick filament. No trypsin-sensitive sites are available but LMM sites (γ) are. (Pepe (3).)

VARIANTS OF THE THEORY INVOLVING FOLDING OF ONE TYPE OF FILAMENT

We may take now examples of alternative theories, each based also on the two types of filament moving past each other, but otherwise visualising quite different mechanisms, including folding up of one type of filament.

The theory of A. G. Szent-Györgyi & Johnson (1) grew out of very interesting work, beginning about 1956,[1] in which attempts were made to locate the fibril proteins in the sarcomere by means of their reaction with antibodies labelled with fluorescent dye. In 1959 Marshall, Holtzer, Finck & Pepe (1) found, besides the staining of the A band by anti-myosin, that anti-actin stained the I band and the M line, while the greater part of the A band was unstained. The LMM and LMM Fraction 1 antigens seemed to be localised in the lateral parts of the A band, and the HMM antigen in the neighbourhood of the M line. This led to the tentative hpothesis that the myosin molecules (very greatly extended to have a length of some 6000 Å instead of the usually reported 1600 Å) were oriented with the H portions

---

[1] Finck, Holtzer & Marshall (1).

pointing inwards to the M line, the fibrous LMM tails towards the Z line; or that the molecule existed in the form of widely separated sub-units.

In 1961 Tunik & Holtzer (1) studied the distribution of antigens in contracted myofibrils. As contraction became greater the anti-myosin became more and more confined to the lateral portions of the A band; the I band antigens (reacting to anti-actin sera) and the H zone antigens (reacting to anti-actin and anti-HMM sera), kept the same relative positions through all stages of contraction – which would not be expected on the sliding model. Szent-Györgyi, Holtzer & Johnson (1) confirmed that the area of low fluorescence at the centre of the A band became progressively wider as the I band shortened, while the width of the fluorescent area at the end of each A band decreased linearly with contraction. They also found that the amount of bound antibody was about the same whether the muscle was at resting length or contracted.

The interpretation was as follows:[1] the myosin-containing structures were interrupted at the centre of the A band, and during contraction these structures became shorter in proportion to the shortening of the sarcomere. The theory of contraction based on the observations needed some further assumptions: 1, that the continuity of the A filaments was maintained by a core of unknown material, though the myosin-containing structure (or myosin lattice) was interrupted at the centre of the band; 2, that connections could be made between the ends of the thick filaments and the contralateral thin filaments in the other half of the sarcomere; in the resting muscle this connection was extensible but in the active muscle was stiffened; 3, that the primary event in contraction was (at A_1) formation of a link between the myosin in the thick filament and actin in a neighbouring thin filament; the myosin lattice was thus anchored to the Z line on its own side of the sarcomere, and its shortening now could draw the Z lines towards one another. The mechanism envisaged is shown in fig. 56.

It must be recorded now that further work by Pepe (1) with the antibody staining method has given another explanation for the apparent removal of myosin from the centre of the A band and has led to his conclusion that antibody results were consistent with the sliding-filament model. Moreover the work of Stephens (1, 2) on selective immobilisation of different parts of the sarcomere by means of narrow-beam ultra-violet irradiation is weighty evidence for this model and against the 'contralateral-filament' model.[2]

In his work of 1966 Pepe (1) paid special attention to the quantitative aspect of the localisation of myosin. He found that uniform staining of the A band in resting muscle (with the exception of the small unstained central area) only happened with a large excess of antibody. With smaller amounts four fluorescent bands were seen – two medial and two at the ends of the

[1] A. G. Szent-Györgyi & Johnson (1).   [2] See p. 261 below.

A band. The reaction of the centrally-situated material with anti-HMM and of the laterally-situated material with anti-LMM was confirmed. Since the appearance, size and method of building up of the A filaments was now known, it was no longer possible to entertain the sort of explanation suggested earlier by Marshall *et al.* Pepe now suggested that the HMM part of

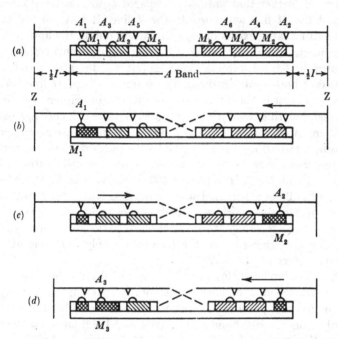

Fig. 56. A theory of contraction. (*a*) A resting sarcomere at equilibrium length: sites which can potentially interact with each other are shown on the actin and on the myosin component. (*b*) On excitation a connection is formed between the myosin lattice on one side and the actin filament on the other side. A and M sites which are within a critical distance of one another will interact, and that portion of the myosin lattice associated with the interacting M site will contract. (*c*) The contralateral actin filament will be pulled in and a new set of A and M sites will interact. (*d*) This interaction will then bring about a third event of the same type. (A. G. Szent-Györgyi, Holtzer & Johnson (1).)

the myosin molecule could only react with anti-myosin or anti-HMM when it was not attached to actin;[1] further that, owing to tight packing of the LMM ends of the myosin molecule the LMM antigenic sites were not available for most of the length of the A filaments. However at the tapering ends of the myosin filaments the packing might be less tight; moreover the tapering would mean a greater distance between the myosin and actin filaments and cross-bridge formation between them might mean a loosening

[1] Compare the suggestion made by Gergely in 1959 (2), p. 549.

of the LMM packing, and consequent exposure of antigenic sites to a degree dependent on the overlap of the filaments. The fact that there was no change in the amount of antibody taken up in the work of Szent-Györgyi *et al.* when the muscle contracted could be explained by decrease in anti-HMM uptake being balanced by increase in anti-LMM uptake.

Pepe showed further that antibody prepared against actin stained the *I* bands and the *M* line and possibly the *Z* line. It was purified (*a*) by absorption with a large excess of tropomyosin to obtain the anti-actin; or (*b*) by absorption with a large excess of F-actin to obtain anti-tropomyosin. In both cases the *I* band staining remained but was diminished, indicating the presence of both proteins there. The staining of the *M* line was undiminished and must be due to some other antigen. In neither case was the *Z* line stained, but this could not be taken as decisive evidence against the presence there of one or both of the proteins. Antibody prepared against tropomyosin, and purified by absorption with purified actin, stained the *I* band, and there were some indications of uneven distribution of the tropomyosin in the band. The strong 400 Å period seen in the electron microscope in *I* filaments treated with unpurified anti-actin could not be due to tropomyosin, since purified anti-tropomyosin did not give this repeat period. Pepe suggested that the interaction of actin and tropomyosin is such that the antigenic sites of the former are only available at these regular 400 Å intervals. (Cf. p. 297.)

Pepe (2, 3) has extended this work by means of electron microscopy and from it has developed a detailed model of the packing of the myosin molecules in different parts of the filament. He also now obtained evidence for a third antigenic site besides the HMM and LMM sites – that of the trypsin-sensitive regions (see fig. 55, p. 257). Antibody for these sites could be prepared from LMM obtained after 15 min tryptic digestion, but not after 60 min digestion; they were available only in approximately the middle third of each half of the *A* band. Study of the effects of sarcomere length on the availability of the trypsin-sensitive sites led to the suggestion that in the closely-packed region of the thick filament much distortion of these sites is needed before the cross-bridges can interact with actin. This distortion makes the sites available for antibody staining, which is not seen at the tapering ends of the thick filaments or under conditions where there is no overlap with actin. The staining is particularly marked in contracted fibrils, as might be expected owing to the greater distance between the filaments. The work with antibody staining has thus resulted in substantial support for the generally accepted sliding mechanism.

Another mechanism depending on the movement past each other of the two types of filament is that of Podolsky (2, 3). Here upon activation the ends of the flexible actin filaments become fixed in relation to the thick filaments, and shortening of the actin filaments[1] takes place as the result of unbalance of electric and entropic forces. Interaction occurs between sites binding substrate on the actin and enzymic sites on the myosin. Certain advantages were claimed for this model in ease of fitting to the thermo-dynamic consequences of Hill's energy relations, but it must be said at once that from the structural point of view it has proved deficient, for it involves constancy of the H zone during contraction and this is contrary to general experience, as we have seen. Podolsky (4) although showing some preference for the folding over the sliding model, did not insist on this attitude and recognised the behaviour of the H zone during shortening as 'a crucial point in scoring various theories of muscle contraction'.

The work of Stephens (1, 2) (in A. G. Szent-Györgyi's laboratory) in 1964 and 1965 was specially designed to test between the three models, of Hanson, the Huxleys and Niedergerke, of Szent-Györgyi and of Podolsky, that we have been considering. He first showed (1) by the fluorescent anti-body method that, when locally contracted sarcomeres could be observed, the shift in fluorescent anti-myosin labelling took place in contracted half sarcomeres, and not (as would be predicted on the Szent-Györgyi–Johnson theory) in the opposite half of the sarcomere. He then developed a method whereby an ultra-violet micro-beam, about $1\,\mu$ in diameter, could be used to inactivate chosen fractions of a sarcomere. The effects of such irradiation on contraction may be summarised as follows (see fig. 57 pl.). Full irradiation of either the I or the A band prevented contraction in the sarcomeres or half sarcomeres involved – a result to be expected on any of the three models. Irradiation of two-thirds of an A band allowed contraction at the non-irradiated third. This result would be consistent with the sliding mechanism but inconsistent with the contralateral-filament mechanism, since one would expect the contralateral connections to be disrupted and the myosin (needed to draw in the actin filaments) to be inactivated. With irradiation of the I band, released parts of actin filaments (overlapping with the A band and so not irradiated) were seen to slide into the A band, with disappearance of the H zone – a result in agreement with both the sliding filament and contralateral-filament theories but not with the theory of actin folding. It was concluded that the results supported the Huxley–Hanson mechanism.

Szent-Györgyi (4) in 1968 recorded his opinion, as the result of the work of Pepe and of Stephens, as well as of further unpublished work in his laboratory, that the Szent-Györgyi–Johnson model must be discarded. He was ready to accept the relevance of the classical sliding-filament theory.

Another such contribution to recent thought on the sliding mechanism is that of G. M. Frank (1) summarising the work of many collaborators. A few

[1] See also Morales (1).

of their findings may be mentioned here. The diffraction of visible light caused by the cross-striations of living muscle was studied. The diffraction picture given by means of a rapidly rotating, slitted disk was scanned on a photomultiplier and the distance was measured between the diffraction bands for successive stages of a single contraction. Thus the course of the curve for the behaviour of the cross-striation could be compared with the curve for the mechanical effect. It was noted that while the mechanical curve was practically monotonic, the change in cross-striation had a step-like character. This was interpreted as meaning that 'the monotonic process of sliding during short time intervals is a step-like pulling up of the thin filaments by the thick ones really shortening several times during a single contraction'. In support of this idea certain conditions are cited in which A-band shortening (without disappearance of the I band) was observed: e.g. in acetylcholine contraction of denervated sartorius.

Further evidence for the sliding-filament theory of H. E. Huxley & Hanson and A. F. Huxley & Niedergerke will be described in the next chapter together with some of the many attempts which have been made, both in theory and by experiment, to arrive at detailed knowledge of the molecular mechanism of the sliding.

12

HOW DOES THE SLIDING MECHANISM WORK?

INTRODUCTION

In the last chapter attention was concentrated mainly on the various types of evidence which contributed to the knowledge of the fine structure of the myofibril at rest and contracted. The generally preferred conception emerged of the interaction of two types of filament, one consisting mainly of myosin, the other mainly of actin, intermittently linked by bridges from the myosin; and of contraction as depending not on shortening of the filaments, but on the degree of overlap (thus on the degree of possibility of interaction) of the two sets of filament. The manner in which such a conception could fit with such well-known facts as the effects of ATP on actin/myosin association, or energy provision by actomyosin-catalysed ATP hydrolysis, was explored in a general way. Now we have two tasks – first that of considering more specific theories attempting to explain *how* the sliding could take place and derive its energy; secondly and mainly that of describing experimentation of recent years which, in various ways, has set out to throw light on possible conformational changes in the proteins during contraction, or under conditions which might obtain during contraction, or which might resemble such conditions. In some cases the possibility that these changes might be relevant is simply stated; in other cases, the results are assembled to support certain detailed ideas concerning the mechanism of sliding and the provision of energy for it.

TWO EARLY THEORIES BASED ON THE SLIDING MECHANISM

One of the first attempts to visualise the sliding process was that of H. H. Weber (8) in 1955,[1] making use exclusively of formation and breakage of covalent linkages. Here ATP first phosphorylates an acid group on the actin, with formation of an energy-rich phosphate bond. Reaction of the acid group with SH on the myosin follows and the phosphate is set free, the bond formed still being energy rich; the energy is liberated in subsequent reactions with phenolic and alcoholic hydroxyls. At this point, by movement

[1] Also H. H. Weber (9).

[263]

along the myosin, the active group on the actin has been brought into the neighbourhood of a second SH group, and the process is repeated. The thermodynamic behaviour of the muscle can be explained in a qualitative way. With small load the filaments move more quickly from State III to State V, and thus the rate of energy release from ATP is greater than with larger loads; on the other hand the greater amount of energy provision for unit distance of shortening with a large load, could depend on individual sites being active more than once during the slower shortening under heavy load. This repeated activity would be consequent upon breaking of the links (under the strain of the load) and their reformation after withdrawal of the actin filaments to their initial position.

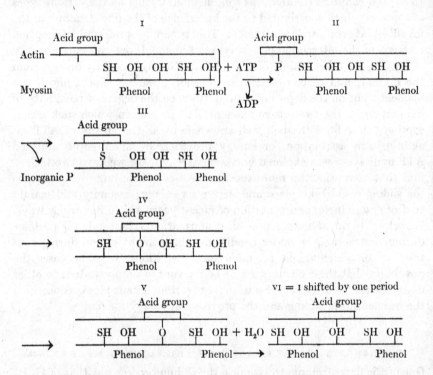

The phosphorylation of actin suggested here was prompted by the findings of Ulbrecht & Ulbrecht (3) about this time in Weber's laboratory. Using many times precipitated natural actomyosin or well-washed myofibrils they found, in presence of Mg, phosphate exchange between added $AD^{32}P$ and ATP. If the myosin was selectively extracted from myofibrils, the remaining 'ghosts' showed the exchange while the myosin did not. These results were interpreted as showing reversible phosphorylation of actin – though actin preparations from muscle acetone-powder did not show the phenomenon. In subsequent work by M. Ulbrecht (1) the extremely large phosphate exchange with cell granules from muscle was discovered; their activity was 30–40 times that of an equal weight of fibrils and the exchange in purified actomyosin or in myofibrils could be accounted for by 2 % contamination with granules. It now seems possible that rather than actin it is myosin that undergoes phosphorylation as an intermediate stage in ATP hydrolysis,[1] though owing to the very transient nature of the phosphorylated form in presence of actin, its existence has been difficult to demonstrate.

A. F. Huxley's very comprehensive study (1) of the contractile properties of muscle in terms of a sliding mechanism was published in 1957. It was postulated here that the myosin filament bore side-pieces capable of oscillating (to an extent limited by an elastic connection) backwards and forwards in Brownian movement about an equilibrium position. This side-piece bears an active site M capable of reacting with a site A on the actin, if the sites are within the right range. Tension can then be exerted by the stretched elastic connection, and as it pulls the side-piece back, the actin filament moves also.

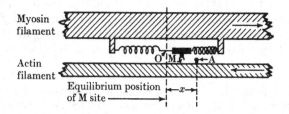

The rate constants (f and g), for the making and breaking respectively of the actomyosin connection, depend on the distance of the site A from the equilibrium position. A biochemical description of the sequence of events could be written thus:

$$1. \quad A + M \longrightarrow AM \text{ (rate constant } f)$$
$$2. \quad AM + XP \longrightarrow AXP + M \text{ (rate constant } g)$$
$$3. \quad AXP \longrightarrow A + X + PO_4$$

where XP is a high-energy phosphate compound, presumably ATP. The potential energy (resulting from the thermal motion of the side-piece) stored in the elastic connections is the immediate source of contraction energy, but this has to be made good by the energy supplied in the ATP breakdown needed to break the actomyosin link. It is supposed that prevention of reaction 3 is the condition for the relaxed state, while for

---

[1] P. 276 below.

onset of the active state reaction 3 followed by 1 is necessary. The activation heat could be associated with reaction 3. When certain values were given to f and g, reasonable agreement with Hill's characteristic equation was found.

A number of phenomena concerned with contraction were discussed in the light of this hypothesis. Thus for the latency relaxation; if we regard the series elastic element as made up of the postulated S filaments running through the H zone and connecting the ends of the actin filaments,[1] a very slight lengthening of the latter would result in shortening of the S filaments with decrease in their tension. The early rise in rigidity would be due to reaction 1.[2]

It is interesting to notice the both Weber's and A. F. Huxley's theories were evolved before the establishment of the actual existence of the cross-bridges.

In his important paper of 1960, H. E. Huxley (9) considered a number of interesting points. For example he recalled the proposal of Hill[3] in 1938 that if the maximum load (P_0) and the actual load (P) are regarded as in some way proportional to the number of linkages generating tension, then P_0-P might be considered a measure of the free linkages. Since P_0-P was taken as proportional to the rate of energy liberation, this statement, that the rate of energy release should be proportional to the number of *free* linkages seems paradoxical. It can be explained however if we regard the re-activation of free linkages as the rate limiting step in the contraction mechanism. As Huxley writes:

It is rather like having a team of men pulling on a long rope. If the rope is being drawn in rather quickly, then at any instant several men will always be in process of shifting their grip on it after completing one cycle of pulling, and the tension will be less than the maximum which could be exerted by all the men simultaneously. The total rate at which work is being done can be ascertained by counting the number of men not pulling at any moment (say, seven), and observing how long it takes them to change their grip (say it was 1 second). Then the total rate of working must be 7 man-cycles per second, for there must be a steady flow of seven men per second into the pool of those who have just completed a cycle of pulling.

It should be mentioned here in parenthesis that in 1964 A. V. Hill (25) returned to the question of shortening heat. With improved methods it now became clear that the shortening heat did vary with the load lifted and that the constant a of the characteristic equation was not the same as the constant (now to be known as α) involved in the thermal study of shortening; however the two constants were clearly connected in some way. Sandow (5) in 1961, assembling values of a and b from a large number of published

[1] Ch. 11, p. 250.   [2] P. 177.   [3] P. 175.

experiments, had expressed doubts as to their constancy, and pointed out that the thermal and mechanical values seemed never to have been determined on the same muscle. The original statement of the characteristic equation contained the hypothesis that the rate of extra energy liberation as work and shortening heat, $(P+a)v$, was proportional to (P_0-P), the difference between maximal isometric tension and actual load lifted. It now seemed necessary to modify this idea, in assuming that the rate of extra energy liberation depends both on (P_0-P) and also on v, the velocity of shortening. This would mean that extra energy liberation is increased during shortening by a greater difference between P_0 and P, but at the same time diminished by a greater velocity v. Podolsky (4) pointed out that this non-linearity of the force–energy relation posed a new question which must be tackled in correlating the behaviour of models based on the double set of filaments with behaviour of the contracting muscle.

Another property of striated muscle only recently realised is the existence of a control mechanism whereby the mechanical conditions can regulate the active state of the excited muscle. This pattern was first brought out by Pringle (1, 2) in work on the rhythmically contracting muscle in certain insects, e.g. in the tymbal concerned in the song of cicadas. Here the rhythm is maintained by deactivation of the myofibrils by their quick-release at the instant of clicking of the tymbal into the IN position; followed by pulling out of the muscle to its initial length (owing to the elasticity of the exoskeleton) and redevelopment of tension. At the moment of quick-release the tension falls from about 18 g to about 0.5 g within 1–2 msec; the redevelopment of tension takes place without further excitation, since a single stimulus can maintain the active state long enough for several clicks to occur. Pringle (5) has considered this control mechanism in relation to the sliding-filament theory and its relevance to contraction of skeletal muscle in general will be considered in chapter 14.[1]

We have already seen what importance Astbury[2] attached to the cross-β configuration for contraction. In 1960, after describing the β-like folds transverse to the fibre axis in the egg-stalk of the fly *Chrysopa*[3] he emphasised that this configuration must be seriously taken into account, in spite of its apparent conflict with conclusions from electron microscopy. He wrote (4a):

This gap between the two approaches must, of course, be bridged before ever we can feel truly confident that we are at last along the right lines. But is there any real gap – any gap, that is, beyond the present inability of the electron microscope to reveal what takes place at the moment of contraction? It shows the structure only before and after contraction (or elongation) but throws no light on the process between.

[1] See p. 360.   [2] Ch. 9.   [3] Parker & Rudall (1).

No evidence for the cross-β configuration in contracting muscle or super-
contracting actomyosin preparations has come to light; indeed Gabelova (1)
specifically states that examination of infra-red absorption after hydrogen-
deuterium exchange[1] gave no evidence for any change from the α to the
cross-β conformation during superprecipitation. Nevertheless it is clear that
transitory and localised changes in the myosin or actin or both must be
contemplated. As we have seen,[2] Huxley, Brown & Holmes (1) as well as
Elliott, Lowy & Millman (1) themselves recognised that they could not be
ruled out. We must now consider possible sites for such changes.

DIRECT EXAMINATION OF CHANGES IN CONFORMATION OF MYOSIN WHEN IT INTERACTS WITH ATP OR ACTIN

Many attempts have been made to determine whether changes in helical
content of myosin follow the addition to it of ATP or of actin;[3] the result
was small or negative. Also treatment of myosin with certain inorganic salt
solutions causing (at high concentration) a very marked decrease in helical
content showed that ATPase activity disappeared almost entirely at low
salt concentrations, where the effect on helical content was small.[4] Thus
only a very limited unfolding round the active centre could be postulated
to account for the loss of enzymic activity. Brahms & Kay (1) in 1962 used
a number of organic solvents and solutions (e.g. of urea) known to cause
important conformational changes in synthetic polypeptides. At low concentrations of many of these reagents there was increased ATPase activity
of the myosin, but no sign of conformational change when this was investigated by hydrodynamic or optical methods. The order of the solvents,
arranged according to their effect on ATPase, was however the same as that
of their ability to cause conformational change in synthetic polypeptides;
again it appeared that any protein unfolding involved in the increase of
enzymic activity was small and restricted to the active site. With 45 %
ethylene glycol, optical rotatory measurements did show a marked decrease
in right-handed helix, while the ATPase activity increased by 200 %.

In 1964 Iyengar, Glauser & Davies (1) examined the effect of ATP upon
the absorbance of heavy meromyosin and acto-HMM at 187–195 mμ. This
procedure has been used for following changes in the α-helix content of
polypeptides and proteins.[5] They found an increase in absorption (to be
interpreted as a fall in the helix content) upon ATP addition to HMM and
more markedly with acto-HMM. The effect which was transitory (no doubt

---

[1] P. 269 below.   [2] P. 255.
[3] Tonomura, Sekiya, Imamura & Tokiwa (1); Kay, Green & Oikawa (1); Tonomura,
   Tokura & Sekiya (1).
[4] Tonomura, Sekiya & Imamura (1).   [5] Rosenbeck & Doty (1).

owing to ATPase activity) was enhanced by the presence of Ca^{2+}, and was not given by ADP. No change in F-actin alone was observed.

Morita & Yagi (1) observed the absorption spectrum of HMM at 288 mμ, the region where aromatic chromophores are markedly effective. A red shift was found when ATP was added; this change was also caused by ADP, inorganic triphosphate and pyrophosphate, but to a less extent. There was no evidence for any change in the difference spectrum of tryptophan at 293 mμ; thus the difference spectrum at 288 mμ was ascribed to side-chain chromophores in tyrosine residues. Doty & Gratzer (1) had observed a blue shift around 280 mμ during the transition from α-helix to random coil: Morita & Yagi therefore interpreted their result to mean no decrease in α-helix, but some other conformational change upon binding of ATP to the active site, involving the burying within the protein of several tyrosine residues.[1]

Another method which has been used with polypeptides to examine the degree of orderliness within the structure has been that of susceptibility to deuterium–hydrogen exchange, detected by changes in the infra-red absorbance. Gabelova (1) worked with this method to examine the effect of superprecipitation of actomyosin by means of ATP, using thin, rapidly dried films of the protein before and after superprecipitation. Her results on the absorbance in the region 2963–3300 mμ led to the conclusion that on superprecipitation there is some transformation of easily exchangeable peptide groups into the hard-to-exchange state. She was inclined to interpret this as meaning an increase in helical content of 10–13 % or an enhancement in stability of part of the existing helices. It may be noticed that Tonomura, Tokura & Sekiya (1) found by the method of optical rotatory dispersion, a decrease in the helical content of myosin solutions when actin was added in the binding ratio of 1:3.7; here of course the conditions were very different – the proteins being in solution and ATP absent.

This is a very difficult field. In 1969 Gratzer & Lowey (1) commented on the variation in results in the literature concerned just with the effect of added ATP upon the conformation of myosin. Their own results on optical rotatory dispersion and on the absorption in the far ultra-violet showed clearly that the presence of ATP caused no change greater than 0.2 % in the α-helix content. Also there was no change in the sedimentation coefficient such as would be expected with a change in shape of the protein. A small conformational change was indicated by a slight increase in binding of bromthymol blue. More such work on the protein, under various conditions relevant to contraction, is needed.

[1] Also Sekiya & Tonomura (1).

A POSSIBLE PART PLAYED BY CHANGES IN PROTEIN HYDRATION

Another method of attack was that of Rainford, Noguchi & Morales (1) who examined the effect of greatly increased pressure on the rates of superprecipitation of actomyosin and of its ATPase activity. At a pressure of 500 atmospheres both these rates were approximately halved; no such effect was found on myosin ATPase. The authors suggested that the site of action of the pressure was the formation of the activated complex of the two proteins, which might involve release of electrostricted water.

The early work of Szent-Györgyi, on adsorption of cations by myosin and shifts in its isoelectric point,[1] is recalled in a paper by Lynn (1), where interest is focused on the consequent changes in hydration of the protein. Experimental evidence was obtained showing the need for ATP or other polyanions in these reversible hydration changes,' and the effect of varying concentrations of Ca and Mg ions – the former ion on the whole favouring dehydration, the latter favouring hydration. Actomyosin showed similar behaviour, but with greater polyanion specificity. These observations are used in the formulation of a theory of contraction not requiring covalent interaction between the A and I filaments. It is supposed that the first event is the change in surface charge and isoelectric point of the K-myosin-ate, brought about by the changing concentration of various cations and of ATP. The Ca-ion increase on stimulation leads in this way to myosin de-hydration, and readier association between the myosin in this form and actin; the result is the drawing in of the actin filaments into the K-myo-sinate gel. The role of the actomyosin ATPase is regarded as that of en-suring rapid changes in the ATP concentration, thus making the system quickly reversible, its presence being necessary for both contraction and relaxation. Energy provision is not discussed. From the effects of disulphide reagents and NEM, it seemed that the same three SH groups essential for ATPase activity[2] were also essential for the hydration and solubilisation of the myosin; it was therefore suggested that the site of changes in hydration was the HMM.

POSSIBLE CONNECTION OF SPECIFIC AMINO-ACID RESIDUES
WITH CONFORMATIONAL CHANGES

The indication in the work of Morita & Yagi has been mentioned that upon addition of ATP some of the tyrosine residues in the HMM molecule showed a somewhat changed absorption spectrum. Tonomura, Sekiya, Imamura & Tokiwa (1) in 1963 found that the number of abnormal tyrosine groups in myosin increased from 3.6 to about 6.5 and 5.5 moles/10^5 g of protein when

---

[1] E.g. A. Szent-Györgyi (10).　　　　　　　　　　[2] P. 286 below.

pyrophosphate and ATP respectively were added. On the other hand Lowey (2) using skeletal muscle myosin, and Verpoorte & Kay (1) using cardiac myosin, have found no evidence for 'buried' tyrosine residues. The position remains unclear.

In 1960 Kubo, Tokura & Tonomura (1) found that two lysine residues (out of the 357 contained in one myosin molecule) were specifically attacked by 2,4,6-trinitrobenzenesulphonate (TBS) giving trinitrophenyl myosin. The ATPase activity was markedly depressed by the binding of the TBS, and its binding was to some extent affected by the presence of SH reagents. These facts suggested that these two lysine groups lie in or near the active centre. In later work the TNP myosin was digested with trypsin and the TNP-containing peptides were examined.[1] These were all contained in the HMM part of the molecule, and were characterised by very high proline content. One peptide contained the sequence Asp-Pro-Pro-TNPlysine. Since it is known that proline does not fit into the α-helix, it was suggested that the high proline content might be involved in making possible the specific and local transconformation at the active site.

Sekine and his colleagues have identified short amino acid sequences in the myosin (a) at the site (S_1) where maleimide interaction leads to activation of Ca-activated ATPase;[2] and (b) at the site (S_2) where further interaction leads to inhibition.[3] These experiments were made with N-(4-dimethylamino-3,5-dinitrophenyl)maleimide, which could be detected in the peptides after trypsin digestion. The sequences were, for S_1, Ile-Cys-Arg; for S_2, Cys-Asp-Gly. Kielley, Kimura & Cooke (1) independently proposed Ile-Cys-Cys-Arg for S_1.

Work concerned with the specificity of myosin SH groups, in ATPase activity on the one hand and actin binding on the other, is discussed below (p. 285ff.).

THE SIGNIFICANCE OF SUBSTRATE SPECIFICITY, AND OF THE ACTION OF CERTAIN MODIFIERS OF ATPASE ACTIVITY, FOR INTERPRETATION OF THE BEHAVIOUR OF THE ACTIVE CENTRE

During the last twelve years much work has been done on the effects of a number of modifiers upon myosin-ATPase activity. It is impossible to describe these exhaustively here, but interesting generalisations emerge, some in connection with conformational changes in the protein, and a brief account of these must be given.

---

[1] Kubo, Tokuyama & Tonomura (1).
[2] Yamashita, Soma, Kobayashi, Sekine, Titani & Narita (1); see pp. 275 and 288 for a discussion of sites S_1 and S_2.
[3] Yamashita, Soma, Kobayashi & Sekine (1).

In 1954 Friess (1) made the interesting discovery that the chelating agent ethylenediaminetetraacetate (EDTA) has a strongly activating effect on myosin B in 0.6 M KCl, in the presence or the absence of Ca ions. Friess, Morales & Bowen (1), having washed the myosin B with EDTA solution, removed the EDTA, and applied a second dose which again caused activation, thought it likely that the effect of EDTA was mediated by its binding to very firmly held Mg.[1] Bowen & Kerwin (2) independently recorded the EDTA activating effect, finding it more pronounced with myosin A than myosin B, and only observable at ionic strength above about 0.17 – thus concerned with myosin, not actomyosin.

In 1955 Greville & Needham (1) observed that, in concentrations about 10 μM, phenylmercuric acetate (PMA) inhibited the Ca-activated ATPase of myosin, while concentrations of only 4 μM caused some 40 % activation at low ionic strength; the activation increased with rising ionic strength. At the same time Chappell & Perry (1), with washed myofibrils, saw similar effects of PMA. These workers also studied the effect of dinitrophenol (DNP), which was of interest because of its uncoupling effect on oxidative phosphorylation[2] and because Webster (1) had reported an activating effect on the ATPase activity of myosin. Greville & Needham with Ca-activated myosin ATPase found activation with 4 mM DNP and no effect on actomyosin at low ionic strength; but with increasing ionic strength activation appeared, being more than 100 % at $I = 0.3$. Similar effects were obtained by Chappell & Perry, on both myosin and myofibrils; in experiments with synthetic actomyosin the ATPase became less sensitive to DNP activation with increasing amounts of actin added. They also made the interesting observation that DNP activation of myofibrillar ATPase was temperature-dependent, being very much less at 0° than at 20°.

Greville & Needham found that with myosin as enzyme and ITP as substrate the dephosphorylation (more rapid than that of ATP under the control conditions used) was not affected by DNP concentrations up to 2×10^{-5}M; above this very low concentration there was inhibition. Greville & Reich (1) went on to show that CTP, containing an NH_2 group in the 6-position of the ring, behaved like ATP, in undergoing increased breakdown in presence of DNP, PMA or EDTA, while with the 6-OH compound GTP, as with ITP, dephosphorylation was inhibited under the same conditions. Similar observations on the specificity of the nucleoside triphosphates were made by Kielley & Bradley (1). Greville & Reich suggested that some interaction between enzyme and substrate, dependent on the NH_2 group,

---

[1] Mühlrad, Fábián & Biro (1) and Offer (4) have since made a good case for regarding the activating action of EDTA as consisting in removal of minute traces of Mg added with the reagents. With Chelex-purified K-activated myosin and reagents, and using EDTA–metal buffer, Offer found myosin ATPase was completely inhibited in presence of 10^{-7}M Mg.   [2] P. 414.

affects the enzymic velocity in the unmodified enzyme, but is disoriented in presence of these modifiers.

Kielley & Bradley using *p*-chlormercuribenzoate (PCMB) and *N*-ethylmaleimide (NEM) independently observed the activating effect of controlled concentrations of these SH reagents, followed at higher concentrations by inhibition. They suggested that two types of SH group were present at the active site; of these, one or more seemed to take part in an interaction preventing the maximal activity of the Ca-activated ATPase. Taking into account the EDTA activation of myosin, and the fact that in presence of EDTA no activation but rather inhibition was observed with SH reagents also added, Kielley & Bradley suggested that SH groups might be involved in binding of metals.

In 1960 Gilmour (1) examined the pH dependence of the ATPase and ITPase activity of Ca-activated myosin. The ATPase curve at 25° had the shape we have already mentioned,[1] with a trough at about pH 7.0, and a peak in the neighbourhood of 6.3. These marked changes were not seen at 0°, nor were they found at either temperature in the ITPase–pH curve (fig. 58). These observations led him to the view that with decreased H^+ concentration at the higher temperature a negatively-charged inhibitory group appeared and played its part in the binding of the 6-amino group of the substrate. The ATPase–pH curve after treatment of the enzyme with PCMB or DNP was similar to that for the 6-OH compound with untreated enzyme. The facts (1) that the ATPase–pH curve at 0° resembled that for ITPase; (2) that PCMB activated ATPase at 25° but only inhibited at 0°; and (3) that ATP was hydrolysed more quickly than ITP at 0°, more slowly at 25° were all explained on the hypothesis that a conformational change necessary for strong 6-NH_2 binding took place above a critical temperature and that the inhibitory group responsible for this binding was blocked by PCMB. Blum (3) had also discussed the behaviour of myosin with its various substrates in presence of various modifiers and under various conditions in terms of the ability of those containing the 6-NH_2 group in the purine ring to induce a conformational change in the neighbourhood of the enzymic centre; he regarded the enzyme acting on 6-OH substrates as a 'simpler enzyme'. He emphasised that other groups on the enzyme beside the SH group must participate in the postulated conformational change.

In 1961 Gilmour & Gellert (1), by means of changes in optical density at 250 mμ, examined the binding of PCMB to myosin. The binding of 3–4 mole/10^{-5} g of myosin took place rapidly and was associated with acceleration of ATPase activity and inhibition of ITPase. Further binding went on, up to a maximum of 7.0–8.0 mole/10^{-5} g of protein, more and more

[1] Ch. 8, p. 152.

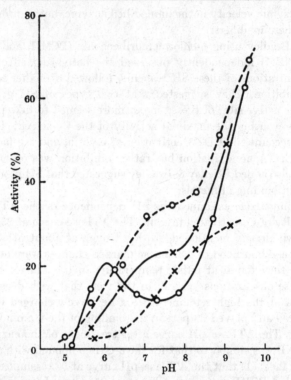

Fig. 58. pH dependence of myosin ATPase and ITPase at 25°. ———, ATPase activity;
- - - - -, ITPase activity; O—O, untreated myosin; ×—×, myosin treated with
p-chlormercuribenzoate. Activity expressed in moles P liberated per g myosin per sec
(after Gilmour (1)).

slowly; this phase was associated with inhibition, which eventually became
complete, of both ATPase and ITPase. A model of the active site to explain
these phenomena was proposed.

Rel. PCMB conc. moles/10^5 g		ATPase activity	ITPase activity
0		100	250
4		200	80
6·5		0	0

x = position blocked by PCMB.

The similarity of action of the modifiers PCMB, PMA and DNP made it tempting to suppose that they all reacted with the same enzymic site, but Levy & Ryan (1) found that there was no competition between PCMB and DNP. They further found that, in presence of Mg^{2+} and absence of Ca^{2+}, ITPase *was* markedly activated by either DNP or PCMB. Thus again metal ions seem to be implicated in this intrinsic inhibition, and it appears not to be restricted under all conditions to substrates having the 6-amino group.

Levy, Sharon, Ryan & Koshland, a little later, studied the temperature dependence of the rate of hydrolysis of ATP and ITP by myosin. ITP hydrolysis gave a biphasic Arrhenius plot, curving at 16°, while that for ATP was linear, unless a modifier such as DNP or PCMB was added, when it became similar to that for ITP with the unmodified enzyme. That ATP and ITP interact with the enzyme at the same site, and not at completely separate ones, was shown with ^{32}P-labelled ATP which completely inhibited ITPase activity. They concluded that at about 16° a conformational change took place in the myosin, involving a change of shape in the enzyme–ITP complex, reflected in the biphasic Arrhenius plot. With ATP and the unmodified enzyme this change did not manifest itself, because the 6-NH_2 group contributed to a tighter fit; but with DNP or PCMB treatment this stabilising effect disappeared and the enzymic behaviour with ATP resembled that with ITP. These observations fit well with those of Gilmour.

Levy, Sharon, Ryan & Koshland (1) have summed up such work as that described above in saying that it 'indicates that the SH groups of the protein, the group at the 6-position in the purine or pyrimidine ring on the substrate, the metal ion and the temperature all played key roles in determining the action of a modifier'. They pointed out that Chappell & Perry's results indicated that actin and DNP competed for a site on the myosin, and it is interesting that Levy, Sharon & Koshland (1) found that the ATPase of actomyosin gave the same transition temperature in the Arrhenius plot as that found with myosin ITPase or with myosin ATPase with modifiers; thus the actin was behaving as a modifier.

Sekine and his collaborators in 1964 drew much of this work together. Thus Sekine, Barnett & Kielley (1) showed that loss of EDTA activation took place when one SH group (out of 15 contained in 2×10^5 g of myosin) reacted with NEM. Sekine & Kielley (1) found that under precisely the same conditions Ca-ATPase (at temperatures above 25°) showed activation. Thus the same SH group (S_1)[1] must be involved in the two cases. Interaction of NEM with this group resulted in an altered pattern of behaviour which now resembled that of ITPase with respect to substrate specificity, and dependence on ionic strength, pH and temperature. Yamaguchi & Sekine (1) have also described the further effects of N-ethylmaleimide (NEM) on

[1] P. 271.

S_1-blocked myosin.[1] The SH groups concerned in the inhibitory phase of the PCMB titration curve reacted very slowly to NEM; but if ATP (or other nucleoside triphosphate substrate) were present during the incubation with NEM, then rapid and almost complete inhibition was obtained. It was suggested that the presence of the substrate induced a conformational change in the active site rendering it sensitive to this inhibitor.

One may perhaps summarise by saying that DNP and the SH-reagents mimic rather closely the effects of actin on myosin ATPase – as has been remarked by Levy, Sharon & Koshland for DNP (1). This is apparent, for example, in the changes which they produce in the relative enzymic activity with different nucleosidetriphosphate substrates. The main interest of the EDTA activation is to emphasise the Mg-inhibition of the myosin ATPase and thus to stress its great difference from the Mg-activated ATPase in presence of actin.

Sekine (1) has designated myosin an 'allosteric protein', and the modifiers considered here 'allosteric effectors'. According to this concept of Monod and Jacob and their collaborators[2] the allosteric site is distinct from the active site; it is complementary in structure to another metabolite (the allosteric effector) which it can bind specifically and reversibly. The formation of this complex does not lead to any reaction involving the effector, but rather to a reversible alteration of molecular structure (the allosteric transition) in this sensitive and flexible area, a change which modifies the properties of the active site. The role of the naturally occurring allosteric effectors (amongst which metal ions may also be included) is thus different from that of co-enzymes, classically regarded as transient reactors and transporters (though co-enzymes might also have an allosteric role).

PHOSPHORYLATION OF MYOSIN AS AN INTERMEDIATE STAGE IN ITS ATPASE ACTIVITY AND AS A POSSIBLE KEY REACTION IN CONTRACTION

Phosphorylation of myosin was early pictured as the obvious means of transfer of energy from ATP to the contractile mechanism.[3] We shall consider here evidence for such phosphorylation.

Brahms & Kakol (1) in 1958 reported some interesting experiments in which, during ATP hydrolysis by myosin, the myosin-bound orthophosphate first increased, then diminished again. Simultaneously there was a decrease, followed by an increase, in the content of free SH groups (measured by amperometric mercurimetric titration). The results suggested transitory phosphorylation of the protein through the SH groups. From experiments

---

[1] Also Sekine & Yamaguchi (1).
[2] Monod, Changeux & Jacob (1); Monod, Wyman & Changeux (1).   [3] Ch. 9.

using ^{32}P-ATP and unlabelled inorganic P, or unlabelled ATP and labelled inorganic P, Gruda, Kakol & Rzyoko (1) concluded that orthophosphate bound to myosin or to HMM, during ATPase activity, originated in the ATP. However the experience of Gergely & Maruyama (1) using similar methods, led them to the conclusion that the source of the protein-bound phosphate was inorganic phosphate in the medium. The binding of this phosphate depended on the *presence* of ATP, but went on unchanged when mercurials were added to prevent ATP splitting. Gruda, Kakol & Nie-mierko (1) later agreed that phosphorylation of myosin could go on in the presence of inhibitors but maintained the position that the protein-bound phosphate came from the ATP. They suggested that the inhibitors prevented not phosphorylation but dephosphorylation of the myosin.

The conditions in such experiments make a clear-cut demonstration of myosin phosphorylation and its connection with ATPase activity very difficult: an added complication being that, as the Polish workers had shown, ATP itself is bound by myosin; for instance it seems that the immediate stopping of the reaction (by cooling, by treatment with acetate buffer at low pH or by $(NH_4)_2SO_4$ addition) without removal of any of the bound P or any breakdown of the bound ATP, presents a problem. Levy & Koshland and their colleagues used a different approach, with the use of ^{18}O-labelled water.

Koshland & Clarke (1) had shown in 1953 that during splitting of ATP catalysed by washed strips of lobster muscle in presence of ^{18}O-labelled water, the label was found in the phosphate and not in the ADP produced. Thus the cleavage of the terminal O—P bond takes place thus:

$$AMP-O-\overset{\overset{O}{\|}}{\underset{\underset{O^-}{|}}{P}}-O-\overset{\overset{O}{\|}}{\underset{\underset{O^-}{|}}{P}}-O \quad \bigg| \quad \overset{\overset{O}{\|}}{\underset{\underset{^{18}OH}{O^-}}{P}}-O^- \longrightarrow ADP + {}^{18}O^- -\overset{\overset{O}{\|}}{\underset{\underset{O^-}{|}}{P}}-O^-$$

The number of atoms of ^{18}O per molecule in the inorganic phosphate formed was greater than one – indicating that an extensive exchange reaction had occurred between this phosphate (presumably during its attachment to the protein) and the water of the medium. Here again then was evidence for enzymic phosphorylation of the myosin. Koshland, Budenstein & Kowalsky (1) went on to test whether such an exchange could be demonstrated with myosin or actomyosin ATPase, activated by Ca. Only one atom of ^{18}O was introduced into each molecule of phosphate, and the reaction seemed to involve a simple displacement mechanism; any phosphorylated intermediate, if formed, being of very transitory nature. But in 1958 Levy & Koshland (1)[1]

[1] Also Levy & Koshland (2).

found that the exchange reaction could be shown with actomyosin or myosin ATPase in the presence of Mg ions but not of Ca ions. (Very long incubation times were of course needed to show ATP splitting with myosin ATPase.) In control tests there was no exchange in the presence of myosin of the ^{18}O of water with inorganic phosphate or with the dissociable phosphate of unhydrolysed ATP remaining when the experiment was stopped; hence the exchange was taken as occurring at some intermediate stage of the ATP hydrolysis. (However, the possibility that the exchange concerned the terminal phosphate of ATP in the enzyme–substrate complex seems not to have been excluded.) Levy and his collaborators[1] also found the interesting fact that there was no correlation between extent of exchange reaction and rate of hydrolysis when the effects were studied of changing the nucleotide or metal ion used, changing the KCl or Mg concentration, or varying the temperature. On the other hand, the nucleotides and metal ions most effective for contraction and maximum tension production gave the highest ^{18}O exchange.[2]

At this point Dempsey & Boyer (1) intervened in 1961 with the finding that, in the *presence* of ATP, actomyosin or myosin *can* catalyse an exchange of inorganic phosphate oxygens with the oxygen of water.[3] This was demonstrated by using ^{18}O labelled H_3PO_4 and showing that the loss of label was equal to the ^{18}O exchange measured under exactly similar conditions by incorporation of ^{18}O from water into inorganic phosphate. Dempsey & Boyer interpreted this phenomenon as fitting in with the concept of structural changes in myosin induced by the presence of ATP making possible the interaction with inorganic P – perhaps the reversible formation of a phosphoryl derivative.

It is satisfactory to be able to record that a considerable measure of agreement has been established between the groups of workers on ^{18}O exchange, since Koshland & Levy (1) now agree that exchange between inorganic P of the medium and the ^{18}O of the water does occur during ATP hydrolysis, but they do not find that it accounts for all this exchange. The exchange at the intermediate stage was always somewhat higher than that at the medium level under their conditions, and in cases of rapid ATP splitting (as in the presence of dinitrophenol) very much higher. Boyer[4] has reported some success in demonstrating the preferential exchange with phosphate released from ATP when using the conditions used by Levy and Koshland. The reasons for the quantitative discrepancy between the two

[1] Levy, Sharon, Lindemann & Koshland (1).
[2] Also R. G. Yount & Koshland (1) for similar results with HMM. They suggest that certain amino acids at the active site are common to the exchange reaction and the contraction process.
[3] Cf. Gergely & Maruyama, p. 277 above.
[4] See p. 101 in discussion on Boyer (1).

groups are not yet entirely clear, though differences in concentration of inorganic P used and difference in length of incubation period account for part of it.[1] The work of Swanson & Yount (1) has analysed carefully the conditions for ^{18}O inorganic phosphate exchange catalysed by HMM. A significant point was that when all possibility of myokinase activity was excluded, it became clear that ADP was a more effective co-factor than ATP. This makes it more plausible that the inorganic phosphate exchanges at the active site (in a state conditioned by the binding of the phosphoryl-ated nucleotide) since it seems unlikely that ATP and inorganic phosphate could bind simultaneously to the same site.

Dempsey, Boyer & Benson (1) using glycerinated fibres contracting with ATP found the rate of incorporation of ^{18}O from $H_2^{18}O$ into H_3PO_4, as well as the ATPase activity to be increased by increasing load. No differentiation was made in these experiments between the two possible types of exchange – at intermediate or medium level.

Some very suggestive experiments have recently been reported by Cheesman & Hilton (1), using intact muscle. Frog sartorius or rectus ab-dominis was kept for 30–60 min in Ringer solution containing ^{32}P H_3PO_4; then at 0° was washed, minced and centrifuged. The residue was washed once with $0.05\,M$ $Na_2CO_3/0.05\,M$ $NaHCO_3$, then many times with water followed by several times with acetone. The same procedure was followed using muscles washed after the incubation with labelled phosphate, then put into KCl or acetyl choline contracture in the absence of labelled phosphate. The acetone powders from both sets of muscle contained phosphate, about $1.85\,\mu g/mg$ dry powder. This could be extracted with trichloracetic acid and in the extract consisted entirely of inorganic phos-phate. The phosphate in the powder from the control had relative specific activity of 61, that from the muscle after contracture of 39. Similar results were obtained with glycerinated fibres contracting with ATP. The label in the minced extracted muscle did not exchange with unlabelled inorganic phosphate ions. Relaxation with return to normal length restored the bound ^{32}P to the control value. Some evidence was obtained that the loss of label accompanied the activation process. Thus with rectus abdominis pre-incubated in Ca-free medium, immersion in isotonic KCl leads to depolarisa-tion but no contraction; with this treatment the labelled P content fell to the same extent as in contracture. These results fit in with a tentative picture drawn by Cheesman & Whitehead (1), where the myosin of resting muscle has some or all of its ATPase sites phosphorylated and is thus prevented from interaction with actin. On activation, this phosphate is released, and the making and breaking of links in contraction then goes on, involving successive phosphorylations and dephosphorylations of the ATPase

[1] Also Dempsey, Boyer & Benson (1).

sites. On relaxation the protein is left in the phosphorylated state, this phosphate having been supplied from the muscle's store of labelled ATP and phosphocreatine.

Quite another line of experimentation has been pursued by Tonomura and his colleagues, leading them to the conclusion of a phosphorylated myosin intermediate during ATP hydrolysis by myosin or actomyosin. This work followed on the discovery made by A. Weber & Hasselbach in 1954 (1) that in hydrolysis of ATP by muscle fibrils, actomyosin flocules or myosin, there was an 'explosive' phase lasting some 15 sec, during which the ATP was split at least twice as fast as during the steady phase which supervened. They suggested that this phenomenon (for which they could offer no satisfactory mechanism) might well have physiological interest since in a tetanus the ATP splitting at each excitation might be regarded as an 'initial splitting'; an explanation might be found here for the much lower splitting rate observed for actomyosin in glycerinated fibres than *in vivo*.

This interesting initial phase was now intensively studied by the Japanese workers using mainly myosin or actomyosin in 0.6M KCl. They showed[1] that the initial rate might be as much as five times as high as that of the succeeding stationary state and, by means of ^{32}P labelled ATP, that the initial burst of phosphate liberation originated in the ATP. It was much larger with Mg than with Ca activation.[2] In 1961 Tonomura, Yagi, Kubo & Kitagawa (1) suggested that the inorganic phosphate appearing in the explosive phase was actually derived from a phosphorylated intermediate of myosin – broken down on fixation with trichloracetic acid. In later work[3] they found that the production of ADP (measured by means of an enzymic system) ran a linear course from the beginning, parallel to the later curve for phosphate production.

A puzzling feature of this work is the difficulty of reconciling the early observations (where in both the German and Japanese work the P_i of the burst phase corresponded to 20–50 mole/mole of myosin) with the later ones by the Japanese workers were a 1:1 ratio is claimed for P_i bound to the active site. In 1965 Kanazawa & Tonomura (1) found that the size of the burst was greatly affected by Mg ion concentration: at concentration > 1 mM, the stoichiometric relation obtained of 1 mole of P_i/mole of myosin, but with lower concentration the extra P_i liberation was much higher – up to 7 mole/mole of myosin. An explanation was suggested in terms of the initial phosphorylation of the active site, followed by transfer of phosphate to

[1] Tonomura & Kitagawa (1, 2).
[2] The ATP splitting in the steady state in presence of Mg was of course very low – of the order of 0.01 μmole P/mg protein min for myosin. Compare the figures given by Hasselbach (1) of about 0.02 μmole/mg min in rather similar circumstances.
[3] Imamura, Tada & Tonomura (1).

some other sites in the molecule and re-phosphorylation of the active site. Similarly in the earlier work much higher uptakes of p-nitrothiophenol (see below) were found, and a similar possible explanation was advanced.

Tonomura, Kitagawa & Yoshimura (1) studied the effect of actin on the initial ATPase activity. Using myosin B, they found the extra initial phosphate under the conditions of the 'clearing phase' (0.075M KCl, 4 mM Mg ions, 2 mM ATP) but it was absent or almost so in the superprecipitated state; in the latter state the ATPase activity remained high, as if the initial burst continued into the steady period. Further evidence for the formation of a phosphorylated intermediate was found in the following facts.[1] (1) The postulated phosphorylated myosin was isolated by means of a Millipore filter. (2) When ATP and myosin were used in equimolecular proportions, a stoichiometric reaction seemed to take place, coming to an end after some 20 sec when all the ATP was used up. No steady-state liberation was seen and this evidence for the stability of the intermediate agreed with results on the isolated protein just mentioned. (3) The phosphorylated intermediate reacted with the nucleophilic reagent p-nitrothiophenol (NTP) in a manner analogous to the reaction of acetylphosphate with hydroxylamine. The reagent, added after the explosive phase, was slowly taken up in amount approximately equivalent to the phosphate of the explosive phase. The rate of NTP binding to actomyosin was much lower in the superprecipitated than in the clear state.[2] The binding of NTP to myosin in approximately 1.5:1 molar ratio (under conditions in which the phosphate burst showed a 1:1 molar ratio with myosin) completely prevented superprecipitation when actin was added, and the actomyosin type of ATPase activity; with myosin it prevented the extra P uptake but did not affect the steady state ATPase of the myosin. It was concluded that the phosphorylation of myosin is both an obligatory intermediate step in actomyosin ATPase and a key reaction in superprecipitation – therefore by analogy in contraction.[3]

Tonomura, Kanazawa & Sekiya (1) have emphasised the need for Mg ions both for contraction and for the rapid phase and also a correlation between the ability of ATP analogues to elicit contraction and their ability to cause the initial burst of phosphate liberation.[4]

The findings of the Japanese group may be summarised (for myosin ATP in presence of Mg ions) in the following scheme.[5] Myosin was regarded as a 'double-headed enzyme' which could break down ATP by either of two routes – by simple hydrolysis (responsible for the steady-state breakdown)

[1] Imamura, Kanazawa, Tada & Tonomura (1).
[2] Also Tonomura, Kitagawa & Yoshimura (1).
[3] Tonomura & Kanazawa (1); but see Tonomura, Kitagawa & Yoshimura (1).
[4] See also Ikehara, Ohtsuka, Kitagawa, Yagi & Tonomura (1).
[5] Tokiya & Tonomura (1).

or through phosphoryl myosin. The breakdown of the phosphorylated protein was very slow for myosin but immensely speeded up with actomyosin, so that the accumulation of $E \sim P$ responsible for the initial phosphate burst no longer took place. In a later paper[1] conditions are described in which the burst phase could be observed with actomyosin.

$$E+S+H^+ \underset{}{\overset{k_{+1}}{\rightleftharpoons}} E_1S+H^+ \xrightarrow{k_{+2}} E+ADP+P_i+H^+$$

Inhibited by EDTA $\quad k_{+3} \downarrow \qquad \uparrow k_{+5}$

$$E_2SH^+ \xrightarrow{k_{+4}} E \sim P+ADP+H^+$$

hydrolysed by TCA

$$k_{+1} \gg k_{+2}, \ k_{+3} > k_{+4} \gg k_{+2}+k_{+5}$$

Tokiwa & Tonomura followed, by a stopped-flow method, the H^+-ion liberation during the initial period. As indicated in the scheme above, they found a phase of H^+ absorption followed by a phase of liberation, with return to the initial level. Finlayson & Taylor (1) have very recently examined the transient phase and found, with Mg- or Ca-ATP, liberation of H ions (in one or two steps) at rates greatly exceeding that of the steady state. Absorption was not seen under any of their conditions. They therefore do not agree with the formulation of the Japanese workers, but suggest only that their own results could be associated with changes in the conformation of free enzyme or of enzyme–substrate complex.

In a paper of 1969 Tonomura and his collaborators have proposed a molecular mechanism of contraction in terms of the following cycle.[2]

1. Resting state: Myosin present as the myosin–phosphate complex – the α-state.

2. Excitation with setting free of Ca ions: conformational change in troponin,[3] followed by conformational change in actin; link formed between actin and myosin, with setting free of phosphate.

3. Contraction: Phosphorylation of myosin by ATP and rapid dephosphorylation of phosphoryl myosin by the action of F-actin; this energy-releasing process being accompanied by conformational change in the myosin (to the β-state) and sliding of the actin filaments.

4. Relaxation: On cessation of the stimulus, ATP forms the myosin–phosphate–ADP complex and the actin–myosin link is broken; spontaneous return of the myosin to the α-state follows.

[1] Kinoshita, Kanazawa, Onishi & Tonomura (1).
[2] Tonomua, Nakamura, Kinoshita, Onishi & Shigekawa (1).
[3] See p. 298.

Bendall (1) made a very careful study of the time course of ATP hydrolysis using myofibrils under different conditions. In the Ca-activated system he found clear evidence for product inhibition, which would account to a large extent for the falling off in enzymic activity. With the Mg-activated system, this inhibition was much less important, being nil at ionic strength less than 0.15 at temperatures up to 18.5° (similar conditions to those usually used by the Japanese group in their work on actomyosin). The initial burst of ATPase activity with myofibrils Mg-activated was much reduced if the fibrils were supercontracted before the experiment by ATP treatment, and it was concluded that in experiments starting with relaxed fibrils much of the decline in ATPase activity was due to reduction in number of the active centres during superprecipitation.[1] An interesting question also studied by Bendall was the energy of activation of ATP hydrolysis – the Ca-activated system giving a normal value of about 11 kcal/mole, the Mg-activated system a high value of 22–28 kcal, and a very high value for E_A/T. This was taken as evidence that more reaction steps are concerned in the breakdown when Mg-activated than when Ca-activated.

Bowen, Stewart & Martin (1) also observed the rapid initial phase of phosphate release with myosin B – indeed in much larger amount than did the Japanese workers. The burst of phosphate also occurred with thin glycerinated fibre bundles, whether their ATPase was activated by Ca or Mg ions. In the latter case of course there was contraction, in the former case not; so that in these experiments the masking of binding sites during shortening seems to be ruled out as the cause of the falling off of phosphate production. No inhibition was found with products, ADP or phosphate, under their conditions. In contrast to the results in Tonomura's laboratory, they found ADP formation to run parallel to phosphate release from the beginning. They concluded that the fast initial phase was a true characteristic of the enzyme but did not agree about the existence of phosphorylated myosin. In the case of glycerinated fibres the reaction could be stopped by transferring the fibres from the reaction medium to water at 26°; after spinning off the fibres and adding the water to the reaction medium, the extra initial phosphate was still seen.

In 1966 Sartorelli and his collaborators[2] were able to confirm the burst of phosphate production, about one mole/mole of myosin, under the same conditions as those used by the Japanese workers. They could not however find evidence for the phosphorylation of the protein. They tried separation of the myosin by phenol extraction near the neutral point, but even after

---

[1] This was not the experience of Weber & Hasselbach, who found the explosive phase unaffected by previous shortening of the fibrils.
[2] Sartorelli, Fromm, Benson & Boyer (1).

this very mild procedure the myosin showed no P_i content. Also, when the burst of P_i production occurs, if P–N or P–S linkages are being broken, water oxygen atoms must appear in the phosphate. This was not found when ^{18}O labelled water was used. Carboxyl phosphate was not excluded but reasons were given showing that its presence was unlikely. Phosphorylated tyrosine, serine or threonine would be expected to be stable under the conditions of separation used. The authors make various suggestions aimed at explaining both the burst of phosphate release and the oxygen exchange results. One possibility is the formation of a monomeric metaphosphate at the active site, with subsequent hydration to inorganic phosphate.

In the later work of Schliselfeld, Conover & Bárány (1)[1] the binding of ^{14}C or ^{32}P labelled ATP to myosin or actomyosin was studied. The K_m of labelling of myosin in presence of Mg was of a similar low value to the K_m of the ATPase activity; the K_m of labelling of actomyosin was much higher and of a similar value to the K_m of actomyosin-ATPase activity. Treatment with PCMB led to similar loss of labelling value and of ATPase activity. The observations of Morita (1) in 1969 had shown that the very rapid formation of the u.v. difference spectrum caused by ATP addition to HMM depended on complex formation at the site where the steady-state ATPase operates. The rate of decay of the difference spectrum in presence of Mg agreed with the maximum velocity of the steady-state ATPase. In presence of Ca decay was much faster. Schliselfeld, Conover & Bárány, from their own results and those of Morita, suggested that the conformational change in the myosin caused by ATP led to blocking of the release of ATP or its hydrolysis products; so the rate of formation of the conformational change is reflected in the initial burst, and its decay in the steady-state Mg-ATPase activity. In the presence of actin there would be increased speed of release of ADP and P_i.

In very recent work Lymn & Taylor (1) and Taylor, Lymn & Moll (1) have made an extensive examination of the transient state with myosin ATPase and of the effect of the myosin-product complex on the steady-state hydrolysis. They discuss the agreements and disagreements of their results with those of Tonomura and his colleagues. They conclude that the hydrolysis mechanism may be indicated thus:

$$\text{M} + \text{S} \underset{k_{-1}}{\overset{k_1}{\rightleftharpoons}} \text{MS} \underset{K_{-2}}{\overset{k_2}{\rightleftharpoons}} \text{MADPP}_i \overset{K_3}{\longrightarrow} \text{M} + \text{ADP} + \text{P}_i$$

and that the phosphate burst depends on the large value of K_2 compared with K_3. Use of radioactive P and column chromatography allowed K_2 to be measured without use of acid, and thus the need to invoke a phosphorylated intermediate was obviated. Lymn & Taylor mention further experiments,

---

[1] Also Schliselfeld & Bárány (1).

still to be published, showing that actin activation concerns acceleration of the product dissociation step. Eisenberg, Zobel & Moos (1) were also of the opinion that the activating effect of actin on the ATPase of HMM and sub-fragment 1 could be accounted for in the same way, from their kinetic results, by assuming increased rate of release of products.

THE INTERACTION OF MYOSIN AND ACTIN, PARTICULARLY WITH REGARD TO SPECIFIC SH GROUPS ON THE TWO PROTEINS

We have already seen much evidence for the association of actin and myosin, and for the dissociation of the complex under the influence of ATP and of raised salt concentration.[1] In 1949 Snellman & Erdös studied the stoichiometric relations between the two proteins and found a combining ratio of 1 actin to 2.5–3 of myosin by weight; when solutions containing different proportions of actin and myosin were ultracentrifuged, the acto-myosin sedimented, leaving either actin or myosin in the supernatant, though the results were not very clear-cut. In ultracentrifuging actomyosin preparations from myosin B, H. H. Weber (10) described a series of actomyosins of increasing sedimentation constant; and Gergely & Kohler (1) later also had evidence from light-scattering studies of a stepwise formation and dissociation. Johnson & Landolt (1) emphasised that the actin/myosin/actomyosin system seemed to be in reversible equilibrium, and in their work with the ultracentrifuge usually found slower components as well as the actomyosin. At 0.6 M KCl they considered the equilibrium to be well towards actomyosin formation, while at 2 M KCl dissociation was practically complete. They also found a combining ratio of 1:2 or 3. A. G. Szent-Györgyi (1) used the method of myosin saturation with actin to give maximum viscosity increase. Using actin purified by Mommaerts' method[2] he obtained a ratio of actin to myosin of 1:4. Spicer & Gergely (1) also obtained approximately this ratio, allowing for the impurities in their actin, when they determined in superprecipitation tests the actin:myosin ratio leaving least soluble protein in the supernatant. In 1957 Gergely & Kohler (2), from this ratio (also found in their light-scattering experiments) and from the molecular weight of myosin (ca. 500000) and of F-actin (ca. 3×10^6) pointed out that one filamentous molecule of F-actin must combine with about 24 molecules of myosin. These ratios, and the molecular weights of myosin and G-actin, also indicate that each myosin molecule is combined with 2 G-actin sub-units (in the same actin filament or in two different filaments) or, an alternative explanation, that only alternate G-actin sub-units combine with myosin.

In contrast to this assessment, the work of M. Young (1) in 1967 indicated,

[1] Ch. 8.   [2] P. 214.

as we have seen,[1] combination of one HMM molecule with one actin monomer, and this agreed with the electron-microscope observations of Huxley.[2] Young suggested that the earlier high values given for actin might be due to impurities in that protein.

In 1952 Laki, Spicer & Carroll (1) had noticed a reversible variation with temperature of the combining ratio, the proportion of actomyosin being greater at higher temperatures than at lower. The formation of actomyosin was thus to be regarded as an endothermic reaction. This effect was confirmed by Tonomura, Tokura & Sekiya (1) using the light-scattering method. They found ΔH and ΔS values (57.5 cal. and 234 cal/deg mole) very high compared with values obtained with antigen–antibody combination. Nanninga (2) however, using the ultracentrifuge method, was unable to confirm these high values and found the temperature dependence of the equilibrium small. In later work Johnson & Rowe (4) were able to prepare at low temperature a highly purified actomyosin of low myosin content, and concluded that the number of myosin-binding sites on the actin fibrils was temperature dependent. It would be interesting to know whether this observation has any relevance to the living conditions of cold-blooded animals. They also presented evidence which suggested that the equilibrium F-actin + myosin $\rightleftharpoons$ actomyosin previously advocated was too simple; under certain conditions formation of G-actomyosin seemed to play an important part.

We turn now to the part played by SH groups. Early work on myosin had shown that the susceptibility of its SH groups to attack varied greatly according to the SH reagent used:[3] Turba & Kuschinsky (1) in 1951 had realised that differentiation could be made between SH groups concerned in actin-combining power and in ATPase activity; with the work of Bárány & Bárány (1) in 1959 it became clear that three categories of SH group were involved. With certain reagents, e.g. dithioglycollate or mercurials, half or more of the SH groups could be blocked without effect on the ATPase activity or actin-combining power. Further blocking affected both properties, though not always to the same extent. But if actomyosin were treated in the same way as the myosin, for example with dithioglycollate, iodoacetamide or N-ethylmaleimide, then a myosin could be prepared from it with no ATPase activity but with undiminished power to combine again with actin. Thus it was shown that certain SH groups, not those concerned in ATPase activity, could be protected by combination with actin and were presumably concerned in this combination. It was also shown with actomyosin that, parallel to its loss of ATPase activity, there was a decrease in its ATP sensitivity – in other words a reduction in the high affinity for ATP at a site concerned in the actin dissociability[4].

[1] P. 209.   [2] P. 249.   [3] Ch. 8, p. 164.   [4] P. 291.

In 1963 Gröschel-Stewart & Turba (1)[1] studied the progressive attack on distinct SH groups in HMM, using ^{14}C-N-ethylmaleimide. On tryptic digestion after increasing periods of ^{14}C-NEM treatment, labelled peptides were found, the production of which could be individually correlated with (1) loss of EDTA activation; (2) loss of Ca-ATPase activity; (3) loss of actin-combining power.

The specificity of the different SH groups of myosin was also clearly shown up by the work of Stracher (1, 2), making use of the disulphide–sulphydryl interchange with dithiopropionate. Treatment of the myosin with this reagent for 96 h at 0° led to exchange of 14 out of the 15 SH groups contained in 200000 g of myosin, and to complete loss of Ca-activated ATPase. Activity could be restored by treatment of the DSEM (disulphide-exchanged myosin) with β-mercapto-ethanol; but this recovery was impossible if the one remaining SH group of the DSEM had been carboxymethylated by means of IANH$_2$. When the disulphide exchange was carried out in the presence of ATP and Mg ions, two SH groups failed to exchange; if in the presence of actin, three residues did not exchange. That the protected SH residues were different in the two cases was shown by the fact that DSEM (actin-protected) could combine with actin, while DSEM (ATP-protected) could not. Further, the disulphide-exchanged derivatives were carboxymethylated by means of iodoacetamide-1-^{14}C, and the resulting labelled proteins were digested with trypsin. The peptide digest of DSEM yielded two labelled peptides, only one sulphydryl, the other histidyl. The peptide pattern obtained from DSEM (actin-protected) showed three carboxymethylated cysteinyl residues, one corresponding to that in DSEM, as well as the carboxymethylated histidyl residue. The localisation of these SH residues may be pictured schematically thus:

```
                                                  SH
        DSEM————————————————————————————————————————
                                        SH        SH
            DSEM (ATP)————————————————————————————————
                          SH        SH            SH
            DSEM (actin)————————————————————————————————
```

An interesting echo of the work described earlier on the S_1 and S_2 sites of myosin[2] is found in Stracher's observation that during IANH$_2$ treatment there was a transitory stimulation of myosin ATPase. This stimulation was seen also with DSEM (ATP-protected) during IANH$_2$ treatment when tested in the following way: samples were removed at intervals of a few minutes from the start of the IANH$_2$ treatment, the protein was precipitated to remove the IANH$_2$ and was then treated with β-mercaptoethanol. After only two minutes' IANH$_2$ treatment, the ATPase activity was not

[1] Also Gröschel-Stewart, Rüdiger & Turba (1).   [2] Pp. 271, 275.

only restored but enhanced; this enhancement was not seen with DSEM or with DSEM (actin-protected) (see fig. 59). Thus it can be suggested that the SH group responsible for the phenomenon of activation with SH reagent (Sekine's S_1) is that reacting with disulphide or ATP, while the SH group not reacting with disulphide is Sekine's S_2. Trotta, Dreizen & Stracher (1) (using labelling with ^{14}C-iodoacetamide) have recently shown that S_1 and S_2 are found in HMM subfragment 1 if the latter is prepared by papain digestion. But if tryptic digestion is used, while subfragment 1 retains all the

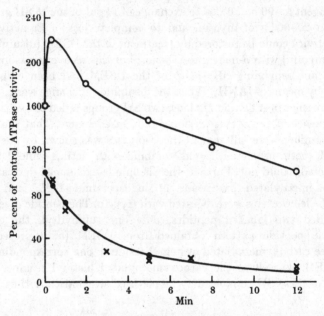

Fig. 59. Effect of reaction time of DSEM derivatives with $IANH_2$ on ATPase activity. ●—●, reaction of DSEM with $IANH_2$; ○—○, reaction of DSEM (ATP) with $IANH_2$; ×—×, reaction of DSEM (actin) with $IANH_2$ (Stracher (1)).

ATPase activity, a low-molecular weight fraction separated from it on Sephadex, contains over 60 % of the S_1 and S_2 sulphydryls. The authors suggest that these sulphydryls are not situated at the active centre, but in a neighbouring allosteric region whence they can influence catalytic events.

Stracher (3) had brought evidence in 1965 for the presence of histidine at the active site of myosin. DSEM was subjected to photo-oxidation, with the result that ATPase activity rapidly disappeared. Histidyl residues were destroyed at two different rates – 8 out of 25 being destroyed at a high rate. It was found that the first-order rate constant for this fast destructive action was the same as the first-order rate constant for loss of ATPase activity, and it was concluded that one out of these eight histidyl residue was involved in the catalytic function. The single SH group remaining in the DSEM was very slowly photo-oxidised, and must have been shielded in the molecule in some way. Stracher suggested that the two SH groups concerned in ATPase activity may function as binding sites for the ATP, while the histidyl residue may act as the catalytic group involved in cleavage of the terminal phosphate.

Perry and his collaborators have recently brought to light a very interesting aspect of the myosin–actin interaction. The activation at low ionic strength of actomyosin ATPase by Mg ions, in contrast to myosin ATPase, is well-known and it has been assumed that this change in enzymic property depended on the association of the two proteins in actomyosin leading also to its increased viscosity, higher sedimentation constant, etc. Working with acto-H-meromyosin, which is soluble at low ionic strength, Leadbeater & Perry (1) first showed that marked activation of the Mg-ATPase was caused by actin addition to H-meromyosin, under conditions in which (as assessed by viscosity measurements) the HMM and actin might be regarded as dissociated. They suggested two types of interaction of HMM (and by analogy of myosin) with actin – one causing the viscosity effects and a more subtle one bringing about the modification of enzymic behaviour. Perry & Cotterill (2, 4) went on to show that treatment of F-actin with low concentrations of phenylmercuriacetate or p-chlormercuribenzoate inhibited its ability to form a viscous complex with HMM; nevertheless the Mg activation of HMM appeared when the poisoned actin was added (see fig. 60). They had earlier shown (3) that with controlled trypsin treatment actin lost its ability to interact (assessed viscosimetrically) while retaining its power to induce Mg activation of the ATPase up to 40 % of the control value. Thus besides recognising two types of SH group in myosin concerned with ATPase activity and actin-combining power respectively, they suggest that two groups should be recognised in the actin – one, an SH group, concerned with combination with myosin, the other of unknown character responsible for some influence on the ATPase site of the myosin.[1]

Eisenberg & Moos (1) have since found that a very high actin: HMM ratio is necessary in order to achieve maximum combination (measured by the effect on ATPase activity). At the ratio ordinarily used (HMM: actin about 3:1) the acto-HMM is quite dissociated, but the effect of the actin on the Mg-activated ATPase, still clearly seen, could be due to cyclical inter-

[1] Bárány & Finkelman (2) have shown that, under certain conditions, mercurials can cause actin depolymerisation. In the present experiments Perry & Cotterill relied on the fact that the viscosity of the acto-H-meromyosin solutions was the same after ATP addition, whether normal or poisoned F-actin had been used. They took this to indicate that the effect of the mercurial was on the viscosity of the complex and not on that of the actin.

action of the two proteins, only a small proportion being present as the combination at a given time. Thus effects of ATP on viscosity might be expected to be less obvious than effects on ATPase activity. Further, Eisenberg & Moos (2) found with actomyosin preparations in the 'cleared' state that, in the interval before superprecipitation, the increase in ATPase activity above that of the myosin without actin was always accompanied by increased viscosity of the actomyosin.

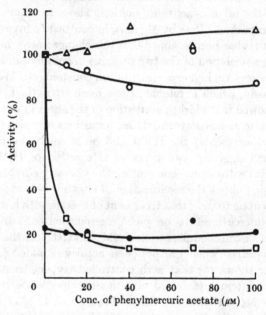

Fig. 60. ATPase activity and viscometric behaviour with ATP of acto-H-meromyosin formed from F-actin treated with phenylmercuric acetate. Activities as percentages of the values obtained with the unpoisoned control. □, fall in relative viscosity on addition of ATP; ○, ATPase activity, Mg-activated; △, ATPase activity, Ca-activated; ●, ATPase activity in absence of Mg or Ca ions. (Perry & Cotterill (2).)

Earlier discussions had often assumed that dissociation of actomyosin by ATP depended on binding of ATP to a dissociation site distinct from the enzymic site. This seemed to be the case in the experiments of Portzehl (2) with glycerinated threads or fibres, where on Salyrgan treatment contraction with ATP was abolished while its ability to cause plasticisation remained. On the other hand, as we have also seen, in the work of Bárány & Bárány (1)[1] on actomyosin sols there were indications that the same SH group was concerned in ATPase activity and in dissociability of the actin from the myosin caused by ATP. In 1968 Eisenberg, Zobel & Moos (1) found

[1] P. 286.

evidence that at low ionic strength acto-subfragment I both hydrolyses ATP and is dissociated by ATP. If only one binding site is present for ATP (compare the results of M. Young)[1] then it would appear that binding of ATP to the enzymic site causes also dissociation of actin from the actin-binding site. Eisenberg & Moos (1) brought further evidence, based on kinetic studies of interaction of HMM, actin and ATP, for this point of view. Also from a kinetic study of the formation and dissociation of actomyosin Finlayson, Lymn & Taylor (1) concluded that, for Mg nucleotide triphosphates, the results could be interpreted on a scheme where dissociation resulted from substrate binding to a site distinct from the actin-binding site. Whether a special dissociation site, apart from the enzymic site, was involved was left undecided.

A POSSIBLE PART PLAYED BY ACTIN IN THE MECHANO-CHEMISTRY OF CONTRACTION

We have seen that one of the possible ways suggested by Hanson & Huxley (2) in 1955 for mediating the sliding past each other of the filaments was alternating increase and decrease in length over short portions of the actin filaments. The binding of actin to the myosin was pictured as causing slight stretching of the former; detachment at one end would allow shortening and attachment at another site. Increase in length again, followed by detachment (and re-attachment) at the other end would lead to passage of the actin by one step along the myosin. Such a step was visualised as small compared with the step (about 132 Å) expected if the sliding depended on the other mechanism – actomyosin connection through the alternately contracting and extending myosin branches. As we have also seen, there is now good evidence for movement of the bridges in contracting muscle. Nevertheless the possibility remains that small changes in the actin filaments may play their part also, and the interesting results and hypotheses of Oosawa and his collaborators will be considered here.

The complex behaviour of actin has already to some extent been described, including the evidence for changes in the intramolecular structure of the monomers on polymerisation and, to a less extent, on nucleotide addition. It became clear that, while the rate of polymerisation of G-actin was greatly influenced by the presence of nucleotide phosphates and of Mg and Ca ions, polymerisation could still take place in the absence of such compounds and cations, though the preservation of certain SH groups seemed to be necessary. By 1962 there was general agreement that ATP dephosphorylation was not essential for polymerisation and that the ADP normally

[1] P. 209.

bound in F-actin was held in the polymer by non-covalent bonds; however, ATP seemed to stabilise G-actin and its dephosphorylation did seem to be a rate-regulating factor in the bond formation of polymerisation. Further, there was evidence, during slow polymerisation in presence of very low salt concentration, for the formation of short actin filaments as an intermediate stage, and for equilibria between the various states of the actin molecule. In 1961 Oosawa, Asakura & Ooi (1) tentatively proposed participation of a linear-helical transition of F-actin in the contraction mechanism although it was not until 1963 that there was direct evidence from the electron microscope of the helical nature of F-actin filaments. This proposition was based mainly on their experience of the steady-state intermediate stages in actin polymerisation. Support for this theory was found in the behaviour of F-actin under sonication[1] when, even at low ionic strength, actin polymerised to a form of moderate size. This was shown by increase in flow birefringence and viscosity, and also by the fact that actin in the sonic field was not denatured by EDTA, as is G-actin. After stopping of the sonic vibration, rapid ATP splitting was observed, the initial rate being proportional to the actin content and the final amount of ATP split being much more than equivalent to the actin present. It was assumed that at this phase short polymers were reforming. This state of affairs was expressed in the following scheme:[1]

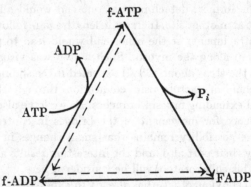

The broken lines denote slow reactions, the solid ones rapid reactions; f-ADP and f-ATP denote the interrupted form of the actin molecule. In applying this picture to isotonic contraction, it is assumed that the F-actin (helical) in the filament is shorter than the interrupted form which would be produced on stretching by interaction with myosin;[2] on release, in the presence of ATP, the f-actin would go rapidly to the helical form with ATP

[1] Asakura, Tanaguchi & Oosawa (1).
[2] Oosawa, Asakura & Ooi (1).

splitting (see fig. 61). In isometric contraction also, each stretched part is actively transformed into the helical state; but the shortening must be slight and accompanied by interruption (elongation) at another part of the filament.[1] So as long as ATP is undergoing dephosphorylation rapid cyclic reaction happens here and there on the filament with tension production.[1]

We have seen[2] that Martonosi, Gouvea & Gergely (2) could find no evidence from experiments *in vivo* for participation in contraction of any exchange of bound and free ADP. The conditions of such experiments are of course very complicated, and it is possible for instance that the ADP of

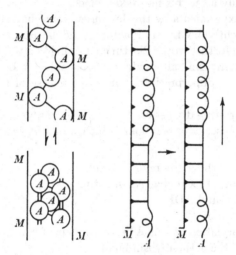

Fig. 61. Models of contraction of the actin filament induced by interaction with the myosin filament (see text). (Oosawa, Asakura & Ooi (1).)

the F-actin equilibrates with some localised ATP pool. They also tested the release of actin-bound ^{14}C-ADP from actomyosin on superprecipitation, and found it amounted to only some 15 %, which they considered negligible. However, in later work Kitagawa, Martonosi & Gergely (1)[3] as well as A. G. Szent-Györgyi & Prior (2) have found evidence for release of up to 60 % of the bound nucleotide during superprecipitation. Szent-Györgyi & Prior showed that the presence of nucleotide in the medium was necessary for this release; and that the release depended on the superprecipitation, being delayed like the latter by low temperature or lack of Mg ions. The exchange reaction continued as long as ATP was present, long after the end of the macroscopic changes of superprecipitation. That the ATP binding

---

[1] Also Oosawa, Asakura, Asai, Kasai, Kobayashi, Mihashi, Ook, Taniguchi & Nakano (1); Oosawa, Asakura, Higashi, Kasai, Kobayashi, Nakano, Ohnishi & Taniguchi(1).
[2] P. 188.   [3] Also Kitagawa, Drabikowski & Gergely (1).

sites could react more than once was shown in the following way. The actomyosin used contained ^{3}H-ADP in its actin while the medium contained ^{14}C-ATP; the loss of the bound ADP was compensated by uptake of ^{14}C-ADP and this was then re-exchanged at the same rate as the ^{3}H-ADP.

Moos, Estes & Eisenberg (1) also observed incorporation of ^{3}H-labelled ATP into actomyosin. They found a certain amount of nucleotide exchange with F-actin alone, but it was greatly increased by addition of myosin. The increased exchange seemed not to be associated with the actual superprecipitation process but, as in the experiments of Szent-Györgyi & Prior, continued steadily after this had taken place.

Preliminary experiments by the Japanese group[1] had suggested that Ca ions, usually tightly bound to F-actin, could be released during superprecipitation. In further work[2] F-actin containing ^{45}Ca was used in preparation of the actomyosin, all easily exchangeable Ca being removed by thorough washing. Superprecipitation led to release of some 30 % of the bound Ca.

In a very interesting discussion A. G. Szent-Györgyi & Prior propose the following possible series of reactions (which would fit with the ideas of the Oosawa group):

1. Formation of the actin–myosin bond.
2. Myosin-induced local changes in actin.
3. Release of bound ADP.
4. Incorporation of ATP.
5. Dephosphorylation of this ATP and repair of the actin structure.
6. Rupture of the actin–myosin bond.

They also envisaged the possibility that in some circumstances the actin-bound ADP might be rephosphorylated. Experiments with myofibrils showed that ADP exchange was much slower than with actomyosin; possibly in the myofibril exchange and rephosphorylation might both play a part, the latter predominating.

Thus although the ADP exchange cannot be quantitatively related to changes in state of the actomyosin, these experiments are indicative of conformational changes in the actin during superprecipitation and possibly during contraction.

[1] Asakura, Hotta, Imai, Ooi & Oosawa (1).
[2] Referred to in Oosawa, Asakura, Higashi, Kasai, Kobayashi, Nakano, Ohnishi & Taniguchi (1) and Oosawa, Asakura, Asai, Kasai, Kobayashi, Mihashi, Ooi, Tanaguchi & Nakano (1).

PROTEIN FACTORS MODIFYING THE INTERACTION OF MYOSIN,
ACTIN AND ATP

It will be necessary to take into account the possible role in contraction of a number of proteins recently detected in small amount in muscle preparations and originating in the myofibrils.

TROPONIN. In 1956 Perry & Grey (1, 2) showed that EDTA inhibited the Mg-activated ATPase of myofibrils or of 'natural actomyosin' when present in concentration only one-tenth or less of that of the $MgCl_2$. In contrast the ATPase activity of 'synthetic actomyosin' (prepared by adding together purified myosin and actin) showed no sensitivity to low EDTA concentrations. Similar effects were seen with EGTA. It thus seemed that in the natural actomyosin traces of some other metal were necessary besides Mg for the enzymic activation. EDTA has a greater affinity for Ca than for Mg ions, and this specificity is much more marked with EGTA.[1] This seems to be the first use of this reagent in an actomyosin system; we shall see how important it later became. The implicit inference from these experiments is that the unidentified metal is calcium.

Nothing further seems to have been done directly to explain this distinction between natural and synthetic actomyosin until the work of Ebashi (5)[2] in 1963. Natural actomyosin, besides showing ATPase inhibition, lost its power of superprecipitation in presence of low EGTA concentrations, while reconstituted actomyosin (or actomyosin after mild trypsin treatment) did not. In the light of all that was now known about the importance of Ca ions for Mg-activated ATPase,[3] he put down this change in the behaviour of the two latter proteins to the loss of some important element controlling responsiveness to Ca ions. He extracted from muscle and from natural actomyosin a protein fraction (now known as the 'EGTA sensitising factor' or ESF) which when added to the synthetic actomyosin restored its sensitivity. This factor had many of the properties of tropomyosin, but tropomyosin prepared by Bailey's method was ineffective.

In 1964 A. M. Katz (1) had noticed delayed superprecipitation of actomyosin and inhibition of its Mg-ATPase activity during the clearing phase, caused by presence of tropomyosin (in this case prepared by the classical Bailey method). Later (2) he prepared from actin (extracted from the acetone-dried powder at 25–37° and therefore containing tropomyosin) a protein fraction which conferred EGTA sensitivity on reconstituted actomyosin. This material, though similar to tropomyosin, had some divergent properties.

[1] Schwarzenbach & Ackermann (1); Raaflaub (1).
[2] Also Ebashi & Ebashi (1).   [3] See ch. 13.

The 'tropomyosin-like protein' was then shown by Ebashi & Kodama (1, 2, 3) to consist of two components – tropomyosin precipitating at pH 6.4 and a smaller part remaining in the supernatant. The latter, which was given the name troponin, was obtained in a form apparently homogeneous in the ultracentrifuge with $S_{20W} = 3$. It had low viscosity, but when added to tropomyosin greatly increased the viscosity of the latter. It had no effect on tropomyosin-free F-actin. Neither tropomyosin prepared by Bailey's method nor troponin alone could sensitise actomyosin to the action of Ca ions – a combination of the two was needed.

About this time Mueller (3) prepared tropomyosin in presence of dithiothreitol to preserve its SH groups. This material had sensitising activity. Later, however, Hartshorne & Mueller (1) were able to separate tropomyosin and troponin by ammonium sulphate fractionation and isoelectric precipitation, and to show that the two proteins together were needed for ESF activity, maximum activity being given at a tropomyosin:troponin ratio of 1:1.3 by weight. The two proteins formed a complex with a single peak in the ultracentrifuge. Perry, Davies & Hayter (1) had rather different experience when purifying by fractionation on DEAE cellulose the ESF contained in the 'extra protein'[1] from well-washed myofibrils: the three fractions obtained all showed comparable factor activity but appreciable amounts of tropomyosin appeared only in one fraction.

A further stage was reached with the finding by Yasui, Fuchs & Briggs (1) that activity of the factor depended on the SH groups in the troponin, and did not correlate with the SH groups of tropomyosin. There was no disadvantage in the use of Bailey tropomyosin, provided troponin was also present. These workers also studied the Ca affinity of troponin. Twenty-four μmoles of high-affinity Ca-binding sites were found per gram; the number was almost doubled when the SH groups reacted with SH reagents, so that these groups must play no part in Ca binding. Ebashi, Ebashi & Kodama (1) had also concluded that Ca-binding depended on troponin and not on tropomyosin.

Wakabayashi & Ebashi (1) further showed that changes in concentration of free Ca ions caused a change in the physical state of troponin. From the sedimentation and disc-electrophoresis patterns it appeared that, at concentrations about 2×10^{-9}M Ca^{2+}, the troponin molecules tended to aggregate; increased Ca-ion concentration (e.g. 10^{-5}M) promoted reversible dissociation.

In 1963 Hanson & Lowy (1) had suggested that the greater part of the tropomyosin is situated in the actin filaments, two strands of this protein possibly running in the two grooves on the surface of the filament; its high α-helix content would fit it for a role of mechanical support, or by masking certain monomers in the filament it might help to determine the sites for

[1] P. 244.

interaction with myosin. Ebashi and his collaborators[1] have studied the localisation of native tropomyosin and of troponin in the fibre. After treatment with a low concentration of trypsin (to which these proteins are both very susceptible) myofibrils were seen to bind only in the *I* band externally applied native tropomyosin labelled with a fluorescent dye. Evidence for the *I* band location of both proteins was found on treatment of myofibrils with fluorescent antibodies.

Evidence for the distribution of troponin along the thin filaments at 400 Å periods was found in the electron microscope after treatment of myofibrils with anti-troponin conjugated with ferritin.[2] This periodicity is significant, since it had already been observed in the *I* filament in muscle by Page & Huxley (1). Hanson (1) has recently shown that, in spite of its similarity to the axial period of F-actin, it cannot be due to the latter protein: sections of aggregates of unpurified F-actin show this transverse striation, but it is absent in sections from the highly purified protein. It is relevant that, as we have seen,[3] paracrystals of tropomyosin show this periodicity in the electron microscope.[4] In view of the Ca-binding power of troponin it is also interesting to recall the early electron-microscope studies of Draper & Hodge (1, 2, 3) in which an axial periodicity (250–450 Å) was put down to localisation of Ca and Mg. Ohtsuki and his collaborators suggest that, since tropomyosin binds to F-actin, it is likely that it runs along the F-actin filaments; and since it also binds troponin it may determine the location of the latter.

According to Kitagawa, Drabikowski & Gergely (1) Bailey-type tropomyosin has an inhibitory effect on the release or exchange of the ADP of F-actin; it would be of interest to enquire into possible effects here of other protein factors.

ATPASE INHIBITORY FACTORS. An interesting finding by Hartshorne, Perry & Schaub (1)[5] in the course of their work on ESF concerns still another myofibrillar protein factor – one which inhibited Mg-ATPase of desensitised actomyosin[6] in the absence of EGTA. Furthermore Schaub, Hartshorne & Perry (1, 2) describe a different factor causing inhibition of the Ca-activated ATPase of desensitised actomyosin, with no effect on myosin ATPase. This latter inhibitory action seemed to be a property of

[1] Endo, Nonomura, Masaki, Ohtsuki & Ebashi (1).
[2] Ohtsuki, Masaki, Nonomura & Ebashi (1). See also p. 260.
[3] Ch. 10.
[4] See Nonomura, Drabikowski & Ebashi (1) for a study of the localisation of troponin in tropomyosin paracrystals.
[5] Also Hartshorne, Perry & Davies (1).
[6] Actomyosin washed repeatedly at low ionic strength and thus rendered insensitive EGTA (see Schaub, Hartshorne & Perry (1)).

tropomyosin itself, since purification in presence or absence of thiol reagents, or denaturation followed by recovery failed to abolish it.

THE TROPONIN COMPLEX. Important steps towards the understanding of the control of actomyosin ATPase have recently been taken in two laboratories.

Hartshorne, Theiner & Mueller (1) separated troponin from tropomyosin by means of ammonium sulphate and isoelectric fractionation. The troponin could then be separated into two components by treatment with acid 1.2 M KCl. Troponin B, in the presence of tropomyosin, inhibited the Mg^{2+} activation of synthetic actomyosin; this inhibition was unaffected by the presence of EGTA or of Ca ions. A mixture of troponins A and B and tropomyosin approximated in its behaviour to the ESF. There was activation of the ATPase in absence of EGTA and inhibition in its presence. Thus, as the authors suggest, the function of troponin A is presumably to confer Ca^{2+} dependence on the inhibitory factor.

Similar results have been obtained by Schaub & Perry (1). They prepared the troponin complex free from tropomyosin by isoelectric precipitation and chromatography on SE-Sephadex. By rechromatography in presence of a dissociating agent such as urea, the troponin could be separated into two factors – the inhibitory factor (troponin B) and the Ca-sensitising factor (troponin A). The inhibitory factor seems to be identical with the factor previously described by Hartshorne, Perry & Schaub (1). It inhibited the MgATPase and superprecipitation of densensitised actomyosin, independently of the level of Ca ions; the presence of tropomyosin was needed. The original biological activity of the actomyosin was restored by adding the Ca-sensitising factor, the optimal proportion to the inhibitory factor being 2:3.

As regards behaviour of these proteins, it is interesting to notice that Schaub & Perry found a change in the electrophoretic mobility of the sensitising factor in presence of EGTA. The calcium-sensitising and the inhibitory factors complex, and Hartshorne et al. also had indications of complex formation between troponin B and tropomyosin.

The mechanism of interaction of these proteins remains to be elucidated. Schaub & Perry have pointed out that, although these purified troponin preparations need tropomyosin for full activity, there is evidence that protein fractions apparently free from tropomyosin can confer ESF activity on desensitised actomyosin;[1] the possibility of still another component in the tropomyosin remains open.[2] Schaub & Perry

[1] Perry, Davies & Hayter (1). See p. 296.
[2] Also Arai & Watanabe (1) for indications of the presence of a third component in the tropomyosin/troponin complex.

have suggested a new nomenclature which is embodied in the following scheme:

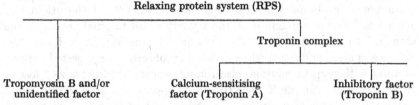

Relaxing protein system (RPS)

| Tropomyosin B and/or unidentified factor | Calcium-sensitising factor (Troponin A) | Inhibitory factor (Troponin B) |

Troponin complex

α-ACTININ. In 1964 Ebashi & Ebashi (2)[1] reported a factor which they partly purified by ammonium sulphate fractionation from the crude extract obtained in the preparation of native tropomyosin. It had the property of greatly promoting the superprecipitation of actomyosin at very low ATP concentration (10^{-5}M) and over an extended range of KCl concentration. They called it α-actinin. Maruyama & Ebashi (1) showed that gelation followed the addition of α-actinin (in amount about 20 % of the actin) to an F-actin solution. They recalled the gel form of F-actin recorded by Johnson, Napper & Rowe, and surmised that this effect was due to α-actinin in their actin.[2]

α-Actinin was purified by Nonomura (1). In the ultracentrifuge it gave three components (S^0_{20w} = 6.2, 10 and 25); all were equally effective in accelerating superprecipitation and all depolymerised to the same sub-unit (S_{20w} = 4) on urea treatment. The $6S$ component was shown to be responsible for the gelation of F-actin,[3] and by antibody staining Masaki, Endo & Ebashi (1) localised this component in the Z line and in the middle of the A band. It was also shown that the M-line protein is a minor constituent of α-actinin.[4]

Maruyama & Ebashi had considered the effect of α-actinin on superprecipitation (measured by turbidity changes) as equivalent to a contraction-promoting effect. Mommaerts and his collaborators did not agree.[5] Using gel synaeresis as a measure of contractility they found no requirement for α-actinin at low ionic strength; further with the turbidimetric test they confirmed an activation by α-actinin at low ionic strength, but found no effect at physiological salt concentrations. Ultracentrifuge tests showed combination between F-actin and α-actinin. They were of the opinion that α-actinin is not a required co-factor for contraction.

Ebashi & Ebashi (3) had remarked on the resistance of α-actinin to

---

[1] Also Ebashi & Ebashi (1, 3).
[2] P. 216.
[3] Drabikowski & Nonomura (1). Drabikowski, Nonomura & Maruyama (1).
[4] P. 250.
[5] Seraydarian, Briskey & Mommaerts (1); Briskey, Seraydarian & Mommaerts (1, 2).

trypsin. Mommaerts and his collaborators[1] later used this attribute in methods of preparing highly-purified α-actinin. After prolonged digestion a fraction was obtained, constituting only about 6 % of the original α-actinin but containing almost all the activity of the original protein in the standard turbidimetric test. The presence of α-actinin in and near the Z line was also confirmed: when isolated myofibrils were digested for two minutes with trypsin solution, electron-microscope studies showed loss of density in and near the Z line; there was apparently some solubilisation process and release of α-actinin. A protein prepared from the supernatant of such fibrils had high activity in α-actinin tests.

It thus seems that the I filaments contain actin, tropomyosin and troponin, while at the Z line (where actin filaments are known to be linked *in vivo*) α-actinin is located. Drabikowski & Nowak (2) have recently found that tropomyosin can prevent the gelation of F-actin by α-actinin or the 6S component. As we have seen,[2] tropomyosin may also be present at the Z line; the physiological significance of possible competition between the two proteins for effects on the actin is not yet clear. Further, one must remark that if α-actinin is confined to the region of the Z line, it is difficult to see how it could take part in the reactions of the overlap region.

We have seen that during contraction a movement of the cross-bridges very clearly is involved. Time will show whether structural changes of the sort we have been considering, both inter- and intramolecular, play any part in the generation and transmission of the driving force for the sliding of the filaments by means of this movement.

FURTHER THEORETICAL CONSIDERATIONS OF THE MECHANISM OF SLIDING

It remains to consider now some recent theories in which attention has been mainly concentrated upon the happenings in the cross-bridges.

A very detailed discussion of a possible contractile mechanism is that of Davies (4) appearing in 1963. In this theory existence of the cross-bridge in two different states is envisaged (fig. 62).

(a) In the resting muscle it consists of an extended asymmetrically arranged polypeptide chain. One ATP molecule is held at the free end of the bridge by means of three positive charges on three Mg ions, and the remaining negative charge on the ATP is repelled by a fixed negative charge in the HMM near the base of the polypeptide chain. Thus the chain is kept extended.

(b) When on stimulation calcium is released from the sarcoplasmic reticulum[3] a Ca ion forms a link between the ADP⁻ on the actin and the

---

[1] Goll, Mommaerts, Reedy & Seraydarian (1).   [2] P. 250.   [3] Ch. 13.

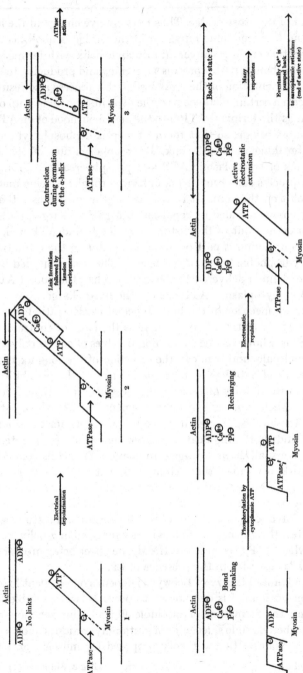

Fig. 62. A scheme for the contraction mechanism. Ca²⁺-dependent contraction during α-helix formation plus ATP-dependent extension of the cross-bridges. (Davies (4).)

ATP⁻ at the end of the cross-bridge. Thus the negative charge in the latter
is neutralised and the coulombic repulsion in the bridge is abolished. The
extended polypeptide chain is converted into an α-helix with formation of
H and hydrophobic bonds, in this process shortening and producing tension.
With appropriate orientation of the cross-bridge this process also results in
pulling of the actin a certain distance past the myosin. The folding up of the
cross-bridge brings the terminal ATP into the neighbourhood of an ATPase
site inside the cross-bridge, so that the next step is dephosphorylation of
the ATP and breaking of the Ca link. Rephosphorylation of the ADP
follows, by means of sarcoplasmic ATP and phosphocreatine. From this
point the whole process can occur again, linkage through Ca being made at
some further point on the actin. It is assumed in this mechanism that, in
the polypeptide chain extended by repulsion of negative charges, potential
energy is stored on account of the distortion of the bound ATP molecule.
The energy for contraction is part of that set free during the formation of
hydrogen and hydrophobic bonds of the α-helix. The energy needed for re-
extension of the chain is delivered to the system when the bound ADP is
phosphorylated by cytoplasmic ATP near to the fixed charge.

In this theory we meet again the importance of electrostatic forces and
the folding of the polypeptide chain familiar as the basis of contraction in
earlier theories;[1] here however the shortening involves of course only a part
of the myosin molecule, and is not in the direction of the long axis of the
filaments. A wealth of data has been assembled to test the possibilities of
the theory – from the affinity of Ca ions for phosphate in various forms to
the properties and behaviour of synthetic polypeptides. The theory stands
up to many of these tests, though in some ways, e.g. in its treatment of
activation heat and shortening heat, it proves inadequate (as pointed out
by Wilkie (1)). We shall discuss it again in chapter 14 in this connection.
Again, recent work[2] has shown that actomyosin made with ADP-free actin
can still superprecipitate normally; this finding and the part now known
to be played by troponin raise doubt about the details of the Ca link
pictured. As we have seen,[3] Davies and his collaborators, in a study which
was planned to test this theory, did find evidence for a fall in α-helix content
of the protein when ATP was added to HMM, the effect being greater with
acto-HMM, and greater still in the presence of Ca.

The recently advanced theory of Loewy (1) depends on covalent linkages
and the reaction postulated is one known to occur between soluble fibrin
molecules during the formation of insoluble fibrin.[4] This reaction takes
place between carboxylamide sidechains of glutaminyl residues and ε-amino
groups of lysine residues. In the theory proposed for muscle (fig. 63) an

---

[1] Chs. 7 and 9.   [2] Tonomura, Tokiwa & Shimada (1).
[3] P. 268 above.   [4] Matačić & Loewy (1).

actin-ADP acyl derivative is first formed and then reacts, by means of an acyl transfer, with an ε-amino group on the myosin. It is proposed that Ca ions activate this last reaction. The suggestion is made that myosin behaves as an acyl-transferase in the presence of the actin–ADP intermediate; myosin, after going through an allosteric transition, might also be responsible for hydrolysis of the isopeptide bond or a hydrolytic site on the actin might be brought into play (as during sonication). This theory is shown to be consonant with a number of well-known observations, such as

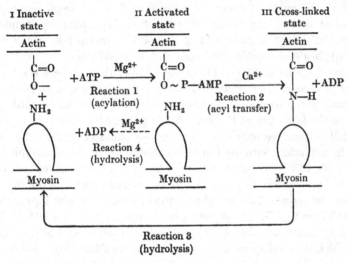

Fig. 68. A proposed cycle of covalent reactions involved in the contraction of striated muscle. Reaction 2 causes a conformational change symbolised as a change in orientation and shape of the myosin cross-bridges. It is suggested that reaction 4 could occur in the absence of troponin. (Loewy (1).)

the release of ADP from F-actin during superprecipitation; and the inhibitory effect of ε-amino groups of polylysine or protamine on contraction of glycerinated fibres.[1] The author outlines a number of ways which could be used to test for the formation of such isopeptide cross-bridges.

Perry & Cotterill (4) have proposed a scheme in which the ordered arrangement *vis-à-vis* one another of the two types of active centre in each of the two proteins[2] might have a part in ensuring the unidirectional aspect of the filament sliding. Perry (7) included the ESF factor in this scheme, suggesting that in the absence of Ca ions this protein could bind specifically to the enzyme site on the myosin or to the centre on the actin interacting with this site. Saturated with calcium, the ESF would have no

[1] See Weber (11).   [2] P. 289 above; also Perry (5).

effect on these centres. Schaub, Hartshorne & Perry (1) emphasised that, with desensitised myosin, superprecipitation took place in presence of EGTA concentrations which ensured a free Ca^{2+} concentration of much less than 10^{-7}M (the critical threshold for the myofibrillar ATPase and for contraction *in vivo*). Thus contraction, in so far as it is represented by superprecipitation, can take place in absence of Ca^{2+} .The involvement of a Ca bridge in the manner suggested by Davies (or of the Ca-activated acyltransferase reaction of Loewy) would thus be contraindicated.[1] Perry (7) thus contemplates the possibility that the regulatory effect of Ca^{2+} is exerted at the enzymic centre of myosin and not through making and breaking of actomyosin links. Thus the first happening on stimulation would be the activation of MgATPase (involving also the relevant centre on the actin); bridge formation and contraction would follow.

Ebashi, Kodama & Ebashi (1) in the light of their further results with troponin envisage a conformational change in the latter due to the Ca ion, this change being mediated via tropomyosin to F-actin and resulting in a change in the structure of F-actin, affecting its interaction with myosin.

Bendall (9) has recently proposed a formulation in which actin is prevented from binding with myosin on account of charge repulsion due to the ATP^{4-} or $MgATP^{2-}$ bound at the active site. After Ca^{2+} has neutralised the negative charge on the troponin closely associated with the actin, the latter can attach to myosin. This would be both via the actin-combining centre and via the ATP or MgATP already attached to the myosin. This latter binding is taken to cause destabilisation of a helical portion of the polypeptide chain, which then collapses and in so doing pulls off the ADP residue of the ATP leaving the phosphate group on the actin. The free energy released in this dephosphorylation may be partly degraded into heat, partly absorbed as internal energy of the polypeptide chain. Part or all of this increased internal energy can be released as free energy during the collapse and used for external work. Since, after this has happened, the internal energy of the shrunken polypeptide chain will have fallen almost to the same value as that of the original helix, reformation of the latter can take place with little energy expenditure; it receives ATP^{4-} or $MgATP^{2-}$ from the sarcoplasm and is ready for another cycle. The analogy is drawn with the heat shrinkage of collagen, during which tension can be developed and work done as the result of the collapse of a helix; here the internal energy has been temporarily increased by the heat energy taken in.

A very different view of the contraction mechanism was expressed by Elliott (2) in 1967. He pointed out how difficult it was for any chemical

---

[1] Loewy doubts that superprecipitation in such circumstances can be taken as analogous to contraction; he points out that tension production by synthetic threads in absence of Ca^{2+} has not been demonstrated.

theory of cross-bridge interaction to account for the fact that the contractile process can go on at varying surface-to-surface distances of the myofilaments (50–130 Å). He saw no experimental evidence for filament swelling to maintain contact or for such variations in cross-bridge length. Elliott wished rather to consider muscle as a colloidal system where balance of electric double-layer effects (long-range repulsive forces), van der Waals attraction and hydration effects control the filament spacing and the constant volume of the filament lattice in resting and contracting muscle. On this view the cross-bridge projections were the sites of charge concentration giving rise to the double-layer effect. The work of Rome (1) on the spacing of the low-angle equatorial X-ray reflections from glycerol-extracted psoas muscle has thrown some interesting light on this subject. In the glycerinated muscle examined in 50 % glycerol/0.01 M phosphate buffer at pH 7.0, the change in the lattice dimensions with sarcomere length was similar to that in living muscle. Changes in pH of the medium resulted in changes in the interfilament distance, fall in pH (which would cause a decrease in charge of the negatively-charged filaments) leading to a corresponding decrease in the distance apart of the filaments. Somewhat similar changes, though much smaller, were observed when the glycerinated fibres were examined in physiological salt solution. There is some support here for the idea that the long-range electrostatic forces play a part in determining the filament lattice. The fact that the changes in lattice dimensions with sarcomere length in the case of the glycerinated muscle in 50 % glycerol agree so much better with the results in the living muscle than do the changes with glycerinated muscle in physiological salt solution suggests, as Elliott pointed out, that in the tissue few free ions are present where the inter-filament forces are acting; or possibly these ions are of a specific nature. Rome (2) has extended these studies on glycerinated muscle and has included experiments on living toad sartorius. Since the membrane of the living cell has only very limited permeability to many ions, it is difficult if not impossible to arrange known changes in the internal ionic composition. However evidence was obtained that the lattice dimensions were affected by pH changes, or entry of water or potassium ions.

Elliott, Rome & Spencer (1) have developed a contraction hypothesis, which they suggest could be applied not only to striated muscle but also to vertebrate smooth muscle and primitive systems such as the slime moulds. Increase of repulsive forces between myosin and actin filaments or between actin and actin filaments, would lead to their increased lateral separation. As constant volume is maintained there must be a longitudinal shortening. The authors consider that viscous resistance to volume change (depending on such factors as the bound state of much of the muscle water, and the restricted availability of channels for fluid flow) may be high enough to

allow tension development in consequence of the constant volume behaviour. On this view the combination of myosin with the actin could act by
increasing the negative charge on the actin filaments; or, by decreasing the
distance between filaments, could lead to increased mutual repulsion.
Several methods for investigation which might throw light on the postulated
phenomena are suggested.

H. E. Huxley (13) in a recent discussion of the sliding mechanism remarks
that the long-range electrostatic forces envisaged by Rome and Elliott may

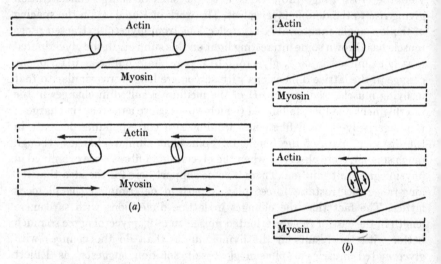

Fig. 64. A diagram illustrating possible mechanisms for producing relative sliding
movement by tilting of cross-bridges. (a) If separation of filaments is maintained by
electrostatic force-balance, tilting must give rise to movement of filaments past each
other. (b) A small relative movement between two subunits of myosin could give rise to
a large change in tilt, by the mechanism shown (H. E. Huxley (13)).

act as a cushion on which the filaments can slide past each other with low
internal friction; in this way they might be essential for efficient working.
He holds the view that the most likely place for the generation of force in
contraction is at the attachment of the globular part of HMM to the actin.
We have already considered the special nature of the 400 Å-long tail of the
HMM – exemplified by its water solubility; and the possibility that, being
free from the LMM filament for most of its length, it could form a flexible
connection between the LMM and the head part of the HMM. The junction
between the heavy and light meromyosins – the point where the tail is
attached – is susceptible to trypsin treatment, while the globular portions
may be removed from the tail by papain treatment. Thus these two points
of attachment might be expected to be flexible. Huxley points out that, on

the other hand, a rather rigid attachment of the HMM heads to the actin is suggested by the highly-ordered arrangement of the 'arrowheads' formed when the HMM reacts with actin filaments. It can be pictured then (fig. 64) that the energy-generating parts of the HMM (the globular sub-units) are rigidly attached to the actin, while being connected with the rest of the myosin molecule by a flexible but inextensible link capable of conveying tension. Changes in tilt of the heads would then cause a relative sliding force between the filaments. The two heads might function independently or their action in conjunction might be necessary. He also suggests that during the ATP splitting the two head subunits might alter their relative positions, thus altering their angle of attachment to the actin. Huxley points out that on such a scheme it would not be surprising if only rather minor changes in configuration accompanied ATPase activity of purified myosin. He emphasises that, for increased knowledge of the mode of attachment of the cross-bridges, much is to be hoped from possible crystallisation of the globular subunit (now isolated in fairly pure form)[1] and its study by crystallographic techniques.

[1] Kominz, Mitchell, Nihei & Kay (1); and Lowey, Slater, Weeds & Baker (1).

13

EXCITATION, EXCITATION–CONTRACTION
COUPLING, AND RELAXATION

PRELUDE TO ELECTRO-PHYSIOLOGY

Let us first consider the ideas of Francis Glisson (1) in the middle of the seventeenth century. Glisson believed that all the 'fibres' of the organism, not only muscle fibres, had a special property of irritability causing them to respond by contraction to external or internal stimuli; by 'fibre' he understood a gossamer-like but strong and elastic thread. He recognised three types of impulse: one intrinsic, due to the fibres' own structure and independent of the nervous system; the second transmitted in the *spiritus vitalis* through the blood; the third in the *spiritus animalis* conveyed through the nerves. We have already mentioned[1] Glisson's acceptance of the standpoint that the cause of contraction could not be *inflation* of the muscle by animal or vital spirits. Bastholm (1)[2] has well described the intricacies of Glisson's theory; besides the *motus naturalis*, dependent on the intrinsic properties of the fibre and quite independent of the nervous system, he postulated the *motus sensitivus externus* due to the nerve stimulation, and the *motus sensitivus internus*, due to the action of the supreme regulatory principle *phantasia* upon the nerve. It should be remembered that William Harvey, from his embryological studies, had a little earlier deduced that irritability was an intrinsic property of living tissue, since the heart and other muscles in a foetus could contract before the brain appeared.[3]

About a century later Albrecht von Haller (1) devoted himself to the study of sensitivity, particularly of all types of muscle. Great numbers of experiments led him to distinguish between irritability (the innate property of muscles to respond by shortening to a variety of stimuli), and sensibility (the property of transmitting to the soul the impression of contact or of pain).[4] He showed experimentally that the muscles had irritability and contracted independently of the nerve supply.[5] Bastholm[6] has paid tribute

---

[1] Ch. 1.
[2] For the account given here of work to the end of the 19th cent. I am also indebted to the article by Brazier (1) and to Liddell (1) ch. 2.
[3] J. Needham (2); Temkin (1); Pagel (2).   [4] Von Haller (1) pp. 658–9.
[5] E.g. pp. 677 and 691.   [6] Bastholm (1) p. 238.

to von Haller in the following words: 'he signifies the introduction to a new period in the history of muscle physiology: Experimental muscle physiology which is not content to be guided by any particular trend but in which detail is added to detail for the building of a structure whose completion it is not granted to the individual to see.'

We have already considered the influential ideas of Descartes on the mechanism of contraction; it is perhaps surprising to find them in sway some 70 years later over such a man as the great teacher Hermann Boerhaave (who numbered von Haller among his pupils). In his Academical Lectures (delivered at the University of Leyden early in the eighteenth century) there is the following description:[1]

The matter from whence the Juices or Spirits of the Brain are prepared, is the viscid and tenacious serum of the Blood, which by passing thro' many Degrees of Attenuation, at length acquires the Subtilty of a Spirit, after its Particles have been moulded or formed by passing frequently thro' the smallest Series of Vessels in the Body...till at last losing the Nature of Lymph it acquires the subtle one of Spirit.

Later he writes:[2]

that in the very instant of time when a Muscle contracts, all the Fibres thereof are pressed or urged from within outwards towards every Point of the Surface...that this Cause must of Necessity come from the Brain, Cerebellum and Origin of the Nerves; and that it is able to overcome the Resistance by which it is strongly opposed in the Muscles. And that we must therefore consider this Cause can be no other than a very thin fluid Body, very easily or quickly moved, and that it must be forcibly thrust into or applied to the Muscle.

But newer ideas were in the air. Isaac Newton (1) in 1718 discussing the all-pervading elastic aether, wrote:[3] 'Is not Animal Motion performed by the Vibrations of this Medium, excited in the Brain by the Power of the Will, and propagated from thence by the solid, pellucid and uniform Capillamenta of the Nerves into the Muscles, for contracting and dilating them?'. This idea was greatly welcomed by some physiologists[4] who, though accepting the notion of a nervous principle running down the nerves, were troubled by various considerations – that it was difficult to imagine its flow as fast enough, and moreover it had never been seen. Again, the work of the early eighteenth century on electrical phenomena and their effects on animals and man (with frictional machines as sources of electricity, gold-leaf electroscopes for its detection and Leyden jar con-

---

[1] Boerhaave (1) II, p. 297.
[2] Boerhaave (1) III, p. 216.
[3] Query 24, p. 328.
[4] B. Robinson (1).

densers) stimulated speculation. Stephen Hales[1] wrote in his 'Statical Essays', published 1726 to 1733,

From this very small Force of the arterial Blood among the muscular Fibres we may with good reason conclude, how short this Force is of producing so great an Effect, as that of muscular Motion, which wonderful and hitherto inexplicable Mystery of Nature, must therefore be owing to some more vigorous and active Energy, whose Force is regulated by the Nerves; but whether it be confined in Canals within the Nerves, or acts along their surfaces like electrical Powers, is not easy to determine.

Others also were attracted by this idea but some, like Monro[2] found stumbling-blocks:

Without stating the difficulty there is in conceiving how the Electrical Fluid can be accumulated by or confined within our Nervous System, we may observe that where the Electrical Fluid, or Fluid resembling that put in motion by the foregoing Experiments, is accumulated by an Animal, such as the *Torpedo* or *Gymnotus*, a proper apparatus is given to the Animal, by means of which it is able to collect and discharge this Fluid...The nervous power is excited by chemical and mechanical Stimuli; and on the other hand is destroyed by Opium and other Poisons, which cannot be imagined to act on the Electrical Fluid.

So the idea of the *succus nerveus* was disappearing. Von Haller (2),[3] dissatisfied with both the intangible ether and electrical phenomena as the basis of nervous action, also emphasised that a ligated nerve, stimulated above the ligature, does not swell. Leeuwenhoek (3) in 1674 had tried to see if the nerve did enclose a cavity, but his results were inconclusive. It was only in 1837 that the renowned microscopist Purkinje (1) described the clear protein-like substance contained within the nerve cylinder. In the report of this lecture it is stated that in this same year Remak also saw the nerve fibre contents in the form of a band or ribbon.

THE EARLY EXPERIMENTAL WORK ON BIO-ELECTRICITY

We come now to the work of Aloysio Galvani in the seventeen eighties and his conception of animal electricity. Galvani carried out numerous experiments showing that muscle, e.g. in the legs of frogs, could be stimulated to contraction by atmospheric electricity. In 1791 he recorded (1) many of these, and also other observations on electrical force and muscle movement. Thus when, using a metal bow (made, for better results as he found, of two different metals) he touched with one end the severed spinal column of a

---

[1] Vol. II, p. 58.
[2] Monro (1). This publication, dated 1793, was posthumous, the paper itself bears no date.
[3] Von Haller (2), paragraphs xv and vii.

frog, with the other the muscle surface, a contraction was elicited in the muscle. Alessandro Volta (1) emphatically denied that these results were evidence of bio-electric currents, maintaining that they were due simply to the juxtaposition of two metals, a phenomenon which he studied in further experiments leading to the development of the voltaic pile and the electric battery. Later results of Galvani, in which only a single metal was used, were similarly explained by Volta as due to impurities in the metal, and the controversy continued. In 1794, some four years before Galvani's death, an anonymous pamphlet appeared,[1] (believed to have been written with his

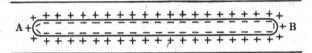

Fig. 65. Uninjured fibre, with electrical double layer, giving no current. (Bernstein (5).)

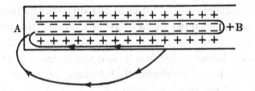

Fig. 66. Injured fibre with electrical double layer, giving injury current. (Bernstein (5).)

collaboration) in which contraction without metals was demonstrated for the first time. This same experiment was described later by Galvani's nephew Aldini (1) in the following words: 'Holding in my hand one limb (of the prepared frog) I let the other hang freely. In this situation, with a glass rod I raised the nerves in such a way that they did not touch the muscles: suddenly I took away the rod; and every time that the nerves and the spinal marrow fell on the muscular parts, a contraction was excited.' He thus claimed the experiment as his own while acknowledging, in a preamble, the direct instigation of Galvani.

The work of Matteucci (1) in 1838 brought the first galvanometric detection of current flow in muscle. He showed that current within the muscle ran from the interior of the muscle mass to the undamaged surface, when the damaged and undamaged parts were connected (figs 65 and 66).

---

[1] '*Dell'arco conduttore nelle contrazione dei muscoli*' and the Supplemento to this tract. For a discussion of the authorship see Fulton & Cushing (1). The manuscript, now lost, was said by du Bois-Reymond to be partly in Galvani's writing, and he attributed the crucial experiment to Galvani himself.

This current became known first as the 'Längsquerschnittstrom' and sometimes as the demarcation current. It has been termed now for many years the current of injury.[1] It is significant as we shall see, that Matteucci observed the effect of strychnine convulsions in diminishing this current. He related his experiments to those of Galvani in showing that when a connecting arc between muscle and nerve elicited contraction, if this arc was in circuit with a galvanometer, there was at the same time a deviation of the needle. In 1842 Matteucci (2) was able to demonstrate to the Paris Academy the classical experiment in which he obtained contraction in the leg muscle of a frog when its nerve was laid on a muscle already excited to contraction in another leg.[2]

About this same time Matteucci's younger contemporary du Bois-Reymond (1) began his work on animal electricity, work which continued for many years and was distinguished by great skill in instrumentation. In 1843 he demonstrated the effect of stimulation upon the current of injury; this was a diminution in current, the 'negative variation', shown by a backward swing of the detecting apparatus. Since the injury current runs in the muscle from the damaged to the undamaged surface, the former (that is to say, the muscle interior) must be negatively charged, the latter positively charged; thus the negative variation meant simply a decrease in the potential difference between the inside and the outside of the muscle. Du Bois-Reymond also found evidence that tetanic stimulation was accompanied by a series of negative variations.[3]

MEMBRANE DEPOLARISATION AND THE ACTION POTENTIAL

By 1871 Bernstein had been able to determine the form and magnitude of the variation corresponding to each stimulus. At the maximum of the variation the current sank to about zero, and he did not consider that a reversal of sign took place.[4] He also studied the propagation along the muscle of the excitatory wave, from the point of excitation, concluding that the change propagated was associated with negativity of the fibres,[5] i.e. a decrease in the external positive charge. But Bernstein is chiefly remembered for his realisation of the importance of the muscle membrane, a realisation inspired by the dictum of Ostwald (1) which he quotes (4): 'not only the currents in muscles and nerves, but also indeed the puzzling activity of electrical fish will find their explanation in the properties of semi-permeable membranes'. Bernstein's conception can be found, with

---

[1] Bernstein (4).   [2] Also Matteucci (3).
[3] Also Biedermann (1) I, pp. 321 ff.   [4] Bernstein (5) pp. 54 ff.; p. 105.
[5] Bernstein (5) p. 39; Biedermann (1) p. 373 ff.; see also Lillie (1) for many early references.

earlier references, in his paper of 1902 (4) and more explicitly in his book of 1912 (5). He suggested that the living sarcoplasmic membrane of the un-injured fibre is responsible for the potential difference between injured and uninjured parts. Suppose such a membrane were impermeable to one ion of an electrolyte, say the negatively charged phosphate ion of KH_2PO_4; then, since the concentration of K ions is high inside, these would seek to wander out, and an electrical double layer would be set up, with the inside negative, the outside positive.[1] This double layer would exist in the uninjured fibre, but would only be detectable when after injury connection was made between inside and outside of the membrane. The negative variation was explained by increase in permeability of the membrane as the result of some chemical alteration on stimulation. As expected from the Nernst equation[2] for a concentration cell, he found (4) that the strength of the muscle injury current was proportional to the temperature between $0°$ and $32°$, the changes being reversible.

The high concentration of K in muscle was already well known at this time. In 1896 J. Katz (1) had published an elaborate quantitative study of the content in many different types of muscle of a number of cations, in-cluding K^+, Na^+ and Ca^{2+}, and of anions including Cl^- and inorganic phosphate. His values for K and Na agree well with later values, and are compared in the table below with the values given by Fenn (3) for the blood plasma content.

μmoles/kilo of fresh tissue or plasma	Muscle	Plasma
Potassium	79	2.5
Sodium	23.9	103.8

But Overton (1) in 1902 was the first to observe the indispensability of Na ions for muscle excitation. Muscles were placed in sucrose solution, isosmotic with blood; if about 0.1 % NaCl were present, the muscle remained normally excitable, but without the Na ions excitability was lost. Overton suggested that there was perhaps some exchange between the intracellular K and the Na of the medium.

Many years later, in 1936, Fenn & Cobb (1) attacked, by means of chemical estimations, the question of electrolyte changes in muscle during activity. In rat and frog, especially with direct stimulation, they found a 50 % decrease in potassium content, and a rather more than compensating

---

[1] Bernstein (5) p. 92 ff.

[2] $E_r = \dfrac{RT}{ZF} \log_e \dfrac{[C]_i}{[C]_o}$ where E_r is the potential difference between inside and outside of the fibre. R is the gas constant, T the absolute temperature, Z the valency of the ion, F the Faraday and $[C]_i$ and $[C]_o$ concentrations of an ion inside and outside.

increase in sodium. A little later it was possible to examine the permeability of the muscle cell membrane *in vivo* to certain cations and anions by the use of labelled isotopes. Thus Joseph, Cohn & Greenberg (1) found evidence of slow uptake of radioactive K into the muscle after its administration by stomach pump; and Hahn, Hevesy & Rebbe (1), after injecting labelled K, Na and inorganic P, found penetration of K into the muscle cells and of Na to a very much smaller extent. Labelled P was also found in the muscle cells, but quantitative interpretation of its rate of entry was complicated by the metabolism of phosphate esters.

In 1941 the important paper of Boyle & Conway (1) appeared. From quantitative determinations, using excised sartorius muscle, they concluded that the cells were permeable to K^+ and cations of the same or smaller diameter, but impermeable to Na^+ and larger cations; permeability was also found to the smaller anions like Cl^-, but not to larger ones such as phosphocreatine and ATP. They were of opinion that the high K content in muscle was held by the electrostatic attraction of indiffusible anions, including negatively charged proteins as well as phosphate esters. The potential difference between the inside and outside was explained in a fairly quantitative way by the differences in K and Cl concentration; and good agreement was found, when the external KCl concentration was varied, between the observed change in potential difference and that calculated from the equations for the Donnan equilibrium.

All this culminated in the work of Hodgkin and his collaborators on the ionic basis of electrical activity studied in single muscle fibres. In these experiments a very small electrode with a tip diameter less than 0.5μ (as first introduced by Gerard) was used to penetrate the fibre and did so without changing the membrane potential. It was already known[1] from work on the giant nerve axon of the squid *Loligo* that during the action potential the potential difference between outside and inside was not only reduced but was reversed in sign, the inside actually becoming momentarily positive to the outside; this decrease in membrane potential propagates itself all the way along the fibre. This phenomenon of reversal of potential difference, seen now in the work of Nastuk & Hodgkin (1) in muscle fibres (fig. 67), could not be explained in terms of the classical membrane theory of Bernstein which postulated simple breakdown of the semipermeable membrane. Nastuk & Hodgkin observed a resting potential difference of about 88 mV; for a membrane moderately permeable to K and Cl but impermeable to Na, the difference expected would be 100 mV, but a slight permeability to Na would, they calculated, explain the observed value. Furthermore they found that the reversed potential difference across the active membrane varied linearly with the logarithm of the Na concentration

[1] For refs. see Hodgkin (1, 2).

with approximately the slope expected for a Na electrode. It was thus possible to account for the action potential by assuming that the membrane becomes, suddenly and specifically, highly permeable to Na. It was suggested[1] that Na enters during the rising phase of the action potential in nerve (and presumably also in muscle) then a delayed increase in permeability to K allows this cation to leave the muscle cell during the falling phase of the potential. It must also be supposed that during recovery Na is actively pumped out of excitable cells, and probably also that K is pumped in; for nerve there is evidence of the metabolic processes involved[2] in the necessary energy provision, but so far not satisfactorily for muscle.[3]

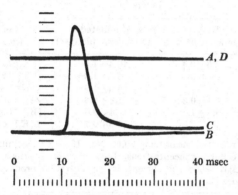

Fig. 67. Resting potential and action potential recorded in a single muscle fibre at 6°. Ordinate scale, 10 mV steps; abscissa, msec. Records (*A*) and (*B*) were obtained with the microelectrode outside the fibre at the beginning and end of the experiment; (*B*) and (*C*) with it inside in the resting and stimulated conditions. (Nastuk & Hodgkin (1).)

In 1959 Hodgkin & Horowicz (1) followed the movements of labelled Na and K during activity of single muscle fibres (table 12). There was a gain in Na and a somewhat smaller loss in K, as in the experiments of Fenn & Cobb. Thus the conduction of impulses turned out to involve changes in membrane permeability which permit movements of these ions down their concentration gradients, and the restoration of the *status in quo ante* depends upon active transport by contragradient pumping mechanisms.

The connection of the action-potential spike with Ca^{2+} provision in the traces needed for tension production[4] will emerge. All that has been said here refers to vertebrate twitch muscle; different behaviour is seen in certain other muscles. Thus in some vertebrate smooth muscles changes in permeability to Ca^{2+} play an important part in actually producing the depolarising currents of the action potential.[5] Then Evans, Schild & Thesleff (1) observed with a number of mammalian smooth muscles that

[1] Hodgkin (1, 2).   [2] Hodgkin (2).   [3] Swan (1).   [4] Cf. p. 162   [5] Ch. 23.

when the tissue was depolarised by immersion in Ringer solution where Na ions were replaced by K ions, contractions were elicited by acetylcholine, but were unaccompanied by membrane potentials. Similar reactions were studied by Keatinge (1) in sheep carotid artery. Durbin & Jenkinson (1) found that such responses were abolished in absence of Ca ions, and suggested that the tension development in these conditions depended on increased permeability of the cell membrane to calcium, this ion then entering from the medium.[1]

TABLE 12

	Fluxes		
	Resting (pmole/cm^2.sed)	Stimulated (pmole/cm^2.impulse)	Concentrations (m-mole/kg H$_2$O)
Na influx	3.5 ± 0.4	19.4 ± 1.5	[Na$_o$] 120
Na efflux	ca. 3.5	3.8 ± 1.0	[Na$_i$] 9.2 ± 1.2
Na entry	.	15.6 ± 1.8	Δ[Na$_i$] $+0.0077 \pm 0.009$
K influx	5.4 ± 0.8	1.8 ± 0.3	[K$_o$] 2.5
K efflux	8.8 ± 1.2	11.4 ± 0.6	[K$_i$] 140 ± 7
K exit	.	9.6 ± 0.7	Δ[K$_i$] -0.0047 ± 0.0003

Mean fibre diameter, 100 μ; mean temp 20.6°. Na$_o$, K$_o$ = extracellular concentrations; [Na$_i$], [K$_i$] = intracellular concentrations. Internal concentrations are in mmole/kg water. Δ[Na$_i$] and Δ[K$_i$] show the changes in concentration produced by one impulse. (Hodgkin & Horowicz (1).)

THE RELAXING FACTOR AND THE IMPORTANCE OF CALCIUM

In 1883 Ringer (1) observed that the contraction of the heart could not be maintained in various solutions of sodium and potassium salts. But if lime salts were added after the heart had ceased to beat, good contraction reappeared and was sustained. Thirty years later Mines (1) confirmed this need of the heart for calcium and recorded also that, though movement ceased in its absence, spontaneous electrical variations, normal in form and magnitude, continued. Mines pictured the contraction mechanism as possibly consisting of the calcium salt of a colloidal material, in the form of strands with the property of shortening in contact with lactic acid, but becoming soluble when the calcium concentration was lowered. Bailey (1) in his detailed pioneer work on the ATPase activity of 'myosin' emphasised the significance of the activation which he observed with this enzyme by means of Ca rather than Mg ions. Paradoxically it now seems that he was probably dealing with myosin rather than actomyosin,[2] in which case the Ca activation would not be relevant to the contraction process;[3] nevertheless there was some cogency in his arguments, and we shall soon be considering

[1] Cf. p. 337.   [2] P. 151.   [3] P. 166.

the need of Mg-activated actomyosin for the presence also of calcium, albeit in traces.

In 1947 Heilbrunn & Wiercinski (1) first showed the importance of intracellular calcium for contraction. They used single muscle fibres of the frog and injected 2×10^{-4} Ca by means of a micropipette. There was shortening by 50 % or more of the muscle length and the effect was not given by Mg or any other physiologically important cations. They suggested that calcium held in combination with cell structure had to be released on excitation. Weise (1) had already shown in 1934 that an ultrafiltrate of fresh rat muscle (well disintegrated in 0.9 % NaCl, then filtered through a collodion filter) contained no detectable calcium. If the rats had worked for some six hours on a treadmill, about 50 % of the total calcium content appeared in the ultrafiltrate. Results suggestive of the release upon stimulation of calcium bound in the muscle were also obtained by Woodward (1) in 1949, with frog sartorii after soaking in Ringer solution containing ^{45}Ca. With resting muscle little calcium was given off on washing; but after electrical stimulation there was an increase of 30–200 % in calcium released to the bathing medium, the increase being greater the greater the stimulation. Analysis of the time curves of release showed that the ^{45}Ca came from the fibres, not the interfibrillar spaces.

By 1952 accumulated data on calcium involvement in muscle behaviour stimulated Sandow (3) to describe his ideas on excitation–contraction coupling, comprising excitation, an inward-acting link and activation of contraction. In this formulation calcium was pictured as released upon stimulation from a bound state in the cell membrane, being then trans-ported to the myosin–ATPase of the fibrillae by some kind of exchange–diffusion, in which the contractile material acted also as a carrier. By this exchange–diffusion mechanism, in which electrostatic attractions might play a part, Sandow hoped to get over the difficulty emphasised by A. V. Hill (28) that simple diffusion of any activating material must be far too slow to account for the extremely rapid involvement of the whole cross-section of the muscle in the activation process.

The work of Marsh has already been mentioned[1] as contributing to the evidence that ATP breakdown (with low ATP concentration at the active enzymic sites) is associated with contraction, while a high maintained ATP concentration is needed for relaxation. The important feature of his work was the discovery of a factor (now known as the relaxing factor) in presence of which ATP breakdown is inhibited and relaxation becomes possible. Marsh (1, 2) used a homogenate of beef muscle in 0.16 M KCl, and with this *brei* of fibre bundles followed contraction and relaxation by measuring the volume of the centrifuged fibre layer. With a fresh homogenate, volume

[1] Ch. 8, p. 168.

increase was observed on ATP addition, followed by a decrease as the ATP disappeared through enzymic action. These changes could be repeated many times on the same sample (fig. 68). Microscopic examination showed that increase in volume on ATP addition was associated with fibre lengthening, while decrease in volume was accompanied by shortening. He considered that the change in volume was the consequence, not the cause, of contraction. Now if the supernatant was removed by centrifuging and the fibre bundles were suspended in 0.16 M KCl, the only effect ever observed on ATP

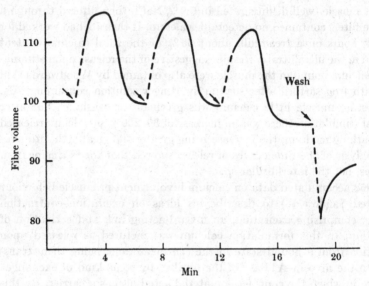

Fig. 68. Effects of ATP addition on homogenised fibres: reversible volume increase with fresh fibres, irreversible synaeresis with washed fibres. Fibre volume expressed as a percentage of the control volume after the same time of centrifuging. 0.76 mg ATP-P added at each arrow. (Marsh (2).)

addition was irreversible volume decrease, and in the same circumstances the ATPase activity of the fibres was greatly increased. The responsible factor in the supernatant was found to be non-dialysable and labile to heat and acid; further its effect was completely prevented by the presence of 0.002 M Ca.

This work of Marsh was quickly followed by attempts in many different laboratories to determine the nature of the factor. Hasselbach & H. H. Weber (1) confirmed its inhibitory effect on actomyosin ATPase, and showed that it inhibited superprecipitation. In its presence there was a lowering of the optimum ATP concentration for these activities. Bendall (3, 4, 5) used the effect on shortening of glycerinated fibres under load as a test for the factor, as well as observing its effect on their ATPase activity. He found the factor potency of soluble muscle protein preparations followed their myokinase content, and he prepared homogeneous myokinase which showed good relaxing activity in his test system. ATP and Mg ions (1–8 mM in each case) were necessary for the relaxing effect, and it was prevented by 0.2 mM Ca. With well-washed fibres this concentration of calcium had no effect on the ATPase activity in presence of 4 mM $MgCl_2$; but with briefly washed fibres, or ones well washed to which factor had been added, 0.2 mM Ca caused a marked rise in ATPase. Thus the effect of the Ca seemed to be an indirect one, via the factor. About the same time Goodall & A. G. Szent-Györgyi (1) found that a combination of two components of muscle extract (one a protein, the other identified as phosphocreatine) could under certain conditions cause relaxation of glycerinated fibres, when added to the medium containing ATP and Mg ions. Lorand (1) then showed that the protein component was creatine kinase, and he suggested that relaxation was equivalent to rephosphorylation of ADP bound to the contractile protein, so that any enzymic mechanism capable of bringing about this reaction might be expected to act as a relaxing factor.

THE PARTICULATE NATURE OF THE FACTOR. The first indication of a particulate relaxing factor was given in 1955 by Kumagai, Ebashi & Takeda (1). They tackled the question of its isolation by means of ammonium sulphate fractionation of muscle extracts. Two fractions, A and B, were obtained, at 10–20 g $(NH_4)_2SO_4/100$ ml of extract and at 30–40 g/ 100 ml respectively. With freshly prepared glycerinated fibres each fraction showed some relaxing activity, but with fibres after some weeks preservation use of the two fractions together was necessary. Fraction A was highly opalescent with strong Mg-activated ATPase activity, and seemed to correspond to Kielley & Meyerhof's particulate Mg-activated ATPase, from which it could not be separated by centrifuging at $18\,000g$.[1] Fraction B contained myokinase and creatine phosphokinase. Kumagai et al. suggested that in the test fibres of earlier workers some of the active A fraction was still contained. Ebashi in 1958 specifically referred to the 'granule-bound' nature of the relaxation-factor (1, 2); he found that the Mg-activated ATPase activity of the particles seemed to parallel their relaxing activity; the latter however could be destroyed without affecting the former.

---

[1] This preparation was made by Kielley & Meyerhof (1) in the following way. An extract of muscle made at 0.5 M KCl was diluted to remove myosin. Precipitation was carried out at 35 % ammonium sulphate saturation and the material obtained was purified by reprecipitations. A suspension was spun at 10 000 g to remove further impurities, then at 18 000 g. The enzyme obtained bound to particulate matter was activated by Mg ions (optimally at 3×10^{-3} M) and inhibited by Ca ions. Later work (see p. 323) showed it to consist of fragmented sarcoplasmic reticulum.

In 1957 Portzehl (4), bearing in mind the importance of ATP concentration for relaxation[1] and for the effect of the relaxing factor, turned to the use of isolated myofibrils for a study of this factor; she pointed out that with the fibre bundles 0.25–0.3 mm in diameter, used by workers up to this time, adequate supply of ATP from the medium to the core of the bundle would be unlikely; moreover it might be difficult or impossible to wash such fibre bundles free from all traces of relaxing factor.[2] She therefore studied the effect of the factor upon ATPase activity and shortening of glycerinated fibrils, only 1 μ in diameter, observed under the microscope. To mitigate any effects of the muscle's own calcium, Portzehl introduced the practice of

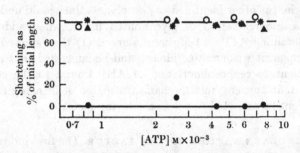

Fig. 69. Influence of the phosphocreatine–creatine kinase system and of the relaxation factor on fibril contraction with ATP. I = ∼ 0.17; pH 6.3; oxalate, 4×10^{-3} M, $MgCl_2$, $2–6 \times 10^{-3}$ M. ▲, with ATP; ✳, ATP + 0.01 M phosphocreatine 3–13 mg creatine-kinase; ●, ATP + factor; ○, ATP + factor + 4.5×10^{-3} M Ca. (Portzehl (4).)

preparing the relaxing factor by homogenising the muscle in medium containing 5 mM K-oxalate. She found this extract, in the presence of ATP and Mg, to be potent in preventing shortening and ATPase activity, while creatine kinase plus phosphocreatine had no effect (fig. 69).[3] Portzehl (5) went on to find that all the factor-containing material was spun down on centrifugation at 35000 g, the resuspended material showing the factor activity, while the supernatant (although it contained myokinase) was inactive (fig. 70). Identification of a granular fraction as the site of the relaxing factor was confirmed by Bendall (6).

The work of Nagai, Makinose & Hasselbach in 1960 (1) introduced a number of new points. Thus they found that before inhibitory action could be shown, incubation of the granules with ATP and Mg was necessary for a few minutes. Although several washings with a medium containing ATP

---

[1] Pp. 161 ff.

[2] The experiments of Makinose & Hasselbach (1) later brought direct evidence for this latter supposition.

[3] Also Briggs & Portzehl (1).

and Mg had no such effect, repeated washing by means of a fibril suspension seemed to exhaust the granules of their relaxing activity, and these observations led to the suggestion that the granules produce and release a labile inhibitor.[1] In this paper an electron micrograph of the particles (by H. E. Huxley) appeared, and they were referred to as 'Teilchen oder Bläschen'.

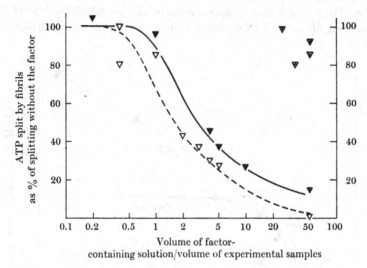

Fig. 70. Binding of the relaxing factor to the granule suspension. ▽—▽, crude extract; ▼—▼, granule suspension; ▽, supernatant. The granules obtained from 1 ml of extract were suspended in 1 ml of solution containing 4.5×10^{-3} M ATP, 0.005 M MgCl$_2$, 0.005 M K-oxalate, 0.02 M histidine, pH 7.0. (Portzehl (5).)

THE CALCIUM PUMP. The realisation of the removal of calcium from the medium by the microsomal particles, indeed of the active uptake and storage of the calcium in the vesicles of which the microsomal fraction is in fact composed, now came quickly. It is interesting to recall that A. Weber in 1959, in the paper (3) to be discussed below on the importance for contraction of free Ca ions, made the suggestion: 'It might be worth while to investigate whether the Marsh–Bendall factor, which is particulate and presumably does not interact with the myofibrils, acts by binding calcium.' In a short note in 1960 (mainly devoted to the relaxing effects of chelating agents) Ebashi (3) refers to the work of himself and Lipmann[2] which had shown that the relaxing material strongly concentrated calcium, removing it from the medium in an ATPase-coupled reaction. Taking into consideration his own results with chelators and the results of A. Weber, Ebashi

[1] Also Makinose & Hasselbach (1).
[2] Ebashi & Lipmann (1), not published until 1962.

proposed that this calcium removal was directly concerned with relaxation
and was the function of the relaxing factor. Independent and a little later
was the work of Hasselbach & Makinose (1) who now used the term 'calcium
pump'. They had observed by means of tests on myofibrillar ATPase that
the inhibition of relaxing-factor activity by small amounts of calcium was
only temporary, and that the effect of repeated small applications was
reversible. They found this disappearance of the Ca effect to depend on
disappearance of the Ca from the medium and its storage within the

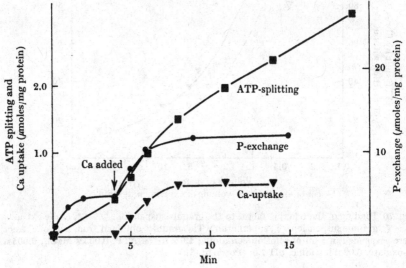

Fig. 71. ATP 'extra' splitting, calcium uptake and phosphate exchange. Mg^{2+} =
oxalate = ATP = 5×10^{-3}M; I = 0.1, histidine = 0.01M; pH = 7.0. (Hasselbach &
Makinose (2).)

particles. In the presence of 5 mM oxalate the Ca accumulated as the
oxalate within the vesicles, and crystals could be seen there; as much as
0.6 mole could be stored in 1 litre of granules. As soon as the Ca uptake
began, the ATPase activity of the particles increased about seven-fold,
returning to the initial value when the added calcium had been removed.
Poisoning of the ATPase by mM Salyrgan led to disappearance of Ca uptake
and of the inhibitory effect of the particles on myofibrillar ATPase. Hasselbach & Makinose (2) observed that the extra ATP splitting stimulated by
Ca (unlike the basic ATP splitting) was accompanied by an ATP–ADP
exchange reaction (fig. 71) implying an intermediate phosphorylation
reaction.[1]

---

[1] P. 327.

Ebashi & Lipmann (1) also brought evidence for active calcium accumulation (more than a thousandfold) in a particulate fraction isolated on centrifuging between 10400 and 38000 g; for the dependence of this concentrating activity on the presence of ATP and ATPase; and for an ATP–ADP exchange reaction catalysed by the same fraction. Their description of electron micrographs of the material is interesting. The main part (about 80 %) consisted of vesicles, 60–200 mμ in diameter, spherical or flattened; and tubules, 30 mμ in diameter. They were similar in shape and size to the tubules and cisternae of the sarcoplasmic reticulum and seemed to consist of healed fragments of the reticulum. Occasionally vesicles were seen with the localised membrane thickening characteristic of the terminal cisternae of the triad system.[1] They proposed an energy-requiring transport of Ca ions against an osmotic gradient into the tubules and vesicles. Thus we know now what the Marsh factor was – microsomal fragments with a voracious appetite for Ca ions.

THE ROLE OF CALCIUM IONS IN ACTIVATION OF THE ATP-ACTOMYOSIN SYSTEM

We must now turn back a little to consider new knowledge of the part played by traces of Ca ions in the activity of Mg-activated ATPase and in contraction.

In 1959 A. Weber (3) considered the evidence from the literature on conditions for actomyosin-ATPase inhibition. This could clearly be caused by high ATP concentration, as in the experiments of Heinz & Holton (1), Geske *et al.* (1) and Perry & Grey (1); and had been variously interpreted as due to substrate inhibition or to competition between free ATP and the true substrate MgATP^{2-}. Both inhibition of the ATPase (at low ionic strength) and relaxation of previously contracted fibres by the metal-chelator EDTA had also been well-attested.[2] Perry & Grey had made the significant observation that the Mg-activated ATPase was inhibited by addition of EDTA in concentration only a small fraction of the concentration of MgCl$_2$ present. Bozler (1), who, as we have seen, first showed a full cycle of contraction and relaxation by varying ATP concentration, found this prevented by 0.1 mM Ca. He later showed that the relaxing effect of EDTA was similarly stopped by Ca addition and therefore postulated that the relaxing substances, including the relaxing factor, acted by combining with Ca^{2+} bound to myosin.

A. Weber now proposed that free Ca in low concentration might be necessary as well as Mg^{2+} for optimal hydrolysis of ATP by actomyosin at

---

[1] Porter & Palade (1); and pp. 330 ff. below.
[2] Perry & Grey (1); Geske, Ulbrecht & Weber (1); Bendall (4); Watanabe & Sleator (1).

low ionic strength, and for contraction; all these observations could be explained by the removal of Ca ions from the medium by combination with the ATP or EDTA. She did indeed find the following results. In the presence of 8 mM ATP or 0.05 mM EDTA, 1–2 μM free Mg^{2+} activated actomyosin ATPase, but at higher concentrations it inhibited unless free Ca^{2+} was present. When the free Ca^{2+} concentration was 6×10^{-6}M, the rate of hydrolysis depended only on the concentration of $MgATP^{2-}$. She concluded (a) that for maximum rate of Mg-activated ATPase the enzyme must be in equilibrium with a very low Ca concentration; (b) that $MgATP^{2-}$ is the substrate for the enzyme. Weber & Winicur (1) in 1961 examined the relation of Ca^{2+} to actomyosin superprecipitation as a model of contraction, precautions being taken to remove the last traces of calcium from the protein and from the reagents. Such actomyosin preparations usually showed little ATPase activity and superprecipitated very slowly unless 0.1 mM Ca was added. Some did show maximum activity without calcium addition and were little inhibited by EGTA. This aberrant behaviour depended on the actin preparation used, and its significance will be apparent from the discussion of troponin in chapter 12. Then Weber & Herz (1) using ^{45}Ca studied binding of exchangeable Ca to actomyosin under different conditions. An amount of more than 1–2 μmoles/g myosin was necessary for superprecipitation. On treatment of the actomyosin with 4 mM MgATP in 0.1 M KCl, more than half the exchangeable Ca was removed. With addition also of 2 mM EGTA most of the bound Ca was removed, and ATPase activity and superprecipitation were lost. Hasselbach (6) has summed up the effects of Mg^{2+}, Ca^{2+} and $MgATP^{2-}$ at various concentrations individually and in presence of each other. He suggested at that time (1964) that Ca ions operated by counteracting a dissociating effect of free Mg ions upon the actin/myosin combination.

Ebashi (3) had already shown that (provided 0.01 M Mg ions were also present) the relative chelating power of a number of Ca-chelating agents correlated well with the relaxing action of the compounds on glycerinated muscle fibres. Weber, Herz & Reiss (1, 3) found that the addition of the relaxing factor (0.1 mg N/ml) lowered the Ca content of myofibrils to the same low level as did addition of 2 mM EGTA (the best Ca-chelating agent). A close correlation existed between the loss of the Ca content of the myofibrils and their power to superprecipitate and to hydrolyse ATP. The relaxing factor could lower the ionized Ca content of the medium to 10^{-8}M, and the extent to which Ca was removed from the fibrils correlated fairly well with the lowering of external Ca concentration. This mechanism was regarded as probably adequate to explain the relaxing factor effects on actomyosin systems.

Although much work has been done on the EGTA sensitising factor, the actual mechanism of the part it plays in regulating the action of Ca ions in the contractile process remains obscure. As we have seen, Ebashi, Kodama & Ebashi (1) suggested in 1968 that Ca attachment to troponin might cause conformational changes which could be passed on via tropomyosin to F-actin. A modification of the interaction of actin and myosin might then result. Weber and her colleagues had shown that lack of calcium, inhibited superprecipitation only if the Mg concentration was greater than 0.01 mM;[1] or if the ATP concentration was greater than 6 μM.[2] Binding of the Mg or the ATP at the higher concentration to inhibitory sites could be postulated, the function of the Ca being concerned with relief of this inhibition. Levy & Ryan (2, 3) later observed inhibition of superprecipitation by ATP concentrations above about 10^{-4}M in the absence of Ca, the inhibition being relieved by Ca addition. Various treatments abolishing SH groups in the system led to a state of affairs in which there was good superprecipitation in absence of Ca. They postulated that high concentrations of MgATP[2-] interfere with superprecipitation by binding to an inhibitory site, and that SH groups on some other protein than the actomyosin (presumably the tropomyosin–troponin complex) regulated the action of the inhibitory site. Then as we have seen, Yasui, Fuchs & Briggs (1) found that addition of PCMB-treated troponin plus tropomyosin had no inhibitory effect on superprecipitation, although the calcium-binding power of the troponin was not diminished – rather, much enhanced. Again it was necessary to invoke transmitted conformational changes.

This brief survey of a very complex field leaves many unanswered questions.[3] We might turn in contrast to the far simpler system of Ca-activated myosin. Here the kinetic analyses of Nanninga (3) in 1959 favoured the view that free ATP is the substrate of the enzyme, while at a neighbouring site presence of Ca^{2+} is necessary for the enzymic activation.

Going back to the muscle fibre itself, we find in 1964 the experiments of Portzehl, Caldwell and Rüegg (1) in which calcium buffers (prepared with EGTA) were injected into the large intact muscle fibres of the crab *Maia squinado*. In this way the concentration of ionised calcium injected should scarcely be influenced by dilution in the muscle, and also probably less affected by removal of calcium by binding than in the case of injection of unbuffered solutions. It is satisfactory to record that the Ca^{2+} concentration needed for isotonic contraction was 0.3–1.5 μM – very similar to the threshold needed in the experiments of Weber and her colleagues (1 μM) for ATPase activation and superprecipitation of actomyosin preparations. Podolsky & Costantin (1) also in 1964 applied calcium by means of a micropipette to isolated frog fibres from which the sarcolemma had been dissected away.

[1] Weber & Winicur (1).   [2] Weber & Herz (1).
[3] See the reviews by Weber (4) and by Ebashi & Endo (1).

Application in amounts greater than $10\,\mu$M caused contraction. Relaxation followed and it was shown that mere dilution of the calcium could not suffice to lower the calcium concentration sufficiently. These experiments also confirm that relaxation can be controlled by a structure within the cell, independently of the cell membrane.

THE QUESTION OF A SOLUBLE RELAXING FACTOR

In the earlier days of research on the relaxing factor the difficulty of visualising close contact between the fibrils and the particles was often emphasised. Several workers brought evidence which they believed showed the existence of a soluble, Ca-sensitive relaxing factor, though its effects were not so striking as those of the vesicular suspension; it was also maintained that a soluble co-factor for the activity of the particles could be demonstrated. Thus Gergely, Briggs and their collaborators[1] found that the effect of the centrifuged cell particles in causing relaxation of glycerinated fibres and inhibition of their ATPase, did not account for the whole effect of the crude muscle extract. Combination of the particles with the supernatant after centrifuging at 35 000 g was needed to bring the relaxing effect of the granules up to that of the original extract. A dialysate of the supernatant served equally well. Parker & Gergely (1) then found that a soluble relaxing substance could be obtained by incubating the cell particles in buffered KCl solution containing sucrose, ATP, Mg and oxalate for 15 min at 25°, followed by centrifugation at 50 000 g. This factor contained no protein and had considerable heat stability.

More recently Seidel & Gergely (1) have put a different interpretation on such experiments. They found that commercial crystalline ATP contained about 1 mole calcium/500 moles ATP. Using ATP freed from Ca by treatment with Chelex 100 or Dosex 50 they found (in confirmation of the results of Weber & Winicur) that 2×10^{-5}M Ca was needed for maximal activation of Mg ATPase and rapid synaeresis of myofibrils; in presence of equimolecular ATP and Mg and absence of Ca the inhibition of ATPase ran parallel with the loss of bound Ca from the myofibrils. They reached the conclusion that the contaminating Ca present in the ATP of the medium before incubation with particles was enough to activate fully the myofibrillar ATPase. The relaxing effect of the medium after incubation with particles was due simply to its being calcium-free: the 'soluble relaxing system' and a solution of Dowex-50 treated ATP had identical effect in inhibiting ATPase. With regard to the co-factors contained in the dialysate of the supernatant and in the dialysed supernatant, Seidel (1) described evidence to show that these

---

[1] Briggs, Kaldor & Gergely (1); Gergely, Kaldor & Briggs (1); Briggs & Fuchs (1); Nagai, Uchida & Yasuda (1).

effects on synaeresis of myofibrils (obtained only in presence of particles) were probably due to presence of inorganic phosphate and of myokinase respectively. We have already discussed the relaxing effect of myokinase under conditions where ATP concentration might be limiting. With regard to the phosphate, in experiments done in absence of oxalate Martonosi & Feretos (1) had observed a potentiating effect on the Ca uptake of the vesicles – no doubt due to precipitation of Ca phosphate within them. In experiments done in presence of oxalate (as were these of Seidel) phosphate alone had no effect but did increase the effect of myokinase; the reason for this is not clear.

It seems then that the activity of the relaxing factor depends primarily on the power of the vesicles contained in it to sequester Ca and on the resultant loss, at a low enough Ca content of the medium, of bound exchangeable Ca from the myofibrils or actomyosin preparations, followed by the dissociation of the actomyosin in the presence of the essential ATP and Mg. Any non-toxic substance capable of binding Ca could of course assist in this process, but the need to postulate a physiological requirement for such a substance seems doubtful. F. N. Briggs (1)[1] later agreed that removal of calcium is the essential function of any relaxing factor. Having shown that the 'soluble relaxing systems' he had used did indeed act in this way, he still suggested however that a soluble chelating factor might have a part to play in regulation of the effect of calcium released from the vesicles.

THE MECHANISM OF THE CALCIUM PUMP

We have seen that Hasselbach & Makinose (2) as well as Ebashi & Lipmann (1) had observed an ATP–ADP exchange reaction of the relaxing factor particles, which appeared on Ca addition and ran parallel with the extra ATP splitting also provoked by the calcium. In both cases the exchange reaction was interpreted as indicating a reversible phosphorylation of the membrane; Hasselbach & Makinose proposed a mechanism for the pump based on current theories of active transport. Thus the carrier mechanism might involve phosphorylation of an unknown substance on the outer surface of the membrane, this substance then acquiring greatly increased affinity for calcium. Upon diffusion of the calcium complex to the inner surface of the membrane, the phosphate would be split off, the calcium affinity of the carrier diminished and the calcium liberated, in spite of the comparatively high concentration of calcium inside.

Hasselbach & Makinose (3) brought evidence in 1963 that during calcium uptake in presence of oxalate, the calcium was indeed contained in the vesicles as calcium oxalate, and thus not bound to the vesicular structure.

[1] Also Briggs & Fleishman (1).

Further, the calcium uptake was strictly correlated, in time and in amount, to the extra ATP breakdown and to the exchange reaction. Both these reactions ceased when the external concentration of free Ca fell to about 10^{-8}M. Vesicles would take up 2 moles of calcium/mole of ATP hydrolysed, though this ratio might be smaller at external concentrations of free calcium less than 10^{-7}M. The final concentration of calcium within the vesicles was some 500 times as high as the concentration outside, implying an affinity of the calcium carrier at the inside of the membrane only one five-hundredth of that at the outside. The properties of the postulated carrier were discussed in the light of these findings.

At the same time and independently Martonosi & Feretos (1, 2) were studying these questions. They also found calcium uptake accompanied by equimolecular oxalate uptake, and under optimal conditions a calcium uptake of 2 moles/mole of ATP split. An interesting observation was the calcium efflux, shown by rapid exchange of the calcium stored with ^{45}Ca in the medium. This happened in absence of oxalate, only slight exchange taking place if oxalate (or phosphate or pyrophosphate) were present (presumably because with calcium precipitation within the vesicles the internal concentration of Ca^{2+} is kept low).

In a further study of calcium transport by the isolated vesicles, in presence of oxalate, Makinose & Hasselbach (2) showed that Ca uptake ceased when a certain maximum capacity of the vesicles was reached, this maximum depending on the free Ca concentration inside (and therefore on the concentration of oxalate supplied). If with a sufficiently concentrated vesicle suspension this maximal capacity was not reached, calcium uptake went on until the external calcium fell so low that the activity of the pump almost ceased. They also observed the efflux of calcium from vesicles loaded with calcium and then placed in a medium of lower calcium content. With inactivation of the pump (by ATP removal or by poisoning with Salyrgan) efflux was markedly increased. It seemed likely that a small activity of the pump was needed after maximal filling to compensate for this efflux which in presence of 5 mM oxalate had only about 0.02 % of the velocity of the maximally activated intake. Weber, Herz & Reiss (2) have also made a very detailed study of the kinetics of the calcium transport. In the absence of oxalate or any calcium-precipitating agent the calcium content of the vesicles at steady-state filling depended on the pCa of the medium; when this steady state was achieved there was a sharp decline in ATPase activity to about 15 % of the initial rate. Calcium influx was dependent on ATP breakdown, with a two to one molar ratio; on the other hand, calcium efflux into medium of low calcium content was independent of ATP hydrolysis and seemed to be due to free diffusion. Two possible schemes for the carrier mechanism were discussed.

Ebashi later somewhat changed his attitude with regard to the mechanism of calcium uptake. While prepared at first, as we have seen, to suggest transportation through the vesicular membrane,[1] he later in work with Ohnishi[2] considered it more plausible that the greater part of the calcium was bound to the external surface of the sarcoplasmic reticulum and that only a small part accumulated inside. The reasons for this view were (1) the great rapidity of the calcium uptake (measured as changes in absorbance of murexide by means of a double beam spectrophotometer and rapid-flow technique) as compared with the rate in known cases of sodium transport; (b) the fact that in his laboratory stoichiometric relations were not found between calcium uptake and ATP breakdown.[3] Ohnishi & Ebashi still agreed however that the effects of oxalate were evidence for transport of some calcium into the interior of the vesicles, but they considered this a minor part. Weber, Herz & Reiss (2) could not confirm the variation in Ca/ATP ratio found by F. Ebashi & Yamanouchi; but they did on other grounds consider that there was strong evidence for the binding of some of the accumulated calcium to the vesicle structure, this binding being more probably in equilibrium with internal than external calcium.

In 1965 Makinose & The (1) showed that calcium accumulation by the sarcoplasmic vesicles could depend on the breakdown of other nucleoside triphosphates besides ATP. Rates for calcium transport and extra NTP hydrolysis varied together and for each of the two processes the compounds could be arranged in the following sequence:

$$ATP > ITP \ (\equiv 0.8 \ ATP) > GTP \ (\equiv 0.7 \ ATP) > CTP \ (\equiv 0.5 \ ATP)$$
$$> UTP \ (\equiv 0.25 \ ATP).$$

These results fit well into the carrier hypothesis, the carrier being phosphorylated to some extent whichever triphosphate is used, and the calcium transport depending only on the affinity for calcium of the phosphorylated carrier and its amount. Rather similar results, but not so clear-cut, were found by Carsten & Mommaerts (2). Makinose (1) went on to show that the vesicles could transfer terminal phosphate of ITP or GTP to ADP and the terminal P of ATP to IDP as well as to ADP. In the presence of Mg or Mn, the addition of Ca ions (2×10^{-7}M) increased the trans-phosphorylation rates 5 to 30-fold. It was shown that this activity was not due to myokinase or nucleoside-diphosphokinase contamination of the vesicles.

Hasselbach & Seraydarian (1) found seven SH equivalents/10^5 g of vesicular protein. Of these, three reacted readily with NEM without loss of ATPase activity or of calcium transport. Blockage of the other four

---

[1] Also Ebashi (4).   [2] Ohnishi & Ebashi (1).
[3] F. Ebashi & Yamanouchi (1).

equivalents led to loss of calcium transport and storage and of extra ATPase splitting. The electron-microscope experiments of Hasselbach & Elfvin (1), using the SH-reagent Hg-phenyl azoferritin, located the SH groups on the outer surface of the transporting membranes. An interesting contrast in the behaviour of the Ca-dependent ATPase activity and the Ca-dependent exchange reaction was seen by Balzer, Makinose & Hasselbach (1) upon treatment of the vesicular fragments with the drugs reserpine, prenylamine or chlorpromazine at a concentration of 3×10^{-5}M. The ATPase was 50 % inhibited while the phosphate exchange was unaffected. The calcium uptake and storage were depressed to the same degree as the ATPase activity. It seems that these drugs do not affect the membrane phosphorylation and the translocation of the calcium, but do inhibit the hydrolysis of the phosphorylated membrane constituent.

THE STRUCTURE AND FUNCTION OF THE SARCOPLASMIC RETICULUM

In 1955 A. F. Huxley & Taylor (1) set out to test the possibility (which had been suggested by Tiegs (4) in 1924) that the Z line, stretching across the fibre, might be the anatomical basis for the link between excitation by membrane depolarisation and contraction in the fibre interior. Huxley & Taylor, using a micropipette with a tip diameter of 2μ, and single striated muscle fibres, applied a current much weaker than that needed to initiate propagated activation of the muscle membrane. Depolarisation of a very small area led to contraction, but only when the tip was opposite the I band, and the contraction was confined to this single band. Further experiments by Huxley (2) confirmed this result for the frog, but showed that the sensitive area in crab muscle was close to the $A-I$ boundary; depolarisation at this point caused shortening of the adjacent half I band. Just at this time the electron microscopy of Porter & Palade (1) drew attention to well-defined structures, which they named 'triads', in the spaces between myofibrils at the level of the Z line in amphibian muscle. We shall have more to say about the triads in a moment. Robertson (1) had shown tubules, resembling the middle element of the triads in lizard muscle (*Anolis carolinensis*), but at the $A-I$ junction. In order to test the possible connection between these and the inward spread of activation, Huxley & Straub (1) performed their micro-stimulation experiments with lizard muscle (*Lacerta viridis*). With local reduction in membrane potential centred at the $A-I$ boundary, contraction resulted in one half I band, the Z line being pulled over into contact with the A band; they also confirmed with the electron microscope Robertson's location of the triads in this muscle. Huxley & Taylor (2) carried these experiments further, and also discussed the mechanism of the inward conduction. The response to the local depolarisation was always graded according to stimulus strength; thus it seemed that the electric conduction concerned was a passive process falling off in intensity with distance.

The early history of the growth of knowledge of the sarcoplasmic reticulum down to the fine and clarifying work of Veratti (1) (published in 1902) has been told by Bennett (1, 2), Porter (2) and D. S. Smith (1). The great difficulties involved in this early work in the interpretation of the images seen in the light microscope comes plainly out of Smith's account. As Veratti wrote 'One of the things that strikes us most forcibly is the disproportion between the complexity of questions that authors have set out to resolve and the imperfection of the methods they have employed'. He chided many of his predecessors for their neglect of the use of thin sections and of selective staining. He himself used silver-impregnated preparations and with this method obtained clear and reproducible results, most of which have been confirmed by electron microscopy. It has amazed the later workers in this field that work so accurate, elegant and interesting could have been forgotten within ten years and not rediscovered for another forty years. It is good to know that Veratti lived to see its complete rehabilitation.

Veratti described the reticulum of the adult mouse-muscle fibre as having 'a geometric regularity. The fibre is crossed at regular intervals by a series of flat transverse reticula made up of the finest fibrils with granules along their length and at nodal points, running in the interstices between the muscle columns; the transverse reticula are joined together by rather scarce longitudinal filaments'. He also saw filaments emerging from the reticular apparatus, crossing the 'sarcoplasmic mantle' of the cell, and reaching the sarcolemma. It is striking that the locations given by Veratti to the transverse reticulum – in some types of muscle at the Z line, in some others at the A–I boundary, in still others at both the Z line and this boundary – correspond to a great extent with the results of modern work, some of which have been discussed. Veratti himself took these variations to be probably artifacts, due simply to imperfections in his staining methods. Of his predecessors he mentioned particularly Fusari[1] and Ramon y Cahal (both of whom used similar methods to his own). Retzius (another predecessor)[2] entertained the idea that the reticular system might function in intracellular transmission of external nervous excitation (2). In support of this he recalled the fact that in a contraction wave the muscle can clearly be seen to contract throughout its whole diameter, one-sided contraction occurring exceptionally only with dying muscle. The question of a possible

[1] For a summary of this work see Fusari (1).
[2] Retzius (1, 2).

relation between the motor nerve endings and the reticular apparatus was mentioned by Veratti as a subject for further investigation.

It was in 1953 that Bennett & Porter (1) made their electron-microscope study of muscle of the fowl, using thin sections instead of the disintegrated material mainly examined by previous workers. They found that in this way something like the natural relations of fibrils, sarcolemma and sarcoplasmic components were preserved. They were able to identify the sarcoplasmic reticulum of Veratti and Retzius, and equated it with the endoplasmic reticulum of cells in general. Its arrangement recalled to them the cross-fibre reticulum described by Thin in 1874 (1). It has a regular relation to the sarcomere bands of the myofibrils, showing a loose lace-like structure around the fibrils at the level of the Z line, crossing the space between the fibrils and reaching the sarcolemma. At that time believing that there was during contraction an increase in density of the A and I bands, Bennett & Porter suggested that the sarcoplasmic reticulum might be concerned in the movement of sarcoplasmic material into the fibrils. In 1955 Bennett (2) and in 1956 Porter (1) each proposed that something like impulse conduction might take place along the reticulum.

Porter (1) and a little later Porter & Palade (1) described in detail the endoplasmic reticulum of striated muscle cells of newt larvae and of rat heart and skeletal muscle (fig. 72 pl.). It consisted of a three-dimensional lattice of membrane-enclosed spaces – tubules and flattened vesicles – in the sarcoplasm surrounding the fibrils. The orientation was largely longitudinal but close to the A–I junction in the rat muscle or at the Z line in the newt larvae a special arrangement was seen. Here the longitudinal reticulum terminated in dilated transversely extended cisternae, which faced similar elements belonging to the domain of the next sarcomere. Between them was a space (about 500 Å across) occupied by another vesicular transverse element. This three-component structure was what they named the triad,[1] and they commented that from reports in the literature it must be widespread (see fig. 73 pl.). Again the possibility of conduction of excitation was canvassed and this point of view was taken up vigorously by Peachey & Porter (1) in 1959. They pointed out that if the excitatory impulse passed from the surface always by the same mechanism, it seemed paradoxical that rapidly contracting muscle consists of large cells, 50–100 μ in diameter, while slow smooth muscle cells are only 6–10 μ wide. But they then discussed the distribution of sarcoplasmic reticulum, pointing out its richness in fast-contracting muscles and its paucity in smooth muscle. By this time a good experimental basis for the idea of the conduction of excitation inwards by the triad system had been provided by the work of Huxley and his collabo-

[1] Amongst arthropods the triad system is replaced by the dyad, in which the tubular element is in contact with sarcoplasmic reticulum only on one side.

rators described above. Peachey & A. F. Huxley (2) in 1962 were also able
to identify in the electron microscope the twitch and slow striated muscle
fibres of the frog. The former always contained triads, but these were absent
from the latter.

Formidable difficulties remained however to be overcome. The requirement was for some connection between the depolarisation at the cell membrane and the activation at the centre of the fibre. The central component
of the triad was in the right place and running in the right direction, but
often it appeared to consist of a row of separate vesicles, and moreover
showed no connection with the sarcolemma. With regard to the first
difficulty it must be remembered that Robertson's tubules (1) appeared to be
continuous, and A. F. Huxley (3) also showed an electron micrograph where
the middle element of the triad seemed to be continuous. Then Andersson-
Cedergren (1) showed in 1959, for mouse skeletal muscle, by means of three-
dimensional reconstructions from serial sections, that the central elements,
for which she proposed the name 'T-system', were in fact highly convoluted
but continuous tubules extending over long distances. They showed signs
of intimate contact with the sarcolemma, but continuity could not be
observed. Similarly Revel (1) in 1962 found that the intermediate element
of the triad in bat cricothyroid muscle could be seen as a slender continuous
tubule coming in contact with the sarcolemma.

Upon the second question – whether or not the membrane of the T-
system was continuous with the sarcolemma and its lumen open to the
extra-cellular space – much information appeared in 1964.[1] Franzini-
Armstrong & Porter (2), using a new glutaraldehyde fixation method, could
see in muscle from the fish *Molliensis* that the central element of the triad
was a long continuous sac, and that sarcolemmal invaginations actually
formed the walls of this T-system (see fig. 74, pl). Thus the T-system tubules
might be regarded as bringing a derivative of the cell membrane to within
$1\,\mu$ of the contractile fibrils. With osmic acid fixation of this same material
the T-system appeared as a row of vesicles and the sarcolemma was always
closed. It was suggested that continuity of the T-system might actually be
of general occurrence. H. E. Huxley (12) came to a similar conclusion from
experiments on frog sartorius, in which the muscle was immersed in a
solution of the protein ferritin. The ferritin molecules (110 Å in diameter)
are easy to recognise in the electron microscope because of their dense core
of ferric hydroxide. After soaking and fixation the central elements of the
triad were seen to contain much ferritin, which was also present in some
longitudinal tubules 'continuous with' the central element. None was
visible in any other part of the muscle fibre. Franzini-Armstrong (1) also

---

[1] For somewhat earlier descriptions of sarcolemmal invaginations in cardiac muscle see
Simpson & Oertelis (1) and D. G. Nelson & Benson (1).

described apparent 'bridges' between the terminal vesicles of the longitudinal reticulum and the T-system.

Page (2) in independent work, saw the central element as a convoluted tube, into which either ferritin molecules or colloidal gold particles (up to 200 Å in diameter) would pass. D. K. Hill (5) studied the space accessible to albumin in toad sartorius by soaking the muscle in a solution of the protein labelled with tritium. With autoradiographs of thin sections ($0.5\,\mu$) examined in the light microscope, he found that albumin had penetrated and was concentrated at the Z line and at the A–I junction, about half at each site. The volume of the albumin-accessible space at the Z line corresponded to that of the T-system, given by Page as about 0.2 % of the fibre volume. There was no evidence as to the morphological identity of the albumin-accessible space at the A–I boundary. Hill referred to Veratti's strong opinion that all muscles contain transverse reticula at both sites.[1] Again Endo (1) used the fluorescence microscope to detect the entry of the fluorescent dye lissamine rhodamine-B-200 into certain locations in single frog muscle fibres. This dye does not penetrate into sarcoplasm, but with the fibres washed after two minutes soaking in the dye, fluorescent striations were seen in a position identified as that of the centre of the I band. It is interesting that also in 1964 Girardier and his collaborators[2] were able to bring electrophysiological data to show that current can flow between the interior and exterior of the cray-fish muscle fibre, through the channels of the transverse tubular system. The membrane of this system they found to be anion permselective, differing in its permselectivity from the plasma membrane of the cell.

From all that has gone before it is to be inferred that the result of the change upon excitation in the membrane of the transverse tubule penetrating the fibre will be release of calcium. There remained now, as H. E. Huxley pointed out, the problem of the exact location of the calcium store which was to be tapped. He calculated that if the central structure of the triad contained calcium in the same concentration as in the extracellular space (about 2 mM) then as the volume of the central structures is not more than 0.5 % of the fibre volume, release of all their calcium would raise the overall cell calcium only by about 10^{-5}M. This seemed too small as the concentration of myosin is about 10^{-4}M. He suggested that depolarisation of the central structure might trigger calcium release from the terminal cisternae. Specialised 'junctional structures' are seen between the central and side structures of the triad, and these may be involved in this trans-

[1] Peachey & A. F. Huxley (1) in 1964 also observed in crab muscle tubular infoldings of the plasma membrane; these occurred both near the A–I junction and at the Z line. The former only were concerned in excitation, and it was suggested that the latter might assist metabolic exchange in these large fibres.

[2] Girardier, Reuben, Brandt & Grundfest (1).

mission process. Peachey (1) later calculated that the terminal cisternae in the frog sartorius make up about 5 % of the fibre volume, so that if they contained 2 mM Ca, release of this could raise the level in the cell to the required degree. Page (2) had also calculated the volume of the terminal cisternae to be 4–8 % of the fibre volume.

Fahrenbach (1) in 1965 made an electron-microscope study of the triadic junction. He concluded that the region of the apposed membranes of the T-system and the lateral cisternae has a similar construction to that of the 'tight junction' or *zona occludens* known in other tissues and characterised by fusion of two membranes.[1] Peachey, considering the question of the mechanism of release of calcium from the cisternae, emphasised that in the triad about 80 % of the transverse tubule is covered by the flattened surface of the terminal cisternae, only 20 % being directly in contact with the sarcoplasm. It thus seemed plausible to suppose that most of the ionic current in the tubule membrane should pass first to the cisternae.

Costantin, Franzini-Armstrong & Podolsky (1) studied localisation of the calcium store, using single frog fibres from which the sarcolemma had been dissected away. These fibres were suspended in paraffin oil, and oxalate (10 mM in 140 mM KCl) was applied. Electron-dense material accumulated in the terminal sacs or cisternae of the reticulum adjacent to the *I* band. It is true, as the authors remark, that translocation of the calcium during the precipitation process cannot be ruled out, but the most direct interpretation would be that the terminal sacs are specialised regions in which the calcium accumulation existed before the oxalate was added.

In the studies of Winegrad (1) about the same time, autoradiographs were made of ultra-thin sections of frog toe muscles after soaking in Ringer solution containing radioactive calcium. In unstimulated muscle the calcium was localised in the centre of the *I* band (where the triads are) and to some extent in the region of overlap of the *A* and *I* filaments. With muscles after tension production the percentage of calcium in the overlap region increased, the increase being larger the greater the tension. It was calculated that the amount of exchangeable calcium in the overlap zone in muscles in isometric contraction was about 4–7 moles per mole of myosin; this ratio is similar to that needed in maximum superprecipitation of myofibrils.[2]

In connection with the localisation of calcium uptake, the question of the localisation of ATPase activity in the sarcoplasmic reticulum is important. It is interesting that Essner, Novikoff & Quintana (1) subsequently reported for rat cardiac muscle that ATPase activity could not be demonstrated in the T-system or in any parts of the sarcoplasmic reticulum except the transverse cisternae. The method of incubating sections with ATP and Mg in lead-containing media was used, and deposits of lead phosphate were

[1] Farquhar & Palade (1).   [2] Weber, Herz & Reiss (6).

looked for. As the authors point out, negative results may only imply inhibition of the enzyme (by lead or some other component of the system)
and its disappearance at less active sites; it might also be necessary to
reckon with translocation of the lead salts. Peachey (1) has suggested that a
function of the longitudinally distributed reticulum might be to recapture
the calcium and return it to the transverse cisternae. The present ATPase
results at their face value would not bear out this idea.

In applying the concept of the calcium pump to the muscle *in vivo*, it is
of course necessary to consider the quantitative aspects. Weber, Herz &
Reiss (3) have calculated that the capacity for storage is adequate: thus
with 10^{-7}M free calcium in the medium, 2 mg of reticulum (a minimal
amount per gram of muscle for the rabbit) can take up 0.2 μmole of calcium,
whereas the amount of calcium which must be removed from the fibrils in
1 gram of muscle is only 0.1 μmole. Hasselbach (6) has similarly calculated
that there is a wide safety margin in the rate of sequestration of the calcium
when compared with the rate of relaxation. The question of the rate of
release of the calcium was more difficult, but this was attacked by Jöbsis &
O'Connor (1) in 1966. They used muscle from toads which had been injected intraperitoneally for several days before death with murexide
solution. The calcium determinations were made by transmitting through
the muscle a monochromatic beam of light, either at the peak or trough
wave-length of the calcium–murexide difference spectrum, and comparing
the optical changes with those at two adjacent wave-lengths. The time
course of decrease in light transmission (corresponding to increase in
calcium ion concentration in the sarcoplasm) is shown in fig. 75 (pl.) in
relation to the curve of tension production. The free calcium concentration
begins to rise within 4 msec of stimulation of the muscle and reaches a
maximum coincident with the beginning of contraction at about 80 msec;
rather surprisingly, the concentration dies away while the tension of the
isometric twitch remains high. It seems likely[1] that interaction of the calcium with the contractile proteins and also its chelation with ATP may
explain this fall. The sequence of events thus seems to be: depolarisation of
the T-system (coincident with the action potential which lasts only a few
milliseconds) causing a transitory change in the cisternal membrane thus
leading to increased permeability and rapid exit of Ca ions; increase of free
Ca concentration in the sarcoplasm followed by uptake of Ca by the contractile system from the moment of the start of contraction; after the contraction is over, release of the Ca from troponin combination and its sequestration by the Ca pump. When the steady state of filling is again reached,
the reactions concerned with the pumping would again fall to the low
resting values. Sandow (6) has recently suggested that the latency of

[1] Jöbsis (1) and the ensuing discussion.

relaxation might find its explanation in the fall in osmotic pressure within the sarcoplasmic reticulum consequent on the Ca release.

Finally it is interesting to consider the explanation of the effect of a number of reagents which increase the rate of shortening and tension development, and prolong the active state. In the case of the alkaloid caffeine, which readily penetrates the cell,[1] it has been shown by Bianchi (2) that presence of this drug increases the rate of both influx and outflux of radiocalcium with muscle fibres of frog sartorius. Herz & Weber (1) found later that this substance inhibits Ca uptake by the isolated sarcoplasmic reticulum, and causes release of calcium provided the vesicles are more than half filled. The relations are not simple. If the calcium filling is maximal but the external calcium concentration is less than 10^{-8}M, release is much reduced, but at 10^{-7}M the release is enough to account for the contracture observed. Sandow (4) has also discussed a number of other potentiators, and in the action of the lyotropic anions it also seems likely that inhibition of the calcium pump may occur.[2]

THE ROLE OF CALCIUM IN THE EXTRACELLULAR MEDIUM

Later work has substantiated the early observations indicating that contraction in muscle was accompanied by calcium efflux. Shanes & Bianchi (1) in 1960 studied the release of ^{45}Ca from frog sartorius and found it to happen only during stimulation. Observations were also made by Bianchi & Shanes (1) on ^{45}Ca influx. For resting muscle this was similar to that going on in nerve axon, but on stimulation of the muscle it greatly outstripped that of the nerve. Much work was also done on the effects of calcium depletion in the external medium. Thus G. B. Frank (1, 2) showed in 1958 that toe muscles of the frog in Ca-free medium do not develop contracture when subjected to K-depolarisation; and Edman & Grieve (1) in 1963 that such muscles do not respond to electrical stimulation either with contracture or with change in membrane potential. It thus appears that here calcium is concerned not only in activation of the contractile fibrillae but also in production of the action potential.[3]

In the past there has been a tendency to regard the external membrane of the cell as a likely site for the calcium store,[3] the influx of calcium from the external medium on stimulation presumably being needed to replace that mobilised and passing into the cell interior. The evidence for calcium storage in the sarcoplasmic reticulum is now, as we have seen, very well established; in any case it has been calculated both by Winegrad (2) and by

---

[1] Bianchi (1).
[2] Ebashi, Otsuka & Endo (1).
[3] These questions are well discussed by Sandow (4).

Frank (3) that the amount of calcium influx would be equivalent to less than 1 % of the myosin contained in the fibrils and thus quite inadequate for its activation. It is important however to remember that if calcium ions are concerned in the polarisation changes in the membrane, the presence of extracellular calcium in the T-system will be essential for the setting free of the stored calcium from the terminal vesicles.

14

HAPPENINGS IN INTACT MUSCLE: THE CHALLENGE OF ADENOSINETRIPHOSPHATE BREAKDOWN

We have already discussed[1] Lundsgaard's realisation in 1934 that phosphocreatine breakdown must be considered a recovery process, leading to the restitution of some unknown energy-rich substance. The evidence from his experiments on whole muscle seemed to rule out ATP as this substance, since ATP breakdown was observed only after stimulation to exhaustion, just before rigor supervened. Lohmann's experiments a little later on muscle extracts, however, bore the clear implication that ATP breakdown must precede phosphocreatine breakdown. The idea of ATP hydrolysis as the energy-yielding reaction closest to the muscle machine was greatly strengthened by Engelhardt & Lyubimova's discovery of the ATPase activity of myosin and the subsequent work on interactions *in vitro* of ATP and actomyosin gels. Since no decrease in ATP content of normally contracting, unexhausted muscle was observed, the confident assumption was generally made that resynthesis of the ATP used was too rapid to permit of measurement of its hydrolysis. This attitude was too complacent for although the assumption turned out to be right in the end, its rigorous proof was attended by extreme difficulties.

We shall consider first in this chapter the various ways and means which experimenters tried hoping to get light on the chemical reactions accompanying contraction, and the evidence which came out of such work for ATP dephosphorylation during a short series of twitches or a short tetanus. I have divided this discussion into two periods – that in the ten years subsequent to Lohmann's discovery and that after 1949 when A. V. Hill (26, 27) made a very telling challenge to biochemists. In the earlier period, with the methods then available, it had rarely been possible to claim even indirect or qualitative evidence for a chemical change during a single twitch. Now Hill pleaded that the concept of dependence of contraction on ATP breakdown should be removed from the limbo of hypotheses eternally awaiting test. Thus the elucidation of the metabolism of a single twitch or a very short tetanic contraction became the urgent goal of a great many workers.

[1] Ch. 5.

INVESTIGATIONS BEFORE 1949

THE IMPLICATIONS OF CHANGES IN VOLUME AND IN TRANS-
PARENCY DURING CONTRACTION. Ernst (1) in 1925, motivated by the
wish to investigate the quantitative applicability of the imbibition theory of
contraction,[1] was the first to look for minute changes in volume in muscle
during contraction. As we have seen, seventeenth-century workers, making
measurements on the whole limb, had satisfied themselves that there was
no gross change in volume under these conditions.[2] Ernst argued that the
shift of water into the 'inotagmata', envisaged in the imbibition theory,
should be accompanied by a decrease in volume, since imbibition of water
by proteins *in vitro* leads to volume decrease of the system.[3] He did indeed
find a volume diminution of about 0.02 cm/g of muscle. The occurrence of
this change in both isometric and isotonic contractions he took as evidence
against the imbibition theory, since in isometric contraction the swelling of
the inotagmata would be prevented.[4] According to his results (2, 3) this
change preceded the contraction, running parallel with the action current;
he proposed that it was due to electrostriction of water in the muscle fluid
consequent upon the setting free of ions in the excitation process.

Some years later Meyerhof confirmed the size and the direction of this
volume change, but found that it only began with contraction. It was much
greater with isometric than with lightly-loaded isotonic contraction, and
increased as the load increased in the latter.[5] Meyerhof & Möhle (2) brought
arguments against new formation of ions, and resulting change in the
ionisation of proteins, as the principal cause of the volume decrease; they
had already argued (1) against compression of water during the contraction
as an explanation. Both in isometric and in isotonic tetanic contractions
part of the volume decrease persisted after the contraction was over.[6] This
residual volume decrease (which might be as much as 80 % of the whole
change) and its variations under different conditions directed their attention
to metabolic processes as a possible cause.

Meyerhof & Möhle (2) then measured the volume changes associated with
the chemical reactions occurring (or expected to occur) during contraction.
The results are shown in table 13. Hartmann (1) made a careful comparison
of volume changes and chemical changes in the same muscle, having the
muscle suspended in paraffin oil during stimulation to avoid imbibition
effects noticed in Ringer solution. His results are shown in fig. 76 *a* and *b*
for both normal and IAA-poisoned muscle. In the latter the volume decrease persisted while in the former, where lactic acid formation could take

---

[1] Ch. 7, p. 141. [2] Ch. 1. [3] E.g. Hofmeister (1). [4] Also Ernst (2).
[5] Meyerhof (22); Meyerhof & Möhle (1); Meyerhof & Hartmann (1).
[6] Meyerhof & Möhle (1); Meyerhof & Hartmann (1).

place, it was succeeded by dilatation. With the poisoned muscle the observed volume change agreed to about 6 % with that expected. With normal muscle the agreement was less good, the final volume being 20 % less than expected.[1]

TABLE 13

	Average change in volume (cm^3/mole)
Glycogen ⟶ lactic acid	+23.3
Hexosediphosphate ⟶ lactic acid	+23.3
Hexosemonophosphate ⟶ lactic acid (in frog muscle extract)	+23.3
Glycogen ⟶ phosphate ester (in extract with fluoride)	+7
AMP ⟶ IMP	−21.4
ATP ⟶ IMP	−42
ITP ⟶ IMP (in rabbit muscle extract)	−21
Hydrolysis of phosphocreatine (in frog and rabbit muscle extract)	−11.5
Hydrolysis of glycerophosphate	−8.4
Hydrolysis of hexosediphosphate	−6–12
Hydrolysis of hexosemonophosphate (with kidney phosphatase)	−0–3.5

In 1935, with improved apparatus, both Ernst & Koczkás (1) and Meyerhof & Möhle (4) could observe in tetanic contractions variations in volume with each stimulation (up to 150/sec) superimposed on the curve for volume contraction. Ernst took this for further evidence that volume change was an expression of the excitatory process. Meyerhof & Möhle on the other hand regarded these small, easily reversible changes as also due to chemical reactions. They pointed out that, when isotonic and isometric single twitches were compared, the easily reversible part of the volume change resembled the slowly reversible part in being much greater for the isometric than for the isotonic twitch especially if the latter was unloaded; they related this to the heat output in the two types of twitch and so to the metabolic rate.[2] Dubuisson (13) in 1939 then threw out the query as to whether this small and rapidly reversible change could be due to ATP breakdown.

In a much later paper Ernst, Tigyi & Örkényi (1) concluded that two processes are concerned in volume changes during tetanic contraction: (1) the large persistent change; (2) the rapid variations superimposed on the

---

[1] Also Meyerhof & Möhle (3). Ernst & Koczkás (2) criticised the method used by Hartmann in correcting for dilatation due to heat production in the muscle. Meyerhof (23) was able to refute these criticisms.

[2] It would of course only be under certain conditions that heat production in the isometric twitch would exceed that in the isotonic. See chapter 3, p. 54.

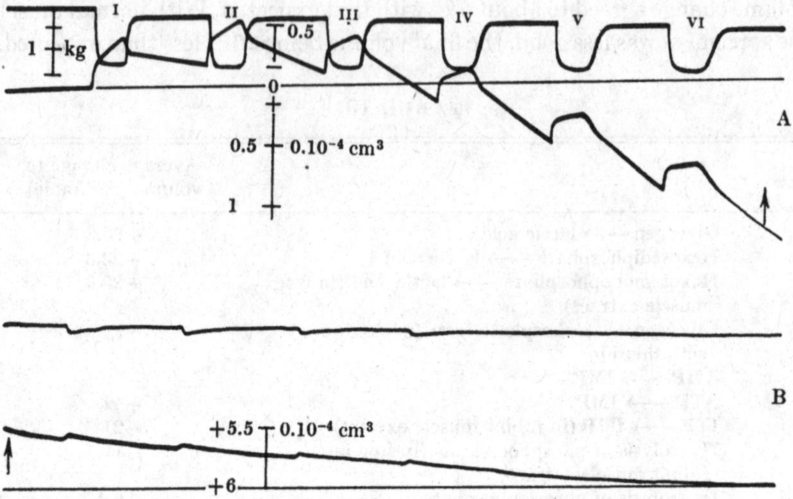

Fig. 76*a*. The volume changes of frog gastrocnemius (in paraffin oil) during a series of tetanic stimuli. Curves (A) and (B): in each case the upper curve shows periods of contraction, the lower shows accompanying volume changes. (A) fresh muscle; (B) exhausted muscle.

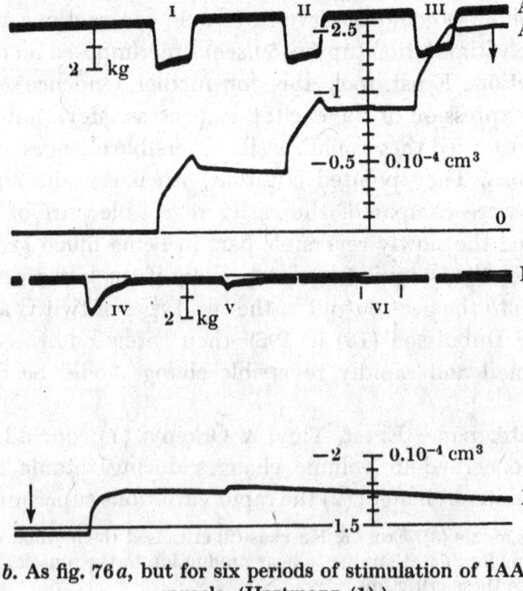

Fig. 76*b*. As fig. 76*a*, but for six periods of stimulation of IAA-poisoned muscle. (Hartmann (1).)

slow change, and in their view concerned with excitation. The slower, more difficultly reversible change they explained as due to 'tension-conditioned crystallisation of myosin'. Ernst, Tigyi & Sebes (1) in 1954 followed the size of the volume decrease between 4° and 24°; they found no temperature dependence – evidence against Meyerhof's explanation. This rather bitter controversy seems never to have been resolved.

A. V. Hill (27) in 1950 pointed out that all experiments had been on muscles of mechanically complex structure, e.g. the gastrocnemius, and expressed the view that the reality of the small reversible changes could not be decided until measurements had been made on muscles with long straight fibres. This suggestion was not taken up by Ernst and his colleagues, but in 1962 Abbott & Baskin (1) did use frog sartorius for such experiments. They found a volume increase within 2 to 3 milliseconds of stimulation (thus at the time of the latency relaxation and increase in transparency)[1] not recorded by the earlier workers. They saw no sign of the very early volume decrements reported by Ernst as coincidental with the action potential. At 0° the volume decrease became maximal before the tension reached its peak, and this was taken to indicate connection of the former with the active state rather than with contraction. Evidence for this connection rather than with excitation was found by Baskin (2) in observations on sartorius muscle in Ringer solution of 2.5 times the normal strength. In these conditions action potentials are still propagated but the ability to produce tension gradually disappears. In parallel with the latter change he found the volume change gradually becoming undetectable. He considered that the increase and decrease in volume were direct indication of protein reorganisation during contraction.

Of interest here also are the observations in 1934 of von Baeyer & von Muralt (1) and von Muralt (4) on changes in scattering of light passing through a muscle. Conditions, such as anaerobic fatigue and various forms of rigor, leading to lactic acid accumulation were accompanied by a decrease in transparency, while breakdown of phosphocreatine in iodo-acetate-poisoned muscle led to a parallel increase in transparency. With a long series of stimuli on normal muscle an immediate decrease in extinction, followed by a very much larger increase in extinction, could be seen (fig. 77). They correlated these effects with the findings in Meyerhof's laboratory[2] that phosphocreatine disappearance is much greater in relation to lactic acid formation on stimulation of fresh muscle than later in an anaerobic contraction series. Fundamentally different processes seemed to be responsible for the extinction changes in the two cases: with lactate formation they were only seen in presence of Ringer solution, and seemed to depend on imbibition of water; the changes with phosphocreatine decrease occurred

[1] Ch. 9, p. 177.   [2] Ch. 5.

as clearly when the muscle was suspended in paraffin oil as in Ringer solution. The relevance of such work to the general muscle problem as seen at that time is well discussed by von Muralt (3).

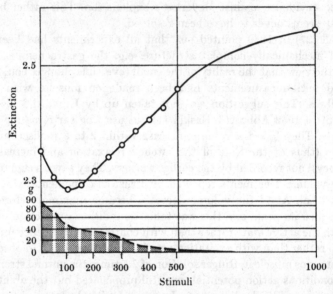

Fig. 77. Changes in transparency of frog sartorius during a stimulation series. Stimulation 4 times/min. Readings taken every 50 stimuli up to 500 stimuli, then finally at 1000 stimuli. Tension production is also marked on the ordinate. (von Baeyer & von Muralt (1).)

THE IMPLICATIONS OF CHANGES OF pH DURING CONTRACTION. We have already considered in chapter 5 manometric experiments by Meyerhof and his collaborators in which pH changes were studied in muscle stimulated under different conditions and ascribed to phosphocreatine breakdown followed by lactic acid formation. In 1937 Dubuisson (9, 10) began a very thorough study of pH changes resulting from stimulation. Using a glass electrode of which the glass membrane, applied to the muscle surface, was only $1-2 \times 10^{-2}$ mm thick, he showed that the delay was about four seconds between the variation of pH in the muscle surface under the glass membrane and its recording via the electrometer and galvanometer. With the slow isometric tetanic contraction of the frog stomach the following sequence could be distinguished: change (b) to the acid side (proportional to the tension developed during the contraction); change (c) to the alkaline side at the beginning of relaxation, or with longer tetani, beginning within the contraction period; change (d), a second acid phase put down to lactic acid formation (see fig. 78). Changes (b) and (c), acid and alkaline respectively, were in the direction to be expected from hydrolysis of ATP followed by hydrolysis of phosphocreatine; moreover the effect upon (b) and (c) of changes in the initial pH of the muscle were such as would fit with this explanation. These pH variations could also be seen with striated muscle (frog gastrocnemius) but the time course could only be followed in slow smooth muscle. Dubuisson & Schulz (1) did indeed verify that there was correspondence between the pH changes in normal and IAA-poisoned contracting muscle and the changes in phosphocreatine and lactic acid content.

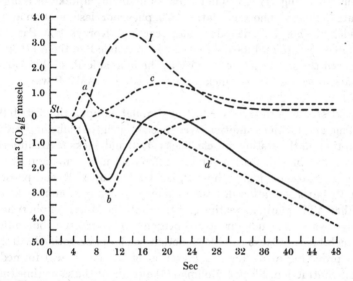

Fig. 78. The continuous line shows variations of pH during contraction of smooth muscle of the frog's stomach (calculated from the variations in CO_2 concentration of the Ringer film covering the muscle). *I*, isometric contraction; *St*, stimulus. The dotted curves (a), (b), (c), (d) show the component phases. (Dubuisson (13).)

A confusing feature of these pH curves in Dubuisson's early work was the appearance of a very early alkaline change (a). This was later found by Dubuisson (11) to be due to slight stretching of the muscle – such alkalisation by stretch had already been observed by Margaria (1), who used frog sartorius muscle stained with bromcresol purple or bromthymol blue by injection of the dye into the living animal.

We shall return to the discussion of the significance of these pH changes.

THE USE OF RADIOACTIVE ISOTOPES IN MUSCLE BIOCHEMISTRY.
In the late nineteen thirties the possibilities of radioactive isotopes as

biological tools drew attention, and Hevesy collaborated in many metabolic probes involving their use. As far as muscle was concerned, the first question attacked was that of the rate of decomposition and rebuilding of phosphocreatine in intact resting muscle. Hevesy & Rebbe (1) could show that at 2°, 3 h after injection of labelled inorganic phosphate into the frog, the phosphocreatine and the ATP each showed 50 % of the specific activity of the inorganic phosphate present in the extracts. Since the P atoms of organic phosphorus compounds had been found not to exchange spontaneously, this indicated the utilisation of the energy-rich P compounds and their rebuilding via enzymic pathways involving uptake of inorganic phosphate (oxidative phosphorylation,[1] the phosphorylase reaction and the glyceraldehydephosphate dehydrogenase reaction). Korzybski & Parnas (1) also injected ^{32}P into rabbits and found that the labile P of the ATP 30 min later showed the same specific activity as the inorganic P, so that isotopic equilibrations seemed to be quickly established. The AMP-P was scarcely labelled even after 2 h.

The next step was to try the effect of stimulation upon the distribution of labelling, and for this conditions were needed such that labelling was not in equilibrium in the resting muscle. Sacks (1, 2) and Sacks & Altshuler (1), using cats, did find in resting muscle differences in specific activities of inorganic P, hexosemonophosphate P, the labile P of ATP, and phosphocreatine P; but stimulation did not produce any changes, and Sacks concluded that ATP, phosphocreatine and the Embden–Meyerhof phosphorylating glycolysis were not directly concerned in contraction but only in resting metabolism. Sacks saw the function of phosphocreatine hydrolysis to lie in providing alkali for neutralisation of the lactic acid formed on anaerobic contraction. Flock & Bollman (1) also about the same time (using rats after ^{32}P injection, in which the specific activity of the inorganic P was much greater than that of the ATP-P) found several cases in which prolonged work had very little or no effect on the ATP labelling. The results were quite variable – in one case an experiment in which 35 % of the ATP was resynthesised three times in one hour (by alternate periods of work and rest) led to 25–50 % increase in the specific radioactivity of Pγ and Pβ groups of the ATP. Bollman & Flock (1) had also found that even with work conditions giving 80 % breakdown and resynthesis of phosphocreatine there was no increase in the labelling. They suggested that for the resynthesis of ATP and phosphocreatine the muscle made use of the same phosphate (unlabelled) split off on hydrolysis, and they remarked that a reason for this might be that much of the labelled inorganic phosphate present might in fact be extracellular – in this, as we shall see, they were right.

Kalckar, Dehlinger & Mehler (1) now made a critical study of the com-

[1] Ch. 17.

plications of technique involved. Clearly the rate of the metabolism bringing about labelling of the phosphate esters in presence of labelled inorganic phosphate can only be calculated if the specific activity of the *intracellular* inorganic P is known. But the extracellular ^{32}P of the muscle, in equilibrium with serum ^{32}P, will remain (since penetration from the extracellular space to the cell interior is slow, as Hevesy & Rebbe (2) had shown) higher for a time than the intracellular value. If the total ^{32}P of the muscle is assumed to be intracellular, spuriously low rates of rejuvenation of the P compounds will be found. Attempts were made[1] to assess the extracellular P by assuming (*a*) that the extracellular space was 12 % of the muscle volume; (*b*) that the specific radioactivity of the inorganic P in this space equalled that of the serum. The labelled intracellular inorganic P could then be found by difference. This method, however, is crude, and could give quite impossible results.[2] Kalckar and his collaborators attempted to overcome this difficulty by perfusion of the muscle with ice-cold Ringer solution, in order to remove extracellular labelled inorganic P. This procedure resulted in a large fall in the specific radioactivity of the inorganic phosphate of the muscle; the results given below were interpreted to indicate a much higher turnover rate in the perfused muscle of the organic P compounds with *intracellular* inorganic P than could be perceived in the unperfused control.

Results as percentage of ^{32}P concentration in the muscle inorganic P
30 min after injection of labelled inorganic P

Inorganic P		Phosphocreatine P		P[γ] and P[β] groups of ATP	
unperf.	perf.	unperf.	perf.	unperf.	perf.
235	100	61	61	58	57

It was not until ten years later that Ennor & Rosenberg (1, 2) showed, by comparison with muscle frozen immediately after the experimental period, that during the period (1 h) which the perfused muscle and its control spent at 0° marked changes went on in the ^{32}P distribution, as a result of continuing metabolism. Their own experiments showed that these changes included breakdown of weakly labelled organic P compounds (with dilution of the specific radioactivity of the intracellular inorganic P pool) as well as changes in the specific radioactivity of the phosphate groups of ATP and phosphocreatine. Bearing in mind also the criticisms to be made of such procedures as those of Sacks & Altshuler, they concluded that with the present state of knowledge it was not possible to determine in muscle the absolute turnover rate of organic phosphorus compounds deriving their

---

[1] E.g. Sacks & Altshuler (1).
[2] See the discussion by Ennor & Rosenberg (1).

phosphate from inorganic phosphate. However it was possible to measure the relative turnover rates of the three P atoms of ATP: 5 min after the injection of labelled inorganic P the specific radioactivity of P^γ was about six times that of P^β, and P^β in turn had specific radioactivity about 16 times that of P^α. As we shall see, much use was made of the isotopic method in later work attempting to detect concealed reactions accompanying contraction.

INVESTIGATIONS AFTER 1949

THE RESPONSE TO THE CHALLENGE. We shall follow now the efforts to establish whether or not ATP breaks down in a single twitch, and indeed whether such breakdown occurs in the contraction or relaxation phase. The formidable nature of such a task is shown by the following considerations. From heat measurements it is known that about 3 mcal are released per gram of muscle per twitch; this would need the breakdown of about 0.3 μmole of energy-rich P compound per gram of muscle – about 0.03 in a whole sartorius containing about 2.5 μmoles.

It will become clear as we proceed that experimentation initiated for this purpose gradually took on a wider aspect: more and more experimenters became interested in following the distribution in time of the cleavage of energy-rich phosphate bonds during a short contraction, and the correlation of such metabolism with heat production and mechanical performance. Some very interesting phenomena came to light in this way.

Amongst the factors contributing to success in this whole field should be mentioned the right choice of method of fixation; the introduction of new and extremely micro-methods, especially chromatographic, of separating and estimating the phosphate esters; and the use of suitable inhibitors. It was necessary of course to use muscles which could be obtained in symmetrical pairs, in order to have proper comparison of control and stimulated metabolism; apart from this it will be seen that much turned on the choice of muscle.

Lundsgaard (9) replying to A. V. Hill at the Royal Society Discussion on Muscle of 1950, described certain experiments he was doing using normal muscles and muscles in IAA-rigor. His experience on freezing these muscles reminded him of Embden's observation in 1922[1] that the introduction of a fresh muscle into liquid air acts as a stimulus to strong contraction. As a preliminary contribution to solving Hill's problem he therefore estimated the ATP content in frog gastrocnemii, one in an excitable state frozen suddenly in liquid air, the other as control frozen down slowly in a deep-freeze at $-10°$ and then, stiff and inexcitable, put into liquid air. An average decrease of 8.5 % (statistically significant) was found in the ATP

[1] Embden & Lawaczeck (1); ch. 4, p. 63.

content in the rapidly frozen muscles. Mommaerts & Rupp (1) also about this time found evidence of ATP breakdown in comparing frog leg muscles frozen in liquid air at the height of contraction with resting controls extracted with ice-cold perchloric acid.

Munch-Petersen (1) in 1953 for arrest of metabolism used immersion in liquid propane cooled to −165° by means of liquid nitrogen. Since the propane is far from its boiling point, this arrangement avoids the formation round the muscle of a vapour layer of low thermal conductivity, as happens in liquid air or nitrogen.[1] She was thus able to get instantaneous freezing of the muscle in a non-stimulated state; test showed that immersion of a fresh muscle into liquid air caused tetanic responses with a series of action potentials. Using separation of ATP and ADP by means of ion exchange resins she was able to show a statistically significant increase of 7 % in the ADP content of various thin muscles from the tortoise (electrically stimulated) during the rising phase of a single twitch, compared with the symmetrical non-stimulated muscle. The ADP content was about 0.9 μmole/g, so that the change would be about 0.06 μmole. From the heat liberation in a twitch/g of muscle ADP formation of about 0.3 μmole would be expected – thus only about 20 % was observed.

Mommaerts[2] in 1954 turned to the use of liquid propane at −180°. But using the gracilis and sartorius muscles of the turtle (where the twitch lasts 2–3 sec) with very rapid arrangements for immersion and crushing, he could find no evidence for formation of expected or possible metabolic products – ADP, AMP, free creatine, creatinine or pyruvate. Yet estimations were made during the rising phase, at the peak of contraction and during relaxation, under both anaerobic and aerobic conditions.

Later experiments showed that some of these difficulties were not unconnected with the types of muscle chosen. Subsequently, in 1962, Mommaerts and his collaborators[3] were able to show phosphocreatine breakdown in frog sartorius with a single short tetanus (0.3 sec). Free creatine and inorganic P were formed in equivalent amounts – 0.44 μmole of creatine and 0.35 μmole of inorganic P/g (the latter estimated by their newly worked out, highly micro-method).[4] ATP and ADP showed no change. They went on to re-investigate[5] the behaviour of the turtle sartorius, but again found no phosphocreatine breakdown during a 10 sec tetanus and the recovery period. They suggested that there might be a reaction within the proteins which could either proceed alone or be coupled to ATP breakdown. Afterwards Mommaerts (13) came to the conclusion that there is

[1] E.g. Bell (1).   [2] Mommaerts (11, 12); Mommaerts & Schilling (1).
[3] Mommaerts, Seraydarian & Wallner (1).
[4] Seraydarian, Mommaerts, Wallner & Guillory (1).
[5] Mommaerts, Olmstedt, Seraydarian & Wallner (1).

probably a much smaller activation metabolism in this muscle and that it was therefore unsuitable for study of metabolism in a single twitch. The rectus femoris of the turtle, a stronger and faster muscle, did split phosphocreatine in quantities comparable to those found with the frog sartorius.

In 1953 the application of chromatography to this problem began.[1] Fleckenstein, Janke, Davies & Krebs (1),[2] using this method with the slow, tonic, rectus abdominis of the frog, and about 1 sec electrical stimulation, found a fall in phosphocreatine (but not in ATP) if the experiment was done at 17°. There was a rise in inorganic P more than equivalent to the phosphocreatine disappearance and a fall in the fraction (total organic P-phosphocreatine P). These changes could not be observed in experiments at 0°. They found that the rate of resynthesis of phosphocreatine during recovery was far too low to explain the smallness of the change in phosphocreatine during contraction and postulated an unknown source of energy which could either re-phosphorylate ATP, or was closer to contraction than ATP. These experiments made use of freezing in liquid nitrogen, so the controls may have been unreliable, although it was reported that no spontaneous shortening was shown.

Fleckenstein, Janke & Davies (1, 2) turned next to the isotopic technique. They soaked the frog rectus abdominis for 30 min at 20° in Ringer solution containing ^{32}P, and found the relative degree of labelling to run as follows:

$$P\alpha : P\beta : P\gamma : \text{phosphocreatine P}$$
$$1.9 : 29 : 100 : \qquad 71$$

They found no effect of stimulation on the ^{32}P taken up.

Fleckenstein and his collaborators[3] recognised the difficulties of interpreting the results of such isotope studies. As in the studies *in vivo* already discussed, the cell membrane is a barrier to diffusion of the ^{32}P. Any changes in permeability (more likely to occur under conditions *in vitro*) will lead to changes in intracellular concentration and so to changes in rate of incorporation. Moreover, the ^{32}P entering may be esterified close to the cell membrane and may not penetrate to all parts of the contractile mechanism. They tried instead to measure turnover of phosphate esters by means of ^{18}O labelled water, since this very readily enters the cell. The fact that the ^{18}O of the chromatographically separated P compounds, upon proton bombardment in the cyclotron, could give by a (p,n)-reaction the highly radioactive, easily measurable ^{18}F allowed estimation of very small quantities of incorporated ^{18}O. The postulated reactions were:

[1] See Caldwell (1); Fleckenstein & Janke (1).
[2] Also Fleckenstein, Janke, Lechner & Bauer (1).
[3] Fleckenstein, Gerlach, Janke & Marmier (1).

$$
\begin{array}{c}
\text{OH} \\
| \\
\text{RO—P—OH} \\
\| \\
\text{O}
\end{array}
+ \text{H}_2{}^{18}\text{O} \longrightarrow \text{ROH} +
\begin{array}{c}
\text{OH} \\
| \\
\text{H}{}^{18}\text{O—P—OH} \\
\| \\
\text{O}
\end{array}
$$

$$
\text{ROH} +
\begin{array}{c}
\text{OH} \\
| \\
\text{HO—P—}{}^{18}\text{OH} \\
\| \\
\text{O}
\end{array}
\longrightarrow
\begin{array}{c}
\text{OH} \\
| \\
\text{RO—P—}{}^{18}\text{OH} \\
\| \\
\text{O}
\end{array}
+ \text{H}_2\text{O}
$$

$$
\begin{array}{c}
\text{OH} \\
| \\
\text{RO—P—}{}^{18}\text{OH} \\
\| \\
\text{O}
\end{array}
+ \text{H}_2{}^{18}\text{O} \longrightarrow \text{ROH} +
\begin{array}{c}
\text{OH} \\
| \\
\text{H}{}^{18}\text{O—P—}{}^{18}\text{OH} \\
\| \\
\text{O}
\end{array}
$$

Thus the rate of labelling of a phosphate ester might be taken as measuring its rate of synthesis; while rate of labelling of the inorganic phosphate would measure the overall rate of splitting of such labelled compounds. It was found that in resting muscle there was very rapid replacement of ^{16}O by ^{18}O atoms in inorganic phosphate, while the ^{18}O was incorporated into ATP and creatine phosphate at about 30 % of this rate. Although increased incorporation into inorganic P could be seen with 10 sec tetani, longer periods of contraction were needed to get increased labelling of the ATP and phosphocreatine. Recovery, on the other hand, was associated with incorporation into all compounds. This work of course depended on the assumption that the exchange reaction

$$
\begin{array}{c}
\text{OH} \\
| \\
\text{HO—P—OH} \\
\| \\
\text{O}
\end{array}
+ \text{H}_2{}^{18}\text{O} \longrightarrow
\begin{array}{c}
\text{OH} \\
| \\
\text{HO—P—}{}^{18}\text{OH} \\
\| \\
\text{O}
\end{array}
+ \text{H}_2\text{O}
$$

was not possible. But evidence that this reaction can occur in the muscle cell was later found, as we have seen.[1] Interpretation is thus made uncertain.

Lange (1), using frog rectus and 60 % alcohol at 0° for fixation, was one of the few workers in this period who described the finding of definite ATP breakdown and ADP increase; this was with contractions 3 sec in duration, induced by K$^+$ or acetylcholine treatment. Phosphocreatine breakdown was only found after 5 sec of stimulation. He also used muscles poisoned with enough IAA (10^{-2}M) to inhibit the creatine kinase: ATP breakdown and no phosphocreatine breakdown upon contraction was reported. The phosphate compounds were separated by high tension paper electrophoresis and paper chromatography.

Wajzer and his collaborators,[2] using isolated muscle fibres from frog semitendinosus, stretched perpendicularly to a monochromatic ultra-violet beam, described a change in absorption at 265 mμ and 240 mμ as the result

---

[1] Dempsey & Boyer (1); ch. 12, p. 278.
[2] Wajzer, Weber, Lerique & Nekhorocheff (1); Wajzer, Nekhorocheff & Dondon (1).

of a single twitch. This they interpreted as due to formation of hypoxan-thine-containing nucleotides from adenine-containing nucleotides, i.e. to reversible dephosphorylation and deamination of ATP. In the later paper they found after a 0.5 sec tetanus of the rectus internus major a decrease of 1.5 μmoles of adenine nucleotide and an increase of 0.5 μmole of inosine nucleotide. However, Cain, Kushmerick & Davies (1) in 1963 were unable to find any evidence for this connection of inosinic acid with contraction, either by means of the isotope technique or by spectrophotometric assays.

μmoles per g (average values)

	ATP	ADP	ITP	IDP
Control	2.75	0.47	0.22	0.07
After contraction	2.69	0.49	0.26	0.09

In 1959, in the course of an interesting review of the work in this field, Davies, Cain & Delluva (1) reported some significant results on inorganic phosphate release. Using retractor penis muscles of the turtle, frozen very rapidly in melting dichlordifluoromethane, they estimated the inorganic P of the resting muscle to be 1.9 μmoles/g, after a short contraction 2.4 μmoles/g. This increase was of the order to be expected, and they concluded that it came from some very labile unidentified compound.

A quite different method of approach to the problem of ADP formation was that used by Chance & Connelly (1) in 1957. Chance & Williams (1) had shown that ADP stimulated the respiration of mitochondria and sarco-somes, and that this stimulation was accompanied by changes in the steady-state levels of oxidation–reduction of certain respiratory enzymes. These changes could be recorded by a special double-beam spectrophoto-meter. Chance & Connelly[1] used a very thin frog sartorius muscle, kept under fully aerobic conditions in flowing Ringer solution at 7°. They found that stimulation once per second led to an abrupt increase in absorbency at 340 mμ relative to 386 mμ; the tracing returned gradually to the original level after cessation of the stimulus. Reasons were given for attributing this change to oxidation of reduced pyridine nucleotide in the sarcosomes, oxidation stimulated by increased ADP concentration in their vicinity.[2] From knowledge of the amount of ADP needed to cause such changes in isolated mitochondria, it was calculated that 0.009 μmole/g twitch would suffice. This of course is only a small fraction of the amount expected if ATP breakdown supplied the energy for the twitch. Chance & Connelly con-sidered that the affinity of the respiratory chain components for ADP was so high as to rule out possibility of its resynthesis from phosphocreatine as a cause of the low amount reacting in the sarcosomes.

[1] Also Connelly & Chance (1).
[2] Ch. 17 for an account of oxidative phosphorylation.

Other workers however were sceptical of this. Thus Carlson & Siger in 1959 (1) used IAA-poisoned muscle, considering it as a closed system where only the coupled equilibria

$$(1) \quad \text{phosphocreatine} + \text{ADP} \rightleftharpoons \text{creatine} + \text{ATP}$$
$$(2) \quad \text{ATP} + \text{AMP} \rightleftharpoons 2\,\text{ADP}$$

functioned to restore ATP hydrolysed on contraction. They estimated ATP, phosphocreatine and free creatine in muscle resting and after contraction; from analysis of their results they calculated an equilibrium constant for reaction (1) in good agreement with that found *in vitro*. The calculated value of ADP concentration at equilibrium was very low – only about 0.035 μmole/g. They pointed out that an active creatine kinase system would account for Chance & Connelly's findings. Jöbsis (3) later also calculated that, assuming the breakdown of ATP/twitch g of muscle amounted to about 0.3 μmole, then the increase in ADP to be expected, in presence of the creatine kinase, would be approximately only 0.002 μmole. He concluded that the results of Chance & Connelly did not bring evidence against ATP as the primary energy source.[1] It should be mentioned that some workers have argued for compartmentalisation of the muscle ATP. Thus Hohorst, Reim & Bartele (1) found in resting muscle a steady-state equilibrium close to the thermodynamic equilibrium. But this was not so in contracting muscle, where great changes take place in phosphocreatine and creatine concentrations, with only slow and small changes in ATP and ADP concentrations. It seems that here in suggesting the existence of some 85 % of the ATP as 'storage ATP' they neglected to take into consideration the activity of myokinase, and the extent to which the ATP and ADP concentrations do change during contraction.[2]

In 1954 Distèche & Dubuisson (1) returned to the measurement of pH changes in muscle, with new techniques which eliminated the delay in recording. They showed that this delay was now of the order of only 50 msec when very fast stepwise changes in pH were made under the electrode. There remained the delay due to diffusion of CO_2 in the muscle to the electrode, and correction was made for this by the method of Hill (28). Distèche (1) in a very thorough presentation of this whole subject, was able to give corrected curves not only for tetanic contractions, but also for isometric and isotonic twitches in frog as well as tortoise muscle. The phenomena observed, with the succession of changes (*b*), (*c*) and (*d*), were fundamentally the same as those described by Dubuisson.[3] It is interesting that in a single twitch only the acid change and its reversal appeared, the alkaline change only becoming apparent after a number of twitches.

---

[1] Also Jöbsis (2).
[2] Also Talke, Arese & Hohorst (1). For a discussion see Maréchal (1) p. 144.   [3] P. 345.

Distèche, influenced by the failure at that time of chemical methods to find ATP breakdown in a short contraction, was prepared to relinquish the idea that phase (b) was the result of such breakdown; he subscribed to the proposal made by a number of workers[1] that the first reaction of contraction was hydrolysis of phosphate groups attached to protein, the latter existing in the phosphorylated state during repose. This hydrolysis would release hydrogen ions, and it was pictured that rephosphorylation of the protein by means of ATP (with no appreciable pH change) followed immediately.

PROPOSALS FOR ALTERNATIVE SOURCES OF ENERGY-RICH PHOSPHATE. Besides this idea of phosphorylated protein as a participant, a number of other possible alternatives to ATP were canvassed from time to time. Thus Goodall (1) in 1956 referred to the earlier experiments of Severin, Georgievskaya & Ivanov (1). These Russian workers had synthesised carnosine diphosphate; they observed that it was hydrolysed enzymically in aqueous extracts of muscle, and that when carnosine was added to minced muscle in the presence of oxygen at 17°, the easily hydrolysable phosphate content increased above that in the control. Goodall now suggested that this might be the unknown donor postulated by Fleckenstein, Janke, Davies & Krebs (1). Goodall observed with carnosine diphosphate a relaxing effect on glycerol-extracted muscle fibres comparable with that of phosphocreatine. But Cain, Delluva & Davies (1) were quickly able to get evidence against this. Retractor penis muscles were used, immersed for 4 h at 25° in physiological salt solution containing ^{32}P. Several periods of stimulation followed by recovery were included in this time, and this treatment should have caused turnover of all phosphate compounds concerned in contraction. To the extract of the frozen muscle carrier quantities of carnosine mono- and diphosphates were added. Chromatographic analysis showed labelled phosphocreatine, but no labelling in the carnosine compounds.

About this time O. Cori and his collaborators[2] thought they had evidence that, in presence of an enzyme from rat muscle, the carbonyl phosphate group of diphosphoglyceric acid could be directly transferred to creatine without intervention of ADP. This work of course opened up still another possibility for the by-passing of ATP breakdown in contraction, but it was short-lived. In 1961 Morrison & Doherty (1) were able to show that enzyme preparations carrying out this 'direct' transfer were always contaminated with nicotinamide-adenine dinucleotide[3] which non-enzymically gave rise to ADP-ribose. The latter could then be hydrolysed to adenylic acid by an enzyme present in muscle extracts. The further interesting observation was

---

[1] E.g. Davies, Cain & Delluva (1).
[2] Cori, Traverso-Cori, Lagarrigue & Marcus.   [3] Also Rogozkin (1).

made that highly purified AMP could be phosphorylated to ATP in presence of myokinase and creatine phosphokinase. Either enzyme alone was without effect, and similar observations concerning phosphorylation of AMP by phosphopyruvic acid were made.[1] It thus appears that, though ADP is the acceptor of choice, AMP can under certain circumstances play this role. A little later Cori, Marcus & Traverso-Cori (1) withdrew their claim for the direct pathway, finding that their enzyme preparation contained minute amounts of ADP or ATP – 0.2–0.6 μmole/g protein. About this time Cain, Kushmerick & Davies (2) examined by the isotopic method and paper electrophoresis the possibility that the phosphoglyceric acids or phospho-pyruvic acid might be directly concerned in contraction. The amounts found were very small – too small in each case to act as the net energy source for a single contraction; little or no change occurred on contraction.

THE BREAKTHROUGH. This came in 1962 with the use of the inhibitor fluorodinitrobenzene (FDNB). Iodoacetate, to prevent glycolysis and re-phosphorylation of ADP from this source, had been used by a number of workers; as we shall see, much useful information was gained in this way about phosphocreatine breakdown, but ATP breakdown could not be observed except at the onset of rigor. In 1959 Kuby & Mahowald (1) reported that 1-fluoro-2,4-dinitrobenzene was a potent inhibitor of ATP-creatine phosphokinase and myokinase. As Davies (1) has recorded, this compound was regarded as so aggressive that it would be unlikely to show specificity; thus it was not till three years later that the behaviour of muscles treated with it was tested. Davies and his collaborators[2] were then astonished to find that, although the number was limited, normal contrac-tions could take place and ATP breakdown could be demonstrated in a single tetanic contraction of the frog rectus abdominis. ATP, ADP and AMP were measured by the firefly luminescence technique and fluorimetrically,[3] and the following values were found after 40 min at 0° in 3.8×10^{-4}M FDNB.

| | μmoles/g | | |
	ATP	ADP	AMP
Rest	1.25	0.64	0.1
One contraction (125 g cm/g)	0.81	0.90	0.24
Rest	1.24	0.61	0.07
Double contraction	0.59	0.88	0.41

ATPase and myokinase were both active. There was no fall in phospho-creatine content, nor rise in free creatine. These experiments on the slow

[1] Also Bücher & Pfleiderer (1).
[2] Cain & Davies (1); Cain, Infante & Davies (1).
[3] Wahl & Kozloff (1); Estabrook & Maitra (1).

rectus abdominis were repeated by Infante & Davies (1) on the frog sartorius, where ATP breakdown could be shown during the rising phase of a single twitch.

Infante & Davies (2) later made a thorough investigation of the effects of FDNB on the energy-producing reactions of striated muscle. No oxygen was taken up but lactate formation went on at a normal rate. In extracts the creatine kinase was completely inhibited, and myokinase about 30 % inhibited, while the Ca-activated ATPase showed somewhat enhanced activity. There was no net synthesis of ATP during rest intervals – the lactic acid formation observed was extremely slow compared with the ATP breakdown during contraction. Dydynska & Wilkie (1) have confirmed the main conclusions of Davies and his collaborators with regard to the effects of FDNB.[1] In their experience however the creatine kinase was not quite completely inhibited. They also made the interesting observation that in poisoned resting muscle there is twice as much inosine as in the unpoisoned muscle, and that poisoned muscle after a long series of twitches contains more than the poisoned resting muscle. They suggest that the deaminase may have a role in disturbing the equilibrium of the myokinase reaction, and thus reducing the concentration of ADP (which is known to inhibit actomyosin ATPase).[2]

The particular and crucial problem of the primary part played by ATP in contraction metabolism was thus solved; we turn now to consider the metabolic changes during a short contraction and immediately after.

RELATION OF THE METABOLISM TO THE HEAT PRODUCTION AND MECHANICAL EVENTS

Some time before the introduction of FDNB as a revelatory inhibitor, muscles treated with IAA or 2,4-dinitrophenol were being used in the study of phosphocreatine hydrolysis in the course of contraction under different conditions. Efforts were made to find the metabolic counterparts of the activation and maintenance heat, the shortening heat and the work performed.

Carlson & Siger (1, 2) in 1960, using IAA-treated frog sartorius (at 0° in order to minimise any oxidative recovery) found that with isometric twitches up to about forty in number, the phosphocreatine dephosphorylation varied in a linear manner with the number of twitches. 0.29 μmole was split per gram per twitch, in good agreement with the heat values to be expected. There was no net change in ATP content up to about seventy twitches. Carlson & Siger suggested that on contraction the creatine kinase system phosphorylated actin-bound ADP, and that the ATP formed

[1] Also Maréchal & Beckers-Bleukx.   [2] E.g. Bendall (1).

remained bound while the free creatine escaped to the sarcoplasm; as the result of ATPase activity free phosphate would be formed and would also escape.

$$CP + \text{F-Actin–ADP} \longrightarrow \text{G-Actin–ATP} + C$$
$$\text{G-Actin–ATP} \longrightarrow \text{F-Actin–ADP} + P_i.$$

This formulation drew support from the work of Yagi & Noda (1) according to which the bound ADP of deoxycholate-treated or glycerinated fibres could be phosphorylated by phosphocreatine in presence of the kinase. Moos (1) however was unable to confirm these findings, showing that small quantities of liberated nucleotide were necessary for the phosphorylation.[1] Noda & Bono (1) later agreed with this, and Carlson (1) admitted that the hypothesis described became less tenable.

In 1960 also Mommaerts and his collaborators[2] began a close study of metabolism connected with activation, shortening and work, following the phosphocreatine breakdown in IAA-treated frog sartorius. In isotonic tetanic contractions they found increase in free creatine and phosphate to run parallel with the work performed. No metabolic counterpart of the shortening heat could be established. The authors were inclined to explain this by the diminution in activation heat with shortening, so that the two effects might cancel each other.[3] Maréchal & Mommaerts went on to examine metabolism in isometric tetanic contractions. Here the maintenance of tension during a time up to 60 sec was accompanied by phosphocreatine breakdown at the constant rate of 0.28 μmole/g sec. This rate agrees well with the maintenance heat if the heat of hydrolysis of phosphocreatine and associated ionic changes is taken as 10000 cal/mole. When several short tetani were studied instead of one long tetanus, an 'extra metabolism' was found which amounted to about 0.33 μmole/g tetanus. The extra metabolism would include that associated with the elastic work and internal shortening during tension generation and also any continuation of heat production during relaxation.[4] The connection of these metabolic findings with the course of the maintenance heat as described by Hill[5] was discussed, though no quantitative correspondence with the early less efficient phase of heat production and the later more efficient phase could be established.

---

[1] Also West, Nagy & Gergely (1).

[2] Mommaerts & Seraydarian (2); Mommaerts, Seraydarian & Maréchal (1); Maréchal & Mommaerts (1).

[3] See Aubert (1); Brown (1). These two papers quoted by Mommaerts & Seraydarian are concerned with dependence of maintenance heat on muscular length. Carlson, Hardy & Wilkie (1) criticised this explanation, pointing out that the situation envisaged by Mommaerts & Seraydarian would have made impossible the original detection of shortening heat.

[4] Maréchal (1) for a thorough discussion of maintenance metabolism and extra metabolism.

[5] Hill (18, 19) ch. 19.

From 1962 Davies and his collaborators[1] described a long series of experiments on frog rectus abdominis. The curve relating phophocreatine breakdown to work done was a straight line passing through the origin, so that no evidence appeared for the chemical breakdown corresponding to activation heat or shortening heat (see fig. 79). The same conclusion was reached as a result of experiments in which the muscles were made to do different quantities of work while shortening the same amount, or the

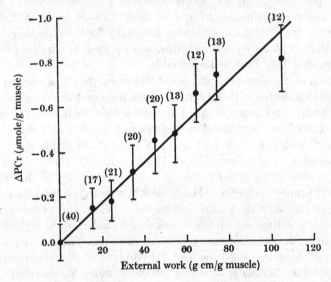

Fig. 79. Relationship of phosphocreatine breakdown to the amount of external work done by DNP-treated frog rectus abdominis muscles contracting against a constant load to different degrees and for different times. The figures in parenthesis refer to the numbers of pairs of muscles used. Slope of line: 2600 cal of external work/mole phosphocreatine. (Cain, Infante & Davies (1).)

same work while shortening to different lengths. Shortening against zero load gave no significant increase in metabolism. In these experiments muscles which had been previously treated with 2,4-dinitrophenol $(2.5 \times 10^{-4}\mathrm{M})$ were used. This had the effect of reducing the phosphocreatine content from 12 μmoles to less than 3 μmoles; the muscles could still perform several contractions, and the low phosphocreatine content made demonstration of changes in its concentration much more accurate. Infante, Klaupiks & Davies (2) also did some experiments with FDNB-treated frog sartorius. They found with isometric tetanic contractions a linear relation between duration of stimulus and metabolism – here measured as change in

[1] Cain, Infante & Davies (1); Cain, Infante, Klaupiks, Eaton & Davies (1); Infante, Klaupiks & Davies (1, 3).

ATP content. The process of tension development seemed to have no important effect.

Carlson, Hardy & Wilkie (1) also studied contraction in nitrogen of IAA-treated frog sartorius. They found no correlation between phosphocreatine hydrolysis and degree of shortening; thus they write '...the energy output

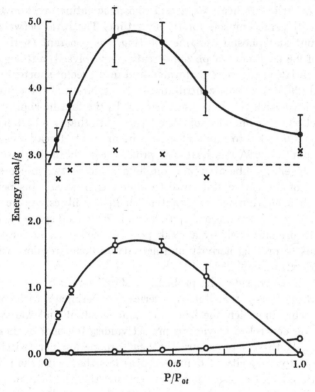

Fig. 80. The variation in energy output with load in isotonic twitches; the load was allowed to fall during relaxation. Abscissa, load P as a fraction of the peak isometric tension P_{0t}. Ordinate, energy output in mcal/g and twitch (mean of 100 twitches, twenty by each of five muscles). Upper curve, ●, total heat. Lower curves, ○, external and internal work; ×, total heat minus total work. (Carlson, Hardy & Wilkie (1).)

in a complete cycle (of contraction and relaxation) consists of a fixed activation heat plus the work...'. Measurements of heat and work output in complete isotonic twitches on unpoisoned muscles confirmed this indication that $F = A + W$ (see fig. 80). Thus both phosphocreatine metabolism and heat production showed a constant activation term, linear dependence on work and no significant dependence on shortening. They pointed out that in the experiments of A. V. Hill in 1949 the shortening heat was demonstrated in the early stages of a twitch. They suggested that with isotonic contraction the heat (exclusive of that generated by the falling load in relaxation) shows different distribution in time from that with isometric twitches. As Wilkie (1)[1] later expressed the situation: 'while it is shortening the muscle produces heat at a faster rate than if it is isometric; but the *total* quantity of heat is similar in the two cases'.

In these experiments heat, work and phosphocreatine breakdown were measured in the same muscle for the first time. The ratio between total energy output and phosphocreatine splitting was constant (with energy liberation of 9.8 kcal/mole of phosphocreatine hydrolysed) although there were great variations in work performed and in degree of shortening.

A. V. Hill (29, 30) in 1964 contributed to the solution of these problems of energy release with two important papers. In the first he compared the heat production in a complete isometric twitch with that in a twitch at first isometric but subjected to rapid release at the moment of peak tension. He concluded that 15–20 % less heat was produced in the latter case. Thus 15–20 % of the energy liberated in a complete isometric twitch comes from continuation of the active state into the relaxation phase, this being associated with a continuing state of tension in the fibres. As he wrote: '...the arrangement reminds one of a system in which a displacement from rest tends to prolong itself by a weak positive feedback; in the muscle, tension tends to prolong activity and activity prolongs tension, with the liberation of extra energy.'[2]

Hill went on to consider the problem of shortening heat raised by the work of Carlson, Hardy & Wilkie. In a series of experiments with complete isotonic twitches in which the load was kept constant and the speed of shortening was controlled at various pre-set values, the heat of shortening was obvious. With this procedure the heat production per twitch, with shortening up to a velocity of 6 mm/sec, was less than with the isometric twitch. With velocity of shortening greater than this the heat production rose in parallel with the speed and so with the degree of shortening, but it never rose above its isometric value. Hill gave the following explanation for the absence of shortening heat at low rates of shortening, envisaging these factors diminishing heat production: (1) the very slow shortening would occur largely at a time when the active state was decaying; (2) elastic energy present in the muscle and connections would be dissipated as heat in the isometric contraction and much of it in the nearly isometric contractions, while with greater shortening this energy would be removed by the ergometer as work; (3) the 'positive feedback' mentioned above would be

---

[1] Also Wilkie (2).

[2] In *Trails and Trials* (p. 346) Hill (11) refers to Pringle's concept (6) of 'activation affected by tension' in explaining the properties of insect fibrillar muscle.

operative in the isometric and in the contractions at low speeds. The real existence of the shortening heat in a complete isotonic twitch was thus demonstrated. Hill explained the negative results of Carlson *et al.* with regard to shortening heat as due to the following state of affairs. In isotonic twitches with small loads allowing rapid and extensive shortening, the heat of shortening would be large, while since the tension produced would be small the 'positive feedback' just discussed would be small. With large loads shortening would be limited, while the extra heat liberated as a result of maintained tension, continuing even into the relaxation phase, would be large. These effects would tend to balance each other. Hill preferred to write $E = A + W + ax + h$, where $h = $ heat due to tension persisting after the conclusion of the work and shortening. In his opinion it would make little difference if one regarded the effect of shortening on heat production as simply a change in the maintenance heat, produced by shortening. (See also Hill (25).) Thus the claim of Carlson, Hardy & Wilkie was justified – that a muscle which shortens never produces more net heat than one which does not shorten.

It is interesting to notice that Baskin (1) recently found with isotonic twitches of frog sartorius that the amount of oxygen consumption depended on the work done. As with the other methods of measuring metabolism, the experiments were inconclusive with regard to any extra uptake associated with shortening. Using isometric twitches of the frog sartorius at different lengths, Baskin & Gaffin (1) give an equivalent for the activation heat in terms of oxygen consumption. They found a close correspondence between the oxygen consumption and the tension produced at different lengths (after subtraction of the 'activation' oxygen). This part of the oxygen consumption would represent the heat of shortening of the contractile elements against the elastic ones, and the work done by the stronger portions in extending the weaker.

In a paper in 1965 Hill (32) made a Further Challenge to Biochemists. He clarified the conditions under which the muscle should be examined in order to establish whether or not ATP (or phosphocreatine) breakdown accompanied shortening heat: the muscles of a pair should be used, each receiving the same stimulus, one shortening the other not. Such conditions up to that time had not been met, but in 1967 Davies, Kushmerick & Larson (1) accepted the challenge. They used pairs of frog sartorii, one muscle contracting isometrically, the other isotonically lightly loaded. Each was stimulated for 170 msec and frozen by 200 msec. ATP breakdown for the isometric contraction was greater by a mean value of 0.128 μmole/g muscle than ATP breakdown for the isotonic contraction. When the heat of the measured shortening was calculated from Hill's data together with its ATP equivalent, the discrepancy became 0.356 μmole.

Davies in his model of the sliding-filament mechanism in 1963 equated the shortening heat with large heat output accompanying transformation of random coils into α-helices in the first contraction of the cross-bridges. The cyclic making and breaking of links would be thermally neutral, but at the final lengthening of the links on relaxation absorption of heat would be expected. Davies (3) has suggested that the absence of shortening heat observed by Carlson, Hardy & Wilkie with complete isotonic twitches might depend on such heat absorption during the relaxation period.

If we try to survey the present position many uncertainties besides those concerned with the shortening heat are encountered. Besides the difficulty for biochemists of the very small quantities of metabolites involved, it has to be remembered that differences between results at different times and in different laboratories may be due to differences in species of frog used and in the season. Then there are the variations in experimental procedure; for example Carlson, Hardy & Wilkie (2) have emphasised the problems involved in the choice of method for accurately and consistently determining the weight of the muscle; these problems of course might be differently solved by different workers.

As we have seen, Davies and his colleagues working with FDNB-poisoned rectus abdominis had found no metabolic equivalent of the activation heat, in contrast to the experience of Mommaerts and his colleagues and of Carlson, Hardy & Wilkie all working with sartorius treated with IAA. This was at first put down to the different muscle used.[1] But later Davies, Kushmerick & Larson (1) reported experiments on frog sartorius in which the muscle was stimulated to unloaded contraction at different frequencies. ATP breakdown in μmoles/g pulse was about 0.08 for slow rate of stimulation (about one/second), only 0.02 for 10 stimuli/sec. The expected change calculated from the activation heat was about 0.12 μmole per twitch. They explained the activation heat as due to release of bound calcium from the sarcoplasmic reticulum without need for ATP hydrolysis, the active state resulting from the increase in free calcium. This is a matter of surmise, since at present there are no figures for heat change associated with the setting free of calcium. The capture of the Ca^{2+} by troponin in the contractile system may also play a part.

With regard to the return of the calcium to the reticulum, Davies, in the account of his theory in 1963, recognised that this would need ATP utilisation and he claimed that the ATP breakdown observed during relaxation[2] was of the right order for this requirement. Mommaerts & Wallner (1) however have recently described experiments specially designed to examine relaxation metabolism with isometric and isotonic contractions. With

---

[1] See the discussion in *Biochemistry of Muscle Contraction* pp. 468 and 474, after the paper of Davies, Cain, Infante, Klaupiks & Eaton.　　　[2] Infante & Davies (1).

paired, FDNB-poisoned sartorius muscles they could find no difference in ATP or phosphocreatine breakdown in the muscle frozen at the height of contraction as compared with that frozen after relaxation. In this connection we may return for a moment to the experiments of A. V. Hill; if indeed tension is prolonging the active state in the muscle relaxing after isometric contraction, we should expect to find here increased ATP breakdown as a result of the retardation of the return of calcium to the reticulum. Clearly it is hard to reconcile this with the very carefully controlled results of Mommaerts & Wallner. However, the latter authors themselves emphasise that circumstances might arise in which the ATP breakdown is delayed.

The immediately post-contractile events in IAA-poisoned muscle have also been the subject of some difference of opinion. As we have seen, Lundsgaard (6) had found at 0° a large 'recovery' utilisation of phosphocreatine.[1] Maréchal & Mommaerts (1) in 1963 reported no increase in free creatine during the 15 min after isometric contraction at 0°, even after 30 sec tetanus; but Spronck (1) confirmed Lundsgaard in finding phosphocreatine breakdown continuing some 30 sec after only five isometric twitches at 2°. Hexosediphosphate content rose during this time by an amount about equivalent to 15 % of the total phosphocreatine disappearance. Thus the ATP resulting from the recovery breakdown of phosphocreatine can either replenish the ATP content of the muscle diminished by contraction, or take part in abortive carbohydrate breakdown.

Carlson, Hardy & Wilkie (2) continued their work on IAA-treated sartorius with an examination of the relation between phosphocreatine utilisation, tension development and heat production in isometric twitches and in 10 or 30 sec tetani at 0°. They considered that the balance of evidence indicated little ester formation under their conditions; thus the whole breakdown, whether during or after contraction, would be taken as representing the ATP used in contraction. They concluded that the heat plus work production associated with phosphocreatine hydrolysis in vivo was about 10.6 kcal/mole – in good agreement with the values for the heat of hydrolysis in vitro in the presence of muscle extract. Wilkie (4) determined this relation for different types of contraction – isometric twitches and tetani, isotonic twitches performing work – and found it always 11 kcal/mole (see fig. 81). He therefore concluded that, whatever the mechanical conditions, the breakdown of phosphocreatine over the whole cycle of contraction and relaxation was proportional to the sum of the heat and work produced. Here we seem to be on firm ground, but it is clear that we have not yet reached the goal of knowledge of the moment-to-moment metabolism and energy release of the contracting muscle.

[1] P. 102.

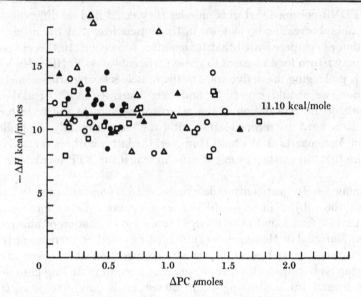

Fig. 81. Ordinate: individual ratios (heat + work, mcal)/phosphocreatine split (μmoles). Each point thus gives an estimate of $-\Delta H$, the *in vivo* enthalpy change of phosphocreatine splitting, in kcal/mole. Abscissa: total phosphocreatine split (μmoles). $\bigcirc$, $\bullet$, isometric twitches; $\square$, isometric tetani; $\blacktriangle$, isotonic twitches, positive work; $\triangle$, stretches, negative work. (Wilkie (4).)

THE METABOLISM OF MUSCLES STRETCHED DURING ACTIVITY

We have already briefly mentioned[1] the experiments of Fenn (2) in which the muscle was stretched by means of a heavy load applied soon after the stimulus. The work done on the muscle might be expected to appear as heat, but Fenn found that the net energy liberated (total heat output minus the work done on the muscle) was less than in an isometric contraction. He took this result to mean a decrease in energy liberated by the muscle under stretch.

This very interesting question was taken up again in 1951 in A. V. Hill's laboratory. Abbott, Aubert & Hill (1) found that when a muscle was stretched during the active phase of contraction, the total heat production was greater than that of an unstretched muscle, but the excess heat production was only about 50 % of the energy of the work done on the muscle. They discussed three possibilities as to the fate of the missing work: (*a*) that it was absorbed in driving backwards chemical processes which had actually occurred in the muscle as a normal part of its contraction. (*b*) That it was absorbed in some other unknown physical or chemical process. (*c*) That it

[1] P. 56.

was all degraded to heat but that chemical processes normally occurring in the stimulated muscle were suppressed as a result of the stretch. Abbott & Aubert (1) found that with slow stretches all the work done on the muscle disappeared; they found it impossible to decide between the possibilities described above. In 1959 Hill & Howarth (1), with improved techniques, were able to measure heat, work and elastic energy throughout the course of the contraction of toad muscle with and without stretch. They found that the total heat appearing up to the end of relaxation in the stretched muscle might be equal only to the work done on the muscle. Thus since the work had disappeared and elastic energy remaining was nil, the net energy liberated by the muscle itself (total heat minus work done on the muscle) was nil. It was concluded that 'the chemical products of the reactions provoked by the stimulus have been wholly returned to their initial state'. This complete disappearance of heat production, other than that due to the work done on the muscle, depended on the amplitude and timing of the stretch. The total heat production could be rather greater than the work done, but total heat produced minus the work done was always much less than the total heat production in the corresponding isometric contraction.

In 1963 Maréchal and his collaborators examined the phosphocreatine metabolism of stimulated IAA-treated frog sartorii during stretch.[1] They found with slow stretching a saving of about 30 % in the breakdown as compared with the unstretched partner, in good accord with the heat measurements. This also was the experience of Wilkie (4). But with rapid stretches Maréchal reported increased phosphocreatine hydrolysis, which finds no parallel in the heat measurements and is still unexplained. Infante, Klaupiks & Davies (4), using FDNB-treated muscle, found even with quick stretches less ATP used, the rate of breakdown being only about 50 % of that in the controls.

Wilkie (3) has emphasised that in interpreting such results it is important to distinguish between actual reversal of energy-yielding reactions and their suppression.[2] The metabolic results as they stand give no evidence for reversal, and can most simply be explained by supposing that ATP breakdown has been prevented in the circumstances of the stretch. This might be pictured on the sliding mechanism in various ways. Thus A. F. Huxley (1) suggested prevention of the reaction $A + M \longrightarrow AM$ in the postulated series

$$A + M \longrightarrow AM$$
$$AM + XP \longrightarrow AXP + M$$
$$AXP \longrightarrow A + X + \text{inorganic phosphate}$$

while Davies and his collaborators, in the context of Davies' theory of contraction,[3] suggest that external work is used to break hydrogen bonds in

---

[1] Maréchal (1).   [2] See also A. V. Hill (31) for discussion.   [3] Ch. 12.

the contracted H-meromyosin side-chains. Since hydrolysis of an ATP molecule is assumed to be responsible normally for this breakage and extension in each link, the breakage by stretch would spare ATP. It cannot of course be decided from the metabolic results that no resynthesis of ATP has gone on – breakdown might actually have been greater than that observed. It is difficult to understand the apparent expectation of Davies[1] that a net resynthesis might be found; as Wilkie (3) has remarked, in resting muscle the unbound ADP is so low that its total phosphorylation could hardly lead to a detectable change in the ATP concentration. In the experiments of Hill & Howarth, study of the variation of CTH – CWD (corrected total heat minus corrected work done) with time throughout the contraction with stretch did show a transient negative phase. But Wilkie pointed out that this analysis, making great demands on the myothermic technique, involved a small difference between two large values; he found the evidence for reversal less compelling than that for suppression. Changes in ATP concentration during the contraction with stretch have not been followed, but this would of course be a very difficult thing to do.

We see then that the challenge has been met, and that we can say with confidence that ATP is the fuel directly used by the muscle machine. We can say too that the energy-rich phosphate bond quantitatively supplies the energy needed for work and for the heat output accompanying the work. The exact details of the energy requirements – in the actual process of tension production or for such needs as Ca^{2+} release and uptake – and their distribution in time still to some extent elude us.

[1] Davies (2) p. 40.

15

RIGOR, AND THE CHEMICAL CHANGES
RESPONSIBLE FOR ITS ONSET

We have already discussed the view of Kühne in 1864 that death rigor had three causes, all contributing to precipitation and coagulation of the muscle proteins – the precipitation of the globulin (myosin) fraction in a manner resembling its precipitation on dilution; the precipitation of the albumin fraction under the influence of lactic acid; and the operation of a clotting activator.[1]

We come now to the experiments of Claude Bernard, already briefly mentioned in chapter 5. Part of the very clear description in his book (2) of 1877 may be quoted here:[2]

It has even been thought possible to lay down a general rule that muscular rigidity and acid reaction were two phenomena essentially bound the one to the other. This is not so; the two phenomena are not bound the one to the other; it is the presence of the glycogenic substance which is the necessary forerunner of the acid reaction of the muscle, this reaction arising essentially from the lactic fermentation of the glycogenic matter. If in fact one kills an animal after having made its muscle glycogen disappear, the rigidity of the corpse comes on although the muscle conserves its alkaline reaction. For this effect, it is enough to kill the animal by starvation, which has the result of completely depriving the muscle tissue of glycogenic matter; on this account the rigid muscles are not acid: thus no glycogen in the muscle, no lactic acid in its tissue. An example is again supplied for us by the muscles of crustaceans in general, and notably of the crayfish, on which it is very easy to make the experiment. These muscles do not contain more than a trace of glycogenic material. After death they also become rigid while preserving their alkaline reaction.

Bernard also noted the more rapid onset of the alkaline rigor.

In 1882 Schipiloff (1) made a very definite claim to have shown that death rigor was due to the precipitation of chemically unaltered myosin out of its half-fluid state in the muscle plasma, this precipitation being caused by the postmortal lactic acid increase. She took the later disappearance of rigor as due to the solubilising effect of greater amounts of acid. These views were based (a) on the behaviour of myosin in salt solutions – precipitation of the myosin by water or a very little acid, resolution of the precipitate in

[1] Ch. 7.   [2] Bernard (2) p. 429.

larger amounts of acid – but no details of concentration or conditions are given; (*b*) on the effects of various solutions used for perfusion of frog muscles – e.g. the immediate appearance of 'death rigor' on perfusing with 0.1–0.25 % lactic acid, resolution of this rigor on perfusion with a solution 0.5 %. This effect was never confirmed and von Fürth could not repeat it.

In the years between 1882 and 1926 the great majority of workers accepted the idea that lactic acid formation was responsible for the hardening and shortening of muscle in death rigor. Indications of this can be seen in the early work of W. M. Fletcher, for instance in his study (1) of the effects of lactic acid solutions on muscles immersed in them. With an acid concentration of 0.05 % a reversible contraction was seen, with higher concentrations a sudden rigor of different character. In 1902 he accounted for the delaying effect of oxygen on rigor formation as due to the prevention of accumulation of metabolic products within the muscle (3); and by 1913, as we have seen[1] he was ready to regard lactic acid as the metabolic product concerned (4). In this 'lactic acid era' of muscle contraction, it was natural that the conception of lactic acid as the *causa movens* in the active muscle should be carried over into explanations of rigor, as may be seen in the Croonian Lecture of Fletcher & Hopkins (2). Von Fürth (3) quotes Winterstein as seeing in ordinary muscle contraction nothing other than a transient rigor, and Hill (5) in 1916 wrote: 'No processes involving marked energy exchange occur in contraction other than those which occur in rigor, and on this ground the heat production of the two processes is intimately linked with lactic acid formation. Thus if we study the chemistry of rigor, we are studying at the same time the chemistry of normal muscle contraction.' Nevertheless it is surprising that in none of the writings of this period have I found any reference to the 'alkaline rigor' of Claude Bernard; the title of his book *Leçons sur le Diabète* may have suggested to workers on muscle that there was nothing in it for them. Even in Michael Foster's chapter on 'Glycogen' in his life of Bernard there is no mention of these striking observations[2] which, if known to Fletcher and Hopkins, might have had a decisive effect on their thinking.

Von Fürth also became interested in rigor. In an early paper (2) he compared the effects of various toxic agents on the precipitation of myogen and myosin solutions with their effects in causing rigor in living animals. He could find no evidence in this way for the attribution of rigor to protein coagulation. As we have seen,[3] von Fürth was an adherent of the *Säurequellungstheorie* of contraction, according to which, under the influence of lactic acid, a water shift took place within the doubly refracting part of the contractile elements; as a result the ultramicroscopic particles swelled at

[1] Ch. 3.   [2] Foster (3).   [3] Ch. 7.

the expense of the circumambient fluid in the fibre. He considered that this same process was the basis of the changes of *rigor mortis*, and that acid-facilitated coagulation of the plasma happened only later, associated with *Entquellung* (loosening of the water binding) and the passing off of rigor. Experimental basis for these views he found in his work with Lenk (1). Here the imbibition of fluid by muscle soaked in water or in dilute lactic acid solutions was followed. There was rapid uptake, followed after about 25 h by a more or less rapid loss of water. This latter phenomenon was interpreted as an *Entquellung* accompanying the resolution of rigor. Support for the assumption that clotting was involved in the latter process was found in the observation that certain factors known to cause clotting of plasma (heat, sodium rhodanate solution) when injected into the muscle vessels also caused water loss rather than uptake.

The Embden school had ideas about the mechanism of contraction at variance with those in which lactic acid played the central role.[1] They were influenced by their demonstration (not at first accepted by other workers) that part of the lactic acid formation was post-contraction, and they regarded some colloid–chemical reaction in the fibril protein as the first source of the muscle's energy. With regard to chemical events, Embden attached chief importance to the formation from glycogen of a phosphate ester (which he called lactacidogen) and to its dephosphorylation. The formation of lactacidogen was normally enhanced in the presence of fluoride ions; Embden and his collaborators[2] found that this 'synthetic activity' was diminished in *brei* made from fatigued or rigor muscle. They interpreted this observation as meaning a deterioration, in both these states, of the colloid-chemical apparatus. As further evidence Deuticke (2) cited his finding that, both in fatigue and death rigor, the solubility of muscle proteins in various phosphate buffer solutions was diminished.

It was in 1926 that Hoet & Marks (1) and Hoet & Kerridge (1) described their results on alkaline rigor. About this time many workers were experimenting with insulin, and more than one group had observed the very rapid onset of *rigor mortis* in animals dying after insulin convulsions;[3] the same was true after death in hypoglycaemia due to prolonged thyroid feeding.[4] Dudley & Marrian (1) found after death from large insulin doses that the glycogen of the muscles had vanished. Kuhn & Baur (1) noted also the very low lactic acid content of such muscles. The importance of the exhaustion of the glycogen store for this precipitate rigor was shown in the following way by Best, Hoet & Marks (1). Insulin was injected into a rabbit in which the sciatic nerve on one side had been cut; the paralysed muscles retained their glycogen

---

[1] P. 76.   [2] Embden & Jost (3); Deuticke (1).
[3] E.g. Baur, Kuhn & Wacker (1). Such convulsions are the result of low glucose concentration in the central nervous system.   [4] Burn & Marks (1).

and remained flaccid and reactive for some hours after the corresponding muscles (which had undergone convulsions and lost their glycogen) had passed into immediate rigor. In similar experiments Hoet & Marks (1) now showed that this rapid rigor had taken place at a pH value as high as 7.0, with no rise in lactic acid. There was however considerably more inorganic phosphate in the rigor muscles than in the denervated muscles. Baur, Kuhn & Wacker had recorded an alkaline reaction in muscles in insulin rigor and also in the rigor of starved animals. Hoet & Marks discussed the early observations of Claude Bernard, and the suggestions of Embden on the importance of lactacidogen synthesis as an indication of the good functional state of the fibril colloids. They now proposed that, in the light of recent work, Embden & Jost's results should be interpreted as showing that rigor supervenes when there is rapid disappearance of hexosephosphate from the muscle, the hexosephosphate perhaps in some way stabilising the colloidal complex. Meyerhof (4) was of the opinion that the increased phosphate content might represent hydrolysis of ATP, since this was known to occur in heat rigor and in some forms of chemical rigor, e.g. after iodoacetate poisoning.

Hoet & Kerridge (1) also took up the matter of Claude Bernard's crustacea. They found indeed, with a number of different species, that soft-shelled specimens (which had just moulted) showed greatly diminished activity. In contrast to the stronger, hard-shelled normal specimens, their muscle contained practically no glycogen. Twenty hours after death, by which time the pH of the muscles of the hard-shelled individuals was 6.5 or below, the pH in those of the soft-shelled was 7.1 or above. These latter muscles became rigid much more quickly.

A few years later E. C. Bate Smith (4, 5, 6) started an important series of papers on rigor mortis. As we have seen,[1] he had reached the conclusion that the myosin in the living muscle must be in the form of a weak gel; during rigor, as the hardness of the muscle increased, there was more and more difficulty in obtaining plasma from it, and this he put down to setting of the gel in the muscle. It could be shown *in vitro* that the process of gelation was quite independent of any change in pH. Bate Smith did not agree that the results of Deuticke (1)[2] showed decrease in solubility of the normally soluble proteins (including myosin) during rigor onset; he interpreted them rather as indicating greater difficulty in extraction of these proteins from behind insoluble protein barriers. He himself found with rigor muscle that, on use of an efficient extractant (7% $LiCl_2$) with very thorough grinding and repeated extractions, more than 90% of the protein normally obtained from fresh muscle could be extracted. In a later paper (5) he developed a method of measuring extensibility of the muscle, with which results much

[1] P. 136.   [2] P. 192.

more quantitative could be obtained than with the old method of measuring hardness. He found that in normal muscles during onset of rigor the decrease in extensibility ran closely parallel to the lactic acid increase, but, bearing in mind the non-acid rigor in abnormal conditions, he suggested that stiffening and acid production were both related to an unknown third change.

The discovery of phosphocreatine in muscle in 1929 and the study of its significance in contraction scarcely impinged on ideas about rigor. But the case was far otherwise with ATP. After the demonstration of the ATPase activity of classical myosin, the Szeged group soon showed the complex nature of the protein, in which actin and myosin were associated. Actin was regarded as the solid structure of the fibril with myosin loosely attached, and it was shown that ATP had a decisive effect on the physical state of the actomyosin – contracting, relaxing or dissolving it according to the conditions. Banga & Szent-Györgyi (1) found that as minced muscle stood and its ATP content decreased, so the 'myosin' contained in it became progressively less extractable with Edsall solution. In 1943 Erdös in Szent-Györgyi's laboratory showed that there was a close *post mortem* parallelism between hardness of the muscle and decrease in the amount of ATP contained in it. As the rigor developed and the ATP disappeared, the actomyosin in the muscle became more and more insoluble. The very small preparation from rigor muscle contained only myosin, no actin.

Bate Smith & Bendall (1, 2, 3) confirmed Erdös' finding; they showed that the decrease in extensibility was correlated with decrease in the ATP content. This correlation held for alkaline rigor after starvation or insulin injection, as well as for acid rigor. They explained the difference in the delay period of rigor onset as depending on the glycogen content at death, but only indirectly, the role of the glycogen being to undergo glycolysis and thus bring about resynthesis of ATP. The glycogen content at time of death was varied in their experiments in several ways: for low values with high pH, by insulin injection; for intermediate values, by 2–3 days starvation; for high values, by ample diet of carrots and oats; for conservation of the glycogen level at death with little lactic acid formation, by injection of the muscle-relaxing drug myanesin before killing. However, even during the lag period, the resynthesis of ATP did not quite keep pace with its breakdown, and when the pH had fallen to about 6.5 a great increase in ATP hydrolysis began; it was at this time that the significant change in extensibility occurred. The connection of phosphocreatine with rigor phenomena was investigated for the first time by Bendall (8). He found with rested muscles that phosphocreatine was broken down from the moment of death of the animal, ATP breakdown starting only when at least 70 % of the phosphocreatine had disappeared. Thus both phosphocreatine breakdown and glycolysis contribute to the ATP upkeep, the former bearing the greater burden early in the process, the latter in the later stages.

Since lactic acid formation is often still going on at the time of rapid onset of rigor, it seems that the sharp change may be due to increased ATP breakdown rather than to decreased resynthesis. This might be explained[1] by the increasing activity of the water-soluble ATPase extracted by Sakov (1) from minced muscle, with pH optimum about 6.0; or by increased activity of the ATPase of the cell particles. But another explanation seems more probable. As Bendall has underlined, it seems that the myofibrillar ATPase itself must be quiescent during the delay period, the gradual ATP hydrolysis being due to these sarcoplasmic enzymes. If the splitting during the delay period were due to actomyosin ATPase, some shortening and performance of work would be expected. Bendall explained this control of the myofibrillar ATPase as due to the 'Marsh factor' (or in the language of seven years later, the activity of the calcium pump).[2] As is well known, a certain concentration of ATP is needed for the activity of the Marsh factor, i.e. of the calcium pump. Therefore if the ATP level fell below this concentration the rapid myofibrillar ATPase could come into play. Further the concentration of ATP needed in this way varies with pH, being lower at low pH values than at high.[3] This fact supplies Bendall's explanation (7) for his observation (8) that the sudden decrease in extensibility happens at an ATP concentration of 4 μmoles/g if the ultimate rigor pH is high; while the critical ATP level is about 2 μmoles/g at ultimate rigor pH of 5.8 (see fig. 82).

Bendall & Davey (1) have followed the fate of the vanishing ATP. After rigor, inosine, hypoxanthine, inosinediphosphate, inosinetriphosphate, ATP, ADP and inosinic acid were all present, the inosinic acid in much the greatest amount. The reactions concerned seem to be:

$$ATP \longrightarrow ADP + inorganic\ phosphate \tag{1}$$
$$2\ ADP \rightleftharpoons ATP + AMP \tag{2}$$
$$AMP \longrightarrow IMP + NH_3\ (at\ 37^\circ). \tag{3}$$

At 17° the following reactions were also found:

$$ADP \longrightarrow IDP + NH_3 \tag{4}$$
$$IDP + ATP \rightleftharpoons ITP + ADP. \tag{5}$$

Reaction 4 was studied by Webster (1) using washed myofibrils; this deaminase seems to be closely associated with actomyosin. Evidence for reaction 5 in muscle suspensions has been reported by Krebs & Hems (1).

The question of the nature of the shortening in rigor is an interesting one. In older work it was considered to be a slow and irreversible contraction, though Bethe (1) showed that the tension produced was only about 15 % of

[1] Bendall (7).　　　[2] Ch. 13.　　　[3] Briggs & Portzehl (1).

that in a tetanic contraction. Bendall (7) in a histological study found that only a fraction of the fibres showed signs of having contracted, while most seemed to have been passively compressed and distorted. Although shortening occurs as a rule in rigor at 37°, it is not usual at room temperature, except in cases where there is little fall of pH. Here again it seems that the

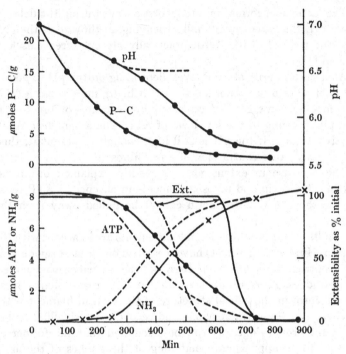

Fig. 82. Time course of the chemical and physical changes during rigor in the psoas muscle of an immobilized animal. (———), well-fed animal; pH change: from 7.0 to 5.6. (- - - - -), starved animal; pH change: from 7.0 to 6.5. The curves labelled Ext. show extensibility changes. P—C = phosphocreatine. (Bendall (7).)

end of control by the Marsh factor is responsible, allowing actomyosin ATPase to become active at a pH where contraction of a few fibres is still possible.

As regards the resolution of rigor, from the experiments of Marsh (3) and of Bendall[1] it seems that this can be staved off indefinitely if putrefaction is avoided, and it is unlikely that the earlier explanations have any basis. The well-known fact that rapid rigor passes off rapidly is probably to be accounted for by the less acid state of the muscle, which would thus afford a more appropriate milieu for bacterial growth.

[1] Bate Smith & Bendall (4).

Summing up we can distinguish the following sequence of events in typical *rigor mortis*.

(1) Cessation of the circulation, preventing oxidative phosphorylation,[1] leaves the resynthesis of ATP (continually undergoing slow dephosphorylation by sarcoplasmic ATPases) dependent on phosphocreatine breakdown and glycolysis.

(2) The lactic acid formation in glycolysis leads during the delay period to falling pH, the extent of the fall depending on the size of the glycogen store at the time of death. With practically glycogen-free muscles there may be no fall.

(3) At a certain critical ATP level, depending on the pH (2 μmoles/g if the pH falls below 6.3; 4 μmoles if the pH is high), more rapid ATP breakdown comes on. This can be explained by cessation of function of the calcium pump owing to the low level of ATP. The actomyosin ATPase is thus activated and, if the pH is still high enough to permit it, this ATP hydrolysis may cause contraction of a few fibres.

(4) The decrease in extensibility is readily explained on the sliding mechanism, the actin and myosin filaments in absence of ATP remaining bound together, so that the actin cannot be pulled away from the surrounding myosin.

The results of X-ray diffraction measurements are in accord with this last conclusion. Huxley & Brown (1) have described on onset of rigor a dramatic change occurring in the low-angle X-ray diffraction pattern associated with the cross-bridges. These results show that there is first a change to a more disordered form in the helical structure of the myosin filament with little or no change in the actin spacing; and secondly a swivelling or bending round of the cross-bridges. It can be calculated that the former change would lead to an approximate matching of the pitches of the actin and myosin helices (not found in living contracting muscle); the movement of the cross-bridges would then allow as many of them as possible to attach to the actin with only small longitudinal movement in the bridge. Huxley & Brown suggest that the retention of the 143 Å meridional repeat in rigor could be explained if this depended just on the points at which a flexible part of the HMM is attached to the LMM firmly built into the thick filament. Such a narrow part of the HMM molecule, joining the LMM and the globular part of the HMM, would contribute little to the X-ray pattern and its movements would allow changes in position of the cross-bridges without alteration in the myosin sub-unit repeat.

[1] Ch. 17.

16

RESPIRATION

Early investigators were deeply impressed by the fact that muscles worked perfectly well without a contemporary oxygen supply;[1] so that for many years the concept of the 'inogen' molecule, with its built-in oxygen, dominated the thought of those concerned with muscle contraction. 'Inogen', as we have seen, did not survive the work of Fletcher; and by the early nineteen twenties the Meyerhof Cycle had taken its place as an explanation of the relationship between anaerobic and aerobic metabolism.[2]

Pasteur (1) was the first to observe that in presence of oxygen much less sugar was used by yeast than under anaerobic conditions – indeed only about one-twentieth, notwithstanding the much greater proliferation aerobically. In 1861 he wrote:

It must be admitted that yeast, so greedy of oxygen that it removes it from atmospheric air with great activity, has no more need of it and does without when one refuses it this gas in the free state, presenting it instead in profusion in the combined form in fermentable material; if one refuses it this gas in the free state, immediately the organism appears as an agent of sugar decomposition. With each respiratory gesture of the cells, there will be molecules of sugar whose equilibrium will be destroyed by the subtraction of a part of their oxygen. A phenomenon of decomposition will follow, and hence the ferment character which on the other hand will be missing when the organism assimilates free oxygen gas.

This concept of anaerobic metabolism which, as Meyerhof[3] has remarked, amounted to that of an intramolecular oxido-reduction process, was much nearer to the viewpoint made familiar by later work on muscular glycolysis than was the inogen theory; but Pasteur's ideas were not considered in relation to muscle until the work of Meyerhof. As we have seen,[4] in 1920 Meyerhof (3) showed that less glycogen was broken down to lactic acid by chopped muscle in oxygen than in nitrogen. His explanation of this Pasteur effect however involved the same breakdown to lactic acid under both conditions, with utilisation of part of the lactic acid for resynthesis aerobically. The process of actual glycogen resynthesis during oxidative recovery from fatigue in intact muscles was correlated with loss of lactic acid of which only

---

[1] Ch. 2.　　　　[2] Ch. 3.　　　　[3] Meyerhof (4) p. 176.　　　　[4] Ch. 3.

a part was oxidised. The energy for the resynthesis was assumed to come from this oxidation and the oxidation quotient:

$$\frac{\text{lactic acid disappearing}}{\text{lactic acid oxidised}}$$

might be as high as 4.7. This point of view was very generally held for the next decade, although it was founded almost entirely on work on frog muscle at low temperatures. The only variant was the suggestion that oxidation of protein or fat might under certain conditions provide the energy for lactic acid resynthesis. This idea was based on circumstantial evidence, e.g. Takane's conclusion (1) that carbohydrate loss could not account for all the energy exchange in respiring rat diaphragm muscle.[1]

About 1925 it began to appear that the liver might play an important part in the recovery of muscle glycogen after exercise. This was the construction put on their results by Jannsen & Jost (1); after infusion of lactic acid into living warm-blooded animals they found the acid first accumulated by the muscles, then rapidly given up again. During this period of release by the muscles the arterial lactic acid content fell very rapidly, indicating that it was taken up by some other organ. A few years later C. F. Cori & G. T. Cori (2) found with lactic acid fed to rats or injected subcutaneously that more than 70 % was retained as liver glycogen. They first definitely stated what is now known as the Cori cycle, in which much the greater burden of lactic acid resynthesis falls on the liver.

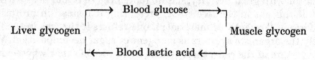

About the same time Himwich, Koskoff & Nahum (1) found by examination of arterial and venous blood that glucose was removed from the blood stream by the muscle, lactic acid by the liver; they too suggested the liver as 'an integral part of muscle recovery'.

In contrast to this ready synthesis of glycogen by the liver, Eggleton & Lovatt Evans (1) in 1930 were not able to show glycogen synthesis in cat muscles when lactate was injected into the animals intra-arterially or intravenously. Indeed they did not find such a synthesis in hind limbs of frogs perfused with lactate solution, although Meyerhof, Lohmann & Meier (1) had recorded some 10 % increase in glycogen under such conditions. Sacks & Sacks (1) could find no glycogen synthesis in the muscle of rabbits recovering from contraction, but observed that the lactic acid was lost from

---

[1] It is interesting that in these experiments with rat diaphragm in oxygenated Ringer-lactate, although the lactate diminished there was no decrease in the carbohydrate loss in the muscle; cf. the results of other workers on mammalian muscle below.

the muscle by diffusion. Similarly Flock & Bollmann (2) found that very
little glycogen had been re-formed in the muscles of rats 20 min after work,
although by this time all the lactic acid excess had been removed from the
blood.

A. V. Hill and his colleagues at first accepted the Meyerhof cycle as the
basis of recovery phenomena;[1] it seemed to fit well in frog muscle with the
time-distribution of heat production and with the 'oxygen debt' – defined
by Hill & Lupton (1) as 'the total amount of oxygen used, after cessation of
exercise, in recovery therefrom'. However the experiments of Martin, Field
& Hall (1) in 1929 on exercise in man showed that there was no quantitative
relationship between excess oxygen used and the lactic acid disappearing
from the blood.

By the time of the very thorough study of human exercise made by
Margaria, Edwards & Dill (1)[2] in 1933, the results of Lundsgaard had
brought in the phosphocreatine era.[3] Like Martin et al. they found that the
lactic acid mechanism alone was inadequate to explain all the processes of
payment of the oxygen debt, but they now had another explanation to put
in its place. They found that no extra lactic acid appeared in the blood up
to a work rate of about 60 % of the maximum, and that the oxygen debt
might be divided into two main parts: (1) the alactacid debt, very rapidly
repaid – to the extent of about 50 % in 0.5 min; (2) the lactacid debt,
oxygen consumption accompanying removal of lactic acid from the blood
and considered to be involved in glycogen synthesis. This part of the debt
was incurred only when the muscle had been under anaerobic conditions, as
during work in the neighbourhood of maximum rate. It was suggested that
the alactacid debt was related to oxidations furnishing energy for phosphocreatine resynthesis (fig. 83). Bang (1) also in 1936, studying muscular
exercise in man and using a new micro-method for following lactic acid
changes, concluded that the dominating conception of 'no muscular
activity without lactic acid formation' could not be substantiated. He found
a rise at first in blood lactic acid, followed by a steady fall during exercise;
he proposed that, in ordinary exertion, there was an initial brief anaerobic
phase followed by a steady state in which there was no lactic acid formation, but in which phosphocreatine was resynthesised by oxidative processes. With very strenuous exercise lactic acid formation might also occur
later.

Relevant here also are the experiments of von Muralt (4) in which he
followed the changes in transparency in frog sartorius muscle during 5 min
after a 2 sec tetanus. Under aerobic conditions the increase in transparency
(associated, as we have seen,[4] with phosphocreatine breakdown) showed
reversal in a two-phase curve. The earlier phase, which occurred also im-

[1] P. 49.   [2] Also Margaria & Edwards (1).   [3] Ch. 5.   [4] Ch. 14.

mediately after anaerobic stimulation, was put down to phosphocreatine resynthesis through lactic acid formation; the second phase, only occurring aerobically, was explained by aerobic resynthesis.

In 1933 R. Hill (1) began the study of the kinetic properties of the haemoglobin of muscle tissue (myoglobin) which had just previously been crystallised by Theorell (1) and definitely shown to be different from blood haemoglobin.[1] This study was continued by R. Hill (2) and by Millikan (1).

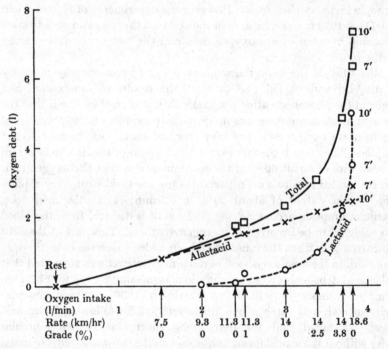

Fig. 83. Amounts of alactacid and lactacid oxygen debts as a function of the metabolic rate: subject Clapham. (Margaria, Edwards & Dill (1).)

Myoglobin has a very high oxygen affinity, three to fifteen times that of haemoglobin, and the rate of combination with oxygen or carbon monoxide is several times as fast as for haemoglobin. R. Hill (1) had realised the importance for oxygen transfer to the muscle of the fact that, in the middle range of the dissociation curve, there was a large difference in oxygen saturation of the two pigments, a difference which would allow the muscle pigment to take up oxygen from the blood pigment. In 1937 Millikan (2) described a photoelectric method whereby the degree of oxygen saturation of the intracellular myoglobin could be measured in the soleus muscle of the

[1] For some account of early work on the muscle pigment see ch. 19.

cat with blood and nerve supply intact. On contraction deoxygenation of the myoglobin took place very rapidly, starting in less than 0.2 sec from the beginning of contraction and reaching the maximum value in less than 1 sec. Calculation showed that the initial rate of oxygen uptake was 1.3–3.5 mm³/g sec, in good agreement with figures in the literature (obtained by conventional methods) for the effects of exercise. It also seemed that some half of the alactacid debt of Margaria *et al.* would not be a true metabolic debt but was needed to replenish oxygen stores used at once when contraction began. Millikan (3) considered the relation of these results to the observations on frog muscle[1] where recovery heat production occurred to so great an extent after contraction was over. The explanation seems to be twofold – the difference in temperature of the two sets of experiments (the mammalian muscle at 37° and the frog muscle at about 4°) and the much higher metabolic rate, even at the same temperature, of mammalian muscle as compared with amphibian. Thus Cattell & Shorr (1) found with dog's scalenus muscle much more rapid oxidative recovery heat than with frog sartorius, even at the same temperature. As another example of variation in metabolic rate, D. K. Hill (1) has remarked that the resting rate of oxygen usage by locust muscle is equal to the maximum possible rate for stimulated frog muscle.

As we have seen,[2] Sacks and his collaborators drew from isotope experiments the conclusion that ATP and phosphocreatine breakdown and lactic acid formation were not directly concerned in contraction. This sceptical attitude had begun with earlier work.[3] In experiments on muscle of anaesthetised cats and rabbits (with nerve and blood supply to the muscle intact) they examined the rate of resynthesis of phosphocreatine and ATP during 20 min of recovery, and concluded that the rates were far too low to permit of the thesis that breakdown of these substances and their oxidative resynthesis were the basis of energy provision for contraction; the function of phosphocreatine breakdown was preservation of neutral pH. But all the work we have just been considering shows how rapid is the response to need of increased oxygen uptake – within the first minute – and at this time phosphocreatine and ATP resynthesis may well go on at many times the rate to be observed 5–20 min later. Meyerhof (24) in discussing these experiments pointed to the work of von Muralt in which, by following changes in transparency, the very rapid resynthesis of phosphocreatine was indicated during the first minute of recovery, the rate being greatly reduced after this point.

Summing up, we may say that it seems likely that the initial stage of a contraction has an anaerobic source of energy – ATP and phosphocreatine breakdown – and that in many muscles resynthesis at the expense of

---

[1] Ch. 3.   [2] Ch. 14.   [3] Sacks & Sacks (1, 2, 3, 4).

energy supplied by oxidation follows with great rapidity. Carbohydrate seems to be the fuel of primary importance in most animals, but fat and possibly protein may be used also. There seems in such muscles no need for lactic acid formation except under special conditions – where a short lapse of time may be needed for circulatory adjustment at the beginning of contraction, and also in exhausting exercise. Muscles with high rate of lactic acid formation also exist.

With regard to chopped muscle later work has shown that there is no reason to suppose that the Pasteur effect depends on the Meyerhof cycle. It now seems likely that the diminution in glycolysis under aerobic conditions is the result of successful competition by the aerobic mechanism for certain factors, particularly ADP,[1] which are necessary for both glycolytic and oxidative processes.

We must turn now to consideration of the mechanism of oxidative metabolism, in preparation for understanding the synthesis of ATP connected with it.

THE TWO ASPECTS OF RESPIRATION

The best place to begin is Wieland's study (1) in 1912 of 'hydrogenation and dehydrogenation'. He showed that in presence of palladium or platinum black many organic compounds could lose hydrogen to the catalyst – e.g. CO_2 was formed from formaldehyde and quinone from hydroquinone. In this oxidation process he emphasised (2) that there was no need to postulate activation of oxygen, since it could go equally well in absence of air if methylene blue or another quinonoid dye were present to accept the hydrogen. He pictured the following steps for oxidation of ethyl alcohol to acetic acid:

$$2CH_3CH_2OH \ + \ O{:}O \longrightarrow 2CH_3CH.O \ + \ 2H_2O$$

$$CH_3CH.O \ + \ H_2O \longrightarrow CH_3C{\underset{OH}{\overset{H}{<}}}OH$$

$$2CH_3C{\underset{OH}{\overset{H}{<}}}OH \ + \ O{:}O \longrightarrow 2CH_3COOH \ + \ 2H_2O$$

The first hydrogenation product of O_2 was taken to be hydrogen peroxide, which vanished rapidly by reaction with active hydrogen, going to H_2O. These ideas Wieland applied to biological oxidation processes (2, 3). He recognised that dehydrogenating enzymes had manifold specificity towards both the dehydrogenated and the hydrogenated molecules.

[1] Ch. 17.

These conceptions were further developed by Thunberg. In the 'methylene blue tube' of his invention, in which the interaction of substrate and methylene blue, in presence of enzyme, could be conveniently studied *in vacuo*, he examined the behaviour of many substrates, starting with succinic acid (1). He found (2) that chopped frog muscle, if sufficiently long extracted with water, lost its power of decolorising methylene blue. This power was restored by adding any one of many different substances. He concluded that sugar, fats and amino acids passed in metabolism through a whole series of intermediate stages by means of dehydrogenations, combined with splitting off of CO_2. For this breakdown a whole series of specific enzymes was necessary. Thus all foodstuffs were hydrogen donators and hydrogen was the general fuel of the body. The work on these lines was summed up by Thunberg in his Sedgwick Memorial Lecture of 1930 (3).

Meanwhile, from about 1907, Batelli & Stern (1, 2, 3, 4) had been working on oxidations in animal tissues. Succinic acid added to chopped tissue, suspended in three times its weight of water, called forth a very vigorous oxygen uptake, and malic acid was formed. Citric, fumaric and malic acids could also be oxidised, though not so rapidly. They also found that all tissues of higher animals have the power to oxidise p-phenylenediamine by means of molecular oxygen;[1] with acid-precipitated preparations from many tissues, the oxygen uptake on addition of succinic acid ran parallel to the oxygen uptake with p-phenylenediamine. Furthermore there was a parallelism between this activity and the respiratory activity of the fresh, undamaged tissue. Again, cyanide, which reversibly inhibited tissue respiration,[2] inhibited both succinic oxidation and indophenol formation. They concluded that the enzymes concerned in these oxidations took part in the '*Hauptatmung*' of the cell. They also made the significant observation that these oxidases could not be extracted from the cell débris but remained attached to washed tissue particles.

Like Batelli & Stern, Warburg was concerned with the part played by oxygen in tissues, rather than with dehydrogenations. From 1908 he was engaged on a study of oxygen transfer. He had evidence that all cells contain iron, and that iron could act catalytically *in vitro* to bring about the oxidation of organic substances – ferrous iron taking up oxygen and ferric iron reacting with these substrates. Since for cells the ratio: $\dfrac{O_2 \text{ used in cmm}}{mg/Fe \times h}$ is 10000–100000, while for the catalytic effect of Fe on autoxidation of

[1] For the history of earlier work on indophenol oxidase see Keilin (1) pp. 129, 177. This was the name given to the enzyme catalysing the oxidative synthesis of indophenol from α-naphthol and dimethyl-p-phenylenediamine. The activity of the enzyme can be measured by the rate of oxidation of p-phenylenediamine or certain other aromatic amines and phenols, or by colorimetric determination of the indophenol formed.
[2] Spitzer (1); Batelli & Stern (5).

cysteine the ratio is 120 000–400 000, he was satisfied that the iron content of the cell was more than sufficient for purposes of respiration. In experiments on unfertilised sea-urchin eggs with membranes damaged so that they could take up iron from the medium surrounding them, he found the respiration to increase in proportion to the Fe taken in. In 1925 in a review (1) of several years work he wrote:

Respiration is a reaction on surfaces and as such is hindered by all substances which, without being combustible, themselves dislodge the combustible substances from the surfaces.

The unspecific surface forces...do not suffice to bring about a reaction between the organic materials and oxygen. The absorbing surfaces are not homogeneous, but contain, stored in them, a substance which transmits the oxygen according to chemical forces. This substance is the respiratory enzyme.

The oxygen-transferring part of the respiratory enzyme is iron.

In the 'unspecific surface forces' Warburg apparently massed together all the dehydrogenases, in spite of the fact that much was already known about their specificity. He made catalytically active iron-containing material by preparing charcoal from crystalline haemin[1] and found its catalytic activity to be inhibited by cyanide, as of course was the respiration. His experience with charcoal made from various materials led him to the opinion that the catalytically active substance was Fe bound to nitrogen.

These two points of view on respiration were considered quite incompatible by their obstinate protagonists,[2] but in 1924 Szent-Györgyi (8) drew attention to the various facts suggesting a double mechanism in cell respiration. There is for example no activation of the hydrogen of p-phenylenediamine in anaerobic presence of tissue and methylene blue, yet hydrogen is readily removed from this molecule in the presence of tissue and oxygen. Therefore this 'is an expression of oxygen activation without hydrogen activation'. In the case of succinic acid, which reacts with either methylene blue or oxygen, activation of hydrogen alone or of both hydrogen and oxygen can occur, only the latter being inhibited by cyanide. Since in oxygen with both succinic acid and p-phenylenediamine present the oxygen uptake is not summed, there is indication of identity of the enzyme systems concerned. Fleisch (1) also considered this question in relation to succinic acid, and pictured one enzyme, succinoxydone, as uniting two functions.

As Szent-Györgyi wrote '...beide Theorien ihre schöne, weite, experimentelle Grundlage haben'.

---

[1] P. 383 below.
[2] E.g. Warburg (4).

THE RESPIRATORY CHAIN

THE CYTOCHROMES. As far back as 1864 MacMunn published a first paper on a new muscle pigment, widespread in the animal kingdom, which he called 'myohaematin'. He studied this on fragments of tissue, such as the thoracic muscle of the fly, compressed between slides and well illuminated, by means of a microspectroscope. It was characterised by four absorption bands. This absorption spectrum belonged to the reduced state of the pigment, and could be made to appear and disappear by treatment of the tissue with ammonium sulphide or hydrogen peroxide. He believed that the pigment had a respiratory function and could undergo oxidation and reduction *in vivo*, but he was not able to demonstrate this. A very similar, if not identical pigment, occurring in many other organs, he called 'histohaematin'. He considered these pigments to be allied to haemochromogen, although no derivative of haemoglobin was known with a four-banded spectrum in the reduced state, and a hardly visible spectrum in the oxidised state. The reader may like to be reminded of the terms for some of the derivatives and components of haemoglobin.

Haemochromogen = haem + nitrogenous base or protein.

Haem = protoporphyrin (the nucleus of haemoglobin) combined with Fe in the ferrous form.

Haematin = haem in the ferric form.

Haemin = haematin chloride.

At the time of MacMunn's work the absorption spectra and some other properties of haemoglobin, haemochromogen, haemin, haematin and porphyrin had already been described.[1] The structure of protoporphyrin was only elucidated much later, between 1910 and 1940, largely by the work of H. Fischer.[2]

Haem

[1] See Keilin (1) p. 96.   [2] E.g. Fischer (2).

Keilin, in the early nineteen twenties, was studying the respiration of *Gasterophilus* during the various stages of its metamorphosis.[1] He found in the thoracic muscles of the adult a pigment with four distinct absorption bands, and a little later saw this again in a small lump of yeast compressed between slides. Moreover in an instance where the yeast had just been vigorously shaken the bands were missing, suddenly to appear again after a few moments. This phenomenon could be repeated again and again, and Keilin (1) wrote many years later '...this visual perception of an intracellular respiratory process was one of the most impressive spectacles I have witnessed in the course of my work'. A study of the literature quickly showed him that he had rediscovered the myohaematin of MacMunn. In

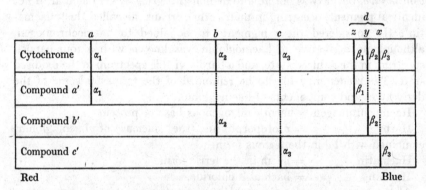

	a	b	c	z	y	x
Cytochrome	α_1	α_2	α_3	β_1	β_2	β_3
Compound a'	α_1			β_1		
Compound b'		α_2			β_2	
Compound c'			α_3			β_3

Red   Blue

Fig. 84. Diagram showing the three haemochromogen components a', b' and c' of cytochrome. (Keilin (1).)

his outstanding intellectual autobiography Keilin (1) has discussed the reasons for the neglect and misunderstanding of MacMunn's work in the intervening 35 years. Since the presence of the pigment in yeast and bacteria established its intracellular location, Keilin named it 'cytochrome'.[2]

By 1925 Keilin (2) had been able to show that cytochrome consisted of three haemochromogens, each with two absorption bands and the unusual absorption spectrum was thus explained; each was responsible for one of the four absorption bands, while the fourth band was composite, made up of the second bands of the three haemochromogens (fig. 84). The three cytochromes are now designated a, b and c, while the bands are known as a, b, c and d. Only cytochrome c is soluble, the others being tightly associated with tissue particles.

In 1926 Warburg found that respiration in yeast was inhibited by carbon monoxide and that the inhibition was relieved by illumination – the latter

[1] *Gasterophilus intestinalis* is a dipteran parasite of equine animals.
[2] Keilin (2).

experiment being suggested by Haldane & Smith's observation in 1896 that the carbon monoxide compound of haemoglobin was light-sensitive. He went on to make observations in which the respiration of a yeast suspension was inhibited by CO, and the suspension was then illuminated by light of different wavelengths, the return of respiration being investigated at each one. In this original way the absorption bands of the CO compound could be deduced and were found to be similar to those of a haematin compound.[1]

In two later papers in 1929 and 1930 (3, 4) Keilin studied the respiratory function of the cytochromes. He came to the conclusion that the enzyme known as indophenol oxidase and Warburg's respiratory enzyme was responsible for their oxidation, since this was inhibited by the same agents (KCN, H_2S, CO in the dark). The cytochrome oxidase was prepared in the form of a colloidal suspension of thoroughly washed heart-muscle particles, almost free from cytochrome c. He showed that all agents (such as narcotics) inhibiting the tissue dehydrogenases delayed the reduction of oxidised cytochrome, but did not affect the oxidase activity. Thus it was demonstrated that the cytochromes acted as carriers between the two types of activation mechanism. It was by virtue of the ability of their haem-bound iron to undergo reversible change between the ferrous and ferric states that electron transfer could take place between the substrate-dehydrogenase systems and the system made up of cytochrome, cytochrome oxidase and O_2. In 1938 Keilin & Hartree (1, 2) showed that cytochrome a consisted not of one but of two components, one of which was autoxidisable. This they designated a_3 and brought evidence that it was indeed cytochrome oxidase. Its identity with Warburg's respiratory enzyme was shown by the close similarity of the absorption spectra in presence of CO.

In the discussion of Keilin & Hartree (1) it now began to appear that the cytochrome oxidase might contain copper as well as iron. There was, for example, the observation of Elvehjem (1) that addition of copper to the nutrient medium produced an increase in cytochrome a content of yeast cells and intensified their indophenol oxidase reaction; similar results were obtained on adding copper to a copper-deficient diet of rats.[2] Much more recently Wainio, van der Welde & Shimp (1) have found, on purifying cytochrome oxidase and using cytochrome c as substrate, a good correlation between copper content, haem content and activity of the oxidase. The copper appears to be attached to the protein part of the molecule, as in the haemocyanin respiratory pigments of many invertebrates.

The existence of another cytochrome (c_1) in heart muscle was reported by Yakushiji & Okunuki (1) in 1940 ,with an absorption band very close to that of cytochrome c. Its existence was confirmed in 1955 by Keilin & Hartree (3)

---

[1] Warburg (2). [2] Cohen & Elvehjem (1).

using their method of examination at the temperature of liquid air, in which condition the absorption bands are greatly sharpened.[1]

In 1938 Ball (1) suggested that a key to the order in which the three haems of the cytochromes reacted in the oxidation–reduction chain might be found in their oxidation–reduction potentials. By observing the change in intensity of the characteristic absorption bands (a at 605 mμ; b at 563 mμ and c at 550 mμ) in media of different potentials he concluded that the three components were half reduced, a at $+0.29$ V, b at -0.04 V and c at $+0.27$ V. The order expected in the chain would thus be a, c, b from the oxygen end. This fits well with the known behaviour of the cytochromes[2] and is generally accepted.

The realisation in 1948 that the respiratory chain was situated in the mitochondrion was of major importance. Green, Loomis & Auerbach (1) were using a washed particulate preparation (which they called 'cyclophorase') prepared from tissues, including pigeon breast and heart muscle; they found this to contain the complete system[3] for oxidation of pyruvate to CO_2 and H_2O. Then Hoogeboom, Schneider & Pallade (1), having for the first time developed a procedure for preparing mitochondria retaining their normal morphology, demonstrated that the succinoxidase activity of rat liver cells resided entirely in the mitochondrial system. Kennedy & Lehninger (1, 2), also using morphologically intact liver mitochrondia, found that these contained all the fatty acid oxidase activity; moreover, these mitochondria also catalysed all the reactions of the tricarboxylic acid cycle and the phosphorylations coupled with it.[4]

THE FLAVOPROTEINS, COENZYME Q AND NON-HAEM IRON. As time went on it became clear that other substances besides the cytochromes were concerned in the chain of electron transport. Let us begin with the flavoproteins.

The first demonstration of participation of such compounds in respiration came from Warburg & Christian (5, 6, 7) in 1932. They obtained, from the press-juice of bottom yeast, what they called a 'yellow enzyme' which was reduced to the leuco-form on addition of hexosemonophosphate with its dehydrogenase and co-enzyme (NADP).[5] The reduced enzyme was oxidised by air to the yellow form and so could bring about oxidation of hexosemonophosphate by atmospheric oxygen, this oxidation being insensitive to CO and cyanide. In 1935 Theorell (2), in Warburg's laboratory, showed that the active group of the yellow enzyme was the monophosphoric ester of a riboflavine.

---

[1] Also Sekuzu, Orii & Okunuki (1).
[2] E.g. Slater (1); Sekuzu et al.; also the further discussion below.
[3] P. 395 below.   [4] P. 400 below; ch. 17.   [5] P. 117

The flavoprotein concerned in this particular case with oxidation of hexosemonophosphate by means of molecular oxygen proved to be the forerunner of many, some containing mono-, some di-nucleotides, later found to be implicated in respiratory processes. Amongst these flavoproteins we shall find the amino-acid oxidases, acyl-coenzyme A dehydrogenases; and several enzymes forming part of the respiratory chain. Thus several flavoproteins must be contained within the mitochondrion, but we shall be concerned with those within the structure of the respiratory chain, especially with the two (NADH dehydrogenase and succinic dehydrogenase) linking the chain to the tricarboxylic acid cycle, through which oxidation of carbohydrate, fatty acids and some amino acids is channelled.

Succinic dehydrogenase was isolated in 1956 from heart muscle independently by Wang, Tsou & Wang (1) and by Singer, Kearney & Bernath (1). It had earlier been supposed that this enzyme might be identical with cytochrome *b*, but Tsou (1) showed in 1951 that this could not be so, since in Keilin–Hartree heart-muscle preparations the absorption spectrum gave evidence that this cytochrome could still be oxidised and reduced after the dehydrogenase had been completely inactivated. It was further noted that Axelrod, Potter & Elvehjem (1) had found that in rats on a flavoprotein-deficient diet the liver was low in succinoxidase activity. The Chinese workers found that this deficiency concerned only the dehydrogenase part of the oxidase complex. Their dehydrogenase, purified from Keilin–Hartree heart-muscle preparations, was almost electrophoretically pure; it consisted of a metalloflavoprotein containing non-haematin iron – 4 g atoms to 1 mole of flavine and about 150000 g of protein. The purified enzyme made at the same time by Singer, Kearney & Bernath from heart mitochondria was of very similar composition.

It was not until the nineteen sixties that NADH dehydrogenase was isolated from highly purified mitochondrial fragments by Ringler, Minekami & Singer (1). The enzyme contained flavine-mononucleotide and about 16 g atoms of non-haem iron per mole of FMN. It catalysed the oxidation of NADH by ferricyanide at a very high rate. The isolation of these two flavoprotein dehydrogenases was an exceedingly difficult task on account of their insolubility – contingent as we shall see, upon their forming part of the respiratory chain structure. Special treatment of the mitochondria or mitochondrial fragments (such as digestion with phospholipin A of snake venom, alkaline extraction of the acetone-dried material, etc.) was needed to free the enzyme, and a careful discussion of the effects of such treatments on the integrity of the native enzymes is given by Singer (1).

We come now to co-enzyme Q. Crane (1) and Morton (1) have told this story. It was early realised that mitochondria contain a large amount of lipid material; this, after extraction and purification by chromatography, showed a strong absorption band at 275 mμ, and could be reduced by various reagents. For a time it was known as Q_{275}. Its reversible reduction. catalysed by mitochondria or mitochondrial fragments, in presence of succinate, malate, pyruvate or NADH was shown.[1] The active material concerned, very widely distributed in plants and animals, was in 1958 identified as consisting of a number of 2,3-dimethyloxy-5-methyl-benzo-quinones with an unsaturated isoprenoid side-chain in the six position. The number (n) of isoprenoid units varied from six to ten. Substances of this kind were also being isolated from the unsaponifiable fraction of liver and many other animal tissues, and called ubiquinone. Q_{275} could be extracted from mitochondrial fragments by means of iso-octane, whereupon the fragments lost their power to oxidise succinic acid. This power was restored by adding Q_{275}[2] and the name coenzyme Q was introduced. Crane (1) has commented on the unusual nature of this case, where it seems that the coenzyme must be pictured as floating in the lipid phase of the functional lipoprotein structure. We shall return below to such problems.

THE STRUCTURE OF THE RESPIRATORY CHAIN

We have seen that the suggested order of the cytochromes in mediating hydrogen transfer was a_3, a, c, b. Keilin many times emphasised his belief that there must be some structural micro-morphological connection between them,[3] and by 1948 it was known that the chain must be situated in the mitochondrion. During recent years intensive work from Green's laboratory, by methods of carefully controlled disruption of mitochondria (including for example treatments with cholate or deoxycholate, sonication, fractional precipitation with ammonium salts, etc.) has resulted in prepara-tion of fragments containing one, two or more constituents of the chain. (See table 14.) The makeup of these fragments is consistent with their derivation from a parent structure of orderly and constant arrangement. Moreover in several instances reconstitution of longer parts of the chain

---

[1] Crane, Hatefi, Lester & Widmer (1).   [2] Crane, Widmer, Lester & Hatefi (1).
[3] See for example Keilin & Hartree (4, 5) on the 'succinic dehydrogenase–cytochrome system'.

with the expected properties has been achieved from such fragments. The effects of certain inhibitors used to differentiate between parts of the chain are also given in the table.

In 1956 Crane, Glenn & Green (1) prepared from heart muscle mitochondria what they called the electron-transferring particle (ETP). It contained flavine, cytochromes a, b, and c_1, non-haem iron and copper in

TABLE 14

I		
Succinate-CoQ reductase	Contains flavine and cyt. b in equivalent amounts, but the haem is not reduced by succinate. Not inhibited by amytal.	Ziegler & Doeg (1), (2). Tisdale, Wharton & Green (1).
Succinate $\longrightarrow$ * CoQ		
IV		
Cytochrome oxidase	High activity and high spectral purity. Increase in oxidase activity parallel to increase in cyt. a. Equimolecular relation between cyt. a and Cu at all levels of purification. Cu reduced by substrates of oxidase and re-oxidation inhibited by cyanide.	Wainio, van der Welde & Shimp (1). Griffiths & Wharton (2).
Reduced cytochrome $c \longrightarrow$ oxygen		
II, III		
NADH-cytochrome c reductase	Contains NADH dehydrogenase flavoprotein, cyt. b and c, CoQ and non-haem iron. Essentially free from succinic dehydrogenase and cytochrome oxidase.	Hatefi, Haavik & Jurtshuk (1).
NADH $\longrightarrow$ cytochrome c.		
II		
NADH-CoQ reductase.	Made from purified NADH-cyt. c reductase. Contains NADH dehydrogenase flavoprotein, CoQ and lipid. Highly specific for substrate and acceptor. Inhibited by amytal.	Hatefi, Haavik & Griffiths (1).
NADH $\longrightarrow$ CoQ		
III		
CoQH$_2$-cytochrome c reductase	Catalyses the reduction of cyt. c by reduced CoQ. Contains cytochromes b and c. Free from cyt. c, flavoprotein and dehydrogenases of citric acid cycle. Inhibited by antimycin A, not by amytal.	Hatefi, Haavik & Griffiths (2).
CoQH$_2$ $\longrightarrow$ cytochrome c		
I, III		
Succinate-cytochrome c reductase.	Contains the same molar ratio of flavine, cyt. c and cyt. b as parent mitochondria.	Tisdale, Wharton & Green (1).
Succinate $\longrightarrow$ cytochrome c		

* The arrows in the first column indicate the transfer of hydrogen and electrons.

constant proportions. Cytochrome c was also present. Both flavine and the cytochromes were reduced by NADH or succinate; if all the soluble cytochrome c had been removed during preparation its addition was necessary for oxidation of NADH by molecular oxygen. Table 14 shows a summary of much further work yielding particles which fell into place in the following scheme :[1]

Succinate $\rightarrow$ [$f_s \rightarrow$ FeNH] $\rightarrow$ Q $\rightarrow$ b

$\rightarrow$ FeNH $\rightarrow c_1$ $\rightarrow$ c $\rightarrow$ $\begin{matrix} a-Cu \\ a_3-Cu \end{matrix}$ $\rightarrow O_2$

Reduced NAD $\rightarrow$ [$f_D \rightarrow$ FeNH] $\rightarrow$ Q $\rightarrow$ b

I   III   IV   II

f_s = succinic dehydrogenase; f_D = NADH dehydrogenase.
a, b, c, c_1 = cytochromes. a_3 = cytochrome oxidase.
Q = coenzyme Q. FeNH = non-haem Fe.

From the work of Tisdale, Wharton & Green (1) it appeared that cytochrome b could exist in more than one part of the chain. In succinate-CoQ reductase it was present in amount equivalent to the flavine, but seemed to take no part in electron transfer. In succinate-cytochrome c reductase there were three moles of cytochrome b per flavine mole, and the cytochrome underwent oxidation and reduction.

In reconstitution experiments[2] the possibilities of re-associating the isolated fragments have been studied. For example, I, II and III have been shown to recombine in the same ratio in which they exist in the mitochondrion, though perfect stoichiometry in the re-association of complex IV has not been achieved. While the separation of this last complex from the rest went easily, the separation of the other three from each other was difficult.[3] Green regards the more soluble cytochrome c, and coenzyme Q, as mobile links between the complexes, shuttling electrons from one complex to the next.

Chance and his colleagues made a different approach to the problem of the respiratory components.[4] Using intact mitochondria, they worked on the hypothesis that the time sequence in a series of spectroscopic changes is the same as the sequence of the chemical reactions. Their suggested sequence on the basis of these kinetic studies runs:

$$O_2 \longleftarrow a_3 \longleftarrow a \longleftarrow c \longleftarrow b \longleftarrow \text{fp} \longleftarrow \text{NAD} \longleftarrow \text{substrate}$$

in good general agreement with the early proposals from spectroscopic study and with the results of the 'dissection' method.

[1] Green (1).   [2] Hatefi, Haavik, Fowler & Griffiths (1); Fowler & Richardson (1).
[3] Green (1); Green (2).   [4] Chance & Williams (1), see p. 410 below.

In a very interesting paper Criddle, Bock, Green & Tisdale (1) showed that some 70 % of the mitochondrion consists of a structural protein which can be obtained in a homogeneous state. At pH 7.0 it forms a water-insoluble polymeric aggregation, but at pH 11.0 it can be converted into a monomeric form of low molecular weight – about 25 000. Cytochromes a, b and c_1 showed an analogous polymer/monomer transition, and the structural protein can combine in a 1:1 ratio with these cytochromes. These associations were shown to be specific, since a large number of other purified proteins tested evinced no tendency to bind to the structural protein. The cytochromes and the structural protein could all bind phospholipid.

Thus the respiratory chain emerges as an ordered collection of specific haemochromogens, held together by specific links with a structural protein and with phospholipid, so arranged that electrons may pass from one active group to the next, ever higher in oxidation–reduction potential. In this arrangement certain flavoproteins, coenzyme Q and certain non-haem iron metalloproteins have their place and play their part. Many uncertainties remain: for instance the role of cytochrome b; the complicated details of certain discrepancies between the behaviour of some components while in the intact chain and after extraction from it, or after particular treatments; and some results of the kinetic method difficult to explain.[1] But the general picture is now well established.

The work of Palade and of Sjöstrand in the early nineteen fifties with the electron microscope showed the mitochondrion to consist of an elongated body bounded by a two-layered membrane; the inner membrane showed regular infoldings forming internal ridges, called by Palade the cristae. Very beautiful electron micrographs of the fine structure of the cristae of heart muscle mitochondria can be seen in the paper, published in 1964, by Fernández-Morán, Oda, Blair & Green (1)[2] (see fig. 85 pl.). All the visible membrane surfaces were covered by attached particles; each of these was made up of a head (80–100 Å in diameter), a stalk and a base piece forming the attachment. The headpiece and stalk could be stripped away from the membrane by sonic irradiation or treatment with bile salts. These workers, as the result of biochemical considerations,[3] tentatively suggested the headpieces as the seat of complex IV, the stalk containing III and the base piece I and II. However, from the simultaneous work of Stasny & Crane (1) it appeared that the fraction with highest cytochrome content and enzyme activity consisted of membranous material deficient in attached sub-units. Chance, Parsons & Williams (1) also found that the cytochrome concentration was increased per mg of protein in the inner membranous structure

---

[1] See e.g. Chance (1); Redfearn (1).
[2] Many references to early work are given in this paper.
[3] Blair, Oda, Green & Fernández-Morán (1).

after the sub-units had been stripped off by exposure of the mitochondria
to sound of high frequency. Green (1) later agreed that the electron-transfer
chain can be isolated as a vesicular structure made up of base pieces entirely
devoid of particles.

PREPARATION OF CARBOHYDRATE, FAT AND PROTEIN FOR ENTRY INTO THE RESPIRATORY CHAIN

A century of biochemical study has made it clear that the large molecules
of these classes of substances, components of the animal body and its food,
cannot be used directly as energy sources. They are broken down enzymically to smaller molecules – to pyruvate in the case of carbohydrates, to
free fatty acids and to amino acids. All these can pass through the mechanism
of the citric acid cycle, but complicated processing is necessary before their
entry into the cycle, especially for the keto and fatty acids. During the
operation of the cycle, CO_2 is produced, and H is transferred to NAD and
NADP. The H from these reduced coenzymes as well as from succinic acid
enters the respiratory chain. At certain stages in the chain, as we shall see,[1]
energy is yielded up and ATP is formed. In the citric acid cycle itself ATP
is formed only at the stage of oxidation of α-ketoglutaric acid to succinic
acid.

THE CITRIC ACID CYCLE. In 1920 Thunberg (2) introducing his 'new
method of investigating intermediate stages in metabolism', and much
influenced by the ideas of Hopkins,[2] emphasised his belief that food stuffs
presented to the cell become labile, not because they become incorporated
into a labile protoplasmic molecule, but because they meet a whole series
of enzymes which act upon them. 'From this modern point of view, one
considers that a nutrient molecule, entering the cell, e.g. a sugar molecule,
passes only through stages very simple from the chemical point of view,
representing substances well known to organic chemistry.' To the suggestion
that certain substances reacted *in statu nascendi* he answered:

that all natural processes, and therefore also all chemical processes, run their
course in time, and that the expression, that a substance already in the moment of
its formation is metabolised, can only mean that the substance has a very short
lifetime, not that it can dispense with having a lifetime at all. So long as a chemical
conversion goes on in a cell, so long must the totality of the intermediate stages
of this process be represented in the cell.

He concluded that intermediate substances would be expected to pass two
tests: they should be shown to undergo very rapid metabolism, and they
should be detectable as normal tissue constituents, even if (indeed probably)

in very small concentration. He found fumaric acid to be a good example. Its rapid metabolism was well known, and Einbeck (1) in 1914 had demonstrated its presence in fresh muscle.

Some years later D. M. Moyle (1)[1] found that succinic acid formation went on in minced muscle (particularly red muscle) anaerobically incubated in phosphate buffer. The content fell in presence of oxygen and oxidation proceeded further than to malic and fumaric acids. Then in 1930 Toenniessen & Brinkmann (1) studying oxidative metabolism of carbohydrate, showed that pyruvic acid perfused through mammalian muscle gave rise to succinic acid. They suggested the following cyclic sequence of reactions:

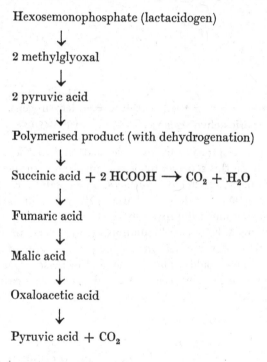

Hexosemonophosphate (lactacidogen)

↓

2 methylglyoxal

↓

2 pyruvic acid

↓

Polymerised product (with dehydrogenation)

↓

Succinic acid + 2 HCOOH → CO_2 + H_2O

↓

Fumaric acid

↓

Malic acid

↓

Oxaloacetic acid

↓

Pyruvic acid + CO_2

This was a very intelligent guess, naming as it did four of the components which were afterwards demonstrated to occur in the citric acid cycle.

In 1934 Gözsy & Szent-Györgyi (1) described the great increase in respiration of minced pigeon-breast muscle caused by addition of succinic or fumaric acid – an effect which they described as catalytic since the dicarboxylic acid was not consumed. In 1935 Szent-Györgyi (9) put together much work from his laboratory to illustrate the very special characteristics of succinic dehydrogenase: (a) It was one of the most active of enzymes

[1] Also D. M. Needham (5).

known at that time, and gave maximum activity with minimal substrate concentration. (*b*) It was unique in his experience in needing no co-enzyme. (*c*) Washed muscle *brei* supplied with succinate took up oxygen in large amounts, while other substrates had no effect.[1] Thus in contrast to other dehydrogenases, succinic dehydrogenase could bring its substrate into direct contact with the cytochrome series. As a result of these and further considerations, Szent-Györgyi suggested a tentative scheme in which succinic acid, continually produced by transfer of hydrogen from metabolites to oxaloacetic acid, was continually oxidised to fumaric acid by means of the respiratory chain.

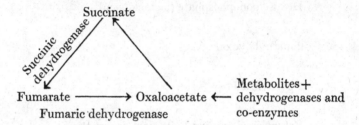

The inhibition of respiration by malonic acid[2] and maleic acid[3] could be explained nicely on this scheme, the first competing with succinic acid, the second with fumaric, on enzymic surfaces. The catalytic effect of fumaric acid, and of substances yielding fumaric acid, was confirmed by Stare & Baumann (1) using much improved methods of estimation.

In 1937 Martius & Knoop (1)[4] found, using an enzyme preparation from liver, that citric acid with methylene blue gave α-ketoglutaric acid, and that cis-aconitic acid could replace citric acid in these experiments in the Thunberg tube, while anaerobically it could give rise to citric acid. They therefore suggested:

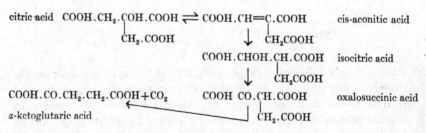

The background was now prepared for the conception of the tricarboxylic or citric acid cycle, which was proposed by Krebs & Johnson (1) the same

---

[1] See the papers following his Introduction.   [2] Gözsy & Szent-Györgyi (1).
[3] Thunberg (2).   [4] Also Martius (1).

year. They found that citric acid has similar catalytic effects on respiration of muscle tissue (especially if carbohydrate were added also) as had been found by Szent-Györgyi and his co-workers for succinic, fumaric, malic and oxalacetic acids. This led them to look for regeneration of citric acid from products of oxidation, and they did find that large amounts of citrate were formed anaerobically from oxalacetate added to the muscle. The observed rate of citric acid synthesis was about 90 % of that to be expected from the respiratory rate. The compound supplying the other two carbon atoms could not be determined. In the light of many experiments on pyruvate oxidation by pigeon-breast muscle, Krebs & Eggleston (1) later suggested pyruvate as the substance reacting with oxalacetate to give citrate and CO_2 with removal of 2H. An intermediate 7-carbon compound was suggested by Krebs (1) in 1943, and he proposed the following scheme:

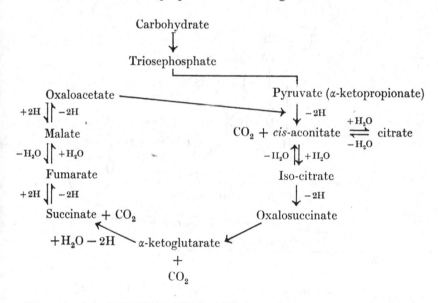

We come now to the elucidation of the condensation mechanism in citrate formation, i.e. the reaction which produces a 6-carbon compound from a 4-carbon one and a 3-carbon one with loss of CO_2. This may be considered an acetylation. In 1945 Lipmann (4), interested in the excretion of the drug sulphonamide in its acetylated form, examined liver homogenates and extracts to probe the enzymic mechanism concerned. He found that the acetylation needed the presence of ATP; there were also indications of the need for a coenzyme, which Lipmann and his colleagues[1] later found to be a pantothenic acid derivative. The isolation of 'coenzyme A' (CoA)

---

[1] Lipmann, Kaplan, Novelli, Tuttle & Guirard (1).

followed in 1950.[1] In contained pantothenic acid[2] linked through a phosphate bridge to a second component, probably an adenylic acid, and a third component – they suggested possibly an amino acid. This work strongly suggested that the actual acetyl donor, or 'active acetate', was acetylated CoA, and in 1951 Lynen & Reichert (1) isolated and partly purified acetyl CoA from yeast which had been actively oxidising glucose. They showed that the CoA reacted with acetate through a sulphydryl group, and a little later (after the work of Snell and his collaborators on the chemical nature of the growth factor of *Lactobacillus bulgaricus*)[3] Lynen, Reichert & Rueff (1) showed that CoA contained β-mercaptoethylamine. The formula can now be written as below.

With regard to the activation of the acetate to enable it to enter into combination with CoA, ATP is necessary. Acetyl phosphate is not an intermediate in animal tissues; but it seems from the work of Whitehouse,

[1] Ibid. (2).
[2] Isolated and named in 1933 by Williams, Lyman, Goodyear, Truesdail & Holaday (1) from extracts of very diverse tissues showing power of stimulating yeast growth. Its constitution was established by Williams & Major (1) in 1940.
[3] Snell, Brown, Peters, Craig, Wittle, Moore, McGlohan & Bird (1).

Mocksi & Gurin (1) with pigeon liver, that an enzyme-bound intermediate adenylacetate is formed with release of pyrophosphate.

$$\text{Acyl-OH} + \text{ATP} + \text{enzyme} \rightleftharpoons \text{enzyme-Acyl-AMP} + \text{pyrophosphate}$$
$$\downarrow + \text{CoA}$$
$$\text{Acyl-CoA} + \text{AMP}$$

The actual production of citrate (and thus the clinching of the cycle mechanism) has been studied by Stern & Ochoa and their collaborators.[1] The reaction was shown to need (besides CoA) ATP and Mg^{2+} or Mn^{2+}. In presence of the enzyme (which was in highest concentration in heart and skeletal muscle of all tissues examined) acetyl-CoA + oxalacetate ⟶ citrate + CoA; the reaction was reversible, with the equilibrium far to the citrate side. It is very interesting that calculations based on the equilibrium constant show the acylmercaptide bond to be energy-rich and comparable with the energy-rich phosphate bond – the free energy released on its breakage being about 12 000 cal.[2] Lynen & Reichert had also concluded that a new type of energy-rich bond was concerned here.

In the oxidation of citric to α-ketoglutaric acid two enzymes are involved. First aconitase establishes, via a common intermediate bound to the enzyme, equilibria between citric, cis-aconitic and isocitric acids. The mechanism of action was studied by Speyer & Dickman (1) on enzyme prepared from heart muscle. Then isocitric dehydrogenase–oxalosuccinic carboxylase activates the oxidative decarboxylation of isocitric acid to α-ketoglutaric. This enzyme was obtained about 95 % pure by J. Moyle & Dixon (1) from heart muscle; throughout purification the ratio of the dehydrogenase to carboxylase activity remained constant and it was concluded that a single enzyme was concerned. This enzyme from heart muscle needs NADP, but in some tissues it depends on NAD.

OXIDATION OF KETO ACIDS. The oxidative decarboxylation of pyruvic (α-ketopropionic) and of α-ketoglutaric acids, providing active acyl for the citric acid cycle, has proved to be extremely complicated.

First Lohmann & Schuster (1) showed in 1937 that the diphosphate ester of vitamin B_1, thiamine-pyrophosphate, functioned as co-carboxylase in the splitting of pyruvic acid by yeast preparations, with formation of acetaldehyde and CO_2; in 1948 O'Kane & Gunsalus (1) found that for *Streptococcus* a factor was needed for pyruvic oxidation and dismutation, not identical with any known accessory factor. About this time Peters[3] and his collabo-

---

[1] Ochoa, Stern & Schneider (1); Stern, Shapiro, Stadtman & Ochoa (1); Stern, Ochoa & Lynen (1).
[2] The free energy of the bond is now given as 8250 cal. (Burton, 2). See ch. 6.
[3] Peters (2).

rators found the pyruvate oxidase of brain was very sensitive to arsenicals and that it could be protected by dithiols but not by monothiols. This suggested combination of the arsenic with an essential dithiol grouping in the enzyme system.

In 1951 it was shown that both pyruvate oxidation (with bacterial preparations)[1] and α-ketoglutarate oxidation (with heart and skeletal muscle preparations)[2] yielded acyl-CoA intermediates – acetyl-CoA and succinyl-CoA respectively. The mode of participation of pyruvate oxidation in the cycle was thus clear; in the case of succinyl-CoA the energy of the actyl-CoA bond could be used for other syntheses while the liberated succinate entered the cycle.[3]

In the same year Reed and his collaborators[4] began the account of their studies of the unknown factor, the existence of which had been realised for several years, associated with pyruvate oxidation. They prepared from liver a crystalline acidic compound, soluble in organic solvents and little soluble in water, which they named α-lipoic acid. This compound was essential for oxidative decarboxylation of pyruvate by certain bacteria. Sanadi, Littlefield & Bock (1) then showed that their highly purified α-ketoglutaric oxidase from pig heart contained both α-lipoic acid and diphosphothiamine. Evidence for the structure of α-lipoic acid was soon provided[5] and the following structure, dithioloctanoic acid, was established.

$$\underset{\underset{S\!-\!-\!-\!S}{|\qquad\quad|}}{H_2C \qquad CH(CH_2)_4COOH} \overset{CH_2}{\diagup\diagdown}$$

Reed (1) has recounted the evidence for the interplay of α-keto acid, thiamine-pyrophosphate, lipoic acid, CoA and flavine-adenine dinucleotide, together with three enzymes, in the enzymic production of acyl-CoA. Much the greater part of this evidence comes from work on *E. coli*; the pyruvate dehydrogenation system of this organism has been separated into the three enzymic components named below.[6] However, enough work has been done on preparations from muscle to indicate that the same structure for the α-ketoacid oxidase exists there. Thus Sanadi and his collaborators[7] have shown that the α-ketoglutaric dehydrogenase complex from pig-heart or pigeon-breast muscle contains tightly bound lipoic acid and catalyses reversible oxidation of NADH by lipoic acid. It was further found[8] that

[1] Korkes, del Camillo, Gunsalus & Ochoa (1).
[2] Sanadi & Littlefield (1).   [3] Kaufman, Gilvarg, Cori & Ochoa (1).
[4] Reed, De Busk, Gunsalus & Hornberger (1).
[5] Bullock, Brockman, Patterson, Pierce & Stokstad (1).
[6] Koike, Reed & Carroll (1).   [7] Sanadi, Langley & Searls (1).
[8] Sanadi, Langley & White (1).

presence of CoA was necessary in order to obtain the arsenite-inhibited form of the enzyme. This was taken to indicate the formation of succinyl-S-dithiol octanoate (reaction 2) followed by transfer of succinyl to CoA (reaction 3), this latter reaction leaving the thiol free for attack by arsenite. Then in 1960 Massey (1) studied the flavoprotein from pig-heart particles

$$\text{R}\overset{\text{O}}{\overset{\|}{\text{C}}}\text{COOH} + \text{TPP--E}_1 \longrightarrow [\text{RCHO--TPP}]\text{--E}_1 + \text{CO}_2 \tag{1}$$

$$[\text{RCHO--TPP}]\text{--E}_1 + \;\overset{}{\underset{\text{S--S}}{\diagup\diagdown}}\text{--(CH}_2)_4\overset{\text{O}}{\overset{\|}{\text{C}}}\text{--E}_2 \longrightarrow \;\overset{}{\underset{\text{HS} \quad \text{SCR}}{\diagup\diagdown}}\text{--(CH}_2)_4\overset{\text{O}}{\overset{\|}{\text{C}}}\text{--E}_2 + \text{TPP--E}_1 \tag{2}$$

$$\overset{}{\underset{\text{HS} \quad \overset{}{\underset{\text{O}}{\overset{\|}{\text{SCR}}}}}{\diagup\diagdown}}\text{--(CH}_2)_4\overset{\text{O}}{\overset{\|}{\text{C}}}\text{--E}_2 + \text{HSCoA} \longrightarrow \;\overset{}{\underset{\text{HS} \quad \text{SH}}{\diagup\diagdown}}\text{--(CH}_2)_4\overset{\text{O}}{\overset{\|}{\text{C}}}\text{--E}_2 + \text{RCSCoA} \tag{3}$$

$$\overset{}{\underset{\text{HS} \quad \text{SH}}{\diagup\diagdown}}\text{--(CH}_2)_4\overset{\text{O}}{\overset{\|}{\text{C}}}\text{--E}_2 + \text{FAD--E}_3 \longrightarrow \;\overset{}{\underset{\text{S--S}}{\diagup\diagdown}}\text{--(CH}_2)_4\overset{\text{O}}{\overset{\|}{\text{C}}}\text{--E}_2 + \text{HFAD--E}_3 \tag{4}$$

$$\text{HFAD--E}_3 + \text{NAD}^+ \longrightarrow \text{FAD--E}_3 + \text{NADH} + \text{H}^+ \tag{5}$$

E_1 = carboxylase. E_2 = lipoic reductase-transacetylase.
E_3 = dihydrolipoic dehydrogenase.
TPP = thiamine pyrophosphate. R = CH_3 or $\text{HOOC(CH}_2)_2$. (Reed (1).)

concerned in the transfer of hydrogen from lipoic acid to NAD. His keto-glutarate dehydrogenase preparation[1] contained protein-bound thiamine-pyrophosphate, lipoic acid and FAD in stoichiometric proportions. The complex could be resolved into two fractions: one colourless, containing all the protein-bound thiamine-pyrophosphate and lipoic acid, which catalysed oxidation of α-ketoglutarate with ferricyanide as H acceptor; the other yellow, containing the flavoprotein. Neither alone could carry out the NAD-linked oxidation of the keto acid, but on mixing the two fractions this activity was obtained.

It is interesting to remember that in 1938 Dewan & Green (1) had noticed

[1] Massey (2).

that NADH reacted very slowly with methylene blue or cytochrome c
while in presence of dehydrogenase preparations it reacted readily; they
suggested that an enzymic 'co-enzyme factor'[1] was involved. Soon afterwards Straub (4) prepared a flavoprotein from heart muscle, containing
FAD, and Corran, Green & Straub (1) found that this flavoprotein fulfilled
the requirements of their coenzyme factor. Massey now found this flavoprotein to be identical with that in his preparations; thus its natural function seems to be removal of H from reduced lipoic acid, rather than from
NADH.

OXIDATION OF FATTY ACIDS. The hypothesis of Knoop (1) of the oxidation of fatty acids at the β-carbon atom, enunciated at the very beginning
of this century and based on experiments on the whole animal, has stood
the test of time. But it was only after some 35 years that investigation
began of the enzyme systems involved. By 1943 the work of Muñoz &
Leloir (1) had shown that washed liver particles would oxidise fatty acids,
if they were provided with inorganic phosphate, fumarate, adenylic acid,
cytochrome c and Mg^{2+} or Mn^{2+}. Lehninger (1, 2) in 1945 found that in such
a particulate system simultaneous oxidation of other metabolites was
needed for fatty acid oxidation, but if ATP were present, this coupled
oxidation was unnecessary. Presumably the function of the coupled oxidation was to provide ATP by oxidative phosphorylation.[2] Heart-muscle
homogenates, if suitably fortified, could also oxidise higher fatty acids.[3]
We have already discussed the activation of acetate – the formation of
acetyl-CoA in presence of ATP. It was shown in 1953 by a number of workers
that higher straight-chain saturated acids can be activated in the same way.
Kornberg & Pricer (1) for example, and Mahler, Wakil & Bock (1) demonstrated acyl-CoA formation with a number of higher fatty acids, adenylic
acid and pyrophosphate being formed in stoichiometric amounts. The
collaboration of Lynen & Ochoa (1) also in this year, showed that four
enzymes were involved in reactions (1), (2), (3) and (4) on p. 401; similar
results were obtained in Green's laboratory[4] with preparations from beef
liver mitochondria. In both cases all the reactions were readily reversible.

These enzymes were isolated and further studied in detail in the early
nineteen fifties, preparations from liver particles or mitochondria being
used: the butyryl coenzyme-A dehydrogenase (of reaction 1) by Green,
Mii, Mahler & Bock (1); the hydratase (of reaction 2) by Wakil & Mahler (1);
the β-hydroxyl-acyl-CoA dehydrogenase (of reaction 3) (requiring NADP)

[1] This became known for a time as diaphorase; see Corran, Green & Straub (1). The term
'diaphorase activity' is now often used for the activity of any enzyme transferring
hydrogen from reduced NAD or NADP to a dye.
[2] Ch. 17.   [3] Lehninger (3).
[4] Beinert, Bock, Goldman, Green, Mahler, Mii, Stansley & Wakil (1).

by Wakil, Green & Mahler (1); and the acyl transferase (of reaction 4) (also demonstrated in heart preparations) by Goldman (1).[1]

$$RCH_2CH_2COSCoA + B \longrightarrow RCH = CHCOSCoA + BH_2 \quad (1)$$

$$RCH = CHCOSCoA + H_2O \longrightarrow R\underset{\underset{OH}{|}}{\overset{\overset{H}{|}}{C}} - CH_2COSCoA \quad (2)$$

$$\underset{\underset{OH}{|}}{\overset{\overset{H}{|}}{R}C} - CH_2COSCoA + B \longrightarrow RCOCH_2COSCoA + BH_2 \quad (3)$$

$$\beta\text{-ketoacyl-CoA} + CoA \longrightarrow \text{acetyl-CoA} + \text{acyl-CoA.} \quad (4)$$

B = hydrogen acceptor. The acyl-CoA (with 2 C atoms less) resulting in the 4th reaction enters again into reaction (1) in what has been called the 'fatty acid spiral'.

Carnitine has important effects in stimulating fatty acid oxidation. In 1950 Lundsgaard (10) had described the production by perfused muscle of a factor which was necessary for the attainment of maximal oxygen consumption in perfused liver. Fritz (1) working in Lundsgaard's laboratory was able to show in 1955 that, out of some twenty constituents of muscle extract, carnitine was the only one giving somewhat similar effects when tested on rat-liver slices, with labelled palmitic acid as substrate. Carnitine (β-OH,γ-trimethyl ammonium butyrate) is widely distributed in nature; it has long been known to occur in muscle and indeed, as Fraenkel (1) showed, its concentration is greater there than in any other tissue.

The oxidation of palmitate by washed-particle preparations from skeletal muscle and liver was enhanced by carnitine.[2] Fritz, from the observation that with water extracts of liver particles (capable of fatty acid oxidation) there was no augmentation with carnitine, deduced that the carnitine effect was concerned in facilitating transfer of long-chain fatty acids across a mitochondrial membrane to the intramitochondrial sites of fatty-acid oxidation. The incorporation of ^{3}H-labelled carnitine or ^{14}C-labelled palmitate into palmitylcarnitine was demonstrated by Fritz & Yue (1), using heart-muscle preparations. This synthesis required the presence of ATP and co-enzyme A; but from palmityl-CoA and carnitine it could take place without these additions. The following sequence of reactions was proposed:

Palmitic acid + ATP + CoA $\rightleftharpoons$ palmityl-CoA + AMP + pyrophosphate (1)
Palmityl-CoA + carnitine $\rightleftharpoons$ palmitylcarnitine + CoA. (2)

The enzyme responsible for reaction (2) had been described by Friedman & Fraenkel (1) in 1955, but its significance was not appreciated for some years. Fritz & Marquis (1) in 1965 pictured reactions (1) and (2) as taking place at a site, perhaps in the mitochondrial cristae space, separated by a

[1] See also the excellent review by Green (3).   [2] Fritz & McEwen (1); Fritz (2).

barrier (impermeable to palmityl-CoA, but permeable to palmitylcarnitine) from the site of fatty acid oxidation. At this latter site the sequence continued:

$$\text{Palmitylcarnitine} + \text{CoA} \rightleftharpoons \text{palmityl-CoA} + \text{carnitine} \qquad (3)$$
$$\text{Palmityl-CoA} \longrightarrow \text{intermediates of } \beta \text{ oxidation} \longrightarrow 8 \text{ acetyl-CoA.} \qquad (4)$$

Fritz's reviews (3, 4) of 1961 and 1967 give many references to work in this field.

The detailed studies of Chappell & Crofts (1) concerning the effects of inhibitors on the oxidation of palmitate and palmitylcarnitine by isolated mitochondia led them (also in 1965) to the similar suggestion that carnitine has the function of transferring acyl groups between two mitochondrial pools of co-enzyme A; the 'outer' pool being accessible to fatty acids or acyl CoA, the 'inner' being in contact with the enzymes of oxidation. The results of extensive assays by Tubbs and his collaborators[1] of the content in carnitine and acylcarnitine, CoA and acyl-CoA in liver and heart under different conditions fitted in with this picture.

Work on rather similar lines to those pursued by Fritz went on independently in Norway in 1962 and 1963. The paper of Bremer (1) gives references. Bremer observed the reversible acetylation of carnitine by mitochondria and the metabolism of fatty acid esters of carnitine by mitochondria without previous hydrolysis. He discussed the activity of carnitine as a carrier. It is interesting that pyruvate inhibited mitochondrial deacylation of acetyl or octanoylcarnitine while α-glycerophosphate promoted extra-mitochondrial deacylation of fatty acylcarnitine. This was seen as a double mechanism for carbohydrate sparing of fatty acid oxidation.

We have seen that earlier work suggested utilisation of fat by muscle in certain circumstances. More direct evidence came from the work of Buchwald & Cori (1) in 1930 on gastrocnemii of spring frogs. Here where the carbohydrate content was low, stimulation did effect a decrease in fat content. Such a result could not be found with mammalian muscle. More recent work has clearly shown that the mammalian heart can remove free fatty acids supplied in the perfusing fluid. In the experiments of Shipp, Opie & Challoner (1) there was marked inhibition of glucose oxidation under these conditions and ^{14}C labelled glucose was shown to give rise to glycogen. Garland, Newsholme & Randle (1) found enhanced respiration of endogenous fatty acids in both heart and diaphragm muscle of the rat in diabetes or starvation, while in the experiments of Williamson & Krebs (1) even normal rat hearts preferred acetoacetate to glucose.

With regard to the human subject, Andres, Cader & Zierler (1) found that glucose taken up by the resting forearm accounted for only about 7 % of the

[1] Tubbs, Pearson & Chase (1); Pearson & Tubbs (1).

oxygen uptake; there was indirect evidence that glycogen usage was unimportant, and since the respiratory quotient (R.Q.) was 0.8 they concluded that the resting muscle used mainly lipids. Havel, Naimark & Borchgrevink (1) reported that with human resting muscle, during fasting, some 25 % of the energy metabolism was supplied by free fatty acids entering the tissue; with carbohydrate feeding, only about 5 % was from the fatty acid source. During exercise the percentages rose to 45 and about 10 respectively. Hultman (1) quotes experiments on men in which the effects of different work loads were tested with regard to the R.Q.[1] With loads corresponding to 29, 53 and 79 % of the maximum oxygen uptake, the R.Q.'s corresponding were 0.87, 0.90 and 0.93 – indicating the taking over by carbohydrate metabolism with increased energy requirement. The effect of diet was also examined[2] and it appeared that with the same load the R.Q. was lowest in work after a fat and protein diet (when the average glycogen content of the muscle was only 0.63 g/100 g). Thus the proportions of carbohydrate and fatty acids used vary, both with the degree of exercise and with the diet.

OXIDATION OF AMINO ACIDS; TRANSAMINATION. Early studies of amino-acid oxidation were made by Neubauer about 1910 with yeast.[3] Using phenylaminoacetic acid he showed that the keto-acid phenylglyoxylic acid was formed and suggested the mechanism as in the figure.

$$\underset{\substack{|\\ \text{COOH}}}{\overset{\substack{\text{R}\\|}}{\text{HC}-\text{NH}_2}} \xrightarrow{+[O]} \underset{\substack{|\\ \text{COOH}}}{\overset{\substack{\text{R}\\|}}{\text{C}=\text{NH}}} + \text{H}_2\text{O} \xrightarrow{+\text{H}_2\text{O}} \underset{\substack{|\\ \text{COOH}}}{\overset{\substack{\text{R}\\|}}{\text{C}}}{\overset{\text{OH}}{\underset{\text{NH}_2}{\big<}}} \longrightarrow \underset{\substack{|\\ \text{COOH}}}{\overset{\substack{\text{R}\\|}}{\text{C}=\text{O}}} + \text{NH}_3$$

In the early nineteenth thirties Krebs examined amino-acid oxidation with animal tissues – kidney and liver slices and homogenates.[4] Two different enzyme systems were present, one concerned with D-amino acids, the other with the naturally occurring L-amino acids. Warburg & Christian (6) showed in 1938 that the former was a flavoprotein, with isoalloxazin-adenine-dinucleotide as prosthetic group. Blanchard, Green, Nocito & Ratner (1, 2), using kidney and liver, later isolated the L-amino acid oxidase; it too is a flavoprotein, in this case with a mononucleotide prosthetic group. Both oxidases carried out the following reaction:

$$\text{RCH(NH}_2)\text{COOH} + \text{O}_2 + \text{H}_2\text{O} \longrightarrow \text{RCO.COOH} + \text{NH}_3 + \text{H}_2\text{O}_2.$$

[1] Hermansen, Hultman & Saltin (1).
[2] Bergstrom, Hermansen, Hultman & Saltin (1).
[3] Neubauer & Fromherz (1).
[4] H. A. Krebs (2).

Dixon and his collaborators have studied the D-amino acid oxidase,[1] as well as the D-aspartate oxidase of kidney.[2] The former oxidises many monocarboxylic acids, but not the two dicarboxylic acids, D-glutamate and D-aspartate. These are both oxidised by the latter enzyme, which has no action on the monocarboxylic acids. Curiously enough, the activity of the L-amino acid oxidase was low or negligible in tissues of several different types of animal; it seems possible that oxidation of these amino acids may often take place indirectly through another system – L-glutamate dehydrogenase and transaminases.

Thunberg (2) first observed reduction of methylene blue with glutamic acid, using finely cut, washed muscle. Nearly twenty years later the enzyme was studied by von Euler and his collaborators in liver preparations.[3] The reaction yielded α-ketoglutaric acid and ammonia; it was reversible and dependent on NADP. Later work has shown that the coenzyme specificity– for NAD or NADP – varies with the source of the enzyme; also that other amino acids may be used, but only at a fraction of the rate with glutamate.[4]

We turn now to transamination. In 1930 Needham (6) had found that glutamic or aspartic acid added to minced muscle under anaerobic conditions gave rise to succinic, malic and fumaric acids without formation of ammonia or any change in the soluble nitrogen fractions. Aerobically these two amino acids increased the oxygen uptake, again with no change in the soluble N fractions, although it was shown that the aspartic acid actually disappeared. She suggested that the amino group entered into combination with some reactive carbohydrate residue and was retained as a new amino acid.

Several years later Braunstein & Kritzmann (1) in the Soviet Union published their first papers on transamination, using muscle tissue; glutamic acid vanished anaerobically while alanine appeared. The disappearance of glutamic acid was increased by adding pyruvate, and it soon become clear that this was a reaction between the glutamic acid and α-keto acids, several of the latter being able to replace pyruvate. The reaction was reversible and a whole series of α-amino acids could transfer their nitrogen to α-ketoglutaric acid. Kritzmann (1) prepared from muscle an extract containing the transaminase specific for glutamic acid and inactive with aspartic acid. Later, realising the participation of a thermostable co-enzyme, she was able to show the presence also of aspartic transaminase in such extracts (2). This activator seemed to be less readily dissociable from the glutamic enzyme.

The development of chromatographic methods enabled Cammarata & Cohen (1) and Rowsell (1, 2) to make extended studies of the scope of these enzymes. It is interesting that the former found heart-muscle extracts to be almost always more active than liver and kidney extracts.

[1] Dixon & Kleppe (1, 2).

[2] Dixon & Kenworthy (1).

[3] Von Euler, Adler, Günther & Das (1).

[4] Frieden (1).

With regard to the coenzyme, Snell (1) in 1945, working on the vitamin B_6 group, observed that pyridoxal reacted with glutamic acid at raised temperatures to give pyridoxamine and α-ketoglutaric acid. He connected this observation with the work of the Braunstein laboratory, and found (in experiments with Schlenk)[1] that the tissues of rats deficient in B_6 had lowered transaminase activity. Shortly afterwards Kritzmann & Samarina (1) were able to show reactivation of glutamic transaminase by pyridoxal phosphate. Schlenk & Fisher (1) then found that highly purified transaminase preparations usually contained pyridoxal phosphate (a form of B_6) while a few contained a pyridoxamine derivative. They were of opinion that the amino group transfer took place by means of the reversible change between pyridoxal and pyridoxamine.

A question must now arise concerning evidence for the utilisation in muscle of energy provided by amino-acid oxidation. We have seen that even in strenuous exercise examination of nitrogen excretion showed that protein could not play more than a very minor part;[2] and increased ammonia in the blood during muscular contraction was traced to adenylic acid, not protein, breakdown.[3] In the early work of Thunberg and of Needham, other amino acids than glutamic and aspartic had little or no effect on the oxygen uptake of muscle *brei*. Later experiments confirm the view that protein is little called upon in normal muscle activity. Thus Cruickshank & McClure (1), Clarke (1) and Williamson & Krebs (1) could find no evidence for amino-acid utilisation by the perfused mammalian heart. Hicks & Kerly (1) examining the perfusate of rat heart supplied with glutamic, aspartic and other amino acids showed that little transamination accompanied its activity, although the amino acids could penetrate, and heart homogenates had good transaminase activity.

As far as one can surmise, then, it seems that such oxidation of amino acids as goes on in muscle is likely to be via glutamic dehydrogenase and transaminase. The latter enzyme may of course have a function (perhaps more important) in amino-acid synthesis.

CONCLUSION

In this account we have dealt only with the mitochondrial respiratory chain. Although in the main oxidative metabolism is carried out in the mitochondria, it seems that the microsomal[4] and the soluble[5] fractions of the cell also play some part. Microsomes have been found to contain a cytochrome b_5, though little cytochrome c; also NADP and co-enzymes Q_9 and Q_{10}. Amongst respiratory enzymes reported there are $NADH_2$ and $NADPH_2$

---

[1] Schlenk & Snell (1).   [2] Ch. 2.   [3] Ch. 4.
[4] Reid (1) p. 340.   [5] Anderson & J. G. Green (1) p. 496.

cytochrome *c* reductase and L-amino acid oxidase. The pentosephosphate oxidation pathway is contained in the soluble phase of the cell of some mammalian tissues but does not operate in normal muscle.[1]

Our present knowledge of the citric acid cycle may be summed up in the accompanying diagram,[2] which indicates also (1) how fatty acids may participate, by means of the acetyl-CoA split off at each β-oxidation step; (2) how amino acids may enter via transaminase activity and the oxidative deamination of glutamic acid. As we have seen the reduced coenzymes are oxidised via the respiratory chain, and the same is true of the oxidation of succinic to fumaric acid, although coenzyme is not involved. It may be observed that almost all the reactions portrayed are reversible, but the significant point has been made by Krebs (3) that, in the case of the oxidative decarboxylation of α-ketoglutaric acid the equilibrium is so far in the direction of succinic acid as to ensure that the operation of the cycle is virtually unidirectional.

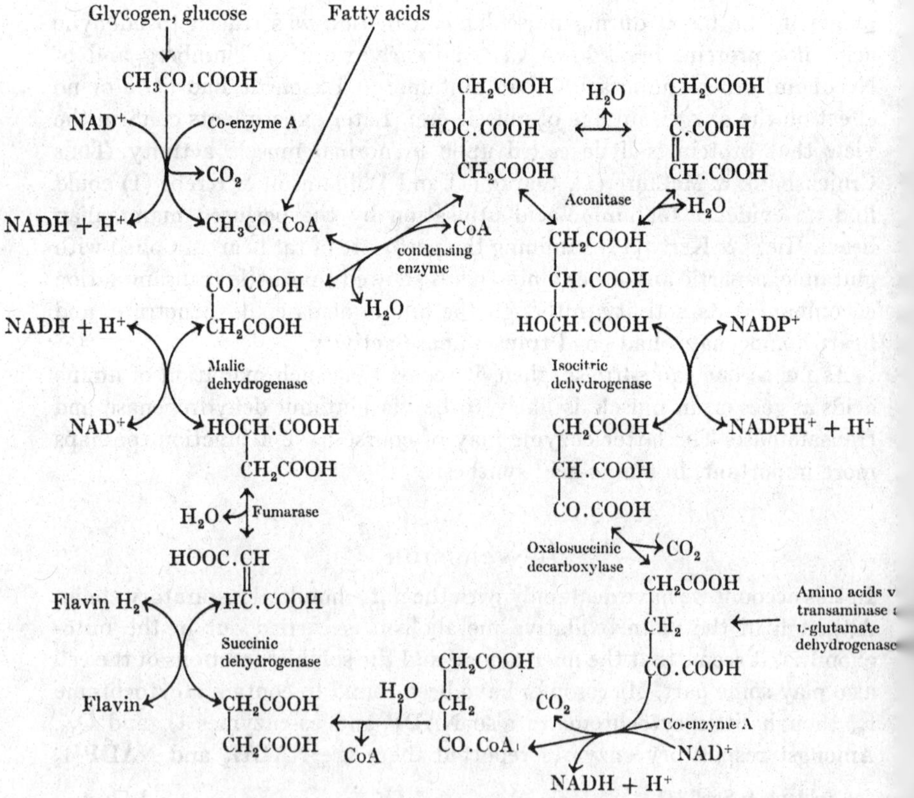

[1] Cf. ch. 19 p. 488; ch. 21, p. 502. [2] After Fruton & Simmonds' *General Biochemistry*, Wiley

17

OXIDATIVE PHOSPHORYLATION

EARLY EVIDENCE FOR OXIDATIVE PHOSPHORYLATION

We have already mentioned Lundsgaard's observation (3) in 1930 that with iodoacetate-poisoned muscle, no longer able to carry on glycolysis, there was clear evidence of aerobic resynthesis of phosphocreatine during or after its breakdown in contraction. About the same time we find Engelhardt's study of oxidative phosphorylation in isolated red blood cells, the forerunner of studies still continuing on the mechanism of this phosphorylation in cell-free preparations and purified enzyme systems. In Engelhardt's experiments (4, 5) erythrocytes from mammals and birds were used in presence of glucose, and of fluoride to prevent glycolysis. In conditions of vigorous respiration of the suspension, inorganic phosphate was taken up with formation of pyrophosphate. In a period of anaerobiosis, pyrophosphate broke down; on re-admission of oxygen there was increased oxygen consumption, pyrophosphate synthesis paralleling the rate of extra oxygen uptake and levelling off with it. It is clear from Engelhardt's discussion (5) that he considered this pyrophosphate to be contained in ATP, and he adumbrated an interesting conception of metabolic cycles in which breakdown products act as stimulants of resynthetic processes.

The next contribution was from Kalckar (2) in 1937, using kidney-cortex extracts.[1] In this tissue there is no phosphate uptake accompanying glycolysis, as there is in muscle extracts, but aerobically in presence of fluoride he found that for every mole of oxygen used, about 1 mole of phosphate was esterified. Adenylic acid was converted to ATP, and glucose, fructose or fructose-6-phosphate could be phosphorylated.

Then in 1939 Belitzer & Tsibakova (1) made a striking observation. They used findly chopped or ground pigeon-breast or rabbit-heart muscle, well washed with a neutral salt solution of low ionic strength. NAD, Mg^{2+}, creatine and substrate were added. With the substrates pyruvate, citrate, ketoglutarate, succinate, fumarate, malate and lactate there was good respiration accompanied by synthesis of phosphocreatine; synthesis was not prevented by IAA or fluoride. The P/O ratios (atoms of P esterified to atoms of oxygen consumed) lay between 1.9 and 2.1. They concluded that phosphorylation was not connected only with the primary acts of dehydro-

---

[1] Also Kalckar (3).

genation of glucose, such as that occurring in glycolysis when phosphorylation accompanies transfer of 2H to NAD with a P/H_2 (P/O) ratio of 1; the preliminary removal of 12 atoms of H from one glucose molecule would lead to consumption of 6 atoms of O, and by analogy to esterification of 6 atoms of P. Therefore subsequent, so far unidentified, reactions of H transport must be invoked. They found that oxidation of succinic acid to fumaric acid (in presence of arsenite to prevent oxidation of fumaric and malic acids) was coupled with phosphorylation with a P/O ratio about 1; they suggested that oxidation in the 'fumaric system' (perhaps with Szent-Györgyi's cycle in mind,[1] though this was not explicitly stated) might be one of the intermediate sites of phosphorylation. Oxidation of p-phenylenediamine, though very vigorous, was not accompanied by phosphorylation, and this led them to rule out the cytochrome system. Belitzer (1), like Engelhardt, laid great stress on the stimulatory effect of breakdown products (in this case free creatine) on respiration in muscle brei, this stimulation depending upon the presence also of free phosphate – thus upon the opportunity for synthesis of phosphocreatine.

In 1940 Colowick, Welch & Cori (1) confirmed Kalckar's findings of aerobic phosphorylation in kidney extracts. They used a number of dicarboxylic acids as substrate and like Belitzer & Tsibakova noted that the step succinate to fumarate was associated with P-esterification; they surmised that this dehydrogenation (according to Szent-Györgyi's view linking a number of substrates to the cytochrome chain) might play an important part in phosphorylation. A little later Ochoa (1, 2), using brain and heart-muscle extracts, reported that while he could find an average P/O ratio of 3 with α-ketoglutarate oxidation to succinate, the oxidation of succinate to fumarate seemed to account for the formation of only one energy-rich phosphate bond. He pointed out that the catalytic function of the succinic to fumaric reaction in the form visualised by Szent-Györgyi was now very doubtful, and he seems to have been the first to suggest (in 1943) that phosphate bonds might be generated during the passage of H from the potential level of flavoprotein to that of oxygen.[2] He failed however to find phosphorylation accompanying the oxidation by molecular oxygen of large amounts of NADH in heart extracts.

Judah (1) in 1951 compared the phosphorylation obtained on oxidation of α-ketoglutarate and β-hydroxybutyrate, finding P/O ratios of nearly 4 and about 3 respectively. F. E. Hunter (1)[2] had already shown in 1949 that the anaerobic oxido-reduction of α-ketoglutarate with oxaloacetate had a P/O ratio of 1,[3] while Lehninger (4) later found that a similar dismuta-

---

[1] Ch. 16.
[2] Also Ochoa (2).
[3] Also Cross, Taggart, Covo & Green (1).

tion of β-hydroxybutyrate with oxaloacetate involved no such phosphorylation. Thus in the oxidation of α-ketoglutarate one energy-rich phosphate bond is formed at the substrate level, and 2 to 3 (just as with β-hydroxybutyrate oxidation) beyond NADH in the chain.

Much further work in a number of laboratories, with continually increasing refinement of methods,[1] has led to general agreement on the following values of P/O ratios:

$$\text{ketoglutarate} + \tfrac{1}{2}O_2 \longrightarrow \text{succinate} + CO_2 \qquad 3 \text{ to } 4$$
$$\text{hydroxybutyrate} + \tfrac{1}{2}O_2 \longrightarrow \text{acetoacetate} + H_2O \quad 2 \text{ to } 3$$
$$\text{succinate} + \tfrac{1}{2}O_2 \longrightarrow \text{fumarate} + H_2O \qquad 1 \text{ to } 2.$$

THE LOCALISATION OF THE PHOSPHORYLATION SITES IN THE RESPIRATORY CHAIN

Friedkin & Lehninger (1) with washed particles were able to show that oxidation of reduced NAD by oxygen did indeed lead to phosphate esterification, and in 1951 Lehninger (5) using highly purified NADH and particles which had been pretreated under hypotonic conditions, found a P/O ratio of 1.89 for this oxidation. This hypotonic treatment probably removed some 'availability barrier'.

In 1950 Slater (3) with heart-muscle particles showed phosphorylation coupled with oxidation of α-ketoglutarate with ferricytochrome c as H acceptor, in presence of cyanide or under anaerobic conditions. The results indicated two phosphorylation sites. He later found (4) that phosphorylation was distinctly greater when oxygen was the acceptor rather than cytochrome c; in the latter case the P/O ratio approached 2, in the former 3. Meanwhile Nielsen & Lehninger (1) in 1954 found phosphorylation upon oxidation of ferrocytochrome c by oxygen, reduction of the oxidised cytochrome c by endogenous substrates being prevented by the presence of antimycin A.[2] The P/O ratio was 0.44–0.86, indicating a single phosphorylation site.

The enormous difficulties inherent in such experiments must be stressed and are well described by Slater (2). These include the marked ATPase activity of mitochondrial preparations, obscuring the ATP synthesis; the possibility that inhibitors added for a particular purpose may act also in other very subtle ways; and the barriers that may exist to penetration of added substances (such as NADH) into the active sites of the chain.

The facts enumerated so far may now be compared with the structure of the respiratory chain:[3]

---

[1] For an account of these see Slater (2).
[2] P. 416 below.
[3] Ch. 16.

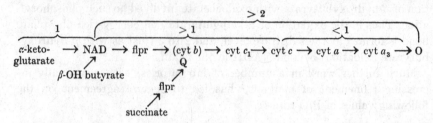

This general picture has been confirmed by means of the entirely different technique used by Chance and his collaborators.[1] With suspensions of fresh mitochondria they followed at various wave-lengths the changes in optical density consequent upon various changes in conditions. They could thus find the state of oxidation–reduction of the individual components of the chain in presence or absence of oxidisable substrate, or of phosphate acceptor or of respiratory inhibitors. After addition of the respiratory inhibitor antimycin A to actively metabolising mitochondria they found a steady state in which there was complete oxidation of cytochromes a_3, a and c while cytochrome b and NAD became completely reduced. Thus the action of this inhibitor was localised between cytochromes b and c, and Chance & Williams named such points, where one carrier becomes more oxidised while its neighbour becomes more reduced, 'crossover points'. As we have seen, both Engelhardt and Belitzer had emphasised the indispensability of phosphorylation for respiration, and Lardy & Wellman (1) found in 1952 with carefully prepared mitochondria that their respiration was indeed dependent on presence of phosphate acceptors; mitochondria in such a state are known as 'tightly coupled'. Chance & Williams now studied the changes when such actively respiring mitochondria were brought into the 'resting' or 'inhibited' state by exhaustion of phosphate acceptors. This transition led to reduction of all the respiratory carriers from NAD to cytochrome c, cytochrome a becoming more oxidised. This crossover point lay between cytochromes a and c. With low concentrations of azide, the crossover point on using up of all ADP lay between cytochromes b and c, while in the presence of cyanide it was shifted again, to lie between NADH and flavoprotein. Thus the lack of acceptor at each of these points was inhibitory, and the normal formation of three energy-rich phosphate bonds was indicated. The uncertainties of interpretation of such results as regards exact localisation of the sites responsible for the $\sim$P bond formation have been discussed by Chance & Williams and by Slater (6). But it seems clear that they indicate three phosphorylation sites, and that they may well accord with the general view that these lie between NAD and flavoprotein, between cytochromes b and c and beyond cytochrome c.

[1] Chance & Williams (1).

What of the thermodynamic aspects of this situation? Lipmann (6) in 1946 pointed out that the available potential span (1.2 V) between oxygen and a pair of average substrate hydrogens is amply sufficient to allow for four energy-rich phosphate bonds to be formed. More recently Slater (6), has considered this question in greater detail. He used the figures of Chance & Williams (2) for the degree of reduction of each component of the chain in mitochondria oxidising β-hydroxybutyrate in presence of ATP and inorganic phosphate; using also the oxidation–reduction potentials of these components it is possible to calculate the change in free energy (ΔG) for each oxidation–reduction step, assuming (a) no phosphorylation (ΔG_1); (b) phosphorylation (ΔG_2).[1] It is clear that phosphorylation can only occur if ΔG_2 has a negative value, and the calculations showed that this occured only at the stages:

$$\text{NADH} + \text{flpr} + \text{H}^+ \longrightarrow \text{NAD} + \text{flprH}_2$$
$$2 \text{ cyt } b^{2+} + 2 \text{ cyt } c^{3+} \longrightarrow 2 \text{ cyt } b^{3+} + 2 \text{ cyt } c^{2+}$$
$$2 \text{ cyt } a^{2+} + \tfrac{1}{2}\text{O} + 4\text{H} \longrightarrow 2 \text{ cyt } a^{3+} + \text{H}_2\text{O}.$$

Slater emphasises that these calculations must be regarded with some reservations, but it appears from them and from what has just been narrated that the most likely sites for $\sim$P generation are shown in the following scheme:

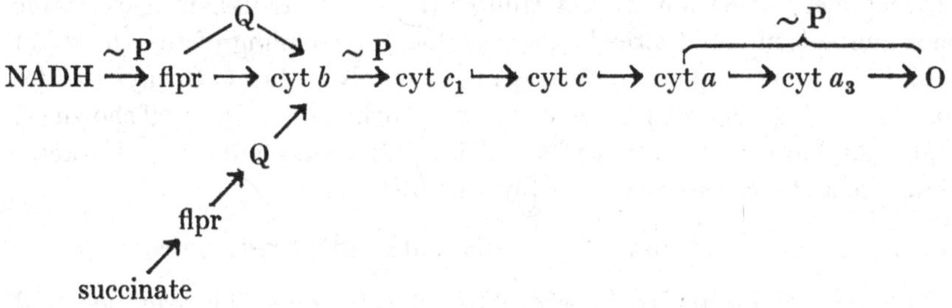

THE MECHANISM OF OXIDATIVE PHOSPHORYLATION AT SUBSTRATE LEVEL

Since the theories of mechanism of oxidative phosphorylation owe much to what was already known of the mechanism of phosphate uptake during oxidoreduction of glyceraldehyde phosphate in glycolysis, and during the

[1]
$$\Delta G_1 = \Delta G_0' + RT \ln \frac{(\text{A})(\text{BH}_2)}{(\text{AH}_2)(\text{B})} \text{ cal/mole}$$

$$\Delta G_2 = \Delta G_1 + 8000 + RT \ln \frac{(\text{ATP})}{(\text{ADP})(\text{P}_i)} \text{ cal/mole}$$

where G_0' = the standard free energy change for the reaction $\text{AH}_2 + \text{B} \rightleftharpoons \text{A} + \text{BH}_2$ and 8000 cal is the value of the standard free energy change for the reaction
$$\text{ADP} + \text{P} \rightleftharpoons \text{ATP} + \text{H}_2\text{O}.$$

oxidation of α-ketoglutarate to succinate and CO_2, it would be well to consider these here.

We have seen[1] that in the oxidation of glyceraldehyde phosphate to diphosphoglyceric acid Warburg & Christian postulated entry of the phosphate into the aldehyde group of the glyceraldehyde phosphate, the energy-rich bond being formed on oxidation of the phosphorylated aldehyde group to the carboxyl phosphate; evidence for this mechanism was sought by Meyerhof and his collaborators, but was not found:

$$CH_2O.CHOH.CHO + H_3PO_4 \longrightarrow CH_2O.CHOH.\underset{\underset{HO-\overset{\|}{\underset{O}{P}}-OH}{|}}{\overset{\overset{OH}{|}}{C}}.\overset{OH}{\underset{H}{O}}-\overset{OH}{\underset{OH}{P}}{=}O$$

$$CH_2O.CHOH.\underset{\underset{HO-\overset{\|}{\underset{O}{P}}-OH}{|}}{\overset{\overset{OH}{|}}{C}}.\overset{OH}{\underset{H}{O}}-\overset{OH}{\underset{OH}{P}}{=}O + NAD \longrightarrow CH_2O.CHOH.\underset{\underset{HO-\overset{\|}{\underset{O}{P}}-OH}{|}}{\overset{\overset{O}{\|}}{C}}-O \sim \overset{OH}{\underset{OH}{P}}{=}O + NADH + H^+.$$

By 1951 the study of acetylations had led to the recognition of the energy-rich acylmercaptide bond,[2] and Hunter (2) in his discussion of oxidative phosphorylation could already suggest that the coupling of the energy of oxidations with synthetic processes might involve other pathways besides that via the high-energy phosphate system. He had in fact himself shown (1) that phosphate esterification was coupled with oxido-reduction of α-keto-glutarate and oxaloacetate probably with the step:

$$\alpha\text{-ketoglutarate} + \text{coenzyme} \longrightarrow \text{succinate} + CO_2 + \text{reduced coenzyme}.$$

Then Kaufman and his collaborators[3] studied the reversible reaction found to be intermediate in this oxido-reduction:

$$\text{succinyl-CoA} + \text{ADP} + P_i \longrightarrow \text{succinate} + \text{CoA} + \text{ATP}$$

and in 1955, using an enzyme system from spinach, Kaufman (1) found evidence (based on experiments with ^{14}C labelled succinate and ^{32}P labelled ADP) for the following series of reactions:

$$\text{Succinyl} \sim \text{CoA} + P_i + E \rightleftharpoons \text{succinate} + E\ (\text{CoA} \sim P)$$
$$E\ (\text{CoA} \sim P) \rightleftharpoons E \sim P + \text{CoA}$$
$$E \sim P + \text{ADP} \rightleftharpoons \text{ATP} + E.$$

Here we have the energy-rich $P-O \sim P$ bond, an $S \sim C$ linkage, and an energy-rich phosphate bond on the enzyme. It remains to add in parenthesis that

---

[1] Ch. 6.　　　　　[2] Ch. 16.　　　　　[3] Kaufman, Gilvary, Cori & Ochoa (1).

Sanadi and his collaborators,[1] with beef heart mitochondria, finding need in this phosphorylation system for a coenzyme besides NAD and ATP, traced the participation of guanosine triphosphate as the primary product of phosphorylation, a transphosphorylase being necessary for the formation of ATP.

In 1952 an important study of glyceraldehyde phosphate dehydrogenase was described by Racker & Krimsky (1). The classical work of Rapkine (1) had shown the sensitivity of this enzyme to SH reagents. Racker & Krimsky now confirmed this and brought evidence for a two-stage mechanism whereby oxidation was first effected by combination of the aldehyde group of the glyceraldehyde phosphate with the enzyme (they suggested through the SH groups) and transfer of H to enzyme-bound NAD; this stage they suggested would be followed by phosphorolysis at the energy-rich $S \sim CO$ linkage, with formation of the energy-rich carboxyl phosphate linkage:

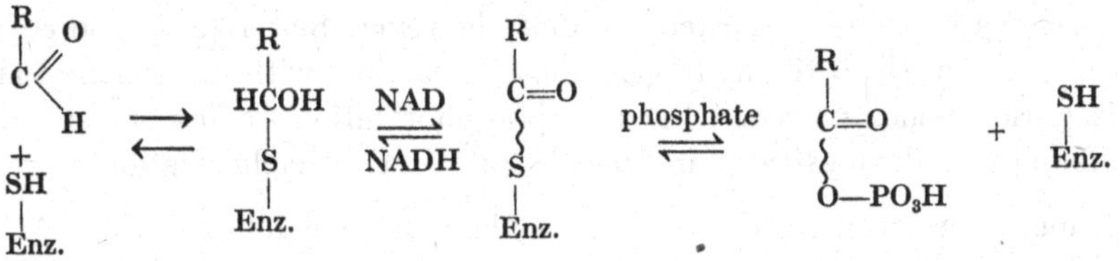

Two main lines of evidence supported this concept. (*a*) It was already known[2] that other aldehydes than glyceraldehyde phosphate are also oxidised by the enzyme, e.g. acetaldehyde and glyceraldehyde. It was now found that when acetaldehyde and coenzyme were added to the pure enzyme (used in large amounts) in absence of phosphate, formation of acyl enzyme could be demonstrated by its power to acetylate glutathione, sulphanilamide or coenzyme A. (*b*) If this hypothesis is correct, the reaction

$$\text{diphosphoglycerate} \rightleftharpoons \text{monophosphoglycerate} + H_3PO_4$$

(the reverse of the third reaction above) should be catalysed in absence of a hydrogen donor. The equilibrium is unfavourable to the demonstration of this reverse reaction, but it runs readily if arsenate is added. Then presumably an unstable acyl arsenate is formed which breaks down at once to release the monophosphoglyceric acid. Under certain conditions this arsenolysis could be shown completely unaccompanied by oxido-reduction. A great deal of later work has been devoted to many facets of the problem of the mechanism of this reaction; evidence has accumulated both for limited oxidation of substrate in absence of phosphate, and for the reality of the postulated reaction of the substrate with SH groups on the enzyme.[3]

[1] Sanadi, Gibson, Ayengar & Ouellet (1); Ayengar, Gibson & Sanadi; Sanadi, Gibson & Ayengar (1).   [2] Harting (1).
[3] E.g. Harting & Velick (1); Segal & Boyer (1); Koeppe, Boyer & Stulberg (1).

Thus here we find again the formation of an energy-rich bond enzyme-bound, in this case in all probability the carbon-sulphur linkage, used to form an energy-rich carboxyl phosphate bond.

THE MECHANISM OF PHOSPHORYLATION IN THE RESPIRATORY CHAIN

CHEMICAL THEORIES. In 1946 Lipmann (6) discussed possible mechanisms for transformation of electrical potential into phosphate bond energy at various potential levels in the respiratory chain. His suggested scheme for the level $E_0 = 0.45$ V involved entry of the phosphate into the molecule followed by oxidation with formation of an energy-rich phosphorylated CO group attached to a double bond (as in phosphoenolpyruvate).[1] The idea of an intermediate in oxidative phosphorylation containing an energy-rich phosphate bond was at first the only one entertained and used in tentative expressions of energy transfer. In 1953, however, Slater (5) proposed a mechanism in which the hydrogen transfer reaction, with formation of an energy-rich bond, preceded the transphosphorylation – a mechanism resembling that demonstrated in the substrate level phosphorylations.

Scheme I representing earlier views, may be written as follows, where AH_2 and B are adjacent members of the respiratory chain.

$$AH_2 + P \rightleftharpoons AH_2P \tag{1}$$
$$AH_2P + B \rightleftharpoons A \sim P + BH_2 \tag{2}$$
$$A \sim P + ADP \rightleftharpoons A + ATP. \tag{3}$$

Scheme II, based on Slater's formulation, has an additional component, C, needed for interaction of AH_2 and B:

$$AH_2 + B + C \rightleftharpoons A \sim C + BH_2 \tag{1}$$
$$A \sim C + P \rightleftharpoons A \sim P + C \tag{2}$$
$$A \sim P + ADP \rightleftharpoons A + ATP \tag{3}$$
$$A \sim C + H_2O \longrightarrow A + C. \tag{4}$$

Slater's mechanism had certain advantages in explanation of observed results. Thus Keilin and Hartree heart-muscle preparations[2] (consisting of disrupted mitochondria) could actively oxidise added substrates without coupled phosphorylation and for such oxidation phosphate was not needed; here the regeneration of A and C was explained by the hydrolysis of reaction 4 in Scheme II, replacing the phosphorolysis of reaction 2. Then we may consider the action of 2,4-dinitrophenol. This agent had a long-established sinister reputation for interference with synthetic processes – nitrogen assimilation, growth and differentiation, formation of adaptive enzymes.[3] As regards muscle, Ronzoni & Ehrenfest (1) in 1936 had shown that it

---

[1] Ch. 6.   [2] Keilin & Hartree (5); Bonner (1).   [3] Loomis & Lipmann (1) for refs.

provoked enormous wastage of energy. In resting frog muscle immersed in salt solution containing 3×10^{-5}M dinitrophenol, oxygen consumption was much accelerated and at the same time there was greatly increased break-down of phosphocreatine; anaerobically lactic acid formation was accelerated 6–7 times, and this continued until complete breakdown of creatine phosphate and ATP had taken place. Lardy & Elvehjem (1) in 1945 suggested that all these effects might be due to speeding up of hydrolysis of compounds containing high-energy phosphate (e.g. A~P in Scheme I) and this explanation was adopted by many workers. However, Loomis & Lipmann (1), who found that dinitrophenol could uncouple phosphorylation from oxidation in cyclophorase preparations, had shown that this agent could 'replace' phosphate in such preparations – that is to say, phosphate was no longer necessary for respiration.[1] Slater now suggested that this effect would be the result of a stimulation by dinitrophenol of reaction 4 in Scheme II. The abolition by dinitrophenol of the phosphate requirement for respiration was confirmed by Borst & Slater (1) using mitochondria.

Later work, particularly on the combined effects of various agents, can be interpreted in terms of Scheme II or extensions of it. Thus in 1958 Lardy and his collaborators introduced the use of the antibiotic oligomycin, which has become very important in such studies.[2] It inhibited respiration of liver mitochondria as well as preventing phosphorylation, and this inhibition of respiration was completely reversed by dinitrophenol. They concluded that oligomycin worked by blocking a reaction involved in the transfer of phosphate, this blockage being responsible for the holding up of respiration. It followed that dinitrophenol must exert its effect at some point prior to the incorporation of phosphate – in confirmation of the formulation of Scheme II.

In 1961 Huijing & Slater (1) published a paper confirming this behaviour of oligomycin, and giving a very interesting résumé of the action of a number of respiratory-chain inhibitors.

(a) Dinitrophenol, while not inhibiting respiration, uncouples all phosphorylation steps in the respiratory chain, and so stimulates the respiration of systems deficient in ADP or phosphate.

(b) Arsenate uncouples oxidative phosphorylation,[3] probably by substituting for phosphate to form unstable arsenic compounds.

(c) Oligomycin inhibits respiration like antimycin and phosphorylation like dinitrophenol, but unlike antimycin has no effect on uncoupled, non-phosphorylating respiration. Its inhibitory effect on respiration is completely released by dinitrophenol but not by arsenate. Thus is it primarily

---

[1] E.g. Loomis & Lipmann (1); Lardy & Wellman (2); Cross, Taggart, Covo & Green (1).
[2] Lardy, Johnson & McMurray (1); Lardy & McMurray (1).
[3] Crane & Lipmann (1).

an inhibitor of phosphate transfer, the inhibition of respiration resulting from this. Lardy & McMurray (1) had shown that oligomycin blocked exchange of inorganic P with ATP, and the loss of ^{18}O from labelled phosphate; thus it inhibited the enzyme involved in P uptake. Huijing & Slater now showed it did not affect the ATP $\rightleftharpoons$ ADP exchange reaction, i.e. the inhibition did not involve the terminal phosphorylated compound. They suggested that it might block formation of $C \sim D$ (where D might be an enzyme).

Huijing & Slater showed that these facts could be interpreted by an extension of Scheme II, postulating a minimum of three energy-rich intermediates (one phosphorylated) between the respiratory chain component concerned and ATP.

$$
\begin{array}{c}
\text{Oligomycin} \\
\lceil - - - - - \rceil \\
\quad +C \qquad\qquad +D \qquad\qquad +P_i \qquad\quad +ADP \\
AH_2 + B \rightleftharpoons A \sim C + BH_2 \rightleftharpoons C \sim D + A \rightleftharpoons D \sim P + C \rightleftharpoons ATP + D \\
\quad \downarrow DNP \qquad\qquad\qquad \downarrow H_2As_3O_4 \\
\quad A + C.DNP \qquad\qquad C + D \\
\quad \downarrow H_2O \\
\quad DNP + C
\end{array}
$$

The respiratory inhibitors antimycin A and amytal are believed by Hulsmann to act in the same way as dinitrophenol,[1] with the difference that the combinations formed are stable and thus prevent respiration as well as phosphorylation. Antimycin A has been shown by Potter & Reif (1) and by Keilin & Hartree (3) to exert its inhibitory influence on respiration by blocking hydrogen transfer between cytochrome b and cytochrome c_1 (phosphorylation site 2). Lehninger (3) in agreement with this found no interference by antimycin with phosphorylation coupled to cytochrome c oxidation. Amytal was shown by Chance & Hollunger (1) to act at the first phosphorylation site in the chain, flavoprotein remaining oxidised, while NAD became reduced.

THE CHEMIOSMOTIC THEORY. This hypothesis was put forward by Mitchell (1) in 1961. It envisages respiration and ATP synthesis as coupled by means of a respiration-dependent proton translocation through a coupling membrane in the mitochondrion. Basic features are: (1) The synchronous electron and proton translocation system which creates the gradient of H^+ and OH^- across the membrane, H^+ accumulating outside

[1] Quoted by Slater (2) p. 365.

and OH⁻ inside. (2) The reversible ATPase situated in the membrane, anisotropic in the sense that it is accessible to OH⁻ but not H⁺ from the inside, and to H⁺ but not OH⁻ from the cytoplasm (fig. 86).

Since it is further assumed that the membrane does not allow passive diffusion of ions, the respiration-coupled proton pump must generate an electric-potential difference across the membrane; it is necessary also to

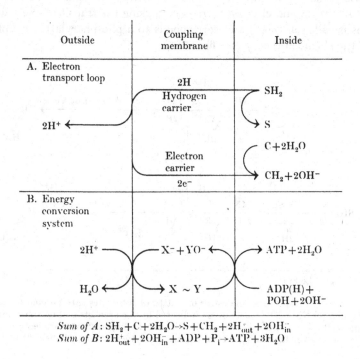

$$\text{Sum of } A: \text{SH}_2 + \text{C} + 2\text{H}_2\text{O} \rightarrow \text{S} + \text{CH}_2 + 2\text{H}^+_{\text{out}} + 2\text{OH}^-_{\text{in}}$$
$$\text{Sum of } B: 2\text{H}^+_{\text{out}} + 2\text{OH}^-_{\text{in}} + \text{ADP} + \text{P}_i \rightarrow \text{ATP} + 3\text{H}_2\text{O}$$

Fig. 86. Separation of H⁺ and OH⁻ by a loop and their recombination in ATP synthesis according to the 'chemiosmotic' hypothesis. S = substrate; C = hydrogen carrier; $X \sim Y$ = high-energy anhydride intermediate. (Pullman & Schatz (1).)

assume that the coupling membrane contains specific exchange–diffusion systems for bringing about the uptake and extrusion of certain ions, thus regulating the pH and osmotic differential, and permitting entry and exit of essential metabolites without collapse of membrane potential.

The ejection of protons as a result of respiratory chain activity gives rise to an electrochemical potential difference across the membrane. This derives from the pH difference and the membrane potential, and the relative importance of these two components varies with the conditions. The hydrolysis of ATP by the reversible ATPase system is supposed to be linked with the outward movement of protons. As a result of the electrochemical potential difference established by the respiration, protons are driven inwards through the ATPase system which is caused to operate in the reverse (dehydration) direction, so that ATP is synthesised from ADP and inorganic phosphate.

The translocation of protons is brought about by the oxidation in the respiratory chain of a reduced hydrogen carrier by an electron carrier. The sum of hydrogen and electron carriers catalysing the transit of one pair of protons is called a 'loop', and corresponds to a phosphorylation coupling site of the chemical theory (see fig. 87).

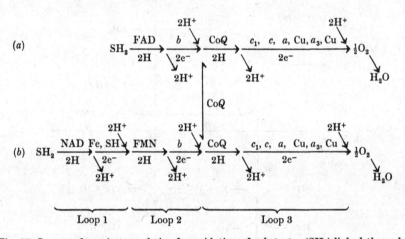

Fig. 87. Suggested respiratory chains for oxidation of substrates (SH_2) linked through: (a) FAD; and (b) NAD. The o/r loops are indicated by the brackets and by the points of entry and exit of H^+ at the junctions between the hydrogen (2H) and electron ($2e^-$) currencies of oxido-reduction. c_1, c, a and a_3 are the cytochromes. (Mitchell (2).)

Mitchell (1) and Mitchell & Moyle (1) brought evidence for this hypothesis in their finding that either oxidation of substrate or ATP hydrolysis in a mitochondrial suspension led to ejection of protons from the mitochondria. For each loop traversed of the respiratory chain, two protons appeared in the suspension medium; with the ATPase activity also two protons were ejected for each ATP molecule hydrolysed, provided that corrections were made so that only the oligomycin-sensitive ATPase was considered.[1] In this work anaerobic weakly-buffered mitochondrial suspensions were used, those for the oxidation experiments containing a substrate. Small amounts

---

[1] Myers & Slater (1) had shown the existence of four ATPases in mitochondrial fragments. Only one of these was appreciably active in fresh mitochondria and Huijing & Slater found it was inhibited by very low oligomycin concentrations.

of oxygen or ATP were injected, and the pH changes in the medium were measured with a sensitive glass electrode. Reid, Moyle & Mitchell (1), using ADP and labelled inorganic P in the medium reported ATP synthesis when a transient pH fall of about 4.6 units was suddenly caused by HCl injection into the medium. This synthesis was inhibited by oligomycin and was therefore explained by the reversed action of the oligomycin-sensitive ATPase.

With regard to the postulated impermeability of the mitochondrial membrane to protons, this was shown by Mitchell (1); he also brought evidence that uncoupling agents (dinitrophenol, azide) catalyse the flow of protons through the membrane, thus abolishing the proton gradient. This was supported by the further work of Mitchell & Moyle (2, 3). They concluded that the permeability of the coupling membrane to hydrogen and hydroxyl ions is low enough to account for the observed tight coupling between respiration and phosphorylation in terms of the chemiosmotic hypothesis, and that uncoupling by such agents as dinitrophenol can be quantitatively accounted for by their observed activity as catalysts of proton translocation. This effect of uncouplers is particularly significant support for the hypothesis, since it has become evident that a number of chemically unrelated uncouplers have this effect on the membrane.[1]

Chappell & Crofts (1) in 1965 described experiments on ion transport in isolated mitochondria which they were able to explain on the assumption that the membrane had limited permeability to all ions, and that specific 'permeases' existed for each type (e.g. for univalent ions, divalent anions) transferred across the membrane. They also were prepared to postulate the proton pump, and to explain their results with uncoupling agents in terms of increased membrane permeability.[2]

Such results are to be expected on the chemiosmotic hypothesis, but this cannot be considered proven, any more than the chemical hypothesis is proven. Several criticisms and alternative explanations have been voiced. Thus Chance & Mela (1) (using a method in which bromthymol blue bound to the mitochondria was considered to act as a sensitive indicator of intramitochondrial pH changes) reported that proton ejection during oxidation of a substrate was very low if the cation concentration (particularly of Ca^{2+}) in the medium was low. If the Ca^{2+} concentration was raised above a limiting level, respiration-dependent proton ejection greatly increased. Thus these proton movements might be connected, not with ATP synthesis, but with an energy-linked Ca^{2+} transport. Similar observations were made by Chappell & Haarhoff (1) who, while pointing out that these considerations meant deprivation of support for the thesis of Mitchell, yet considered that this thesis might still be correct if, for example, ejected

---

[1] See also Mitchell (3).   [2] Also Chappell & Haarhoff (1).

protons remained unestimated on account of their retention close to the coupling membrane. Again, Christie & his collaborators[1] have studied the active transport of K^+ by mitochondria when an energy source is supplied. A 1:1 exchange of K^+ with H^+ was observed. Cockrell, Harris & Pressman (1) have suggested that the pH decrease observed in the experiments of Mitchell & Moyle (where K^+ was present in the medium) might arise from an energy-dependent H^+ for K^+ exchange.

Mitchell & Moyle (4) have discussed the question of low ratio of proton ejection to oxygen uptake sometimes found in the experiments already mentioned; they indicated the difficulties involved and the precautions necessary for obtaining accurate measurements of the proton translocation.

The objection has also been made by Slater (7) that the Mitchell loops (requiring transfer of 2 H atoms followed by the transfer of 2 electrons) do not fit well into the present picture of the respiratory chain. However, he points out that very little is known about possible changes in the number of dissociated groups in the protein of the oxidised and reduced forms; thus it is at present impossible to say that the oxido-reduction equilibria rule out the chemiosmotic theory. Mitchell (4) has discussed this question.

As we have seen, this hypothesis allows for passage also of other ions than H^+, on an exchange–diffusion basis; this means that the driving force does not depend only on the pH gradient, but on the sum of this gradient and the membrane potential. Slater (8) has recently raised thermodynamic difficulties with calculations showing the need for a pH value as high as 13.5 in the cristae space, if the system is to have the needed efficiency.[2] However, as Pullman & Schatz (1) have pointed out, this objection is not necessarily valid, since the thermodynamics characteristic of extremely small volumes and the energy-requirement for ATP formation in membranes are still matters of dispute.

THE HYPOTHESIS INVOLVING CONFORMATIONAL CHANGES. Boyer (2)[3] has recently suggested that, since intensive search has not revealed the presence of phosphorylated intermediates in oxidative phosphorylation, or indeed of any marked amount of non-phosphorylated precursors,[4] consideration should be given to the possibility that oxidations bring about a 'high-energy state' by conformational change in some protein component of the system. This high-energy state would be present in amount stoichiometric with the respiratory chain components, and ATP itself would be the first

[1] Christie, Ahmed, McLean & Judah (1).
[2] Also Cockrell, Harris & Pressman.
[3] Also Boyer, Bieber, R. A. Mitchell & Szabolcsi (1).
[4] P. 422 below.

phosphorylated product formed from inorganic P. This conception was represented thus:

$$AH_2 + B + Y \rightleftharpoons A + BH_2 + \sim Y$$
$$\sim Y + P_i + ADP \rightleftharpoons Y + ATP + HOH.$$

Green and his collaborators support such a theory.[1] Their close study in the electron microscope of conformational changes in the inner membrane of mitochondria led them to recognise three states of the membrane – non-energised, energised and energised-twisted. Conditions for electron transfer or for ATP hydrolysis generated the energised state, and this was prevented by reagents which inhibited electron transfer or ATP hydrolysis. The speed of conformational changes in the cristae was described as consistent with the idea that these changes were the primary event in energy transfer. It is interesting to note (as pointed out by Harris *et al.*) the analogy with myofibrils, where ATP hydrolysis is accompanied by conformational change.

COUPLING FACTORS

A number of workers have found with sub-mitochondrial particles capable of carrying out oxidation of added substrates without coupled phosphorylation that it has been possible to restore phosphorylating power by adding material from the supernatant fraction. For example Linnane, in Green's laboratory, in 1958 found a soluble, heat labile, non-dialysable factor active in this way, contained in the supernatant after preparation of submitochondrial particles from heart muscle. Similar results were obtained by Penefsky, Pullman, Datta & Racker (1), who obtained P/O ratios up to 0.8 with the recombined system. Beyer (1) in Green's laboratory, also working with heart-muscle preparations, found a coupling factor which restored phosphorylation with succinate or NADH as substrate, but not with reduced cytochrome *c*. This suggested coupling activity specifically at site II of the chain. Again Racker's group has prepared five factors; one of these (F4), described in 1965 by Zalkin & Racker (1), is very insoluble and similar in its properties to the structural protein of Criddle *et al.* (1)[2]. The structural protein as previously prepared showed no coupling factor activity, but when made without use of dodecyl sulphate, it behaved in a strikingly similar way to the factor, both restoring oxidative phosphorylation.

Pullman & Schatz (1) and Schatz (1) have discussed (with many references) the multiplicity of coupling factors reported and their intricate relationships. Racker and his collaborators have identified and partly purified three – F_1, F_2 and F_3. F_4 appears to contain F_2 and F_3 bound to structural

---

[1] Penniston, Harris, Asai & Green (1); Harris, Penniston, Asai & Green (1); Green, Asai, Harris & Penniston (1).
[2] Ch. 16.

protein. Soluble coupling factors presumably can be taken to represent isolated portions of the series of components needed (according to such schemes as that on p. 416) in energy transfer between the respiratory chain and ATP synthesis. As Schatz has remarked, there are also indications that F_1, for example, may exert at least part of its effect by inducing a change in the structure of the mitochondrial membrane.

DISCUSSION

The nature of the intermediates schematically designated $A \sim C$, $C \sim D$, $D \sim P$ is naturally a matter of great concern. Clark, Kirby & Todd (1) have discussed the theoretical possibilities of quinones in the chain as bearers of an energy-rich phosphate bond, and Ernster (1) of flavoproteins as the site of energy-rich bonds not involving phosphate. Moreover, efforts have been made actually to isolate such compounds from mitochondria. Interesting compounds have been obtained[1] but so far there is no convincing evidence that any of these is an actual intermediate. The interest in sulphur-containing energy-rich bonds, which we discussed at substrate level of phosphorylation, is echoed in the work of Fluharty & Sanadi (1) who found arsenite and α-(p-arseno-phenyl)-n-butyrate to uncouple oxidative phosphorylation.

Nevertheless, there seems to be good evidence for accumulation under certain conditions of a small amount of energy-rich material in the respiratory chain. This was first realised by Chance & Hollunger (1) in 1957. They found that if succinate was added to mitochondria in absence of ADP there was rapid and marked reduction of NAD, which was inhibited by dinitrophenol or amytal. They suggested that here was a reversal of electron flow in the respiratory chain, the energy of succinate oxidation being used for reduction of NAD instead of passing on through the cytochromes (where its passage was blocked by lack of P acceptor.) These observations were confirmed and extended in several laboratories, for example by Klingenberg and his colleagues,[2] by Ernster and his colleagues[3] and by Snoswell (1). Ernster and Snoswell both remarked in 1961 that oligomycin did not inhibit such endergonic reactions, and therefore it could be inferred that this energy-linked reversal of electron flow did not involve a phosphate compound. Lee, Azzone & Ernster in 1964 found that submitochondrial particles, prepared in various ways from mitochondria, showed respiration but no coupled phosphorylation. These particles still showed the energy-generating reaction needed for coenzyme reduction and oligomycin did not

[1] Slater (2); Griffiths (1) for refs.
[2] Klingenberg & Slenczka (1); Klingenberg, Slenczka & Ritt (1).
[3] Ernster (1); Lee, Azzone & Ernster (1).

inhibit, indeed it stimulated the reaction; here was further indication of the non-phosphorylated nature of the high-energy intermediates. Such findings could be explained on any of the theories we have considered.

The chemiosmotic and configurational hypotheses have the advantage that postulation of intermediate chemical compounds, whether phosphorylated or not, is unnecessary. The chemiosmotic theory provides a better explanation than does either of the others for the action of diverse chemical uncouplers, in their effect on membrane permeability. As we have seen, this theory is at present open to criticism as regards interpretation of the proton movements and on thermodynamic grounds. The configurational hypothesis is attractive, but suffers from the disadvantage of the difficulty of finding approaches for its rigid investigation.[1]

So far as the problems of this book are concerned, the aim of our enquiry into the complex happenings of oxidative phosphorylation has been to trace the paths whereby ATP is synthesised ready for its interaction with actomyosin. Now it has been found in the last few years that energy produced in oligomycin-resistant respiration (as we have seen, unassociated with ATP formation) can be used for a number of energy-requiring processes – Ca uptake by mitochondria;[2] cation transport in rat-liver slices;[3] phosphate transport in rat-kidney slices.[4] Is it then possible that contracting myofilaments might draw energy from the same source? This question was first raised by Slater & Hülsmann (1) in 1959, in the context of the failure of many experiments in the nineteen fifties designed to show ATP breakdown during contraction;[5] Harary & Slater (1), by means of experiments on isolated heart cells, have recently answered it in the negative. In 1960 Harary & Farley (1) used single heart cells from new-born rats, isolated by trypsin treatment and cultured attached to glass in a liquid medium. The rate of beat, normally about 150/minute, could be affected by a number of agents. With 5×10^{-5}M dinitrophenol it fell to less than 10/minute, being restored again by 5×10^{-6}M ATP. In 1965 Harary & Slater made a more extended study with the same type of tissue preparation. The results are shown in the table following:

	Effect on beat
0.5×10^{-3}M dinitrophenol	Immediate cessation
0.5×10^{-3} dinitrophenol $+ 5 \times 10^{-3}$M ATP	Restored
15 g/ml oligomycin	No effect
mM iodoacetate	No effect
mM iodoacetate + oligomycin, 15 g/ml	Complete cessation
0.5×10^{-3}M dinitrophenol + oligomycin 15 g/ml	Restored.

[1] For reviews discussing these theories see Schatz (1) and Pullman & Schatz (1).
[2] DeLuca & Engstrom (1).   [3] Van Rossum (1).
[4] Wu (1).   [5] Ch. 14.

The interpretation was as follows. Dinitrophenol prevents contraction by inhibition of ATP formation in the respiratory chain on the one hand and stimulation of its breakdown on the other. Presumably ATP formed in other ways, for example by glycolysis, would also fall a victim to this hydrolytic process. Oligomycin and iodoacetate have no effect separately, since either oxidative phosphorylation or glycolysis can provide enough ATP for contraction at the low temperature used (22°) for these experiments but with both together the possibility of ATP formation is abolished and contraction ceases. The effect of oligomycin in counteracting dinitrophenol inhibition of contraction was considered to be due to the fact that oligomycin not only uncouples respiration but also stimulates ATPase activity, this DNP-stimulated ATPase being inhibited by oligomycin.[1] ATP produced by glycolysis would thus be spared. Thus the necessity of a supply of ATP for the myofibrils was upheld by the evidence of this work that it is essential for the contraction of the beating heart.

[1] See Huÿing & Slater (1)

18

THE REGULATION OF CARBOHYDRATE
METABOLISM FOR ENERGY SUPPLY TO
THE MUSCLE MACHINE

Some aspects of the regulation of metabolism have already been discussed. Thus it became clear, e.g. from the work of Engelhardt and of Chance that the presence of a supply of inorganic phosphate and of phosphate acceptor was a pre-requisite for respiration.[1] Such necessity for the supply of reactant and removal of end product is readily understood; but as Lardy (1) has remarked, we know little about the properties of individual enzymes of the oxidative phosphorylation machinery, and more subtle types of regulation may well also be at work there. On the other hand with glycolysis, whether of glycogen or of glucose, we have the chance to study a variety of other controlling factors which operate in response to the call for change in energy provision – for example, rapid changes in enzyme concentration, allosteric effects and in one case (surprisingly enough) marked activation by product.

We have seen in chapter 6 how in 1935 Parnas & Baranowski (1) realised that the initial stage of glycolysis in muscle extract, the formation of hexose ester from glycogen, took place at the expense of inorganic phosphate; and how in the next year Cori & Cori showed that the first product was not glucose-6-phosphate but glucose-1-phosphate, which in dialysed muscle extract was readily converted into the former ester:

$$\text{Glycogen} + H_3PO_4 \rightleftharpoons \text{glucose-1-phosphate} \rightleftharpoons \text{glucose-6-phosphate.}$$

The enzyme forming glucose-1-phosphate became known as phosphorylase after a suggestion of Parnas (5), since it was pictured as breaking down glycogen step by step, by means of introduction of phosphate instead of water. From this opening we pass on to find in phosphorylase exceptionally interesting traits of enzymic mechanism, traits which enable it to play a key role in the regulation of glycogen breakdown upon stimulation, or in anaerobiosis or under the influence of certain hormones. The second key enzyme is phosphofructokinase (bringing about the phosphorylation of fructose-6-phosphate by means of ATP).[2]

[1] Ch. 16. [2] Ch. 4.

Glucose-1-phosphate $\xrightleftharpoons{\qquad\qquad}$ glucose-6-phosphate
$\qquad\qquad$ (phosphoglucomutase)
$\qquad\qquad\qquad\qquad\qquad\qquad$ $\downarrow\uparrow$ $\qquad$ (phosphoglucoisomerase)
$\qquad\qquad\qquad\qquad$ +ATP $\qquad$ $\downarrow$
Fructose-1,6-diphosphate $\longleftarrow$ $\qquad\qquad$ fructose-6-phosphate
$\qquad\qquad\qquad$ (phosphofructokinase)
$\quad$ +ADP

Where the utilisation of glucose (brought to the muscle in the bloodstream from the liver) is concerned, there is an important control of its entry through the cell membrane, as we shall see.

As for the resynthesis of glycogen from glucose in muscle, it is not mediated, as was for a long time supposed, by the reverse action of phosphorylase, although this reaction is reversible; it was not for some twenty years that the true synthetic path, via uridine diphosphate glucose, was elucidated.

THE STRUCTURE OF GLYCOGEN AND THE MECHANISM OF ITS BREAKDOWN BY PHOSPHORYLASE

C. F. Cori & G. T. Cori (1) had observed in their earliest work that the formation of hexosemonophosphate from glycogen in washed, minced frog muscle was enhanced by addition of adenylic compounds. G. T. Cori, Colowick & C. F. Cori (1) found with dialysed muscle extracts that the essential compound was adenylic acid, ATP and ADP[1] being inactive in absence of pyrophosphatase and myokinase respectively.

Investigation of the kinetics of the reaction at pH 7.0 showed that at the equilibrium point 23 % of the glycogen had been converted into glucose-1-phosphate.[2] An important condition for synthesis was the presence of a trace of glycogen: it was observed by G. T. Cori & C. F. Cori (3) that most tissue preparations, including those from muscle, showed a lag period, the greater the purer the enzyme, for glycogen synthesis from glucose-1-phosphate; with liver preparations, which always contained some glycogen, this lag was never seen. It could be overcome in preparations from other tissues by adding a few milligrams % of glycogen. C. F. Cori, G. T. Cori & Green (1) surmised in 1943 that glucose-1-phosphate must react with terminal units in the glycogen, lengthening the chains by addition of maltosidic (i.e. 1, 4) chain units. It later became clear that two enzymes (phosphorylase and a 'branching' enzyme or transglucosylase)[3] took part in synthesis; for breakdown phosphorylase, a transglucosylase and a 'debranching' enzyme were needed. Thus G. T. Cori & C. F. Cori (4) showed that, by the action of phosphorylase alone on glucose-1-phosphate, there is

[1] C. F. Cori, G. T. Cori & Green (1).
[2] G. T. Cori & C. F. Cori (2); Sutherland, Colowick & C. F. Cori (1).
[3] See pp. 428 and 443 below for the branching enzyme.

produced not glycogen but a substance 'limit dextrin' resembling the amylose fraction of starch. The action together of phosphorylase plus a supplementary enzyme, obtained from heart or liver, resulted in the formation of a typical glycogen.

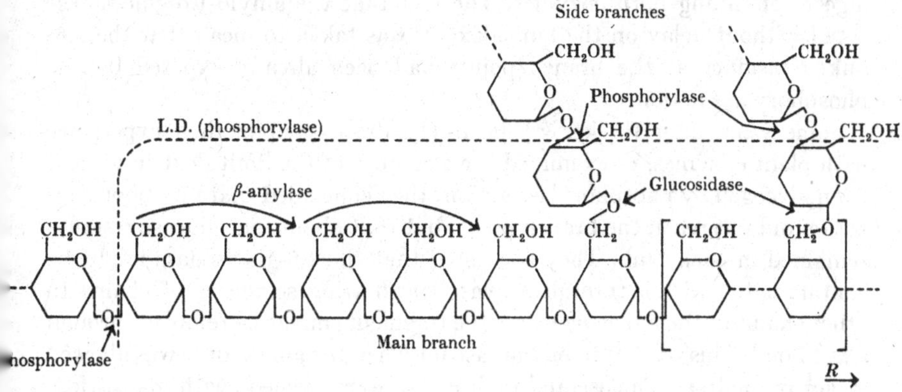

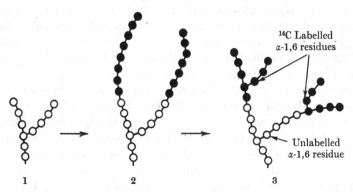

Fig. 88. Structural model of a portion of a branched polysaccharide showing the sites of enzymatic action. L.D. corresponds to the limit dextrin formed by exhaustive action of phosphorylase. *R* refers to the reducing end. (G. T. Cori & Larner (1).)

Fig. 89. Branch point synthesis by radioactive labelling technique. (1) Incompletely degraded glycogen segment (non-reducing end), (2) Glycogen segment with labelled outer chains, (3) Glycogen segment after action of branching enzyme. ○, unlabelled glucose residue; ●, ^{14}C-labelled glucose residue. (Larner (1).) For discussion see p. 443.

By 1950 much more was known of the constitution of glycogen. Its highly ramified structure,[1] consisting of straight-chain arrays of glucose units in α-1,4-linkage, with α-1,6-linkages at the branch points, is indicated in figs 88 and 89. G. T. Cori & Larner (1) showed that in the presence of phosphorylase plus inorganic phosphate glucose units (as glucose-1-

[1] See for references G. T. Cori & Larner (1); Illingworth, Larner & G. T. Cori (1).

phosphate) were split off from the outer chains, and on addition of the specific glucosidase (also found in the muscle) the 1,6-linkages at the branch points were broken. The degree of branching was gauged by the relative amounts of glucose and glucose-1-phosphate appearing in the digest containing both enzymes. The fact that the amylo-1,6-glucosidase acted without delay on the limit dextrin was taken to mean that the 1,6-linked residues at the branch points had been already exposed by the phosphorylase action.

Some years later Walker & Whelan (1), drawing upon their experience with plant enzymes, re-examined the structure of the limit dextrin formed from glycogen by phosphorylase action; they concluded that phosphorylase action did not reach the branch point, but that about four glucose residues remained in each limb. They suggested that the 1,6-glucosidase might be contaminated with a transglucoslyase transferring segments of chains to other chains in the molecule where the fragments might be rejoined through a 1–4 bond. This proved to be the case for when the glucosidase was allowed to act on maltose, glucose and maltotriose were formed; with maltotriose as substrate, glucose, maltose, maltotetraose and higher oligosaccharides appeared. This suggestion was further borne out by the finding of Brown & Illingworth (1) that their most highly purified 1,6-glucosidase contained an oligo-1,4 ⟶ 1,4-glucantransferase. This had no effect on maltose or maltotriose but was able to convey a chain of three or two maltosidically linked glucose units from donor to acceptor; the acceptor might be glucose, an oligosaccharide or another part of the same molecule.

Further aspects of glycogen breakdown, which could only be explained by transglucosylase activity in their best 1,6-glucosidase preparations were described by Abdullah & Whelan (1). With action on glycogen of these glucosidase preparations (which were free from amylase) they observed a slow but marked increase in intensity of iodine staining – to be interpreted as due to lengthening of some of the chains in the polymer. At the same time the susceptibility of the glycogen to β-amylase attack was increased, indicating removal of 1,6-linkages by the glucosidase. Brown, Illingworth & Cori (1) also encountered this increase in iodine colour given by glycogen on treatment with highly purified 1,6-glucosidase, and they explained it in the same way. Their experiments showed the greater availability of the structure to phosphorylase when this enzyme was added after the incubation with the glucosidase.

It seems then that, in the breakdown of glycogen, the action of this transglucosylase must intervene between the action of the phosphorylase and the 1,6-glucosidase, removing three glucose units from one side of the branch point to the other. Thus one side becomes seven units long, while on the other a single 1,6-linked glucosyl residue is prepared for the glucosidase action. These questions are discussed in detail in a Ciba Foundation Symposium of 1964.[1] The great difficulty met with in attempts to separate the transferase and the glucosidase was emphasised. Cori[2] suggested the possibility that a two-headed enzyme was involved.

ACTIVATION AND INACTIVATION OF PHOSPHORYLASE

We come now to the complicated processes by which the activity of phosphorylase is governed.

ADENYLIC ACID. It was first observed in 1943 by G. T. Cori & Green (1) that this enzyme can exist in two forms – one, form a, having about 70 % of its maximum activity without adenylic acid addition; the other, form b, entirely inactive in the absence of adenylic acid. An enzyme (known at that time as the 'prosthetic-group removing' or PR enzyme) which could be separated from phosphorylase, brought about the change of the a form into the b form; it was supposed that this change was due to removal of adenylic acid from the a form. However, G. T. Cori & C. F. Cori (5) could find no convincing evidence for this removal. Keller & G. T. Cori (1) in 1953 found that under the influence of the PR enzyme the molecular weight of phosphorylase was halved, falling from about 495000 for the a form to about 242000 for the b form.

Madsen & C. F. Cori (1) in 1957 described the binding of 4 moles of adenylic acid per mole of protein by phosphorylase a, while phosphorylase b bound only 2; the affinity for the nucleotide was low and it was easily lost from the protein. They had previously found (2) that the energy of activation of phosphorylase a was lowered by presence of adenylic acid and they speculated on the possibility of its attachment in the neighbourhood of the active centre. Cohn & G. T. Cori (1) in 1948 had looked for interchange between labelled inorganic phosphate and adenylic acid phosphate but none was found.

PYRIDOXAL-5-PHOSPHATE. Another facet of the complex make-up of phosphorylase was first described by C. F. Cori and his collaborators[3] in 1957. They found with highly purified preparations (dialysed, Noritetreated and recrystallised) that 8 atoms of phosphorus were bound, per mole of phosphorylase a, and that of these 4 could be extracted by precipitation of the protein with trichloracetic acid. The phosphorus compound in the extract was identified by its spectrum, paper electrophoresis and

---

[1] Abdullah, Taylor & Whelan (1); Brown & Illingworth (2); Hers, Verhue & Mathieu (1).
[2] See discussion of the paper by Hers et al. p. 169.
[3] Baranowski, Illingworth, Brown & C. F. Cori (1).

enzymic tests as pyridoxal-5-phosphate. Further work by Cori & Illingworth (1) showed that phosphorylase b contained 2 moles of this compound per mole of enzyme. In both cases the pyridoxal-phosphate could be removed by precipitation of the enzyme on acid treatment (to pH 3.6) in presence of ammonium sulphate. The protein, after washing with alkaline ammonium sulphate was inactive, but activity was recovered on short incubation with pyridoxal-5-phosphate (and, in the case of phosphorylase b, of adenylic acid also). As with the adenylic acid associated with the enzyme, no evidence could be found, on use of inorganic ^{32}P or ^{32}P-labelled glucose-1-phosphate, for any direct participation of the phosphate group of the pyridoxal-phosphate in the enzymic mechanism.[1]

PHOSPHORYLASE b KINASE AND THE FORMATION OF PHOSPHORYLSERINE. In 1955 Fischer & E. G. Krebs (1)[2] showed convincingly that mainly phosphorylase b is contained in resting muscle, and that conversion to the a form goes on readily in muscle extracts, provided ATP, a protein fraction of the extract and divalent metal ions are present. If contraction was minimised during preparation of the extract, and the extraction fluid contained EDTA, practically only the b form was found. It had previously been thought that the evidence pointed to the presence mainly of the a form in resting muscle and of the b form in stimulated muscle.[3]

Krebs, Kent & Fischer (1) in 1958, using crystalline phosphorylase b and ^{32}P-labelled ATP, showed that four phosphate groups were introduced per molecule of phosphorylase a formed in the conversion process and four molecules of ADP appeared. They gave the name phosphorylase b kinase to the enzyme. About this time Wosilait (1) was studying the acid-soluble material released by trypsin treatment of ^{32}-labelled phosphorylase obtained by this kinase action. All the soluble ^{32}P was contained in phosphorylserine. The same observation was made independently by Fischer and his collaborators,[4] who also showed the amino-acid sequence in which the phosphorylserine was contained: Lys., GluNH$_2$.Ileu.SerP.Val.Arg. Krebs, Kent & Fischer could find no exchange of phosphorylserine phosphate with inorganic phosphate during the action of phosphorylase on glycogen.

Just about the same time as the discovery of the phosphorylation needed in its formation from the b form, Sutherland & Wosilait (1) found release of inorganic phosphate from liver phosphorylase a during its inactivation by the PR enzyme, in this case also prepared from liver. They reported a close parallelism between the phosphate release and the degree of inactivation.

---

[1] Illingworth, Jansz, Brown &. C. F. Cori (1).
[2] Also Krebs & Fischer (1).   [3] G. T. Cori & Green (1).
[4] Fischer, Graves, Snyder, Crittenden & Krebs (1).

ACTIVATION OF PHOSPHORYLASE *b* KINASE; THE ROLE OF CYCLIC
AMP. Evidence was next brought forward that the phosphorylase *b* kinase
could itself be activated in ways which seemed relevant to conditions in the
muscle. Rall, Sutherland & Berthet (1) in 1957 observed that cell-free
homogenates from liver could greatly increase the formation of phosphorylase *a* from phosphorylase *b*, if the hormone adrenaline were present.
An active, heat-stable and dialysable factor was formed in the particulate
fraction, and this stimulated phosphorylase *a* formation in the supernatant
fraction. Rall & Sutherland (1, 2) showed that the factor concerned was
3′,5′-adenosine monophosphate, and that it was formed from ATP (with
release of inorganic pyrophosphate) in the presence of Mg ions; particulate
preparations from heart and skeletal muscle also brought about formation
of this cyclic adenylic acid, especially in the presence of adrenaline. They
named the enzyme concerned adenyl cyclase.

Posner, Stern & Krebs (1) linked these results with happenings *in vivo* by
showing that adrenaline injection into intact animals led in the muscles to
increased cyclic adenylic acid levels, activation of phosphorylase *b* kinase
and phosphorylase *a* formation. It is significant also for regulation that
Rall & Sutherland and their collaborators have detected and purified a
phosphodiesterase which catalyses hydrolysis of the cyclic adenylic acid
at the 3′ position, yielding 5′-adenylic acid.[1] Butcher & Sutherland (1)
showed that all tissues producing the cyclic adenylic acid also contain this
diesterase.

Still more complicated relationships were disclosed in 1959 by E. G.
Krebs, Graves & Fischer (1) who extracted phosphorylase *b* kinase from
muscle in a form almost inactive at pH 7.0. It could be activated by preincubation for a short time with either 10^{-3}M Ca ions or with ATP and Mg
ions. Later work[2] showed that the latter effect seemed to involve phosphorylation of the kinase; it was enhanced by adding a trace of cyclic
adenylic acid, which had no action alone. Then in 1967 Ozawa and his
collaborators[3] found that this activation of phosphorylase *b* kinase by
means of cyclic AMP, ATP and Mg ions needed also the presence of Ca ions
in very low concentration – about 10^{-6}M for maximum effect. This calcium
activation was to some extent reversible. The stimulation of the kinase by
preincubation with 10^{-3}M Ca ions is again a complex matter. Meyer, Fischer
& Krebs (1) showed in 1964 the need for a separate protein factor ('kinase-
activating factor') which was separated from the kinase and purified; there
was no evidence that it was an enzyme. The kinase was inhibited by certain
metal-chelating agents, notably EDTA, and the inhibition was relieved by

---

[1] Butcher & Sutherland (1) for references.
[2] Krebs, Love, Bratvold, Trayser, Meyer & Fischer (1).
[3] Ozawa, Hosoi & Ebashi (1); Ozawa & Ebashi (1).

adding Ca^{2+}; on the other hand, once activation by calcium had been accomplished chelating agents could not reverse it.

In 1962 Sutherland and his collaborators[1] found that the relative potencies of adrenaline and other catecholamines in stimulating formation of cyclic AMP in dog heart were very similar to their relative effects (observed by Mayer & Moran (1)) in increasing contractile force in the heart and its phosphorylase a content. With further work however it became clear that the situation could not be explained by the train of events: cyclic AMP formation; increased phosphorylase a formation; increased glycogen breakdown; increased muscular activity. Drummond, Valaderes & Duncan (1) in 1964 noticed, both with perfused rat hearts and with dog hearts *in situ*, that small doses of adrenaline could cause increase in contractile force without any rise in concentration of phosphorylase a. The activation of the phosphorylase came only later. The separation in time (by some 10–15 sec) of the glycogenolytic and inotropic responses was also found by Williamson & Jamieson (1). Sutherland and his collaborators[2] in 1965 agreed that the adrenaline (probably via cyclic AMP) was exerting its effect on some unknown system involved in the heightened performance, a system operating prior to the increase in energy supply from glycogen breakdown. Indeed, as they suggest, it is possible that the increased glycogen breakdown is an effect, not the cause, of the increased work done.[3] More recently Drummond, Duncan & Hertzman (1) have found activation of phosphorylase b kinase in perfused rat hearts within 1 sec after injection of the maximum physiological dose; the rise in cyclic AMP levels followed an almost identical course, while the inotropic peak was at about 10 sec. In commenting on the earlier work showing glycogen breakdown lagging behind the contractile response, they suggest that there may be structural separation of the phosphorylase b kinase and the phosphorylase. For example, the phosphorylase b kinase might be situated in the cell membrane, and so be readily accessible to the adenyl cyclase, believed to be membrane-located;[4] while phosphorylase a may be free in the cytoplasm.

Sutherland & Robison (1) found no effect of cyclic AMP on the rate of actomyosin superprecipitation or on the ATPase activity of various heart-muscle fractions. Williamson (1) has tentatively suggested that cyclic AMP might increase the amount of ionised calcium in the heart, thus of course affecting actomyosin ATPase activity. This increase would presumably depend on a change in permeability of the membrane of the sarcoplasmic reticulum or of the cell.[5]

---

[1] Murad, Chi, Rall & Sutherland (1).
[2] Robison, Butcher, Øye, Morgan & Sutherland (1).
[3] Also Namm & Mayer (1).   [4] Davoren & Sutherland (1).
[5] Ch. 19. Also compare what is known of the effects of adrenaline in smooth mammalian muscle (ch. 23).

ALLOSTERIC EFFECTS IN PHOSPHORYLASE ACTIVATION. Parmeggiani & Morgan (1) were the first to emphasise, in 1962, that changes in concentration of adenine nucleotides might be important in controlling phosphorylase activity in other ways as well as effecting the b to a transformation. This was suggested by the observation that in anoxia there was more rapid and extensive glycogen breakdown than seemed explicable by the phosphorylase a formation observed. Cornblath and his collaborators[1] also found evidence for other factors involved in glycogenolysis besides the b to a transformation. Comparing the effects of anoxia and glucagon, they showed that with anoxia there is only some 20 % conversion to the a form, with the hormone 50 % conversion; yet glycogen breakdown was more rapid in the former case, and preliminary experiments indicated that here increased tissue levels of AMP and inorganic P were responsible. Then, after estimation of the levels of ATP, AMP, ADP and inorganic phosphate present in the aerobic and anaerobic states of perfused rat hearts, Morgan & Parmeggiani (1, 2) tested these levels in vitro on solutions of crystalline rabbit muscle phosphorylase a and b. With tests corresponding to aerobic conditions, ATP strongly inhibited phosphorylase b with little effect on phosphorylase a; AMP stimulated both forms while ADP had little effect on either in presence of aerobic AMP concentrations. With levels corresponding to anaerobic conditions, the effects of the changes in ATP, AMP and inorganic P concentration were additive with both forms of the enzyme. They concluded that in the resting muscle, where phosphorylase is mainly in the b form, its activity is kept at a very low level by ATP inhibition and very low concentration of the substrate inorganic P. In anaerobiosis, the ATP inhibition is counteracted by marked rise in AMP and inorganic P, these changes re-inforcing the effect of some change (about 14 %) of the phosphorylase b into a. AMP increased the activity of the b form by increasing the apparent affinity of the enzyme for its substrates; they also showed (3) that ATP inhibited phosphorylase b not directly, but by raising the K_m for AMP.

Helmreich & Cori (1), concerned with the puzzle of the mechanism of AMP activation of phosphorylase b, also showed at this time that in the presence of this nucleotide the affinity of the muscle enzyme for both its substrates, glycogen and inorganic P, was increased. They were prepared to follow Monod, Changeux & Jacob in seeing in phosphorylase b a typical allosteric enzyme, the activity of which was dependent on configurational changes brought about by the effector AMP. As the result of his independent experiments, Madsen (1) reached a similar point of view.

In table 15 an attempt has been made to summarise these intricate

---

[1] Cornblath, Randle, Parmeggiani & Morgan (1).

relationships.[1] As we shall see presently, they play their part in a delicate and complex machinery for the control of the glycogen breakdown in skeletal as well as in heart muscle.

TABLE 15

Phosphorylase *b*	Phosphorylase *a*
Inactive without AMP	70 % active without AMP
Inhibited by ATP	
M.W. 242000	M.W. 495000
Loosely binds 2 moles AMP	Loosely binds 4 moles AMP
Contains 2 moles pyridoxalphosphate	Contains 4 moles pyridoxalphosphate

$$\text{Phosphorylase } b \xrightarrow[\text{ATP, Mg}]{\text{phosphorylase } b \text{ kinase}} \text{Phosphorylase } a + \text{ADP}$$
(4 moles serine $\longrightarrow$ phosphorylserine)

$$\text{Phosphorylase } b + 4H_3PO_4 \xleftarrow[\text{(PR enzyme)}]{\text{phosphatase}} \text{Phosphorylase } a$$

$$\text{Inactive phosphorylase } b \text{ kinase} \xrightarrow[\substack{\text{or preincubation with} \\ \text{Ca+protein factor}}]{\substack{\text{Preincubation with} \\ \text{ATP, Mg and cyclic AMP}}} \text{Active phosphorylase } b \text{ kinase}$$

$$\text{ATP} \xrightarrow[\substack{\text{Mg} \\ \text{stim. by adrenaline}}]{\text{adenyl cyclase}} \text{Cyclic AMP} + \text{pyrophosphate}$$

$$\text{Cyclic 3,5-AMP} \xrightarrow{\text{Cyclic phosphodiesterase}} \text{5-AMP}$$

ACTIVATION AND INHIBITION OF PHOSPHOFRUCTOKINASE

In 1941 C. F. Cori (1) remarked that both glucose- and fructose-6-phosphates are normally found in muscle, and that the amount of these esters can increase markedly under certain conditions, without any accompanying increase in lactic acid. He took this as an indication that phosphofructokinase can act as a limiting factor in glycolysis in intact muscle. Much later work on means of inhibiting and activating this enzyme has borne out this suggestion. Lardy & Parks (1) in 1956 showed that beef-liver phosphofructokinase is strongly inhibited by ATP if the latter is present in excess of the Mg ions. An important step was taken when Passonneau & Lowry (1) with muscle phosphofructokinase found that this inhibition could be overcome by addition of cyclic AMP, AMP or inorganic P.[2] In other experiments, to simulate anoxic conditions, inorganic P concentration was raised and ATP concentration lowered, AMP and ADP being added at low levels. The result was a thirtyfold increase in the phosphofructokinase activity.

---

[1] Also table 17 below.   [2] Also Mansour, Clague & Beernink (1); Mansour (1).

The reaction product, fructosediphosphate, was also a potent activator. They suggested that the enzyme has two sites for ATP, one an active site, the second inhibitory;[1] de-inhibitors or stimulators such as AMP and fructosediphosphate might compete for this second site and occupy it without being inhibitory. There is evidence in this work, and in that of Pogson & Randle (1) that increasing concentrations of ATP increase the K_m for fructose-6-phosphate.

About this time Randle and his collaborators were engaged on a study of glucose uptake by rat heart and diaphragm under a variety of conditions. They found glycolysis to be accelerated by anoxia and in particular observed decrease in hexosemonophosphate content and increase in fructosediphosphate content in hearts under anaerobic conditions.[2] It seemed that the rate of conversion of fructose-6-phosphate into the diphosphate, on change from aerobic to anaerobic conditions, increased to a much greater degree than either formation of the monophosphates or conversion of the diphosphate to lactic acid. They concluded that the phosphofructokinase step was rate-limiting under aerobic conditions, and later connected these findings with the results we have just discussed on the effects of the adenine nucleotides and inorganic phosphate on this step.[3] In further work they found that the glycolysis rate and phosphofructokinase activity were diminished in hearts from diabetic or starved rats, although here no consistent decrease in concentration of AMP or inorganic P could be found.[4] Characteristic of these conditions was the release of fatty acids for oxidation, and similar depressant effects were found on perfusion of the hearts with fatty acids, ketone bodies or pyruvate. A search for some intermediate in fatty acid oxidation which might inhibit phosphofructokinase led to the discovery that citrate has this property. It was shown that citrate concentration in the tissue was indeed raised to an inhibitory level under the various conditions where phosphofructokinase inhibition could be recognised. No inhibition was found on testing several other members of the citric acid cycle. Simultaneously and independently this inhibitory effect of citrate on phosphofructokinase was observed by Passonneau & Lowry (2) using brain and liver, and by Parmeggiani & Bowman (1) using hearts in alloxan diabetes or perfused with fatty acids. Passonneau & Lowry (3) summed up the situation in the following comment:

From the facts available, phosphofructokinase is the valve which controls the flow from fructose-6-phosphate to pyruvate. The slow flow into the citrate cycle

---

[1] Also Lowry & Passoneau (1).
[2] Newsholme & Randle (1). For later work see Garland & Randle (2) with many references.   [3] Garland, Randle & Newsholme (1).
[4] Newsholme, Randle & Manchester (1); Garland & Randle (1); Garland, Randle & Newsholme (1).

under oxidative conditions can be regulated by the balancing effects of fructose-6-phosphate and citrate concentration; in anoxia or other emergency, increase in P_i, AMP or NH_4^+ can turn the valve on full to maintain or restore ATP levels.

Mansour and his collaborators[1] in their work on purification of sheep-heart phosphofructokinase encountered very interesting behaviour indicating the great lability of the enzyme. With hearts from freshly killed guinea-pigs the enzyme activity was found almost entirely in the supernatant after centrifuging at 24000g; but with guinea-pig hearts kept for a short time, or with sheep hearts from the slaughter-house, the enzyme was spun down; it seemed that it became insoluble, or perhaps adhered to some heavy cell component. In this state it had comparatively little activity, but active soluble enzyme could be obtained from it by incubation with 0.1 M $MgSO_4$, or (rather better) with 0.01 M ATP+0.01 M $MgSO_4$.[2] The stabilising effect here was found to be due to the sulphate and not to the magnesium ion. Other substances which could stabilise and activate the enzyme included ADP, AMP, cyclic AMP and inorganic phosphate. Mansour, Wakid & Sprouse (2) had evidence that the enzyme can exist in a number of different polymeric forms, depending on its concentration and on the solvent. But whether the degree of polymerisation affects the enzymic activity and whether changes in molecular weight can take place *in vivo* is not yet known.

The recent detailed study by Kemp & Krebs (1) of the nature of the binding of a number of metabolites to phosphofructokinase (prepared from skeletal muscle) gave results consistent with those of the kinetic studies we have already discussed. They observed, amongst many other relevant facts, that increased pH, AMP, ammonium ions and phosphate ions can all increase the affinity of the enzyme for fructose-6-phosphate.

The effect of gross fall in pH in limiting glycolysis has long been known. Ui (1) has recently shown that very small changes even within the alkaline range (e.g. from 7.6 to 7.3) can cause decreased phosphofructokinase activity in presence of 3 mM or higher ATP. The effect of ATP concentration, between pH values of 7.0 and 9.0, was very marked and it was suggested that the inhibitory ATP site was pH dependent. Since any glucose-6-phosphate accumulation resulting from phosphofructokinase inhibition would be inhibitory to hexokinase,[3] any slight pH rise relieving this phosphofructokinase inhibition would lead to a disproportionately great increase in glycolysis of glucose.

Some of these effects have been put together in summary form in table 16.

---

[1] Mansour, Wakid & Sprouse (1); Wakid & Mansour (1); Mansour, Wakid & Sprouse (2).
[2] Garland, Newsholme & Randle (1) were the first to observe the activating effect of sulphate ions, which in their experiments acted by preventing citrate inhibition.
[3] P. 438 below.

TABLE 16

Inhibitors	Activators or de-inhibitors
ATP. This inhibition is pH-dependent, being greater at lower pH values. Citrate.	ADP, AMP, cyclic AMP, inorg. P overcome ATP inhibition. Increased pH, NH$_4$ ions, and phosphate ions increase the affinity of the enzyme for fructose-6-phosphate. Fructose diphosphate. SO$_4$ ions.

CONTROL OF GLUCOSE METABOLISM IN MUSCLE

Evidence has accumulated from the results of many workers that the entry of glucose through the cell membrane from the blood stream is the rate-limiting step in the utilisation of free glucose by muscle, and further that insulin increases the rate of penetration.[1] Although insulin was discovered in the early nineteen twenties this understanding of the mechanism of its facilitation of carbohydrate metabolism was not hinted at for nearly twenty years. Direct or indirect stimulation of the enzymic formation of glucose-6-phosphate was a favoured explanation for the observed effect that insulin injection brings about reduction in blood-sugar concentration. In 1939 Lundsgaard (11) published a paper in an Uppsala medical journal in which he showed that the intracellular glucose concentration of muscle is extremely low. He argued that it was therefore clear that the explanation of insulin action, in causing some three times the rate of entry of glucose into the tissue at constant blood-sugar concentration, could not be due to increased concentration gradient between the blood and tissue. He suggested that the increased influx of glucose depended on the influence of insulin on the active process which must be assumed to be a factor in its transfer. This paper remained entirely unknown, and the idea had to be rediscovered by Levine and her co-workers[2] in the late nineteen forties. Then it was clearly demonstrated with eviscerated animals that insulin increased the permeability of muscle tissue to the non-metabolised sugar D-galactose. In 1956 Park and his collaborators[3] summarised their results, which confirmed this effect of insulin with non-utilisable sugars, and also showed that the hormone activated membrane transport of glucose itself. The work soon afterwards of Randle & Smith (1, 2) and of Morgan, Randle & Regen (1) is

---

[1] For an account of the growth of knowledge of the effects of insulin in the muscle since the pioneer experiments of Burn & Dale (1) in 1924, see Krahl (1).
[2] Levine & Goldstein (1) for an account with full references.
[3] Park, Post, Kalman, Wright, Johnson & Morgan (1).

relevant here. With isolated rat diaphragm and perfused heart, anoxia or the presence of uncouplers of oxidative phosphorylation activated membrane transport of glucose and other sugars; these effects were not due to membrane damage. Thus the entry of glucose into the muscle cell seems to be restrained by availability of high-energy phosphate, and insulin may play its part by in some way overcoming this restraint.[1] Morgan and his collaborators in 1961 found that inward transport of glucose, as a function of external glucose concentration, conformed to Michaelis–Menten kinetics and they regarded the transport as depending on glucose combination at specific sites in the membrane, thus constituting the first reaction in glucose metabolism.[2] When insulin was added, phosphorylation became the limiting step at all external concentrations of glucose above 25 mg/100 ml. Insulin could cause as much as a thirtyfold rise in transport capacity.

Once glucose has entered the cell, its phosphorylation depends on the enzyme hexokinase. Randle & Smith (1) showed with diaphragm that xylose from the medium would accumulate in the tissue under conditions where glucose accumulation was not seen. This was explained as due to glucose phosphorylation being more rapid than its entry, while xylose is not metabolised by muscle. Inhibition of hexokinase by glucose-6-phosphate, the product of the reaction, had been observed by Crane & Sols (1) in 1955, using skeletal muscle extracts. Kipnis, Helmreich & Cori (1) observed that injection of adrenaline into rats led to intracellular accumulation of glucose and rise in glucose-6-phosphate concentration in both diaphragm and gastracnemius. They explained the glucose accumulation as probably due to this inhibitory effect of glucose-6-phosphate, the rise in concentration of this ester being a consequence of increased glycogen breakdown under the influence of adrenaline. Later Karpatkin, Helmreich & Cori (2) extended their experiments on frog sartorius to compare the effects of anaerobiosis and stimulation, with and without insulin on the penetration and phosphorylation of tritiated deoxyglucose. Both anaerobiosis and stimulation could increase the rate of phosphorylation independently of effects on penetration. A large increase in hexokinase activity thus seemed to be involved.

All the factors we have already discussed in connection with the activity of phosphofructokinase will of course also affect the further breakdown of glucose. Furthermore, a change in phosphofructokinase activity by leading to a change in concentration of its substrate glucose-6-phosphate, will affect the activity of the hexokinase.

[1] Randle (1) suggested that the insulin role may be activation of a phosphatase. Friedman & Larner (2) later made the interesting point that this would be a similar mechanism for the hormone action to that they suggest in its activation of glycogen synthetase. See p. 444 below.

[2] Morgan, Henderson, Regen & Park (1); Post, Morgan & Park (1).

REGULATION OF CARBOHYDRATE BREAKDOWN DURING
ANAEROBIC CONTRACTION AND RETURN TO REST

During the last few years Cori and his collaborators have made a close study
of changes taking place in frog sartorius muscle under strictly anaerobic
conditions, relating them to the regulation of glycogen breakdown on contraction and return to rest. Danforth, Helmreich & Cori (1) showed in 1962
that not more than 5 % of phosphorylase is present in resting frog sartorius
in the *a* form; on stimulation to sustained isometric contraction there was a
very rapid increase to about 50 % with a 1 sec stimulation, and to more
than 80 % with a 2.5 sec stimulation. About 60 sec after cessation of
stimulation, the concentration had returned to the resting value. These
changes seemed to depend entirely on variations in activity of the phosphorylase *b* kinase, the phosphorylase *a* phosphatase remaining constant in
activity. Danforth & Helmreich (1) used frog sartorius contracting isotonically. After a lag period, there was an increase in phosphorylase *a* up to
a steady state. The phosphorylase *a* content rose with the frequency of
stimulation up to about 8 shocks/sec, changes in the work done at a given
rate of stimulation having no effect on the lag period or on the steady state
level at each frequency. They considered the possibility that some event in
the excitation-contraction sequence might be important in triggering off
the increase in phosphorylase *a*, and suggested that the stimulatory effect
of Ca ions on phosphorylase *b* kinase (observed *in vitro*) might operate at
this point *in vivo*.

An illuminating comparison is made in these two sets of experiments
between the effect of adrenaline treatment and of stimulation. The increase
in phosphorylase activity on stimulation was both much greater and much
more rapid as the result of stimulation than after treatment of the muscle
with adrenaline. Moreover the adrenaline effect was inhibited by dichloro-
isoproterenol, which did not inhibit the increase in phosphorylase *a* on
stimulation. Within 60 sec of adrenaline treatment increase in the muscle
of the phosphorylase *b* activator, cyclic AMP, could be detected; it was not
known at that time whether stimulation caused an increase in this substance, but in any case it seems that two different mechanisms must be
involved in phosphorylase activation with stimulation on the one hand and
adrenaline treatment on the other. Posner, Stern & Krebs (2) later testing
this point found no increase in cyclic AMP on electrical stimulation of frog
or rat muscle. Drummond, Harwood & Powell (1) also have recently shown
that, upon electrical stimulation, phosphorylase *b* $\longrightarrow$ *a* conversion could
take place in the absence of any detectable increase in cyclic AMP or
activation of phosphorylase *b* kinase.

The importance of the phosphorylase *b* to *a* transformation was underlined by the experiments of Danforth & Lyon (1) on contraction in two strains of mice, one normal, the other unable to form *a* from *b*. Glycogen breakdown was considerably greater on contraction, and the increased rate of breakdown came on more quickly, in the normal strain, though rapid breakdown could also occur in the abnormal one.

Karpatkin, Helmreich & Cori (1) examined the effect of stimulation under anaerobic conditions on the lactic acid and phosphate ester content of frog sartorius.[1] The amount of lactic acid formed in a 30 min period was proportional to the rate of stimulation, up to 48 shocks/min, with an increase in rate more than 100-fold. At stimulation rates of 18 shocks/min there was no rise in hexosemonophosphate and presumably phosphofructokinase as well as phosphorylase was activated. With frequencies about 24/min, glucose-6-phosphate began to rise and it would seem that phosphofructokinase became rate-limiting; inorganic P also increased at this frequency. Helmreich & Cori (2) then followed the changes in ATP and AMP over the range of stimulation up to 48 shocks/min. There was a gradual fall in concentration for the former and rise for the latter; it was calculated that, judging from known effects *in vitro*, the change in ratio would not cause more than a doubling of the rate of phosphorylase *b* activity. Özand & Narahara (1), in Cori's laboratory, using muscles which had been exposed to adrenaline, found that electrical stimulation greatly increased the fructosediphosphate content – probably an indication of the stimulation of phosphofructokinase activity mediated through the increased supply of glucose-6-phosphate, itself a consequence of increased phosphorylase activity. The ratio of glucose-6-phosphate to fructose-6-phosphate did not change, so that phosphoglucoisomerase seemed not to be rate-limiting.

Muscle metabolism can also adjust itself very rapidly in returning after contraction to the resting stage. Helmreich & Cori (2) record that within 5 min from the beginning of rest the rate of lactic acid production had fallen to 5 % of the value during stimulation.

From all that has been said it emerges that experiments *in vitro* suggest many mechanisms for regulation of energy provision from carbohydrate breakdown *in vivo*.[2] Helmreich & Cori have discussed the possibility that the release of Ca ions on membrane depolarisation[3] could, besides activating actomyosin ATPase, also mediate activation of the two key enzymes phosphorylase and phosphofructokinase. Removal of Ca by the Ca pump would (besides inhibiting the ATPase) abolish this activation.[4] Some

---

[1] Also Helmreich & Cori (2).
[2] For regulatory mechanisms in insect flight muscle see ch. 22.   [3] Ch. 13.
[4] Vaughan & Newsholme (1) however, in very recent work, could find no effect of Ca ions (0.001 to 10 μM) upon the activity of phosphofructokinase, fructosediphosphatase or hexokinase, prepared from a variety of muscles.

reversible alteration in the state of the two enzymes on stimulation might be involved, enabling them readily to interact with substrates, co-factors or both. A difficulty has been the failure to inactivate *in vitro* the Ca-activated phosphorylase *b* kinase by means of chelating agents; this difficulty seems to have been at any rate partly overcome by the work of Ozawa *et al.*[1] As we have seen, changes in ATP and AMP in the conditions of contraction used by the Cori group, seem to have little relevance to the large changes in glycolysis; but, as Helmreich & Cori have emphasised, it must always be borne in mind that estimations of the total concentration of constituents in the muscle take no account of possible compartmentation within the cell and consequent variations in distribution. Again the work of D. K. Hill (6) should be recalled here.[2] Electron-microscope examination suggested that the labelled adenine nucleotides, concentrated in the I bands of the fibres, were situated between the fibrils; and stained sections indicated the presence of vesicles of the transverse sarcoplasmic reticulum at the appropriate sites.

Helmreich & Cori also suggest that the extremely rapid lactic acid formation, with little or no accumulation of intermediates, implies some special structural organisation of the glycolytic series of enzymes. A relevant paper is that of Margreth, Muscatello & Andersson-Cedergren (1) who examined the localisation of certain enzymes by means of biochemical estimations combined with differential centrifugation and microscopic examination. They found, for instance, that the only fraction to show UDPG-glycogen synthetase activity[3] was the sarcotubular fraction spun down between 75000 and 105000*g*. This fraction also had much higher phosphofructokinase activity/mg protein than the other fractions. Some caution is needed in interpreting the latter findings, in view of the results of Mansour and his collaborators on the ready changes in solubility of this enzyme.

THE PATHWAY OF GLYCOGEN SYNTHESIS FROM GLUCOSE AND ITS CONTROL

The question of glycogen synthesis from glucose in muscle was early considered by the Coris and their collaborators. Sutherland, Colowick & C. F. Cori (1) showed in 1941 that at pH 7.0, at the equilibrium point of the phosphoglucomutase reaction, 94% of glucose-6-phosphate was present and only 6% of the glucose-1-phosphate. The position of this equilibrium was thus unfavourable for synthesis. At the equilibrium point of the phosphorylase, the ratio of inorganic P to glucose-1-phosphate was about 3; with a higher ratio breakdown would occur, with a lower ratio synthesis.

[1] P. 431 above. [2] Ch. 9. [3] P. 442.

In tissues the ratio is always much higher than this,[1] and it was realised that for synthesis to take place a drastic fall in inorganic P would be needed. This they accomplished *in vitro* by addition of barium ions, and it was assumed that in resting muscle *in vivo* various reactions, including oxidative phosphorylation, would bring about the necessary low concentration.[2] A disturbing feature, however, was the effect of adrenaline which, in activating phosphorylase, always called forth glycogen breakdown, never glycogen synthesis.[3]

A new chapter opened with the work of Leloir and his collaborators[4] in 1949 when they showed that the transformation (in enzymic preparations from *Saccharomyces fragilis*) of galactose-1-phosphate into glucose-1-phosphate needed a thermostable co-factor, found also in mammalian tissues. The isolation of this co-factor soon followed and the elucidation of its structure as uridinediphosphate-glucose (UDPG), uridine-5-phosphate and glucose-1-phosphate being joined by a pyrophosphate bridge. In 1957 Leloir & Cardini (1)[5] discovered that UDPG can act in presence of a liver enzyme (with a trace of glycogen as primer) as glucose donor in the synthesis of glycogen, UDP being formed in amount equivalent to the glycogen. The enzyme concerned (which has become known as UDPG-α-glucan transglucosylase, or more simply glycogen synthetase) was greatly activated by glucose-6-phosphate. When UDPG labelled in the glucose carbon was incubated with the enzyme and glucose-6-phosphate, the radioactivity was transferred to the glycogen formed, from which it could be removed as glucose-1-phosphate by means of phosphorylase. As was shown later[6] the glucose-6-phosphate takes no part in the reaction, since incubation of UDPG and labelled glucose-6-phosphate led to no labelling of the glycogen formed.

THE MECHANISM OF THE SYNTHESIS. Light was first thrown on the mechanism of this synthesis by Villar-Palasi & Larner (1) using rat diaphragms and skeletal muscle. They studied two enzymic activities, one bringing about pyrophosphorylysis of UDPG, the other converting UDPG to glycogen.

$$\text{UTP} + \text{G-1-P} \rightleftharpoons \text{UDPG} + \text{pyrophosphate} \tag{1}$$
$$\text{UDPG} + \text{glycogen} \longrightarrow [\text{Glycogen} + 1 \text{ glucose}] + \text{UDP}. \tag{2}$$

The uridinediphosphate glucose 'pyrophosphorylase' reaction (reverse of reaction (1)) had already been recognised by Munch-Petersen and her colleagues in 1953 in yeast extracts,[7] and the enzyme was purified by

---

[1] Larner, Villar-Palasi & Richman (1) later gave a value of about 300 in rat diaphragm.
[2] Cori (1).   [3] Sutherland & Cori (1); Leloir (1).
[4] Caputto, Leloir, Trucco, Cardini & Paladini (2).
[5] See also the work a little later of Leloir, Olavarria, Goldenberg & Carminatti (1).
[6] Leloir (1).   [7] Munch-Petersen, Kalckar, Cutolo & Smith (1).

Munch-Petersen (2). Presumably UDP can be rephosphorylated by ATP; the work of Trucco (1) indicated this, at any rate in yeast extracts.

Villar-Palasi & Larner (1, 2) came to the conclusion that this was a system in which glycogen synthesis was greatly favoured under physiological conditions. Reversal of reaction (1) did not take place under conditions where breakdown of glycogen by phosphorylase could be shown. Moreover the affinity of the UDPG pyrophosphorylase for pyrophosphate was low and pyrophosphate is readily removed by pyrophosphatase in the tissues. The high phosphoglucomutase which they observed assumes importance in view of the need to provide glucose-1-phosphate against an unfavourable equilibrium.[1] The synthetic role of the UDPG system, as contrasted with the glycolytic role of phosphorylase, also comes out of the work of Robbins, Traut & Lipmann (1). With homogenates or pigeon-breast muscle, good incorporation of radioactive glucose or glucose-6-phosphate into glycogen was found, and even more rapidly with UDPG; but addition of phosphorylase b kinase, increasing the concentration of phosphorylase a, strongly inhibited the incorporation of glucose and glucose-6-phosphate. Larner, Villar-Palasi & Richman (2) also observed with rat diaphragms in presence of insulin that glycogen formation could go on when the ratio of inorganic P to glucose-1-phosphate was over 300 – a situation far removed from the equilibrium of the phosphorylase reaction and one which would greatly favour glycogenolysis by that enzyme.

Larner (1) in 1953 had studied the participation of a branching enzyme necessary, during the synthetic activity of phosphorylase, for the formation of the highly-branched glycogen structure. Outer chains of glycogen were first acted upon by phosphorylase plus inorganic phosphate, and a polysaccharide was isolated having outer chains of only 4–5 glucose units. These were then lengthened by the synthetic action of phosphorylase in presence of ^{14}C-labelled glucose-1-phosphate. When they had reached a certain length of 6–11 glucose units, they could be acted upon by the branching enzyme, amylo-1,4 $\longrightarrow$ 1,6-transglucosylase. By activity of this enzyme short segments of 1,4-linked chain were transferred to the 6-OH of glucose units in the chain;[2] this was shown by the fact that 1,6-glucosidase in presence of phosphorylase now released labelled glucose units. In 1965 Brown, Illingworth & Kornfeld (1) showed that such branching activity is needed also with the UDPG synthetase to carry the synthesis beyond the stage of long linear chains giving a blue iodine reaction.

CONTROL OF GLYCOGEN SYNTHETASE ACTIVITY. Villar-Palasi & Larner (3) in 1961 observed that in extracts of rat diaphragms, which had been incubated with insulin, phosphoglucomutase, UDPG pyrophosphoryl-

[1] P. 116 above.   [2] See p. 427.

ase and phosphorylase all showed unchanged activity while the glycogen synthetase had been stimulated some 35 % and its activity approached the maximum rate of glycogen synthesis. In the presence of glucose-6-phosphate the total synthetase activity was the same whether the extracts were made from control or from insulin-treated diaphragms. It was therefore suggested that the insulin caused increased activity of an enzyme active without glucose-6-phosphate and correspondingly decreased activity of an enzyme responsive to glucose-6-phosphate.[1]

Friedman & Larner (1) followed this up by showing that two types of activity of the glycogen synthetase could indeed be detected in muscle homogenates, one type (called D) being dependent on the presence of glucose-6-phosphate, the other (called I) being independent. The two types of activity seemed to be interconvertible, for when ATP and Mg ions were added, there was a decrease in the I type while the total activity, in the presence of glucose-6-phosphate, was constant. It then appeared[2] that two enzymic reactions were concerned:

$$\text{Synthetase I} + n\text{ATP} \longrightarrow \text{synthetase D} + n\text{ADP} \qquad (1)$$
$$\text{Synthetase D} \longrightarrow \text{synthetase I} + n\text{P}_i. \qquad (2)$$

Evidence for reaction (1) was found in the incorporation of ^{32}P from ATP into the enzyme; while the second reaction could be demonstrated by using the purified, labelled D enzyme and measuring the release of ^{32}P from it in presence of muscle extract. The analogy with the $a–b$ transformation of phosphorylase springs to the mind: in both cases one form of the enzyme is phosphorylated, but in the case of the synthetase the form dependent on the effector glucose-6-phosphate is phosphorylated while in the case of phosphorylase the form dependent on the effector adenylic acid is dephosphorylated. But as we shall see, the analogy is rather between the synthetase and phosphorylase b kinase.[3] Friedman & Larner (2) suggest that the effect of insulin lies in inhibition of the phosphorylation of synthetase I rather than in stimulation of the phosphatase. This would accord with the action of the hormone in promoting glucose entry into the cell.[4]

The two types of the synthetase could be clearly differentiated by means of their different stability under certain conditions, and they were purified by Rosell-Perez, Villar-Palasi & Larner (1, 2). Glucose-6-phosphate had an effect on the 'independent' form, in that it increased the affinity of the enzyme for the substrate UDPG without increasing the maximum velocity; with the dependent form the major effect was an increase in maximum velocity – as much as 50-fold with 10^{-2}M ester. Traut & Lipmann (1) also were interested in the change in affinity for substrate brought about by glucose-6-phosphate; from the change in the pH-activity curve of the

---

[1] Also Craig & Larner (1).   [2] Friedman & Larner (2).   [3] P. 447.   [4] P. 438, fn.

enzyme which accompanied this effect, and also from the fact that glucose-6-phosphate counteracted inhibition by p-chlormercuribenzoate, they deduced a conformational change in the enzyme.

Belocopitow (1) had shown in 1961 that adrenaline treatment of rat diaphragms caused a decrease in synthetase activity. With homogenates of rat skeletal muscle he tested the supernatants for phosphorylase and synthetase activity after preincubation with ATP and cyclic AMP; the two activities were affected in opposite ways, the phosphorylase being activated and the synthetase inhibited. This effect on the synthetase was confirmed by Rosell-Perez & Larner (1) when they showed that phosphorylation of the synthetase by ATP was enhanced by cyclic AMP. Appleman, Birnbaumer & Torres (1) made a close study of the requirements in ATP, Mg and cyclic AMP for inhibition of synthetase and of the enzyme activating phosphorylase b kinase, and found these the same for the two enzymes.

Belocopitow, Appleman & Torres[1] (1) then showed further ways (besides the ATP–AMP effect) in which the system converting the I to the D form of synthetase (later called by Friedman & Larner (3) the synthetase I kinase) resembled the system activating phosphorylase b kinase. In both cases preincubation with Ca ions, in presence of a protein factor which seemed to be very similar for both (Ca-activating factor or CAF) caused increased activity, and a third similarity was activation by trypsin treatment. There was no evidence in either case that the protein factor had enzymic properties. Attempts to separate by chromatographic methods the protein factor operative with the phosphorylase b kinase from that operative with the synthetase I kinase had no success, although both were purified tenfold. Thus the idea was strengthened that in fact only one system was concerned. Friedmann & Larner (3) on the other hand, working on the conversion of the I to the D synthetase in presence of ATP and cyclic AMP were able to make a clear separation of phosphorylase b kinase from synthetase I kinase. The existence of a 'phosphorylase kinase kinase' has recently been suggested by Krebs and his colleagues.[2] The phosphorylation of phosphorylase b kinase was shown to be autocatalytic, the enzyme catalysing its own phosphorylation. But the involvement of a second enzyme was indicated by the fact that the activating agent, cyclic AMP, does not bind to phosphorylase b kinase.

These relationships may be summarised thus (table 17).

Belocopitow, Appleman & Torres suggested that this Ca-mediated inhibition of glycogen synthesis might be the counterpart of a Ca-mediated increase in glycolysis on stimulation; they pointed out however that the Ca ions set free for excitation are at a concentration of only about

[1] Also Appleman, Birnbaumer, Belocopitow & Torres (1).
[2] De Lange, Kemp, Riley, Cooper & Krebs (1).

TABLE 17

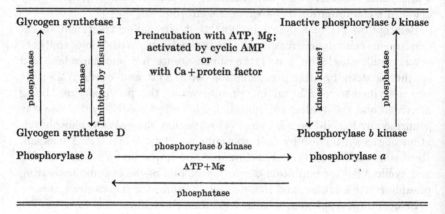

10⁻⁶M, while for these effects on glycogen metabolism about 10⁻⁴M is
needed.[1]

It will be remembered that, while both adrenaline treatment and
stimulation elicited increase in phosphorylase a concentration, stimulation
caused no formation of cyclic AMP; the suggestion was made that on
stimulation the activation of phosphorylase b kinase by Ca was operative.
Danforth (1) more recently tested the effect of stimulation on concentration
of synthetase D and could find no evidence for an increase. This is perhaps
surprising in view of the results of Belocopitow and his collaborators on the
very similar behaviour *in vitro* of phosphorylase b kinase and synthetase-I
kinase. Danforth describes on the other hand an increase in the I form of
the synthetase from 20 to 35 % of the total synthetase activity; this took
place during a 20 sec tetanus, the increase going up to 90 % by the end of
3 min. The proportion of the I form then fell progressively over the next
hour. He emphasises that this increase is very slow in comparison with the
response of the phosphorylase system, which is a matter of very few seconds.
Further light is thrown here by the work of Piras and his collaborators.
Piras, Rothman & Cabib (1) showed in 1968 that synthetase D is more
markedly inhibited at low glucose 6-phosphate concentrations by certain
metabolites (for example inorganic phosphate, ATP, ADP) than is synthetase I. Piras & Staneloni (1) then followed the rate of glycogen synthesis
in vivo after a 10 sec stimulus. This rate increased during the first 4 min of
recovery, and the highest fraction of the synthetase in the I form was
present at 4 min after cessation of the stimulus. Their comparisons of rates
of glycogen synthesis *in vitro* under various conditions with the rate *in vivo*

---

[1] But see Ozawa *et al.*, p. 431 above.

led them to the conclusion that glycogen synthetase activity during contraction and recovery can best be explained by interconversion of the I and D forms and by action of effectors on the two forms of the enzyme.

Recognition of still another method of regulation of glycogen synthesis came out of the work of Danforth, when he observed that in mouse muscle *in vivo* as well as in intact rat diaphragms the I form of the synthetase increased when the glycogen content was lowered.[1] Recently this effect has been confirmed by Larner and his collaborators who, using a series of rat hearts with a wide range of glycogen content, reported a close inverse correlation between this content and the synthetase I activity.[2] Villar-Palasi & Larner (4), using extracts of skeletal muscle, showed that the phosphatase converting the D form into the I form was markedly inhibited by physiological concentrations of glycogen. There was no direct effect on the synthetase I kinase activity. They also confirmed earlier reports that glycogen stimulates phosphorylase *b* kinase.[3] Thus glycogen acts in a double feedback mechanism to control its own synthesis and degradation.

It seems clear that adrenaline could have an inhibitory effect on glycogen synthesis via the cyclic AMP activation of the synthetase I kinase, assuming of course that the concentration of glucose-6-phosphate is below optimal. On the other hand insulin, encouraging glycogen synthesis, increases the I form at the expense of the D form. It is not known whether an inhibition of the kinase or an activation of the phosphatase is concerned. Much of course hinges on the concentration of glucose-6-phosphate available to the enzyme. Larner, Villar-Palasi & Richman (1) give a value of 5×10^{-4}M for the content in diaphragm after insulin treatment, and showed that such a concentration increased synthetase activity in muscle extracts some twenty times. The figure given by Helmreich & Cori for resting sartorius muscle is of the same order. A fall in glucose-6-phosphate content might thus be important in preventing synthesis, and might result from increased phosphofructokinase activity. It must also be remembered that the ester may be compartmentalised in the cell. Danforth's value of 3.4×10^{-4}M in the intracellular water of diaphragms after incubation with glucose and insulin was, he considered, high enough to activate the D form to 20 % of maximal activity, while the actual synthetase activity was much lower. He suggested physical separation of the enzyme from the activator as a possible explanation. A similar construction might be put on the results of Özand & Narahara (1), who found the ratio of glucose-6-phosphate to fructose-6-phosphate to be nearly twice as high in sartorius muscle as the equilibrium ratio obtained *in vitro* with phosphoglucoisomerase.

[1] Also Bär & Blanchaer (1); Hultman & Bergström (1).
[2] Larner, Villar-Palasi, Goldberg, Bishop, Huijing, Wenger, Sasko & Brown (1).
[3] Krebs, Love, Bratvold, Trayser, Meyer & Fischer (1).

GLUCONEOGENESIS

We have seen[1] that there is little or no evidence for resynthesis of lactic acid to carbohydrate within the mammalian muscle. In recent years much work has been done on the pathway of such synthesis in some other tissues, notably liver and kidney cortex, and considerations arising from such studies have turned out to be important for muscle also.

In his Croonian Lecture of 1963[2] Krebs discussed the reversibility of the glycolytic pathway, and emphasised that, though most of the reactions are readily reversible, there are three energy barriers to be overcome in its utilisation for synthesis: (a) The formation of glycogen from glucose-1-phosphate. (b) The formation of fructosemonophosphate from fructosediphosphate. (c) The formation of phosphopyruvate from pyruvate. The first difficulty is surmounted by the uridinediphosphate pathway; the second by the presence of a specific fructosediphosphate phosphatase, hydrolysing the 1,6-diphosphate to the 6-phosphate; and the third by means of the pyruvate carboxylase reaction,[3] converting pyruvate to oxalacetate, followed by the phosphopyruvate carboxykinase reaction,[4] yielding phosphopyruvate:

$$\text{Pyruvate} + CO_2 + ATP \rightleftharpoons \text{oxalacetate} + ADP + H_3PO_4$$
$$\text{Oxalacetate} + GTP \rightleftharpoons \text{phosphopyruvate} + CO_2 + GDP.$$

In 1965 Krebs & Woodford (1) reported the presence of fructosediphosphatase in a wide range of skeletal muscles from man to frog. Since Keech & Utter (2) had found no phosphopyruvate carboxykinase and practically no pyruvate carboxylase in skeletal muscle, Krebs & Woodford interpreted this to mean that carbohydrate synthesis could take place in these muscles but only from phosphorylated 3-carbon intermediates, not from lactate or pyruvate. They gave reasons for believing that α-glycerolphosphate is likely to be the phosphorylated intermediate concerned, and discussed its possible significance in vertebrate muscle on the lines of its well-known importance in insect flight muscle.[5]

A complication affecting this picture of the pathway of carbohydrate synthesis in muscle must be borne in mind. This is the finding by Opie & Newsholme (1) that the overall intracellular content of AMP exceeds the concentration needed for complete inhibition of fructosediphosphatase. They discuss however the possibility that the AMP may be unevenly distributed in the cell owing to 'compartmentation'; furthermore the AMP inhibition was sensitive to Mg and Mn ions. The effect of the former in relieving the inhibition could be largely explained by chelation of AMP, but

---

[1] Ch. 16.
[2] Krebs (4), where many references are given to work of his own and other laboratories.
[3] Utter & Keech (1, 2); Keech & Utter (1).
[4] Utter & Kurahashi (1); Utter, Keech & Scrutton (1).   [5] Ch. 22.

Mn ions seemed to have a specific effect, perhaps reducing the affinity of the enzyme for AMP.

Krebs & Woodford questioned the validity of Meyerhof's glycogen synthesis from lactate in frog muscle. But recently Meyerhof's work has been completely vindicated by Bendall & Taylor (1). They were able, of course, to use much more dependable methods of analysis of the small amounts involved; this meant opportunity to shorten the duration of the experiments and to do the different estimations on the same muscle. With muscle suspended in lactate solution during some hours an oxidation quotient of 6.2 was found, lactate disappearing, glycogen formation and oxygen uptake all being estimated. They calculated that with the pyruvate carboxylase route for phosphopyruvate generation, 7 moles of ATP are needed for synthesis of one glucose unit from 2 moles of lactate. With a P/O ratio of 3, oxidation of 2 moles of lactate would give 34 moles of ATP from ADP, allowing thus for an oxidation quotient of 4.86, close to the observed value. With regard to the difficulty of the very low activity reported by Opie & Newsholme for pyruvate carboxylase even in frog muscle, Bendall & Taylor suggest that the widely distributed 'malic enzyme'[1] present in all the muscles tested by Opie & Newsholme, might also be important in providing malate from pyruvate, the malate then being oxidised to oxaloacetate:

$$CH_3CO.COOH + H_2CO_3 \underset{NADP}{\overset{NADPH}{\rightleftharpoons}} COOH.CH_2CHOH.COOH$$

$$COOH.CH_2CHOH.COOH \underset{NADH}{\overset{NAD}{\rightleftharpoons}} COOH.CH_2CO.COOH.$$

Since formation of oxaloacetate by this route from pyruvate does not need ATP, the requirement for the glycogen synthesis from pyruvate would be only 5 moles per glucose unit. An oxidation quotient of 6.8 would thus be possible. For synthesis to take place by either of these routes conditions must be such that the lactate dehydrogenase can function to produce pyruvate: high pH and high oxygen tension are necessary. Further work on the enzymes concerned in this synthesis and on their different distribution in red and white muscle is discussed in chapter 19.

DISCUSSION

The interlocking controls of glycogen breakdown on the one hand and glycogen synthesis on the other form a truly amazing pattern, integrated in the muscle cell with the needs of contraction and relaxation. As we have seen, a dependence of this pattern on some postulated structural basis has

[1] Ochoa, Mehler & Kornberg (1).

been from time to time suggested. This is one striking example of the regulation of cell metabolism with which we have been dealing in many aspects.

The nature of the regulation has been the subject of careful conjecture and deduction over many years. Hopkins (1) in 1913 in his inspiring plea for the study of 'the dynamic side of biochemistry',[1] spoke of the control of cell reactions by specific catalysts, a state of affairs the existence of which had been realised only since about the turn of the century. As these catalysts are of colloidal nature, he, pictured the reactions catalysed as occurring in a medium not strictly homogeneous. This point of view is described in the memorable phrases: 'On ultimate analysis we can hardly speak at all of living matter in the cell...Its life is the expression of a particular dynamic equilibrium which obtains in a polyphasic system'; and 'life, as we instinctively define it, is the property of the cell as a whole, because it depends on the organisation of processes, upon the equilibrium displayed by the totality of the co-existing phases'.

Peters (3) took up the theme again in the Harben Lectures of 1929. He felt now that the 'idea of the cell as a heterogeneous colloid system with specific enzymes', useful in the past, could not be so much longer. He laid great stress on the knowledge which had accumulated from the work, for example, of Hardy, Haskins, Langmuir, Leathes and himself concerning the chemical aspects of adsorption, and the orientation of molecules at interfaces; this gave rise to the conception of a cell surface 'made up of molecules so anchored that they constitute a chemical mosaic'. Organisation he pictured as a directive force radiating to a distance of several microns from the mosaic pattern of protein and other surfaces within the cell. Such questions were also considered in the Terry Lectures of J. Needham (5) in 1936, who spoke of the 'extension of morphology into biochemistry'. By this time it was possible for him to emphasise the hopes for the future which lay in the method of X-ray diffraction, as evinced by the work for example of Astbury and Bernal.

As we have seen, it was not until the 1940's that the electron microscope began to make its direct visual contributions to the solution of such problems. The need for much further work on the localisation of enzymes, substrates and activators becomes obvious again and again when one attempts to comprehend the interaction of regulatory mechanisms. Fortunately such work becomes ever more possible with the introduction of new techniques.

[1] Address to the Physiological Section of the British Association for the Advancement of Science.

19

A COMPARATIVE STUDY OF THE
STRIATED MUSCLE OF VERTEBRATES

RED AND WHITE MUSCLE; EARLY WORK

PIGMENTATION. The variations in colour of skeletal muscle, from the deep crimson of the pigeon breast to the whiteness of fish muscle, have long been a matter of discussion; in 1678 Lorenzini[1] commented on the striking differences of colour in certain muscles of the rabbit. It was at first supposed that the red colour was due to a greater supply of blood, but Kölliker (1) in 1850 considered from his histological studies that the pigment was contained actually within the contractile substance of the fibres. Kühne (4) in 1865 showed that this was indeed the case, by perfusing muscles to wash out all blood, and finding that a haemoglobin-like substance remained in the muscle plasma. In spectral analyses of the reduced and oxygenated forms of the substance, as well as of the carbon monoxide compound and the haematin formed from it, he found no significant difference from blood haemoglobin; but the work of Mörner (1) in 1896 made it clear that in all these cases the absorption bands lie nearer the red end of the spectrum with muscle haemoglobin than with blood haemoglobin. He suggested the name myochrome, and the designation myoglobin now in general use was introduced by Günther (1), who confirmed the results of Mörner, only in 1921. Keilin[2] has described the confusion in the literature during the first quarter of this century between myochrome and MacMunn's myohaematin, confusion only cleared up by Keilin's own work.

PHYSIOLOGICAL BEHAVIOUR. From 1870 for the rest of the century much work was done on the distribution, the histological structure and the physiological behaviour of red and white muscles. From a wide survey of the animal kingdom Lankester (1) concluded that the most active and strongest muscles were provided with this pigment. About the same time Ranvier (1) began experiments on the physiological behaviour of red and white muscles. He found with the rabbit that, on electrical stimulation, the red semitendinosus contracted and relaxed slowly while the white vastus internus acted quickly. The muscles of the ray behaved similarly. Fewer

---

[1] Cited by Paukul (1).   [2] Keilin (1) pp. 109 ff.

stimuli were needed with the red muscles to obtain a smooth tetanus. Kronecker & Stirling (1) confirmed these findings and also showed that, though the single twitch with white muscle was higher than that with red muscle, in incomplete tetanus the curve for red muscle rose much higher. A number of other workers, e.g. Rollett (4), Grützner (1) and H. Fischer (1), provided results of the same kind. Thus Fischer in 1908 showed with cat gastrocnemius that the rapid twitch could reach a tension of 600 g and tetanus a tension of 700 g for a very short time; while with the red soleus the twitch tension was only 200 g, but the tetanus tension was 900 g maintainable for a long period. Paukul (1) emphasised the very much longer relaxation time for a twitch in red muscle of the rabbit.

HISTOLOGICAL STUDIES. The histological appearance[1] of the red and white fibres was much discussed, especially as regards their diameter and the amount and nature of the sarcoplasm associated with them. Much variation in these properties was reported; the fibres of white muscles were often larger and clearer and of the red muscles finer and more granular. The difference in granularity was the more marked and the more constantly found; indeed Knoll always referred to *protoplasma-armen* and *protoplasma-reichen* fibres. As we shall see, this difference has turned out to be of great significance. Knoll (1) reached the conclusion that the *protoplasma-reichen* fibres were more numerous in the most active muscles, and was inclined to the view that the protoplasm filled a nutritional role for the recovery of the fibrillar substance after activity.

Grützner (1), partly from histological studies and partly from the results of experiments showing differences in response according to intensity of stimulus, came to the conclusion that the muscles of higher mammals were composed of an intimate mixture of fibres of the clear and opaque types, between them determining the speed of contraction and relaxation in any particular muscle. He regarded the red fibres as acting as an 'inner support', holding the resting muscle firmly at a definite length. Denny-Brown (1) however, in a careful study in 1929 of a wide range of mammalian muscles, concluded that Grützner's hypothesis could not be sustained. In his experience fibres of similar speed of contraction formed a group, sharply delineated from other such groups within a given muscle. In more complex muscles the groups of slowly contracting fibres were usually deeply situated. Within any such group he found histological differences in fibre diameter and granule content and concluded that these had no direct relation to the speed and nature of contraction.

[1] Ranvier (1); Grützner (1); Knoll (1); Knoll & Hauer (1); Paukul (1). Also D. M. Needham (7) for a review.

METABOLISM. The interesting paper of Gleiss (1) in 1887 was the first to tackle the question of differences in metabolism between the slow and fast muscles. Comparing the muscles of frogs and toads, he found that the latter (in spite of equally long and intense work) developed less acid than the former. He showed this by crushing the muscles in saturated NaCl solution containing litmus, and also by testing the extract with neutral iron chloride solution which gives a yellow colour with lactic acid. The results were similar in comparisons of the red and white muscles of rat, cat and rabbit. Fletcher (4) twenty-five years later also was interested in lactic acid production in such muscles. Using the new quantitative method of Fletcher & Hopkins (1) he found that this production went on much more rapidly in the uninjured gastrocnemius of the rabbit during onset of death rigor at 38° than in the red soleus; in contrast, rigor with great shortening came on at a much earlier time in the red muscle.

As we have seen,[1] Embden in the early nineteen twenties attached great importance to lactacidogen (later shown to be hexosemonophosphate) in the metabolism of contraction. The quickly-contracting pale biceps femoris of the rabbit had a lactacidogen P content of 0.3 %, while the slow semitendinosus contained only about half this amount.[2] On the other hand (a more important point, as could only be appreciated much later) the slow muscles had a higher content of residual P (the total organic P minus the lactacidogen P).[3] Lyding (1) working in Embden's laboratory showed that the pale chicken-breast muscle contained only some 20 % of the amount contained in pigeon-breast muscle.

In 1912 Batelli & Stern (1), studying oxidation of p-phenylenediamine and of succinic acid, found greater oxidation power in the red than in the white muscles of pigeon, hen and calf. Observations of Ahlgren (1) and of Holden[4] confirmed this relation. We thus arrive at the first clues to the metabolic characteristics of red as compared with pale muscle.

RED AND WHITE MUSCLE; LATER WORK

Between 1920 and 1950 little was done on comparisons of red and white muscle. The early work had led to the general view that red muscle was associated with strenuous, prolonged contraction, the white muscle with rapid but less sustained and vigorous contraction. This point of view is well described by Ff. Roberts (1) in 1916. He pointed out that in the falcon, for example, the proximal segments of the wing stay permanently extended by prolonged contractions of the red proximal muscles, the white distal seg-

[1] Ch. 4.   [2] Embden & Adler (1).
[3] For discussion of the importance of phospholipids in mitochondria see Lehninger (7).
[4] Quoted in D. M. Needham (7).

ments performing the necessary rapid movements. So the pectoral muscles are characterised by both slow movement and great endurance. Further, white muscles often pass over two or more joints, while the more deeply placed red muscles usually unite adjacent segments of a limb. A white muscle can cause movement at either of the joints over which it passes, but clearly it can only exert full effect on one joint if the other is held firm. Such fixation would be a function of red muscle. It is interesting that during this intervening period there seems to have been a tendency to doubt the importance of the pigment itself. Thus Denny-Brown wrote: 'the red pigmentation of the slow, less differentiated muscles does not appear to be essential to the slow type of contraction process, and is probably the outward sign of some function not closely related to contraction.' Then as we have seen[1] between 1933 and 1939 R. Hill and Millikan described its special oxygen-combining characteristics, and its function in providing an oxygen store became clear. Interest in these questions was started again by the work of Lawrie (1) and of Paul & Sperling (1) in 1952; since then much has been done.

METABOLISM. Lawrie (1, 2) gave figures for the myoglobin content and the succinic oxidase and cytochrome oxidase activity, for different muscles of different animals; he found a correlation between redness and activity of the cytochrome system. In skeletal muscle he found the following values to have an inverse relation to the myoglobin content:

(a) The initial level of ATP and creatine phosphate.

(b) $\dfrac{\text{Rate of fall of pH aerobically}}{\text{Rate of fall of pH anaerobically}}$.

(c) Rate of fall of pH anaerobically.

(d) $\dfrac{\text{Rate of aerobic breakdown of energy-rich P compounds}}{\text{Rate of anaerobic breakdown of energy-rich P compounds}}$.

In other words, the greater the myoglobin content of the muscle, the greater its power of respiratory metabolism and the less its glycolytic activity. The work of Paul & Sperling (1) added evidence for the great importance of respiration in certain red muscles. They measured the 'cyclophorase quotient';[2]

$$\frac{\text{Dry wt of cyclophorase in mg} \times O_2 \text{ uptake per mg dry wt (corr. for salt content)}}{\text{Initial wet wt}}.$$

Using α-ketoglutarate and succinate as substrates they found the quotient to be about 20 times as high for breast muscle in the pigeon as in the

---

[1] Ch. 16.

[2] Cyclophorase is a mitochondrial preparation, see ch. 16.

chicken. They contrasted muscle cells called on for strong rhythmic and prolonged exercise with those active sporadically.

	Cyclophorase quotient	Mitochondrial content by phase-contrast microscopy
Chicken breast	110	Few
Pigeon breast	1830	+ + + +
Rabbit gastrocnemius	110	0
Rabbit soleus	85	+
Rabbit diaphragm	545	
Heart	1110	

Lawrie (3) went on to show that red muscles had higher capacity for aerobic resynthesis of energy-rich P compounds, while white muscles were more efficient at anaerobic resynthesis of such compounds and contained a greater store of them. ATPase activity was higher in the white muscle.

About this time Chappell & Perry (2) made a more careful study of skeletal-muscle mitochondria. They pointed out that cyclophorase preparations when made from muscle might be seriously contaminated by myofibrillar material. By an improved method they made granule preparations very similar in properties to those of mitochondrial preparations from liver or kidney. With pigeon-breast muscle this granule fraction (sedimented between 600 and 3500g) made up about 20% of the total N. The yields from other muscles fell off in the following order: pigeon breast > hamster leg > rat leg > rabbit back > rabbit leg.

HISTOCHEMICAL STUDIES. A little later J. C. George and his collaborators in India published a series of papers on the fibre constitution of pigeon pectoralis major. They had observed that this muscle contained two types of fibre – one broad and clear, the other narrow and dense with lipid inclusions. They had also noticed that after exercise the content in fat globules appeared less, and they suggested possible utilisation of fat as the fuel here for muscular activity.[1] George & Naik (1) then found, using Sudan black B, that the narrow fibres were deeply stained, the broad fibres unstained (see fig. 90 a pl.). On the other hand, a high content of glycogen was shown in the broad fibres, by means of the Best carmine stain, with haematoxylin counterstaining (see fig. 90 b pl.). In a histological study by George & Naik (2) on mitochondrial distribution between the two types of fibres in this same muscle, high mitochondrial content characterised the narrow fibres; this fitted in with the conclusions drawn from the biochemical experiments described above. George & Scaria (2) also made a histochemical demonstration of much greater lipase content in the narrow fibres.

[1] George & Jyoti (1, 2); George & Scaria (1).

Ogata's histochemical observations (1, 2) on striated muscle of frog, fish, birds and mammals led him to the conclusion that three types of fibre exist, in varying degrees, in different muscles: the red fibres distinguished by their high succinic dehydrogenase and cytochrome oxidase content and their ready stainability with Sudan black; and the white fibres, low in these enzymic activities and in reaction with the stain; and an intermediate type.[1] Ogata (3) also obtained evidence by the method of Farber, Sternberg & Dunlop (1), that the red fibres contained more NADH and NADPH diaphorase activity than white fibres. The method depends on the formation of the insoluble blue diformazan upon reduction of ditetrazolium chloride. The test was made in washed slices supplied with NAD- or NADP-dependent substrates and co-factors. Since with two cases of the former a single common pattern of staining was found, while with three cases of the latter a different common staining pattern appeared, it was deduced that the actual reduction took place at the site of the diaphorase activity in each case (see chapter 16).

FURTHER BIOCHEMICAL STUDIES. The intensive histochemical work of the nineteen fifties was followed by much renewed biochemical activity in this field.

In 1960 Ogata (4) using practically pure red fibre preparations (from the central part of the soleus) and purely white fibre preparations from the posterior edge of the adductor magnus, found the white muscle to contain 3.7 times as much glycogen as the red. It also showed 1.7 times the glycolytic activity of the red and contained twice as much phosphocreatine and 7 min hydrolysable P.[2]

The experiments of Domonkos & Latzkovits (1) indicate the greater activity of the citric acid cycle in red than in white muscle. Thus with thin slices (cut parallel to the fibres) of rabbit soleus and semitendinosus much less lactate was formed in air from added pyruvate by the red muscle. Pyruvate utilisation, on the other hand, was much greater in the red muscle; the effect of added fumarate in increasing pyruvate utilisation (the classical Szent-Györgyi effect) was seen only with the white muscle.

George & Talesara (1) described a quantitative study of distribution patterns of cytochrome oxidase, malic oxidase, lactic dehydrogenase and α-lipase in the red and white fibres of pigeon-breast muscle. It was known that the red fibres appeared predominantly in the interior of the muscle, while the white fibres formed a higher proportion towards the periphery. One side of a breast muscle was used for histological control of the distribution, and pieces of muscle from exactly similar locations on the other side

---

[1] Also Nachmias & Padykula (1).
[2] Also Beatty, Petersen & Bocek (1).

were used for homogenisation and enzyme estimation. The earlier qualitative conclusions were confirmed.

Light from an entirely new angle was thrown on metabolic differences between muscles of different functional type by the work in 1963 of Kaplan and his collaborators on the lactate dehydrogenase isoenzymes.[1] They showed that two distinct forms of this enzyme (differing in amino-acid composition, electrophoretic mobility, immunological reactions, etc.) occur in the tissues of most animals. One form (H) predominates in the heart, the other (M) in many skeletal muscles. From our present point of view the important difference is that with the former enzyme (H) reduction of

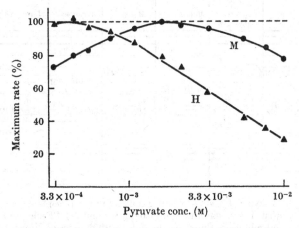

Fig. 91. Effect of pyruvate concentration on the activity of the pure H and M lactic dehydrogenases of the domestic fowl. (Wilson, Cahn & Kaplan (1).)

pyruvate to lactate is inhibited at low substrate concentrations (3.3×10^{-4}M); while with the latter (M) inhibition occurs only at much higher substrate concentration (10^{-2}M). Thus it seems that in the heart, a characteristically aerobic organ, rapid lactate production from pyruvate would not be possible, and complete oxidation of glucose via the Krebs cycle would be favoured. The M enzyme in contrast would allow rapid lactic acid formation and accumulation of an oxygen debt. In breast muscles from a number of birds a high ratio of the H/M enzymes was associated with capacity for great activity in prolonged flight; while in birds such as the domestic fowl, flying only occasionally for short times, pyruvate did not inhibit the breast muscle enzyme (see fig. 91). The ratio was measured both by substrate inhibition studies and by electrophoretic mobilities.

Bücher and his collaborators also about this time applied to red and white

[1] See Wilson, Cahn & Kaplan (1); Kaplan & Goodfriend (1).

muscles their methods of the thorough biochemical study of patterns of enzymic activity (see fig. 92).[1] Pette & Bücher (1),[2] using white and red muscle of the rabbit, found in 1963 that, although in both the Embden–Meyerhof pathway showed comparable proportions of the different enzyme activities, the pathway functioned at very different levels in the two cases. Thus for the group of enzymes triosephosphate isomerase, glyceraldehyde-phosphate dehydrogenase, phosphoglyceric kinase, phosphoglyceromutase and enolase the activity ratio for white to red was 6.7; for the group phosphoglucomutase, hexosephosphate isomerase and phosphofructokinase the

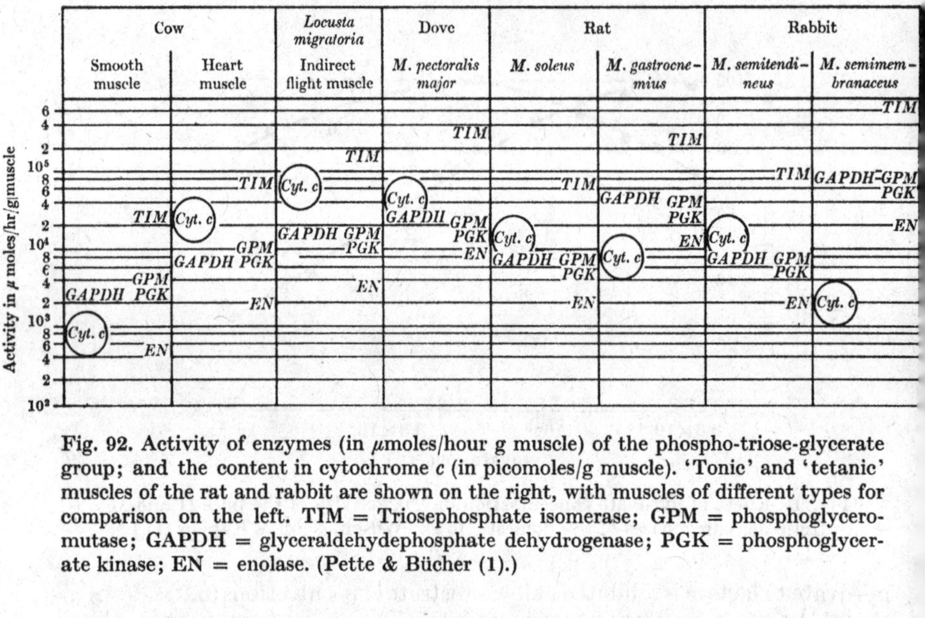

Fig. 92. Activity of enzymes (in μmoles/hour g muscle) of the phospho-triose-glycerate group; and the content in cytochrome c (in picomoles/g muscle). 'Tonic' and 'tetanic' muscles of the rat and rabbit are shown on the right, with muscles of different types for comparison on the left. TIM = Triosephosphate isomerase; GPM = phosphoglyceromutase; GAPDH = glyceraldehydephosphate dehydrogenase; PGK = phosphoglycerate kinase; EN = enolase. (Pette & Bücher (1).)

ratio was 10.5. In accordance with the higher mitochondrial content of red muscles, the cytochrome c content of these was many times as great as that of white muscles; certain enzymes, such as glutamic dehydrogenase and isocitric dehydrogenase, localised in the mitochondria, also showed much higher activity in the red muscle. Creatinekinase was about twice as active in the white muscle. A particularly interesting feature of the enzymic pattern was the much greater activity (some 30-fold) of the intramitochondrial (as well as the cytoplasmic) α-glycerophosphate dehydrogenase in white muscle. This 'compensation phenomenon' as the authors termed it, they considered to be of functional significance for the glycerolphosphate cycle.[3]

[1] Compare ch. 22.
[2] This paper summarises much work of this group and gives many references.
[3] Ch. 22.

Blanchaer (1), using mitochondria from white muscle of the guinea pig, found that oxygen uptake with α-glycerophosphate as substrate was greater than with most other substrates including NADH. With mitochondria from red muscle, on the other hand, there was a larger uptake with added NADH than with α-glycerophosphate. These differences persisted after permeability of the mitochondria to NADH had been increased by means of osmotic swelling; thus they seemed to depend on differences in enzymic constitution of the mitochondria and not on differences in their permeability in the native state.

Recently Opie & Newsholme (2) have carried further Krebs' hypothesis of the part played by fructosediphosphatase in muscle.[1] The activity of this enzyme was high in white muscle, low in red muscle; none was found in heart muscle. Moreover there was in white muscle a good correlation between the activities of this enzyme and of phosphoenolpyruvate carboxykinase, while the latter enzyme was absent from red muscle. No pyruvate carboxylase was found in the white muscle. They suggest that, during the active glycolysis in white muscle, special need arises for the oxidation of extramitochondrial NADH; this might be achieved by the operation of the α-glycerophosphate cycle (as suggested by Pette & Bücher) and possibly also by the malate/oxaloacetate cycle.[2] Excess of metabolites accumulating in these cycles (especially during exercise) might be re-converted to carbohydrate, by the glycolytic chain in reverse, with the assistance of the fructosediphosphate phosphatase and phosphoenolpyruvate carboxykinase. Opie & Newsholme also found that the lactate dehydrogenase isoenzyme ratio, defined as the ratio of activities measured at low and high pyruvate concentration, was high for red muscle (3.9–6.8) and low for white muscle (*ca.* 1.0). This is in agreement with what would be expected from the results of Kaplan and his collaborators. Thus once again in these different ways the importance of glycolytic metabolism in white muscle is underlined, both anaerobically and aerobically.

The greater oxidative activity of red muscle is also apparent in the work of Bär and Blanchaer (1), using rat muscle (diaphragm as red, external oblique muscle of the diaphragm as white, identified histochemically and chosen because thin enough for ready penetration of substrates and oxygen in manometric experiments). They found higher transformation of ^{14}C-labelled lactate and glucose to CO_2 in the red muscle. This effect with labelled glucose was also observed by Bocek, Petersen & Beatty (1), using the adductor muscles from the paws of rats, and selecting groups of fibres predominantly red or white.[3] Their results also gave biochemical support for the preferential utilisation of fat in red muscle, since with labelled

---

[1] Ch. 18, p. 448.   [2] Ch. 22.

[3] Beatty, Petersen & Bocek (1).

acetoacetate there was greater oxygen uptake, and greater labelling of the CO_2 formed, with this type of fibre. In the work of Bilinski (1) on the dark and white muscle of the trout this last point was also made. Oxidation of ^{14}C labelled long-chain fatty acids was much greater in tissue slices of the former. A little later Masoro, Rowell & McDonald found triglycerides and phospholipids present in higher concentration in the soleus than in the gastrocnemius of the monkey. Froberg (1) also found a similar differential concentration using white muscle from the rat gastrocnemius and red fibres from the heads of the same muscle.

Study of the enzymic reactions concerned with glycogen itself revealed some interesting differences between the two types of muscle. Already in 1958 Jinnai (1) brought evidence for greater phosphorylase activity in white than in red muscle fibres of the cat.[1] Then Hess & Pearse (1) found with individual fibres of the rat that phosphorylase activity and activity of the branching enzyme were greater in white muscle, while synthetase activity was greater in the red. These demonstrations were all made on fresh frozen sections by the histochemical methods of Takeuchi,[2] whereby glycogen formation was followed by the iodine reaction, glucose-1-phosphate and uridine diphosphoglucose respectively being used as substrates for phosphorylase and synthetase; the activity of the branching enzyme was gauged by difference in tint produced in the iodine reaction. In biochemical experiments Stubbs & Blanchaer (1) in 1965 reported greater synthetase activity in red than in white muscle, while the converse was true for phosphorylase. Stimulation caused a significant conversion of phosphorylase b to phosphorylase a in white but not in red muscle. In these experiments total synthetase (the glucose-6-phosphate independent form + the dependent form) was measured, and the effect of stimulation on the I/D ratio was not tested.[3]

Bocek and her collaborators[4] did indeed find greater glycogen labelling in the red fibres during aerobic incubation with [U-^{14}C] glucose. They also did interesting experiments[5] in which the muscle glycogen was labelled by aerobic incubation of the fibres with [U-^{14}C] glucose, the fibres being then transferred to aerobic glucose-free medium. Under these conditions the glycogen breakdown was similar in the two types of muscle, but the decrease in the specific activity of the remaining glycogen was 24 % for the red, 67 % for the white. Thus it seems that the newly added units were more labile in the white muscle.

In much of the enzymic work the distinction has been made rather

---

[1] Also Dubowitz & Pearse (1) for human muscle.
[2] Takeuchi (1, 2); Takeuchi & Glenner (1).   [3] Cf. Danforth, ch. 18.
[4] Beatty, Petersen & Bocek (1); Bocek, Petersen & Beatty (1).
[5] Bocek, Basinger & Beatty (1).

crudely between red and white muscle. But it is noticeable that certain workers have pointed out the occurrence of groups of fibres with different characteristics within the same muscle, and also that at least three different types of fibre could be distinguished.[1] In 1964 Romanul and Dawson[2] made a thorough study of this question. In the rat it was found that the content in redness of fibres could be classed in the following way: heart > soleus > plantaris > most parts of the gastrocnemius – but some deeper parts of the plantaris and gastrocnemius were redder than the soleus. In the plantaris and gastrocnemius, in fact, eight types of fibre were distinguished, three only in the soleus; these types fell into three main groups. At one end of the spectrum were fibres with a high capacity to use glycogen, moderate lipid metabolism and low oxidative metabolism with low myoglobin content (Group I); in the middle those with moderate glycogenolytic ability, low lipid metabolism and high oxidative power with high myoglobin content (Group II); at the other end, those with low glycogen breakdown, and very high lipid metabolism together with high oxidative power and high myoglobin content (Group III). These results were obtained both by histochemical methods and by quantitative enzymic assays; in each case comparison was made with heart muscle. It is significant that the fibres of Group III when found in predominantly white muscle are situated in positions where the contraction required of them would be slower and more prolonged than in the rest of the muscle. The question was posed, but is not yet answered, as to whether the fibres of Group II have different contraction characteristics from those of the other two groups.

MUSCLE PROTEINS. This was begun by Meyer & Weber (1) as long ago as 1933. They found similar amounts of myosin in the two kinds of muscle, more stroma protein in the red and more water-soluble protein in the white. This distribution was confirmed by the Báránys and their collaborators in 1965, who also found the actin content similar in the two cases.[3] With the homogeneous myosin preparations which they made from white and red muscles of the rabbit they found ATPase activity to be two or three times as high in the white muscle as in the red. This was true for Ca or EDTA activated myosin, and also for actomyosin made from it. Gergely and his colleagues also carried out a thorough comparison of the ATPase activity and some other properties of the contractile protein of red and white muscle from the rabbit. Using myofibrils, natural actomyosin and myosin they found the low ATPase activity of the protein from red muscle; the ATP sensitivity of the actomyosin from the red muscle was only about half that

---

[1] E.g. Denny-Brown (1); Nachmias & Padykula (1); Ogata (1, 2).
[2] Romanul (1); Dawson & Romanul (1).
[3] Bárány, Bárány, Reckard & Volpe (1). See also M. Bárány (1).

of actomyosin from the white, although the final viscosity was the same for both.[1] Structural differences between myosins from the two sources were suggested by the following observations:

(1) Red-muscle myosin ATPase (Ca-activated) has a higher apparent activation energy.[2]

(2) The activation phenomena by N-ethylmaleimide[3] differ somewhat in the two cases.[4]

(3) Red-muscle myosin in more resistant to tryptic digesion.[5]

Maddox & Perry (1), comparing myosins from pigeon-breast muscle and rabbit skeletal muscle, found that the values for sedimentation and diffusion coefficients were identical in the two cases.

Varga and his collaborators[6] recorded no demonstrable difference in antigenic properties between the myosin of white muscle of rabbit semitendinosus and that of the soleus of the cat. Gröschel-Stewart, however (1),[7] has described experiments showing the immunological difference between the myosins of red and white muscle. The γ-globulin fraction of immune sera prepared from the sternal part of the pectoralis (mainly red) of the human was conjugated with fluoresceinisothiocynate; when this labelled antibody was applied to frozen sections of muscle, it was seen that reaction took place with the fibres of red muscle, not with those of white muscle. At least two types of fibre could be found in the same muscle: in the psoas few fibres reacted; in the diaphragm about half the fibres became strongly fluorescent. Heart muscle behaved as red muscle.

Gutmann & Syrový (1) have drawn attention to the greater proteolytic activity and higher RNA content in slow muscles (including those with Feldstruktur)[8] than in white muscle. They suggest that greater protein turnover is involved in the metabolism of slow muscles. It has also been shown that in the chicken skeletal muscle, diaphragm and heart, in this order, there is increasing proteolytic activity and nucleic acid content.[9] The metabolism of frog muscle fits into this pattern[10] but an exception is the slightly higher proteolytic activity (by about 40%) in the fast extensor digitalis longus of the rat as compared with the soleus. Biron, Dreyfus & Schapira (1) also found greater incorporation of labelled, mixed amino acids into certain proteins of the red muscles of the rabbit than into those of the while muscles. Myosin, aldolase and lactic dehydrogenase preparations

---

[1] Seidel, Sréter & Gergely (1). It is to be noted that Ermini & Schaub (1) found with adult rat muscle similar ATPase activity in red and white myofibrils.

[2] Sréter, Seidel & Gergely (1).

[3] Ch. 12.   [4] Sréter, Seidel & Gergely (1).

[5] Gergely, Pragay, Scholz, Seidel, Sréter & Thompson (1).

[6] Varga, Köver, Kovaćs, Jókay & Szilagyi (1).

[7] Also Gröschel-Stewart & Doniach (1).

[8] P. 464 below.   [9] Cited by Gutmann & Syrový.

[10] Personal communication from Professor Gutmann.

were made from the muscles in both cases, and in those from the red muscle about 50 % higher incorporation was found. To what extent greater enzymic activity on the one hand and increased permeability of the muscle membrane to amino acids on the other hand play a part is still uncertain. Dreyfus (1) measured permeability of red and white muscles by following the entry of the non-metabolised α-aminoisobutyric acid after intravenous injection; the content was much higher in the red muscle. It was suggested that this greater permeability was connected with the greater state of tension in the red muscle, since the difference was much less in animals under anaesthesia.

A very interesting study has been made by Hamoir and his collaborators[1] on the myogen proteins of carp white and red muscle. The electrophoretic patterns were very different, a number of peaks (designated 2, 3 and 5) appearing in white muscle preparations but not in those from red muscle. The sedimentation diagrams were similar and in both cases showed about one-third of the material sedimenting very slowly at a rate of 1.7S. This slowly sedimenting portion was however different in the two types of muscle – with white-muscle myogen it consisted mainly of the electrophoretic components 2, 3 and 5, while with red-muscle myogen it was made up almost entirely of myoglobin. The proteins of components 2, 3 and 5 were of very low molecular weight – 9000–13000; evidence for their abnormal amino-acid composition was found, including absence of tryptophan and methionine and presence of much phenylalanine. The low molecular weight fraction failed to catalyse production of fructose-6-phosphate from glycogen or glucose-1-phosphate, or of lactic acid from fructose-1,6-diphosphate. Although these results show that this fraction cannot be responsible for the whole glycolytic pathway, they do not exclude the possibility that enzymes catalysing individual stages might be contained therein. From an extensive survey of the literature Hamoir et al. suggested that the group of proteins described is characteristic of fish white muscle. At present its function is obscure.

As early as 1953 Connell (2) had described markedly different electrophoretic patterns of the muscle extracts (made with buffered 0.05 M KCl) from 20 different species of fish. This species specificity has been confirmed in much recent work briefly reviewed by Hamoir (5).[2] Comparisons between aquatic and terrestrial amphibians suggest a correlation between the aquatic life and presence of these low molecular weight albumins. Possibly their high negative charges at neutral pH could prevent loss of cations into the external medium; but this would not explain their absence in fish red muscle.

[1] Hamoir & Konosu (1); Konosu, Hamoir & Pechère (1); Pechère & Focant (1).
[2] See also Bushana Rao, Focant, Gerday & Hamoir (1).

THE SARCOPLASMIC RETICULUM. Pellegrino & Franzini (1) in 1963 found in electron micrographs of red and white muscle from the rat a greater amount of sarcoplasmic reticulum in the latter. Both showed two triads per sarcomere, but the cisternae at the A band level were much more developed in the white fibres. They also noted the large size of the mitochondria in the soleus.

Soon afterwards Sréter & Gergely (1)[1] compared the Ca uptake by particulate preparations (sedimentating at 8000–30000g) from red and white muscle of the rabbit. There was an astonishing difference in activity, the uptake by particles from the white muscle being some twenty times as great per mg of protein. The ATPase of the red-muscle preparation showed considerable inhibition by azide, indicating much mitochondrial contamination, which was absent from the white-muscle fraction. Marked differences were shown in electron micrographs of the particulate preparations from the two sources. That from white muscle consisted mainly of tubular elements, while in the red-muscle preparations rounded vesicles predominated. The relationship of these fragments to the intact reticulum in the cell is still a matter of conjecture. Sréter et al.[1] have discussed the possibility that the red-muscle reticulum really does operate a Ca pump but that some difference in the elements in this tissue leads to their readier inactivation during preparation of the particulate fraction in vitro. This seems very likely in view of the later results of Samaha & Gergely (1) with human tissue, where the particulate preparations from red muscle were 50 % as active as preparations from white muscle. The experience of several workers with the Ca-binding particles of heart muscle,[2] where great preparative difficulties have been met and overcome, points in the same direction.

SLOW FIBRES OF VERTEBRATE STRIATED MUSCLE

Krüger and his collaborators have accumulated evidence for the presence, in certain frog and bird muscles, of a special type of fibre showing different structure ('Feldstruktur') from the familiar 'Fibrillenstruktur' e.g. of the frog sartorius.[3] The Feldstruktur is seen in cross-section to consist of larger ribbon-like areas; such fibres are characterised by a capability of maintaining long-lasting contractures. Kuffler & Vaughan Williams (1, 2) in 1953 showed that frog skeletal muscle contains two types of nerve connections. One is associated with large nerve fibres, stimulation of which leads to propagated muscle impulses and rapid contractile response; the other is

---

[1] See also Sréter (1); Gergely, Pragay, Scholz, Seidel, Sréter & Thompson (1).
[2] See below.
[3] Krüger (1) for an account of some thirty years' work.

associated with small nerve fibres, stimulation of which causes local activation at numerous points around multiple small-nerve junctions and slow contraction. Schaechtelin (1) also found histological and physiological evidence for these two types in the rectus abdominis of the frog. Then in 1961 Hess (1) showed conclusively for the first time that *Feldstruktur* occurs also in certain mammalian muscles. Using the extrinsic muscles of the guinea-pig eye he found, by phase contrast and electron microscopy, *Fibrillen-* and *Feldstruktur* with two kinds of nerve endings: single '*en plaque*' endings on each fibre of the former, and in the latter several '*en grappe*' endings on one fibre. Hess & Pilar (1), using intracellular micro-electrodes, could distinguish the two fibre types with the two types of innervation in the extraocular muscles of the cat; these corresponded to the twitch and slow muscles of the frog, the former giving propagated action potentials, the latter only non-propagated junction potentials.

Peachey & A. F. Huxley (2) at this same time isolated twitch and slow muscle fibres from the frog ileofibularis. Each was tested for type of contraction by means of direct electric shocks and then examined in the electron microscope. The fibres which responded to single shocks with propagated twitches contained fibres less than 1μ across with clearly defined sarcous elements. The fibres which did not twitch but responded to repeated shocks with local slow contractions were composed of large ribbon-like fibres fused together. Peachey & Huxley found triads in the fast fibres but none in the slow fibres. Page (3) however later found triads in the latter, but only at every fifth or sixth sarcomere, while in the fast fibres they were present at every Z line.

A. F. Huxley (4) has reviewed the question of slow fibres. He points out that, though it is now clear that slow tonic fibres equivalent to those of the frog do exist in mammals, these are not to be confused with the slow phasic fibres of slow muscles such as the soleus which 'are slow only in the sense of giving twitches of longer time course than some other muscles.' He envisages the possibility of a whole spectrum of fibres varying continuously as regards structure, speed of contraction, type of electric response and innervation. An example of this would be the reaction (described by Gutmann & Hanzlikova (1)) of the slow-phasic soleus of the rat to give contracture with acetylcholine or caffeine – a reaction resembling that of slow-tonic muscles.

Gutmann and his collaborators have made a study of metabolic differences between fast muscle (in the posterior latissimus dorsi) and slow muscle of the *Feldstruktur* type (in the anterior latissimus dorsi) of the chicken.[1] They found greater proteolytic activity and a higher content of RNA in the latter; these differences were not apparent in the newly-hatched bird, but were established after about 30 days. Gutmann &

[1] Syrový, Hájek & Gutmann (1, 2); Hajek, Gutmann & Syrový (1); Syrový & Gutmann (1).

/

Syrový (1) also found that radioactive methionine was incorporated into the slow muscle about twice as quickly as into the fast muscle, although the amino-acid pool was of the same size in both. These differences between the *Feldstruktur* fibres and the fast fibres seem to agree with those observed between red muscles, e.g. the soleus, and fast muscles. The glycogen content and aldolase activity were also low in the slow tonic muscle.

Costantin, Podolsky & Tice (1) have recently used fibres from the tonus bundle of the frog iliofibularis (identified in the electron microscope as slow fibres) for the study of the effects of Ca application after removal of the sarcolemma. With solutions containing 1–10 mM Ca the contractile response was always slow and spread more extensively than in fast fibres; relaxation was very slow. When oxalate was added, opaque deposits formed in the lumen of the internal membrane system; it thus appeared that a Ca pump was at work, though less efficiently than in fast muscles. Moreover it can be concluded that the slow shortening rate of these fibres is an inherent property of the contractile mechanism, not depending on excitation-contraction coupling or changed by provision of a large excess of Ca.

HEART MUSCLE

STRUCTURE. The structure of the muscle cells of the heart is complicated and for many years the histological picture was interpreted as meaning that the cells formed a syncytium, being continuous without boundaries both laterally and longitudinally.[1] Schäfer (1) however in 1910 portrayed broad striated cells (each with its nucleus) joined end to end at clearly defined junctions to form fibres. These cells, it was believed, were characterised by their capacity to branch, the branches (disposed in many planes) joining up with cells of neighbouring fibres.[2]

The electron microscope in the nineteen fifties[3] brought evidence in favour of the cellular rather than the syncytical arrangement. The junctional lines of Schäfer (now known as the intercalated discs) were clearly shown to be cell borders. These discs, consisting of the interdigitating plasma membranes of two cells, cross the fibre normally to the long axis. Between the plasma membranes is a space of some 140 Å containing a dense material which seems to fasten the fibrils terminating there. According to Moore & Ruska (1) the myofibrils formed a 'synfibrillar system' consisting of branches composed of 200–1000 myofilaments; but Spiro (2) later considered that the branched appearance was due to deep invaginations of the sarcolemma. The structure of the myofibrils was essentially the same as in skeletal muscle, and all the classical bands were visible.

---

[1] E.g. Kölliker (2) II, p. 282.   [2] Also Ranvier (2).
[3] E.g. Sjöstrand (1); Moore & Ruska (1); Ham (1); Spiro (1).

Porter & Palade (1) in 1957 described the sarcoplasmic reticulum in cardiac muscle as small in amount and unevenly distributed. Components consisting of two profiles (representing sections through small vesicles or narrow tubules) could sometimes be made out; these often ran transversely opposite the *I* band. Nelson & Benson (1) in 1963 observed that the transverse tubular system seemed to be continuous with the sarcolemma, though this was not proved at the time.[1] Rostgaard & Behnke (1), using a glutaraldehyde perfusion–fixation method to minimise translocation of substances, were able to locate ATPase activity in the lateral elements of triads and in the flattened cisternae of intercalated discs.[2]

PROTEINS OF THE CONTRACTILE MECHANISM. There has been much difference of opinion about the molecular weight of cardiac myosin. Thus while Gergely & Kohler (2) in 1957 reported a value of 500000 ($\pm$ 10 %), similar to that of skeletal myosin, Ellenbogen and collaborators in 1960 gave 225000.[3] Brahms & Kay (2) on the other hand in 1962 found the much higher value of 758000. Kay (3) still maintained this value in 1965, but in the meantime the Pittsburgh group had revised their figure to 534000. Luchi, Kritcher & Conn (1) also in 1965 gave 500000. This figure was obtained after purification by ammonium sulphate precipitation in presence of 2M LlCl, which was believed to dissociate impurities and did indeed enhance the ATPase activity. There has been much discussion concerning the reasons for these discrepancies; knowledge is continually increasing of the pitfalls to be avoided in the methods used, and of the precautions necessary to preserve the properties of the native myosin. The balance of evidence at present indicates a similar molecular weight, about 500000, for myosin from all types of mammalian striated muscle, including heart muscle.

Iyengar & Olson (1) found a close resemblance in the amino-acid composition of dog cardiac myosin to that of rabbit skeletal myosin reported in the literature. Bárány and his collaborators[4] however found fewer cysteine residues in cardiac myosin and heavy meromyosin than in the proteins from skeletal muscle. Differences between the proteins from the two sources include greater resistance to proteolysis[5] and lower ATPase activity of the cardiac myosin.

The lower ATPase activity of the cardiac protein was noticed by Bailey (1) in 1942, and has been found by many workers since, though there is some variation in the ratios given for the activities. For example, in the

---

[1] Ch. 18.
[2] They took this to indicate the presence of an ATPase-driven calcium pump.
[3] Ellenbogen, Iyengar, Stern & Olson (1).
[4] Bárány, Gaetjens, Bárány & Karp (1).   [5] Mueller, Theiner & Olson (1).

experiments of Bárány *et al.* Ca-activated ATPase of skeletal myosin, actomyosin and myofibrils was about three times as high as for the preparations from the heart. When Mg-activated actomyosin was compared the ratio varied between three and eight. Luchi *et al.* found with myosin prepared by their method that the skeletal protein was only some 40% more active than the heart protein. Katz, Repke & Cohen (1) with highly purified actomyosin from heart and skeletal muscle showed that the curves relating ATPase activity (in presence of Mg) to the Ca ion concentration were very similar in shape, although the heart protein had less than one-tenth the activity of the skeletal muscle protein. In both cases half-maximal activity was reached at 10^{-6} M Ca. Thus the low activity of heart actomyosin could not be explained by low sensitivity to Ca.

Ebashi and his colleagues[1] have found both tropomyosin and actinin in cardiac muscle.

METABOLISM. As we have seen, heart muscle was often used as a 'red' muscle in making comparisons with the metabolism of white muscle although, as Lawrie (1, 2) pointed out, the heart myoglobin content is low. Nevertheless its blood supply is so good that it functions practically entirely aerobically; according to Lawrie even under anaerobic conditions there is no detectable fall in pH. Bing & Michal (1) found no evidence for any glycolysis in the normally beating heart *in situ*, and the heart can use all substrates circulating in the blood; under physiological conditions of feeding and fasting it can change from using predominantly carbohydrate to almost exclusive utilisation of fatty acids.[2]

EXCITATION-CONTRACTION COUPLING. Much difficulty has been experienced in demonstrating the presence of a calcium pump in heart sarcoplasmic reticulum comparable to that in skeletal muscle. Finkel & Gergely (1) in 1961 could find no relaxing effect of microsomal preparations from the heart, tested on myofibrillar preparations from both heart and skeletal muscle; moreover relaxing factor from skeletal muscle, active with skeletal myofibrils, had no effect on cardiac myofibrils. However in 1964 Fanburg, Finkel & Martonosi (1) found that myofibrils prepared from the heart in the usual way contained an ATPase not inhibited by EGTA. Freed from this contaminating ATPase by much washing, glycerinated cardiac fibres showed inhibition of synaeresis and ATPase activity on addition of either EGTA or cardiac microsomal fraction. There was good correlation between this inhibition and the accompanying lowering of Ca concentration. This correlation was also seen by A. Weber, Herz & Reiss (4). Fanburg & Gergely (1) further showed clearly that microsomal particles from the heart

---

[1] Ebashi, Iwakura, Nakajima, Nakamura & Ooi (1)　　[2] Olson & Piatnek (1).

do accumulate Ca in presence of ATP and Mg ions. The ratio of Ca removed to ATP hydrolysed however was much lower, even after removal or inhibition of non-microsomal ATPase, than with skeletal-muscle particles. Carsten (3) by purification of the microsomal fraction by means of gradient centrifugation in sucrose, succeeded in obtaining a preparation showing an initial rate of activity of Ca uptake twenty times that described by Fanburg and Gergely. She emphasised the great lability of the heart particles, and overcame this by adding ascorbic acid in the early stages of the preparation and by the purification procedure. Lee (1) also in 1965 prepared cardiac relaxing factor as potent as its skeletal-muscle counterpart in Ca uptake; to achieve this he found it necessary to add to the heart preparation an ATP regenerating system (phosphocreatine and creatine kinase) as well as oxalate.

It was from time to time suggested that the cardiac sarcoplasmic reticulum caused relaxation not by Ca uptake but by producing a soluble relaxing factor.[1] Weber, Herz & Reiss (5) could find no evidence for this.

It seems then that the heart operates a Ca pump, and that the difficulties in its demonstration have now been overcome.

It has been known since the time of Ringer that the strength of the heart's beat is dependent on both Ca^{2+} and Na^+ concentrations in the bathing medium; Wilbrandt & Koller (1) in 1948 showed that the strength of beat remained roughly constant if the ratio Ca/Na was kept constant and depended on the Ca/Na^2 ratio in the Ringer's fluid. This finding was confirmed by Lüttgau & Niedergerke (1) in 1958 and explained by them as due to competition between Ca and Na ions for hypothetical negatively-charged molecules in the cell membrane.[2] Thus:

$$Ca + Na_2R \longrightarrow CaR + 2Na$$

where CaR was a compound in some way activating tension on entry into the cell while the compound Na_2R was inactive. Niedergerke (2) using ^{45}Ca went on to show that more Ca ions are taken up by the heart made to contract by bathing in media containing reduced Na^+ concentration or high K^+ concentration at constant Na^+ concentration.[3] The rate of Ca^{2+} efflux was also increased and there was a marked net efflux when the heart was returned to normal Ringer solution. The strength of contracture and the Ca uptake both depended on the degree of Na depletion. The rate of attainment of maximum contraction was however much higher than the rate of attainment of maximum Ca concentration in the muscle; similarly on return to normal Ringer solution the rate of relaxation was much greater than that

[1] E.g. Honig & Stamm (1).
[2] For the Na carrier hypothesis arising out of work on nerve axon see Hodgkin, Huxley & Katz (1); Frankenhaeuser & Hodgkin (1); Frankenhaeuser (1).
[3] Also Niedergerke (1).

of the Ca efflux. Niedergerke suggested that only a fraction of the exchangeable Ca is released at the inner membrane and involved in starting
and maintaining the contraction; after cessation of contraction it would be
rapidly inactivated and stored in this form before expulsion.

Both Niedergerke and Katz remarked on the apparent sparseness of the
sarcoplasmic reticulum in heart muscle as compared with skeletal muscle,
which would make the former very dependent on externally supplied Ca.
Since 1963 it has become clear, as we have seen, that the heart does contain
a calcium pump; it is nevertheless very possible that the limited amount of
sarcoplasmic reticulum does not contain enough Ca for maximum activation of the contractile process and that any factor which increased Ca entry
into the fibres would help to increase size and strength of contractions and
perhaps also to replenish Ca stores.

A. M. Katz (3) has also considered the situation *in vivo*. Increased Ca
uptake has often been observed in situations involving increased performance of the heart; thus Langer (1), using the papillary muscle of the
dog's ventricle, had found that 40 % increased rate of contraction was accompanied by increased Ca uptake from the perfusing fluid, in amount
2×10^{-5} moles/kg wet wt. Katz calculated that this increase could give an
increase in actomyosin ATPase activity of 40 %. The effect of adrenaline on
the heart, and suggestions concerning consequent increase in its ionized
calcium, have been considered in chapter 18.

DEVELOPING MUSCLE

HISTOLOGICAL AND PHYSIOLOGICAL OBSERVATIONS. We have been
considering the histological characteristics of the slow red and the fast white
muscles. Knoll (1) already in 1891 suggested that the protoplasm-poor
fibres were more highly developed, and Denny-Brown in 1924 made observations on kittens which showed that at birth all the fibres were small and
densely packed with granules, while after two weeks the gastrocnemius had
20 % of large clear fibres. This muscle still contracted slowly and it only
began to acquire rapidity of movement after another two or three weeks.
In much later work Dubowitz (1) besides noting the size of the fibres,
diagnosed their degree of differentiation by examining histochemically their
content in phosphorylase on the one hand and in certain oxidising enzymes
on the other. With human, guinea pig, rabbit and hamster the enzymic
differentiation (increase in the former enzyme and decrease in the latter
enzymes) had already taken place by the time of birth. With the rat, less
active and mature at birth, differentiation did not occur till some ten days
later. Goldspink (1, 2) making fibre-diameter measurements in the mouse
biceps brachi found that in the adult muscle this showed two peaks –

around $20\,\mu$ and $40\,\mu$. In the period of growth of the animal from 8 g to 20 g there was no increase in number of fibres, but a rapid increase in the proportion of large ones. All gave a positive histological reaction for succinic dehydrogenase, more marked in the smaller ones. An interesting point was that, during the early period after birth, the newly formed and differentiating fibres seemed to be freely permeable to inulin; this high permeability may be significant at this time of rapid conversion of small to large fibres.

The greater differentiation of the white muscles is borne out by physiological studies such as that of Close & Hoh (1),[1] who compared the intrinsic speed of shortening of the soleus and the flexor digitalis longus of 1- to 2-day old kittens with that of the adult. The intrinsic speed of the red muscle did not change but that of the white increased. In this case even at birth the white muscle was faster but earlier work of Close (1, 2) on the rat and mouse showed that in early stages both muscles have similar intrinsic speeds; possibly this is true also of the kitten in intra-uterine life.

ATPASE ACTIVITY OF THE CONTRACTILE PROTEINS. The finding in 1942 of the ATPase activity of the structural protein myosin naturally stimulated interest in the behaviour of this enzyme during embryonic development and in the first days of post-natal life. The earliest papers were those of Moog (1) with chick muscles and of Herrmann & Nicholas (1, 2) with rat muscles. Little can be gleaned from these concerning myosin ATPase, since the amount of 'total apyrase' of muscle homogenates was followed, and the 'myosin' fractions prepared in order to follow the increase in this protein must have been very highly contaminated with cell particles especially in the early stages of embryonic development.

D. S. Robinson in 1952 (1, 2, 3) with chick embryos went carefully into these two questions – the presence of other ATPases and the proportion of 'myofibrillar' protein present in the fraction prepared by extraction at high salt concentration followed by dilution to $I = 0.12$ in the presence of 15 % ethanol. He devised a method for assessing the protein content of the nucleoprotein present; then, by estimations of nucleic acid, could correct the dilution precipitate for its nucleoprotein content. He found that after fourteen days incubation of the eggs the myofibrillar fraction was only some 7 % of the total muscle protein; and that at this stage the dilution precipitate contained more than 60 % of nucleoprotein. The myofibrillar content rose to about 15 % of the total muscle protein by the time of hatching. It is interesting that the breast muscle would respond to electrical stimulation after only 8 days' incubation, so that some sort of contractile apparatus must have been already functional notwithstanding the low actomyosin

[1] Also Buller, Eccles & Eccles (2); Buller & Lewis (1).

content. With the rat de Villafranca (2) confirmed an earlier observation of Herrmann, Nicholas & Vosgian (1) that the myosin ATPase increased in activity just at the time contractions began – about the sixteenth day of prenatal life. From this time the ATPase activity rose rapidly, the rate falling off about three weeks after birth to reach the adult level at about fifteen weeks. With the chicken on the other hand, the specific activity of the myofibrillar fraction was at the adult level at the time of hatching.[1]

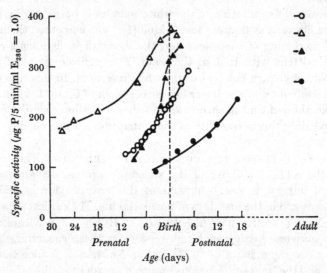

Fig. 93. Change in specific ATPase activity of purified myosins from skeletal muscle during development. Adult values indicated by symbols in top right-hand corner of figure. ○, rabbit; △, guinea pig; ▲, fowl; ●, rat. (Trayer & Perry (1).)

More recent work has concentrated on the study of highly purified myosin prepared at different stages of development. Thus Bárány and his collaborators with 2-day old rabbits prepared (by means of ammonium sulphate fractionation) myosin homogeneous in the ultracentrifuge and free from nucleic acid.[2] The Ca-activated ATPase was about 60 % of the adult value. Trayer & Perry (1) used foetuses of the rabbit, guinea pig and rat, and chick embryos.[3] Again each myosin preparation was highly purified, by chromatographic procedures. The increase in ATPase activity with development could be clearly seen; in the earliest foetuses used the activity was only one-third of the adult value. The rat again lagged behind the other three animals in reaching activities characteristic of the adult (fig. 93). The

[1] Robinson (2); Ohshima, Maruyama & Noda (1).
[2] Bárány, Tucci, Bárány, Volpe & Reckard (1).
[3] Also Perry & Hartshorne (1).

differences in specific activity were maintained throughout changes in pH or other environmental conditions.

The increasing sensitivity during development of the myofibrillar ATPase to EDTA has been noticed by Ohshima and his collaborators for the chick. Full sensitivity was reached about the time of hatching. Holland (1) with rabbit myofibrils similarly found increasing sensitivity to EGTA, continuing until the adult response was given about 25 days after birth. As he remarks, these effects are probably due to increasing amounts of EGTA sensitising factor (containing tropomyosin + troponin) associated with the myofibrils.[1] Ermini & Schaub (1) with rat myofibrils found inhibition with EGTA to increase from 40 to 60 % over the six weeks after birth.

OTHER PROPERTIES OF FOETAL MYOSIN AND ACTOMYOSIN. The earliest observation I have found on foetal myosin is that of Ballantyne (1) in 1895, who concluded that the 5-months human foetus must contain myosin, since from this point of development *rigor mortis* was usually observed after death. In the work we have just been considering the presence of myosin at different stages was inferred from ATPase activity and solubility characteristics, but efforts have also been made to detect the developing myosin by more specific methods. Early work in this field was that of Csapo & Herrmann (1) in 1951, using the viscosimetric method. This measures actomyosin by means of the fall in viscosity of solutions on ATP addition. They could detect actomyosin in chick muscle before the tenth day of development and by the time of hatching it amounted to about 30 mg/g of muscle and reached the adult level (about 60 mg/g muscle) by the thirtieth day after hatching.

Immunological methods have been much used. Thus Holtzer & Abbott (1) tested for myosin by means of fluorescein-labelled antimyosin. They found that the first cross-striation could be detected in this way in chick myoblasts about the third day. From this time glycerinated material would contract on addition of ATP. Ogawa (1) recorded that by immunological methods actin and myosin could each be detected in chick embryos after four days' incubation, the appearance of the actin preceding that of the myosin. Nass (1) prepared actomyosin precipitated several times in the classical way from muscle of frog embryos. She then used adult anti-actomyosin to purify the embryo actomyosin by precipitation. The amount precipitable increased linearly from the tail-bud stage to the time of hatching; thereafter the increase was more rapid. She further found that the contraction of threads made (by the method of Hayashi (1)) from this actomyosin was at the hatching stage some 60 % of the adult contraction.

Pinaev (1) examined the flow birefringence and the intrinsic viscosity of

[1] Ch. 12.

the actomyosin solutions prepared from skeletal muscle of 28-day embryos, newborn, 10-day old and adult rabbits. There was a progressive increase in these values, from which change in shape of the actomyosin particles was deduced. Trayer & Perry (1) on the other hand, with their highly purified myosin, found no difference in sedimentation and diffusion coefficients between the foetal and adult samples from the rabbit. Ohshima, Maruyama & Noda (1) also reported increase in particle length and in sedimentation coefficient of actomyosin during chick development, while these properties remained unchanged for myosin.

Further differences between adult and foetal myosin lie in the quantum yield of the tryptophan fluorescence[1] and in the nature of the light sub-units obtained on dissociation of myosin.[2] In 1967 also Johnson, Harris & Perry (1) showed the presence of 3-methylhistidine in adult myosin as a constituent of the primary structure of HMM; in the work of Trayer, Harris & Perry (1) on foetal myosin a much smaller content was found. It is relevant that red and heart muscle myosin, both as we have seen showing lower ATPase activity than white muscle myosin, contain less 3-methylhistidine than the latter, when compared within the same species. Thus the extent of methylation seems to be in some way related to the functional activity of the muscle, but how is not yet apparent. Perry (7) has recently suggested[3] that myosin exists in two isoenzyme forms; from their content of 3-methylhistidine (1.6 residues per molecule of HMM) the usual preparations of adult muscle myosin may contain 80 % of the adult form (with 2 residues/HMM) and 20 % of the foetal form lacking this methylated amino acid.

DEVELOPMENT OF THE MYOFIBRIL. A recent advance in our knowledge of the development of myosin has been the identification of the polyribosomes synthesising this protein. Heywood, Dowben & Rich (1) used the supernatant obtained after gentle homogenisation of leg muscle from 14-day chick embryos; by layering on a sucrose gradient they obtained fractions containing ribosomes of various sizes. Four fractions were separated, Fraction A containing the largest polysomes, fraction D the smallest. Each was incubated in vitro with the necessary ingredients for protein synthesis[4] – Mg ions, buffer, mercaptoethanol, ATP, GTP, labelled ^{14}C amino-acid mixture, chicken-liver tRNA and pH 5-enzyme from muscle. The reaction was ended by adding ribonuclease. Fractions A and B were very active in protein synthesis. After incubation, unlabelled carrier myosin was added and the myosin was then isolated by three precipitations at low ionic strength. Protein from fraction A had the highest radioactivity/mg of

[1] Trayer, Perry & Teale (1).   [2] Perrie, Stone & Perry, quoted in Perry (9).
[3] Also Trayer & Perry (1); Perry (9).
[4] E.g. Protein Biosynthesis, ed. R. J. C. Harris.

ribosomal RNA. This protein from fraction A also migrated on acrylamide gel in 12 M urea concurrently with the myosin marker. On examination of the polysomes in the electron microscope it was seen that those in fraction A contained approximately 50–60 ribosomes in each cluster. Since it was known from the earlier work of Warner, Rich & Hall (1) that the polysomes synthesising haemoglobin (molecular weight 17 000) are pentamers, the involvement of these large polysomes in the synthesis of myosin fits well with what is known of the molecular weight of myosin sub-units.

Heywood & Rich (1) later showed that fraction A was small in amount until about the fourteenth day of incubation. Fraction B was important at the tenth day, while fraction C like fraction A increased later. Isolation and purification of the proteins formed in each case showed: (a) that with fraction A only one peak – that of myosin – appeared on the acrylamide gel; (b) that with fraction B there were two or three peaks, one with the mobility of actin; (c) that with fraction C the more rapid of two peaks corresponded to tropomyosin (see fig. 94). This evidence of the early formation of actin is in agreement the immunological evidence and with electron microscope observations made by Allen & Pepe (1), who found in sections of embryo chick muscle that thin filaments randomly dispersed in the cytoplasm appeared before any thick filaments; the latter were first seen about the end of the third day. The thin and thick filaments became aligned by some unknown mechanism, forming unstriated fibrils. Then a banding pattern supervened, caused by the occurrence of tubules at $1.5\,\mu$ intervals along the filaments; later (about the sixth day) dense Z lines were seen associated with the filaments at the level of the tubules. I bands, H zones and M lines were only apparent after the Z lines were formed.

There has been much controversy concerning the site of formation of the myofibrils and many workers have suggested their origin in the microscomes and mitochondria. To take only two examples: Godlewski (1) in 1902 concluded that they were formed by fusion or aggregation of cytoplasmic particles; Levi & Chèvremont (1) in 1941 that they arose from transformation of elongated mitochondria. These ideas have been discarded, but the extremely active protein synthesis which can be demonstrated in the particulate fractions of the muscle cell of very young chicks has been emphasised by Winnick & Winnick (1) in 1960. Perry & Zydowo (1, 2) have shown that washed myofibrils of adult rabbit or hen skeletal muscle contain some 60 % of the total muscle RNA, and they have prepared a ribonucleoprotein from these fibrils. They give good reason for discarding contamination with cell particles as the explanation, and suggest that this RNA protein may have a role in synthesis of the contractile proteins. It is certainly difficult to visualise how proteins formed in the sarcoplasm could reach the fibril interior to support the turnover known to take place there. These observations led Winnick & Winnick to prepare an RNA-protein fraction from myofibrils labelled after injection of ^{14}C valine. They found the specific activity there four times as high as in the original myofibrils. If such a mechanism is used in the adult, it might well be a continuation of one going on during development.

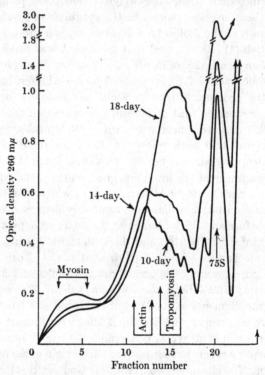

Fig. 94. Sucrose gradient analyses of cytoplasmic extracts from 0.7 g of embryonic chick leg muscle at different ages of embryological development. The positions labelled myosin, actin and tropomyosin correspond to fractions A, B and C mentioned in the text. Single ribosomes are labelled 75S. (Heywood & Rich (1).)

It may be significant that Robinson (3) in the early stages of development found a high percentage of the Ca-activated ATPase was not precipitable on dilution of an extract made with buffered 0.6 M KCl. He made the tentative suggestion that a watersoluble prototype of myosin might be present. This still remains a possibility and the fact that Holtzer[1] did not find myosin–antibody binding in the sarcoplasm does not rule it out. Holtzer describes the lateral growth of single myofibrils as taking place by apposition of new material (presumably synthesised *in situ*) while growth in

[1] See H. Holtzer (1).

length went on by meristematic addition of new sarcomeres. Allen & Pepe could find no evidence for any relation between other cellular elements and myofibrillogenesis. Their electron micrographs show many free polyribosomes in the cytoplasm, the long chains containing 70–75 ribosomes. Amongst the fine electron micrographs of Heuson-Stiennon (1) showing myofilaments mixed with polysomes in the cytoplasm of rat embryonic muscle about the seventeenth day, there is one showing a myofilament attached to the end of an aggregate of ribosomes (see fig. 95, pl.).

OTHER STUDIES OF THE ENZYMIC EQUIPMENT. In 1956 Kasavina & Torchinsky (1) applied paper micro-electrophoresis to the problem of estimating certain enzymes in muscle sarcoplasm during development of rabbits, rats, guinea pigs and hens. Aldolase and glyceraldehydephosphate dehydrogenase were found to accumulate as development progressed. Again a distinction was found between the animals born more mature – guinea pigs and hens – and those born less mature – rabbits and rats; the increase was perceptible by the end of the embryonic period in the former, but with the latter came on only gradually during the post-embryonic period.

In 1962 Hartshorne & Perry (1) described a very thorough study of the sarcoplasmic proteins in adult and foetal skeletal muscle of the rabbit, using chromatography on diethylaminoethylcellulose and starch-gel electrophoresis. The patterns given were very different; in particular the more positively charged protein components were low in the foetal-muscle sarcoplasm and their amount increased rapidly after birth. One of these components was the enzyme aldolase. In view of what has been said earlier about the relatively great importance of glycolysis in white muscle, it is interesting to find that Hartshorne & Perry, in a few experiments using these methods on adult heart and on separated red and white adult muscle, found the aldolase content much higher in the white muscle: e.g. the activity was ten times as great there as in the adult heart, and the activity of the latter was comparable with that of foetal muscle.

Stave (1) in 1964 examined in the rabbit the activity of certain enzymes of the glycolytic chain – phosphoglucomutase, phosphoglucoisomerase, glyceraldehydephosphate dehydrogenase, glycerophosphate dehydrogenase, and lactic dehydrogenase; he found all very low in foetal life but rising after birth to become near the adult values in about 30 days. Moreover Dawson, Goodfriend & Kaplan (1) found that there were changes in the molecular form of lactic dehydrogenase during development. In the rabbit gastrocnemius the foetal muscle contained a relatively high proportion of the H isoenzyme; later the M form increased at a striking rate.[1] Grillo (1)

_____
[1] Also Fine, Kaplan & Kuftinec (1).

found glycogen in the striated muscle of the chick embryo as early as the fourth day, before the appearance of phosphorylase; Grillo & Ozone (1) showed that glycogen synthetase was present then also. Rinaudo & Bruno (1) later measured increase in phosphorylase activity which went on at a moderate pace from the fifth day of incubation then much more rapidly in the days after hatching. These findings agree well with those we have just discussed and also fit in with the observations from many sources reviewed by Shelley (1) where the glycogen content of the mammalian foetus rose during the latter half of pregnancy, to fall steeply during the first day or two after birth. It should be remembered that Claude Bernard[1] in 1859 followed (by means of staining with acid tincture of iodine) the laying down of glycogen in the developing muscle cells of the calf; he also recorded the complete disappearacne of glycogen from the muscle of a kitten one day after birth. The importance of muscle glycogen in the exacting period immediately after birth is discussed by Shelley. It seems to be needed, not only for energy provision but also for keeping up the body temperature. The glycogen of the heart seems to be particularly concerned in combating anoxia.

An interesting point was made by Beatty, Basinger & Bocek (1) when they showed that certain enzymes of the pentose pathway were more important in foetal than in infant monkey muscle. It still made up only about 0.4% of the total carbohydrate metabolism. They considered that this higher content was concerned, not with energy provision, but with production of the ribose for RNA synthesis.

In 1959 Read & Johnson (1) followed the creatine kinase activity in heart and skeletal muscle of chick embryos from the fourteenth day and up to 90 days after birth. None of this enzyme could be detected in the gastrocnemius till the twenty-third day; at the twentieth day this muscle contained creatine, but no phosphocreatine and no creatine kinase; it did not react to electrical stimulation. By the twenty-third day all these constituents were present and the response to electrical stimulation was a slow wormlike contraction followed by very slow relaxation. The kinase appeared earlier in the heart, about the sixteenth day. At the time of birth the activities of the enzyme in the gastrocnemius and the heart were about one-third and two-thirds respectively of the adult values. A few years later Reporter, Konigsberg & Strehler (1) using the very delicate firefly-luminescence method for estimation of the kinase, found this enzyme to increase in the chick from as early as the eleventh day onwards.[2] In tissue cultures of the embryonic muscle they saw spontaneous contractions about the seventh day, and the earliest appearance of creatine kinase was about the same

---

[1] Bernard (2) p. 502.
[2] See also Eppenberger, Eppenberger, Richterich & Aebi (1).

time. Kendrick-Jones & Perry (1) in 1967 described the changes in activity during muscle development of a number of enzymes concerned with nucleotide metabolism – adenylic deaminase, myokinase and creatine kinase. They used a number of different animals and in all cases the activity was low (as had earlier been noticed by Stave (1) for the rabbit) in the early half of the prenatal period; it rose either a few days before birth or after birth about the fourth or fifth day. This later development was characteristic of the rabbit and the rat, while, as we have seen already in the work of Dubowitz and of Trayer & Perry, the fowl and the guinea pig tended to show mature characteristics at birth. Kendrick-Jones & Perry found that by encouraging young rabbits to move about (by taking away the nest from their cage) the formation of creatine kinase was hastened. At 16 days after birth the activity of the enzyme might be as much as double that in the muscles of undisturbed animals.

The concentration of 'energy-rich' phosphate esters in developing muscle has also been investigated. In 1935 Koschtojanz & Rjabinowskaya (1) found that phosphocreatine only appeared in foetal rabbits about the twenty-third day and then only at the level of 4 mg P%. Twelve hours after birth it had risen to about 15 mg% – about a quarter of the adult value reached after 6 weeks. Movements were elicited on the nineteenth or twentieth day by electrical stimulation or pricking, but no phosphocreatine could be detected at that time. In the work of Herrmann & Cox (1) in 1951 precautions were taken to minimise muscular movement by inducing anaesthesia in embryos more than 18 days old. Both phosphocreatine and ATP were present, albeit in small amount, about halfway through embryonic life; this of course still seems late in comparison with the reports of movement at about the eighth day. The major increase in concentration did not come until about the fourth day after hatching.

The work of Eppenberger and his collaborators[1] on the isoenzymes of creatine kinase and their behaviour during development is of great interest. Creatine kinase is found only in two organs – muscle and brain. Three isoenzymes exist – the brain enzyme (I) which migrates towards the anode, the muscle enzyme (III) migrating towards the cathode and an intermediate form (II). Foetal muscle in rat and chick contained I, which slowly disappeared during development and was replaced by II, finally by III; at an intermediate stage a mixture of isoenzymes was present. In the heart the change was to a final mixture of II and III, while in the brain I remained as the only form. Dance & Watts (1) had already suggested (from the number of peptides obtained after tryptic digestion of the purified enzyme from rabbit muscle) that the protein had a dimeric structure. On

[1] Eppenberger, Eppenberger, Richterich & Aebi (1); see also Burger, Richterich & Aebi (1).

this basis it could be suggested that the isoenzymes should be designated MM, BB and MB. Later work[1] showed that the muscle and brain enzymes from chicken and rabbit had the same molecular weight (80000) but that there were differences in amino-acid composition and in peptide maps.

An important question recently investigated is the development of the calcium pump, as we have seen essential in the regulation of muscle activity.[2] With the chick, as shown by Ohshima, Maruyama & Noda, at the fourteenth day of incubation the inhibitory effect upon adult myofibrillar ATPase of microsomal particles prepared from the embryo was much less than at the end of incubation, when the adult level was reached. The same phenomenon was observed for the rabbit microsomal fraction by Holland (1), but here low inhibitory power persisted after birth and the adult situation was only slowly attained. In agreement with these observations Fanburg, Drachman & Moll (1) showed that the rate of Ca uptake/mg of microsomal protein was very low in chick embryos till the seventeenth day; then there was a rapid rise till hatching time. With rabbits, as Holland found, the Ca-uptake capacity of the microsomes was low at birth and rose rapidly from the seventh postnatal day. Holland[3] also described a rise in MgATPase activity of the sarcoplasmic reticulum fraction (corresponding to the 'basal ATPase' of Hasselbach)[4] to a sharp peak occurring in the rabbit and rat at 8–10 days after birth, in the chicken and guinea pig just before birth. The 'extra' ATPase of Hasselbach (brought into play by the presence of Ca ions) on the other hand increased steadily and the total ATPase almost reached its maximum value in the rabbit *longissimus dorsi* at 8–10 days after birth, continuing thereafter to rise slowly to the adult value (table 18). These results were interpreted to mean that the Ca

TABLE 18. *ATPase activity and $^{45}Ca^{2+}$ uptake in the microsomal fractions from developing rabbit longissimus dorsi muscle*

Age of rabbits (days after birth)	ATPase activities (μmoles of P_i/min./mg. of N)			Rate of Ca^{2+} transport (μmoles/min/ mg of N)	Efficiency (moles of Ca^{2+}/ mole of ATP)
	'Basal'	'Extra'	'Basal' + 'extra'		
1 day	1.5 ± 0.1 (6)	0.6 ± 0.05 (6)	2.1	0.18 ± 0.01 (6)	0.3 ± 0.02
8 day	2.4 ± 0.2 (6)	1.2 ± 0.09 (6)	3.6	0.54 ± 0.01 (5)	0.45 ± 0.03
Adult	0.9 ± 0.01 (9)	3.0 ± 0.2 (9)	3.9	4.8 ± 0.4 (7)	1.6 ± 0.2

Results are expressed as the means of initial rates ± S.E.M. of the numbers of experiments given in parentheses. (Holland & Perry (1).)

---

[1] Eppenberger, Dawson & Kaplan (1); Dawson, Eppenberger & Kaplan (1).
[2] Ch. 13.   [3] See also Holland & Perry (1).   [4] P. 322.

transport system in the sarcoplasmic reticulum is made up of two components – the ATPase itself and a system which couples this enzymic activity to Ca uptake. Thus as the coupling system takes up its function the 'basal ATPase' falls sharply and the 'extra ATPase' rises.

The general opinion amongst workers in this field seems to be that the normal means of energy provision, both enzymes and substrates, are available about the time that contractions are first seen. In cases where this was not apparent, it seems likely that the explanation lies in a failure of biochemical detection. No doubt in the weak contractions at these early times very little energy expenditure would suffice.

The study of muscle development is intensely interesting in itself, and it has thrown light in a number of directions – e.g. on the *raison d'être* of the isoenzymes, and the relationship between the fast and slow adult muscle fibres. Much remains to be done, but during the last two decades much progress has been made.

DISCUSSION

After work stretching over some hundred years the conclusion may be considered established that in vertebrate striated muscle two main types may be distinguished – those contracting rapidly but lacking staying power; and those contracting slowly and able to maintain powerful activity over long periods. We can recognise also a number of intermediate types. The fact has emerged that the rapid muscles depend largely on glycolytic activity and make use predominantly of carbohydrate for their energy supply; while the slower muscles are characterised by oxidative activity and fat metabolism is more important in them. The high capacity for oxidation involves the presence of numerous mitochondria, often of large size; this explains the dense appearance of the red fibres, and also probably accounts for the high 'residual phosphate' content first observed by Embden. The greater diameter of the white fibres may perhaps be correlated with their greater differentiation, as appears from ontogenetic studies. The red colour which characterises the slower muscles is due partly to the content of myoglobin (which constitutes an oxygen store) and partly to cytochrome *c*. There has been much discussion of the existence of a foetal myoglobin but Wolfson and his collaborators,[1] using criteria of gel filtration, antigenicity and absorption spectra, recently concluded that in the foetus besides the foetal form of haemoglobin only the usual adult form of myoglobin is present.

The question arises as to the suitability of the mode of energy provision to function. It would seem that in the case of the rapidly acting muscle, the glycolytic system can provide immediately for the muscle needs, regardless

[1] Wolfson, Yakulis, Coleman & Heller (1).

of oxygen supply; the activity cannot however be prolonged, since accumulation of lactic acid will cut down enzymic activity. In this connection it is interesting that, as shown by Davey (3, 4), the distribution of carnosine[1] and anserine[2] is low in muscles with great oxidative activity and high in muscles with marked glycolytic activity. The important buffering power of these two dipeptides in the physiological range was first pointed out by Bate Smith (6) in 1938 and has been confirmed by Davey.

It must be remembered that another function has long been assigned to carnosine and anserine by Severin and his collaborators.[3] They found increased contraction amplitude in neuromuscular preparations *in vitro* when the medium contained one of these dipeptides. They also described experiments in which, during aerobic incubation of muscle *brei*, more labile P was found if these substances were present; more phosphocreatine was formed if creatine were added. They suggested that the dipeptides in some way protected ATP from cleavage by ATPase and that their imidazol group might take part in reaction leading to formation of energy-rich P.[4] All these effects of course might be explicable in terms of maintenance of optimum pH, but Meshkina & Karyavkina (1) have recently repeated such experiments, using media well buffered before addition of carnosine and showing no change in pH by the end of the experimental period. They confirmed the earlier results. They invoked the hypothesis of Watts & Rabin (1) as possibly relevant – that a hydrogen-bonded thiolimidazol pair is present in the active site of creatine kinase. Bowen (4, 5) also observed in 1965 with glycerinated rabbit fibres that addition of 0.1 M carnosine to the medium caused a great increase in speed and degree of shortening, a specific effect not due to buffering. He suggested that the imidazole moiety of the histidine might increase the number of binding sites on myosin for ATP. Stracher's evidence for histidine at the active site of myosin may be relevant here;[5] and also the recent finding by Johnson, Harris & Perry of the presence of 3-methylhistidine as a constituent of the primary structure of HMM and subfragment I, as well as in hydrolysates of actin.[6]

Thus it seems that carnosine and anserine may have more than one function in muscle – the buffering effect and this effect of unknown mechanism promoting formation and utilisation of energy-rich P. With regard to the latter function there seems no information at present comparing the behaviour of red and white muscle.

It is not surprising that certain anomalies appear when one attempts to generalise over so wide a field. Thus the heart is justly regarded as the

---

[1] β-alanyl-L-histidine.　　　　　　　　　[2] β-alanyl-1-methyl-L-histidine.
[3] For references see Severin (1) and Meshkina & Karyavkina (1).
[4] Histidine is α-amino-β-imidazoylpropionic acid.
[5] Ch. 12, p. 288.　　　　　　　　　　　　[6] P. 474.

example par excellence of the strongly-working untiring red-muscle type. Yet, as Lawrie (2) showed, it contains little myoglobin in comparison with its supply of cytochrome oxidase; presumably this is because of its particularly good blood supply. On the other hand, Lawrie found the blue whale psoas muscle to have an exceptionally low content of cytochrome oxidase compared with its myoglobin; it seems likely that this relationship is an adaptation helping during diving to prolong the time during which the myoglobin-bound oxygen can be effective. The diaphragm, usually regarded as a red muscle, has been studied by Gauthier & Padykula (1) over a range of thirty different mammals; it is interesting as showing the variations which may occur within one muscle called upon to perform differently in different animals. If it is assumed that the rate of contraction of the diaphragm varies with the rate of breathing, it would be required to contract about seven times as fast in the mouse as in man. Thin fibres rich in mitochondira were found in the small mammals, while large fibres with relatively few mitochondria predominated in the large mammals.

Finally we may consider the important myosin ATPase. We have seen that with red and white muscle and with developing muscle there is a correlation between speed of contraction and ATPase activity of myosin and actomyosin. Bárány (1) also has made a survey of muscles from a number of different animals, vertebrates and invertebrates, for which there existed in the literature figures for the contraction times. He has thus been able to extend and to make more quantitative the correlation between speed of shortening and ATPase activity of the contractile protein.

It is interesting to remember here the recent preliminary results of Davies, Goldspink & Larson (1). Using hamster fast and slow muscles, poisoned with fluorodinitrobenzene, they found a linear relationship between the amounts of tension developed in a single isometric contraction and the cost of maintenance in terms of ATP breakdown. Thus for the biceps brachii 30.8×10^3 g sec corresponded to the use of 1 μmole of ATP; while for the soleus the value was 191×10^3 g sec. With the anterior latissimus dorsi of the chick as much as 540×10^3 g sec tension was developed; but when isotonic contraction, with carrying out of external work was considered, this muscle was only one-third as efficient as the hamster sartorius. It would seem that in the fast and slow muscles the energy from the ATP breakdown is applied in some way differently to the muscle machine. In view of such considerations as those brought forward by Rüegg (8) visualising different possibilities for cross-bridge formation in series or in parallel,[1] one would like to have information about the sarcomere lengths of the different muscles. Further work in this field will be of much interest.

[1] Ch. 22, p. 533.

20

ENZYMIC AND OTHER EFFECTS OF DENERVATION, CROSS-INNERVATION AND REPEATED STIMULATION

The effects of severance of the nerve to a muscle, whether accidentally in man or in experiments in animals, are striking enough. Besides the loss of power to move, except upon electrical stimulation, a state of atrophy with gradual wasting away of the fibres ensues.[1] About the middle of last century there was much discussion about the need for special trophic nerves. S. Mayer (1) in Hermann's 'Handbuch der Physiologie' of 1879 considered this question at length and concluded that there was no proof of the existence of trophic nerves, but that in special cases (e.g. muscle and glands) the nerve supply was concerned in a trophic process of a special nature. Michael Foster (2) also, writing at this same time, was of the opinion that much of the effect of loss of nerve supply could be explained by absence of the usual functional activity. Nevertheless he expressed the view that 'some more or less direct influence of the nervous system on metabolic actions, and so on nutrition, will be established by future inquiries.' Langley (1, 2) some forty years later specifically studied the part played by lack of movement in atrophy after nerve section and found that neither passive movement nor active movement caused by electrical stimulation was effective in delaying atrophy. He emphasised however that denervated muscle may be in constant fibrillation, so that it remained possible that the atrophy was caused by fatigue. He considered that this fibrillation theory was in harmony with some of the known facts, but not with all. Tower (1) in 1937 made a histological study of denervated muscle compared with muscle innervated from an isolated and quiescent region of the spinal cord. She concluded that both contractile activity and some unidentified trophic agent from the nerve must influence the muscle beyond the motor end-plates.

During the last twenty years much work has been done in the attempt to find out what biochemical changes take place in the denervated muscle, with the aim of discovering the primary cause of the atrophy.

---

[1] Though, as we shall see, in exceptional cases muscle tissue may hypertrophy.

BIOCHEMICAL EFFECTS OF DENERVATION

EFFECTS UPON THE PROTEINS AND NUCLEIC ACIDS. Amongst the
first papers in this field were those of E. Fischer & Ramsey in 1945 (1, 2).
They found a decrease of about 14% in total protein three weeks after
denervation of rabbit muscle; the amount of 'precipitable myosin' had
decreased by over 70%. The properties of the myosin they prepared seemed
unchanged: it had normal refractive index and double refraction of flow,
and could be spun into doubly-refracting threads. Daily electrical stimulation was effective in slowing the loss of weight and the decrease in myosin
content. Kohn (1) much later, using rats found 50% loss of protein in two
weeks, all fractions being equally affected; the isolated myosin resembled
the normal in viscosity, ATPase activity and susceptibility to digestion.

Weinstock, Epstein & Milhorat (1) had reported in 1958 an increase in
cathepsin activity in a number of wasting conditions. In 1966 Syrový, Hajek & Gutmann (1) described a thorough study of the enzymes which might
be responsible for the protein wastage in denervated muscle, using the
mixed muscles of rat leg. First they demonstrated, by use of dialysed
extracts, that the increased proteolytic activity was not due to activators
of low molecular weight. Since lysosomes contain cathepsin, the question
was asked whether the rise in proteolysis was due to release of the enzyme
from these particles.[1] By measuring proteolytic activity in presence and
absence of the surface-active substance Triton-X 100 (which liberates
lysosomal enzymes) it was found that both bound and free cathepsins were
increased, particularly the latter. There was little difference in the degree
of activation (about 80%) of the cathepsin whether haemoglobin or the
muscle protein was used as substrate; this indicated that the increased
breakdown was not due to some structural change in the muscle proteins
themselves. Peptidase activity was also increased.

The effect of denervation on incorporation of labelled methionine was
also tested on a number of rat muscles, including the gastrocnemius.[2] Decreasing incorporation was first seen, but later there was increase, apparently depending on an increase in the free amino-acid pool. The complexity of these questions of protein turnover is further realised when one
remembers that the slow *Feldstruktur* muscle, the anterior latissimus dorsi,
undergoes *hypertrophy* after denervation. As we have seen, this muscle
contains high cathepsin activity, but Hajek, Gutmann & Syrový (1) found
that, unlike the fast posterior latissimus dorsi, it showed little increase in
this enzymic activity after denervation. As we have also seen, the slow
muscles tested have been characterised, even in the normal state, by a
higher rate of amino-acid incorporation than the fast muscles. Certain red

[1] Ch. 21, p. 500.   [2] Hayek, Gutmann, Klicpera & Syrový (1).

muscles, e.g. the hemidiaphragm, are known to show this denervation hypertrophy, but of a more transient nature, lasting only days instead of weeks. Gutmann & his collaborators showed that increased incorporation of methionine into the proteins in the denervated diaphragm *in vivo* went

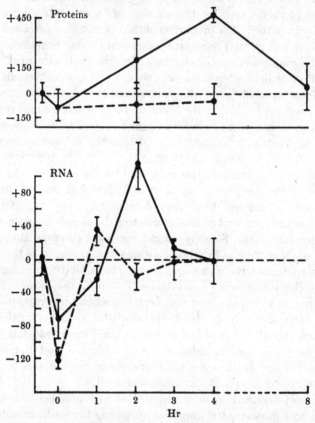

Fig. 96. Changes of non-collagen proteins and ribonucleic acid after stimulation of the normal (full line) and denervated (interrupted line) muscle. Zero on ordinate represents values immediately after stimulation. Proteins expressed as µg non-collagenous N/ muscle. Ribonucleic acid expressed as µg ribonucleic acid P/100 mg protein N. Stimulation at frequency of 300/mm for 6 min. (Zak & Gutmann (1).)

on for several days after denervation and seemed not to depend on increase in the free amino-acid pool.[1] Bajusz (1) has discussed the different effects of denervation on the red granular and the white agranular fibres in the triceps surae of the mouse. The white fibres began to atrophy after a few days, the red only after a much longer time; thus a prolonged stage was

[1] Gutmann, Hanikova, Hájek, Klicpera & Syrový (1).

reached in which the red fibres were of larger diameter than the white fibres.[1]

It can be seen that stimulation of a normal muscle leads to a transient increase in the RNA content,[2] followed by increased protein synthesis. Gutmann & Žak showed that in denervated muscle the increase in RNA could still be observed but was not followed by the increase in protein synthesis (fig. 96).[3] So the question arose whether the disturbance in protein metabolism in denervated muscle is the result of disturbed nucleic acid metabolism, and further work showed increase in both DNA and RNA (per mg of non-collagenous protein) following denervation in the tibialis anterior of the rat; the ratio RNA/DNA rose continuously. At later stages the content of each nucleic acid fell. Intensive electrical stimulation reduced the loss of protein but had much less effect on the nucleic acid. Gutmann (1) suggests that this dislocation may be a characteristic feature of metabolic disturbance in muscles lacking some 'precursor' normally supplied by the nerve. McCaman & McCaman (1) also observed the rise in DNA in denervated mouse muscle – to twice the normal value per gram of protein after 28 days.

EFFECTS UPON THE CELL MEMBRANE AND THE SARCOPLASMIC RETICULUM. The possible part played by increased membrane permeabilty in protein loss during atrophy must also be considered. Pellegrino & Bibbiani (1) incubated control and denervated rat muscles in Krebs Ringer–glucose solution, and found in the effluent from the denervated muscle a greater amount of aldolase (expressed as percentage of the total aldolase). Since the amount of aldolase passing out from the control could also be affected by various factors (such as anoxia) they considered the permeability change as not the primary change in denervation atrophy but rather as dependent on some modification of metabolism. Graaf, Hudson & Strickland (1) also found a very marked fall in aldolase in rat muscle of the hind limb after unilateral denervation; after 30 days there remained only 7 % of that present in the control muscle. They considered that this change was likely to be due to increased membrane permeability, and that aldolase, a soluble enzyme, would be particularly susceptible to loss in this way. The aldolase content of serum, is however not raised in neurogenic atrophy as it is in muscular dystrophy.

Quite a different aspect of membrane change is that described by Axelsson & Thesleff (1) in 1957. While the normal muscle is sensitive to acetylcholine only in the region of the end-plates, the whole muscle membrane becomes sensitive a week or two after denervation. (This state is

---

[1] Also Bajusz (2).   [2] See also p. 496.
[3] Žak & Gutmann (1); Gutmann & Žak (2).

characteristic of foetal muscle as was shown by Diamond & Miledi (1)). Gutmann & Sandow (1) in 1964 found another change in sensitivity resulting from denervation. Mammalian fast muscle normally does not react to caffeine with a contracture but shortly after denervation the extensor digitorum longus acquired this susceptibility; it was suggested that in some way the sarcoplasmic reticulum was affected so that it gave up its calcium under the influence of caffeine.

EFFECTS UPON THE ENZYMIC CONSTITUTION. Studies of the effects of denervation on the enzymes concerned in energy supply show that whether red or white muscle is used, the enzymic equipment is cut down. These studies have been mainly histochemical. Thus Nachmias & Padykula (1) found a loss of some 50 % of the succinic dehydrogenase activity in the soleus 14 days after denervation. B. Smith (1) used the rat tibialis anterior, which normally has a 'checkerboard' appearance owing to the presence in it of both red and white fibres with their different histochemical reactions. A week after section of the sciatic nerve, the checkerboard began to disappear, the red fibres losing succinic dehydrogenase and NADPH diaphorase activity, the white fibres losing phosphorylase, NAD-dependent glycerophosphate dehydrogenase and lactic dehydrogenase. There was increased activity in hydroxybutyric dehydrogenase and a non-specific esterase. She suggested that the diminished energy needs were met by metabolism of ketones and fatty acids. The results of Hogenhuis & Engel (1) on guinea-pig muscle agree in showing on denervation decrease in examples of both glycolytic and oxidative enzymes; they also observed a slight fall in myofibrillar ATPase after some weeks.

George and his collaborators[1] reported a very interesting fact in the increase, after denervation, of glucose-6-phosphate dehydrogenase and 6-phosphogluconic dehydrogenase in red muscle fibres and later in white fibres of pigeon breast. Accumulation of fat was observed and it was suggested that these enzymes of the pentose phosphate pathway were concerned in fatty acid synthesis and cholesterol formation. That the NADPH needed for such synthesis was supplied by oxidative reactions in this pathway had been shown by Siperstein (1) in 1958. Garcia-Buñuel & Garcia-Buñuel (1) also found increase in these two dehydrogenases by estimations on homogenates of normal and denervated rabbit muscle, both soleus and gastrocnemius, the increase being faster in the latter. They were inclined to seek the explanation in infiltration of connective tissue, where activity of these enzymes is normally much greater than in muscle. Aloisi (1) however, who found such increase in denervated frog-muscle, thought this explanation unlikely.

[1] Cherian, Bokdawala, Vallyathan & George (1).

Dawson, Goodfriend & Kaplan (1) followed the changes in lactic dehydrogenase isoenzymes after denervation of red and white muscles of the chick.[1] In the gastrocnemius the total lactic dehydrogenase was reduced to about 20 % of normal after one month, the loss falling almost entirely on the M type. In the soleus also there was a large fall in total enzyme; in this case the loss in the H form was greater, an unusual instance of lability in this form of the enzyme. McCaman & McCaman (1) studied some enzymes concerned with nucleotides. Adenylic deaminase activity fell rapidly, while 5-adenylic nucleotidase increased on a protein basis, to about 4 times the normal activity.

We shall return to discussion of denervation effects after considering the biochemical impact of cross-innervation.

BIOCHEMICAL EFFECTS OF CROSS-INNERVATION OF FAST AND SLOW MUSCLES

Remarkable results were published by Buller, Eccles & Eccles (1) in 1960. With both kittens and cats, the motor nerve supplying a fast muscle (the flexor digitorum longus) was cut and its proximal end was sutured to the distal end of the cut nerve supplying a slow muscle (the soleus). The same procedure of cross-innervation was carried out to provide the fast muscle with innervation normally supplying the slow muscle. Re-innervation went on gradually and after some time (one to eight months) the behaviour of the muscles had changed. The slow muscles tended towards the properties of the fast in the time course of the rising and falling phases of the twitch, while changes in the opposite direction were seen in the fast muscle innervated by the nerve of the slow muscle (see fig. 97). Various possible explanations were considered and rejected as unlikely, and the authors were left with the suggestion that some substance passes down the nerve axon, crosses the neuromuscular junction and directly affects the muscle fibres. Biochemical investigation of the fibres in the changed muscles was clearly needed, and soon followed.

Thus Drahota & Gutmann (1) found that the low content in glycogen and in potassium characteristic of the soleus as compared with white muscle was changed eight months after denervation and re-inervation by the peroneal nerve. Both values rose, to equal those in the normal fast extensor digitorum (see fig. 98). Romanul & van der Meulen (1) in 1966 after performance of the cross-innervation on rats and cats got physiological results in accord with those of Buller, Eccles & Eccles. Histological examination showed reversal of the enzyme profiles (see fig. 99 a and b pl.): the high activity of the phosphorylase and of the two glycerophosphate dehydro-

[1] P. 457 above.

genases (characteristic of fast muscle) was now found in the soleus and was largely lost in the fast flexor hallucis longus. The activity of esterase and β-hydroxybutyrate dehydrogenase (roughly paralleled by that of cytochrome oxidase and succinic and citric dehydrogenases) was inversely proportional to that of the glycolytic enzymes. Guth & Watson (1) recently used the supernatants of aqueous homogenates of rat muscle and separated

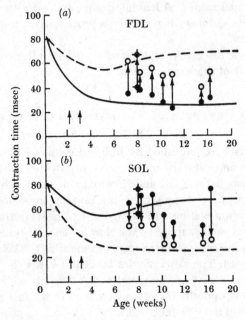

Fig. 97. Contraction times to show changes produced on cross-union between soleus and flexor digitorum longus nerves. The kittens used were operated on at 16–22 days of age except for the one indicated by crosses, for which the operation was at 4½ weeks. The standard curves for contraction time–age are shown as continuous lines for the muscle itself and as broken lines for the muscle belonging to the nerve that now provides its innervation. (Buller, Eccles & Eccles (1).)

the sarcoplasmic proteins by acrylamide gel electrophoresis. Characteristic patterns were obtained with consistent differences between the soleus and plantaris. Denervation of the plantaris caused the protein pattern gradually to become similar to that of the soleus; on re-innervation the normal pattern was recovered. De- and re-innervation of the soleus had no effect but re-innervation by white-muscle nerve caused the soleus pattern to resemble that of the plantaris. Prewitt & Salafsky (1) have also confirmed the physiological effects of cross-innervation using the soleus and flexor digitorum longus of young cats. With regard to enzymic changes, pyruvic kinase and aldolase increased in the soleus while malic and isocitric dehydrogenases decreased. The converse changes went on in the flexor digitorum longus. Then Dubowitz (2) performed cross-innervation operations on the soleus and the fast flexor hallucis longus or flexor digitorum longus of cats and rabbits. After some weeks (five to thirteen) a dramatic change in the histochemical pattern of the fast muscle was seen, parts taking on the

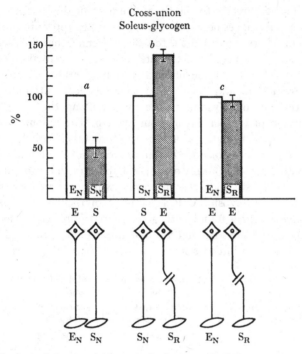

Fig. 98. Content of glycogen (a) in normal soleus muscle (S_N) expressed in % of glycogen content of normal extensor digitorum (E_N); (b) in cross-innervated soleus muscle (S_R) expressed in % of normal soleus muscle; (c) in cross-innervated soleus muscle expressed in % of normal extensor muscle. (Drahota & Gutmann (1).)

appearance of normal soleus muscle. These areas became weaker in ATPase and phosphorylase, and stronger in NADH dehydrogenase. Myosin prepared from these muscles showed no change in ATPase activity; it was suggested that the changed activity observed histochemically was not extensive enough to be apparent by biochemical techniques. Converse changes were seen in the cross-innervated soleus, but these were less consistent.

Thus it seems that the preferred energy supply for ATP resynthesis – whether glycolytic or oxidative – is determined by the nerve supply. The mechanism remains a matter of conjecture. Guth & Watson (1) suggest that the rate of production or transmission or release of some trophic factor is related to the extent and frequency of the nerve impulses. Possibly the nerve releases messenger RNA.

With regard to the nature of substances which may be transmitted by the nerve, the recent experiments of Korr, Wilkinson & Chornock (1) may have significance. They applied inorganic phosphate, ^{32}P-labelled, and ^{14}C-labelled amino acids directly to the tip of the floor of the fourth ventricle in rabbits. These substances began after several days to appear in the muscle cells (and the muscle cells only) of the tongue. It seemed that they travelled down the hypoglossal nerve, for if this delivery was prevented on one side the tongue was labelled on the other side only. Kerkut (1) also, using preparations from snail and frog, found that labelled glutamate passed from the central nervous system to the muscle along the nerve trunks during stimulation of the central nervous system. The amount was proportional to the number of stimuli and to the time taken in its passage.

It is important now to consider the actomyosin/ATP mechanism itself in cross-innervated muscle. Bárány and his collaborators have found similar concentrations of myosin and actin in normal white and red muscles, but less Mg-activated actomyosin ATPase activity in the latter.[1] Sexton & Gersten (1) have recently found with glycerinated fibres from the soleus and gastrocnemius of the rat, that the red fibres developed significantly higher isometric tension but at a lower rate than the white fibres.

	Isometric tension g/mm^2	Rate of development mg/min
Gastrocnemius	18.43 ± 1.04	230.0 ± 27.7
Soleus	27.78 ± 1.96	140.6 ± 18.7
	$p = <0.01$	$p = <0.001$

Here one supposes that only the actomyosin/ATP system can be concerned; ATP is provided, so that the differences in rate of resynthesis in the two muscle types (important *in vivo*) do not play a part in the difference of performance. These results fit in with those of Bárány *et al.* and indicate that in the two types of muscle some differences in enzymic centre must be involved – perhaps also differences in timing of cross-bridge linking.[2] Little work has been done on the actomyosin/ATPase mechanism of cross-innervated muscles, but we have mentioned the observations of Dubowitz, who prepared myosin from cross-innervated flexor hallucis longus and flexor digitorum longus of cats and rabbits. No change in ATPase activity, whether Ca-, EDTA- or actin-activated was seen. It is difficult to reconcile

[1] Bárány, Bárány, Reckard & Volpe (1).
[2] Cf. ch. 22, p. 533.

these results with those just mentioned for the normally innervated muscle and further work in this field would be of great interest.[1]

The matter of the degree of nervous control of metabolic processes exercised in the different types of muscle has also had attention. It seems clear that after denervation red and white muscles become much less different from each other, the changes taking place mainly in the white muscle, both as regards weight loss and enzymic constitution. This dedifferentiation has naturally been connected in discussion with a return towards the foetal type; however, regulatory influences of some sort seem still to be at work. Thus Drahota & Gutmann (1) noticed that the equalisation of potassium content which goes on after denervation of red and white muscle (rat soleus as compared with the white extensor digitorum longus) was achieved entirely by increase in the concentration in the red muscle. Again, Schapira and his collaborators showed that during denervation atrophy the myoglobin content of the soleus fell to less than 50 % of its normal value;[2] at the same time the content in the gastrocnemius rose, so that the latter muscle, ordinarily containing only one-sixth of that in the soleus, now contained about the same concentration. It is interesting to recall that Langley & Hashimoto (1) already in 1918 had recorded corresponding changes in colour of such denervated muscles. Thus although the state of the denervated muscle is in many characters closer to the foetal type, it would be more true to say that an intermediate type is reached.

ENZYMIC ADAPTATION TO CONTRACTILE ACTIVITY

We have seen that during development increased muscular activity may lead to increased content in the muscle of certain enzymes; and that changes in innervation, leading to changed physiological response, can also result in changed enzyme content. The question as to the direct effect of contractile activity of adult muscle upon the concentration of certain enzymes contained in it has also been attacked. Kendrick-Jones & Perry (2) quote a number of papers by Russian workers between 1930 and 1952, reporting

---

[1] Buller, Mommaerts & Seraydarian (1) have recently examined the myosin and myofibril ATPase activity in cross-innervated flexor digitorum longus and soleus of the cat. The enzymic activity fell in the former, becoming nearly equal to that in the normal soleus. There was only slight increase in the ATPase activity of the cross-innervated soleus. Mommaerts, Buller & Seraydarian (1) also demonstrated a reciprocal alteration in the calcium-transporting activity of sarcoplasmic vesicle preparations from the two types of muscle after cross-innervation. It still remains to be tested whether there is an alteration in amount of sarcoplasmic reticulum present; and also whether the changes in activity depend to any extent on greater sensitivity of one type of vesicle to isolation procedures. As the authors point out, the changes in ATPase activity could explain the neurogenically altered contraction velocity, while changes in power of calcium sequestration could explain the neurogenically determined duration of the contraction cycle.

[2] Maleknia, Ebersolt, Schapira & Dreyfus (1).

increase in glycerophosphate, lactic and succinic dehydrogenases and hexokinase in the muscle of rats as the result of training over several days by short periods of strenuous exercise; but some more recent work did not bring confirmation. Thus Hearn and his collaborators[1] (using rats swimming for 30 minutes daily for 35 days and killed 24 hours after the last exercise) found no increase in aldolase, succinic, lactic or malic dehydrogenases, or phosphorylase in skeletal muscle per unit of protein; in the heart some increase in lactic and succinic dehydrogenases was found. Gould and Rawlinson obtained similar negative results for dehydrogenases and phosphorylase, and also for creatine kinase and total Ca-activated ATPase, estimated in muscle homogenate.[2] However Gutmann and his collaborators reported in 1956 an activation of glycogen synthesis after electrical stimulation for some minutes of the muscles of the rat.[3] The glycogen content, 4 hours after stimulation, might be as much as 30 % greater than in the unstimulated muscle. This metabolic activation was dependent on nerve supply: in denervated muscle, although stimulation caused normal glycogen breakdown, resynthesis went on very slowly, only to the control unstimulated level. Kendrick-Jones & Perry moreover have themselves recently found with rats that 4–12 hours exercise could lead to increase in creatine kinase activity of 30–50 % (table 19). Similar increase was found in aldolase activity. With regard to the kinase they emphasise that the changes were more marked if the animals were killed for the enzymic assays as soon as the exercise was over; and also that the experiment started from a lower baseline if the animals had been restricted in movement for some days beforehand. Such unsuspected fluctuations in enzyme content may have played a part in the variability of the results obtained by earlier workers. Increase in creatine kinase, depending on the length of the stimulation, and in prolonged experiments reaching as much as 50 %, was also obtained on stimulation of frog sartorius in vitro (see fig. 100). Particularly interesting were some experiments on desert locusts. These had been grown from larvae in captivity and had never flown before; no creatine kinase or adenylic deaminase seemed to be present in their muscles. But after periods of flight, totally three hours, in a current of air, homogenates of the flight muscle showed significant increase in myokinase and aldolase.

In an attempt to decide whether the increase in creatine kinase activity was due to activation of an inactive enzyme or to synthesis of new enzyme protein, Kendrick-Jones & Perry (3) tried the effect of a number of inhibitors on frog sartorius. The amino-acid analogues ethionine and

[1] Hearn & Wainio (1, 2); Gollnick & Hearn (1).
[2] Gould & Rawlinson (1); Rawlinson & Gould (1).
[3] Vrbová & Gutmann (1); Gutmann, Bass, Vodička & Vrbová (1); Bass, Gutmann & Vodička (1); Žak & Gutmann (1); Gutmann (1).

TABLE 19. *Effect of exercise on the creatine phosphokinase and aldolase activities of rat hind-leg muscle*

Expt. no.		Age (days)	Body-wt. (g)	Creatine phosphokinase (μg creatine/ mg N of extract/ min)	Aldolase (μg triose P/mg N of extract/ 15 min)
1	Controls	100	121.3 ±4.1	497 ± 12.9 (3)	
	20 hr exercise immediately before death		126.4 ±5.2	659 ± 6.5 (3)	
2	Controls, 100 days normal activity, 21 days confined	121	178 ± 6.2	283 ± 7.0 (3)	
	100 days normal, 20 days confined, 1 day exercise		179 ±5.7	410 ± 13.9 (3)	
3	Controls		34.8 ±0.7	406 392 387	100 98 94
	4 hr exercise, 4 hr normal	21	33.5 32.0	439 445	108 111
	8 hr exercise, 4 hr normal		31.5 33.1	490 502	139 142
4	Controls	28	38.4 39.1	441 439	67.2 65.0
	12 hr exercise immediately before death		37.5 38.8	570 640	106 107
5	Controls	21	35.8 36.5	421 437	
	12 hr exercise immediately before death		34.2 34.9	568 587	
6	Controls	21	35.7 36.2	434 439	
	12 hr exercise, 72 hr normal		35.8 35.7	460 446	

Control and exercised animals were litter-mates of comparable weight. The animals used in expts 1 and 2 were all from the same litter; likewise for expts 5 and 6. Animals exercised in a treadmill usually performed 5000 to 6000 revolutions in a 12-hour period. (Kendrick-Jones & Perry (2).)

p-fluorophenylalanine as well as the protein-synthesis inhibitor puromycin prevented the rise in kinase. But the inhibitors of RNA synthesis, actinomycin D, 5-bromouracil and 8-azaguanine were without effect on this rise; in these cases the sartorii (stimulated once every 5 seconds for 6 hours) showed on contraction normal tension and heat development in presence of the inhibitors. The results seem to indicate enzyme increase depending on new protein synthesis, controlled by messenger RNA which is stable

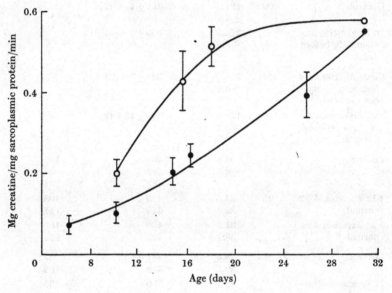

Fig. 100. Effect of activity on the creatine kinase activity of the leg muscles of young rabbits. ○, litters disturbed at 5 days *post partum*; ●, undisturbed control litters. (Perry (8).)

during the six hours of the experiment. It was further shown that, after long activity, labelled leucine was incorporated more rapidly into fibrillar and sarcoplasmic proteins (particularly into the sarcoplasmic granule fraction) than in unstimulated controls; after only two hours stimulation incorporation was not different from normal. An interesting comparison may be made between these results and those of Gutmann and his colleagues who found with rats that, after a preliminary fall, there was an increase in both RNA and non-collagenous protein after stimulation, the former reaching a maximum about 2 hours after the stimulation, the latter about four hours after.[1] In the Czech experiments the contraction period was short, 300

[1] Žak, Gutmann & Vrbová (1); Gutmann & Žak (2). Cf. also fig. 96.

stimuli per minute being applied for 6 minutes. The fall in RNA content, immediately after cessation of strenuous exercise, was also observed by Shchesno in rabbits (1). It was greater in white muscle than in red.

The experiments of Hultman & Bergström (1) on themselves are important in this connection.[1] One leg performed strenuous exercise while the other remained at rest. Biopsies showed that (if a high carbohydrate diet was taken) glycogen resynthesis in the exercised limb was rapid and a 'rebound' effect (as in the earlier-mentioned work of Gutmann et al. with rats) was seen, so that the glycogen content rose some 100 % above that in the control limb, the increased synthesis lasting several days. The authors emphasise that the factor stimulating resynthesis must be local, since only one limb was affected; and that effects due to shift of synthetase D to I or to increased blood flow would be of short duration. They consider the possibility of a local effect of the high insulin output consequent upon the high carbohydrate diet. Another question to be taken into account, in view of the experiments described above, is the possibility of actual synthesis of enzyme.

In 1969 Lamb and his collaborators observed the 'rebound' effect in both red and white muscles of the guinea pig.[2] After extensive training, the glycogen content was about 75 % higher than in the resting controls. As regards enzymes: hexokinase activity rose in the trained muscles to about twice the control value and total synthetase activity was enhanced in the hearts of such animals. No changes of this sort occurred in the liver. It seems likely that these two enzymes are involved in the mechanism of the glycogen supercompensation after exercise.

Holloszy (1) in 1967 observed increased activity in a whole series of oxidative enzymes in the mitochondrial fraction from the gastrocnemius of rats trained by strenuous running on a treadmill. Cytochrome c increased in concentration and the total mitochondrial protein content was about 60 % higher than in the controls. Mild exercise was ineffective. It is important to notice that the mitochondria from the exercised muscles had high respiratory activity and tightly coupled oxidative phosphorylation. Thus the increased respiratory activity was associated with increased power to produce ATP. Pattengale & Holloszy (1) also found the myoglobin content to increase by about 80 %. In the much earlier experiments of Lawrie (4) long-continued training of rats (two hours exercise daily for several weeks) led to increased myoglobin content of the muscles, while short, severe exercise had no such effect. Fowl muscle gave a similar response when the myoglobin of free-range and closely confined birds was compared.

It seems then that conditions have been found in which stimulation to contraction is accompanied or followed by a definite increase in the activity

---

[1] Also Bergström & Hultman (1).
[2] Lamb, Peter, Jeffress and Wallace (1).

of certain enzymes – this increased activity, from the results of Kendrick-Jones & Perry, probably in some cases at any rate, to be attributed to actual increase in the amount of the enzymes concerned. Further work along these lines, and on the influence of the nerve supply on this effect of stimulation (suggested by the work of Gutmann and his collaborators) will be of great interest.

21

SOME ASPECTS OF MUSCLE DISEASE

The subject of muscle diseases is a very wide and intricate one, and here we shall deal only with biochemical research on two forms – progressive muscular dystrophy and glycogen storage disease. This chapter came to be written because the muscle biochemist is so often confronted with the question 'Does all this research on muscle help in the cure of muscle diseases?' The short answer at present is 'No' as far as these two types are concerned. Both are hereditary, and in the case of glycogen storage disease the defect has been traced in different types to lack of a particular enzyme. In the case of progressive muscular dystrophy it has been considered clear that the primary cause is in the muscle itself and not in any deterioration of its nerve supply, but in spite of much experimentation this cause has not been found. Nevertheless much ground has been cleared, and we can say a little about the various views held as to possible causes. It must, however, always be remembered in what follows that the work of the last few years on the effects of cross-innervation on the enzyme patterns of red and white muscles suggests that the nerve supply exerts a more subtle influence on muscle metabolism than had previously been realised.[1] The possibility thus remains that in muscular dystrophy some specific change occurs in the nerve without visible degeneration.

PROGRESSIVE MUSCULAR DYSTROPHY

Descriptions of this disease were first published about the middle of the nineteenth century. Erb (1) in 1884 showed clearly that diseases existed in which muscle degeneration seemed to be the primary event, not consequent upon damage in the central nervous system. Walton (1) has given a brief historical account of great interest. The fact that diseases of this sort ran in families naturally led to the idea of their hereditary nature, but the genetics could only be worked upon in recent times.[2] The muscular dystrophy with which we are concerned here has been defined by Walton (2) as 'a genetically determined, primary, degenerative myopathy'. In a study dependent on human material much variation in the results of different observers is

---

[1] Ch. 20.
[2] Walton (2) for a survey.

inevitable. Nevertheless it has been possible to group the forms into three categories:[1]

(a) Childhood or Duchenne type (first described by Duchenne in 1868).

(b) The facio–scapulo–humeral type.

(c) The limb–girdle type.

Of considerable importance for research was the discovery in 1955 by Michelson, Russell & Harman (1) of a primary, hereditary myopathy in an inbred strain of laboratory mice, although it is still uncertain how far the mouse disease can be equated with any category of the human dystrophy. Since that time hereditary diseases of skeletal muscle have been described in chickens, hamsters, lambs and the white Peking duck.

STUDY OF THE MUSCLE PROTEINS. A fall in muscle protein is very obvious in muscular dystrophy. Using mice, Weinstock, Oppenheimer and Simon and their collaborators have all observed loss in non-collagenous protein amounting to some 25 % of the total protein.[2] Oppenheimer, Bárány & Milhorat showed that myosin was chiefly affected, falling by about 50 %; Simon, Gross & Lessell also found that the loss was almost entirely due to myosin. Weinstock, Epstein & Milhorat and Pennington (2) observed that catheptic activity in dystrophic mice was markedly increased. Later work has shown that the proteinase concerned here is cathepsin D;[3] haemoglobin is ordinarily used as the substrate for this enzyme in tests in vitro; but as Weinstock (1) has reported, the highly purified enzyme will increase autolysis several-fold when added to muscle extracts. The presence of cathepsin A in increased concentration in the muscle of dystrophic chickens has also been shown by Iodice & Weinstock (2). They pointed out that cathepsin D is the predominant proteinase in lysosomes,[4] and that cathepsin A is also known to be associated with liver lysosomes.

Another group of workers had also emphasised the important part which may be played in hereditary muscular dystrophy by the increase in lysosomal hydrolytic enzymes.[5] Besides cathepsin, they found RNAase, β-glucuronidase and aryl sulphatase to be increased in the muscles of affected mice. They considered the possibility that increase of such lysosomal enzymes might be the cause of the dystrophy, and pointed to the suggestive fact that lysosomal β-glucuronidase is known to be under the control of a single gene in the mouse.

---

[1] E.g. Walton (2); Tyler (1).
[2] Weinstock, Epstein & Milhorat (1); Oppenheimer, Bárány & Milhorat (1); Simon, Gross & Lessell (1).
[3] Iodice & Weinstock (1); Iodice, Leong & Weinstock (1).
[4] For a discussion of lysosomes see p. 507 below.
[5] Zalkin, Tappel, Desai, Caldwell & Peterson (1); Tappel, Zalkin, Caldwell, Desai & Shibko (1).

As in the atrophy due to denervation, the question of the increased permeability of the muscle membrane is important. Sibley & Lehninger (1) in 1949 observed two cases of muscular dystrophy with very high serum aldolase content. In 1954 Dreyfus, Schapira & Schapira (1) found that aldolase was almost absent in muscle biopsies from patients with progressive muscular dystrophy. Dreyfus and his colleagues also reported increased phosphohexoisomerase and transaminase in the serum in human myopathies; they stressed the diagnostic value of such observations since the rise in these three enzymes takes place at a very early stage in the progress of the disease, and is not found in cases of neurogenic atrophy. Shortly afterwards they reported raised serum aldolase content – up to 30 times the normal values – in 93 % of the human cases tested.[1] Zierler (1), using segments of mouse diaphragm or peroneus longus (free from transected fibres), tested their permeability for aldolase *in vitro*. More of this enzyme was found in the effluent of the dystrophic muscles than in the effluent of normal muscles, despite their lower content of the enzyme. Ebashi and his colleagues were the first to observe an abnormally high serum creatine kinase activity in muscular dystrophy – raised in some 70 % of the patients they examined.[2] This was not typical of any other disease in their experience.

Coleman & Ashworth (1) in 1959 compared the rate of incorporation of isotopically labelled glycine into proteins and nucleic acid of normal and dystrophic mice. The rate was much higher for the latter. A little later Kruh, Dreyfus, Schapira & Gey (1) also described such experiments. The labelling picture given by myosin in the normal muscle (a rise in radioactivity in the first day followed by a steady plateau) suggested a life-span for this protein of about 20 days. In the myosin of dystrophic animals the labelling rose as rapidly and continued to rise for about five days, then fell rapidly. Results similar but of a rather different time pattern were found for the water-soluble protein. These phenomena (which experiment showed could not be due to increased permeability to glycine) were interpreted as meaning accelerated protein turnover in the dystrophic muscle. The experiments of Simon, Gross & Lessell (1) using labelled DL-leucine, also showed this increased turnover with dystrophic mice. They concluded that the muscle-protein wastage was not due to defective synthesis but to a rate of catabolism exceeding the synthesis. This imbalance seemed chiefly to affect the myofibrils.

The nature of the myosin present in dystrophic muscle has naturally attracted attention. Oppenheimer, Bárány & Milhorat (1), using mice, found that the ATPase activity and ability to combine with actin were not affected. The ultracentrifuge patterns were very similar, but light-scattering

[1] Schapira, Dreyfus, Schapira & Kruh (1).
[2] Ebashi, Toyokura, Momoi & Sugita (1).

experiments showed that the dystrophic myosin was more polydisperse, apparently containing, besides the normal component, larger and heavier particles probably myosin aggregates. Smoller & Fineberg (1), also using mice, confirmed the tendency of the dystrophic myosin to aggregation; and they reported differences in the amino-acid composition of the normal and dystrophic myosins. They suggested that a genetic abnormality in the myosin might be a primary defect in this dystrophy. However, Bárány, Gaetjens & Bárány (1) could not agree with this latter finding. Their analyses showed no difference in amino-acid content between the myosins from normal and dystrophic mice; or between the myosins from normal and dystrophic chickens (although there were definite differences between the normal myosins from chicken and mouse). They confirmed that no change in ATPase activity or actin-combining power was to be found in the two dystrophic myosins. Further assurance of the normality of the myosin from dystrophic chickens has recently been given by Morey and his collaborators[1] – with regard to elution pattern from DEAE-Sephadex, ATPase activity, sedimentation coefficient and ultraviolet spectrum.

Monckton & Nehei (1) described in 1966 preliminary results indicating that after sucrose-gradient centrifugation, the ribosomal fraction of Duchenne dystrophic patients lacked the heavier particles found in similar tests in normal muscle and believed to be concerned there in myosin synthesis.[2]

EFFECTS ON ENZYMES CONCERNED IN GLYCOLYSIS AND RESPIRATION. In 1954 Dreyfus, Schapira & Schapira (1), using biopsy material from human cases, found that phosphorylase as well as aldolase was very markedly decreased as compared with normal muscle; then it was shown that phosphoglucomutase was also lowered and that overall glycolysis fell by about 75 %.[3] In all this work the reference base used was non-collagen protein. Dreyfus, Schapira, Schapira & Demos (1) made the further interesting observation that, though the activity of a number of enzymes concerned in oxidation (cytochrome oxidase, succinic oxidase, succinic dehydrogenase, aconitase and fumarase) was diminished in the myopathic muscle if the results were expressed on a wet weight basis, this effect disappeared when non-collagen protein was used as the basis; no significant difference was then found between normal and diseased groups. With dystrophic mice Pennington (1) similarly found no change in the succinic-tetrazolium reductase of the muscle mitochondria.

McCaman (1) on the other hand found that certain enzymes of the pento-sephosphate pathway were markedly increased in activity; she commented

---

[1] Morey, Tarczy-Hornoch, Richards & Duane-Brown (1).
[2] Ch. 19.   [3] Schapira, Dreyfus, Schapira & Kruh (1).

on the similarity of this behaviour to that shown by normal gastrocnemius after denervation.[1] Pennington (2) also recorded increased activity in glucose-6-phosphate dehydrogenase, and in acid phosphatase.

The elaborate enzyme profiles of Laudahn & Heyck (1) give results collected from 145 cases of progressive muscular dystrophy, including 123 Duchenne cases. They correlated the enzymic changes per gram of protein extracted with the stage of severity of the illness judged histologically. The results are in substantial agreement with those described above. The authors noted certain enzymes, e.g. isocitric dehydrogenase and phosphoglucoisomerase which showed a preliminary increase followed at a later stage by a decrease in activity. They were inclined to connect the sustained increase in glucosephosphate and phosphogluconic acid dehydrogenases of the pentosephosphate pathway with increased fat synthesis in the dystrophic muscle. Such increased synthesis is known to occur in dystrophic mice and would need the NADPH provided by these two dehydrogenases.[2] Vignos & Lefkowitz (1) saw the striking fall in glycolysis in muscle of patients with childhood (Duchenne) dystrophy, but not in adult forms.

EFFECTS ON CREATINE KINASE, AND ON NUCLEOTIDE METABOLISM. Vignos & Lefkowitz also found that creatine kinase activity (per unit of non-collagen N) was reduced in human muscle in cases of childhood dystrophy, but this did not happen in the adult dystrophies they studied. Vignos & Warner (1) showed that total creatine was less than normal in both forms and the ratio phosphocreatine/creatine was also diminished. In the childhood dystrophy the ratio ATP/ADP was only 1.57 as compared with 2.5 in the adult dystrophic and normal muscles. They discussed the question whether the defect is in the contractile system itself or in the energy supply; for instance, they found the decrease in myosin to be markedly less than the decrease in glycolytic activity and rather less than that in creatine kinase activity. Kar & Pearson (1) also found reduction in creatine kinase in some very severe human myopathies, and Laudahn & Heyck (1) in childhood dystrophy. A striking fall in adenylic deaminase (by about 75 % on the non-collagen N basis) was observed by Pennington (2, 3) and by McCaman & McCaman (1) in dystrophic mice. There was no increase of this enzyme in the serum.[3] Pennington (2) found a fall by about 50 % in adenylate kinase, a change also recorded by Laudahn & Heyck (1) in childhood dystrophy. Mitochondrial ATPase activity was slightly higher in the dystrophic than in normal mouse muscle in Pennington's experiments (1).

It is interesting that Olson and his colleagues (using either non-collagen nitrogen or mitochondrial N as the basis of comparison) showed that normal

[1] Ch. 20, p. 488. [2] Young, Young & Edelman (1). [3] Pennington (4).

P/O ratios can be obtained with human dystrophic muscle.[1] As the disease progressed, oxidation of glutamic acid (the substrate used) fell off, but the P/O ratio remained normal. They emphasised that, though these results suggest that a defect in oxidative phosphorylation is not the primary one, it must be remembered that other substrates might follow differing enzymic courses in the mitochondria; further investigation, especially of the capacity to oxidise fat, would be important. On the other hand, according to Lochner & Brink (1), the mitochondrial phosphorylation/oxidation ratio was much depressed in the muscles of hamsters suffering from hereditary muscular dystrophy, while the oxygen uptake was normal.

Another line of research was that of Bourne and his collaborators on the increase (observed histochemically) in a number of phosphatases in the connective tissue (endomysium) of human dystrophic muscle.[2] This increased dephosphorylation affected phosphate esters of adenosine, uridine, guanosine and cytidine, as well as NAD and NADP. They pointed out that some of these substances are concerned in the synthesis of sulphated mucopolysaccharides which, bound to protein in connective tissue, form mucoprotein complexes. Failure to produce this connective-tissue ground substance (owing to lack of the essential nucleotides) might result physiologically in failure of the transport mechanism between the capillaries and the muscle fibres. McCaman & McCaman also recorded a fourfold rise in adenine-5-nucleotidase/unit of protein in dystrophic mice. They stressed the great damage which would supervene on the continual removal of adenylic compounds from the system.

An interesting paper by Bing and his collaborators is concerned with myocardial metabolism in patients with progressive muscular dystrophy.[3] The concentrations of oxygen, glucose, inorganic P, pyruvate and lactate were measured in arterial blood and in blood drawn simultaneously from the coronary sinus. The molar ratio of lactate to pyruvate reflects the ratio NADH/NAD, determining the oxidation–reduction potential. Thus the positive difference in this potential found between the arterial and coronary sinus blood indicates a relative increase in the muscle of the lactate/pyruvate ratio, therefore a shift to anaerobic metabolism. This was in spite of the presence of plenty of oxygen, enough to deal with all the sugar extracted from the blood. The authors suggest the possibility of uncoupling of oxidative phosphorylation in the myocardium.

The results of Nichol (1) on thiol compounds may have significance for the activity of a number of enzymic reactions. In dialysates of homogenates of normal and dystrophic muscle the ratio of thiol to disulphide was 0.55 in

---

[1] Olson, Vignos, Woodlock & Perry (1).
[2] Golarz & Bourne (1) where references to earlier work are given.
[3] Sundermeyer, Gudbjarnson, Wendt, den Bakker & Bing (1).

the former, 0.053 in the latter. The actual amount of disulphide (in μmoles/ mg N in the dialysate) was usually higher in the dystrophic extract.

ISOENZYMES. In 1962 Wieme & Lauryssens (1) reported a change which they considered as a dedifferentiation in the lactic dehydrogenase isoenzyme pattern in muscular dystrophy. Kaplan & Cahn (1) about the same time observed in the muscle of chickens with hereditary muscular dystrophy that there was (besides the decrease in total lactic dehydrogenase) an increase in percentage of the H form present.[1] Similar observations were made independently by Schapira and his colleagues on human dystrophic muscle.[2] Both groups suggested that the disease involved a block in the normal developmental pattern, since in the chicken and in man the H type is the embryonic form. Kaplan & Cahn indeed showed that this change in the isoenzyme composition did not apply to mice, where the M type is the embryonic form. Wieme & Herpol (1) also, examining the serum of patients with the disease, found the M type markedly decreased and the H type increased.

Also involved in the changes accompanying dystrophy is creatine kinase. No abnormalities in the patterns for the isoenzymes on starch-gel electrophoresis were detected in human myopathies by Kar & Pearson (1) or in the dystrophic mouse and chicken by F. Schapira (1). Schapira associated this finding with the observation that in the chicken the adult form of the enzyme is arrived at before the end of incubation, while in the mouse there appears to be no characteristic embryonic form. On the other hand Miyasaki and his collaborators with their dystrophic patients did find a picture resembling that from human foetal muscle, peak III being clearly lower than in normal muscle.[3] Only the isoenzyme responsible for this peak was found in the serum.

An interesting new development has been the finding by Hooton & Watts (1) that the creatine kinase of dystrophic mouse muscle, though indistinguishable from the normal by physical criteria (e.g. having identical mobility on starch-gel electrophoresis) yet had only 50 % of the specific activity. It contained only one essential thiol group/molecule instead of two, and 'finger-printing' results suggested a difference in only one peptide in kinases from the two types of muscle. The authors discuss the question whether the defect in this enzyme is hereditary and would alone be likely to account for the pathological state of the muscle; under conditions *in vivo* normal working of both catalytic sites might be essential, so that loss of one might have a disproportionately great effect. Their later work (2) showed (by estimations of the cysteine and cystine content) that the soluble

[1] Ch. 19, p. 457.   [2] Dreyfus, Demos, Schapira & Schapira (1).
[3] Miyasaki, Toyoda, Tomino, Yoshimatsu, Saijo, Katsunuma & Fujino (1). See p. 379.

proteins of dystrophic muscle were more oxidised than normally; this suggests that the change in the kinase SH might be due to oxidation rather than to replacement by some other amino acid.[1]

DISCUSSION. Several motifs concerning the primary cause will have been noticed running through the discussions by biochemists engaged in the study of hereditary muscular dystrophy.

Consider first the idea of the dystrophy as a 'membrane illness'. This aspect has been carefully analysed by Richterich (1). It can be used to explain the high concentration of certain muscle enzymes in the serum and the great decrease in their amount in the affected muscles. It is also well known that dystrophic muscle is deficient in potassium – Young, Young & Edelman (1) for example found about 20 % less potassium and up to 100 % more sodium in dystrophic mouse muscle. Zierler (2) using ^{42}K followed the K$^+$ influx and efflux in isolated mouse muscle from normal and dystrophic animals. Movement of K$^+$ in both directions was increased in the latter, the effect upon efflux greatly exceeding that upon influx. Tyler (1) has mentioned the preliminary observation of a substance in the serum of Duchenne disease patients which changes the permeability of rat skeletal muscle *in vitro*. This concept of increased permeability of the muscle-cell membrane as the primary lesion of course meets many difficulties: the preferential loss of some enzymes more than others is not explained, and concomitant changes in the fibre membrane have not so far been observable by electron microscopy. Aloisi & Margreth (1) have emphasised that in rat psoas muscle about 80 % of the extra-mitochondrial sarcoplasm is occupied by sarcoplasmic reticulum. They assemble evidence that a number of soluble glycolytic enzymes (e.g. phosphofructokinase, glyceraldehydephosphate dehydrogenase) are situated in the lumina of the sarcoplasmic reticulum, and they therefore suggest that structural abnormalities in the sarcoplasmic reticulum might be implicated with escape of enzymes via the T system. Signs of damage in this reticulum in dystrophic muscle (swelling and vacuolation) have been observed.[2]

In this connection it is interesting to remember the experiments described by Brust (1) showing that, while caffeine has a similar enhancing effect on twitch and tetanus tension in normal and dystrophic mice, nitrate affects the latter more strongly than the former. In view of the observations of Bianchi & Shanes (1) it would seem likely that the permeability of membranes to Ca ions has been altered in the dystrophic muscles: they had found that when nitrate replaced chloride in the medium, the enhanced twitch in normal muscle was accompanied by rise in Ca influx. Again, Sreter, Ikemoto & Gergely (1) found that the microsomal fraction of

[1] Cf. the results of Nichol.   [2] E.g. Pearce (1).

dystrophic muscle had greater ATPase activity but considerably less Ca-sequestering power than this fraction from normal muscle. They could not yet decide whether these changes seen *in vivo* reflect similar changes *in vivo*, or are due simply to greater fragility of the dystrophic sarcoplasmic reticulum.

Perry (8) has drawn attention to the changes in enzyme pattern in normal muscle, consequent upon lack of activity and has suggested that these might play a part in the results with bedridden dystrophic patients.

Increase in lysosomal enzymes in the muscle cell has often been invoked as a possible important cause. We have already considered the grounds for the stress laid by Tappel and Weinstock and their collaborators upon lysosomal activity in the phenomena of dystrophy. Until recently there was no good evidence that the muscle cell contained such particles, well-attested though they are in the liver cell, full of hydrolytic enzymes. Pearce (2) however in 1963 described an electron-microscope study on biopsy muscle material from Duchenne-type dystrophy cases. He was impressed by the frequent association of certain vesicles or corpuscles (0.5–1μ in diameter) with areas where fibrils or mitochondria showed pathological changes. Their characteristic appearance suggested a similarity with lysosomes. They appeared to be structurally related to specific parts of the sarcoplasmic reticulum, and it was suggested that the changes in hydrolytic enzymes in the diseased muscle could be dependent on lysosomal changes within the muscle cell. A secondary part might be played by invading macrophages. In 1968 Fitzpatrick and collaborators have given a preliminary report indicating biochemical evidence for the existence of lysosomes in mouse muscle.[1] Thus certain acid hydrolyases in this tissue showed the structure-linked latency characteristic of lysosome-contained enzymes; moreover by differential centrifugation three particle fractions could be obtained, one containing a higher percentage of acid hydrolases than either of the other two – the mitochondrial and microsomal fractions. Very similar results were independently reported by Stagni & de Bernard (1) for rat and beef skeletal muscle. Further simultaneous biochemical and electron-microscope studies on such separated particles will clearly be of great interest. It has of course to be remembered that increase in lysosomal enzymes occurs also in myopathy due to vitamin E deficiency. This makes it unlikely that the primary cause in inherited muscular dystrophy is connected with lysosomal changes.

The idea put forward by Bourne and his collaborators lays the main burden of the muscle wastage upon phosphorolytic enzymes contained in invading connective tissue. Golarz & Bourne (1) attached importance to the specificity of the increased phosphatase activity, nucleotide compounds

[1] Fitzpatrick, Park, Pennington, Robinson & Worsfold (1).

being attacked while phosphocreatine and phosphorylated intermediates
of the glycolytic cycle escaped. They suggest (2) that consideration should
be given to the possibility that the primary lesion in these dystrophies may
be in the connective tissue rather than in the muscle fibres.

No convincing case can be made at present for any specific enzyme defect
as the primary cause, although changes such as those found in creatine
kinase and adenylic deaminase could have very far-reaching effects.
Comments are often made upon the reversion to foetal type in some aspects
of the diseased muscle, but this may well be an effect rather than a cause,
and indeed all along the line it is difficult to distinguish cause from effect.

Progressive muscular dystrophy is accompanied by creatinuria which is
more pronounced in this than in other muscle diseases. Experiments by
Benedict and his colleagues with injection of labelled glycine showed that
the urinary creatine of dystrophic patients was far more richly labelled
than the creatinine excreted.[1] This was taken to mean that the creatine was
being formed as usual in the liver but, failing to enter the muscle in the
normal way, was not converted to creatinine. The inability of the muscle
to fix creatine could be due to the diminished muscle mass, but a specific
defect might also play a part.[2]

Comparisons between the biochemistry of neurogenic and dystrophic
atrophies may be useful. McCaman & McCaman are inclined to believe that
the similarity of the spectrum of enzymes affected in the two cases may
indicate a closer relationship between the dystrophic and denervated states
than has hitherto been appreciated. It is curious that the characteristic
leakage of aldolase and creatine kinase into the serum in muscular dystrophy
is not found with neurogenic atrophy.[3]

Little can be said about therapy based on these biochemical findings.
The realisation of the destruction of neucleotides in the dystrophic muscle
prompted treatment by means of administratiion of nucleosides and
nucleotides. Thus Thomson & Guest (1) gave these substances to a series of
patients mainly of the Duchenne type, at the same time following the
disease by means of systematic testing of muscle power and by serial
assays of serum aldolase. A sudden decrease in the latter (accompanied by
increase in muscular power) was looked for as evidence of amelioration and
these signs were indeed found. Other treatments which have been recom-
mended are administration of digitalis glucosides with the purpose of
changing fibre permeability in a non-specific way; and of anabolic steroids
to promote accumulation of lost constituents.[4]

[1] Roche, Benedict, Yu, Bien & Stetten (1); Benedict, Kalinsky, Scarrone, Wertheim &
Stetten (1).
[2] Schapira & Dreyfus (1). For evidence concerning the normal (probably non-enzymic)
formation of creatinine from phosphocreatine in muscle see Borsook & Dubnoff (1).
[3] Schapira, Schapira & Dreyfus (1); Vignos & Warner (1).   [4] Tyler (1); Walton (1).

GLYCOGEN STORAGE DISEASES

Several diseases of this type, all probably congenital, are known; each affects in its own way the glycogen metabolism of one or more organs, the pathological condition being traceable in most cases to the loss of a single enzyme.

One of the earliest descriptions of a disease involving derangement of carbohydrate metabolism is that given by Parnas & Wagner (2) in 1922 of a child with a large liver tumour. The blood sugar was very low but could be raised by ingestion of carbohydrate or protein; there was no reaction to adrenaline. Parnas & Wagner remarked on the lack of control of carbohydrate metabolism and the unavailability of the carbohydrate reserves. The nature of the tumour was not investigated. In 1929 von Gierke (1) gave a graphic account of a case – 'without analogue in the literature' – of a child with enormous increase in size of liver and kidneys. Heavy deposition of glycogen in the epithelium was shown histologically, and chemical and biochemical tests were also made by Schönheimer (1). No change in the glycogen content took place when the pathological liver was kept for days but the glycogen prepared from it and added to normal liver tissue was readily broken down. In 1933 Pompe (1) recorded a case in which the heart of a 7-months old child had five times the normal weight and was loaded with glycogen. Later work has shown that in this disease all tissues are affected, not only the heart, but heart failure is usually the cause of death. Nearly twenty years later McArdle (1) studied in a young man another myopathy affecting chiefly, perhaps only, the skeletal muscles. He observed during exercise a fall in blood lactate and pyruvate (in contrast to the normal rise) and very high oxygen requirement. On adrenaline injection rise in blood sugar and lactate was slow and only about 50 % of the normal. Contracture – shortening and stiffening of the muscle – was often seen.

G. T. Cori (1) in her Harvey Lecture of 1953 described the results of enzymic examination of biopsy and autopsy specimens from a number of patients with storage disease. With von Gierke's disease of the liver and kidneys she and C. F. Cori (6) had shown that glucose-6-phosphatase (the enzyme essential for glucose formation there) was absent or very much below the normal level of concentration. Illingworth & G. T. Cori (1) had also encountered two isolated cases, in one of which the glycogen deposits in the liver and muscles had very short and numerous outer branches; in the other the outer branches were very long and thin. Cori now suggested that the former state was due to lack of debranching enzyme, amylo-1,6-glucosidase, while the latter could be accounted for by the absence of the branching enzyme, 1,4 $\longrightarrow$ 1,6-transglucosylase.[1,2] She thus recognised at

[1] Also Illingworth & Brown (1).   [2] Ch. 18.

this time four types of storage disease: Type 1 (von Gierke's, in liver and kidneys) due to lack of glucose-6-phosphatase, with normal glycogen structure; Type 2 (Pompe's disease, of generalised distribution) where the enzyme defect was unknown; Type 3 (found in liver and muscle) tentatively explained as due to lack of debranching enzyme; Type 4 (in liver and probably other organs) tentatively explained as due to lack of branching enzyme. In 1956 Illingworth, Cori & Cori (1) were able to examine two cases of generalised glycogen storage disease characterised by glycogen with abnormally short branches; they could detect no amylo-1,6-glucosidase.

Further biochemical study of McArdle's disease was carried on in 1959 by several groups of workers, who traced the defect to lack of phosphorylase in the muscle. Schmid & Mahler (1) found five times the normal glycogen content, but little lactic acid was formed on incubation, and adenylic acid had no stimulating effect. When crystalline phosphorylase was added, or with glucose-1-phosphate added as substrate, lactic acid formation was normal. Schmid & Mahler found that, while adrenaline had no effect upon glycogen breakdown during incubation of excised muscle, injection of glucagon (which activates liver phosphorylase) did bring relief to one patient.

Mommaerts and his colleagues similarly found virtual absence of phosphorylase a and b;[1] there was no lactic acid formation from the large amount of endogenous glycogen but ready formation from a number of phosphate esters. Phosphoglucomutase, PR enzyme, phosphokinase and UDPG-glycogen synthetase were all present. The glycogen laid down had the normal structure.

Schmid, Robbins & Traut (1) in the case of a young man lacking muscle phosphorylase found both UDPG-glycogen synthetase and phosphoglucomutase somewhat reduced; they put this down to general degeneration of the muscle tissue. Phosphorylase phosphokinase was normal in activity. With biopsies from the same patient Larner & Villar-Palasi (1) reported normal amyl-1,6-glucosidase and UDPG pyrophosphorylase. At this time the idea of a special pathway (via UDPG) for glycogen synthesis had only recently been substantiated.[2] These workers as well as Mommaerts and his collaborators welcomed such results as bringing strong evidence for glycogen synthesis by the newly-realised pathway.

The interpretation of the defect in Type 3 became more difficult with the realisation, originating in the work of Walker & Whelan (1) that a trans-α-glucosylase was concerned in debranching as well as the amylo-1,6-glucosidase.[2] This point was first made by Manners & Wright (1); they suggested that three sorts of limit dextrinosis may exist: that depending on deficiency in trans-α-glucosylase (Type 3a); that depending on deficiency in amylo-1, 6-glucosidase (Type 3b); and a third in which both enzymes might be

[1] Mommaerts, Illingworth, Pearson, Guillory & Seraydarian (1).   [2] Ch. 18.

inadequate (Type 3c). They cited two cases of liver storage disease, of which one might be Type 3a or 3c, the other Type 3b. In 1964 Illingworth & Brown (2) summarised results from a number of cases of Type 3. In assaying the two enzymes concerned they made use of a ^{14}C-labelled branched pentasaccharide. Glucosidase activity was shown by formation of glucose and maltotetraose from the pentasaccharide; transglucosylase activity by the formation of branched 7- and 8-unit compounds by transfer from glycogen to the pentasaccharide. They found that tissues from Type 3 disease usually showed neither glucosidase nor transferase activity, although a single case did occur where the transferase persisted. They referred also to another case of the rare Type 4, but no enzymic evidence for the postulated deficiency of the branching enzyme is yet available.

In 1963 Hers (1, 2) described his attack on the enzymology of Type 2 or Pompe's disease, the severest form of glycogenosis. He showed that in normal human liver and heart muscle an acid maltase (pH optimum 4.0) was present. This enzyme corresponded to the one in rat liver acting upon α-1,4-glucosidic linkages, which was shown by Lejeune, Thinès-Sempoux & Hers (1) to be contained (like several other hydrolases) in the liver lysosomes. Hers found this enzyme also in skeletal muscle, though its activity there was only about one-third of that in the heart and one-eighth of that in the liver. In liver, heart and muscle of children with Pompe's disease this enzyme was missing. Since acid maltase can transfer glucose from maltose to glycogen, it was at first thought that this enzyme might be the transglucosylase postulated by Walker & Whelan. That however could not be the case, since the glycogen deposited in Pompe's disease is normal and not a form of limit dextrin. Hers has suggested an explanation, for appreciation of which a certain background is necessary.

In 1955 de Duve and his collaborators brought circumstantial evidence (depending on differential centrifugation and the release of enzymes under certain conditions) for the existence in liver cells of granules containing a number of hydrolytic enzymes.[1] Since that time there has been substantial validation of the presence of such granules in other tissues also. This support comes not only from tissue fractionation and enzymic studies, but also from electron microscopy. Thus Ashford & Porter (1) have described the formation of liver lysosomes as taking place by the segregation of packets of cytoplasm within membranes. With regard to muscle, we have seen how Pearce also observed certain unfamiliar vesicles in areas where pathological change was evident in fibres and mitochondria in a number of myopathies.[2] These he tentatively likened to lysosomes. The question now arises as to the function of lysosomes in normal tissue. De Duve (1) discusses

[1] De Duve, Pressman, Gianetto, Wattiaux & Appelmans (1).
[2] P. 506 above.

'physiological autolysis' and cites such necessary occasions for hydrolysis of tissue constituents as the involution of the uterus after birth and the metamorphosis of insects; he points out also that the observed turnover of the main constituents of normal tissues indicates some degree of autolysis there also; such an arrangement would of course be beneficial under starvation conditions. It may be surmised that the lysosomal vesicles engulf small portions of cytoplasm which are then digested in this segregated area inside a membrane, the products of digestion being afterwards released. Returning now to Pompe's disease, Hers postulated that in the ordinary course of events some glycogen would be contained in the cytoplasm engulfed; in the absence of the acid glucosidase (normally contained in the lysosomes) glycogen would remain undigested within the lysosomes, where its amount would increase progressively. The lysosomes would swell to enormous size, at length reaching such proportions that the muscle fibres become distorted and disrupted. There is no interference with the cytoplasmic breakdown of glycogen by means of phosphorylase and the other enzymes of the glycolytic cycle, but any of these taken into the lysosomes would be digested there by the cathepsin.

In 1965 Tarui *et al.* investigated three cases (in a single Japanese family) characterised by accumulation of hexosemonophosphate as well as glycogen in the muscles and by failure of the lactic acid to rise in the venous blood during exercise.[1] The muscles were found to be almost completely lacking in phosphofructokinase, while pyruvate kinase was significantly increased. Phosphoglucomutase and phosphoglucoisomerase were normal. With fructosediphosphate as substrate, lactic acid formation *in vitro* was twice as fast as with normal muscle homogenates. Further work by Okuno, Hizukuri & Nishikawa (1) revealed that two other enzymes were affected, at any rate in the muscles of one patient. Here both glycogen synthetase and UDGP pyrophosphorylase showed 2–3 times the normal activity. The pathological muscle contained some glucose-6-phosphate independent synthetase whilst in the two normals tested only the dependent form was present. Its greater activity was therefore not entirely due to the accumulating hexosephosphate. Thus the increased deposition of glycogen in this case was due to two causes – interference with the glycolytic chain and enhanced synthesis.

Still another disease of carbohydrate metabolism (in this case a 'glycogen storage deficiency') probably also congenital, was discovered by Lewis, Steward & Spencer-Peet (1) in 1962. Two identical twins showed profound hypoglycaemia when food was withheld for twelve hours; liver biopsies gave only about 0.5 % of glycogen and complete absence of glycogen synthetase. The activity of phosphorylase, UDGP pyrophosphorylase and

---

[1] Tarui, Okuno, Ikura, Tanaka, Suda & Nishikawa (1). Another case of this rare disease has been reported by Layzer, Roland and Ranney (1).

glucose-6-phosphatase was normal. It is not known whether this disease attacks the muscles.

Mahler (1) has recently reported high glycogen content in the muscles of patients suffering from von Gierke's disease of the liver. The blood sugar is low, but blood lactate and pyruvate are high. That these may be the source of the glycogen was suggested by the fact that much more radioactivity was incorporated into glycogen from labelled pyruvate in such muscles. Also their fructosediphosphatase content was much higher than normal.[1]

DISCUSSION. Except in the case of Type 3 and of the new type investigated by Tarui *et al.*, each of these glycogen storage diseases seems to result from the disappearance of a single enzyme; this defect can be interpreted as the failure in operation of a single gene in each case. Investigation of the mechanism of the significantly increased activity of certain enzymes in the disease occasioned by phosphofructokinase loss will be of great interest; the same is true of the suggested increase in glycogen synthesis from pyruvate in the muscles in von Gierke's disease.

The cramp or contracture which is often mentioned is readily understood in terms of falling off of ATP synthesis, and consequent binding of the actin and myosin filaments together in a form of rigor.

Only a few words can be said about therapy. In a case of McArdle's disease Pearson & Rimer (1) reported that the young man's capacity for work on a treadmill was increased 20-fold by infusion of glucose or fructose. Tobin & Coleman (1) also found that glucose and fructose feeding could alleviate symptoms in this disease, but they warned that its use must be limited owing to the danger of even greater glycogen deposition. Tobin & Coleman, as well as Schmid & Mahler (1) found that some relief was given by injection of glucagon, to mobilise liver glycogen.

[1] Ch. 18, p. 448.

22

CONTRACTION IN MUSCLES OF INVERTEBRATES

Here we shall consider on the one hand the nature and arrangement of the proteins making up the muscle machine, and on the other hand the chemical substances and enzymic reactions providing the energy. A vast number of observations has been made on examples of the different phyla, but knowledge is still very incomplete and it is often difficult to generalise, particularly because important differences may turn up between closely related species. It will not be possible to deal in a comparative way with more than a part of the data. In the case of two important types, however, the adductor muscle in molluscs (responsible for the proverbial closing mechanism e.g. in the clam) and the fibrillar muscle of certain flying insects (the fastest and most active muscle known) very thorough investigations have been made. Formulation of detailed suggestions for the *modus operandi* of these two types has been possible.

It is necessary to say a few words about structure, which is very varied. Setting out from histological and electron-microscope observations, Hanson & Lowy (8) have distinguished three kinds of invertebrate muscle;[1] (a) Striated, showing the *A* and *I* bands familiar in vertebrate striated muscle; in invertebrates these are found e.g. in insect muscle and in the phasic adductor of *Pecten*. (b) Muscles such as those in certain cephalopods and annelids showing 'double-oblique striation', which depends on the helical arrangement of thick and thin filaments. The latter are attached to dense bodies, to be regarded as comparable to the *Z* lines. A less regular structure of this sort occurs, e.g. in the adductor muscles of some lamellibranch molluscs. (c) Muscles in which no striation can be detected, although two types of filament may be present, as in the opaque adductor of the oyster, *Pecten* and *Mytilus*. In some cases included under (b) and (c), the thick filaments are known to consist mainly of paramyosin and this may apply rather generally to these two groups. It seems certain that the thinner filaments always consist of actin.

[1] Also Hanson & Lowy (9).

THE PROTEINS OF THE CONTRACTILE MECHANISM

ACTOMYOSIN. The earliest studies of actomyosin from invertebrate muscle were those of Dubuisson and his collaborators in 1947,[1] who obtained electrophoretic patterns from the muscle of the gastropod mollusc *Murex brandaris* (after extraction at ionic strengths of 0.15 or 0.5) very similar to those obtained earlier with frog muscle. The components α, β, and γ were all present in such extracts at high ionic strength.[2] About the same time Amberson[3] and his collaborators extracted from intact king-crab (*Limulus*) muscle by means of pyrophosphate a viscous protein giving flow bire-fringence. Histological examination showed that the birefringence of the *A* bands had disappeared after this treatment.

The first work on invertebrate actomyosin ATPase was that of Gilmour & Calaby (1) in 1953. They worked with the orthopteran insect *Locusta migrans*, using both the femur muscles (resembling most vertebrate skeletal muscle in structure) and the indirect muscles of the thorax, which are intermediate in structure between normal skeletal muscle and the specialised fibrillar flight muscles of Diptera etc. Their preparations hydrolysed only the terminal phosphate of ATP and the activities fell within the range given by Bailey for rabbit actomyosin – Q_P 1600 for the thorax muscle and 2800 for the femur preparation, the latter also having greater ATP sensitivity. During the next few years Maruyama (3, 4) made an extended study of actomyosin preparations in a number of phyla. That made from the body-wall of the echiuroid *Urechis unicinctus* hydrolysed only the end phosphate of ATP, but showed low activity – Q_P only 300 to 800. That made from the contractile protein of a sea-anemone hydrolysed both ATP and ADP at a low rate; its relative viscosity and ATP sensitivity were very low. The actomyosin of the bee (*Apis mellifera*) was examined in more detail (5). It showed similar properties to rabbit actomyosin in its salting-out and sedimentation behaviour, and its high enzymic activity – Q_P 4000 to 5000. A characteristic property was its very high double refraction of flow, even at low velocity gradients, suggesting a particle length of more than 3 μm – actually near that of the sarcomeres themselves.

Tonomura, Yagi & Matsumiya (1) in 1955, examining actomyosin pre-parations from the fast phasic and slow tonic muscles of the adductor of *Pecten*, found actomyosin in greater quantity in the former than in the latter, with higher Q_P (about 700 and 250 respectively) and higher ATP sensitivity. Rüegg (1, 2) in 1957, also using the tonic muscles of *Pecten*, obtained much higher ATPase activity after purification by means of repeated precipitation at ionic strength of 0.25 instead of 0.1–0.05 as in all

---

[1] Dubuisson & Pezeu (1); Dubuisson & Roubert (1).   [2] Ch. 10.
[3] Amberson, Dale Smith, Chinn, Himmelfarb & Metcalf (1).

the earlier work.[1] He also prepared purified myosin from the actomyosin.
He went on to show (3, 10) a correlation between the tonic properties of the
three different portions of *Pecten* adductor muscles and their paramyosin
content.[2] Thus the actomyosin : paramyosin ratio for the striated phasic
adductor was 9:1; for the translucent portion of the 'smooth' adductor 2:1;
for the opaque portion of the latter, 1:2. Admixture with paramyosin might
well explain the very low enzymic activity of the preparation made by the
Japanese workers from the tonic muscles of *Pecten*.

Connell (1) in 1961 made the interesting observation that skeletal-muscle
myosins could be divided into two distinct groups – unstable myosins
prepared from a variety of fish, and more stable prepared from rabbit, ox
and chicken. The criteria used included rate of heat denaturation, effects of
increasing urea concentrations and rate of digestion with trypsin. Bárány &
Bárány (2) later confirmed and extended this classification, concluding that
myosins from mammals and birds were much more stable than myosins
from fish and invertebrates.[3] They considered that the most likely basis for
the behaviour of the unstable group lay in a higher content of anionic
residues; in a comparison of the amino-acid content of rabbit and scallop
myosins there was a difference which would confer on the *Pecten* myosin an
excess of about 100 negatively charged groups per mole. The involvement of
SH groups in this instability was also shown: *Pecten* myosin was completely
devoid of ATPase activity if prepared in absence of cystein or β-mercapto-
ethanol. Protection was afforded by combination with actin, since actomyosin did not lose activity under preparative conditions. Connell saw
a connection between the stability of their myosins and the temperature at
which the animals were accustomed to live, and this seems very likely.

Carsten & Katz (1) made a wide survey of species (including *Pecten*, fish,
frog, bird and mammal) in studying the nature of their actin. There were
differences in amino-acid composition and peptide map, the differences
decreasing with closer relationship. The sedimentation coefficient was the
same for all, and the patterns with starch gel electrophoresis were similar,
except in the cases of *Pecten* and the octopus.

PARAMYOSIN. A new form of tropomyosin was discovered by Bailey (7)
just ten years after his discovery of tropomyosin in vertebrates. We have
already mentioned the Type I fibrils, giving a characteristic X-ray diffraction pattern, first observed by Bear. Such fibrils from clam adductors were
also studied with the electron microscope by Hall, Jakus & Schmitt (1, 2) in

---

[1] Compare the purification of actomyosin from vertebrate smooth muscle, ch. 23.
[2] See below and ch. 10, p. 234.
[3] It is interesting to remember here the *greater* thermostability and resistance to trypsin
shown by vertebrate smooth-muscle myosin, at any rate from certain organs (ch. 23).

1945. They named the responsible protein paramyosin (1). Bailey (8, 9) in 1956 set out to characterise paramyosin chemically. It was known to be specially abundant in the slow parts of lamellibranch adductor muscles, and he extracted it (at ionic strength 0.5) in good yield (30 % of the muscle protein) from this muscle in *Pinna nobilis*. The solution was very viscous, and needle-shaped crystals were formed on lowering the ionic strength to 0.3. Its amino-acid composition was similar to that of vertebrate tropomyosin – it contained no tryptophan or proline, but lysine had been partly replaced by arginine and glutamic partly by aspartic acid. Its high viscosity and double refraction of flow showed it to be a very asymmetric molecule, and it gave a well-defined α X-ray diffraction pattern. Soon afterwards Kominz, Saad & Laki (1) isolated the two forms of tropomyosin from the smooth adductor of the clam. They were separated by ammonium sulphate fractionation, the new invertebrate tropomyosin coming down at 28–40 % saturation, the well-known form at 45–65 % saturation. At the same time a group of workers in England collaborated to show that the Type I or paramyosin filaments did indeed contain the water-insoluble tropomyosin.[1] Needle-shaped crystals of this protein were prepared from the white part of the adductor of the oyster and of *Pinna nobilis*, and these were examined in the electron microscope. A pattern of banding, repeated every 725 Å was seen, closely resembling that of paramyosin filaments isolated from the same muscles. Kay (2), studying the protein by light-scattering, sedimentation, diffusion and viscosity methods, found a molecular weight of 134000, and evidence for dimensions of 1400 Å in length and 19 Å in width. The length of the molecule would thus be about twice that of the period in the paramyosin fibrils and crystals. Rüegg (1) showed that it had no ATPase activity and did not combine with actin.

A few years later Cohen & Holmes (1) obtained X-ray diffraction diagrams from native muscle (the anterior byssal retractor) of *Mytilus edulis*. This pattern, no doubt due to the paramyosin present in large amount, could be accounted for by a coiled-coil α-helical structure containing two chains; a high helical content would agree with the very low percentage of proline found earlier by A. G. Szent-Györgyi & Cohen (1). Lowey, Kucera & Holtzer (1), using paramyosin solutions from *Venus mercenaria*, confirmed the molecular dimensions given by Kay, but found a molecular weight of 220000. This molecular weight and the other known data are consistent with the double-helix model.

An electron-microscope study has recently been made by Tsao and his collaborators.[2] In general agreement with previous work, the crystals showed band patterns which could be represented by $m \times 140$ Å (where

[1] Hanson, Lowy, Huxley, Bailey, Kay & Rüegg (1).
[2] Tsao, Kung, Peng, Chang & Tsou (1); Peng, Kung, Hsiung & Tsao (1).

$m = 5, 10...$). Hall's mica-replica technique was used to investigate the size and shape of the molecules. The majority were 1400 Å long, with a minority 700 Å and a very few 2100 Å. These results suggested the existence of paramyosin subunits 700 Å in length. The diameter of the molecules was estimated to be 20–30 Å.

Kominz, Saàd & Laki had suggested the designation tropomyosin A for the water-insoluble protein, tropomyosin B for the water-soluble one; but we shall rather retain the terms paramyosin and tropomyosin.

ENERGY PROVISION

THE DIVERSITY OF PHOSPHAGENS AND THEIR DISTRIBUTION. Mention has already been made of Kutscher's attribution of arginine to the invertebrates and of creatine to the vertebrates; and of Meyerhof & Lohmann's successful search in 1928 amongst invertebrates for a phosphoarginine analogue of phosphocreatine – the example they chose being crustacean muscle.[1] Meyerhof (21) also found phosphoarginine in a number of molluscs and in the gephyrean worm *Sipunculus nudus*. In 1926 Kutscher & Ackermann (1) had wished to investigate the cephalochordate *Amphioxus*, thinking that there one might find creatine, a chemical aspect of the transition from invertebrates to vertebrates. But *Amphioxus* was not available in great enough supply, and they had to be content with finding creatine in selachian fish muscle. In 1928, however, both Meyerhof (21) and the Eggletons (3) did find phosphocreatine and no phosphoarginine in *Amphioxus*. Meyerhof (4) in his book of 1930 interpreted this 'development of phosphocreatine out of phosphoarginine' as a chemical mutation.[2]

Then in 1932 Needham, Needham, Baldwin & J. Yudkin (1) described the results of a rather wide survey of phosphagen distribution in invertebrate muscle. They found phosphoarginine in representatives of all the phyla they used (Coelenterates, Platyhelminths, Nemertines, Annelids, Podaxonia, Cephalopoda, Urochorda) and unexpectedly *both* phosphagens in the jaw muscles ('Aristotle's lantern') of the echinoderm *Strongylocentrotus*; also in the few available specimens of the enteropneust *Balanoglossus salmoneus*, another protochordate, both phosphagens were demonstrated. This last finding was of special interest in that it offered support to the evolutionary theory of Bateson (1) (based on the similarity of the larval forms of enteropneusts and echinoderms[3] which saw in these two phyla stages in the transition from invertebrates to vertebrates. A little later Baldwin & Needham (1) examined two more echinoderms, finding only

---

[1] Ch. 5.   [2] Meyerhof (4) p. 95.
[3] *Balanoglossus* has a notochord in the adult state, but a larval form very similar to those of echinoderms.

phosphoarginine in the Crinoid specimens, and only phosphocreatine in the Ophiuroids. Needham *et al.* found only phosphoarginine in the holothurian *Synapta*, but Verbinskaya, Borsuk & Kreps (1) reported the presence of both in the closely related *Cucumaria frondosa*. In another group of Protochordata, the tunicates, Needham, Needham, Baldwin & Yudkin were unable to find phosphocreatine, and only uncertain small amounts of phosphoarginine by the methods of that time. However in 1956 Morrison, Griffiths & Ennor (1) showed conclusively by chromatographic methods that phosphocreatine but no phosphoarginine was present. In 1950 Baldwin & W. H. Yudkin (1) had the opportunity of using two more enteropneusts – *Saccoglossus horsti* and *Saccoglossus kowalevskyi*. In these, as in *Balanoglossus clavigerus*[1] only phosphocreatine was found. W. H. Yudkin (1) has pointed out that the early results of Needham, Needham, Baldwin & Yudkin showing both have never been confirmed and this is important. An observation disconcerting in the context of evolutionary theories was that of Roche & Robin (1) in 1934 that phosphocreatine actually occurred in a species as primitive as *Thetia lyncurium*, one of the sponges, while phosphoarginine was present in the related *Hymenaicedon caruncula*.

Re-examining the annelids in 1950, Baldwin & W. H. Yudkin (1) found a number of Polychaeta containing a phosphagen which they considered identical with phosphocreatine, while others contained a 'new arginine phosphate-like phosphagen' – either alone or together with the phosphocreatine. In the Gephyrea (generally held to have close affinities with the Annelida) no phosphocreatine was found, but again this phosphoarginine-like material. In the case of the only Oligochaete tested, *Lumbricus terrestris*, the common earthworm, they could find no trace of either of the known phosphagens, although free arginine is known to be present.

Then came the discovery, from 1953 onwards, of a whole gamut of new phosphagens – N-phosphoguanidines, mono- or disubstituted. This was mainly the work of a group in Paris – Thoai, Roche, Robin and their collaborators. As early as 1932 Arnold & Luck (1) had, to their surprise, been unable to find arginine in a number of marine representatives of the Annelida, including polychaete and gephyrean species. Also in the case of the oligochaete *Lumbricus* they expressed doubt as to whether the phosphagen present could be phosphoarginine on account of its unusual chemical behaviour; and Kurtz & Luck (1) in 1937 decided that the phosphagen of the polychaete *Nereis brandti* could not be either phosphoarginine or phosphocreatine.

In 1953 the Paris group described no less than three new guanidine compounds – glycocyamine, from muscles of *Nereis diversicolor*,[2] taurocyamine from muscles of *Arenicola*[2] (one or other of these also being shown by

---

[1] K. R. Rees, personal communication.   [2] Thoai, Roche, Robin & Thiem (1).

chromatographic methods in a number of marine worms) and guanidino-
ethylserylphosphate (lumbricine) from the earthworm *Lumbricus terrestris*.[1]
In each case Thoai and his collaborators established the presence of the
phosphorylated form and considered that these fulfilled the role of phosphagens. Thus the suspicions of the earlier workers were fully justified.
Hobson & Rees (1) confirmed the presence of phosphoglycocyamine in
Nereis and of phosphotaurocyamine in *Arenicola*; they also established that
the phosphocreatine – like phosphagen described by Baldwin & Yudkin in
certain annelids was indeed this compound. They remarked that this
occurrence of phosphocreatine could hardly be indicative of a close relationship between this phylum and the Chordata, and took it as a case of
convergent evolution.

$$
\begin{array}{lll}
\mathrm{HN\!-\!\!-\!P\!-\!\!-\!(OH)_2} & \mathrm{HN\!-\!\!-\!P\!-\!\!-\!(OH)_2} & \mathrm{HN\!-\!\!-\!P\!-\!\!-\!(OH)_2 \quad COOH} \\
\qquad\;\; \| & \qquad\;\; \| & \qquad\;\; \| \\
\qquad\;\; \mathrm{O} & \qquad\;\; \mathrm{O} & \qquad\;\; \mathrm{O} \\
\mathrm{C\!=\!NH} & \mathrm{C\!=\!NH} & \mathrm{C\!=\!NH} \qquad \mathrm{H_2NCH} \\
 & & \qquad\qquad\qquad\quad \mathrm{O} \\
 & & \qquad\qquad\qquad\quad \| \\
\mathrm{HN.CH_2.COOH} & \mathrm{HN.CH_2CH_2.SO_2OH} & \mathrm{HN.CH_2.CH_2.OP\!-\!O\!-\!CH_2} \\
 & & \qquad\qquad\qquad\quad \mathrm{OH} \\
\;\; \text{Glycocyamine} & \quad \text{Taurocyamine} & \qquad \text{Lumbricine}
\end{array}
$$

In all these compounds one H of the amino group is replaced by the phosphate group, with formation of the energy-rich phosphoamide linkage.

A few years later two more phosphagens were identified. Hypotaurocyamine and the corresponding phosphagen were found by Robin & Thoai
(1) in the muscles of the gephyreans *Phascolosoma Blainvilli* and *elongatum*.
This phosphagen was also found in much smaller amount in *Arenicola
marina* together with much larger amounts of phosphotaurocyamine.[2]
Finally Thoai, di Jeso & Robin (1) found guanidoethylmethyl phosphoric
acid (opheline) and its N phospho derivative in the Polychaete *Ophelia
neglecta*. Two closely related species contained taurocyamine and lumbricine
respectively.

$$
\begin{array}{ll}
\mathrm{HN\!-\!\!-\!P\!-\!\!-\!(OH)_2} & \mathrm{HN\!-\!\!-\!P\!-\!\!-\!(OH)_2} \\
\qquad\;\; \| & \qquad\;\; \| \\
\qquad\;\; \mathrm{O} & \qquad\;\; \mathrm{O} \\
\mathrm{C\!=\!NH} & \mathrm{C\!=\!NH} \\
 & \qquad\qquad\qquad \mathrm{O} \\
 & \qquad\qquad\qquad \| \\
\mathrm{NH.CH_2CH_2SO_2H} & \mathrm{NH.CH_2CH_2\!-\!O\!-\!P\!-\!O\!-\!CH_3} \\
 & \qquad\qquad\qquad \mathrm{OH} \\
\;\; \text{Hypotaurocyamine} & \text{Phosphate of guanidoethylmethyl} \\
 & \text{phosphoric acid.}
\end{array}
$$

[1] Ibid. (2); Thoai & Robin (1).
[2] Roche, Robin, di Jeso & Thoai (1).

THE PHOSPHOKINASES AND PHOSPHAGEN METABOLISM. The presence in invertebrate muscle of these N phospho-guanidine compounds could not in itself guarantee their phosphagen role of providing phosphate for ATP resynthesis and thus taking part in energy provision for contraction. In 1933 Baldwin & Needham (2) showed that fly muscle contained ATP, in amount comparable with that in frog muscle, and in 1935 Lohmann (17, 18) found that extract of crab and of octopus muscle, especially in the presence of added ATP, caused splitting of phosphoarginine but not of phosphocreatine. Lehmann (2) reported synthesis of phosphoarginine by crab-muscle extract, while Baldwin & Needham (1) found synthesis of both phosphagens in extracts of echinoid jaw muscle, if phosphoglyceric acid and ATP were added; also of phosphoarginine only, in extracts of *Holothuria tubulosa*.[1] Much later Griffiths, Morrison & Ennor (1) observed in the echnioid *Heliocidarus erythrogramma* both arginine and creatine, both phosphagens and two specific phosphokinases which could be readily separated. The wider surveys undertaken by W. H. Yudkin (1) in 1954 and by Yanagisawa (1) in 1960 may be summarised thus:

	Yudkin	Yanagisawa
Crinoidea		A (1)
Holothuroidea	A (1)	A (3)
Asteroidea	A (1)	A (5)
Ophiuroidea	C (3)	C (3)
Camerodonta		A+C (5)
Irregular Echinoidea		A (3)
Echinoidea	A (2); A+C (3)	

(where A indicates the arginine kinase, C the creatine kinase, and the number in brackets the number of species tested).

These results fit well with what we have already seen of the distribution of these two phosphagens and illustrate again the specificity of the phosphate-transferring enzymes. Like Baldwin & Needham, Yanagisawa observed that when the two phosphokinases occurred together, the arginine phosphokinase was the more active.

In 1957 Thoai (1) obtained from *Arenicola marina* an enzyme specific for taurocyamine, while that from *Nereis diversicolor* was specific for glycocyamine. Similar findings were reported by Hobson & Rees (2). Thoai & Pradel (1) later made a much purified preparation from *Arenicola* and studied its specificity. Curiously enough, lumbricine was phosphorylated as well as taurocyamine, but creatine and arginine were not. Thoai, Robin & Pradel (1) found the hypotaurocyamine kinase to be more specific.

Very little work seems to have been done on the breakdown of phosphagens in invertebrate muscle actually during contraction, probably owing

[1] See ch. 6 for the reversible action of such kinases.

to the difficulty of obtaining suitable uninjured muscles from many of the creatures studied. The need for such evidence was realised by M. G. Eggleton (1) who in 1934 made a study of the retractor of the foot of *Mytilus edulis* from this point of view. She found that fatigue caused by prolonged electrical stimulation reduced the phosphoarginine content, and that rest under aerobic conditions allowed its restoration. Twenty years later Rey (1) also showed a great decrease in phospholumbricine of the earthworm after tetanic stimulation.

Very little work seems to have been done on glycolysis in invertebrate muscle, though, as we have seen, in a number of cases formation of phosphagen at the expense of phosphorylated intermediates in glycolysis was studied. Lohmann (18) found surprisingly low capacity for lactic acid formation in *Octopus* muscle, and correlated with this very little synthesis of phosphoarginine by means of phosphopyruvic acid in the presence of ATP.

DISCUSSION. The attempt briefly to describe work in so wide a field must lead to presentation of a very fragmentary picture. Difficulties arise from the need to compare results from different varieties in a given species, since the material available to different workers varied greatly.

It is clear that the simple distinction between arginine and creatine as characteristic of invertebrates and vertebrates respectively has to be abandoned, and also any simple conception of the distribution of the phosphagens as a pointer to lines of evolutionary development. Thoai & Roche (1) went so far as to say in 1964 that 'the distribution of the phosphagens does not respect zoological classification.'

The diversity and the distribution of the phosphagens and their specific phosphokinases are indeed puzzling. Baldwin & Yudkin in their discussion of phosphagens in the Polychaeta and Oligochaeta could find no correlation between possession of phosphoarginine, phosphocreatine or both on the one hand, and overall muscular activity, habitat or any other environmental feature on the other. Only in the isolated case is it yet possible to suspect some connection of the type of phosphagen with other factors; Robin & Thoai have suggested that since the gephyreans containing hypotaurocyamine live at great depths in the sea, difficulty in oxidation of hypotaurocyamine to taurocyamine might be such a factor. Neither Ennor & Morrison (1) nor Thoai & Roche (1) in their most interesting reviews were able to suggest any general solution to this problem.

More recently further work has been done on the phylogenetic significance of biochemical evidence. Thus Bolker (1) in 1967 took into account the nature of the predominant steroids in echinoderms as well as of the phosphagens, and worked out suggested relationships between certain classes of the Echinodermata. About the same time Moreland, Watts & Virden (1) drew attention to the interesting fact (already known for some years though disregarded) that while the unfertilized eggs of many Echinodermata contain arginine kinase, the spermatozoa contain creatine kinase. They suggested that this might be an adaptation to protect the phosphagen, since synthesis of nuclear histones (particularly active in the testis) involves a drain on arginine; further, the appearance of creatine kinase in adult muscle could result from the operation of a genetic release mechanism for the synthesis of this kinase. Rather than indicating direct phylogenetic relationship between echinoderms, tunicates and hemichordates,[1] this appearance may indicate a parallel evolution of groups long separated but coming from a common ancestor which contained the genetic potential for creatine kinase formation. Moreland and her collaborators also emphasise the great similarity in amino-acid composition of the two enzymes and in the behaviour of their active centres. This makes it possible to picture that the mutation essentially involved an alteration of the binding site for arginine to one for creatine. It may also be significant for the more primitive nature of the arginine kinase that this enzyme (as has been shown largely by the work of this group) exists as a monomer or dimer, while the creatine kinase is found as a dimer or in a few cases amongst the Echniodermata as a tetramer. Taking into account the properties of the enzymes and their distribution, so far as this is known, in the spermatozoa and the adult, Moreland and her collaborators suggest that the evolutionary sequence might be represented thus:

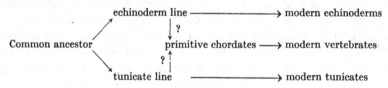

THE HOLDING MECHANISM

EARLY EXPLANATIONS. The power of lamellibranch molluscs, such as the scallop *Pecten*, to keep their shells tightly closed for days against all efforts to open them has been known for centuries; it forms for example the subject of one of La Fontaine's fables, appearing in 1668 – *Le Rat et l'Huitre*.

Coutance recognised in 1878 that the closing-and-opening muscle consists of two parts, together forming a cord of uniform diameter;[2] one part is cross-striated and contracts rapidly; the other is smooth, contracts more

---

[1] For occurrence in the tunicates of arginine kinase, creatine kinase or both enzymes together see Virden & Watts (1) and unpublished work by Watts & Watts quoted by Moreland, Watts & Virden (1).   [2] Quoted by Parnas (6).

slowly, and then takes over the burden of maintaining tension. Very early in the present century much experiment and still more cogitation were devoted to attempts to understand this phenomenon of long-continued, apparently fatigueless contraction; as Parnas (6) described it, an oyster, scarcely breathing, can remain 20–30 days tightly closed, opposing a tension of some 500 g. The amount of energy needed by tetanically contracting striated muscle would be enormous. Bethe (2) in 1903 came to the conclusion, on grounds of the otherwise impossibly great energy expenditure needed, that tonus muscle must obey other laws than those governing muscles concerned in movement. He saw tonus as 'another form of true rest'.

Biedermann (2), chiefly concerned with the behaviour of the earthworm, expressed in 1904 the view that certain types of tonus depend on a coagulation process which hinders or completely prevents the lengthening of the muscle once contracted. He was clearly influenced by the analogies so often drawn at that period between the processes leading to contraction and those concerned in death rigor.[1] Biedermann also discussed the nervous excitation leading to tonus, and held that excitation of inhibitory nerves was responsible for its relaxation – a view which subsequent work has in some cases upheld.

At this time O. Frank, in a long review (1) on the thermodynamics of muscle, quoted Biedermann (2) as recording that, for a steady tetanus, 300 stimuli/sec are necessary for insect muscle, 20–30/sec for rabbit white muscle, and only 2/sec for toad muscle. Frank drew the conclusion that in smooth muscle in general one might get tetanic contraction with stimuli spaced several seconds apart; in this way much less energy would be used in a given time interval. However, he was not prepared to apply this explanation to the closing muscle of the lamellibranchs, believing that tonus here was maintained without any repetition of stimulation. In any case, he did not accept that there was the slightest evidence for any type of contraction with no greater heat production than in rest.

Grützner (2) was perhaps the first to use the rack-and-pinion metaphor, which came to enjoy such popularity, to explain tonus. He wrote in 1904: 'One imagines a stretched elastic rubber thread, which on shortening lifts a weight provided with hooks up to a rack provided with corresponding hooks; and if the shortening is finished, it can stop at any place and hook on at any place.' Marceau (1) in 1909, likened the action of the opaque adductor muscle in maintaining the closure of the shell to 'the action of the ratchet on the moving wheel of a crane, which hinders it from returning backwards, when it has advanced by one tooth.' von Uexkull (1) also used the catch analogy, but emphasised that its mechanism must be capable of loosening

[1] Ch. 15.

and again coming into action. As an alternative he suggested considering the muscle as a sack with contents capable of existing in a coagulated or fluid state, the coagulation being ordered in such a way as to prevent lengthening but not shortening. He brought evidence to show that long-maintained tonus in *Pecten* (which he believed needed no extra metabolism) did depend on continuing nervous stimulation. We may notice that with leech muscle (material no doubt easier to deal with than the lamellibranch muscle) Cohnheim & von Uexkull (1) did find in 1911 substantially greater O_2 uptake on maintenance of 13–70 g weights over half-hour periods.

Parnas (6) in 1910 was the first to face squarely the need for experimental evidence concerning metabolism during tonus, using a number of muscle representatives – *Venus*, *Cytheraea* and *Pecten*. Three creatures were suspended horizontally in an airtight cylinder filled with sea water, and a heavy lead weight was attached to them by means of a very small hole bored in each shell. The change in oxygen content of the sea water during a 3–5 h period was determined by the colorimetric method of Winkler, said to be accurate to 10–3 mg of oxygen. Parnas found no increase in oxygen uptake when the mussels were loaded; he calculated that for maximum contraction of the weight of muscle concerned wth a burden of 3000 g/cm^2 only 8×10^{-3} mg of oxygen was used by the smooth holding muscle. He concluded that the metabolism associated with tonus could not be more than 1/10000 or 1/100000 of the amount which would be needed with similar performance of striated muscle; and that the energy consumption of these muscles in different conditions of length and loading was independent of these conditions. Ritchie (1) in 1928 criticised this work, both as regards adequacy of experimental arrangement and as regards interpretation. If, as he believed, tonus depended on fusion of slow twitches, then the energy requirement of *Pecten* closing muscle should be roughly assessable by considering the known energy requirement per twitch for the frog sartorius and the relaxation time in the two cases. He concluded that this energy requirement was very low and that Parnas' figures, though not dependable, were yet not lower than would be expected and certainly did not prove the point he believed he had made – that the holding mechanism required no energy supply.

About this time Bozler became interested in these problems. In this early work (2, 3) he considered that he had evidence for the participation of two kinds of fibril within the single fibre – one responsible for tonic, the other for tetanic contraction. In later work however (4) he described such fibres as peculiar to the muscle of the chromatophores of cephalopods with which he had been working. By 1930 his ideas on the mechanism of contraction of smooth muscle had undergone a great change. Working in the laboratories of A. V. Hill and of Fenn, he made the first measurements on the heat production of such muscle (5) (using the retractor pharynx of the snail).[1] The rise of tension was rapid, but the relaxation time lasted several seconds. The 'coefficient of economy' $Tl\,dt/H$ (where T = tension, H = heat, t = time and l = length) was much greater in contractions of long duration and increased greatly as contraction continued; in these circumstances the relaxation time increased. CO_2 in low concentration also caused an increase in the coefficient of economy, and Bozler suggested that this effect and the effect of prolonged tension could be explained by a fall in pH of the muscle. Bozler went on to study the mechanical properties. Using the foot muscle of *Pecten* he showed (7) that with single induction shocks there was summation to a high maximum tension and very slow relaxation. But muscle which had held the shell firmly closed for a long time could relax surprisingly fast on receiving another stimulus. Like Biedermann, he took this to mean that the rate of relaxation was under nervous control, and he suggested production of a chemical substance (as with vagus stimulation) capable of altering the physical properties of the muscle. Experiments on relaxed smooth muscle of the snail retractor and the adductor of *Pecten* showed (8, 9) that elastic tension was produced when the muscle was stretched; as soon as the length was left constant, the tension dropped and disappeared on an exponential curve. The time course of this release of tension was the same as the time course of relaxation from an isometric contraction and this was true under various conditions. Bozler concluded that the same viscous factors controlled the tension disappearance in both cases and that these were situated in the contractile elements.

About this time also (1936) Glaister & Kerly (1) estimated lactic acid production in the retractor muscle of the foot of the mussel *Mytilus*. Although the glycogen content (2 g/100 g) was about twice that in frog muscle, lactic acid formation was small, only about 50 mg/100 g (to be compared with 200–300 mg in the frog) after long-lasting tetanus.

In a later paper (10), discussing chiefly the tonus of mammalian visceral muscles but also referring to snail retractor pharynx, Bozler considered an explanation of viscous resistance to extension in terms of bonds between active groups on contractile proteins.

RECENT EVIDENCE. Twenty years later, about the time when the nature of the protein of the paramyosin filaments was established, there was a revival of great interest in the holding mechanism. Thus Hanson & Lowy (4, 5), examining by electron microscopy the funnel retractor of the squid (*Loligo forbesii*) and the pharynx retractor of the snail, observed a double array of filaments, linked by transverse bridges. There was good evidence from much work on molluscan muscle that the larger filaments contained

---

[1] Also Bozler (6).

paramyosin.[1] The location of the actomyosin, known to be present from the work of Rüegg (1) on *Pecten*, was not apparent; Hanson & Lowy (6) with electron-microscope evidence from glycerinated lamellibranch muscles also (see fig. 101 pl), went on to propose in 1959 a sliding-filament mechanism for contraction in tonic smooth muscle.[2] As evidence for this point of view in 1964 Lowy, Millman & Hanson (1) adduced the following points:[3] (*a*) the double array of filaments of two kinds; (*b*) the presence of actin in the thin filaments;[4] (*c*) the cross-linkages between thin and thick filaments; (*d*) the discontinuity of the thick filaments (suggested by the symmetrical, tapered shape of isolated filaments);[5] (*e*) the constancy in length of the filaments during contraction, indicated by the constant diameter of the thick filaments[5] and the lack of change in axial periodicities.[6] Lowy, Millman & Hanson suggested that myosin is associated with paramyosin in mollusc muscle,[7] the myosin providing the bridges and the paramyosin playing a supporting role, particularly important on account of the great length of the thick filaments, compared with those e.g. in frog muscle. The power either to perform phasic contraction or to maintain tension with low energy expenditure was explained by the hypothesis of control of the rate of breakage of actomyosin links. It was suggested that this breaking rate during isometric contraction was controlled by the level of concentration of a relaxant.

We shall return to this linkage hypothesis of tonus, but first a description of a modern version of the catch hypothesis should be given. Johnson & A. G. Szent-Györgyi (1) in 1959 observed that the solubility of paramyosin is critically dependent on pH and ionic strength. Glycerinated muscle from the anterior byssus retractor of *Mytilus* was stretched and the tension measured under conditions where actomyosin linkages could contribute little – e.g. in 10^{-2}M pyrophosphate or 10^{-4}M Salyrgan. It was then found that, at an ionic strength of 0.07, the fibres were relatively stiff at pH values below 6.5 (a range in which paramyosin crystallises out); at higher pH values the fibres were relatively plastic. Such increased stiffness at lower pH values was also seen by Hayashi, Rosenbluth & Lamont (1), using acto-myosin-paramyosin threads prepared from lamellibranch muscle, if the percentage of paramyosin was high. Johnson, Kahn & Szent-Györgyi (1) also noted that at values of pH and ionic strength at which paramyosin crystallises, isotonic shortening of glycerinated preparations of mollusc 'catch' muscles was inhibited; but isometric tension was hardly altered.

[1] Hanson, Lowy, Huxley, Bailey, Kay & Rüegg (1); Elliott, Hanson & Lowy (1).
[2] Also Lowy & Millman (1) for evidence from mechanical properties.
[3] Also Hanson & Lowy (7); Lowy & Hanson (1).
[4] Hanson & Lowy (1).   [5] Lowy & Hanson (1).   [6] Selby & Bear (1).
[7] The recent work of Heumann (quoted by Heumann & Zebe (1)) has shown that the thick filaments have ATPase activity.

They suggested that the tension developed by the actomyosin system might be preserved in the paramyosin system for an indefinite time without continuous energy expenditure, the two systems being activated independently *in vivo*. Kahn & Johnson (1) tried to localise the myosin in the myofilaments of the byssus retractor of *Mytilus*. In the electron microscope cross-sections of glycerinated fibres showed a fine layer of small filaments surrounding the thick ones. Experiments with differential extraction indicated that these fine filaments were removed by myosin-extracting solutions, with a decrease in filament cross-section of some 35%; paramyosin-extraction solutions were only effective after removal of the myosin. This suggested localisation would agree with that proposed as we have seen, by Lowy, Millman & Hanson.

Rüegg independently came to conclusions rather similar to those of Johnson & his collaborators. In further work (2, 3, 10), he had correlated the differing behaviour of the different parts of the smooth adductor of *Pecten* with their differing actomyosin/paramyosin content. He also investigated (4) the rigidity of glycerinated fibres of the paramyosin-rich muscle by means of quick-release and stretch experiments. Rigidity seemed not to depend on actomyosin, which could be put out of action by means of Salyrgan, ethanol denaturation, or pyrophosphate plasticisation, but on the paramyosin system. This rigidity was modified by pH changes or by changes in the free ATP/MgATP ratio present. Increase in this ratio was shown to increase solubility of and electrophoretic mobility of paramyosin *in vitro*, and to plasticise artificial paramyosin threads.

So far we have been considering mainly electron-microscope and biochemical evidence concerning the nature of the tonus mechanism. Further insight derived from experiments on excitation, and on the mechanical performance of the living muscles under different conditions, remains to be discussed.

In 1954 Twarog (1) showed that acetylcholine caused depolarisation and tonic contraction of the anterior byssus retractor of *Mytilus*;[1] if the acetylcholine was washed away, repolarisation followed but the contraction persisted. Addition of 5-hydroxytryptamine (10^{-7}M) caused rapid relaxation of the tonic contraction. If acetylcholine was used in presence of 5-hydroxytryptamine, tension developed but was not sustained. It seems that a mechanism specific for tension maintenance must exist. That the muscle normally contained both acetylcholine and 5-hydroxytryptamine was shown by biological assay. Johnson & Twarog (1) in 1960 described single contractions (stimulated by direct current) as relaxing completely in 20–30 min; for contractions lasting many hours reactivation at low frequency was necessary. Fast relaxation was caused by 5-hydroxytryptamine

[1] Also Twarog (2).

or by stimulus of short duration. They therefore postulated that *in vivo* the response to reflex activation is both contraction and change in extensibility, little energy being needed beyond that setting up the forces initially. Relaxation could be controlled by a second set of nerves altering the 'catch' mechanism, perhaps by secretion of 5-hydroxytryptamine. Blaschko & Hope (1) showed in 1957 that these muscles contain an amine oxidase, which can oxidise 5-hydroxytryptamine.

Jewell in 1959 examined (1) the mechanical state of the living anterior byssus retractor in different conditions. When quick releases were made during phasic contraction, the mechanical properties resembled those of vetebrate skeletal muscle; when they were made during tonic response, the mechanical properties during stimulation resembled those in phasic contraction, but during tonus the quick releases produced only 25 % of the active shortening and redevelopment of tension obtained in skeletal muscle. His experimentation on effects of different types of stimulus led him to the view that excitatory and inhibitory systems were both involved, activation of the former causing contraction in which the active state decays quickly when stimulation ends, leaving what he called the 'fused state'; stimuli which activated the inhibitory mechanism abolished this state.

Lowy & Millman (1)[1] at this time also studied the mechanical properties of the living anterior byssus retractor, under isotonic and isometric conditions. They described two types of response – the phasic, produced by repetitive stimulation, and the tonic, produced by direct current or acetylcholine stimulation. These were characterised by very different time constants of tension decay – 1–7 sec for the phasic response, 5–10 min for the tonic response. On application of 5-hydroxytryptamine the rate of decay with tonic contraction increased to that characteristic of the phasic contraction (see fig. 102). From such results, and from the results of rapid stretch experiments on the muscle during stimulation and during the period of tonic tension decay, Lowy & Millman put forward the 'linkage' hypothesis – that both phasic and tonic contractions depend on the sliding mechanism involving actin–myosin linkages, but that the rate of breakage of these linkages may vary according to the concentration of a relaxant.

Rüegg and his collaborators a few years later studied in detail with this same muscle the effects on the holding mechanism of agents known to inactivate actomyosin. Thus 0.5 M thiourea caused complete inhibition of both ATPase activity and of contractile response to acetylcholine or electrical stimulation.[2] Rüegg (6) however could show that at low temperatures (10°) and in presence of a high tension of CO_2 (up to 100 mm Hg) a passive tension (sensitive to 5-hydroxytryptamine) could be maintained in

[1] Also Lowy & Millman (2).   [2] Rüegg, Straub & Twarog (1); Rüegg (5).

the living muscle in sea-water containing thiourea. The effects of various conditions upon this thiourea-resistant passive tension (for example, of 5-hydroxytryptamine treatment) were similar to the effects upon passive tension remaining in unpoisoned muscles in the tonic response. This tension remainder was also seen in glycerol-extracted fibre bundles contracted

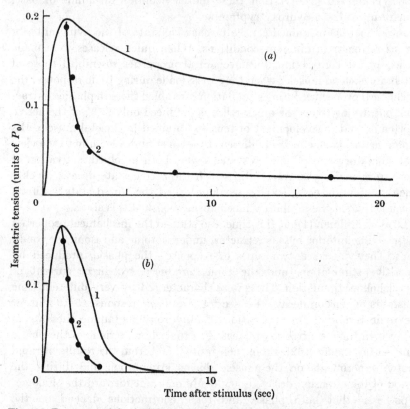

Fig. 102. Responses of the ABRM of *Mytilus* to single stimuli: (a) in sea-water, (b) in sea-water with 10^{-5} g/ml 5-hydroxytryptamine. Curves 1: isometric twitch. Curves 2: active state, determined by measuring the tension redeveloped following releases of 0.04 cm at a speed of 1.0 cm/sec at different times after the stimulus. (After Lowy & Millman (2).)

by ATP then treated with thiourea (see fig. 103). Such passive tension Rüegg put down to paramyosin, since threads of this protein are scarcely plasticised by thiourea plus ATP. In the poisoned fibres the stretch-resistance showed the same dependence on pH, temperature and unchelated ATP as did the stretch-resistance of thiourea-treated paramyosin threads.[1] According to this hypothesis the thick filaments themselves must be able

[1] Also Rüegg & Weber (1).

to interact in some way, to form a continuous stretch-resistant system. There is no conclusive evidence for this from electron microscopy, but Rüegg (7) has drawn attention to a number of possibly significant observations in this field, including a great increase in thick filament counts/fibre cross-section in the contracted state. Some very suggestive results have recently been described by Heumann & Zebe (1) using fresh relaxed and contracted byssus retractor of *Mytilus edulis*. For the contracted muscle very varied and complicated pictures were seen and the authors considered that, with the experimental arrangement so far used, phasic and tonic

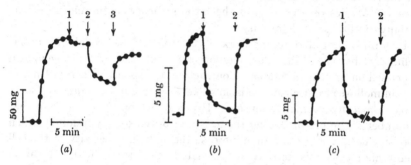

Fig. 103. Incomplete relaxation of extracted ABRM fibre bundles (*a*) and complete relaxation of artificial ABRM actomyosin threads (*b*) and taenia coli (*c*) under thiourea after a 90 % isometric ATP contraction. (*a*) Extracted ABRM fibre bundle contracting in ATP bath. 1st arrow: small relaxing effect of 0.2 M thiourea. 2nd arrow: partial relaxation followed by contraction remainder after addition of 0.5 M thiourea. 3rd arrow: reversal of inhibition after removal of thiourea. (*b*) Artificial actomyosin thread from ABRM contracting in ATP bath. 1st arrow: complete relaxation after addition of 0.5 M thiourea. 2nd arrow: reversal of relaxation after washing with fresh ATP. (*c*) Extracted fibre bundle of taenia coli contracting in ATP bath. 1st arrow: relaxation after addition of 0.5 M thiourea. 2nd arrow: reversal of relaxation after washing with fresh ATP. (Rüegg (5).)

contractions might be occurring in different parts of the same muscle. It is interesting that in contracted muscle they did observe in some fibres a lateral joining up of the thick filaments. Furthermore, in some fibres of a contracted muscle, greatly increased packing of the filaments could be seen; it seemed probable that this depended not only on parallel movement of the thick and thin filaments but also on movement of the thick filaments in relation to each other.

Recent biochemical work has emphasised the low energy expenditure during the operation of the catch. Baguet & Gillis (1) found that the oxygen consumption of the anterior byssus retractor of *Mytilus* during tonic contraction amounted to only about 9 % of the consumption during phasic contraction maintaining the same tension. This could mean a very slow turnover of actomyosin links, but it is not ruled out that the paramyo-

sin catch mechanism may need a small energy supply. Minihan Nauss &
Davies (1) followed the changes in free phosphate and arginine during
phasic succeeded by tonic contraction in this same muscle. In the catch
state there was no phosphoarginine breakdown, but a partial resynthesis
(up to 63 % in 15 min) of that broken down during the phasic period.
A breakdown of phosphoarginine corresponding to the small oxygen uptake
found by Baguet & Gillis could have occurred and been masked by the
recovery reactions. During release of the catch by 5-hydroxytryptamine
there was an increase in free phosphate and arginine, and also liberation of
Ca ions into the bathing medium – phenomena interpreted by Minihan
Nauss & Davies on the basis of the latter's theory of the nature of the
actomyosin link in skeletal muscle.

It is important that recent work of Twarog and her collaborators has
definitely disproved the older hypothesis that the catch mechanism
depended on prolonged tetanus.[1] Contraction in response to nerve stimulation in molluscan muscle depends on spike discharge in the same way as in
typical smooth mammalian muscles.[2] It was shown on the other hand that
a number of factors (including the action of 5-hydroxytryptamine) which
abolish catch, at the same time increase the spike-like activity of the cell
membrane; thus this activity is associated in this case with relaxation
rather than with tension maintenance.

Thus during the last ten years a wealth of knowledge has accumulated
concerning the physiology, biochemistry and fine structure of the holding
mechanism. Nevertheless a complete understanding has not been reached.

RECENT CONSIDERATIONS OF THE PHASIC RESPONSE

As we have seen, the suggestion was often made that the economical maintenance of contraction depended on the length of relaxation time in a series
of tetanic contractions. Rüegg (7, 8) has recently pointed out that a more
significant correlation would be with the rate of shortening. He emphasises
that, though in a tetanus the cell membrane is intermittently excited, quick-
release recovery tests seem to show that the fibres are continuously in an
active state – this state depending on the sarcoplasmic concentration of
free Ca ions. Bárány (1) in 1967 made an extensive study of the relationship
between myosin ATPase activity and maximum velocity of shortening.
For fourteen different types of muscle (including frog, tortoise, *Pecten*,
Mytilus and skeletal and smooth muscles of man) the expression:

$$\frac{\text{muscle length contracted/second}}{\text{ATPase activity/g myosin}}$$

---

[1] Twarog (3, 4, 5); Hikada, Osa & Twarog (1).   [2] Chs 13 and 23.

was approximately constant. This can be interpreted to mean, on the sliding-filament mechanism with hydrolysis of one ATP molecule at each cross-bridge cycle, that the rate of contraction is proportional to the turnover number of the cross-bridges.

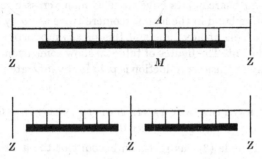

Fig. 104. Diagram illustrating the influence of filament length or sarcomere length on the tension production and shortening speed obtained with a given number of cross-bridges and with a given rate of sliding. Muscles with long filaments have more cross-bridges acting in parallel than short-filament muscles, but any gain in holding economy is offset by loss of speed. (A, actin; M, myosin; Z, Z line.) (Rüegg (8).)

TABLE 20. *Structure and function relationships in various muscles*

Muscle	Length of thick filaments (μ)	Speed of shortening (lengths/ sec)	Force (kg/cm²)	No. of actin fila- ments/μ^2	Economy (sec)
Frog sartorius	1.6	6–10	2.3	1335	1
Oyster yellow adductor	5–8	1.5	5	750	.
ABRM	30	0.25	14	300	200–400*

* During tetanic stimulation. 'Economy' expresses the time during which a muscle of length L will maintain a force P with energy expenditure H. Structural data: Page & Huxley (1); Hanson & Lowy (7); Lowy & Hanson (1); Lowy, Millman & Hanson (1). Mechanical measurements: Hill (15); Jewell (1); Lowy & Millman (2). Energetics: Hill & Woledge (1); Baguet (1); Baguet & Gilles (2). (Rüegg (8).)

A similar point of view was expressed by Baguet & Gillis (2) who measured the oxygen consumption of the *Mytilus* ABRM at rest and after phasic contraction. The value obtained after a contraction could be divided into a constant component and a component depending on the duration of stimulation. The time-dependent component was about 250 times less in this muscle than for frog sartorius, while the rise of tension to its maximum needed 10 sec in the ABRM, only 50 msec in the sartorius. It was suggested that linkage turnover proceeded at only 1/250th of the rate of that in the sartorius.

Rüegg (8) has also discussed another factor affecting achievement of economy in tension development and maintenance – the length of the sarcomeres. As is illustrated in fig. 104, the two short sarcomeres for a given rate of cross-bridge turnover can shorten twice as fast as the long one; this is because the short sarcomeres have twice as many cross-bridges acting in series. On the other hand in the long sarcomere twice as many cross-bridges are acting in parallel at any time and this means proportionately greater tension development. The figures of table 20 show some examples in which greater economy in tension production is paid for by low rate of shortening.

THE CONTRACTILE MECHANISM OF INSECT FIBRILLAR OR ASYNCHRONOUS MUSCLE

STRUCTURE. Pringle (3) has given an account of the anatomy and histology of these very special muscles, found only in certain orders of Insecta – particularly the Diptera, Hymenoptera, Coleoptera and Hemiptera. They comprise the power-producing flight muscles of such insects and the sound-producing tymbal muscles e.g. of certain cicadas. Von Siebold (1) studied these muscles in 1848, and was struck by the great ease with which they broke up, at the slightest pressure, into the very coarse elementary fibrils; he compared this behaviour with that of other voluntary muscles, where the fibrils are firmly bound in bundles. The designation 'fibrillar' muscles thus originated. Holmgren (1) in 1910 described their histology; we may note particularly the very short I bands and the presence of regularly-arranged very large cell particles, some of which he considered particularly characteristic of very active muscles.

These fibrillar muscles *in vivo* contract almost isometrically – by only about 1.5 % in the tymbal muscles of the cicada,[1] 2 % in the indirect flight muscles of the bee.[2] Hanson (4) observed in 1956, by means of phase contrast and polarised light microscopy, the band pattern of blowfly myofibrils, under conditions of contraction and extension, and after extraction with myosin- or actin-removing solutions. She correlated the very short I bands with the small changes in length which the muscle needs to undergo, and concluded that these muscles probably work by the same sliding mechanism as in vertebrate striated muscle.

The supply of air to insect organs by means of the tracheae was early recognised and was discussed e.g. by von Siebold (1) in 1848. Leydig (1) in 1859 was the first to describe the extraordinarily fine ramification of tracheoles inside the flight muscles of the housefly and the bee.

A discovery which seems to be of considerable importance in explaining some of the unique features of fibrillar muscle was made by Auber &

[1] Pringle (2).   [2] Boettiger & Furshpan (1).

Couteaux (1, 2) in an electron-microscope study of Diptera in 1962.[1] This was the attachment of the thick A filaments to the Z line by means of thin prolongations passing between the thin filaments, which are symmetrically arranged around them. Garamvölgyi (1, 2) confirmed these observations, and extended them by means of stretch experiments. On moderate stretch it seemed that the thin I-band portions of the thick filaments could be much elongated at the expense of material actually in the Z line. It should be noticed however that Ashhurst (2) could not find these connecting filaments in the flight muscles of belostomatid water-bugs.

Finally in this brief survey we must mention the observations of Reedy, Holmes & Tregear (1) on the orientation of cross-bridges in the glycerinated muscle of the giant tropical water-bug *Lethocerus maximus*. In very thin sections of relaxed fibres (in presence of ATP) the cross-bridges were seen attached to the myosin at regular intervals of 146 Å, mostly projecting at right angles. In sections of fibres in rigor (due to absence of ATP) the cross-bridges, in symmetrical pairs were slanted at an angle of about 45 degrees, and connected the myosin and actin filaments at intervals of 380 Å; the actin-centred chevrons formed in this way all pointed away from the Z line.[2] The existence of these two states was confirmed by the X-ray diffraction patterns also observed.

METABOLISM. The extraordinary performance of which the fibrillar flight muscles are capable is shown by the work in 1953 of Sotavalta (1),[3] who measured the frequency of wing stroke in midges. The frequency of vibration was determined both by the pitch of flight-tone, and by use of a microphone and high-fidelity tape-recording, transferred to recording film by means of a double-beam cathode-ray oscillograph. He found for *Chironomus* a frequency in free flight of 600–650/sec; for *Forcipomyia* 800–890/sec.

From figures in the literature Weis-Fogh (2) has compared the metabolic rates for the muscles of various types of animal. The astonishing result became apparent that, whereas the maximum metabolic rate for the leg muscles of man is 50–60 kcal/kg/h, the locust wing muscle can average 400–800, *Drosophila* 650, the blowfly *Lucilia* 1700, and the bee 2400. The nature of the fuel used, whether mainly carbohydrate or fat under different conditions of flight, has been reviewed by both Weis-Fogh and Pringle (3). Ashhurst & Luke (1) have described the numerous lipid droplets occurring in the flight muscle of giant water-bugs, in close proximity to the mitochondria.

We turn now to some very interesting characteristics of the respiration

[1] Pp. 251 above and 541 below.
[2] Compare the chevrons formed from actin and H-meromyosin, ch. 10.
[3] References to earlier papers in Finnish are given here.

process in fibrillar muscle. Watanabe & Williams (1) showed in 1951 that the large cell particles (sarcosomes) described in these muscles were in fact mitochondria, containing the cytochromes and many oxidising enzymes.[1] In 1956 Zebe (1) noticed that locust muscle oxidised glycerophosphate at a far greater rate than does vertebrate muscle; this was true also of insects depending on carbohydrate metabolism, so that it could not be concerned only with the high fat metabolism characteristic of the locust. Zebe & McShan (1) then found that a very low lactate dehydrogenase activity was characteristic of insect flight muscle as well as the very high glycerophosphate dehydrogenase activity. Much of the latter was contained in cell particles. Zebe and a group of workers in Marburg continued the investigation,[2] measuring the activity of a number of enzymes concerned in energy-producing metabolism in leg and flight muscles of the locust. The enzyme patterns of the two types of muscle differed greatly (particularly as regards the glycerophosphate and lactate dehydrogenase activities); the pattern of the leg muscle was similar to that of vertebrate skeletal muscle.

Just at the same time and independently, Sacktor and his collaborators[3] were working on this same phenomenon of preferential oxidation of glycerophosphate in insect flight muscle. Thus Chance & Sacktor (1) in a spectroscopic study of the mitochondria[4] of flight muscles of *Musca* found the maximum rate with α-glycerophosphate to be ten times that with succinic acid, the most active of the Krebs cycle intermediates. The two groups of workers outlined similiar suggestions for the important role of glycerophosphate[5] in what is now known as the α-glycerophosphate shuttle. Bearing in mind (1) the evidence for two distinct glycerophosphate dehydrogenases, one readily soluble and requiring NAD,[6] the other mitochondrial and independent of the coenzyme;[7] (2) the difficulty of entry of NADH into the mitochondrion while the α-glycerophosphate entered readily,[8] the following scheme was prepared:

(a) Dihydroxyacetonephosphate + NADH $\rightleftharpoons$ α-glycerophosphate + NAD
    (from glycolysis)   (soluble enzyme)

(b) α-Glycerophosphate + $\frac{1}{2}O_2$ $\rightleftharpoons$ dihydroxyactonephosphate + H_2O
                                              (mitochondrial enzyme)

(c)   NADH + $\frac{1}{2}O_2 \longrightarrow$ NAD + H_2O.

[1] Also Sacktor (1).
[2] Zebe, Delbrück & Bücher (1); Delbrück, Zebe & Bücher (1).
[3] Sacktor & Cochran (1); Chance & Saktor (1).
[4] Also Klingenberg & Bücher (1); Klingenberg & Slenczka (1) for mitochondrial studies.
[5] Bücher & Klingenberg (1); Estabrook & Sacktor (1); Zebe, Delbrück & Bücher (2).
[6] This easily extractable enzyme had been known since the time of von Euler, Adler & Günther (1); it was crystallised and studied by Baranowski (1).
[7] The first evidence for a glycerophosphate dehydrogenase in muscle not dependent on NAD was given by Green in 1936 (4).   [8] Lehninger (2).

In this way the hydrogen from various metabolites (perhaps most important glyceraldehydephosphate in the oxidation of carbohydrate) could be conveyed to the respiratory chain via externally situated NAD and di-hydroxyacetonephosphate. In so active a muscle the benefit of avoiding the oxidation by pyruvate (with formation of lactate and possible dele-terious effects of fall in pH) is obvious. The pyruvate can be oxidised by the citric acid cycle, without its absence affecting the rate of carbohydrate utilisation, NAD needed in this utilisation being supplied by way of the glycerophosphate shuttle.

Estabrook & Sacktor (1) showed that the equilibrium in reactions (a) and (b) is far to the right, as had indeed been shown by Green for the mitochon-drial enzyme of skeletal muscle. They also found that the mitochondrial enzyme of *Musca domestica* was sensitive to versene, its activity being re-stored by addition of Mg ions. They suggested that flight muscle might contain an endogenous chelator, and that perhaps as the result of nervous stimulation, changes in Mg ion concentration might control the oxidation rate; respiration at rest might be through the Krebs cycle, while at the onset of flight increased Mg ion concentration might allow the enormous increase observed in oxidative metabolism. In these mitochondria they found no respiratory control by ADP, but Bücher & Klingenberg obtained a normal activation of respiration by ADP in *Locusta migratoria*. Very recently Hansford & Chappell (1), using *Calliphora vomitaria*, have shown that both ADP and Ca ions can affect the activity of the mitochondrial glycerophos-phate dehydrogenase. The enzyme was markedly stimulated by low levels of Ca^{2+} (10^{-6}M), which acted by increasing the affinity for substrate. In the presence of Ca buffers, at concentrations of Ca^{2+} between 10^{-8} and 10^{-6}, a progressive increase in ADP stimulation of the rate of oxidation was found with rise in Ca concentration. Thus Ca ions, which are essential at such con-centration levels for the MgATPase activity and the oscillatory contraction of insect-flight muscle, also on their release in the muscle cell ensure maximum rate of ADP-stimulated glycerophosphate oxidation.

Mention must also be made of the malate–oxaloacetate cycle, shown to be important in insect-flight muscle, since Bücher & Klingenberg (1) found that two distinct enzymes, one mitochondrial, the other cytoplasmic, were present there, and of high activity. These enzymes differed in pH optimum and in substrate affinity; the cytoplasmic enzyme was oriented towards reduction of oxalate, the mitochondrial enzyme towards oxidation of malate. By means of the cycle shown on p. 538,[1] oxidation of cytoplasmic NADH could again be mediated, as with the glycerophosphate cycle. Bücher & Klingenberg have also considered how the operation of this cycle could stabilise the operation of the citric acid cycle.

[1] Cf. Kaplan (1).

Oxaloacetate + $NADH_2$ ⇌ NAD + malate (cytoplasmic, inhibited by malate)

Malate + NAD ⇌ oxaloacetate + $NADH_2$ (mitochondrial, inhibited by oxaloacetate)

Sacktor & his collaborators later enquired into the regulation of metabolism in the flight muscles of the blowfly, *Phormia regina*.[1] With the beginning of flight there was a hundredfold increase in glycolytic flux, with no accumulation of lactate or citrate. Three sites of regulation could be detected – the phosphorylation of fructose-6-phosphate, the cleavage of trehalose and phosphorylation of glycogen, with other possible controls at the sites of proline and α-glycerophosphate oxidation.

MECHANISM OF THE ASYNCHRONOUS BEHAVIOUR OF INSECT FLIGHT MUSCLE. Pringle (1) in 1949 first made it clear that the rapid oscillations of insect flight muscle are not reflex phenomena; they are independent of the excitation process, electrical records showing only small effects at the wing-beat frequency, with action potentials at a much lower frequency;[2] but nerve impulses above a certain low frequency must be received. This behaviour gave rise to the alternative designation 'asynchronous' for this type of muscle. Pringle formed the opinion that a nervous impulse sets up an excitatory process, altering the state of the contractile apparatus so that it becomes sensitive to a stretch stimulus, responding with a twitch. This contraction would stretch the antagonist which in turn would contract, the cycle continuing as long as the active state lasted. Later Pringle (2) studied the tymbal muscle of the cicada (*Platypleura*). The myogenic rhythm is here also quite asynchronous with the electrical changes. It was shown to be maintained by deactivation of the muscle fibres caused by quick release at the IN click, the muscle being restored to the initial length by the elasticity of the exoskeleton, and thereafter redeveloping tension at a rate depending on the intensity of the active state. A single nerve impulse could maintain the active state long enough for several clicks. Pringle now considered that, both in flight muscle and tymbal muscle, deactivation by release was the most significant happening, though he did not rule out activation by stretch as playing a part. Boettiger & Furshpan (1, 2) using the bumble bee, showed that activation by stretch does indeed enter in, and the importance of both deactivation on quick release and activation on stretch is now agreed upon.[3]

---

[1] Sacktor & Wormser-Shavit (1); Sacktor & Hurlbu t (1).
[2] Also Roeder (1).
[3] For discussion see Pringle (3, 4); Boettiger (1, 2).

Boettiger (2) was the first to show the oscillatory behaviour of flight muscle stimulated under an inertial load. Machin & Pringle (1, 2) have studied in detail the mechanical properties in beetle flight muscle using an electronic feedback to control the very small length changes. When an inertial load was applied to the muscle with damping below a critical value, stimulation caused an oscillatory contraction. The relations found between tension and length are shown in fig. 105. From such curves certain characteristics of this kind of muscle become apparent.[1] Thus the anticlockwise

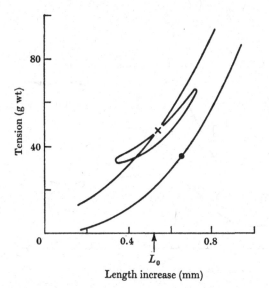

Fig. 105. A typical oscillatory contraction of *Oryctes* basalar muscle. The two sloping lines give the mean tension/length relationship from all experiments with this muscle. ●, unstimulated working point; ×, stimulated but non-oscillatory working point. (Machin & Pringle (1).)

rotation of the oscillatory loop indicates that tension changes lag behind length changes; further that tension is greater during shortening than during lengthening – a state of affairs the reverse of that in active vertebrate muscle.

In 1964 Pringle and his collaborators turned to glycerol-extracted muscle (from the giant water-bugs *Lethocerus* and *Hydrocyrius*) in order to answer the question whether the mechanism controlling oscillatory activity resides in some intracellular membrane system concerned in excitation-contraction coupling, or in the contractile proteins themselves. Jewell, Pringle & Rüegg (1) used a bundle of a very few fibres fixed to a weak torsion band at a

[1] See also Pringle (6).

tension of about 1 mg per fibre. In a medium containing 5 mM ATP and 5 mM Mg ions at pH 7.0 and ionic strength 0.08, oscillations began when the Ca ion concentration was raised to about 8×10^{-8} M. This work was carried further by Jewell & Rüegg (1) who showed that length changes in the fibre preparations resulted in delayed changes in tension; they concluded that this delay between length change and tension change, essential for oscillatory behaviour, is a property of the contractile protein system itself. They confirmed that Ca activation was necessary and made the further very

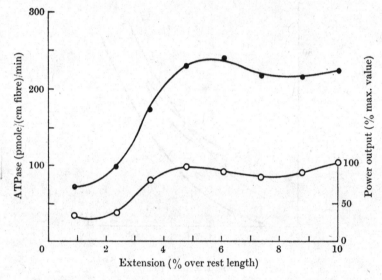

Fig. 106. The ATP split by a fibre bundle (from the fibrillar flight muscle of *Lethocerus cordofanus*) (●), and its power output in response to a 5 cycle/sec, 0.3 % extension test oscillation (○) related to the extension of the muscle. *p*Ca 7.4. (Rüegg & Tregear (1).)

interesting observation that sensitivity to Ca was increased by stretching the fibres; this suggested that stretch might lead to increase in the amount of Ca bound. Rüegg & Tregear (1) measured the ATPase activity of these glycerol-extracted fibres; the presence of Ca ions at concentrations above 10^{-8} M was necessary for enzymic activity of the unstretched muscle, and the ATP splitting increased on stretching (see fig. 106). If oscillation of the stretched muscle was induced, there was further ATP breakdown, apparently proportional to the work done. Jewell & Rüegg gave reasons for concluding that residual mitochondria and sarcoplasmic reticulum in their glycerinated fibre preparations were playing no part in the phenomena recorded. This was borne out by the special study made by Abbott & Chaplain (1) who showed conclusively that this was the case, using glycerin-

ated fibres after treatment with the detergent Tween 80 and with oligomycin.

Chaplain, Abbott & White (1), in Pringle's laboratory, went on to examine
the effect of ADP upon tension changes and ATPase activity in these
glycerinated fibres. They found marked inhibition of the Ca-stimulated
ATPase with about 1 mM ADP and at the same time great increase in
tension, either in the absence of ATP or in the presence of ATP plus Ca ions
– conditions under which bridge formation between the myosin and actin
would be expected. Chaplain (1), using radioactive ^{45}Ca, concluded that the
higher ATPase activity on stretch was accompanied by increased Ca
binding, as Jewell and Rüegg had suggested. He discussed the probable
importance in this connection of the attachment in these muscles of the
A filaments to the Z line; this could mean that applied stretch would
directly stress the myosin filaments themselves, with resultant increase in
Ca binding sites. Thus during the extension phase in the oscillatory cycle,
the sequence could be visualised: stretch; increased ATPase activity;
increase in ADP concentration and resultant delayed increase in tension.
Later work of Rüegg & Stumpf (1), however, did not agree with this effect
of stretch on Ca binding. They found that the ATPase of glycerinated
fibrillar muscle was activated by stretch even in saturating concentration
of Ca (10^{-5}M) and that the Ca ion concentration causing 50% increase in
ATPase activity was not affected by stretch.

Chaplain (2) also described the conditions for substrate (MgATP) inhibition of the actomyosin ATPase of flight muscle. Making use of the kinetics
of this inhibition and the ADP inhibition, and of other characteristics of
the kinetic behaviour of the enzyme as well as of light-scattering studies he
has suggested (3, 4) an allosteric model for the behaviour of the actin-
myosin complex during rest, activation and contraction. The effects of
ADP and Ca on tension production of glycerinated fibres could be fitted to
this model. It must however be noticed that Maruyama & Pringle (1) have
recently found that ADP effects upon insect actomyosin are more complicated and variable than was previously supposed, and a more detailed
analysis is still needed.

In 1968 Maruyama, Pringle & Tregear (1) (confirming and extending
earlier work of vom Brocke (1)) made a special study of the calcium sensitivity of the ATPase activity of actomyosin and myofibrils from insect
flight and leg muscle. In insects having fibrillar flight muscle (*Lethocerus
maximus, Apis mellifera* and *Oryctes rhinoceros* were used) half-activation
of the ATPase was attained with only about 10^{-7}M Ca, while five to fifteen
times this concentration was needed for half-activation of the leg muscle
ATPase (see fig. 107). The maximum activity of the leg muscle ATPase (at
10^{-5}M Ca) was however twice as great as the activity of the flight muscle

enzyme at the same concentration of Ca ions. It is interesting that the actomyosin ATPase of the non-fibrillar flight muscle of *Locusta migratoria* behaved like the ATPase of the leg muscle.

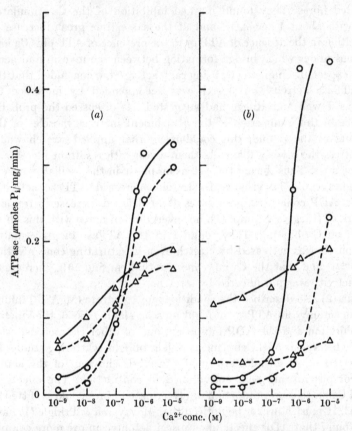

Fig. 107. Effect of free Ca^{2+} concentration on the ATPase activity of water-bug myofibrils (*a*) and actomyosin (*b*). Triangular symbols refer to flight muscle and circles to leg muscle. Medium: 0.02 M tris buffer, pH 7.2; 3.3 mM MgATP (myofibrils) or 1.6 mM MgATP (actomyosin); 0.02 M KCl (solid lines) or 0.06 M KCl (dashed lines). (Maruyama, Pringle & Tregear (1).)

The authors linked the physiological behaviour with this lowered sensitivity to calcium concentration in that it would leave the muscle more free to respond to stretch stimuli with the characteristic oscillatory behaviour. The problems of the pathway of excitation into the interior of the very large fibres of fibrillar muscles was investigated by D. S. Smith (2) in 1961. Electron micrographs of the beetle *Tenebrio* showed that the invading tracheoles drew with them a sheath of plasma membrane from the surface to all depths of the fibres. No sarcoplasmic reticulum could be seen, but certain isolated vesicles occurred in close association with flattened smooth-membraned profiles, shown to be continuous with the peripheral plasma membrane of the fibre; these associations he termed dyads, and considered them to be homologous with the triads of the twitch muscles of vertebrates. Ashhurst (1) has shown that in the flight muscles of *Lethocerus* the sarco-plasmic reticulum was comparatively well-developed; she and Smith (3) have both discussed the general paucity of this reticulum in fibrillar muscle though the transverse tubules are well developed.[1] This fits in with the probability that in oscillatory muscle the role of the sarcoplasmic reticulum is confined to the release of a low concentration of Ca at the onset of a period of activity (renewed release at each contraction not being necessary) and its removal on return to rest.

Maruyama, Pringle & Tregear also obtained indications of the effect of troponin in conferring calcium sensitivity on these insect actomyosins. After digestion with very low concentrations of trypsin, these preparations showed good ATPase activity in absence of calcium, and calcium sensitivity could be restored by addition of native tropomyosin.[2]

Meinrenken (1) has found that after treatment with 0.05 M KCl at pH 7.8, glycerinated fibrillar muscle showed maximum tension development and ATPase activity at neutral pH with or without Ca ions. In this extracted muscle deactivation by release and stretch activation are impossible. He made the suggestion that an alkali-labile factor is present in fibrillar muscle, preventing shortening beyond 90% of rest length, and responsible for the 'contraction prevention by shortening'. Troponin was contained in the extract, as was shown by its power of restoring Ca sensitivity to synthetic skeletal-muscle actomyosin, but other proteins may also have been present. Troponin could not be extracted under the same conditions from insect leg muscle or from rabbit muscle.

Steiger & Rüegg (1), using glycerinated fibre bundles from *Lethocerus maximus*, have recently been able to obtain oscillations at 30° at a frequency of 20–25 cycles/sec – a rate about the same as the wing-beat frequency of the living insect. The rate of ATP splitting and the rate of doing work were both optimal at this frequency. The efficiency (the measured power output/moles ATP split) was about 50% at a frequency of 10 cycles/sec, and fell off at higher and lower frequencies.

It is interesting that Armstrong, Huxley & Julian (1) have found conditions in which oscillatory behaviour, reminiscent of the behaviour of fibrillar flight muscle, can be elicited from frog skeletal-muscle fibres. Thus

[1] As had been noticed much earlier by Edwards, Ruska, Santos & Vallejo-Freire (1).
[2] Ch. 12.

in contractions where the load was very carefully controlled at values near the isometric, sudden small changes of 5–10 % produced damped oscillations in the fibre length.

Pringle (5) and Rüegg (9) in recent penetrating reviews have suggested points of mechanism to explain the characteristics of contraction in insect flight muscle in terms of the sliding-filament model; thus the model differs in some details from that proposed by earlier writers. Only a few remarks concerning them can be made here. For example, Pringle postulates that binding of Ca initiates cyclic movement of a bridge, while binding of ADP stabilises attachment of a bridge to the I filament. Further it is suggested that the number of possible bridges is increased when the A filaments are under tension – a characteristic depending largely on their structural connection to the Z line – and it is shown how important a part this could play in oscillation. Rüegg has speculated that activation by stretch *in vivo* might be due to removal or displacement of an inhibition such as that studied by Meinrenken in glycerinated insect fibres. The point is made by Pringle that insect fibrillar muscle may provide a material from which new details might be learnt about the contractile activity of muscle in general.

23

VERTEBRATE SMOOTH MUSCLE

INTRODUCTION

The pioneering studies of Kölliker (3) in 1849 were the first to tackle this difficult tissue.[1] He recognised its cellular structure and observed its widespread occurrence in the vertebrate organism – in the intestinal canal, glands, liver, uterus, spleen, arteries, bladder, sense organs, etc. Here it is often present in the form of bands embedded in other tissue, and these bands may run in more than one direction; their dissection therefore often presents much difficulty. Then the small diameter of the cells, only $2-5\mu$ at the widest part, and the invisibility of most of their contents in the light microscope must be remembered.

Moreover, the variability in behaviour of smooth muscles from different organs makes generalisations uncertain. This variability is seen, for example, in responses to pharmacological reagents, which may be quite different from those familiar with striated muscles, and which differ from one kind of smooth muscle to another. Again, though some smooth muscles have a complicated system of innervation in close association with the cell, in others (e.g. intestinal muscle) transmission of excitation directly from cell to cell leads to automatic rhythmic tension production. The phasic contraction is always slow; thus McSwiney & Robson (1) gave values for gastric muscle at $37°$ of 0.77 sec for the latent period, 1.7–2.7 sec for the rising phase and 2.8 sec for half relaxation time. This is to be compared with a frog sartorius twitch complete in 0.1 sec at $25°$. The metabolism connected with activity, whether phasic or tonic, is very low and at first was completely missed. Further, in much of the earlier work it was assumed that the smooth muscle of invertebrates could be taken as analogous to the smooth muscle of vertebrates, and it is only in recent times that the important part played by paramyosin in many invertebrate smooth muscles has been realised. Paramyosin has not been found in vertebrate smooth muscle. Thus it may be seen that the reasons are many which make vertebrate smooth muscle a complex subject for research.

We have already considered Grützner's conception of the intimate mixture of two types of fibres in striated muscle,[2] the more slowly con-

[1] See Grützner (2) for a review of experimental work and theory-spinning before 1904.
[2] Ch. 19.

tracting acting as an 'inner support' for the more rapidly contracting. Botazzi (1, 2) went on from this to suggest that it is the sarcoplasm which can contract slowly and in the cell itself provide intrinsic support for the independent rapid contraction of the anisotropic substance. He applied this hypothesis to the function of sarcoplasm in heart and smooth muscle. He pictured the contractile behaviour of the sarcoplasm as similar to the slow rhythmic contraction and expansion of the pseudopodium of an amoeba, and drew evidence in favour of this idea from the sensitivity (reminiscent of the irritability of unicellular organisms) of vertebrate smooth muscle to mechanical and thermal stimuli. He laid much emphasis on the two-peaked curve seen under certain conditions in striated muscle as the result of a single stimulus, relating the first peak to the fibre contraction, the second to the sarcoplasmic. Grützner (2), while referring with interest to Botazzi's conception, was, as we have seen, himself a pioneer in the 'rack and pinion' explanation of tonus.[1]

Although Botazzi felt able to speak firmly about the fibrillar component of the smooth muscle cell, there was much controversy about the validity of the appearance of longitudinal fibrils seen for example by Rouget (1) in 1863 and by Schultz (1) in 1895, in fixed and stained material. It was only with the application of the electron microscope (from about 1955) that the thin, longitudinally arranged filaments, only 50–80 Å in diameter, filling the cells, could be clearly seen. The fibrils observed in the light microscope must have been artificial aggregates of these filaments. Von Ebner (2) knew of the double refraction of smooth muscle cells and of its increase on stretching; Schultz (3) remarked on its fall during contraction and discussed these phenomena in terms of the imbibition theory of contraction.[2] E. Fischer (3) in 1944 with *retractor penis* and intestinal muscle found that the whole length of the cell exhibited both form and intrinsic birefringence, in this way indicating a micellar pattern resembling the *A* band of striated muscle. We shall return to later work in this field.[3]

We have seen that Frank[4] (some twenty years ahead of his time) wished to explain tonus by tetanic fusion of slow responses. In 1928 L. E. Bayliss (1) considered the energetics of smooth muscle. He regarded the very much slower relaxation time, compared with striated muscle, as of primary importance. He calculated that, if the tonus was the result of fusion of very slow phasic contractions, then the low oxygen consumption observed in the experiments of Lovatt Evans (2) with uterine and intestinal muscles would be ample to supply the necessary energy, assuming that heat production was the same per contraction for the smooth muscle as in the experiments of Hill and Hartree for frog sartorius. Ritchie (1) also writing at this time

[1] Ch. 22.
[3] P. 570 below.
[2] Ch. 7.
[4] Ch. 3.

agreed with the view that there was no necessity to postulate a mechanism other than tetanus to explain tonus, if the time relations were remembered.

Bozler has worked for some forty years on problems of smooth muscle contraction. The development of his ideas of the mechanism (ideas which he applied to vertebrate smooth muscle as well as to invertebrate) has been described in chapter 22. In 1928 (2, 3) he held the view that two morphologically distinct kinds of fibre were necessary, the one for tetanus, the other for tonus. But this he quickly gave up as a general theory. In 1948 he wrote in a paper (4) mainly devoted to mammalian smooth muscle: The ability of some smooth muscle to remain contracted for long periods without fatigue...does not require the assumption of any mechanism different from that of tetanic contraction.' 'The peculiarities of smooth muscle are largely due to the slowness of the response and the variability of excitation.' His evidence concerning viscous resistance to relaxation has aleady been mentioned. A number of workers have suggested involvement of various soluble proteins in the tonic function and we shall discuss these later.

In the late nineteenth and earlier twentieth centuries experiments were reported indicating ability of some vertebrate smooth muscles to contract spontaneously in the absence of any nerve supply, and belief in their syncytial nature was fostered.[1] Use of the light microscope gave the impression that protoplasmic bridges connected the cells. Tiegs (1, 2) on the other hand, writing in 1925, placed great emphasis on the 'nerve nets' occurring in close association with smooth muscle cells and not seen with striated muscle fibres.[2] He held that this path served to conduct stimuli to the individual cells, in the absence of the membrane of Krause (the Z line), which he took to be the specialised means of transmission in striated muscle fibres. Bozler (11) in 1941 threw light on these vexed questions when he divided smooth muscles into two groups, the multi-unit and the visceral (or unitary); as described by Burnstock, Holman & Prosser (1) and by Lane & Rhodin (1) evidence has been accumulating in recent years in favour of this classification. In the multi-unit type (of which the vas deferens is an example) there is no spontaneous activity but dependence on abundant innervation.[3] Intestinal muscle, on the other hand, is autorhythmic, responsive to stretch and functional in the absence of nerve.[4] Very interesting differences in the ultrastructure of the two types of muscle have been shown. In the vas deferens Lane & Rhodin (1) found that nerves were widely distributed and closely apposed to the muscle cell membranes at vesicle-bearing areas. In the intestinal muscle the cell axons were few, in large aggregates, and did not have vesicle-bearing zones close to the cell

---

[1] See the review by Lovatt Evans (1).
[2] E. Fischer (4) and Rhodin (1) for reviews of more recent work on this subject.
[3] Burnstock & Holman (1).   [4] P. 566 below.

membranes; broad planar areas of contact between muscle cells were seen, occupying about 5 % of the cell area. It was suggested that they were points of low resistance acting as pathways of transmission. No evidence for syncytial transmission was found. In the vas deferens, though there were numerous points of cell contact, these were entirely different in structure. Taxi (1) independently made an electron-microscope study of the innervation of these two muscles as well as of others belonging to the multi-unit and to an intermediate type. His results are in good agreement with those of Lane & Rhodin. The muscle itself and its membrane are very similar in the two types in that both have cable properties, i.e. close cell-to-cell apposition. This was shown by Tomita (1) and Abe & Tomita (1). Current spread from cell to cell is almost the same in taenia coli as in the vas deferens. The facts that in the multi-unit type the spread physiologically is along nerve fibres and that the excitatory junction potential has the same amplitude in each cell leads to perfect synchronisation in this type.

PROTEINS OF THE CONTRACTILE MECHANISM

ACTOMYOSIN AND MYOSIN. It is interesting to recall that Mehl (2) in 1938 extracted from beef intestine a protein, showing double refraction of flow, which precipitated on dilution. He compared it to 'myosin'. The presence of actomyosin in vertebrate smooth muscle was first demonstrated by Csapo in 1948–50 (2, 3)[1] in his work on uterus extracts (made with buffered 0.5M KCl), using the viscosimetric method. He and his collaborators also found that such extracts of pregnant uterus contained a protein with a sedimentation rate similar to that of actomyosin;[2] this component disappeared when ATP was added and a component appeared resembling myosin in sedimentation rate.[3] Naeslund & Snellman (1) described in extracts of pregnant myometrium a protein of electrophoretic mobility resembling that of actomyosin.

Later Needham & Williams (1, 2) prepared purified actomyosin from uterus by extraction with 0.6M KCl and repeated precipitation in 0.3M KCl. The protein was electrophoretically homogeneous and had viscosity numbers of 0.34–0.4 before ATP addition and 0.17 after – values very similar to those obtained by Portzehl, Schramm & Weber (1) for the skeletal protein.

Myosin with low actin content cannot readily be prepared from smooth muscle by extraction with salt solution of ionic strength 0.35 and pH of 6.5, as is customary with skeletal muscle; this is probably due to the easier extraction of actin from the stroma of the smooth muscle. Uterine myosin

---

[1] Also Csapo (1).   [2] Csapo, Erdös, Naeslund & Snellman (1).
[3] Also Kominz & Saad (1); Ledermair (1).

has been prepared by Cohen, Lowey & Kucera (1) and by Needham & Williams (2) from actomyosin by the method of A. Weber (1) – centrifugal separation of the actin and myosin in the presence of ATP and Mg ions. The purest preparation of Needham & Williams had a viscosity number of 0.13–0.2 (to be compared with the value for skeletal myosin of 0.17–0.2).

Several workers have shown that skeletal-muscle actin and myosin combine in a ratio by weight of about 1:4. This would mean, as in the experiments of A. G. Szent-Györgyi (1), a fall of about 0.11 in the logarithm of the relative viscosity when ATP is added to a solution of pure actomyosin, 1 mg per ml. This same fall was found by Needham & Williams (2) with their purified actomyosin and by Huys (1), an indication that the two smooth muscle proteins combine *in vitro* in about this same ratio.

Smooth mammalian muscle contains surprisingly little actomyosin. Schultz (2) in 1897 remarked that from such muscle no one had prepared a plasma giving spontaneous formation of a clot which would dissolve in salt solution. He associated the presumed absence of myosin with the failure of smooth muscle to show normal death rigor.

The protein obtained by Needham & Williams (3) by dilution of a 0.5 M KCl extract of uterus to 0.1 M (even after several reprecipitations) contained much other protein besides actomyosin; in extracts of pregnant uterus about 50% was actomyosin, with non-pregnant uterus only about 25%. Rüegg, Strassner & Schirmer (1) similarly found that only about 25% of their artery dilution precipitates consisted of actomyosin. Values in the literature for the protein content of dilution precipitates from uterus, artery and stomach muscle usually lie between 6 and 16 mg/gm wet weight, of which probably much less than half is actomyosin, except in the case of the pregnant uterus.[1] Csapo (4) from his viscosity measurements calculated rather higher values for the actomyosin of pregnant uterus – 7–13 mg/gm. Bárány and his collaborators isolated 5 mg of purified actomyosin per gm of gizzard muscle.[2] It should be mentioned that F. R. Goodall (1) found with uterus muscle that, as pregnancy advanced, the content in proteinase greatly increased and extracts of pregnant uterus could cause disintegration of skeletal-muscle actomyosin. This proteinase is probably situated in the lysosomes and has a part to play at involution of the uterus. It is possible that actomyosin values so far given for pregnant uterus are too low, but there seems no reason to suppose that this destructive effect on the contractile protein would be occurring during extraction of other smooth muscles.

---

[1] Schwalm & Cretius (1); Ledermair (1); Ivanov, Mirovich, Moissejeva, Parshina, Tukachinsky, Yuriev, Zhakhova & Zinovieva (1); Ivanov, Mirovich, Jakhova & Tukachinsky (1); Needham & Williams (3); Rüegg, Strassner & Schirmer (1); Rüegg & Strassner (1).   [2] Bárány, Bárány, Gaetjens & Bailin (1).

The slowness of interaction of ATP and smooth-muscle actomyosin was shown in the pioneer experiments of Csapo (2, 4) on viscosity changes in extracts and contraction of actomyosin threads prepared from uterus. Values of ATPase activity of dilution precipitates and of actomyosin and myosin are shown in table 21. It can be seen that the ATPase activity of the smooth muscle preparations is very low compared with that of the skeletal actomyosin and myosin. Activation by Mg is particularly low. Another difference in behaviour of the actomyosin from the two types of muscle is the greater activity of the smooth muscle enzyme at higher ionic strength.

TABLE 21. *ATPase activity (Ca^{2+} activated) at neutral pH of actomyosin, myosin and HMM from vertebrate smooth muscle and from skeletal muscle*

	Ionic strength of medium	
	0.05–0.1	0.5–0.6
Uterus actomyosin (a)	0.57	1.28
Artery actomyosin (b)	0.17	0.51
Rabbit skeletal actomyosin (c)	5.28	3.6
Uterus actomyosin, trypsin-treated (d)	3.29	1.9
Uterus HMM (d)	.	4.6
Skeletal muscle HMM (d)	11.6	6.8
Uterus myosin (a)	0.29	0.64
Uterus myosin, trypsin-treated (a)	0.84	0.72
Chicken-gizzard myosin (e)	0.48	0.66
Chicken-gizzard myosin, trypsin-treated (e)	1.32	0.96
Chicken skeletal myosin (e)	.	2.58
Cow-carotid myosin (f)	0.54	1.04

The results are expressed in μmoles of inorganic P set free per mg N per min.

(a) Needham & Williams (2).   (d) Needham & Williams (1).
(b) Filo, Rüegg & Bohr (1).   (e) Bárány, Bárány, Gaetjens & Bailin (1).
(c) Greville & Needham (1).   (f) Gaspard-Godfroid (unpublished).

The ATPase activity can also be greatly increased by controlled trypsin treatment. Meromyosins very similar to those from the skeletal protein have been prepared from uterus myosin – H-meromyosin by Needham & Williams (1) and L-meromyosin by Cohen, Lowey & Kucera (1). Needham & Williams observed that the Ca-activated ATPase activity, measured at low ionic strength rose by about 200 % during the trypsin treatment; the purified H-actomeromyosin was about half as active as the skeletal-muscle protein. Bárány and his collaborators later also found the activating effect of trypsin treatment with myosin Ca-ATPase prepared from chicken gizzard and made a thorough study of this enzyme. Activation was very much more rapid than the conversion of the myosin into the two meromyosins, this myosin being more resistant to trypsin than the skeletal myosin; moreover a threefold activation was given by the treatment of the enzyme with a number of reagents such as 2M urea, 0.5M guanidine-HCl or 20 vol. % of ethylene glycol. The non-specific nature of the effect suggested a configurational change and this was borne out by the increased accessibility of SH groups to iodoacetic acid in the urea-modified enzyme. Bárány and his collaborators also remarked on the negligible Mg-ATPase activity of gizzard actomyosin and found no way of increasing it.

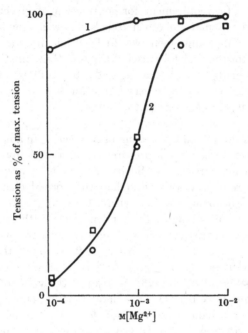

Fig. 108. Dependence of tension development on magnesium concentration. Curve 1: fibre preparation from skeletal muscle; curve 2: fibre preparation from uterus muscle; pregnant uterus ○, non-pregnant uterus □. (Hasselbach & Ledermair (1).)

This lack of Mg activation in actomyosin preparations from smooth muscle is puzzling in view of the requirement shown by skeletal actomyosin for both Mg (10^{-3}M) and Ca (10^{-5}M) if maximal contraction is to be produced. With glycerinated fibres from cow uterus Hasselbach & Ledermair (1) found Mg activation of contraction, but while 10^{-3}M Mg gave good tension production, 10^{-2}M was needed for maximal effect (see fig. 108); no contraction was called forth by 10^{-3}M Ca.[1] Needham & Williams (1), with the idea in mind that a change in the enzymic properties might have been

[1] Also Briggs (1).

effected by the solution and reprecipitation of the protein, tried to prepare native myofilaments from myometrium. This they did by homogenisation at low ionic strength and differential centrifugation. The actomyosin content of the preparation was estimated and its ATPase activity was measured after correction for particle contamination. As far as the results of these exploratory experiments went, it seemed that the filament ATPase was activated as well by Mg as by Ca and as well in 0.1 M KCl as in 0.5 M, but the activity found would still be only about a quarter of that of skeletal actomyosin. In 1963 Filo, Bohr & Rüegg (1) (using EGTA–CaEGTA buffers) compared the free Ca requirement for contraction of glycerinated fibres from skeletal and vascular smooth muscle.[1] This was very similar in the two cases, but the Mg requirement was at least ten times as great for the artery muscle as for the skeletal muscle. Briggs & Hannah (1) mention unpublished experiments in which a Mg-activated ATPase was found in glycerinated fibres. This would be very important if it could be shown that contaminating cell particles were not involved.

ACTIN. Actin (contaminated by tropomyosin) reacting in a normal manner with skeletal myosin has been prepared from uterus by Needham & Williams (2). Rüegg, Strassner & Schirmer (1) with a preparation from artery muscle showed that this actin had the normal characteristics of depolymerisation on homogenisation with 0.1 mM ATP and repolymerisation on raising the ionic strength. The tendency seen with skeletal actin to form associations with tropomyosin was very evident and was studied by Schirmer (1). It is interesting that Huys (1, 2, 3) found that the faster component of uterus actomyosin preparations in presence of ATP had a sedimentation coefficient of only 26.7 instead of the value of 64 given by Portzehl, Schramm & Weber (1) for skeletal F-actin. The value given by Shoenberg and her collaborators was 34S for artery-muscle actin.[2] This smooth-muscle actin according to Huys seemed not to be depolymerised in presence of 0.3 M KCNS, under conditions which with skeletal actin gave conversion to G-actin.

Carsten (4) has prepared from uterus highly purified actin which on starch-gel electrophoresis showed only one band moving the same distance as actin made from skeletal muscle. It resembled the latter in its polymerisation and depolymerisation characteristics (although these changes took place more slowly), its amino-acid composition, its peptide maps and the sedimentation constant of the G form. The two actins behaved similarly in activating myosin ATPase.

Later studies by Gaspar-Godfroid, Hamoir & Laszt (1, 2) showed that,

---

[1] Also Briggs & Hannah (1).
[2] Shoenberg, Rüegg, Needham, Schirmer & Nemetchek-Gansler (1).

by use of methods involving acetone treatment of the tissue, artery muscle gave F-actin preparations with two sedimentation gradients of 70S and 32S. The more rapid would correspond to the F-actin of skeletal muscle, while the less rapid might correspond to a complex of tropomyosin and F-actin. It seems that the native actin of smooth muscle, responsible for the lower sedimentation gradients seen in actomyosin solutions in presence of ATP, is in a different state of aggregation. It was indeed found that the actin prepared from artery actomyosin by fractionation with ammonium sulphate sedimented at the rate of 20–25S. These workers found with artery actin (as Carsten had done with uterus actin) a slowness of polymerisation and depolymerisation as compared with skeletal-muscle actin.

TONOACTOMYOSIN. In 1961 Laszt (1) and Laszt & Hamoir (1) prepared from the muscles of bovine arteries an extract of quite unusual properties. It was made with salt solution (1.5 ml per g of tissue) of ionic strength less than 0.075, pH 7.4; it had marked birefringence and viscosity, the latter being reduced to some extent by ATP addition though the ATP sensitivity was low. When the ionic strength was raised to 0.35 the birefringence, viscosity and ATP sensitivity all increased. In the ultracentrifuge at low ionic strength two principal components having S values of 20–30 and 12 were found; at ionic strength of 0.35 and in presence of 3×10^{-3}M ATP, the ultracentrifuge findings were in agreement with those of Huys (1) for uterine actomyosin in presence of ATP ($S_{20,w} = 27$ and 6.2). Similarly the electrophoretic behaviour at low ionic strength was different from that of striated muscle extracts; but at $I = 0.35$ a component behaving like actomyosin appeared. The viscous behaviour now also agreed with that of striated muscle actomyosin. Laszt & Hamoir concluded that a new form of actomyosin was concerned, particularly involved in tonus, and for this reason they gave it the name of tonoactomyosin. They interpreted their results as meaning that the tonoactomyosin as extracted at low ionic strength was only partly dissociated in the presence of low ATP concentrations, the gradient 10–12S probably being due to side-to-side aggregation of myosin. They suggested that the protein was present in the sarcoplasm (where the ionic strength is about 0.2 and the ATP concentration about 1.5×10^{-3}M) as a labile network; and that under conditions of arterial tonus a more rigid network could be formed, its formation possibly bringing about elimination of water and decrease in cell volume.[1]

Huys (2) very soon examined uterus muscle for this new form of actomyosin and found it there also, confirming the observations of Laszt & Hamoir. Both Huys (2, 3) and Hamoir, Gaspar-Godfroid & Laszt (1) made further careful studies of the effect of ionic strength and of ATP concentra-

---

[1] Also Laszt (2).

tion on the state of aggregation of the protein. The latter workers pointed out that calculations in the literature supported the idea that their material sedimenting at 10–12S was composed of side-to-side dimers of myosin.[1] Needham & Williams (2) also extracted tonoactomyosin from uterus muscle. In one experiment (see table 22) they divided muscle from the same cow uterus into two parts, extracting one part at $I = 0.6$, the other at $I = 0.05$.

TABLE 22. *Preparation and viscous behaviour of tonoactomyosin*

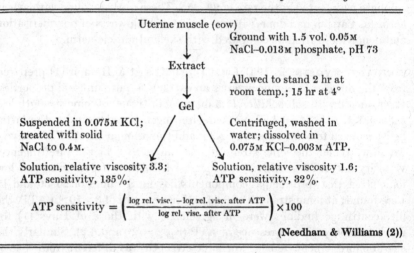

Uterine muscle (cow)

Ground with 1.5 vol. 0.05 M NaCl–0.013 M phosphate, pH 73

Extract

Allowed to stand 2 hr at room temp.; 15 hr at 4°

Gel

Suspended in 0.075 M KCl; treated with solid NaCl to 0.4 M.

Solution, relative viscosity 3.3; ATP sensitivity, 135%.

Centrifuged, washed in water; dissolved in 0.075 M KCl–0.003 M ATP.

Solution, relative viscosity 1.6; ATP sensitivity, 32%.

$$\text{ATP sensitivity} = \left(\frac{\log \text{rel. visc.} - \log \text{rel. visc. after ATP}}{\log \text{rel. visc. after ATP}}\right) \times 100$$

(Needham & Williams (2))

In the latter case the ionic strength of the extract was afterwards raised to 0.45; it appeared from its ATP sensitivity at this ionic strength that about as much actomyosin (as tonoactomyosin) had been extracted per g of muscle at low as at high ionic strength. They concluded that the greater part, if not all, of the uterus actomyosin was tonoactomyosin. This same conclusion was reached by Huys (3) (working with human myometrium) who found that after four extractions at low ionic strength no actomyosin remained to be extracted by Weber–Edsall solution. In a similar manner Rüegg, Strassner & Schirmer (1) showed the identity of the actomyosin and tonoactomyosin from artery muscle. The ATP sensitivity values of successive extracts led Huys to conclude that the tonoactomyosin could not simply be dissolved in the cytoplasm but that some of it took part in cell structure and so came out with difficulty. Rüegg, Strassner & Schirmer, on the grounds that the solubility of tonoactomyosin *in vitro* is about 5 mg/ml while the cell content (allowing for the extracellular space and the high connective tissue content of the artery) must be at least 20 mg/ml, suggested

---

[1] Lowey & Holtzer (1).

some random arrangement *in situ* of myosin aggregates in a dispersed but not completely soluble state.

Studies of conditions for superprecipitation have added to our knowledge of actomyosin.[1] As would be expected, there was no superprecipitation under the conditions effective for skeletal actomyosin; moreover the actomyosin freshly extracted at high ionic strength was only precipitable when the ionic strength was reduced to 0.1 if ATP was absent. Thus it seems that changes taking place under the influence of the high ionic strength are reversible at first. However Schirmer (1) found that ageing of a tonactomyosin gel or lowering of the pH to 6.0 led to a condition of the protein in which ordinary superprecipitation did take place.

In 1962 Hamoir & Laszt (1) prepared the myosin partner from cow-carotid tonoactomyosin. This they did by fractional precipitation with ammonium sulphate. The myosin fraction (with sedimentation coefficient of 6.25) precipitated between 47 and 60 % saturation – a salting-out range higher than that of skeletal myosin. When its amino-acid composition was compared with values in the literature for rabbit skeletal myosin certain differences were found,[2] large compared with the differences between skeletal or heart myosin of different species, as has been shown by Hamoir (4). The comparatively low proline, valine and isoleucine content and high glutamic acid and leucine content indicated a larger α-helicity and this probably underlies the greater stability of this myosin. Gaspar-Godfroid (1) and Hamoir & Gaspar-Godfroid (1)[3] had shown the strikingly greater thermostability of tonomyosin and tonoactomyosin compared with the skeletal muscle proteins. For example artery actomyosin retained nearly 60 % of its ATPase activity after heating to 100°, and remained soluble; while all activity of skeletal actomyosin was lost at 60° and the protein precipitated (see fig. 109). It is tempting to relate these findings to the old observations of Schultz (4) that, with vertebrate smooth muscle the height of contraction increased as the temperature was raised to 39°, thereafter decreasing. He remarked that this maximum temperature for function was considerably higher than with striated muscle. As we have seen, gizzard tonomyosin is more resistant to trypsin than is skeletal myosin; this is true also for artery tonomyosin as was found by Huriaux (1). The L-meromyosin formed showed certain differences from the normal. For instance, in the presence of alcohol it gave rise only to Fraction I, the second fraction characterised by high SH and proline content being absent.[4] Both LMM and HMM had higher salting-out ranges than the normal. Gaspar-Godfroid (2) has recently shown a remarkable effect of the presence of β-mercapto-

---

[1] Filo, Rüegg & Bohr (1); Rüegg, Strassner & Schirmer (1); Schirmer (1).

[2] Huriaux, Pechère & Hamoir (1).

[3] See also Huriaux, Pechère & Hamoir (1).   [4] Ch. 10.

ethanol during the purification of carotid myosin; the Ca-ATPase activity was greatly enhanced and aggregate formation was reduced. Thus curiously enough in this respect the vertebrate smooth-muscle myosin has affinity with the unstable myosins from fish and invertebrate muscle.[1]

Bárány and his collaborators, comparing chicken gizzard and skeletal myosins, also found some differences in amino-acid composition.[2] Finck (1) observed that chicken-gizzard myosin was immunologically distinct from

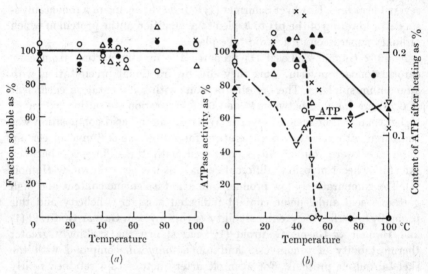

Fig. 109. Influence of temperature of heating on the solubility (*a*) and on the ATPase activity (*b*) of the tonoactomyosin of cow carotid arteries dissolved in a medium of ionic strength 0.35, pH 7.4. No ATP, ●, ▲, × ; 0.18 % ATP initially present ○, △, ▽. In (*b*) the full and interrupted lines refer respectively to activity after heating in absence and presence of ATP, indicating the protective effect of actin combination. The line –·–·–·–· shows the ATP content after heating (right-hand ordinate). (Hamoir & Gaspar-Godfroid (1).)

chicken skeletal and heart myosin, provided the proteins were sufficiently purified. Gröschel-Stewart (1) has recently found immunological specificity with human striated-muscle myosin and uterus myosin.

Hamoir considers that at present there is no evidence to warrant the use of a special name for the myosin of smooth muscle, which can be considered as a third type of the protein, differing from skeletal and heart myosin as they differ from each other. He suggests the very interesting possibility that the exceptional properties of tonoactomyosin may be due to the presence of a third component interacting with the myosin.[3]

[1] Ch. 22.   [2] Bárány, Bárány, Gaetjens & Bailin (1).
[3] Personal communication.

Careful studies of tonomyosin and tonoactomyosin ATPase from artery have been made by Filo, Rüegg & Bohr (1), Gaspar-Godfroid (1) and Schirmer (1). The activity, Ca-activated at pH 7.0 was very similar to that already given above for the proteins extracted from various smooth muscles in the ordinary way. Mg-activation was very low.

In summary then we may say that vertebrate smooth muscle contains only one actomyosin; this shows several properties different from those of the classical actomyosin. After extraction at low ionic strength it may acquire (reversibly or irreversibly) as the result of certain treatments, behaviour indistinguishable from that of skeletal actomyosin. The special characteristics of tonoactomyosin seem to belong mainly to the myosin component. One special character which would seem to be connected with the slowness of the phasic contraction and hence with tonus is the low ATPase activity. How the other biochemical characteristics fit into the picture of the fine structure and contractile mechanism we shall be considering below. Much is still unclear.

TROPOMYOSIN. The yield of this protein in myometrium is about 1.5 mg/g wet wt,[1] to be compared with Bailey's yield of about 4.5 from striated muscle. Thus the ratio tropomyosin:actomyosin is much higher in smooth than in striated muscle – ca, 1:4 in the former, 1:17 in the latter. Tsao, Tan & Peng (1) in 1956 found that tropomyosin from various sources had the same electrophoretic mobility and readiness to crystallise, but there were some well-marked differences. Gizzard tropomyosin had a much higher molecular weight (150000) than pig-heart tropomyosin (89000) or than tropomyosin from rabbit skeletal muscle (53000). Gizzard tropomyosin showed the highest axial ratio and highest polymerising ability of all those tested. Electron micrographs of crystals of this tropomyosin by the Chinese workers and by Cohen & Longley (1) in 1966 showed a repeat period of 800 Å as well as the 400 Å period found in tropomyosin from other sources. Kominz and his collaborators in 1957 had reported differences in amino-acid composition between smooth-muscle tropomyosin (from uterus and bladder) and skeletal-muscle tropomyosin.[2] Cohen & Longley now observed a different band pattern in electron micrographs of gizzard tropomyosin, which would be explained by different amino-acid content. As we have already seen the Chinese workers traced a correlation between polymerisability and arginine content.[3]

Recently Carsten (5) has prepared highly purified tropomyosin from human and from pregnant sheep uterus. The proteins from the smooth muscle of these two mammals were very similar, but on comparison with

---

[1] Needham & Williams (2); Sheng & Tsao (1).
[2] Kominz, Saad, Gladner & Laki (1).   [3] Ch. 9, p. 232

skeletal tropomyosin significant differences were found, e.g. in the amounts present of ten out of the seventeen amino acids estimated. The uterus protein had eight more negative charges per molecule. The peptide maps also differed, only 27–29 peptide spots appearing, instead of the 35 found for the skeletal-muscle protein. Thus smooth-muscle tropomyosin shows greater variability from the skeletal tropomyosin than does smooth-muscle actin from skeletal actin.

Ebashi and his collaborators prepared from gizzard muscle both α-actinin and a protein resembling native tropomyosin;[1] in superprecipitation experiments the latter conferred calcium sensitivity upon synthetic acto-myosin from the rabbit.

Hamoir & Laszt (2) found that about one-third of the tropomyosin was easily extracted from arterial muscle at low ionic strength, while the rest could only be obtained from the residue by the procedure used by Bailey for skeletal muscle. They concluded that part of this protein was contained in the sarcoplasm while part was associated with structural elements.

POSSIBLE IMPLICATION OF SOLUBLE PROTEINS. Sheng & Tsao (1), impressed by the very high content of tropomyosin B in chick gizzard and by the higher content in percentage of the structural proteins in uterus muscle than in skeletal muscle of the same animals, suggested that this protein is the 'substrate for the holding mechanism of the muscle'. Ivanov and his collaborators also implicated soluble proteins in the tonic function.[2] Homogenised uterus muscle was thoroughly washed with large volumes of very dilute salt solution, and the insoluble residue was then extracted with buffered 0.5 M KCl.[3] When this extract was diluted a precipitate containing actomyosin was obtained, but much protein remained in solution. This result was confirmed by Needham & Williams (3). Ivanov and his collabo-rators suggested that the fact that these soluble proteins (which they designated the T fraction) were not removed from the insoluble ones until the latter had been dissolved in salt solution of high ionic strength, showed that the soluble proteins also are part of the contractile mechanism. They found that soluble proteins behaving in this way did also occur in striated muscle but in much smaller proportion to the actomyosin, the actomyosin: T fraction ratios being 80:20 for striated muscle, 40:60 for stomach muscle and 25:75 for uterus muscle. They suggested that this higher proportion in smooth muscle might be connected with a role in the mechanism of tonus. The T fraction was heterogeneous, containing besides other proteins,

[1] Ebashi, Iwakura, Nakajima, Nakamura & Ooi (1).
[2] Ivanov, Mirovich, Moissejeva, Parshina, Tukachinsky, Yuriev, Zhakova & Zinovieva (1); Ivanov, Mirovich, Jakova & Tukachinsky (1).
[3] That little tonoactomyosin was extracted in washing was of course due to the use of very large volumes of solution, diluting the salt and ATP content of the muscle itself.

tropomyosin and the water-soluble myofibrillar protein isolated by Tsao and his collaborators in 1958.[1] No mechanism for this tonic role is obvious. The only known property which could be relevant is the capacity of tropomyosin to change its state of aggregation with changes in ionic strength. The water-insoluble paramyosin is important in the holding mechanism in some invertebrate muscles, as we have seen; this protein has been looked for in vertebrate smooth muscles (in uterus by Needham & Williams (2) and in artery by Rüegg, Strassner & Schirmer (1)) but has not been found.

ENERGY SUPPLY

RESPIRATION AND CARBOHYDRATE METABOLISM. We have referred above to the low oxygen uptake of uterus and intestinal muscle; in these experiments of Lovatt Evans (2) there was decrease rather than increase of oxygen uptake on contraction. This seems at first surprising; but the muscle was already in tone and the contraction was isotonic under constant tension: thus, on analogy with results on skeletal muscle (as discussed also by Bayliss (1)) less metabolism is to be expected at the shorter length. A little later Lovatt Evans (3) estimated the lactic acid changes in uterus, stomach, intestine, etc. The resting value was very low; it rose on contraction of the muscle but the maximum was much lower than for striated muscle. More recently Bülbring (1) has shown clearly with taenia coli that the oxygen uptake was directly proportional to the tension developed and maintained. The lack of suitably-shaped muscles has prevented the measurement of heat production in mammalian smooth muscle.[2]

From a detailed study of glycolytic reactions in uterus extract Hollmann (1) concluded in 1949 that the classical enzymic reactions were involved, but at a very low level of activity even when coenzymes were added. In 1962 Lundholm & Mohme-Lundholm (1) examined, under anaerobic conditions, the effects of electrical stimulation and of certain drugs (adrenaline, noradrenaline, histamine, acetylcholine), in causing lactic acid formation accompanying contraction of the smooth muscle of cow mesenteric artery. They reported a rough correlation between lactic acid production and degree of shortening. However they later found (2, 3) that stimulation of lactic acid formation and of contraction could be dissociated from one another. Thus presence of dihydroergotamine selectively inhibited the contraction effect of adrenaline but left unchanged the stimulatory effect on lactic acid production. With 6 % dextran instead of Tyrode solution as the bathing fluid, there was inhibition of contraction without any interference with lactic acid production. Such observations of course throw doubt on

---

[1] Tsao, Hsu, Jen, Pan, Tan, Tao, Wen & Niu (1).
[2] Personal communication from Prof. A. V. Hill.

the validity of lactic acid production as a measure of the energy requirement of contraction,[1] and might account for the very high requirement calculated in the earlier paper.

TABLE 23

Muscle	Phospho-creatine (μmoles/g wet wt)	ATP+ADP (μ atoms of 7 min hydrolysable P/g wet wt).	Reference
Hen's stomach	1.0	6.8	Dworaczek & Barren-Scheen (1)
Non-pregnant cow uterus	1.9	1.7	Hollmann (1)
Pregnant rat uterus	0.46	1.1	Walaas & Walaas (1)
Pseudo-pregnant rabbit uterus	1.39	2.9	Menkes & Csapo (1)
Pregnant rabbit uterus (term)		2.1	Menkes & Csapo (1)
Rabbit uterus, anoestrus	0.93	1.2	Menkes & Csapo (1)
Non-pregnant rat uterus	1.0	2.5	Volfin, Clausor & Gautheron (1)
		ATP and ADP μmoles/g wet wt (by chromatographic methods)	
Rabbit stomach	0.91	ATP, 0.63; ADP 0.47	Lange (2)
Guinea-pig taenia coli	1.9	ATP 1.8*	Born (1)
Human uterus at term	2.7	ATP 1.25; ADP 1.64	Cretius (1, 2)
Cow mesenteric artery	0.56	ATP 0.68; ADP 0.87	Lundholme & Mohme-Lundholm (1)
Cow carotid artery	0.78	ATP 0.91; ADP 0.63	Daemers-Lambert (3)
Human skeletal muscle	15.3	ATP 3.12; ADP 2.4	Cretius (1, 2)

* By the luciferase method.

Great difficulties have been experienced in demonstrating oxidative phosphorylation in mammalian smooth muscle. Wakid (1) could get only uncertain indications of this with rat-uterus particles but Gautheron, Gaudemer & Zaydela (1) did succeed. Compared with the heart mitochondria, which they also used, the P/O ratios and O_2 uptake for uterus

[1] See also Kulbertus (1) for evidence of the great glycogenolytic stimulation during potassium tonus.

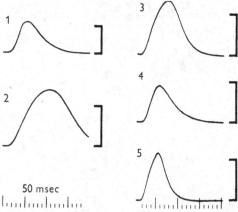

50 msec

Fig. 99 (a) (see p. 489). Isometric recording of twitch contraction to supramaximal stimulation of (1) soleus muscle cross-innervated by flexor digitorum longus nerve; (2) intact soleus muscle on contralateral side; (3) flexor hallucis longus muscle cross-innervated by soleus nerve; (4) same muscle as (3) but stimulus applied to original flexor hallucis longus nerve, some of the fibres of which had re-innervated the original muscle (see text); (5) intact flexor hallucis longus muscle of contralateral side. Recordings made in kittens 6.5 months of age following cross-innervation operation at 2 weeks of age. The upright bar to the right of each contraction represents 550 g (1, 2, 3), 750 g (4), and 600 g (5). (Romanul & van der Meulen (1).)

(b) (see p. 489). Cross-sections of muscles incubated for the histochemical demonstration of α-glycerophosphate dehydrogenase activity, using menadione and MTT. (1) soleus muscle cross-innervated by the nerve to flexor digitorum longus; (2) normal soleus; (3) flexor hallucis cross-innervated by the soleus nerve and also partially re-innervated by its original nerve; (4) normal flexor hallucis muscle. (Romanul & van der Meulen (1).)

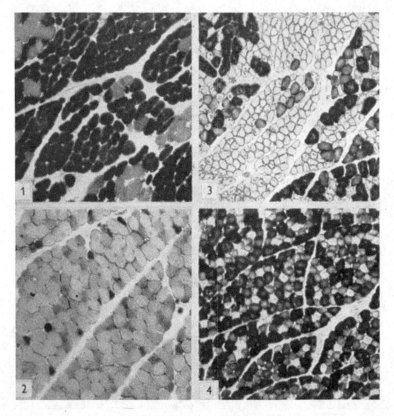

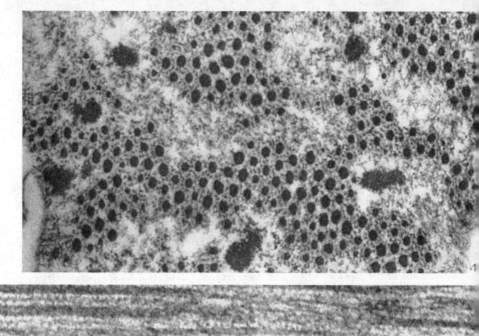

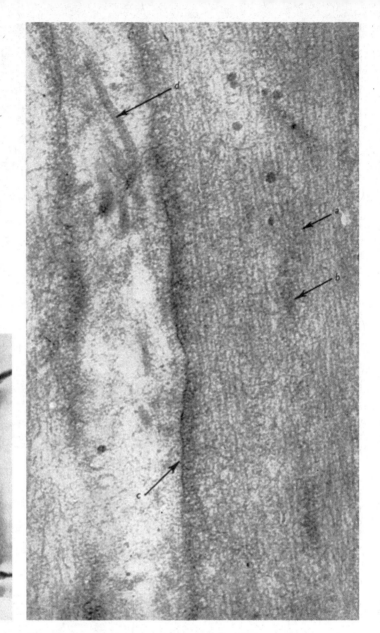

above left Fig. 111 (see p. 586). Behaviour of a motile stalk of *Vorticella* prepared as described in the text. (*a*) In EDTA solution; (*b*) after removal of the EDTA and addition of 5×10^{-5} M Ca^{2+}. (Hoffmann-Berling (6).)

above right Fig. 113 (see p. 569). Electron micrograph of smooth muscle cell of guinea-pig taenia coli. Fixed contracted to half-resting length by means of 2×10^{-4} M acetylcholine. Arrows (*a*) myofilament, (*b*) dense body, (*c*) cell membrane, (*d*) collagen. $100\,000 \times$. (Shoenberg, unpublished.)

opposite page Fig. 101 (see p. 527). (1) Transverse section of glycerol-extracted oyster muscle (*Crassostrea angulata*). Note the cross-bridges and the varying diameter of paramyosin filaments, probably due to tapering. Magnification $100\,000 \times$. (Hanson & Lowy (7).)

　　(2) Longitudinal section of glycerol-extracted oyster muscle showing regularly arranged transverse bridges. Magnification $57\,500 \times$. (Hanson & Lowy (7).)

　　(3) Longitudinal section of *Mytilus* anterior byssus retractor. Magnification $123\,750 \times$. (Hanson & Lowy, unpublished.)

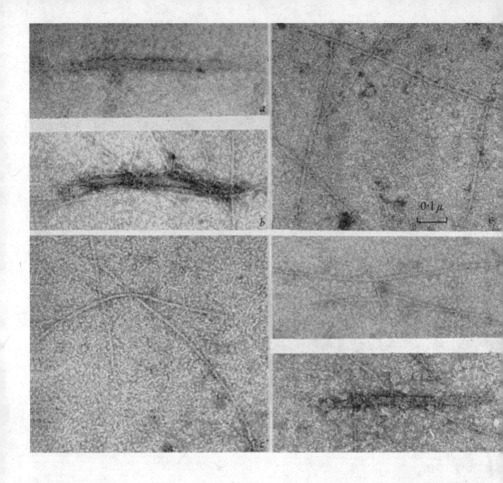

Fig. 114 (see p. 574). Electron micrographs to illustrate the conditions for the finding of myosin filaments in the supernatant from homogenate of chicken gizzard muscle a–f: conditions as in table 26 with diluents 2–7. Actin filaments are seen in all cases, myosin filaments only in a, b and f. (Shoenberg (5).)

Fig. 115 (see p. 588). Cytoplasmic movements of glycerol-water extracted fibroblast cells at the telophase stage. (a) before ATP addition; (b) 5 min after ATP addition (2.5×10^{-3} M); (c) 12 min after ATP addition. (Hoffmann-Berling (5).)

mitochondria were very low. They commented on the fragility of these mitochondria and this creates special difficulties in view of the toughness of the connective tissue present in large amount in most organs containing smooth muscle.

The question of the method of synthesis of glycogen in smooth muscle was for a time puzzling, since Bo & Smith (1) could not demonstrate glycogen synthetase in uterus at a time when much synthesis was going on and with a method which gave good results in tongue muscle. However this difficulty has been overcome; Rubilis, Jacobs & Hughes (1) using guinea-pig and rabbit uterus and Bo, Maraspin & Smith (1) using rat uterus have been able to show the synthetase activity.

It is interesting that indications are found in mammalian smooth muscle of the same sort of metabolic differentiation as exists between red and white skeletal muscle. Thus Dorogan and his collaborators find in the intestine of the dog that the circular muscles regularly show much higher muscle haemoglobin content than the longitudinal muscles, while the latter have much greater aldolase activity.[1] They discuss these results in terms of the different type of function carried out.

METABOLISM OF PHOSPHOCREATINE AND ATP. The earliest figures for the phosphate and ATP+ADP content of smooth muscle seem to be those of Dworaczek & Barrenscheen (1) in 1937 for stomach muscle. They are very low, especially for phosphocreatine, compared with those for skeletal muscle, and this has been the experience of later workers also (see table 23). Csapo & Gergely (1) found about one-seventh as much ATP and phosphocreatine in uterine strips as in striated muscle. These strips could perform, anaerobically in the presence of iodoacetate, about ten contractions instead of seventy to one hundred with skeletal muscles under the same conditions.

Born (1) using taenia coli found that lack of glucose in the oxygenated medium led to diminution of spontaneous active tension and to a parallel fall in phosphocreatine content. In this case ATP was not affected. Anoxia caused more rapid fall in both tension and phosphocreatine content, the ATP now diminishing also to some extent.[2]

In 1964 Daemers-Lambert (1) made a very careful study of the phosphate esters in the coronary, carotid and mesenteric arteries of the cow. Fig. 110 shows the distribution of the total P amongst these compounds, and the

---

[1] Dorogan, Zerner-Faighelis & Pieptu (1).

[2] Ch. 14; Honig has recently drawn attention to an inhibitory effect of inorganic phosphate and 5'-AMP on the ATPase and synaeresis of smooth-muscle myofibrils and actomyosin. When the two substances were added together their effect was synergistic. Since these metabolites accumulate under anaerobic conditions, they could constitute an oxygen-linked feed-back control of ATP utilisation and so of contraction.

very striking difference from the distribution in skeletal muscle. The unusual feature of the presence of guanosine and uridine triphosphates, accounting for about 10% of the total phosphorus, must be mentioned. This fraction fell markedly during contraction.[1]

Tension lasting 30 min, caused aerobically by 60 mM KCl, resulted in little if any change in the P distribution; presumably the level of energy-rich P had been maintained by glycogenolysis. Daemers-Lambert then turned to the use of carotid muscle poisoned by 1 mM iodoacetate. Here with 60 mM KCl tonus was accompanied by a total breakdown of the ATP and phosphocreatine, which were exhausted in about 3 min. The question arose as to the share of the ATP hydrolysis concerned, not with the contraction, but with the transport of the K ions into the cells. To solve this problem she made comparison of the ester breakdown in the muscle kept in 25 mM KCl (where no contraction is induced) with that in muscle kept in 30 mM KCl (where rise of tone is seen after a long latent period). In the former the action of an ATPase and phosphocreatine breakdown could be clearly followed; in the latter, a similar breakdown of energy-rich phosphate occurred during the latent period, with an increased metabolism accompanying the rise of tension. Calculation showed that after correction for the ATP used in production of hexosediphosphate (always found to accumulate) contraction itself was associated with utilisation of ATP amounting to about 0.5 μmole/g of muscle, and occurring to a very great extent within the first 3 min of tonus. The isolation of the ATP breakdown concerned with contraction could be well seen in artery muscle poisoned with fluorodinitrobenzene (4). Lundholm and his collaborators,[2] with aerobic, isometric contraction of unpoisoned mesenteric artery found adrenaline contraction to be associated with a rather similar breakdown of ATP + phosphocreatine – 0.38 μmole/g after 1 min, 0.77 μmole/g after 7 min. Lundholm & Mohme-Lundholm (4) stressed the much greater energy required for the production of tension than for its maintenance.

Daemers-Lambert & Roland (1) used electrical stimulation of IAA-poisoned bovine carotids, in this way avoiding any stimulation of glycogenolysis or ATPase activity not associated with contraction. They found that development of tension (in 20 sec) needed 0.35 μmole ATP/g, while during maintenance energy was consumed at a lower rate – 0.22 μmoles ATP/g min. Phosphocreatine breakdown became important only later, accompanied by a slowing down in ATP disappearance. The variations in ADP and AMP content gave clear evidence of the importance of adenylate kinase as well as creatine kinase in providing the necessary ATP. Kulbertus

[1] See also Mandel & Kempf (1); and Volfin, Clauser, Gautheron & Eboué (1) for such nucleoside triphosphates in rat uterus.
[2] Beviz, Lundholm, Mohme-Lundholm & Vamos (1).

(1) using the method of Dubuisson and Distèche[1] was able to show, during contraction of unpoisoned carotid muscle on electrical stimulation, a sequence of pH changes in accord with the above results. There was first an

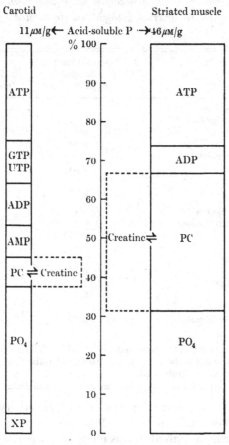

Fig. 110. Comparison of the phosphate balance in carotid muscle and striated muscle. The surface of each column represents the total acid-soluble phosphorus of the tissue. Each ester is expressed in percentage of this total. The free creatine is given in relation to the phosphocreatine present. (Daemers-Lambert (1).)

acid change, interpreted as due to ATP hydrolysis; then an alkaline change due to phosphocreatine breakdown in the Lohmann reaction; and finally the second acid change due to lactic acid formation. Daemers-Lambert (2) has recently studied the P ester metabolism during the contractile response to electrical stimuli of carotid artery poisoned with fluorodinitrobenzene.

[1] Ch. 9.

The rate of establishment of maximum tension was reduced and there was a diminution in the intensity of metabolism of ATP and ADP. No phospho-creatine was used (see table 24).

TABLE 24. *Variations in ATP breakdown associated with development and maintenance of tension in arterial smooth muscle excited by different vaso-constrictor agents*

	Development of tension μM of ATP/g	Maintenance of tension μM of ATP/g/min
Toxic agents		
Potassium chloride*		
MIA	0.36	•
FDNB	0.38	0.0088
Caesium chloride*		
MIA	0.34	0.005
Histamine*		
MIA	0.32	0.009
Phasic agents		
Electrical stimulation*		
MIA	0.355	0.220
FDNB	0.32	0.055
Hypertensin*		
MIA	0.35	0.175
Noradrenaline*		
MIA	0.32	0.185
Adrenaline†		
Without toxic agent	0.35	0.11

* Bovine carotid artery.
† Bovine mesenteric artery. MIA: monoiodoacetic acid. FDNB: fluorodinitrobenzene. Results from: Daemers-Lambert (1, 4, 2); Daemers-Lambert & Roland (1); Beviz *et al.* (1). (After Daemers-Lambert (3).)

Both Cretius (1) using uterus and Daemers-Lambert using artery have remarked on the very high content of ADP in smooth muscle. The latter suggested that the high content of this phosphate acceptor may be impor-tant in regulating the glycolysis and oxidative phosphorylation on which the energy provision so much depends in the absence of a large phospho-creatine store. Luh & Henkel (1) have shown the very low activity of both adenylate kinase and creatine kinase in the myometrium even at the end of pregnancy; the activity is about 4% of the skeletal value in the non-pregnant human uterus with only a small rise, not more than 50%, at the end of pregnancy. Briggs (1) also noticed the absence of adenylate kinase in glycerinated uterine strips, which could not use ADP as an energy source for contraction. Gaspar-Godfroid (1) found that her actomyosin prepara-tions from carotid artery muscle were free from both myokinase and adenylic deaminase. Since it is very difficult to purify skeletal actomyosin from these two enzymes, this probably means that the deaminase also is in low concentration in smooth muscle. This might explain the marked AMP component in smooth muscle adenine nucleotides; though absent in striated muscle, it is often present in smooth muscle in as great amount as the ADP. It has to be remembered (see p. 563 above) that the results of Daemers-Lambert & Roland show that enough myokinase is present in carotid muscle to compete with the creatine kinase in bringing about re-phosphorylation of ADP. The high content of free creatine in proportion to the phosphorylated form was noticed by Dworaczek & Barrenscheen (1); they thought that it was due to extensive breakdown of phosphocreatine - but later workers have not succeeded in much increasing the yield of the latter, and the significance of the free creatine is unknown.

Comparison of the effects of different vasoconstrictor agents has led to the suggestion that two categories exist – those causing phasic contraction (e.g. electrical stimulation, hypertension, adrenaline) and those causing tonic contraction (e.g. KCl, CsCl, histamine).[1] The tonic contraction is charac-terised by its lasting nature and by greater increase in the modulus of elasticity.[2] Daemers-Lambert (3) has discussed these two categories in the light of the biochemical evidence for the energy consumption provoked. Table 24 shows that for tension development the energy requirement is similar for contraction with the two types of agent under a variety of condi-tions. For maintenance of tension there is a marked difference; with the phasic agents there is a requirement of about 0.15 μmole ATP/g min, while with the tonic agents this requirement is negligible. As Daemers-Lambert remarks, the similarity in ATP breakdown for tension production in the two cases suggests that both are using the same actomyosin system; the almost energy-free maintenance of the tonic tension could be the result of some form of reversible stabilisation of the A–M linkages.

Summing up we may say that, as in striated muscle, ATP breakdown is directly concerned in contraction; rephosphorylation of the ADP (by phos-phate transfer in the Lohmann reaction and in glycogenolysis) takes place here too. In the smooth muscle myokinase also plays a part. Since re-phosphorylation is much slower in smooth muscle and the phosphocreatine store is very small, it is possible there to observe the ATP breakdown in iodoacetate-poisoned muscle. This is in contrast to the state of affairs with striated muscle where, as we have seen, fluorodinitrobenzene poisoning is necessary before the ATP breakdown can be made manifest.[3]

The uterus is unique in the enormous increase in size and change of function which the adult organ can undergo during pregnancy. Under the influence of the hormone oestrogen it develops in a spectacular way; there

[1] Laszt (2); Kulbertus (2).   [2] Kulbertus (2).   [3] Ch. 14.

is great increase in size of the muscle cells, with synthesis of the contractile mechanism and of the enzymes concerned in energy provision. We cannot go into details here;[1] it must suffice to say that during the last thirty years many workers have followed the increase in stores of energy-providing material and of the enzymes dealing with these stores.

EXCITATION–CONTRACTION COUPLING

In connection with the fine structure of smooth muscle we have already mentioned the two muscle types, of which the multi-unit type normally responds only to stimulation of motor nerves. Bülbring (2) has given an excellent account of the behaviour of the single-unit type as typified by intestinal muscle, based largely on the work of herself and her colleagues using intracellular electrodes on taenia coli. Many earlier observations of Bozler, using an extra-cellular recording method, were confirmed.

The intestinal muscle cells produce all-or-none propagated action potentials, spontaneously or as the result of various stimuli such as stretch; they also can produce local graded potentials to which the action-potential spikes usually have a rhythmic relation. As we have seen, these cells have a mechanism for conduction of excitation from cell to cell. The pathways concerned are probably variable and the total tension produced will be affected by the degree of synchronisation of cell activity.[2] The rate of transmembrane flux of Na^+ is very rapid, while that of K^+ is slower than in skeletal muscle. This makes for great instability of the membrane potential and the stabilising effect of Ca in limiting Na ingress becomes very important. Goodford (1) studying Ca^{2+} entry into taenia coli under various conditions, observed behaviour compatible with competition between Ca^{2+} and Na^+ (as well as other cations) for fixed negatively-charged sites in the cell membrane. In contrast to skeletal muscle, smooth muscle shows only transient excitation when Ca^{2+} is removed from the medium, and membrane activity soon ceases. The Ca^{2+} inflow may thus take a significant part in the depolarising current across the cell membrane. Observations strongly suggesting that the spike in taenia coli is due to Ca^{2+} entry are recorded by Brading, Bülbring & Tomita (1). They also assemble much evidence in the literature pointing in the same direction. On the other hand, Keatinge (2) has shown clearly that the spike in sheep carotid artery is sodium-based.

Little sarcoplasmic reticulum is present in these muscles; Goodford has pointed out that, if the available evidence for actomyosin content of smooth muscle applies to taenia coli, then only one-tenth as much Ca^{2+} would be needed to stimulate the contractile mechanism as in skeletal

---

[1] For a review of the biochemistry of the uterus see Needham & Shoenberg (1).
[2] Greven (1).

muscle. This would amount to only about 0.02 mmoles Ca/kg wet wt and Goodford suggests that each action potential may be associated with movement of Ca^{2+} from some of the superficial anionic sites, the Ca flow not only helping to depolarise the membrane but also bringing about the contractile response.

As an example of the great variability in smooth muscle reactions, it is interesting to consider the effects of adrenaline. As we have seen in the case of vascular muscle, contraction is caused in some cases, and this is accompanied by depolarisation of the membrane and initiation or increase of spikes as shown by Nakajima & Horn (1) for the mesenteric vein. With intestinal muscle application of adrenaline (about 10^{-9} g/ml) causes immediate relaxation and cessation of spontaneous electrical activity with increase in membrane potential.[1] Increase in membrane potential is also caused by excess Ca^{2+} and it was found in earlier experiments that increased phosphorylase activity quickly followed the adrenaline application; it was therefore suggested that the relaxation effect was due to an increase in the energy supply for processes at the cell membrane, particularly for holding Ca^{2+} there.[2] Later work has substantiated this hypothesis, but certain important new information has led to changes in detail. It was first made clear that there was no phosphorylase activation during the muscle relaxation;[3] on the other hand an increase in ATP and phosphocreatine took place.[4] It was further shown more recently that anaerobic depletion of the glycogen stores (by keeping the tissue in substrate-free medium) abolished both the relaxation response to adrenaline and the metabolic activation concerned in generation of energy stores.[5] Further it could be shown that when β-hydroxybutyrate was supplied aerobically to this glycogen-depleted tissue, relaxation ability and energy-rich phosphate synthesis were both restored, just as well as if carbohydrate were supplied. The finding also that in glycogen-depleted muscle the levels of energy-rich phosphate compounds can be maintained during aerobic but not during anaerobic incubation, suggested that oxidation of an endogenous substrate, perhaps lipid, could be involved. It is interesting to remember that Bozler's experiments had led him to the conclusion (4) that, in muscles showing automaticity, spontaneous fluctuations in chemical processes underlay the changes in membrane potential.

Bueding and his collaborators compare their findings with those on the heart,[6] where also it was shown that the physiological effect of adrenaline

[1] For an account of this work with many references see Bueding & Bülbring (1); Bueding, Bülbring, Gercken, Hawkins & Kuriyama (1).
[2] Axelsson, Bueding & Bülbring (1).
[3] Bueding, Bülbring, Gercken & Kuriyama (1).
[4] Bueding, Bülbring, Gercken & Kuriyama (2).
[5] Bueding, Bülbring, Gercken, Hawkins & Kuriyama (1).   [6] Ch. 18.

(in that case increased contractile activity) preceded increased phosphorylase activity. In the taenia coli, as in the heart, increase in cyclic AMP had already been shown;[1] since this nucleotide affects many enzyme systems, the localisation of its effect here has not yet been possible.[2] Imidazole, which activates the phosphodiesterase catalysing dephosphorylation of cyclic AMP, prevents these actions of adrenaline on smooth muscle, thus fitting into the picture.[3] It has long been known that adenosine derivatives bring about relaxation of smooth muscle,[4] and very low concentrations of ATP (one one-thousandth of the total tissue concentration) applied externally were found to have this effect; it was now suggested that in such circumstances the ATP might be used at the membrane as substrate for cyclic AMP synthesis.[5] Davoren & Sutherland (1) have provided evidence for the location of adenyl cyclase in the membrane of erythrocytes and liver cells.

Later work of Bülbring & Tomita (1, 2, 3) has carried further the analysis of membrane effects of adrenaline and other catecholamines on the taenia coli. They found that adrenaline has two mechanisms of action – the α-action in which membrane permeability to K and Cl ions is increased with resulting hyperpolarisation; and the β-action consisting of the suppression of the slow depolarisation involved in the spontaneous spike discharge. Ca^{2+} in the medium is necessary for the effects of adrenaline and indeed, as we have seen, Ca^{2+} itself produces similar effects to those of adrenaline. The experimental results of Bülbring & Tomita have led them to the following view. With the α-effect there is sudden accumulation of Ca^{2+} (by uptake from the medium and from the cell interior) in a site in the cell membrane where its presence increases K^+ permeability. The membrane potential is thus raised, and since it is then too far away from the 'firing' level, the spontaneous spike discharge ceases. The β-effect is more important with some other catecholamines than with adrenaline; it seems to consist in the suppression of Ca^{2+} removal from the binding site. This action, though not causing hyperpolarisation (since the Ca-concentration in the membrane is probably not increased) is enough to prevent the slow depolarisation or 'pacemaker potential', and in this way to abolish the spontaneous spike discharge. Bülbring & Tomita (2) showed that in presence of imidazole, adrenaline did not cause hyperpolarisation, but had still an effect in slowing spontaneous spike activity. Thus imidazole blocked the α-effect but not the β-effect of catecholamines. They suggest (3) that the processes

---

[1] Bueding, Butcher, Hawkins, Timms & Sutherland (1).
[2] Sutherland & Robison (1).
[3] For experiments on the effects of imidazole in abolishing the inhibition by adrenaline of spontaneous spike discharge and on membrane conductance see Bülbring & Tomita (2).
[4] See the review by Drury (1).   [5] Ch. 18.

described of Ca^{2+} binding to the membrane and suppression of removal from it may need supply of metabolic energy.

Thus for this type of muscle very interesting correlations have been made between metabolic behaviour, electrical happenings at the membrane and physiological response.

MECHANISM OF CONTRACTION

As the result of electron-microscope studies[1] combined with the biochemical results we have been considering, it has recently become possible to formulate tentative hypotheses for the mechanism of contraction of vertebrate smooth muscle.

The difficulties characteristic of smooth muscle studies still beset electron-microscope work: special treatment is necessary for successful fixation of the filaments and here good progress has been made; in examination of contracted muscle there is the complication of unequal contraction of the individual cells even under load, owing to their irregular alignment. Casteels (1) however has been able to record good tension in isometric contraction of electrically-stimulated taenia coli.

In longitudinal section (see fig. 113 pl.) the spindle-shaped cells are seen filled with filaments 50–80 Å in diameter and resembling actin filaments of striated muscle; these are about 100 Å apart, usually parallel to the long axis of the cell but not very regularly aligned. In the studies referred to, between 1955 and 1962, no myosin filaments like those of the A band of striated muscle could be seen; Z lines were also missing but it becomes likely that the dense bodies, irregularly arranged amongst the filaments and with filaments passing through them, are points of attachment;[2] such dense patches were also seen by Panner & Honig at intervals under the plasma membrane and actin filaments, maintaining their roughly parallel orientation to the cell surface, terminated there. This function for the dense bodies was first suggested by Hanson & Lowy (7) as a result of their observations on the structure of oyster muscle fibres.[3]

Thus for several years the myosin, known of course from biochemical studies to be present, remained elusive. Elliott (1) in high- and low-angle X-ray diffraction observations on resting taenia coli and during peristalsis of this muscle (2) found only the actin pattern. In assessing this negative evidence one has to remember that, even with skeletal muscle, the X-ray diffraction pattern of myosin is easily disturbed and may become fainter while the appearance in the electron microscope remains normal.[4] Another

---

[1] Csapo (1); Gansler (1); Mark (1); Caesar, Edwards & Ruska (1); Shoenberg (1); Rhodin (1).   [2] Shoenberg (1); Panner & Honig (1).   [3] Ch. 22.
[4] Personal communication from Dr H. E. Huxley. Cf. ch. 11.

piece of evidence against the existence in regular axial arrangement of myosin filaments with densely packed molecules is provided by the results of Seidel & Weber (1). As we have seen, the whole length of the smooth muscle cell is birefringent. But it was now found that the intrinsic birefringence remained unchanged on complete extraction of actomyosin. Moreover the form birefringence was much higher than that to be expected from the amount of actomyosin present; it also changed little if at all on extraction of actomyosin (fig. 112*a*, *b* and *c*). The presence of this high proportion of doubly refracting material of unknown nature made it impossible to decide whether a small proportion of the form-birefringence might be due to actomyosin. Thus very thin filaments or single molecules lying parallel to the axis might be present.[1]

The discovery of the special solubility properties of smooth-muscle actomyosin cast a new light on the problem of the nature and location of the myosin in the cell. It became possible to speculate that this protein, in dispersed form in the resting muscle, might in the changed conditions resulting from stimulation, undergo some form of aggregation. About this time Choi (1) (1962) observed in electron micrographs of chicken-gizzard muscle a few filaments of thicker and darker type,[2] and Lowy in discussion suggested that these might consist of myosin.[3] Also at this time Shoenberg (2) did filament counts on transverse sections of taenia coli in which the cell orientation is usually parallel to the long axis of the muscle. Two points emerge consistent with a sliding mechanism. First, on contraction the filaments (which remained parallel to the length of the cell) did not change in diameter;[4] secondly, while the cross-sectional area was much greater at

[1] Two important recent contributions come from Lowy and his collaborators (Lowy, Poulson & Vibert (1); Lowy & Small (1)). Evidence is at last provided from low-angle X-ray diffraction patterns for the probable existence of myosin in well-oriented state in living, resting taenia coli. Also in these muscles, provided certain conditions were observed (keeping at 0° holding a 10 g weight for 4 hours before fixation) they saw in the electron microscope ribbon-like structures as well as regular arrays of actin filaments. In longitudinal sections all these were in arrangement parallel to the long axis of the cell. Mean values for the ribbon dimensions were 80 Å thick, 350 Å wide and more than 3 μ long. Evidence is adduced for their myosin nature and for the lability of their structure. Projections which could be cross-bridges were seen from the edges of the ribbons. Comparisons were made of the cross-sectional area of the ribbons with that of the actin filaments; although the ribbons are numerous and it is difficult to see how more could be fitted in to the actin arrays, the ribbon area/actin filament area ratio (about 1 to 2.5–5) seems very low in comparison with biochemical evidence for the relative amounts of the two proteins present in smooth muscle. Lowy (personal communication) estimates that inclusion of the cross-bridges would about double the myosin value. Recent results of Schirmer (personal communication) (based on preparative and analytical centrifugation) are 4.5 mg myosin and 1.1 mg actin/g of tissue. The calculated concentrations in the muscle cell were 15 mg and 3.5 mg/g.

[2] Also Lane (1); Fawcett (1).

[3] For further discussion of this possibility see Shoenberg in Needham & Shoenberg (2); Needham & Shoenberg (1).   [4] Also Conti, Haenni, Laszt & Rouiller (1).

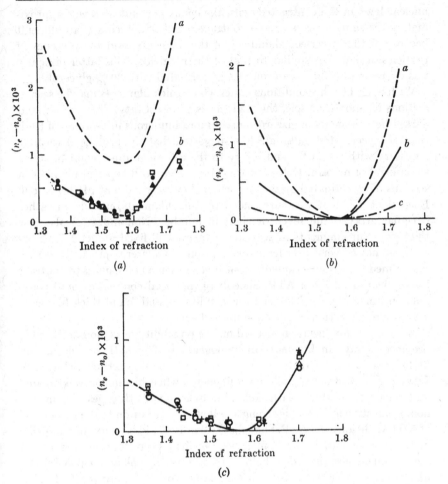

Fig. 112. (a) The dependence of the double refraction of skeletal muscle and vertebrate smooth muscle on the refractive index of the imbibition fluid. Curve a: Skeletal muscle; curve b: □, guinea-pig taenia coli; ●, rabbit taenia coli; ▲, dog retractor penis.

(b) Dependence of the rod double refraction on the refractive index of the imbibition fluid. Curve a: cut. pect. of the frog; curve b: vertebrate smooth muscle (taenia coli of guinea-pig and rabbit and retractor penis of dog). Curve c: rod double refraction, calculated from curve a according to the Wiener formula, which could be due to the actomyosin content of smooth muscle.

(c) Dependence of the rod double refraction of vertebrate smooth muscle on the refractive index of the imbibition fluid after different treatments of the fibres. ○ = Formol fixation of the fresh fibre. Formol fixation: after extraction in 0.6 M KCl, 2×10^{-3} M MgCl$_2$, 10^{-2} ATP, ●; after digestion with trypsin solution, △; after acetone treatment, drying and subsequent extraction with water, +; after extraction in 0.6 M KI $\times 10^{-3}$ M sodium thiosulphate at pH 7.0, □. (After Seidel & Weber (1).)

nuclear level in the contracted cells, the count per unit area was approximately the same for resting and contracted cells. Such counts are difficult because of the imperfect alignment of the filaments, and the presence of particles of diameter similar to that of the filaments. This latter difficulty was overcome by the use of muscles glycerinated with 30 % glycerol.[1]

With regard to the conditions under which thick filaments may be seen in sections of smooth muscle, the evidence is rather scanty. Shoenberg found dark, thick filaments in gizzard muscle more numerous in sections of fresh than of glycerinated material.[2] She suggested that the prolonged washing necessary with the latter and injury to the cell membrane might have led to removal of myosin. Kelly & Rice (1) reported both types of filament in sections of glycerinated gizzard muscle if they were fixed at pH below 6.6. Each thick filament was surrounded by thin filaments, usually in rather irregular array, some seven or eight in number. Nonomura (2) also saw thick and thin filaments in sections of taenia coli fixed at pH 7.2. Cross-sections showed very disordered distribution.[3] In a later paper Kelly & Rice (2) claimed that thick filaments appeared in taenia coli caused to contract by application of 5 mM ATP. Shoenberg (personal communication) could not confirm contraction under these conditions, and found thick filaments only where dissection might have injured the fibres.

Much attention has been focused on the possibility of demonstrating the presence of myosin filaments in homogenates of smooth muscle and of finding conditions for their formation *in vitro* in muscle extracts. Hanson & Lowy (1) in 1963 found only actin filaments when taenia coli was treated under conditions which, with skeletal muscle, gave both types of filament: homogenisation at low ionic strength followed by addition to the suspension of EDTA, Mg ions and ATP. Shoenberg (3) in 1965 did find myosin filaments in these conditions if the homogenisation had been done in presence of Sigma collagenase (added to help disruption of the collagen network). No filaments however were seen when in later work a highly purified collagenase preparation was used,[4] and this fact led to the suggestion that the effect of the Sigma collagenase depended on presence of traces of a protease. Needham & Williams (1) had earlier shown that collagenase treatment (like controlled trypsin treatment) leads to increased ATPase activity of uterine actomyosin and myosin preparations;[5] this increased enzymic activity might be associated with a configurational change in the protein favouring filamentous aggregation. The filaments formed (of dimensions 0.3–0.6 μ in length and 160–240 Å in diameter) were similar in appearance to those

---

[1] Needham & Shoenberg (1) p. 339.   [2] Needham & Shoenberg (2).
[3] Also Hama & Porter (1).
[4] Shoenberg (4); Shoenberg quoted in Needham & Shoenberg (1).
[5] A preparation from Agricultural Biologicals was used.

obtained by Huxley with skeletal muscle. They formed more readily at pH about 6.0 than at pH 7.0 or above, though at pH 7.0–7.4 the numbers increased on standing. Nonomura later described myosin filaments in homogenates of guinea-pigs taenia coli prepared in a medium containing 0.15 M KCl, 1 mM NaHCO$_3$, pH 7.8, 4 mM ATP and 5 mM MnCl$_2$ (for a possible interpretation of these last results see p. 575).

In 1964 Hanson & Lowy (8) were successful in obtaining myosin filaments in taenia coli extracts. They extracted with 0.65 M KCl and separated the myosin from the actin by ultracentrifugation in presence of ATP and Mg ions. The myosin-containing supernatant was used for electron-microscope examination after dialysis against phosphate-buffered 0.1 M KCl. Kaminer (1) also obtained filaments from gizzard myosin. Actomyosin was extracted at high ionic strength, purified and stored in 50 % glycerol. Myosin solutions were prepared from the actomyosin and on rapid dilution from 0.6 M to 0.3, 0.2 or 0.1 M KCl, filaments were formed. Values of pH between 8.0 and 6.0 were used, and the length of the filaments was greater at low ionic strength and low pH. In evaluating such experiments it has to be remembered that both treatment with solutions of high ionic strength and ageing lead to a change in the properties of tonoactomyosin towards those of skeletal actomyosin.[1] Shoenberg and her collaborators could find no myosin filaments in extracts of artery muscle made at low ionic strength and dialysed against buffered 0.1 M KCl, the resultant gel being dispersed in relaxing medium containing 5 mM ATP, 5 mM MgCl$_2$ and 4 mM EDTA; sedimentation and viscosity measurements showed good content of myosin in such extracts.[2]

Shoenberg (4, 5) went on to make tonoactomyosin preparations from chicken-gizzard muscle and to examine them in the electron microscope after treatment in a number of different ways (table 25). This work indicated three factors favourable to filament formation: ageing of the extracts, lowering of pH and presence of Mg ions. The first two findings fit well with the results of Schirmer on superprecipitation; the third is of special interest in view of the high Mg^{2+} requirement for smooth muscle contraction, and of some recent observations of Cohen & Longley (1) on the effects of divalent ions on tropomyosins including that from gizzard muscle. In the presence of 0.01 M MgCl$_2$ or CaCl$_2$ at pH 7.0, fibre formation took place; while at lower concentrations and pH crystals were formed. As is discussed later, a direct effect of the Mg ions on myosin aggregation is a possibility; it is also possible that there may be an indirect effect via the tropomyosin always contained in these preparations, the fibre formation in the latter protein assisting filamentous aggregation of the myosin.

[1] P. 553 above.
[2] Shoenberg, Rüegg, Needham, Schirmer & Nemetchek-Gansler (1).

TABLE 25

Supernatant (or gel separating from it) finally diluted with Huxley's relaxing medium* at pH 7 or brought to pH 6	Filament formation	
	At pH 7	At pH 6
Examined immediately	0	+
Dialysed 24 hours against buffered 0.1 M KCl	0	+ +
Standing overnight	+ +	+ +
Dialysed overnight against 0.1 M KCl containing 0.01 M Mg	+ +	+ +

* This relaxing medium contains (besides the buffered 0.05 to 0.1 M KCl) 4 mM EDTA, 10 mM MgCl₂, 2 mM ATP – thus less ATP and more Mg than the relaxing medium used in the unsuccessful experiments just mentioned.

TABLE 26. *Homogenate of chicken gizzard consisting of 1 part muscle : 1 part 0.1 M KCl buffered with 0.0067 M phosphate (pH 6.87) diluted with 9 volumes of:*

No.	Diluent	Myosin filaments
1	0.1 M KCl buffered with 0.0067 M phosphate with additional 2 mM ATP (pH 6.87)	.
2	0.07 M KCl buffered with 0.0067 M phosphate with additional 10 mM MgCl₂ (pH 6.87)	? +
3	0.07 M KCl buffered with 0.0067 M phosphate with additional 10 mM MgCl₂ + 2 mM ATP (pH 6.87)	+ + +
4	0.06 M KCl buffered with 0.0067 M phosphate with additional 10 mM MgCl₂ + 2 mM ATP + 4 mM EGTA (neutralised with KOH) final pH 6.87	.
5	0.07 M KCl buffered with 0.0067 M phosphate with additional 0.1 or 0.5 mM CaCl₂ (pH 6.87)	.
6	0.07 M KCl buffered with 0.0067 M phosphate with additional 0.1 or 0.5 mM CaCl₂ + 2 mM ATP (pH 6.87)	.
7	0.06 M KCl buffered with 0.0067 M phosphate with additional 10 mM MgCl₂ + 2 mM ATP + 4 mM EGTA (neutralised with KOH) + 0.5 mM CaCl₂ (pH 6.87) Shoenberg (5)	+ + +

These experiments suggested the possibility that previous failure to find filaments in homogenates might have been due to lowering of the divalent ion concentration as the tissues were broken up in the medium added. This idea was borne out by experiments on homogenates which showed the need of added Mg ions for demonstration of myosin filaments (see table 26 and fig. 114 pl.). Occasionally experiments were also successful in showing myosin filaments when homogenisation was done in very small volumes of buffered 0.06 M KCl containing 2 mM ATP (only an equal volume instead of 10 volumes) to avoid excessive dilution of cell contents. The homogenates were negatively stained for electron-microscope examination within 2 min of homogenisation. Experiments such as those summarised in table 26 showed the requirement for a very low Ca ion concentration as well as for Mg ions and ATP. It seems from the results with solutions 3, 4 and 7 that the Ca content of the homogenised tissue itself even in 10-fold dilution is sufficient but a small amount is essential. It should be realised that effects of ageing on filament aggregation must play a much greater part in experiments with extracts, which take up to 48 h to prepare, than in experiments with homogenates. The latter can be used within an hour from the death of the animal.

It should be recalled that in Nonomura's homogenisation experiments, mentioned above, 5 mM MnCl$_2$ was present. This was added in order to prevent contraction, since Nonomura and his collaborators had shown that presence of Mn ions abolished the electrical activity and tension caused by putting the muscle in isotonic KCl. It is possible that these divalent ions had also another effect – like that of Mg ions in facilitating filament formation. Since no EDTA was added, it can be assumed that traces of Ca ions were also available.

We may consider now some of the theories advocated at the present time to explain smooth muscle contraction.

Nemetchek-Gansler,[1] as the result of her electron-microscope studies of stomach, uterus and artery muscle, lays great emphasis on diminution in volume of contracted cells, with loss of water and changes in state of polymerisation of the contractile proteins. Conti, Haenne, Laszt & Rouiller (1) have also described diminution in volume of cells in artery muscle contracted by KCl application. This change was not seen on electrical stimulation. Dark, dense cells such as those described by Gansler and taken to be contracted cells have been seen by Shoenberg in taenia coli contracted under load to 80 % of its resting length; evidence that the cells are really smaller has not yet been provided in Nemetchek-Gansler's published work, though she has recently in a personal communication supplied micrographs showing that the transverse area at the level of the nucleus is less for the dark cells in contracted tissue than in resting tissue.

In the work quoted above, Laszt and his collaborators concluded that the same smooth muscle cell could appear under two different forms characteristic respectively of the state of relaxation or contraction. More recently they have suggested that two types of cell are always present in the muscle of the bovine carotid, whether relaxed or contracted.[2] One type,

[1] Gansler (2, 3). See also Nemetchek-Gansler (1).   [2] Aita, Conti, Laszt & Mandi (1).

the 'oval' cells are described as light and clear; elongated in form with a greater diameter at the level of the nucleus – it is on cross-section at this level that their oval shape appears. In the electron microscope the filaments contained in them are seen to produce a loose network; they show no staining with acid dyes. In contracted tissue shortening of these cells is indicated by the enlargement of the diameter in cross-section and by the folds appearing in the nucleus. The other darker type of cell shows a smaller star-shaped cross-section and stains readily with acid dyes. During contraction the area of the cross-section may diminish; this observation, together with the increased depth of staining that occurs, is taken to mean that cells of this type lose water in contracting tissue. These results are of much interest, especially when one remembers the two types of cell population in many skeletal muscles.[1] Definitive evidence for the characteristics of the relaxed and contracted cells would necessitate counts in relaxed and contracted muscle. This would be difficult to carry out in arterial muscle, because of the non-alignment of the cells, which means that the angle of section cannot be controlled; also the cells may not all contract simultaneously.

Recently Panner & Honig (1) have reported observations on myosin molecules attached to actin filaments in chicken-gizzard muscle homogenised at $I = 0.6$, pH 7.0. To obtain this result it was necessary to stain the preparations within 2 min of homogenisation. The myosin molecules were attached, at regular intervals of about 350 Å, in pairs opposite one another. Occasionally what appeared to be dimers were seen – two myosin molecules associated tail to tail and apparently connecting two actin filaments. In sectioned material Panner & Honig described thin lateral processes (about 25 Å in diameter) between the longitudinal filaments. They suggest a sliding model in which actin filaments from one dense body interdigitate with actin filaments of the opposite sense coming from one or more other dense bodies. Connection between them and their driving past each other are pictured as dependent on myosin dimers with active sites at opposite ends. They point out that this model would be consistent with the force–velocity and length–tension relationships of smooth muscle, which Csapo (5) showed are qualitatively similar to those of skeletal muscle. This is a very interesting contribution, but it remains difficult to see how the myosin dimers are kept in register.

The results of Shoenberg, showing the presence, within a very few minutes, of myosin filaments in homogenates made with small volumes of medium raise the possibility that some such filaments may actually exist *in vivo* but may be lost upon too great dilution of the tissue contents. Whether this is so or not, her work provides good evidence for the view she has put forward: that myosin filaments, capable of entering into a sliding mechanism may be

[1] Ch. 19.

formed *in vivo* under the physiological conditions obtaining when stimulation has taken place.[1] Slight decrease in pH and in ATP concentration might have a role but perhaps more important might be changes in concentration of free Mg and Ca. The concentration of Mg ions used in her experiments was very high; but it must be remembered that the Mg requirement for maximal contraction of glycerinated smooth muscle is ten times as great as that for skeletal muscle. Filo, Bohr & Rüegg (1) argued, from the steep slope of the curve relating Mg concentration in the medium and tension development in the artery-muscle strips, that changes in Mg concentration would make a good regulatory mechanism. We have also already considered the evidence for entry of Ca ions on stimulation of smooth muscle. Since both Mg and traces of Ca are needed for filament formation, Shoenberg suggests that possibly the two ions (known to act together in stimulating skeletal actomyosin ATPase) may function in smooth muscle by stimulating the ATPase activity of its actomyosin *in vivo*, and in so doing may cause configurational changes in the myosin favourable to its aggregation. The slowness of production of important tension, characteristic of vertebrate smooth muscle would then depend, not only on the low activity of the actomyosin ATPase providing the energy[2]; another factor might be this need to assemble the machinery before the sliding mechanism can work.

[1] Shoenberg (5).   [2] Cf. Bárány (1).

24

ENERGY PROVISION AND CONTRACTILE
PROTEINS IN NON-MUSCULAR FUNCTIONS

Let us turn now for a while, in closing this story of the development of ideas concerning the nature of muscle contraction and the pathways of energy provision for it, to survey certain wider horizons which have come into view. In 1933 Hopkins (4), in his presidential address to the British Association, after briefly referring to the sequence of chemical events (as understood at that time) which led up to the mechanical response, continued:

It may be noted as an illustration of the unity of life that the processes which occur in the living yeast cell in its dealings with sugars are closely similar to those which proceed in living muscle. In the earlier stages they are identical and we know now where they part company...I have chosen the case of muscle, and it must serve as my only example, but many such related and ordered reactions have been studied in other tissues, from bacteria to the brain. Some prove general, some more special. Although we are far from possessing a complete picture in any one case we are beginning in thought to fit not a few pieces together. We are on a line safe for progress.

And in conversation he was wont to emphasise the outstanding suitability of muscle as the material for studies on energy relationships, just because in this tissue it was easier to make quantitative measurements of the energy changes involved; he was confident that the relevance of such results to the behaviour of other tissues would in time emerge.

This was also the opinion of Lipmann and of Engelhardt. In 1941 Lipmann (2) wrote: 'More and more clearly it appears that in all cells a tendency exists to convert the major part of available oxidation–reduction energy into phosphate bond energy.' Engelhardt (2) in 1946 considered that he spoke for most research workers in the field when he described ATP as 'the ultimate bearer of chemical energy for the physiological functions of the cell, not restricted to muscle'. He went on to say that at the time evidence was almost completely lacking. We wish now very briefly to indicate some of the knowledge which has accumulated in the last 25 years, concerned with the presence of ATP and ATPase in very diverse types of cell showing motility and indeed other types of energy utilisation (e.g. the production of electric current and osmotic work). Moreover it even appears that in some cases the involvement of proteins analogous to myosin and actin can be discerned.

MOTILITY AND LOCOMOTION

PLASMODIA; AMOEBOID PSEUDOPODIA; CELL MOVEMENTS. The movements concerned here are the flow of cytoplasm (often accompanied by sol/gel and gel/sol transformations) and the change of cell shape, including extension and retraction of pseudopodia. Migration of the cell can result.

Loewy (2) in 1952 was the first to prepare an actomyosin-like protein from a myxomycete – *Physarum polycephalum*. This he did by extracting with 1.2 M KCl/0.1 M K_2HPO_4. On addition of ATP the centrifuged extract showed decrease in viscosity, followed by a slow rise assumed to be due to ATPase activity. Adenylic acid addition also caused a rise in viscosity, reversed by ATP. These nucleotide effects showed considerable specificity. It was suggested that, with a steady supply of ATP, a protein system with these properties would be able to undergo cyclical alterations in structure which would play their part in production of mechanical work.

A few years later Ts'o and his collaborators,[1] using 1.4 M KCl, extracted from *Physarum* a protein of similar properties, including the viscosity change with AMP. The recovery of viscosity after the fall caused by ATP addition was accompanied by parallel release of phosphate. They associated these changes in viscosity with the increased streaming seen in the intact plasmodia on ATP injection and the conversion of gel to sol at the site of injection. The protein extracted was purified by fractional precipitation with $(NH_4)_2SO_4$ and the resulting 'myxomyosin' had enhanced ATPase sensitivity. Viscosity, sedimentation, birefringence and electron-microscope studies showed that it consisted of rigid rods 4000–5000 Å long and about 70 Å wide. There was no change in size or shape of the particles when ATP was added, and the viscosity effect was explained in terms of diminution of interaction between the molecules. Thus this material seems to show some considerable differences from muscle actomyosin. However in 1962 Nakajima (1), extracting plasmodia with Weber–Edsall solution and purifying by means of dilution precipitation, obtained a protein more analogous to the muscle actomyosin. There was no effect of increased viscosity with AMP, and there were many similarities with skeletal actomyosin in superprecipitation and in the enzyme kinetics of the ATPase.

Just about this time also Hatano and his collaborators isolated an actin

[1] Ts'o, Bonner, Eggman & Vinograd (1); Ts'o, Eggman & Vinograd (1, 2).

from acetone-dried *Physarum polycephalum*.[1] This actin existed in two forms – the G-form, with molecular weight of 57000 and sedimentation coefficient of 3.3S, combining with 1 mole of ATP/mole, and being converted into the F-form (30S) on addition of monovalent cations. The F-actin microfilament was 60 Å in diameter and consisted of two strands of helical structure with a half-pitch of 290 Å. The amino-acid composition of the material was very similar to that of skeletal-muscle actin. In presence of Mg ions a second polymer (Mg-polymer) was formed which appeared in the electron microscope as globular aggregates 100–600 Å in diameter; its viscosity was comparatively low, and it seemed to have ATPase activity. F- and Mg-actin each depolymerised to give G-actin, which could be polymerised again to either and the two polymers could be transformed directly one into the other without passing through the G-form. It was suggested that the change might be regulated by divalent ions and might be associated with sol–gel transformation in the protoplasm. Electron micrographs showed bundles of filaments, 70 Å in diameter, in the protoplasmic gel layer; these would seem to be actin, and the question was asked (as with vertebrate smooth muscle) 'Where is the myosin?'

Hatano & Tazawa (1) went on to prepare purified actomyosin from *Physarum* plasmodia by extraction with 0.4M KCl followed by two dilution precipitations. The actomyosin obtained had ATPase activity and ATP sensitivity as high as those of preparations from skeletal muscle. Myosin was isolated from the actomyosin by centrifuging in presence of ATP and Mg ions; it was soluble at low ionic strength (0.03M) and its existence as the dimer form in such solutions was suggested by the single peak at $S = 12.7$ on centrifugation. Electron micrographs of actomyosin in presence of ATP and Mg ions showed filaments of typical F-actin appearance. In absence of ATP the filaments were of greater diameter (100 Å) and their surface was covered by adhering granular particles, considered to be myosin; in the plasmodia *in vivo* also electron-dense material could be seen on and between the microfilaments. Such microfilaments have been shown by Wohlfarth-Bottermann with histochemical methods to have ATPase activity.[2] It thus seems likely that, according to conditions in the protoplasm, the small myosin aggregates may be free or combined with actin filaments. Wohlfarth-Bottermann has also observed that the filaments themselves may appear and disappear regularly with change in direction of streaming. Here interchange between F- and Mg-actin may be concerned. It seems possible that the myxomyosin filaments seen by Ts'o and his collaborators may consist mainly of actin filaments from which the soluble myosin has been washed away. The filaments seen by Wohlfarth-Bottermann (1) in some

[1] Hatano & Oosawa (1); Hatano, Totsuka & Oosawa (1).
[2] Quoted by Hatano & Tazawa (1).

experiments increased greatly in number on addition of ATP and changed their shape.

Kamiya and his collaborators have described experiments in which a thin strand of *Physarum* plasmodium, suspended vertically, after some time began to perform oscillatory contractions and twists.[1] In further preliminary experiments glycerinated preparations of such strips showed oscillatory twisting motion when treated with 10^{-3}M ATP and 10^{-7} Ca ions. It seems that some feedback mechanism may be operative here, as e.g. in glycerinated insect flight muscle.[2]

Theories put forward over the last forty-five years attempting to explain the mechanism of amoeboid movement have been described and considered by Wolpert (1).[3] Several observers have associated ATP with increased mobility or protoplasmic streaming. For example Goldacre & Lorch (1) in 1950 found that injection of ATP could increase the rate and force of movement; whether the ATP was acting as an energy source or as a plasticiser is not known. Thompson & Wolpert (1) attacked the problem in a different way, making a preparation of cytoplasm from *Amoeba proteus* by means of differential centrifugation. Motility in this cytoplasm could be judged by the behaviour of the granules contained in it; addition of ATP led to their active streaming which gradually ceased as gel formation took place, followed by contraction. Regions of partially-oriented fibres could be seen. Simard-Duquesne & Couillard (1) in 1962 extracted an ATPase from amoeba packed by centrifugation and ground with 0.6M KCl; much of the ATPase was water-soluble and activated only by Mg, but a Ca-activated enzyme could be demonstrated soluble only at higher ionic strength, and of activity comparable to that of skeletal actomyosin. It was not purified further.

Hoffmann-Berling, a pioneer in preparation of glycerinated non-muscle tissues, studied many examples of such material during the nineteen-fifties.[4] With epithelial or mesenchymatous cells in tissue culture, for instance, treated in this way (1), ATP caused contraction in conditions similar to those needed for contraction of glycerinated muscle. The same was true for the glycerinated amoebae used later by Simard-Duquesne & Couillard (2). Moreover, in tissue cultures, strips could be prepared from glycerinated hen embryos and Jensen tumour tissue; when these were attached to a tensimeter, ATP evoked tension and subsequent addition of Salyrgan caused relaxation.[5]

FLAGELLA; CILIA; CONTRACTILE TAILS OF BACTERIOPHAGE. The idea of the association of ATP with sperm motility was first voiced in 1945.

---

[1] Kamiya (1).   [2] Ch. 22.   [3] Also Rinaldi & Baker (1).
[4] See Hoffmann-Berling (2) for a review.   [5] Hoffmann-Berling (7).

Mann (1) observed that decrease in ATP content, whether under aerobic or anaerobic conditions, was associated with loss of motility of surviving mammalian sperm, and Engelhardt (2) reported ATPase activity in such sperm.[1] Lardy, Hansen & Phillips (1) also described phosphate esterification in spermatozoa both aerobically and anaerobically; the concentration of ATP was about the same as in muscle. Nelson (1, 2) further showed the interesting distribution of ATPase in bull sperm, the tail containing about three times as much as the midpiece, while practically none was present in the head. Engelhardt & Burnasheva (1) found the ATPase exclusively in the tails, and they prepared from this region a protein with ATPase activity which they called 'spermosin'. More recently Burnasheva & Karansheva (1) have studied the close connection between ATP breakdown and the motility of the infusoria *Tetrahymena*; while Brokaw (1) has found increased ATP dephosphorylation with increased beat frequency of sperm flagella. Reference to several other papers is made in the review by Nelson (3) in 1962. Investigations into the phosphagen and phosphokinase content of sperm have led to a rather confused picture. For example, White & Griffiths (1) could find neither in spermatozoa of the bull, rabbit or ram; but both phosphocreatine and creatine kinase were present in guinea-pig testis. Lehmann & Griffiths (1) have remarked on the extremely high creatine kinase activity of human semen. The presence of this enzyme in echinoderm sperm is well attested,[2] and Yanagisawa (2) followed the changes in phosphocreatine content associated with activity and recovery of echinoderm sperm.

Hoffman-Berling (3) used the particularly large single sperm flagella of tropical locusts after glycerination. Addition of ATP and Mg led to rhythmical activity, and the preparation reacted to various changed conditions in a manner very similar to the behaviour of muscle fibres. An interesting point was the effect of a number of agents – pyrophosphate, urea, increased ionic strength – in increasing the frequency of beat, provided a concentration of ATP at least 10^{-5}M was also present. This can be interpreted to mean that ATP is essential as energy source while the other agents can take the place of higher ATP concentrations in increasing the plasticity of the contractile elements. Hoffman-Berling (2) has emphasised that, since one cannot tell from looking at a motionless flagellum whether it is fully contracted or relaxed, it is not possible to say whether ATP energises the relaxation or the contraction stage. Machin (1) has concluded, from calculation based on observations of Gray (1), that the contractile elements must be distributed along the flagellum; he compared the probable activation of the rhythmical

---

[1] Engelhardt cited unpublished work of Burnasheva, and his own work in *Bull. Acad. Sci. S.S.R.S.* **182**, 195, 1945.
[2] Ch. 22.

waves by local bending with the delayed elastic effect in insect fibrillar muscle.[1]

Much has been done on the proteins concerned in flagellate and ciliary movement.

In 1958 Burnasheva (1) further described spermosin;[2] this protein made up about 20 % of sperm nitrogen, had ATPase activity and was able to complex with skeletal-muscle actin. The complex, of increased viscosity, was dissociated by ATP. Nelson (3) in 1954 had found some immunological indications of the presence of a myosin-like protein in the outer longitudinal fibres of rat sperm flagella.

Gibbons[3] (1) in 1963 studied the isolated cilia of *Tetrahymena pyriformis*. Very little protein could be extracted unless the membrane enclosing them was first broken down by digitonin treatment; the cilia then became gradually soluble in 0.6 M KCl. A fraction was obtained (soluble in mM tris/ mM EDTA at pH 8.3) which contained nearly all the ATPase activity and constituted about 30 % of the cilia protein. The specific ATPase activity was comparable with that of muscle actomyosin. This fraction, named dynein, contained two components having sedimentation coefficients of 30 S and 14 S, the latter having nearly three times the ATPase activity of the former. Electron-microscope studies showed the 30 S fraction to be made up of rodlike particles, up to 4000 Å in length, often with a repeating globular structure of 140 Å. Fraction 14 S consisted of particles, about 85 by 150 Å. It was suggested that conversion of one form to the other might occur, and that ATPase activity might be modulated in this way during bending movements.

In 1968 an actin-like protein was obtained by Renaud, Rowe & Gibbons (1) in the aqueous extract of acetoned cilia of *Tetrahymena* and flagella of sea-urchin sperm. This protein had amino-acid composition resembling that of muscle actin, the differences being no greater than those occurring between muscle actins from different species. After reduction with mercaptoethanol and alkylation with iodoacetate in 8 M urea it migrated as a single band on electrophoresis in polyacrylamide gel, with a velocity the same as that of similarly treated muscle actin. A guanine nucleotide – and no adenine nucleotide – was contained in it, one mole/mole of protein subunit.[4] On polymerisation the classical type of double helix was not given. Stephens (3, 4) found this protein from sea-urchin sperm flagella to consist of globular subunits of molecular weight about 55 000. If dissociation of the

---

[1] P. 538.
[2] Also Burnasheva, Efremenko & Lyubimova (1); Burnasheva, Efremenko, Chumakova & Zueva (1).
[3] Also Gibbons & Rowe (1); Gibbons (2).
[4] Stevens, Renaud & Gibbons (1); Shelanski & Taylor (1) prepared a similar protein from sea-urchin flagella.

protein was carried out by very mild treatment (with the detergent Sarcosyl) subsequent dilution and addition of salt led to re-association into the fibrous form. This process was aided by the presence of GTP, but hydrolysis of the triphosphate during polymerisation has not yet been demonstrated.

The formation of arrowheads on application of heavy meromyosin to sectioned material was shown by Ishikawa, Bischoff & Holtzer (1) for a number of cell types, and its use as a tracer for actin-like filaments was suggested. A non-specific protein-protein interaction seemed to be counter-indicated, since no arrowheads were formed with e.g. collagen, elastic fibres, mitochondria, membranes of the endoplasmic reticulum. However it is possible that conformational changes could prevent interaction with HMM if the monomers, though the same, were differently packed. Holtzer had earlier found that actin itself was immunologically distinct from the material of the mitotic apparatus.[1] Mazia & Ruby (1) who prepared an actin-like protein from erythrocyte membranes, coined the name tektin for a class of similar but not identical actin-like proteins adapted to assembly of a variety of cell structures.

Young & Nelson (1) also at this time prepared from mammalian sperm tails a protein which they called flactospermin. This was achieved by the use of Weber–Edsall solution in the presence of digitonin to solubilise bonds hindering extraction, and the protein was purified by repeated dilution precipitation. On addition of ATP and Mg ions the solution showed a viscosity drop, but considerably less than that given by actomyosin. At low ionic strength superprecipitation could be obtained. By extraction of tails in the absence of digitonin a myosin-like protein (spermosin) was prepared; and by extraction with KI according to the method of A. G. Szent-Györgyi (1) an actin-like protein (flactin). Spermosin interacted with skeletal actin, and flactin (which contained stoichoimetric amounts of bound nucleotide) with skeletal myosin. Thus the presence was demonstrated of a complex deformable in presence of ATP and Mg ions. Nelson considered the Machin model preferable here to the sliding model, recalling that only 3 % shortening is needed to bring about the radius of curvature of the sinusoidal wave.

Another organism in which presence of a contractile protein and ATPase activity have been shown is the bacteriophage T2. Kozloff & Lute (1) showed in electron-microscope studies that the tail of the phage contracted upon contact with the wall of the cell to be invaded and relaxed upon addition of EDTA or of ATP (0.01 M). By the use of highly purified phage particles labelled with ^{32}P, the ATP content could be assessed and it could be demonstrated that the contraction was associated with ATP breakdown. Nevertheless the hydrolysis seems not to be concerned with energy provision but with removal of the ATP: for it was found that, after certain

[1] Cf. Forsheit & Hayashi (1) below.

treatments, tails were obtained completely free from ATP, but still able to contract. This was observed also by Wahl & Kozloff (1) for other similar bacteriophages (T4 and T6).[1] These workers concluded that contraction involved interaction between protein units, and that the function of the ATP was to keep the protein structure extended, its removal being necessary before contraction could take place. The energy source remains unknown.

A different situation is found with bacterial flagella, where only one protein seems to be present and where ATP breakdown and ATP-induced contraction have never been satisfactorily demonstrated.[2] As early as 1948 Weibull (1) brought evidence that flagella of *Proteus vulgaris*, separated from the cells by mechanical shaking followed by repeated centrifugation, and finally disintegrated in acid medium (pH 3.0–4.0), contained only one protein. In the ultracentrifuge only one protein appeared, of molecular weight 41000, $S_0 = 2.4$; it was homogeneous and contained all or most of the flagellar substance. The electron microscope showed that it consisted of particles of various forms. This flagellin has been much studied;[3] the evidence of Weibull that it is a single protein has recently been confirmed and extended by Martinez, Brown & Glazer (1). That it can exist also in a monomer form (M.W. 20000), depending on ionic strength, pH and temperature was found by Erlander, Koeffler & Foster (1). Other interesting papers concern re-aggregation. Thus Abram & Koeffler (1), using flagellin prepared from *Bacillus pumilus* in a similar way to that described above, found that re-aggregation could take place under a variety of conditions; with correct adjustments filaments closely resembling flagella were obtained. Also Asakura and his colleagues[4] studied this question with *Salmonella* protein; under physiological conditions of pH and ionic strength the presence of flagellar fragments as 'seeds' was necessary; growth was unidirectional, from one end only.

Already in 1955 Astbury, Beighton & Weibull (1) described the X-ray diffraction diagram of dried films of flagella. They made the significant finding of a 410 Å periodicity (similar to that in the actin-containing filaments of muscle). They termed the flagella 'monomolecular muscles' and stressed the probable importance of an actin-like protein. More recently Lowy & Hanson (2)[5] have made a detailed electron-microscope study of flagella from a number of bacteria, and have prepared models to express the structure (differing somewhat from one species to another), made up of helically arranged globules similar in size to those of G-actin. They dis-

[1] Also Sarkar, Sarkar & Kozloff (1).   [2] See Newton & Kerridge (1).
[3] Newton & Kerridge (1) and Lowy & Spencer (1) for references.
[4] Asakura, Eguchi & Iino, where references to earlier work are given.
[5] Also Lowy & Spencer (1).

cussed the meaning of the 410 Å periodicity in the light of newer knowledge about the *I* filaments of muscle,[1] with the suggestion of its possible
dependence upon the presence of a second component.[2] Astbury, Beighton
& Weibull discussed the helical undulation of bacterial flagella, and took up
the suggestion (made some fifty years earlier by Reichert) that a helical line
of contraction, continuously displaced along its length, would be needed.
Klug (1) has recently described a model consisting of the protein sub-units
packed into a helical array, but in which two sets of slightly incompatible
bonds operate at different radii in the particles. This coiled-coil structure,
by progressive re-arrangement of the bond–strain pattern and of the
relative displacement of inner and outer parts of the sub-units, could
propagate helical waves in absence of energy supply along its length.
Energy would be generated in the basal body. Doetsch & Hageage have also
discussed this function of the flagellar 'basal bulb' and point out that,
according to the results of Barlow & Blum (1) and of Enomoto (1) flagella
themselves (from *Proteus* and *Salmonella*) contain neither ATP nor ATPase.
Barlow & Blum could find no evidence for their contraction in presence of
ATP.

Still another variation on contraction phenomena is that illustrated by
the motile stalks of the ciliate *Vorticella* (see fig. 111 pl.). Hoffmann-
Berling (6) used stalks separated from cells which had been killed by saponin
treatment. In these, contraction was caused when Ca ions alone were added
(or certain other divalent cations); relaxation was caused by ATP. With
both ATP and Ca present spontaneous, rhythmical movements were seen.
The relaxation, but not the contraction, was inhibited by Salyrgan. It thus
appeared that contraction depended on blocking of excess negative charges
on the contractile protein. The function of the ATP would be to restore
these charges and by its breakdown (assumed but not actually shown) to
provide energy for relaxation. This mechanism is reminiscent of a model
much favoured at one time for muscle contraction. The results of Hoffmann-
Berling have recently been confirmed by Seravin (1), who used cells of the
colonial vorticellid *Carchesium polypinum*, killed by freezing or by heating
to 50°. The stalks did not contract on mechanical stimulation but did
respond to Ca with contraction and to ATP with relaxation, giving rhythmical activity at a certain ratio of the two substances. The important point
was shown histochemically that ATPase activity was as great as in the
living stalks. Townes & Brown (1) have recently studied the complex effects
of pH changes on the reactions of the glycerinated stalk to varying ATP,
Ca and Mg concentrations. They concluded that these three ions react with

---

[1] Also Lowy & Spencer (1).
[2] For a comprehensive review of work on bacterial flagella see Doetsch & Hageage
(1).

the contractile protein, forming a complex which is relaxed at pH 6.8. Addition of $3\,\mu\text{M}$ Ca or raising the pH to 7.0 caused the change in the complex necessary for maximal contraction. The conditions for relaxation with ATP and Mg were studied as well as the special conditions under which their addition could give maximal contraction. ATPase activity was not discussed in this work.

CONTRACTILE HAPPENINGS IN ORGANELLES. The capacity of mito-chondria to undergo swelling and to contract under various conditions on addition of ATP has attracted much interest from biochemists and the possible presence of an actomyosin-like protein has been investigated. For instance Ohnishi & Ohnishi (1) in 1962 extracted from rat-liver mito-chondria by classical methods proteins having some properties resembling those of myosin and actin. Thus on adding ATP to the solution of a mixture of these proteins a drop in viscosity was seen, and extrusion of such a solu-tion into a medium of low ionic strength gave filaments which would shorten with ATP. Ohnishi (1) suggested that these proteins might play a part in the active transport of cations. Vignais and his colleagues[1] however could only partly confirm these findings. In further work with mitochondria which had lost the capacity to contract (through ageing or extraction with $0.6\,\text{M}$ KCl) they observed that this capacity could be restored by addition of either the 'actomyosin-like' fraction or of the supernatant from this frac-tion; but this activity was lost on extraction of lipid from the two fractions. It was then found that phosphatidyl inositol was able to restore the con-tractile ability, and presumably was responsible for the effect of the protein fractions extracted from fresh mitochondria.[2]

Conover & Bárány (1) in a detailed study in 1966 could find no evidence for the presence of an actomyosin-like protein in liver mitochondria, and considered that the fraction described earlier could arise from contamina-tion with vascular tissue. Uchida, Mommaerts & Meretsky (1) also empha-sised the great difficulty they had experienced in obtaining from muscle sarcotubular preparations free from myosin; such preparations were how-ever obtained and functioned as well or better in Ca transport in the absence of myosin. If some contractile protein participates in mitochondrial behaviour, it is so far unidentified.

The mode of participation of platelets (non-nucleated fragments of cytoplasm produced during the maturation of megakaryocytes) in forming 'haemostatic plugs' led to the idea that they contained a contractile protein and an energy source. A good account, with many references, of their function in the complicated processes of haemostasis is given by Bettex-Galland & Lüscher (1); the observations of Sokal (quoted there)

---

[1] Vignais, Vignais, Rossi & Lehninger (1).   [2] Vignais, Vignais & Lehninger (1).

have shown the part played in the formation of the dense aggregate by active contraction of protusions put out from them. Born (2) in 1956 showed that platelets contained a high ATP concentration and that during plasma clotting this ATP disappeared rapidly. In 1959 Bettex-Galland & Lüscher (2) prepared from them, by extraction with 0.6 M KCl, an actomyosin-like protein which they called thrombosthenin.[1] This showed superprecipitation and ATP sensitivity; threads prepared from it by the method of Szent-Györgyi contracted on addition of ATP. It had ATPase activity, the release of only one phosphate group being shown chromatographically; this activity was very low – only 1-2 % of that with muscle actomyosin. Thrombosthenin myosin was obtained by treatment of the protein at low ionic strength with the dissociating agent polyethensulphonate (10^{-6} M) in presence of 10^{-3} M ATP. The actin analogue was prepared from acetone-dried powder of the washed platelets. It is interesting that thrombosthenin actin or muscle actin had effects on the thrombosthenin myosin ATPase activity analogous to those seen on addition of muscle actin to muscle myosin (though on a much lower scale). Thrombosthenin actin could also affect the ATPase activity of muscle myosin.

As we have seen, skeletal and smooth-muscle myosins are immunologically distinct from one another.[2] A very interesting observation was made by Gröschel-Stewart (1) when she found that antisera against human smooth-muscle myosin also reacted with thrombosthenin, while negative results were obtained with antisera against skeletal-muscle myosin. Earlier results of Finck & Holtzer (1), negative when antisera against chicken skeletal-muscle myosin and actin were tested on chicken ciliated epithelium and sperm, fit in with this differentiation of the myosins.

Next there is the fascinating problem of cell-division. Hoffmann-Berling (2) and Roberts (1) have reviewed our knowledge of the phenomena of mitosis, and the many theoretical suggestions for explanation of the mechanisms concerned in the various phases. Considerable light has been shed here by the work of Hoffmann-Berling (1, 4, 5) on glycerinated fibroblast cells in anaphase and telophase (see fig. 115 pl.). He points out (2) that fibres, visible in fixed cells and indicated in living cells by double refraction of the spindle, connect the chromosomes to the spindle poles. The decrease in double refraction as the chromosomes move towards the poles suggests contraction of these fibres. On the other hand, the elongation of the spindle in the later part of anaphase was brought about in glycerinated cells by application of ATP at concentrations high enough to inhibit contraction and increase plasticity, or by other conditions (increased ionic strength, addition of urea) which would have the same effect.

---

[1] The properties of the protein are also described in further papers, e.g. Bettex-Galland & Lüscher (3); Bettex-Galland, Portzehl & Lüscher (1).   [2] Ch. 23.

The happenings at telophase – constriction at the cell equator with simultaneous cell elongation – are particularly interesting. The results of Hoffmann-Berling (5) in 1954 showed that the constriction in glycerinated cells was brought about by ATP, under conditions similar to those needed for muscle contraction. He suggested, from observations with varying ATP concentrations, that the cell poles were more sensitive to the relaxing effect of ATP, possibly because of localisation there of some ATPase inhibitory factor – perhaps resembling the Marsh-Bendall factor.[1]

Then in 1961 Mazia, Chaffee & Iverson (1) succeeded in isolating the mitotic apparatus of sea-urchin eggs from other cell constituents. A solution in 0.5 M KCl, after two hours centrifuging at 105000g to remove particles, showed the presence of an ATPase – about one-third as active as that of muscle actomyosin.

Hoffmann-Berling and his collaborators,[2] returning to these questions in 1964, observed that glycerinated fibroblasts in a state of contraction could be caused to relax by addition of particles from rabbit skeletal muscle in presence of ATP. This relaxation was prevented by addition of 10^{-4} M CaCl$_2$. They compared these findings with the effect in muscle of the particulate relaxing factor, by this time understood as the action of a calcium pump. They pictured that a relaxing system was active in the cell at the poles during telophase and absent in the equatorial area. Later Kinoshita & Yakaki (1) brought some evidence for this. They prepared an anti-serum by injecting relaxing granules from sea-urchin eggs into the rabbit and, after fluorescent labelling, the antibodies were applied to glycerinated fibroblasts. In the cell the material responding seemed to show directed migration during cleavage, migrating towards the centres of the mitotic apparatus at metaphase and anaphase, while the presumptive furrowing region became temporarily denuded of it.

Facts bearing on the actin-like nature of the spindle fibres are supplied by Forsheit & Hayashi (1). Colchicine which inhibits polymerisation of actin also inhibits (at far lower concentration) the polymerisation of the spindle fibres in the mitotic cell. It is suggested that the cell can accumulate colchicine. References are given for evidence on the similarity of action of several agents and conditions on actin and on the mitotic spindle.

Miki-Nomura & Oosawa (1) have recently made a thorough investigation of an actin-like protein obtained from sea-urchin eggs – its reactions with rabbit-muscle myosin to give an actomyosin of classical properties and its behaviour on polyacrylamide-gel electrophoresis to give two main bands identical with those of muscle G-actin. They speculate on its role in cell division, envisaging either a function of actin alone in a monomer–polymer

---

[1] Hoffmann-Berling (2).
[2] Kinoshita, Andoh & Hoffmann-Berling (1); Kinoshita & Hoffmann-Berling (1).

equilibrium; or a part played by an actomyosin in elongation of the spindle and contraction of the cell cortex.

MOVEMENT IN PLANTS. Yen and his collaborators[1] have recently described a detailed study of a protein resembling actomyosin, prepared from carefully isolated vascular bundles of higher plants (the main leaf veins of *Nicotiana*, the petioles of *Cucurbita*, the mid-ribs of *Hydrilla*

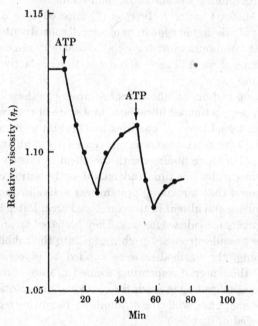

Fig. 116. Change in viscosity of the protein extracted from *Hydrilla verticillata* leaves after addition of ATP. (Yen & Shih (1).)

leaves). The material was frozen at $-10°$ and extracted with Weber–Edsall solution. The extract after centrifugation was concentrated by freeze-drying. Such extracts showed ATPase activity and ATP sensitivity (see fig. 116). The fall in viscosity with ATP was reversed, and the course of this reversal kept pace with liberation of inorganic phosphate. The viscosity response was markedly dependent upon the developmental state of the plants, being most striking at times of vigorous growth. Curiously enough, these extracts showed a slow increase in viscosity with AMP, behaviour reminiscent of that of some myxomyosin preparations. Further purification by ammonium sulphate precipitation and gel filtration gave a preparation

[1] Yen & Shih (1); Yen, Han & Shih (1).

showing a single symmetrical peak on electrophoresis; in presence of ATP a second peak appeared (see fig. 117). The ATPase activity of this preparation was greatly enhanced and was comparable to that of highly purified actomyosin preparations from skeletal muscle. Yen & Shih discussed the probable importance of this protein in the protoplasmic streaming observable especially in the phloem, the avenue of transportation of organic and inorganic solutes, in the vascular bundles.

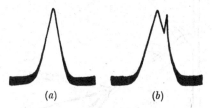

(a)   (b)

Fig. 117. Effect of ATP on the electrophoretic pattern of the plant contractile protein. (a) without ATP; (b) with ATP (1.5×10^{-2}M). (Yen, Han & Shih (1).)

Lyubimova and her collaborators[1] have found that macroscopic motion in plants – e.g. the movement of the leaf of *Mimosa pudica* – can be accompanied by a sharp drop in ATP content. The plant was anaesthetised with ether during removal of the leaves and these were rapidly frozen in liquid nitrogen. The ATP content of purified preparations was measured by the luciferin/luciferase method. It was most abundant in the folding regions of the leaf. After repeated stimulation to drooping it fell to about 30 % of the original, while return of the leaf to its original position was accompanied by restoration of the ATP (see fig. 118, a and b and c).

ACTIVE TRANSPORT; BIOLUMINESCENCE; ELECTRIC DISCHARGE

This survey of possible comparisons between the mechanisms of muscular activity and of movement in other types of cell is necessarily brief, and has been able to touch on only a small part of the experimental and theoretical effort devoted to this problem. The next one which presents itself is that of ATP utilisation in energy provision for work of other types, not involving any obvious movements. The part played by phosphate-bond energy in synthesis of biological materials has already been discussed.[2] It was in seeking the method of resynthesis of the acetylcholine hydrolysed at the nerve end-plates that Nachmansohn & Machado (1) in 1943 demonstrated an early instance of utilisation of the energy of ATP breakdown in bio-

---

[1] Lyubimova, Demyanovskaya, Pedorovich & Itomlenskite (1).
[2] P. 124.

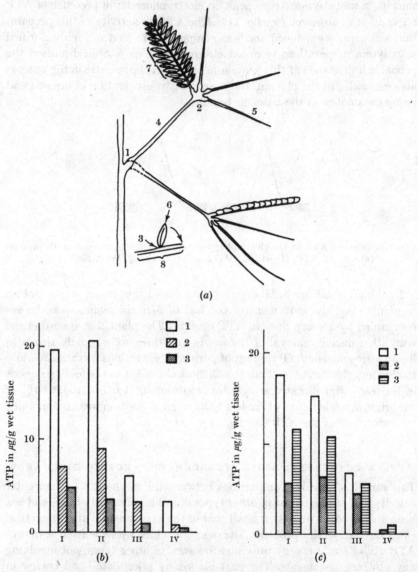

Fig. 118. (*a*) Scheme of the morphological components of the leaf. (1) Basal cushion; (2) secondary cushion; (3) little cushion; (4) basal pedicle; (5) secondary pedicle; (6) vein of the leaf platelet; (7) leaf platelets; (8) folded leaf platelets.

(*b*) Distribution of ATP in various morphological components of the leaf. (I) Basal cushion; (II) secondary cushion; (III) basal pedicle; (IV) leaf platelets. (1) Under ether anaesthesia; (2) cutting of leaf without ether; (3) exhaustion.

(*c*) Resynthesis of ATP in various morphological components of the leaf. I–IV as in fig. 117*b*. (1) Under ether narcosis; (2) exhaustion; (3) after recovery. (Lyubimova (1).)

synthesis in the manner envisaged by Lipmann (2). They found this synthesis to go on in extracts of brain (and of electric organ) in presence of Na acetate, choline cholride, eserine (to inhibit the esterase), fluoride (to inhibit the ATPase) and ATP. The mechanism of such acylations with the participation of co-enzyme A, was later elucidated[1] by work in many laboratories. Another such instance is in methylation reactions, e.g. of guanidinoacetic acid to give creatine in the liver, by means of L-methionine.[2] Pyrophosphate and inorganic phosphate are split off from the ATP and S-adenosyl-L-methionine is formed containing an energy-rich methyl sulphonium bond. The methyl group is then transferred, leaving S-adenosyl-L-homocysteine which is hydrolysed to give homocysteine and adenosine.

ACTIVE TRANSPORT. If one takes the case of active transport of solutes, especially ions, into and out of living cells, reviews such as those of Andersen & Ussing (1) and of Albers (1) show how widespread is the participation of ATP. An instance we have already encountered is the close correlation of ATP breakdown with the activity of the Ca pump in muscle. A particularly clear demonstration is also found in the work of Caldwell & Keynes (1) on the giant axons of the squid after cyanide poisoning; Na efflux was blocked by this treatment, but was restored by injection of ATP into the axons. Caldwell (2), using both ATP and argininephosphate injections, further showed that the change in Na efflux ran parallel to the changes in concentration of high-energy phosphate provided.[3] The mechanism by which the energy of the phosphate bond is utilised in such osmotic work is still a matter for discussion.[4] Roseman & his collaborators[5] in 1964 reported an unprecedented reaction in *Escherichia coli*, involving utilisation of phosphate from phosphopyruvate to form a hexose ester; intermediate formation of phosphohistidine combined in a heat-stable protein takes place:

Phosphoenolpyruvate + histidine–protein $\longrightarrow$ Phosphohistidine–protein + pyruvate
Phosphohistidine–protein + hexose $\longrightarrow$ hexose-6-phosphate + histidine–protein.

The two enzymes concerned have been purified, and it was ascertained that none of the nucleotide phosphates (or other compounds containing energy-rich phosphate bonds) tested could replace phosphoenolpyruvate. This system, concerned in the active transport of sugars, is then specific for phosphoenolpyruvate, but no doubt ATP is responsible for keeping up the supply of this compound.

BIOLUMINESCENCE. The phenomena of luminescence, both of inorganic material and of living things, have excited wonder and awe over many

[1] Ch. 16; also Nachmansohn (4).   [2] Cantoni (1, 2); Cantoni & Vignos (1).
[3] Also Caldwell, Hodgkin, Keynes & Shaw (1).
[4] Albers (1).   [5] Kundig, Ghosh & Roseman (1).

centuries, as recounted so well in E. N. Harvey's 'History of Luminescence' (1). Descriptions of careful observations are found in very early Chinese,[1] as well as Indian and Greek literature. Robert Boyle (2, 3) seems to have been the first, in 1667, to make what may be called chemical experiments on bioluminescence. He had specimens of 'shining wood and fish' and found that both lost their light when kept in a vacuum; luminosity was recovered, even after three days, on admission of air. The light of shining veal was quenched by rectified spirit of wine, but not by water. Subsequent work has confirmed Boyle's finding of the need for air. In a summary prepared by Harvey (2) we find that, out of 57 genera tested, 48 showed necessity for oxygen. It is striking that in only seven cases was ATP required, and in each oxygen was essential also. This requirement for ATP applies to the lampyrid fireflies, *Photinus* and *Photuris*, which have been intensively studied during the last twenty years.

Harvey (2) has summed up the essential chemistry of light production in the following words: 'Bioluminescence is a chemiluminescence in which the energy of a chemical reaction goes to excite some molecule whose electron is raised to a higher energy level. A quantum of light is emitted on return of the electron to its original ground level.' The molecule 'luciferin' undergoing the excitation is very different in different cases; e.g. reduced flavine mononucleotide plus an aldehyde are essential for its formation in bacterial luminescence;[2] while in the ostracod *Cypridina* a chromopolypeptide of cyclic nature is concerned.[3] The structure of firefly luciferin is given as:

$$\text{HO} \quad \overset{N}{\underset{S}{\diamond}} C - C \overset{N}{\underset{S-CH_2}{\diamond}} \overset{H}{\underset{}{C}} - COOH$$

and the sequence of reactions in presence of enzyme (luciferase), luciferin, ATP, Mg and oxygen seems to run as follows:[4]

$$E + ATP + LH_2 \rightleftharpoons E{-}LH_2AMP + PP \qquad\qquad I$$

$$E + LH_2AMP + O_2 \longrightarrow (E.\overset{O}{\overset{\|}{L}}.AMP)^* + H_2O \qquad\qquad II$$

$$(E.\overset{O}{\overset{\|}{L}}.AMP)^* \longrightarrow E.\overset{O}{\overset{\|}{L}}.AMP + light \qquad\qquad III$$

where E = enzyme; LH_2 = luciferin; PP = inorganic pyrophosphate; and * indicates the excited state.

Rhodes & McElroy (1) in 1958 found indication of the linking of the phosphate group of AMP to the carboxyl group of the luciferin. Thus the energy-

---

[1] Needham (1), vol. 4, pt. 1, pp. 72 ff.   [2] Cormier & Totter (1).   [3] Harvey (2).
[4] See McElroy & Seliger (1) as well as Cormier & Totter (1) for many references.

rich bond is retained and it is unlikely that ATP breakdown contributes energy for the light emission (as has often been assumed). This is indicated by the ready reversibility of Reaction I, and by the readiness with which co-enzyme A can replace AMP in the luciferin molecule. Seliger & McElroy (1) showed that for the peak emission of the bioluminescence at 562 mμ the quantum yield was one per mole of luciferin oxidised, which means an energy requirement of at least 57 k cal/mole. McElroy & Seliger (1) propose for the energy-yielding process (producing the excited intermediate) first addition of óxygen to the luciferyl adenylate to form an organic peroxide, followed by a dehydration process:

$$\begin{array}{c} \text{H} \\ | \\ \text{LH}_2 + \text{O}_2 \longrightarrow \text{L} - \text{O} - \text{O} - \text{H} \longrightarrow [\text{L}{=}\text{O}]^* + \text{H}_2\text{O} \\ [\text{L}{=}\text{O}]^* \longrightarrow \text{L}{=}\text{O} + \text{light} \end{array}$$

(the enzyme and adenylic acid are omitted in this scheme).

The adenyl group in the molecule seems to function in some unknown way to change the state of excitation of the luciferin.

ELECTRIC DISCHARGE. The production of electric current by certain fishes is another instance of the utilisation of energy for special biological purposes. It has been shown to involve ion movements similar to those encountered in muscle and nerve. In 1898 Biedermann (1)[1] gave a historical account of our knowledge of this phenomenon and described how such fish were from earliest times known and dreaded in the Mediterranean, as also in the Nile and other African rivers. Many observations were recorded from the seventeenth century onwards; indeed Volta in 1800 (1) defined his pile as an artificial electric organ. We may also recall the triumphant letter of Walsh (1) to Benjamin Franklin in 1773: 'It is with particular satisfaction that I make my first communication, that the effect of *Torpedo* appears to be absolutely electrical.' And he went on to make a close comparison with the Leyden Phial. Evidence (histological, anatomical and embryological) was also assembled by Biedermann to show the derivations of electric organs from muscle, an exception being *Malapterurus*. However, later work (e.g. of Johnels (1) who made a histological study of very young specimens) has tended to indicate that here also the origin is muscular not glandular. The persistence of the actomyosin seems to vary. Bailey (1) found no myosin in *Torpedo* and *Raia* extracts, while Citterio, Maldacea & Ranzi (1) later reported evidence for the extraction of actomyosin and actin in the case of *Torpedo*. They could suggest no physiological reason for their presence.[2]

[1] Vol. II, pp. 356–469.
[2] Also Arioso & Orlando (1) cited by Citterio *et al.* (1).

The offensive and defensive function of the electric discharge was early appreciated;[1] then in 1951 the work of Lissmann (1) showed that in certain genera (*Gymnarchus, Mormyrops*, and *Gymnotis*) the electric organ acts as a direction-finding device like radar, detecting changes in the pattern of flow in the surrounding water of current which it has itself generated.

The first enquiry into the chemical sources of energy for the electric discharge was made by Kisch (1) in 1930. He found a surprisingly high concentration of creatine in the electric organ of *Torpedo*, and showed that 'phosphagen' (phosphocreatine, as his experiments indicated) made up 77 % of the acid-soluble P present. On anaerobic activity of the organ the phosphagen content fell rapidly, to be restored during rest with normal air supply. A few years later Baldwin & Needham (1) and Baldwin (1) showed that a number of phosphate transfer reactions, familiar in muscle, took place in extracts of the electric organs of *Torpedo* and *Raia clavata* – from phosphoglyceric acid to adenylic acid and from phosphoglyceric acid to creatine in presence of adenylic acid. In 1943 Nachmansohn and his collaborators[2] compared the energy available from phosphocreatine breakdown and lactic acid formation as a result of discharge (32×10^{-6} and 16×10^{-6} g cal/gm tissue per impulse respectively) with the electrical energy dissipated – about 8.2×10^{-6} g cal/gm tissue per impulse. Thus ample energy was available.

It was not for another ten years that the origin and localization of the electromotive force was elucidated. The electroplates are single, multinucleated cells, arranged in columns; one side only is innervated. Bernstein (4) in 1912 suggested that at rest both faces (innervated and non-innervated) of each electroplate were polarised; and that the discharge resulted from a momentary disappearance of the potential across the innervated face. Until the advent of micro-electrodes it was not possible to test this hypothesis, but then in 1953 Keynes & Martins-Ferreira (1)[3] found with *Electrophorus* that it must be modified in the sense that the potential across the innervated face was actually reversed by 60 mV or more. They obtained evidence that the response of the innervated face was similar to that in the muscle fibre on spike production; the electroplate could be regarded as a modified muscle fibre, having a normal membrane on its nervous face while on the opposite side there was an inexcitable, specially adapted membrane with very low electrical resistance. The discharge would be triggered by release of acetylcholine at nerve endings on the nervous face. They concluded that the immediate source of electrical energy was probably exchange of sodium and potassium ions down their concentration gradients across the two faces of the electroplate (see fig. 119). An ionic pump would

[1] See Keynes (1) for references.
[2] Nachmansohn, Cox, Coates & Machedo (1).   [3] Also Keynes (1).

be needed for recovery. Evidence for this came with the preparation by Albers & Koval (1) in 1962 of a sodium- and potassium-activated ATPase from electric-organ tissue. Glynn (1) also obtained from this organ a microsomal fraction which contained a powerful ATPase needing Na and K as well as Mg ions, and which had other characteristics of the 'transport ATPase' of red blood cells. Bonting & Caravaggio (1) about the same time showed Na-K-ATPase activity in the non-innervated membrane, and found

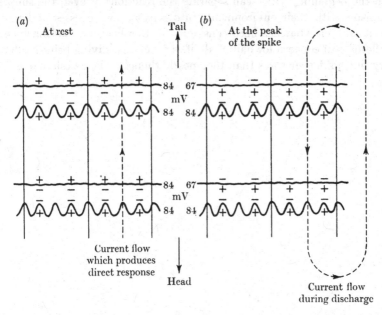

Fig. 119. Diagram illustrating the additive discharge of the electroplates. At rest (*a*) there is no net potential, but at the peak of the spike (*b*) all the potentials are in series, and the head of the eel becomes positive with respect to the tail. The figures are the overall averages obtained for electroplates in the organ of Sachs. (Keynes & Martins-Ferreira (1).)

that cation transport was associated with ATP hydrolysis, about three ions being transported per molecule of ATP used. Keynes & Aubert (1) followed the thermal changes resulting from the electric discharge in *Electrophorus* and *Malapterurus*. There was first a cooling effect, to be explained by entropy changes accompanying work done at the expense of the ionic gradients; then a heat output, interpreted as accompanying the ionic pump driven by $\sim$P energy.

The somewhat fragmentary account in this chapter must suffice to indicate the many possibilities which remain to be explored and the rich harvest that can be expected. One ends with the deep impression that muscle cells are only one special case of a general pattern of contractile proteins, and energy-providing machinery for them, which exists in quite varied forms in almost all manifestations of life. It must also be extremely ancient in evolutionary development, perhaps as much so as nuclear-genetic information storage and replication, for after all if cells were ever going to do as well as to be, mechanisms of the kind considered in this book were needed from the beginning. If we can separate the reactions of even the simplest organisms with their environments into sensory awarenesses and motor activities, then what has been considered in this book as the 'machine of the flesh' is of concern for half of all life's functions. Living beings differ in glory but, in deeper ways than the apostle thought, their flesh is one.

THE PERSPECTIVE SURVEYED

In the preparation of this book an attempt has been made to follow the lines of thought – like Ariadne's threads in a maze – leading to our present conception of the mechanism of muscle excitation and contraction, and of the energy provision for it.

It was only in the third century B.C. that Greek observations led to the realisation – rather dimly by Herophilus, quite definitely by Erasistratus – that the muscles were the organs of movement. Erasistratus even suggested a mechanism: the filling of the muscles with *pneuma* coming through the nerves, and their consequent increase in breadth and diminution in length. We have studied the variants of this theory, still the most important one, in the seventeenth century. We have seen the emphasis laid by Gabriel Fallopius in the early sixteenth century on the significance for movement of the muscle's fibrous structure. Francis Glisson also took a great step forward with his realisation that muscles had an intrinsic irritability independent of the nervous system.

The intensive work in the nineteenth century on the organic chemistry of muscle components and upon catalytic processes had little impact on theories of contraction mechanism; more influential in this direction were the studies of microscopists such as Bowman, Krause and Engelmann. This is apparent in the theories then developed by Pflüger and by Fick. In the first half of the present century we find a number of formulations in which contraction depended on electrostatic forces set up between the contractile protein and ions in its neighbourhood; here the production of hydrogen ions by lactic acid formation played an important part.

In all these theories there was no hint of a sliding mechanism. Indeed the first achievements of X-ray crystallography were interpreted in quite a different direction, the contraction of a muscle being analogised with the supercontraction of keratin, and thought of as a configurational change of fibre-molecules in accordion-pleating fashion. It was the finding of the second fibrillar protein actin, and some knowledge of its interaction with the long-known protein myosin as well as of the interaction of the complex with ATP, that were prerequisites for the discovery of the sliding mechanism. This began to emerge in 1951, and by 1953 stood forth (even before the two types of filament had actually been seen) on impressive evidence. It has continued to gather strength and is now almost universally accepted. If it was always evident that the intrinsic muscle mechanism would be found to

[599]

anticipate some form of human technique, it is satisfying to know now that the piston and cylinder was foreshadowed in it rather than levers and linkwork.

Concerning excitation, we have on the one hand all the work on the endoplasmic reticulum of the histologists before and after Veratti (ca. 1902) and of the subsequent electron microscopists; on the other hand the contributions of those interested in the ions needed for contraction, from Ringer (ca. 1883) onwards. These two lines of work came together only about 1960, in the understanding of the details of the segregation of calcium ions in the sarcoplasmic reticulum and their release on activation.

In turning to the third aspect, that of energy provision for muscle activity and regulation of this supply I may, as in private duty bound, stress the great contribution of Frederick Gowland Hopkins. He was the most consistent and influential advocate, during the early years of this century, of the thesis that 'in the study of the intermediate processes of metabolism we have to deal, not with complex substances which elude ordinary chemical methods, but with simple substances undergoing comprehensible reactions'. Again he said: 'I intend also to emphasise the fact that it is not alone with the separation and identification of products from the animal that our present studies deal; but with their reactions in the body; with the dynamic side of biochemical phenomena.' It has to be remembered that at the time he spoke these words (1913)[1] the idea of the unit of living matter as a very large and labile molecule was still widely current – although, as we have seen, the particular aspect of it concerned in lactic acid formation in muscle had been discarded:

Such assumptions became unnecessary as soon as we learnt that a stable substance may exhibit lability, not because it loses its chemical identity, and the chemical properties inherent in its own molecular structure, by being built into an unstable complex, but because in the cell it meets with agents (the intracellular enzymes) which catalyse certain reactions of which its molecule is normally incapable.

In a later address (in 1933) Hopkins spoke of discrimination among materials needed for the cell's maintenance as being in the first place determined by permeability relationships.[2] How he would have rejoiced if he could have known, for example, the facts recorded earlier in this book concerning the subtle interplay of enzymic interconversion and of enzyme activation and inhibition during the reaction series concerned in carbohydrate utilisation or resynthesis in muscle; or the suspected part played by new enzyme formation as a result of long-continued activity; or the role of permeability not only of cell membranes but of the membranes of the sarcoplasmic reticulum and the mitochondrion.

¹ Hopkins (1).   ² Hopkins (4).

In some contrast with this clear and forthright attitude of Hopkins we may mention the rather curious ideas held in the early nineteen thirties by some workers on fermentation and glycolysis. For example there were Meyerhof's views on the 'physiological' stability stages of hexose esters; and Harden's conception of a coupled reaction in which phosphate esterification of one sugar molecule could induce decomposition of another. To be sure, it is not obvious that this way of looking at things affected the brilliant progress of those days in the complex field of fermentation and glycolysis. These workers expressed themselves as believing in the essential similarity of chemical processes inside and outside the living cell. Thus Harden (2) wrote in 1930: 'I have assumed up to now that the processes in the living cell are essentially of the same kind as those which occur in the various preparations made from the dead cell, but differ from these mainly in the relative intensity of some of the reactions, and I know no valid argument against this assumption.' And Meyerhof (25) explained the absence of hexosephosphate accumulation in the intact yeast cell or in living muscle as 'probably because, as the result of an accurate co-ordination, the whole of the labile ester produced is fermented, and not only about one-half of it, as in the extracts'. Nevertheless, it does seem that the rather facile assumption made by Meyerhof and his collaborators of 'destabilisation' and 'restabilisation' of phosphocreatine as an explanation of certain biochemical facts which did not fit into the metabolic picture at the time, may have held up the understanding of the part played by this compound in energy relations.

Finally one must say a word about the energy-rich phosphate bond and its most important bearer, ATP. It has been suggested[1] that the initial appearance of this compound may have been the decisive event in the transition from the inanimate to the animate state. We have seen and pondered its ubiquity, and also, with reference to the mobilisation of its energy for locomotion, the great similarity in enzyme equipment for dealing with it found in the most diverse forms of living organisms. Very striking are two outstanding exceptions – the case of the utilisation of the phosphate bond energy of phosphopyruvate directly (without ATP intervention) to bring about the active transport of hexose; and the case of bioluminescence where ATP is usually not needed, and even in those types where it is needed its phosphate bond energy is not tapped. But broadly speaking, the insight of Hopkins has been abundantly justified. His conviction was, as we saw, that if only the energy machinery of muscle, with its readily measurable mechanical responses, could be understood, all other forms of cellular work would then be understood more easily, for the machinery would probably turn out to be the same.

[1] By H. F. Blum, quoted by Huennekens & Whiteley (1).

Looking back over the whole road which we have travelled, it is clear that muscle contraction is a phenomenon that was destined to be hid in deep darkness from the most earnest enquirers until long after the rise of modern science at the Renaissance. When our generation was beginning its work after the first world war, hardly anything more was understood of the intrinsic chemical mechanism than had been guessed by Paracelsus or imagined by Alfonso Borelli three or four centuries earlier. One could say that only a couple of stones of the arch over the abyss were laid in place: the recognition of the rodlike character of myosin particles, and the formation of lactic acid from the breakdown of carbohydrate. Now a great many of the stones are well and truly laid in their positions over the half-centering, yet still the keystone may be said to be lacking, the full explanation of exactly how the sliding occurs. We may believe with some confidence that this will not elude us for another twenty centuries.

REFERENCES

Note: *Pflügers Archiv. für gesante Physiologie* is abbreviated to *Arch. f. ges. Physiol.* and *Hoppe-Seyler Zeitschrift f. physiologische Chemie.* to *Z. physiol. Chem.*

Abbott, B. C. (1). The heat production associated with the maintenance of a prolonged contraction and the extra heat produced during large shortening. *J. Physiol.*, **112**, 438. 1951.

Abbott, B. C. & Aubert, X. M. (1). Changes of energy in a muscle during very slow stretches. *Proc. Roy. Soc.* B, **139**, 104. 1951–2.

Abbott, B. C., Aubert, X. M. & Hill, A. V. (1). The absorption of work by a muscle stretched during a single twitch or a short tetanus. *Proc. Roy. Soc.* B, **139**, 86. 1951–2.

Abbott, B. C. & Baskin, R. J. (1). Volume changes in frog muscle during contraction. *J. Physiol.*, **161**, 379. 1962.

Abbott, B. C. & Mommaerts, W. F. H. M. (1). A study of inotropic mechanisms in the papillary muscle preparation. *J. gen. Physiol.*, **42**, 533. 1959.

Abbott, R. H. & Chaplain, R. A. (1). Preparation and properties of the contractile element of insect fibrillar muscle. *J. Cell Sci.*, **1**, 311. 1966.

Abdullah, M., Taylor, P. M. & Whelan, W. J. (1). The enzymic debranching of glycogen and the role of transferase. In Ciba Foundation Symposium *Control of glycogen metabolism*, pp. 123–138. Ed. W. J. Whelan & M. P. Cameron. London, J. & A. Churchill. 1964.

Abdullah, M. & Whelan, W. J. (1). Enzymatic debranching of glycogen. A new pathway in rabbit muscle for the enzymatic debranching of glycogen. *Nature*, **197**, 979. 1963.

Abe, Y. & Tomita, T. (1). Cable properties of smooth muscle. *J. Physiol.*, **196**, 87. 1968.

Abel, K. (1). Die Lehre vom Blutkreislauf im Corpus Hippocraticum. *Hermes*, **86**, 192. 1958.

Abram, D. & Koeffler, H. (1). In vitro formation of flagella-like filaments and other structures from flagellin. *J. mol. Biol.*, **9**, 168. 1964.

Abrams, R. & Bentley, M. (1). Transformation of inosinic acid to adenylic and guanylic acids in a soluble enzyme system. *J. Amer. chem. Soc.*, **77**, 4179. 1955.

Acs, G., Biro, K. S. & Straub F. B. (1). The interaction between actomyosin and polyphosphates. *Hung. Acta. Physiol.*, **2**, 84. 1949.

Adams, F. (1) tr. *Genuine Works of Hippocrates*. Sydenham Soc. publ. no. 15. London. 1849.

Adelstein, R. S., Godfrey, J. E. & Kielley, W. W. (1). G-actin: preparation by gel filtration and evidence for a double-stranded structure. *Biophys. biochem. Res. Com.*, **12**, 34. 1963.

Adler, E. & Günther, G. (1). Über die Komponenten der Dehydrase-systeme. XX. Zur Kenntnis der enzymatischen Triosephosphorsäure Dehydrierung. *Z. f. physiol. Chem.*, **253**, 143. 1938.

Ahlgren, G. (1). Einwirkung der Muskulatur auf Methylenblau. *Skand. Arch. Physiol.*, **41**, 1. 1921.

Aita, M., Conti, G., Laszt, L. & Mandi, B. (1). Sur l'existence de deux sortes de cellules musculaires lisses dans la paroi carotidienne de bovidés. *Angiologica*, 5, 310. 1968.

Albers, R. W. (1). Biochemical aspects of active transport. *Ann. Rev. Biochem.*, 36, 727. 1967.

Albers, R. W. & Koval, G. J. (1). Properties of the sodium-dependent ATPase of *Electrophorus electricus*. *Life Sciences*, 1, 219. 1962.

Aldini, G. (1). Essai théoretique et expérimental sur le galvanisme. Paris, Fournier fils. 1804.

Allen, E. R. & Pepe, F. A. (1). Ultrastructure of developing muscle cells in the chick embryo. *Am. J. Anat.*, 116, 115. 1965.

Aloisi, M. (1). In discussion in *Myopathien*, p. 175, ed. R. Beckmann. Stuttgart, Georg Thieme. 1965.

Aloisi, M. & Margreth, A. (1). The question of glycolytic enzymes localisation in the skeletal muscle fiber and its bearing upon certain aspects of muscle diseases. *Excerpta Medica Foundation*, International Congress Series No. 147. *Exploratory concepts in muscular dystrophy and related disorders*, p. 305. 1967.

Alving, R. E. & Laki, K. (1). *N*-terminal sequence of actin. *Biochemistry*, 5, 2597. 1966.

Alving, R. E., Moczar, E. & Laki, K. (1). An *N*-acetylated peptide chain in tropomyosin. *Biochem. biophys. Res. Com.*, 23, 540. 1966.

Amberson, W., Dale Smith, R., Chinn, B., Himmelfarb, S. & Metcalf, J. On the source of birefringence within the striated muscle fibre. *Biol. Bull.*, 97, 231. 1949.

Andersen, B. & Ussing, H. H. (1). Active transport. In *Comparative Biochemistry*, ed. by M. Florkin & H. S. Mason. Vol. II, pp. 371–402. 1960.

Anderson, N. G. & Green, J. G. (1). The soluble phase of the cell. In *Enzyme Cytology*, p. 475. Ed. D. B. Roodyn. London & New York, Academic Press. 1967.

Andersson-Cedergren, E. (1). Ultrastructure of motor endplate and sarcoplasmic components of mouse skeletal muscle fiber as revealed by three-dimensional reconstructions from serial sections. *J. Ultrastr. Res. Suppl. I*, 1–191. 1951.

Andres, R., Cader, G. & Zierler, K. L. (1). The quantitatively minor role of carbohydrate in oxidative metabolism of skeletal muscle in intact man in the basal state. Measurements of oxygen and glucose uptake and carbon dioxide and lactate production in the forearm. *J. Clin. Invest.*, 35, 671. 1956.

Appleman, M. M., Birnbaumer, L., Belocopitow, E. & Torres, H. N. (1). Complementary factors in the regulation of glycogen synthetase and phosphorylase *b* kinase. *Fed. Proc.*, 24, 537. 1965.

Appleman, M. M., Birnbaumer, L. & Torres, H. N. (1). Factors affecting the activity of muscle glycogen synthetase. III. The reaction with adenosinetriphosphate, Mg^{2+}, and cyclic 3',5'-adenosine monophosphate. *Arch. Biochem. Biophys.*, 116, 39. 1966.

Arai, K-I & Watanabe, S. (1). A study of troponin, a myofibrillar protein from rabbit skeletal muscle. *J. biol. Chem.*, 243, 5670. 1968.

Ardenne, M. von & Weber, H. H. (1). Elektronmikroscopische Untersuchung des Muskeleiweisskörpers 'Myosin'. *Koll. Zeitsch.*, 97, 322. 1941.

Arioso, R. & Orlando, A. (1). Sulle proteine dell' organo ellectrico di Torpedine. *Rend. Acc. Naz. Lincie (Sc. Fiz.)*, 10, 422. 1951.

Aristotle (1). *De Partibus Animalium*. See Ogle.

Aristotle (2). *De Motu Animalium*. See Farquharson.

Armstrong, C. M., Huxley, A. F. & Julian, F. J. (1). Oscillatory responses in frog skeletal muscle fibres. *J. Physiol.*, 186, 26P. 1966.

Arnold, A. & Luck, J. M. (1). Studies on arginine. III. The arginine content of vertebrate and invertebrate muscle. *J. biol. Chem.*, **99**, 677. 1932–3.

Arnold, V. (1). Eine Farbereaktion von Eiweisskörpern mit Nitroprussidnatrium. *Z. phys. Chem.*, **70**, 300. 1910–11.

Asai, H. (1). Electric birefringence of rabbit tropomyosin. *J. Biochem.*, **50**, 182. 1961.

Asakura, S. (1). The interaction between G-actin and adenosine triphosphate. *Arch. Biochem. Biophys.*, **92**, 140. 1961.

Asakura, S. (2). F-actin adenosine triphosphatase activated under sonic vibration. *Biochim. biophys. Acta*, **52**, 65. 1961.

Asakura, S., Eguchi, E. & Iino, T. (1). Undirectional growth in *Salmonella* flagella *in vitro. J. mol. Biol.*, **35**, 227. 1968.

Asakura, S., Hotta, K., Imai, N., Ooi, T. & Oosawa, F. (1). G-F transformation of actin by divalent ions. Conference on the chemistry of muscular contraction, pp. 56–65. Osaka, Igaku Shoin. 1957.

Asakura, S. & Oosawa, F. (1). Dephosphorylation of adenosine triphosphate in actin solutions at low concentrations of magnesium. *Arch. Biochem. Biophys.*, **87**, 273. 1960.

Asakura, S., Taniguchi, M. & Oosawa, F. (1). Mechanochemical behaviour of F-actin. *J. mol. Biol.*, **7**, 55. 1963.

Ashford, T. P. & Porter, K. R. (1). Cytoplasmic components in hepatic cell lysosomes. *J. Cell. Biol.*, **12**, 198. 1962.

Ashhurst, D. E. (1). The fibrillar flight muscle of giant water-bugs: an electron-microscope study. *J. Cell. Sci.*, **2**, 435. 1967.

Ashhurst, D. E. (2). Z line of the flight muscle of belostomatid water-bugs. *J. mol. Biol.*, **27**, 385. 1967.

Ashhurst, D. E. & Luke, B. M. (1). Lipid inclusions in the flight muscles of belostomatid water-bugs. *Zeitsch. f. Zellforsch.*, **92**, 270. 1968.

Ashley, C. A., Avasimavicius, A. & Hass, G. M. (1). Chemical studies of isolated myofibrils in vitro. I. The role of magnesium, calcium and nitrogen-phosphorus partition in contraction induced with adenosinetriphosphate. *Exp. Cell. Research*, **10**, 1. 1956.

Ashley, C. A., Porter, R. K., Philpott, D. E. & Hass, G. M. (1). Observations by electron microscopy on contraction of skeletal myofibrils induced with adenosinetriphosphate. *J. exp. Med.*, **94**, 9. 1951.

Ashmarin, I. P. (1). Enzymic decomposition of adenosinetriphosphoric acid and contraction of actomyosin. *Biokhimiya*, **18**, 71. 1953.

Astbury, W. T. (1). X-ray studies on protein structure. *Cold Spring Harbour Symp. on Quant. Biol.*, **2**, 15. 1934.

Astbury, W. T. (2). X-ray studies of the structure of compounds of biological interest. *Ann. Rev. Biochem.*, **8**, 113. 1939.

Astbury, W. T. (3). Croonian Lecture. On the structure of biological fibres and the problem of muscle. *Proc. Roy. Soc.* B, **134**, 303. 1947.

Astbury, W. T. (4). X-ray study of actin. *Nature*, **160**, 388. 1947.

Astbury, W. T. (4a). Indications on muscle contraction from X-ray diffraction studies of the fibrous proteins. In *Contractile Polymers*, pp. 78–89. Oxford, Pergamon Press. 1960.

Astbury, W. T. (5). X-ray studies of muscle. *Proc. Roy. Soc.* B, **137**, 58. 1950.

Astbury, W. T. Beighton, E. & Weibull, C. (1). The structure of bacterial flagella. *Symp. Soc. exp. Biol.*, **9**, 282. 1955.

Astbury, W. T. & Bell, F. O. (1). Nature of the intramolecular fold in α-keratin and α-myosin. *Nature*, **147**, 696. 1941.

Astbury, W. T. & Dickinson, S. (1). α-β Intramolecular transformation of myosin. *Nature*, **135**, 95. 1935.

Astbury, W. T. & Dickinson, S. (2). α-β Transformation of muscle proteins in situ. *Nature*, **135**, 765. 1935.

Astbury, W. T. & Dickinson, S. (3). X-ray studies of the molecular structure of myosin. *Proc. Roy. Soc.* B, **129**, 307. 1940.

Astbury, W. T., Perry, S. V., Reed, R. & Spark, L. C. (1). An electron microscope and X-ray study of actin. *Biochim. biophys. Acta*, **1**, 379. 1947.

Astbury, W. T., Reed, R. & Spark, L. C. (1). An X-ray and electron microscope study of tropomyosin. *Biochem. J.*, **43**, 282. 1948.

Astbury, W. T. & Woods, H. J. (1). The X-ray interpretation of the structure and elastic properties of hair keratin. *Nature*, **126**, 913. 1930.

Astbury, W. T. & Woods, H. J. (2). X-ray studies of the structure of hair, wool and related fibres. II. The molecular structure and elastic properties of hair keratin. *Phil. Trans. Roy. Soc. Lond.*, A **232**, 333. 1933.

Aubel, E. & Simon, E. (1). Sur la fermentation lactique. *Compt. rend. Soc. Biol.*, **114**, 905. 1933.

Auber, J. & Couteaux, R. (1). L'attache des myofilaments secondaires au niveau de la strie Z dans les muscles de Diptères. *Comp. rend. heb. Acad. Sci.*, **254**, 3425. 1962.

Auber, J. & Couteaux, R. (2). Ultrastructure de la strie Z dans les muscles de Diptères. *Journ. de. Micr.*, **2**, 309, 1963.

Aubert, X. (1). A propos de l'inversion de l'effet Fenn. *Comp. rend. Soc. Biol.*, **140**, 571. 1946.

Avena, R. M. & Bowen, W. J. (1). Effects of carnosine and anserine on muscle adenosinetriphosphatase. *J. biol. Chem.* **244**, 1600. 1969.

Axelrod, A. E., Potter, V. R. & Elvehjem, C. A. (1). The succinoxidase system in riboflavin-deficient rats. *J. biol. Chem.*, **142**, 85. 1942.

Axelsson, J., Bueding, E. & Bülbring, E. (1). The inhibitory action of adrenaline on intestinal smooth muscle in relation to its action on phosphorylase activity. *J. Physiol.*, **156**, 357. 1961.

Axelsson, J. & Thesleff, S. (1). A study of supersensitivity in denervated mammalian muscle. *J. Physiol.*, **149**, 178. 1957.

Ayengar, P., Gibson, D. M. & Sanada, D. R. (1). A new coenzyme for phosphorylation. *Biochim. biophys. Acta*, **13**, 309. 1954.

Azzone, G. F. (1). The isolation and properties of γ-myosin from dystrophic and normal muscles. *Biochim. biophys. Acta*, **30**, 367. 1958.

Baeyer, E. von & Muralt, A. von (1). Lichtdurchlässigkeit und Tätigkeitsstoffwechsel des Muskels. *Arch. ges. Physiol.*, **234**, 233. 1934.

Baguet, F. & Gillis, J. M. (1). Energy cost of tonic contraction in a lamellibrauch catch muscle. *J. Physiol.*, **198**, 127. 1968.

Baguet, F. & Gillis, J. M. (2). The respiration of the anterior byssus retractor muscle of *Mytilus edulis* (ABRM) after a phasic contraction. *J. Physiol.*, **188**, 67. 1967.

Bailey, K. (1). Myosin and adenosinetriphosphatase. *Biochem. Journ.* **36**, 121. 1942.

Bailey, K. (2). Tropomyosin: a new asymmetrical protein component of the muscle fibril. *Biochem. J.*, **43**, 271. 1948.

Bailey, K. (3). Structure proteins II. Muscle. In *The Proteins*, ed. H. Neurath & K. Bailey, p. 951. 1954.

Bailey, K. (4). Some features of the amino acid composition of proteins. *Chem. and Industry*. p. 243. 1950.

Bailey, K. (5). End group assay in some proteins of the keratin-myosin group. *Biochem. J.*, 49, 23. 1951.

Bailey, K. (6). Tropomyosin: a new asymmetric protein component of muscle. *Nature*, 157, 368. 1946.

Bailey, K. (7). The proteins of adductor muscles. *Pubbl. Staz. Zool. Napoli*, 29, 96. 1956.

Bailey, K. (8). Invertebrate tropomyosin. *Biochem. J.*, 64, 9P. 1956.

Bailey, K. (9). Invertebrate tropomyosin. *Biochim. biophys. Acta*, 24, 612. 1957.

Bailey, K. (10). The proteins of electrical tissue. *Biochem. J.*, 33, 255. 1939.

Bailey, K. & Perry, S. V. (1). The role of sulphydryl groups in the interaction of myosin and actin. *Biochim. biophys. Acta*, 1, 506. 1947.

Bajusz, E. (1). Experimental myopathies: Influence of various factors as reflected by histochemical and morphologic studies. In *Myopathien*, p. 121, ed. R. Beckmann, Stuttgart, Georg Thieme, 1965.

Bajusz, E. (2). 'Red' skeletal muscle fibers: relative independence of neural control. *Science*, 145, 938. 1964.

Baldwin, E. (1). Biochemistry of the electric organs of *Raia clavata*. *Biochem. J.*, 32, 888. 1938.

Baldwin, E. & Needham, D. M. (1). A contribution to the comparative biochemistry of muscular and electrical tissue. *Proc. Roy. Soc. B*, 122, 197. 1937.

Baldwin, E. & Needham, D. M. (2). The phosphate distribution in resting fly muscle. *J. Physiol.*, 80, 221. 1933.

Baldwin, E. & Yudkin, W. H. (1). The annelid phosphagen: with a note on phosphagen in the Echinodermata and Protochordata. *Proc. Roy. Soc. B*, 136, 614. 1950.

Balenović, K. & Straub, F. B. (1). Über das Aktomyosin des Kaninchen muskels. *Studies Inst. Chem. Univ. Szeged.*, 2, 17. 1943.

Bálint, M., Szilágyi, L., Fekete, Gy., Blazsó, M. & Biró, N. A. (1). Studies on proteins and protein complexes of muscle by means of proteolysis. *J. mol. Biol.*, 37, 317. 1968.

Ball, E. G. (1). Über die Oxydation und Reduktion der drei Cytochrom-Komponenten. *Biochem. Z.*, 295, 262. 1937–38.

Ballantyne, J. W. (1) Rigor mortis in the foetus. With notes of a case. *Teratologia*, 2, 96. 1895.

Balzer, H., Makinose, M. & Hasselbach, W. (1). The inhibition of the sarcoplasmic calcium pump by phrenylamine, reserpine, chloropromazine and imipramime. *Arch. Pharmak. u. exp. Path.* 260, 444. 1968.

Bamann, E., Fischler, F. & Trapmann, H. (1). Verhalten und Spezifität von Cer, Lanthan, Eisen und Aluminium als Phosphatasemodelle gegenüber physiologisch wichtigen Phosphorsäureverbindungen wie Zuckerphosphorsäure, Adenylsauren, Adenosinetriphosphorsäure und anderen. *Biochem. Z.*, 325, 413. 1954.

Bang, O. (1). The lactate content of the blood during and after muscular exercise in man. *Skand. Arch. Physiol.*, 74, Suppl. 10, 51. 1936.

Banga, I. (1). The phosphatase activity of myosin. *Studies from the Institute of medical Chemistry, University of Szeged*, 1, 27. 1943.

Banga, I. & Szent-Györgyi, A. (1). Preparation and properties of myosin A and B. *Studies from the Institute of medical Chemistry, University of Szeged*, 1, 5. 1943.

Bär, U. & Blanchaer, M. C. (1). Glycogen and carbon dioxide production from glucose and lactate by red and white muscle. *Amer. J. Physiol.*, 209, 905. 1965.

Baranowski, T. (1). Crystalline glycerophosphate dehydrogenase from rabbit muscle. *J. biol. Chem.*, 180, 535. 1949.

Baranowski, T., Illingworth, B., Brown, D. H. & Cori, C. F. (1). The isolation of pyridoxal-5-phosphate from crystalline muscle phosphorylase. *Biochim. biophys. Acta*, 25, 16. 1957.

Bárány, E. H., Edman, K. A. P. & Palis, A. (1). The influence of electrolytes on the rate of viscosity drop in ATP-actomyosin mixtures. *Acta physiol. Scand.*, 24, 361. 1951.

Bárány, K. & Oppenheimer, H. Succinylated meromyosins. *Nature, Lond.*, 213, 626.

Bárány, M. (1). ATPase activity of myosin correlated with speed of muscle shortening. *J. gen. Physiol.*, 50, Pt. 2, 197. 1967.

Bárány, M. & Bárány, K. (1). Studies on 'active centres' of L-myosin. *Biochim. biophys. Acta*, 35, 293. 1959.

Bárány, M. & Bárány, K. (2). Myosin from the striated adductor muscle of the scallop, (*Pecten arradians*). *Biochem. Z.*, 345, 37. 1966.

Bárány, M., Bárány, K., Gaetjens, E. & Bailin, G. (1). Chicken gizzard myosin. *Arch. Biochem. Biophys.*, 113, 205. 1965.

Bárány, M., Bárány, K., Reckard, T. & Volpe, A. (1). Myosin of fast and slow muscles of rabbit. *Arch. Biochem. Biophys.*, 109, 185. 1965.

Bárány, M., Biró, N. A. & Molnár, J. (1). Über die Reaktion zwischen Aktin und zweiwertigen Kationen. *Acta physiol. Acad. Sci. Hung.*, 5, 63. 1954.

Bárány, M., Biró, N. A., Molnár, J. & Straub, F. B. (1). Darstellung enzymfreien Aktin durch Umfällung mit Magnesium. *Acta physiol. Acad. Sci. Hung.*, 5, 369. 1954.

Bárány, M. & Chrambach, A. (1). Exchange of bound calcium of actin. *J. gen. Physiol.*, 45, 589A. 1962.

Bárány, M. & Finkelman, F. (1). The lability of the F-actin bound calcium under ultrasonic vibration. *Biochim. biophys. Acta*, 63, 98. 1962.

Bárány, M. & Finkelman, F. (2). Exchange of F-actin-bound adenosine diphosphate. *Biochim. biophys. Acta*, 78, 175. 1963.

Bárány, M., Finkelman, F. & Therattil-Anthony, T. (1). Studies on the bound calcium of actin. *Arch. Biochem. Biophys.*, 98, 28. 1962.

Bárány, M., Gaetjens, E. & Bárány, K. (1). Myosin in hereditary muscular dystrophy of chickens. *Ann. N.Y. Acad. Sci.*, 138, 360. 1966.

Bárány, M., Gaetjens, E., Bárány, K. & Karp, E. (1). Comparative studies of rabbit cardiac and skeletal myosins. *Arch. Biochem. Biophys.*, 106, 280. 1964.

Bárány, M. & Jaisle, F. (1). Kontraktionszyklus und Interaktion zwischen Aktin und L-Myosin unter der Wirkung spezifischer Interaktions-Inhibitoren. *Biochim. biophys. Acta*, 41, 192. 1960.

Bárány, M., Koshland, D. E., Springhorn, S. S., Finkelman, F. & Therattil-Anthony, T. (1). Adenosine triphosphate cleavage during the G-actin to F-actin transformation and the binding of adenosine diphosphate to F-actin. *J. biol. Chem.*, 239, 1917. 1964.

Bárány, M., Nagy, B., Finkelman, F. & Chrambach, A. (1). Studies on the removal of the bound nucleotide of actin. *J. biol. Chem.*, 236, 2917. 1961.

Bárány, M., Spiró, J., Köteles, G. & Nagy, E. (1). Untersuchung der Aktin-Aktin-bindung. II. Die Schutzwirkung von ATP gegenüber depolymerisierenden Mitteln. *Acta physiol. Acad. Sci. Hung.*, **10**, 159. 1956.

Bárány, M., Spiró, J., Köteles, G. & Nagy, E. (2). Untersuchung der Aktin-Aktin-bindung. I. Die Rolle der SH and NH₂ Gruppen. *Acta physiol. Acad. Sci. Hung.*, **10**, 145. 1956.

Bárány, M., Tucci, A. F., Bárány, K., Volpe, A. & Reckard, T. (1). Myosin of newborn rabbits. *Arch. Biochem. Biophys.*, **111**, 727. 1965.

Barer, R. (1). The structure of the striated muscle fibre. *Biol. Rev.*, **23**, 159. 1948.

Barlow, G. H. & Blum, J. J. (1). On the contractibility of bacterial flagella. *Science*, **116**, 572. 1952.

Barrenscheen, H. K. & Filz, W. (1). Untersuchungen zur Frage der Co-Ferment-wirkung. II Mitteilung: Zur Chemie der Adenosinetriphosphorsäure. *Biochem. Z.*, **250**, 281. 1932.

Baskin, R. J. (1). The variation of muscle oxygen consumption with load. *J. Physiol.*, **181**, 270. 1965.

Baskin, R. J. (2). Relationship between changes in muscle volume and muscle action potential. *Nature*, **195**, 290. 1962.

Baskin, R. J. & Gaffin, S. (1). Oxygen consumption in frog sartorius muscle. I. The isometric twitch. *J. cell. comp. Physiol.*, **65**, 19. 1965.

Bass, A., Gutmann, E. & Vodička, Z. (1). The resynthesis of glycogen in muscle after stimulation. *Physiol. Bohemoslov.*, **4**, 267. 1955.

Bastholm, E. (1). The History of Muscle Physiology. *Acta Historica Scientiarum naturalium et medicinalium.* Vol. VII. Ed. by Bib. Univ. Hauniensis. Tr. W. E. Calvert. Copenhagen, Einar Munksgaard. 1950.

Batelli, F. & Stern, L. (1). L'oxydation de l'acide succinique comme mesure du pouvoir oxydant dans la respiration principale des tissus animaux. *Compt. rend. heb. Soc. Biol.*, **69**, 554. 1910.

Batelli, F. & Stern, L. (2). Die Oxydation der Bernsteinsäure durch Tiergewebe. *Biochem. Z.*, **30**, 172. 1911.

Batelli, F. & Stern, L. (3). Die Oxydation der Citronen-, Apfel-und Fumarsäure durch Tiergewebe. *Biochem. Z.*, **31**, 478. 1911.

Batelli, F. & Stern, L. (4). Oxydation des p- Phenylen-diamins durch die Tierge-webe. *Biochem. Z.*, **46**, 317. 1912.

Batelli, F. & Stern, L. (5). Action des quelques substances sur l'activité respira-toire des tissus isolés. *J. Physiol. Path. gén.*, **9**, 228. 1907.

Bate Smith, E. C. (1). On the coagulation of muscle plasma. *Proc. Roy. Soc.* B, **105**, 579. 1930.

Bate Smith, E. C. (2). On the coagulation of muscle plasma. II. The solubility of myosin. *Proc. Roy. Soc.* B, **114**, 494. 1934.

Bate Smith, E. C. (3). Native and denatured muscle proteins. *Proc. Roy. Soc.* B, **124**, 136. 1937–38.

Bate Smith, E. C. (4). The proteins of meat. *J. Soc. chem. Ind.*, **54**, 152 T. 1935.

Bate Smith, E. C. (4a). Cataphoretic behaviour of zymohexase. *Biochem. J.*, **34**, 1122. 1940.

Bate Smith, E. C. (5). Changes in elasticity of mammalian muscle undergoing rigor mortis. *J. Physiol.*, **96**, 176. 1939.

Bate Smith, E. C. (6). The buffering of muscle in rigor; protein, phosphate and carnosine. *J. Physiol.*, **92**, 336. 1938.

Bate Smith, E. C. & Bendall, J. R. (1). Rigor mortis and adenosine-triphosphate. *J. Physiol.*, **106**, 177. 1947.

Bate Smith, E. C. & Bendall, J. R. (2). Delayed onset of rigor mortis after administration of myanesin. *J. Physiol.*, **107**, 2P. 1947.

Bate Smith, E. C. & Bendall, J. R. (3). Factors determining the time course of rigor mortis. *J. Physiol.*, **110**, 47. 1949.

Bate Smith, E. C. & Bendall, J. R. (4). Changes in muscle after death. *Brit. Med. Bull.*, **12**, 230. 1956.

Bateson, W. (1). The ancestry of the Chordata. *Quart. J. micr. Sci.*, **26**, 535. 1886.

Baudouin, P., Hers, M. G. & Loeb, H. (1). Electron microscopic and biochemical study of Type II glycogenosis. *Lab. Invest.*, **18**, 1139. 1964.

Baur, H., Kuhn, R. & Wacker, L. (1). Insulinwirkung und Totenstarre. *Muenchener Med. Wochensch.*, **71**, 169. 1924.

Bayliss, L. E. (1). The energetics of plain muscle. *J. Physiol.*, **65**, i. 1928.

Bear, R. S. (1). Small-angle X-ray diffraction studies on muscle. *J. Am. chem. Soc.*, **67**, 1625. 1945.

Beatty, C. H., Basinger, G. & Bocek, R. Pentose cycle activity in muscle from fetal, neonatal and infant rhesus monkeys'. *Arch. Biochem. Biophys.*, **117**, 275. 1966.

Beatty, C. H., Petersen, R. D. & Bocek, R. M. (1). Metabolism of red and white muscle from fibre groups. *Amer. J. Physiol.*, **304**, 939. 1963.

Beck, T. (1). Herons (des älteren) Automatentheater. *Beitr. z. Gesch. d. Techn. u. Ind.*, **1**, 182. 1909.

Beinert, H., Bocek, R. M., Goldman, D. S., Green, D. E., Mahler, H. R., Mii, S., Stansley, P. G. & Wakil, S. J. (1). The reconstruction of the fatty acid oxidising system of animal tissues. *J. Amer. chem. Soc.*, **75**, 4111. 1953.

Belágyi, J. & Garamvölgyi, N. (1). Mechanical properties of the flight muscle of the bee, II. Active isometric tension at different muscle lengths. *Acta Biochim. Biophys. Acad. Sci. Hung.*, **3**, 293. 1968.

Belitzer, V. A. (1). La régulation de la respiration musculaire par les transformations du phosphagène. *Enzymologia*, **6**, 1. 1939.

Belitzer, V. A. & Tsibakova, E. T. (1). On the mechanism of phosphorylation coupled with respiration. *Biokhimiya*, **4**, 516. 1939.

Bell, L. G. E. (1). Freezing and drying techniques in cytology. *Int. Rev. Cytol.*, **1**, 35. 1952.

Bell, R. D. & Doisy, E. A. (1). Rapid colorimetric methods for the determination of phosphorus in urine and blood. *J. biol. Chem.*, **44**, 55. 1920.

Belocopitow, E. (1) The action of epinephrine on glycogen synthetase. *Arch. Biochem. Biophys.* **93**, 457. 1961.

Belocopitow, E., Appleman, M. M. & Torres, H. N. (1). Factors affecting the activity of muscle glycogen synthetase. *J. biol. Chem.* **240**, 3473. 1965.

Bendall, J. R. (1). A study of the kinetics of the fibrillar adenosine triphosphatase of rabbit skeletal muscle. *Biochem. J.* **81**, 520. 1961.

Bendall, J. R. (2). The myofibrillar ATPase activity of various animals in relation to ionic strength and temperature. *Biochemistry of Muscle Contraction*, p. 448. Ed. J. Gergely. London, J. A. Churchill Ltd. 1964.

Bendall, J. R. (3). Effect of the 'Marsh factor' on the shortening of muscle fibre models in the presence of adenosine triphosphate. *Nature*, **170**, 1058. 1952.

Bendall, J. R. (4). Further observations on a factor (the 'Marsh' factor) effecting relaxation of ATP-shortened muscle-fibre models, and the effect of Ca and Mg ions upon it. *J. Physiol.*, **121**, 232. 1953.

Bendall, J. R. (5). The relaxing effect of myokinase on muscle fibres; its identity with the 'Marsh' factor. *Proc. Roy. Soc.* B, **142**, 409. 1954.

Bendall, J. R. (6). Muscle-relaxing factors. *Nature*, **181**, 1188. 1958.

Bendall, J. R. (7). Post mortem changes in muscle. In *Structure and Function of muscle*. Vol. III, p. 227. Ed. G. H. Bourne. New York and London, Academic Press, 1960.

Bendall, J. R. (8). The shortening of rabbit muscles during rigor mortis: its relation to the breakdown of adenosine triphosphate and creatine phosphate and to muscular contraction. *J. Physiol.*, **114**, 71. 1951.

Bendall, J. R. (9). *Muscles, Molecules and Movement.* London, Heinemann Educational Books Ltd., London. 1969.

Bendall, J. R. & Davey, C. L. (1). Ammonia liberation during rigor mortis and its relation to changes in the adenine and inosine nucleotides of rabbit muscle. *Biochim. biophys. Acta*, **26**, 93. 1957.

Bendall, J. R. & Taylor, A. A. (1). The Meyerhof quotient and the synthesis of glycogen from lactate in frog and rabbit muscle. A reinvestigation. *Biochem. J.*, **118**, 887. 1970.

Benedict, J. D., Kalinsky, H. J., Scarrone, L. A., Wertheim, A. R. & Stetten, D. (1). The origin of urinary creatine in progressive muscular dystrophy. *J. Clin. Invest.* **34**, 141. 1955.

Bennett, H. S. (1). The sarcoplasmic reticulum of striped muscle. *J. biophys. biochem. Cyt.* **2**, Suppl., 171. 1956.

Bennett, H. S. (2). Modern concepts of structure of striated muscle. *Amer. J. physical Med.*, **34**, 46. 1955.

Bennett, H. S. & Porter, K. R. (1). An electron microscope study of the sectioned breast muscle of the domestic fowl. *Am. J. Anat.* **93**, 61. 1953.

Benzinger, T., Hems, R., Burton, K. & Kitzinger, C. (1). Free energy changes of the glutaminase reaction and the hydrolysis of the terminal pyrophosphate bond of adenosine triphosphate. *Biochem. J.* **71**, 400. 1959.

Bergström, J., Hermansen, L., Hultman, E. & Saltin, B. (1). Diet, muscle glycogen and physical performance. *Acta physiol. Scand.*, **71**, 140. 1967.

Bergström, J. & Hultman, E. (1). Muscle glycogen synthesis after exercise: an enhancing factor localized. *Nature, Lond.*, **210**, 309. 1966.

Bernard, C. (1). *Leçons de Physiologie. Cours du semestre d'hiver 1854–55.* Paris, J.-B. Ballière, 1855.

Bernard, C. (2). De la matière glycogène considerée comme condition de developpement de certains tissus, chez le foetus, avant l'apparition de la fonction glycogénique du foie. In *Leçons sur le diabète.* p. 492. Paris, Ballière. 1877. (Extrait de *Compt. rend. des. séances de l'Acad. des Sci.*, **48**, 1859.)

Bernstein, J. (1). Die Energie des Muskels als Oberflächen-energie, *Arch. f. ges. Physiol.*, **85**, 271. 1901.

Bernstein, J. (2). Zur Thermodynamik der Muskelkontraktion. *Arch. f. ges. Physiol.*, **122**, 129, 1908.

Bernstein, J. (2a). *Untersuchungen über den Erregungsvorgang in Nerven- und Muskelsysteme.* Heidelberg, 1871.

Bernstein, J. (3). Experimentelles und kritischen Theorie des Muskelkontraktion. *Arch. f. ges. Physiol.*, **162**, 1. 1915.

Bernstein, J. (4). Untersuchungen zur Thermodynamik der bioelektrischen Ströme. *Arch. ges. Physiol.*, **92**, 521. 1902.

Bernstein, J. (5). *Elektrobiologie.* Braunschweig, Vieweg & Sohn. 1912.

Berthollet, C. L. (1). Précis d'observations sur l'analyse animale comparée à l'analyse végétale. *Observations sur la physique,* **28,** 272. 1786.

Berzelius, J. (1). *Jahres-Bericht über die Fortschritte der physischen Wissenschaften.* Tr. from the Swedish by F. Wöhler, 1836.

Berzelius, J. (2). *Lehrbuch der Chemie.* Vol. III. Dresden. 1827.

Berzelius, J. (3). *Jahres-Bericht über die Fortschritte der Chemie und Mineralogie.* Vol. **27.** 1848.

Best, C. H., Hoet, J. P. & Marks, H. P. (1). The fate of the sugar disappearing under the action of insulin. *Proc. Roy. Soc.* B, **100,** 32. 1926.

Bethe, A. (1). Spannung und Verkürzung des Muskels bei contracturerzeugenden Eingriffen im Vergleich zur Tetanusspannung und Tetanusverkürzung. *Arch. f. ges. Physiol.,* **199,** 491. 1923.

Bethe, A. (2). *Allgemeine Anatomie und Physiologie des Nervensystems.* Leipzig, Georg Thieme. 1903.

Bettex-Galland, M. & Lüscher, E. F. (1). Thrombosthenin, the contractile protein from blood platelets and its relation to other contractile proteins. *Adv. in Protein Chem.,* **20,** 1. 1965.

Bettex-Galland, M. & Lüscher, E. F. (2). Extraction of an actomyosin-like protein from human thrombocytes. *Nature,* **184,** 276. 1959.

Bettex-Galland, M. & Lüscher, E. F. (3). Thrombosthenin – a contractile protein from thrombocytes. *Biochim. biophys. Acta,* **49,** 536. 1961.

Bettex-Galland, M., Portzehl, H. & Lüscher, E. F. (1). Dissoziation des Thrombosthenins in seine zwei Komponenten. Untersuchung ihrer Adenosintriphosphatase Aktivität. *Helv. chim. Acta,* **46,** 1595. 1963.

Beviz, A., Lundholm, L., Mohme-Lundholm, E. & Vamos, N. (1) Hydrolysis of adenosinetriphosphate and creatinephosphate on isometric contraction of vascular smooth muscle. *Acta physiol. Scand.,* **65,** 268. 1965.

Beyer, R. E. (1). A protein factor required for phosphorylation coupled to electron flow between reduced coenzyme Q and cytochrome C in the electron transfer chain. *Biochem. biophys. Res. Com.,* **16,** 460. 1964.

Bianchi, C. P. (1). Kinetics of radiocaffeine uptake and reléase in frog sartorius. *J. Pharmacol.,* **138,** 41. 1962.

Bianchi, C. P. (2). The effect of caffeine on radiocalcium movement in frog sartorius. *J. gen. Physiol.,* **44,** 845. 1960.

Bianchi, C. P. & Shanes, A. M. (1). Calcium influx in skeletal muscle at rest, during activity, and during potassium contracture. *J. gen. Physiol.,* **42,** 803. 1959.

Biedermann, W. (1). *Electrophysiology.* Trans, F. A. Welby. London, MacMillan & Co. 1898.

Biedermann, W. (2). Studien zur vergleichenden Physiologie der peristaltischen Bewegungen. I. Die peristaltische Bewegungen der Wärmer und der Tonus glatter Muskeln. *Arch. f. ges. Physiol.,* **102,** 475. 1904.

Biedermann, W. (3). Histochemie der quergestreiften Muskelfasern. *Erg. d. Biol.,* **2,** 416. 1927.

Bilinski, E. (1). Utilisation of lipids by fish. I. Fatty acid oxidation by tissue slices from dark and white muscle of rainbow trout (*Salmo gairdmerii*). *Can. J. Biochem. Physiol.,* **41,** 107. 1963.

Bing, R. J. & Michal, G. (1). Myocardial efficiency. *Ann. N.Y. Acad. Sci.,* **72,** 555. 1958–59.

Biörck, G. (1). On myoglobin and its occurrence in man. *Acta med. Scand. Suppl.* **226.** 1949.

Birch, T. (1). *The history of the Royal Society of London.* Vol. II. London, A. Millar. 1756.

Biró, N. A. & Szent-Györgyi, A. (1). The effect of actin and physico-chemical changes on the myosin-ATP-ase system, and on washed muscle. *Acta Hung. Physiol.*, **2**, 120. 1949.

Biron, P., Dreyfus, J. C. & Schapira, F. (1). Différences métaboliques contre les muscles rouges et blancs chez le lapin. *Compt. rend. Soc. Biol.*, **158**, 1841. 1964.

Blair, P. V., Oda, T., Green, D. E. & Fernández-Morán, H. (1). Studies on the electron transfer system. LIV. Isolation of the unit of electron transfer. *Biochemistry (Amer. chem. Soc.),* **2**, 756. 1963.

Blanchaer, M. C. (1). Respiration of mitochondria of red and white muscle. *Amer. J. Physiol.*, **206**, 1015. 1964.

Blanchard, M., Green, D. E., Nocito, V. & Ratner, S. (1). l-Amino acid oxidase of animal tissue. *J. biol. Chem.*, **155**, 421. 1944.

Blanchard, M., Green, D. E., Nocito, V. & Ratner, S. (2). Isolation of l-amino acid oxidase. *J. biol. Chem.*, **161**, 583. 1945.

Blaschko, H. (1). The anaerobic delayed heat-production of stimulated muscle. *J. Physiol.*, **70**, 96, 1930.

Blaschko, H. & Hope, D. B. (1). Observations on the distribution of amine oxidase in invertebrates. *Arch. Biochem. Biophys.*, **69**, 10. 1957,

Blix, M. (1). Studien über Muskelwärme. *Skand, Arch. f. Physiol.*, **12**, 52. 1902.

Blum, J. J. (1). Approximate treatment of diffusion into a cylindrical enzyme system obeying Michaelis-Menten kinetics. *Biochem. biophys. Acta*, **21**, 158. 1956.

Blum, J. J. (2). The enzymatic interaction between myosin and nucleotides. *Arch. Biochem. Biophys.*, **55**, 486. 1955.

Blum, J. J. (3). Interaction between myosin and its substrates. *Arch. Biochem. Biophys.*, **87**, 104. 1960.

Blum, J. J., Kerwin, T. D. & Bowen, W. J. (1). Dependence of length of muscle fibres on ATP concentration. *Arch. Biochem. Biophys.*, **66**, 100. 1957.

Blum, J. J. & Morales, M. F. (1). The interaction of myosin with adenosine triphosphate. *Arch. Biochem. Biophys.*, **43**, 208. 1953.

Bo, W. J., Maraspin, L. E. & Smith M. S. (1). Glycogen synthetase activity in the rat uterus. *J. Endocrin.*, **38**, 33. 1967.

Bo, W. J. & Smith, M. S. (1). Comparison of UDGP-glycogen synthetase activity of the tongue and uterus. *Proc. Soc. exp. Biol. Med.*, **113**, 812. 1963.

Boas, M. (1). *Boyle and seventeenth century Chemistry.* Cambridge University Press. 1957.

Bocek, R. M., Basinger, G. M. & Beatty, C. H. (1). Comparison of glucose uptake and carbohydrate utilization in red and white muscle. *Amer. J. Physiol.*, **210**, 1108. 1966.

Bocek, R. M., Petersen, R. D. & Beatty, C. H. (1). Glycogen metabolism in red and white muscle. *Amer. J. Physiol.*, **210**, 1101. 1966.

Boehm, G. (1). Kurzzeitige Interferenzaufnahmen als neue physiologische Untersuchungsmethode. *Z. f. Biol.*, **91**, 203. 1931.

Boehm, G. & Weber, H. H. (1). Das Röntgendiagramm von gedehnten Myosinfäden. *Koll. Z.*, **61**, 269. 1932.

Boerhaave, H. (1). *Academical lectures on the theory of physic.* (University of Leyden.) London, W. Innys. 1743.

Boettiger, E. G. (1). Triggering of the contractile process in insect fibrillar muscle. In *Physiological Triggers*. Ed. T. H. Bullock. pp. 103–116. Washington D.C., Amer. Physiol. Soc. 1957.

Boettiger, E. G. (2). The machinery of insect flight. In *Recent Advances in Insect Physiology*. pp. 117–142. Ed. B. T. Scheer. Eugene, Oregon, Univ. of Oregon Publications. 1957.

Boettiger, E. G. & Furshpan, E. (1). The response of fibrillar flight muscle to rapid release and stretch. *Biol. Bull.*, **107**, 305. 1953.

Boettiger, E. G. & Furshpan, E. (2). Mechanical properties of insect flight muscle. *J. cell. comp. Physiol.*, **44**, 340. 1954.

Bois-Reymond, E. du. (1). *Untersuchungen über tierische Elektricität*. Berlin, G. Reimer. 1848.

Bolker, H. I. (1). Phylogenetic relationships of the echinoderms: biochemical evidence. *Nature*, **213**, 204. 1967.

Bollman, J. L. & Flock, E. V. (1). Phosphocreatine and inorganic phosphate in working and resting muscles of rats, studied with radioactive phosphorus. *J. biol. Chem.*, **147**, 155. 1943.

Bonner, W. D. (1). Activation of the heart-muscle succinic oxidase system. *Biochem. J.*, **49**, viii. 1951.

Bonting, S. L. & Caravaggio, L. L. (1). Studies on sodium-potassium-activated adenosine triphosphatase. V. Correlation of enzyme activity with cation flux in six tissues. *Arch. Biochem. Biophys.*, **101**, 37. 1963.

Borelli, A. (1). *De motu animalium*. Rome. 1681.

Born, G. V. R. (1). The relation between the tension and the high-energy phosphate content of smooth muscle. *J. Physiol.*, **131**, 704. 1956.

Born, G. V. R. (2). The breakdown of adenosine triphosphate in blood platelets during clotting. *J. Physiol.*, **133**, 61. 1956.

Borsook, H. & Dubnoff, J. W. (1). The hydrolysis of phosphocreatine and the origin of urinary creatinine. *J. biol. Chem.* **168**, 493. 1967.

Borst, P. & Slater, E. C. (1). The site of action of 2,4-dinitrophenol on oxidative phosphorylation. *Biochim. biophys. Acta*, **48**, 362. 1961.

Bosch, M. W. (1). Quantitative electrophoretic study of rabbit muscle proteins soluble in dilute salt solutions. *Biochim. biophys. Acta*, **7**, 61. 1951.

Botazzi, F. (or P.) & Quagliariello, G. (1). Recherches sur la constitution physique et les propriétés chimico-physiques du suc des muscles lisses et des muscle striées. *Arch. internat. Physiol.*, **12**, 234. 289, 409. 1912.

Botazzi, P. (or F). (1). The oscillations of the auricular tonus in the batrachian heart with a theory on the function of sarcoplasm in muscular tissues. *J. Physiol.*, **21**, 1. 1897.

Botazzi, P. (or F.) (2). Uber die Wirkung des Veratrins und anderer Stoffe auf die Quergestreifte, atriale und glatte Musculatur. (Beiträge zur Physiologie des Sarkoplasmas). *Arch. f. Anat. Physiol.*, p. 377. 1901.

Botts, J. & Morales, M. F. (1). The elastic mechanism and hydrogen bonding in actomyosin threads. *J. cell. comp. Physiol.*, **37**, 27. 1951.

Bowen, W. J. (1). Phosphorylysis of adenosine triphosphate and rate of contraction of myosin B threads. *Amer. J. physiol.*, **165**, 10. 1951.

Bowen, W. J. (2). Effect of calcium, magnesium and pH on adenosine triphosphatase of myosin B threads. *Amer. J. Physiol.*, **169**, 221. 1952.

Bowen, W. J. (3). Adenosinetriphosphatase and the shortening of muscular models. *J. Cell. comp. Physiol.*, **49**, Suppl. 1, 267. 1957.

Bowen, W. J. (4). Carnosine and tetramethyl ammonium ion and ATP induced shortening of glycerol-treated muscle fibres. *Biochem. biophys. Res. Com.*, **19**, 427. 1965.

Bowen, W. J. (5). Effects of pH, buffers, carnosine, histidine, and β-alanine on the shortening of glycerol-treated muscle fibres. *Biochim. biophys. Acta*, **112**, 436. 1965.

Bowen, W. J. & Kerwin, T. D. (1). The rate of dephosphorylation of adenosine triphosphate and the shortening of glycerol-washed muscle fibres. *Biochim. biophys. Acta*, **18**, 83. 1955.

Bowen, W. J. & Kerwin, T. D. (2). A study of the effects of ethylenediamine-tetracetic acid on myosin adenosinetriphosphatase. *J. biol. Chem.*, **211**, 237. 1954.

Bowen, W. J., Stewart, L. C. & Martin, H. L. (1). Studies of the fast intitial rate of adenosine triphosphatase of actomyosin and glycerol-treated muscle. *J. biol. Chem.*, **238**, 2926. 1963.

Bowman, W. (1). On the minute structure and movements of voluntary muscle. *Phil. Trans. Roy. Soc. Lond.*, p. 457. 1840.

Boyer, P. D. (1). Oxygen exchange and oxidative phosphorylation studies as related to possible phosphorylations accompanying muscle contraction. In *Biochemistry of muscle contraction*, p. 94. Ed. J. Gergely. London, J. & A. Churchill Ltd. 1964.

Boyer, P. D. (2). Oxidative phosphorylation. In *Biological oxidation*, p. 193. Ed. T. P. Singer. New York, Wiley, 1967.

Boyer, P. D., Bieber, L. L., Mitchell, R. A. & Szabolcsi, G. (1). The apparent independence of the phosphorylation and water formation reactions from the oxidation reations of oxidative phosphorylation. *J. biol. Chem.*, **241**, 5384. 1966.

Boyer, P. D., Lardy, H. A. & Phillips, P. H. (1). Further studies on the role of potassium and other ions in the phosphorylation of the adenylic system. *J. biol. Chem.*, **149**, 529. 1943.

Boyle, P. J. & Conway, E. J. (1). Potassium accumulation in muscle and associated changes. *J. Physiol.*, **100**, 1. 1941.

Boyle, Robert. (1). *The Sceptical Chymist.* 1679.

Boyle, Robert (2). New experiments concerning the relation between light and air (in shining wood and fish). *Phil. Trans. Roy. Soc.*, **2** (Numb. 31), 581. 1667–8.

Boyle, Robert. (3). Some observations about shining flesh. *Phil. Trans. Roy. Soc.* **7** (Numb. 89), 5108. 1672.

Bozler, E. (1). Mechanism of relaxation in extracted muscle fibres. *Amer. J. Physiol.*, **167**, 277. 1951.

Bozler, E. (2). Ueber die Frage des Tonussubstrates. *Z. f. vergl. Physiol.* Abt. C, **7**, 379. 1928.

Bozler, E. (3). Ueber die Frage des Tonussubstrates. *Z. f. vergl. Physiol.*, Abt. C., **7**, 407. 1928.

Bozler, E. (4). Conduction, automaticity, and tonus of visceral muscles. *Experientia*, **4**, 213. 1948.

Bozler, E. (5). The heat production of smooth muscle. *J. Physiol.*, **69**, 442. 1930.

Bozler, E. (6). The energy changes of smooth muscle during relaxation. *J. cell. comp. Physiol.*, **8**, 419. 1936.

Bozler, E. (7). Untersuchungen zur Physiologie der Tonusmuskeln. *Zeitsch. f. vergl. Physiol.*, **12**, 579. 1930.

Bozler, E. (8). Die mechanischen Eigenschaften des ruhenden Muskels, ihre experimentelle Beeinflussung und physiologische Bedeutung. *Z. vergl. Physiol.*, 14, 429. 1931.

Bozler, E. (9). An analysis of the properties of smooth muscle. *Cold Spring Harbour Symposia on quantitative Biology*, 4, 260. 1936.

Bozler, E. (10). The mechanism of muscular relaxation. *Experientia*, 9, 1. 1953.

Bozler, E. (11). Action potentials and conduction of excitation in muscle. *Biological Symposia*, 3, 95. 1941.

Bozler, E. (12). Relaxation of extracted muscle fibres. *J. gen. Physiol.*, 38, 149. 1954–55.

Brading, A., Bülbring, E. & Tomita, T. (1). The effect of sodium and calcium on the action potential of the smooth muscle of the guinea-pig taenia coli. *J. Physiol.*, 200, 637. 1969.

Bragg, L., Kendrew, J. C. & Perutz, M. F. (1). Polypeptide chain configurations in crystalline proteins. *Proc. Roy. Soc.* A, 203, 321. 1950.

Brahms, J. & Kakol, I. (1). Interaction of myosin sulfhydryl groups and phosphorus compounds during cleavage of adenosine triphosphate. *Acta biol. exptl.*, 18, 195. 1958.

Brahms, J. & Kay, C. M. (1). The role of solvent induced conformational changes in the enzymatic ATPase activity of cardiac myosin A. *J. biol. Chem.*, 237, 3449. 1962.

Brahms, J. & Kay, C. M. (2). Molecular and enzymatic studies of cardiac myosin. *J. mol. Biol.*, 5, 132. 1962.

Braunstein, A. E. & Kritzmann, M. G. (1). Ueber den Ab- und Aufbau von Aminosäuren durch Umaminierung. *Enzymologia*, 2, 129. 1937–8.

Brazier, M. A. B. (1). The historical development of neurophysiology. In *Handbook of Physiology*. Vol. I, 1. Ed. J. Field. Washington D.C., Am. Physiol. Soc. 1957.

Bremer, J. (1). Carnitine in intermediary metabolism. The biosynthesis of palmitylcarnitine by cell subfractions. *J. biol. Chem.*, 238, 2775. 1963.

Briggs, A. H. (1). Characteristics of contraction in glycerinated uterine smooth muscle. *Amer. J. Physiol.*, 204, 739. 1963.

Briggs, A. H. & Hannah, H. B. (1). Calcium, magnesium and ATP interaction in glycerinated uterine smooth muscle. In *Muscle*. Ed. by W. M. Paul, E. E. Daniel, C. M. Kay & G. Monckton. pp. 287–294. Oxford, Pergamon Press. 1965.

Briggs, A. P. (1). A modification of the Bell-Doisy phosphate method. *J. biol. Chem.*, 53, 13. 1922.

Briggs, F. N. (1). Current thoughts on the soluble muscle relaxing factor. *Fed. Proc.*, 23, 903. 1964.

Briggs, F. N. & Fleishman, F. (1). Detection of calcium sequestration in a soluble muscle fraction with muscle relaxing activity. *Fed. Proc.*, 24, 208. 1965.

Briggs, F. N. & Fuchs, F. (1). The biosynthesis of a muscle-relaxing factor. *Biochim. biophys. Acta*, 42, 519, 1960.

Briggs, F. N., Kaldor, G. & Gergely, J. (1). Participation of a dialyzable cofactor in the relaxing factor system of muscle. I. Studies with single glycerinated fibres. *Biochim. biophys. Acta*, 34, 211. 1959.

Briggs, F. N. & Portzehl, H. (1). The influence of relaxing factor on the pH dependence of the contraction of muscle models. *Biochim. biophys. Acta*, 24, 482. 1957.

References   617

Briskey, E. J., Seraydarian, K. & Mommaerts, W. F. H. M. (1). The modification of actomyosin by α-actinin. II. The effect of α-actinin upon contractility. *Biochim. biophys. Acta*, **133**, 412. 1967.

Briskey, E. J., Seraydarian, K. & Mommaerts, W. F. H. M. (2). The modification of actomyosin by α-actinin. III. The interaction between α-actinin and actin. *Biochim. biophys. Acta*, **133**, 424. 1967.

Brocke, H. H. vom. (1). The activating effects of calcium ions on the contractile systems of insect fibrillar flight muscle. *Arch. f. ges. Physiol.*, **290**, 70. 1966.

Brody, I. A. (1). Effect of denervation on the lactate dehydrogenase isozymes of skeletal muscle. *Nature*, **205**, 196. 1965.

Brokaw, C. J. (1). Adenosine triphosphate usage by flagella. *Science*, **156**, 76. 1967.

Brown, D. (1). The regulation of energy exchange in contracting muscle. *Biological Symposia*, **3**, 161. Ed. W. O. Fenn. Lancaster, Pa., Jaques Cattell Press. 1941.

Brown, D. H. & Illingworth, B. (1). The properties of an oligo 1,4 →1,4-glucantransferase from animal tissues. *Proc. nat. Acad. Sci. U.S.A.*, **48**, 1783. 1962.

Brown, D. H. & Illingworth, B. (2). The role of oligo 1,4 →1,4-glucantransferase and amylo-1,6-glucosidase in the debranching of glycogen. In Ciba Foundation Symposium *Control of glycogen metabolism*, pp. 139. Ed. W. J. Whelan & M. P. Cameron. London, J. & A. Churchill. 1964.

Brown, D. H., Illingworth, B. & Cori, C. F. (1). Enzymatic debranching of glycogen. Combined action of oligo-1,4 →1,4-glucantransferase and amylo-1,6-glucosidase in debranching glycogen. *Nature*, **197**, 980. 1963.

Brown, D. H., Illingworth, B. & Kornfeld, R. (1). Transfer of glucosyl units to oligosaccharides and polysaccharides by the action of uridine diphosphoglucose-α-glucan transglucosylase. *Biochemistry (Amer. Chem. Soc.)*, **4**, 486. 1965.

Brown, D. M., Fasman, G. D., Magrath, D. I. & Todd, A. R. (1). The structure of adenylic acids *a* and *b*. *J. chem. Soc.*, p. 1448. 1954.

Brücke, E. (1). Untersuchungen über den Bau der Muskelfasern mit Hülfe des polarisirten Lichtes. *Denksch. d. kais. Akad. d. Wissensch. Wien, Math.-nat. Kl.*, **15**, 69. 1858.

Brust, M. (1). Contraction enhancement in skeletal muscles of normal and dystrophic mice. *Amer. J. Physiol.*, **208**, 425. 1965.

Bücher, T. (1). Ueber ein phosphatübertragendes Gärungsferment. *Biophys. biochim. Acta*, **1**, 292. 1947.

Bücher, T. & Klingenberg, M. (1). Wege des Wasserstoffs in der lebendigen Organisation. *Angew. Chem.*, **70**, 552. 1958.

Bücher, T. & Pette (1). Proportionskonstante Gruppen in Beziehung zur Differenzierung der Enzymaktivitätsmuster von Skelett-Muskeln des Kaninchens. *Z. f. physiol. Chem.*, **331**, 180. 1963.

Bücher, T. & Pfleiderer, G. (1). Pyruvate kinase from muscle. In *Methods in Enzymology*, pp. 435–440. Ed. S. P. Colowick & N. O. Kaplan. New York, Academic Press. 1955.

Buchner, E. (1). Alkoholische Gärung ohne Hefezellen. *Ber. d. Deutsch chem. Gesellsch.*, **30**, 117. 1897.

Buchtal, F., Deutsch, A., Knappeis, G. G. & Munch-Petersen, A. (1). On the effect of adenosine triphosphate on myosin threads. *Acta physiol. skand.*, **13**, 167. 1947.

Buchtal, F., Knappeis, G. G. & Lindhard, J. (1). Die Struktur der quergestreiften, lebenden Muskelfaser des Frosches in Ruhe und während der Kontraktion. *Skand. Arch. f. Physiol.*, **73**, 163. 1936.

Buchwald, K. W. & Cori, C. F. (1). Influence of repeated contractions of muscle on its lipid content. *Proc. Soc. exp. Biol. Med.*, **28**, 737. 1930–31.

Bueding, E. & Bülbring, E. (1). The inhibitory action of adrenaline. Biochemical and biophysical observations. In *Pharmacology of Smooth Muscle*, pp. 37–54. Ed. E. Bülbring. Prague, Czechoslovak Medical Press, and Oxford, Pergamon Press. 1964.

Bueding, E., Bülbring, E., Gercken, G., Hawkins, J. T. & Kuriyama, H. (1). The effect of adrenaline on the adenosinetriphosphate and creatine phosphate content of intestinal smooth muscle. *J. Physiol.*, **193**, 187. 1967.

Bueding, E., Bülbring, E., Gercken, G. & Kuriyama, H. (1). Lack of activation of phosphorylase by adrenaline during its physiological action on smooth muscle. *Nature*, **196**, 944. 1962.

Bueding, E., Bülbring, E., Gercken, G. & Kuriyama, H. (2). The effect of adrenaline on the adenosine triphosphate and creatine phosphate content of intestinal smooth muscle. *J. Physiol.*, **166**, 8. 1962.

Bueding, E., Butcher, R. W., Hawkins, W., Timms, A. R. & Sutherland, E. W. (1). Effect of epinephrine on cyclic adenosine 3′,5′-phosphate and hexose phosphates in intestinal smooth muscle. *Biochim. biophys. Acta*, **115**, 173. 196.

Bugnard, L. (1). The relation between total and initial heat in single muscle twitches. *J. Physiol.*, **82**, 509. 1934.

Bülbring, E. (1). Measurements of oxygen consumption in smooth muscle. *J. Physiol.*, **122**, 111. 1953.

Bülbring, E. (2). Electrical activity in intestinal smooth muscle. *Physiol. Rev. Suppl.* **5**, **42**, 160. 1962.

Bülbring, E. & Tomita, T. (1). Effect of calcium, barium and manganese on the action of adrenaline in the smooth muscle of the guinea-pig taenia coli. *Proc. Roy. Soc.* B, **172**, 121. 1969.

Bülbring, E. & Tomita, T. (2). Suppression of spontaneous spike generation by catecholamines in the smooth muscle of guinea-pig taenia coli. *Proc. Roy. Soc.* B, **172**, 103. 1969.

Bülbring, E. & Tomita, T. (3). Increase of membrane conductance by adrenaline in the smooth muscle of guinea-pig taenia coli. *Proc. Roy. Soc.* B, **172**, 89. 1969.

Buller, A. J., Eccles, J. C. & Eccles, R. M. (1). Interactions between motoneurons and muscles in respect of characteristic speeds of their responses. *J. Physiol.*, **150**, 417. 1960.

Buller, A. J., Eccles, J. C. & Eccles, R. M. (2). Differences of fast and slow muscles in the cat hind limb. *J. Physiol.*, **150**, 399. 1960.

Buller, A. J. & Lewis, D. M. (1). Further observations on the differences of skeletal muscles in the kitten hind limb. *J. Physiol.*, **176**, 355. 1965.

Buller, A. J., Mommaerts, W. H. F. M. & Seraydarian, K. (1). Enzymic properties of myosin in fast and slow twitch muscles of the cat following cross-innervation. *J. Physiol.*, **205**, 581. 1969.

Bullock, M. W., Brockman, J. A., Patterson, E. L., Pierce, J. V. & Stokstad, E. L. R. (1). Synthesis of compounds in the thioctic acid series. *J. Amer. Chem. Soc.*, **74**, 3455. 1952.

Burger, A., Richterich, R. & Aebi, H. (1). Die Heterogenität der Kreatin-Kinase. *Biochem. Z.*, **339**, 305. 1964.

Burk, D. (1). The free energy of glycogen-lactic acid breakdown in muscle. *Proc. Roy. Soc.* B, **104**, 153. 1928–29.

Burn, J. H. & Dale, H. H. (1). On the location and nature of the action of insulin. *J. Physiol.*, **59**, 164. 1924.

Burn, J. H. & Marks, H. P. (1). The relation of the thyroid gland to the action of insulin. *J. Physiol.*, **60**, 131. 1925.

Burnasheva, S. A. (1). On spermosin, the contractile protein of sperm cells. *Biokhimiya*, **23**, 558. 1958.

Burnasheva, S. A., Efremenko, M. V., Chumakova, L. P. & Zueva, L. V. (1). Isolation of the contractile proteins from the cilia of *Tetrahymena pyriformis* and investigation of their properties. *Biokhimiya* (Eng. vers.), **30**, 656. 1965.

Burnasheva, S. A., Efremenko, M. V. & Lyubimova, M. N. (1). A study of adenosine triphosphatase activity of isolated cilia of the infusoria *Tetrahymena pyriformis* and the isolation of adenosine triphosphatase from them. *Biokhimiya* (Eng. vers.), **28**, 442. 1963.

Burnasheva, S. A. & Karansheva, T. P. (1). The participation of ATP in the motor function of the infusoria *Tetrahymena pyriformis*. *Biokhimiya* (Eng. vers.) **32**, 222. 1967.

Burnstock, G. & Holman, M. E. (1). The transmission of excitation from autonomic nerve to smooth muscle. *J. Physiol.*, **155**, 115. 1961.

Burnstock, G., Holman, M. E., & Prosser, C. L. (1). Electrophysiology of smooth muscle. *Physiol. Rev.*, **43**, 482. 1963.

Burton, K. (1). Formation constants for the complexes of adenosine di- or tri-phosphate with magnesium or calcium ions. *Biochem. J.*, **71**, 388. 1959.

Burton, K. (2). The free energy change associated with the hydrolysis of the thiol ester bond of acetyl coenzyme A. *Biochem. J.*, **59**, 44. 1955.

Burton, K. (3). Energy of adenosine triphosphate. *Nature*, **181**, 1594. 1958.

Bushana Rao, K. S. P., Focant, B., Gerday, C. & Hamoir, G. (1). Low molecular weight albumins of cod white muscle (*Gadus callarius* L.). *Comp. Biochem. Physiol.*, **30**, 33. 1969.

Busk, G. & Huxley, T. (1) (tr.) [Kolliker's] *Manual of Human Histology*. London. 1853.

Butcher, R. W. & Sutherland, E. W. (1). Adenosine 3′,5′-phosphate in biological materials. 1. Purification and properties of cyclic 3′,5′-nucleotide phosphodiesterase and use of this enzyme to characterize adenosine 3,′5′-phosphate in human urine. *J. biol. Chem.*, **237**, 1244. 1962.

Caesar, R., Edwards, G. A. & Ruska, H. (1). Architecture and nerve supply of mammalian smooth muscle tissue. *J. biophys. biochem. Cyt.*, **3**, 867. 1957.

Cagniard-Latour, C. (1). Memoire sur la fermentation vineuse. *Ann. Chim. (Phys.)*, **68** (2ᵐᵉ série), 206. 1838.

Cain, D. F. & Davies, R. E. (1). Breakdown of adenosine triphosphate during a single contraction of working muscle. *Biochem. biophys. Res. Comm.*, **8**, 361. 1962.

Cain, D. F., Delluva, A. M. & Davies, R. E. (1). Carnosine phosphate as phosphate donor in muscular contraction. *Nature*, **182**, 720. 1958.

Cain, D. F., Infante, A. A. & Davies, R. E. (1). Adenosine triphosphate and phosphorylcreatine as energy suppliers for single contractions of working muscle. *Nature*, **196**, 214. 1962.

Cain, D. F., Infante, A., Klaupiks, D., Eaton, W. A. & Davies, R. E. (1). Chemical changes during muscular contraction. *Fed. Proc.*, **21**, 319. 1962.

Cain, D. F., Kushmerick, M. J. & Davies, R. E. (1). Hypoxanthine nucleotides and muscular contraction. *Biochim. biophys. Acta*, **74**, 735. 1963.

Cain, D. F., Kushmerick, M. J. & Davies, R. E. (2). Phosphoenolpyruvate, the phosphoglyceric acids and muscular contraction. *Biochim. biophys. Acta*, 86, 81. 1964.

Caldwell, P. C. (1). The separation of the phosphate esters of muscle by paper chromatography. *Biochem. J.*, 55, 458. 1953.

Caldwell, P. C. (2). The phosphorus metabolism of squid axons and its relationship to the active transport of sodium. *J. Physiol.*, 152, 545. 1960.

Caldwell, P. C. (3). Factors governing movement and distribution of inorganic ions in nerve and muscle. *Physiol. Rev.*, 48, 1. 1967.

Caldwell, P. C., Hodgkin, A. L., Keynes, R. D. & Shaw, T. I. (1). The effects of injecting 'energy-rich' phosphate compounds on the active transport of ions in the giant axons of *Loligo*. *J. Physiol.*, 152, 561. 1960.

Caldwell, R. C. & Keynes, R. D. (1). The utilisation of phosphate bond energy for sodium extrusion from giant axons. *J. Physiol.*, 137, 12. 1957.

Cammarata, P. S. & Cohen, P. P. (1). The scope of the transamination reaction in animal tissues. *J. biol. Chem.*, 187, 439. 1950.

Cantoni, G. L. (1). On the role of high energy phosphate in transmethylation. In *Phosphorus metabolism*, p. 641. Baltimore, Ed. W. D. McElroy & B. Glass, John Hopkins Press. 1951.

Cantoni, G. L. (2). S-adenosylmethionine; a new intermediate formed enzymatically from L-methionine and adenosinetriphosphate. *J. biol. Chem.*, 204, 403. 1953.

Cantoni, G. L. & Vignos, P. J. (1). Enzymatic mechanism of creatine synthesis. *J. biol. Chem.*, 209, 647. 1954.

Caputto, R., Leloir, L. F., Cardini, C. E. & Paladini, A. C. (1). Isolation of the coenzyme of the galactose phosphate-glucose phosphate transformation. *J. biol. Chem.*, 184, 333. 1950.

Caputto, R., Leloir, L. F., Trucco, R. E., Cardini, C. E. & Paladini, A. C. (1). A coenzyme for phosphoglucomutase. *Arch. Biochem. Biophys.*, 18, 201. 1948.

Caputto, R., Leloir, L. F., Trucco, R. E., Cardini, C. E. & Paladini, A. C. (2). The enzymatic transformation of galactose into glucose derivatives. *J. biol. Chem.*, 179, 497. 1949.

Cardini, C. E., Paladini, A. C., Caputto, R., Leloir, L. F. & Trucco, R. E. (1). The isolation of the co-enzyme of phosphoglucomutase. *Arch. Biochem. Biophys.*, 22, 87. 1949.

Carlsen, F., Knappeis, G. G. & Buchthal, F. (1). Ultrastructure of the resting and contracted striated muscle fibre at different degrees of stretch. *J. biophys. biochem. Cytol.*, 11, 95. 1961.

Carlson, F. D. (1). The mechanochemistry of muscular contraction, a critical revaluation of *in vivo* studies. *Progr. Bioph. mol. Biol.*, 13, 262. 1963.

Carlson, F. D., Hardy, D. J. & Wilkie, D. R. (1). Total energy production and creatine phosphate hydrolysis in the isotonic twitch. *J. gen. Physiol.*, 46, 851. 1963.

Carlson, F. D., Hardy, D. J. & Wilkie, D. R. (2). The relation between heat produced and phosphorylcreatine split during isometric contraction of frog's muscle. *J. Physiol.*, 189, 209. 1967.

Carlson, F. D. & Siger, A. (1). The creatine phosphoryl-transfer reaction in iodoacetate-poisoned muscle. *J. gen. Physiol.*, 43, 301. 1959–60.

Carlson, F. D. & Siger, A. (2). The mechanochemistry of muscular contraction. I. The isometric twitch. *J. gen. Physiol.*, 44, 33. 1960.

Carsten, M. E. (1). Actin, its amino acid composition and its reaction with iodoacetate. *Biochemistry (Amer. chem. Soc.)*, **2**, 32. 1963.

Carsten, M. E. (2). Actin. Its thiol groups. *Biochemistry (Am. chem. Soc.)*, **5**, 297. 1966.

Carsten, M. E. (3). The cardiac calcium pump. *Proc. nat. Acad. Sci. U.S.A.*, **52**, 1456. 1964.

Carsten, M. E. (4). A study of uterine actin. *Biochemistry (Amer. chem. Soc.)*, **4**, 1049. 1965.

Carsten, M. E. (5). Tropomyosin from smooth muscle of the uterus. *Biochemistry (Amer. chem. Soc.)*, **7**, 960. 1968.

Carsten, M. E. & Katz, A. M. (1). Actin: a comparative study. *Biochim. biophys. Acta.*, **90**, 534. 1964.

Carsten, M. E. & Mommaerts, W. F. H. M. (1). A study of actin by means of starch gel electrophoresis. *Biochemistry (Amer. chem. Soc.)*, **2**, 28. 1963.

Carsten, M. E. & Mommaerts, W. F. H. M. (2). The accumulation of calcium ions by sarcotubular vesicles. *J. gen. Physiol.*, **48**, 183. 1964.

Carter, C. E. & Cohen, L. H. (1). Enzymatic synthesis of adenylosuccinic acid. *J. Amer. chem. Soc.*, **77**, 499. 1955.

Carter, C. E. & Cohen, L. H. (2). The preparation and properties of adenylo-succinase and adenylosuccinic acid. *J. biol. Chem.*, **222**, 17. 1956.

Carter, C. E. & Cohen, L. H. (3). Enzymatic synthesis of adenylosuccinic acid. *Fed. Proc.*, **14**, 189. 1955.

Case, E. M. (1). The origin of pyruvic acid in muscle. *Biochem. J.*, **26**, 759. 1932.

Case, E. M. & Cook, R. P. (1). The occurrence of pyruvic acid and methyl glyoxal in muscle metabolism. *Biochem. J.*, **25**, 1318. 1931.

Caspersson, T. & Thorell, B. (1). The localization of the adenylic acids in striated muscle-fibres. *Acta physiol. Scand.*, **4**, 97. 1942.

Casteels, R. (1). Tension development in smooth muscle in response to A.C. field stimulation. *J. Physiol.*, **173**, 12. 1964.

Castiglioni, A. (1). *A History of Medicine.* New York, Knopf. 1947.

Cathcart, E. P. (1). The influence of muscle work on protein metabolism. *Physiol. Rev.*, **5**, 225. 1925.

Cathcart, E. P. & Burnett, W. A. (1). The influence of work on metabolism in varying conditions of diet. *Proc. Roy. Soc.* B, **99**, 405. 1925–26.

Cattell, M. & Hartree, W. (1). The delayed anaerobic heat production of stimulated muscle. *J. Physiol.*, **74**, 221. 1932.

Cattell, McK. & Shorr, E. (1). The recovery heat production of mammalian muscle. *Amer. J. Physiol.* **101**, 18. 1932.

Chance, B. (1). Direct spectroscopic observations of the oxidation-reduction reactions of ubiquinone in heart and kidney mitochondria. In the *CIBA Symposium on Quinones in electron transport*, pp. 327–340. London, J. & A. Churchill. 1961.

Chance, B. & Connelly, C. M. (1). A method for the estimation of the increase in concentration of adenosine diphosphate in muscle sarcosomes following a contraction. *Nature*, **179**, 1235. 1957.

Chance, B. & Hollunger, C. (1). Inhibition of electron and energy transfer in mitochondria. 1. Effects of amytal, rotenone, progesterone and methylene glycol. *J. biol. Chem.*, **238**, 418. 1963.

Chance, B. & Mela, L. (1). Proton movements in mitochondrial membranes. *Nature*, **212**, 372. 1966.

Chance, B., Parsons, D. F. & Williams, G. R. (1). Cytochrome content of mitochondria stripped of inner membrane structure. *Science*, **143**, 136. 1964.

Chance, B. & Sacktor, B. (1). Respiratory metabolism of insect flight muscle. II. Kinetics of respiratory enzymes in flight muscle sarcosomes. *Arch. Biochem. Biophys.*, **76**, 509. 1958.

Chance, B. & Williams, G. R. (1). The respiratory chain and oxidative phosphorylation. *Adv. in Enzym.*, **17**, 65. 1956.

Chance, B. & Williams, G. R. (2). Respiratory enzymes in oxidative phosphorylation. III. The steady state. *J. biol. Chem.*, **217**, 409. 1955.

Chang, Y.-S. & Tsao, T.-C. (1). Conformational changes of rabbit tropomyosin in different solvents. *Scientia Sinica*, **11**, 1353. 1962.

Chaplain, R. A. (1). The effect of Ca^{2+} and fibre elongation on the activation of the contractile mechanism of insect fibrillar flight muscle. *Biochim. biophys. Acta*, **131**, 385. 1967.

Chaplain, R. A. (2). The allosteric nature of substrate inhibition of insect actomyosin ATPase in presence of magnesium. *Biochem. biophys. Res. Com.*, **22**, 248. 1966.

Chaplain, R. A. (3). Indication for an allosteric effect of adenosine diphosphate in actomyosin gels from insect fibrillar flight muscle. *Arch. Biochem. Biophys.*, **115**, 450. 1966.

Chaplain, R. A. (4). Tension development of glycerinated insect muscle fibres as a measure of the conformational state of the myosin. *Biochem. biophys. Res. Com.*, **24**, 526. 1966.

Chaplain, R. A., Abbott, R. H. & White, D. C. S. (1). Indication for an allosteric effect of ADP on actomyosin gels and glycerinated fibres from insect fibrillar flight muscle. *Biochem. biophys. Res. Com.*, **21**, 89. 1965.

Chappell, J. B. & Crofts, A. R. (1). Gramicidin and ion transport in isolated liver mitochondria. *Biochem. J.*, **95**, 393. 1965.

Chappell, J. B. & Crofts, A. R. (2). The effect of atractylate and oligomycin on the behaviour of mitochondria towards adenine nucleotides. *Biochem. J.*, **95**, 707. 1965.

Chappell, J. B. & Haarhoff, K. N. (1). The penetration of the mitochondrial membrane by anions and cations. In *Biochemistry of Mitochondria*, p. 75. Ed. E. C. Slater, Z. Kaniuga & L. Wojtczak. London and New York, Academic Press, 1967.

Chappell, J. B. & Perry, S. V. (1). The stimulation of adenosinetriphosphatase activities of myofibrils and L-myosin by 2:4-dinitrophenol. *Biochim. biophys. Acta*, **16**, 258. 1955.

Chappell, J. B. & Perry, S. V. (2). The respiratory and adenosinetriphosphatase activities of skeletal muscle mitochondria. *Biochem. J.*, **55**, 586. 1953.

Chapuis, A. (1). Les Jeux d'Eaux et les Automates Hydrauliques du Parc d'Hellbrunn près Salzburg. *La Suisse Horlogère*, **67** (no. 4), 39. 1952.

Charleton, W. (1). *Oeconomia Animalis*. London, Daniel & Redman, 1654.

Cheesman, D. F. & Hilton, E. (1). Exchange of structurally bound phosphate in muscular activity. *J. Physiol.*, **183**, 675. 1966.

Cheesman, D. F. & Whitehead, A. (1). Possible role in contraction of structurally bound phosphate of muscle. *Nature*, **221**, 736. 1969.

Cherian, K. M., Bokdawala, F. D., Vallyathan, N. V. & George, J. C. (1). Effect of denervation on the red and white fibres of the pectoralis muscle of the pigeon. *J. Neurol. Neurosurg. & Psychiat.*, **29**, 299. 1966.

Chevreul E. C., (1). Untersuchungen über die chemische Zusammensetzung des Fleischbrühe. *J. prakt. Chem.*, **6**, 120. 1835.

Chibnall, A. C. & Spahr, P. F. (1). Determination of the N-terminal residues in proteins with methoxycarbonyl chloride. *Biochem. J.*, **68**, 135. 1958.

Choi, J. K. (1). Fine structure of the smooth muscle of chicken gizzard. In *Intern. Congr. Electron microscopy*. (Philadelphia). p.M.9. Ed. by S. S. Breeze. New York and London, Academic Press, 1962.

Christie, G. S., Ahmed, K., McLean, A. E. M. & Judah, J. D. (1). Active transport of potassium by mitochondria. I. Exchange of K+ and H+. *Biochim. biophys. Acta*, **94**, 432. 1965.

Citterio, P., Maldacea, L. & Ranzi, S. (1). On the presence of actin and myosin in the electric organ of the *Torpedo*. *Pubbl. Staz. Napoli*, **29**, 434. 1957.

Clark, V. M., Kirby, G. W. & Todd, A. (1). Oxidative phosphorylation: a chemical approach using quinol phosphates. *Nature*, **181**, 1650. 1958.

Clarke, E. W. A simplified heart-oxygenator preparation, suitable for isotopic experiments: with some observations on the metabolism of acetate pyruvate and four amino acids. *J. Physiol.*, **136**, 380.

Clausius, R. (1). Ueber die bewegende Kraft der Wärme. 1850. Ostwald's Klassiker der exakten Wissenschaften, 99. Ed. M. Planck. Engelmann. Leipzig 1898.

Close, R. (1). Dynamic properties of the fast and slow skeletal muscles of the rat during development. *J. Physiol.*, **173**, 74. 1964.

Close, R. (2). Force:velocity properties of mouse muscles. *Nature*, **206**, 718. 1965.

Close, R. & Hoh, F. J. Y. (1). Force:velocity properties of kitten muscles. *J. Physiol.*, **192**, 815. 1967.

Cockrell, R. S., Harris, E. J. & Pressman, B. C. (1). Energetics of potassium transport in mitochondria induced by valinomycin. *Biochemistry (Amer. chem. Soc.)*, **5**, 2326. 1966.

Cohen, C. (1). In discussion in *Symposium on microstructure of proteins. J. polym. Sci.*, **49**, 144. 1961.

Cohen, C. & Hanson, J. (1). An X-ray diffraction study of F-actin. *Biochim. biophys. Acta*, **21**, 177. 1956.

Cohen, C. & Holmes, K. C. (1). X-ray diffraction evidence for α-helical coiled-coils in native muscle. *J. mol. Biol.*, **6**, 423. 1963.

Cohen, C. & Longley, W. (1). Tropomyosin paracrystals formed by divalent cations. *Science*, **152**, 794. 1966.

Cohen, C., Lowey, S. & Kucera, J. (1). Structural studies on uterine myosin. *J. biol. Chem.*, **236**, PC23. 1961.

Cohen, C. & Szent-Györgyi, A. G. (1). Optical rotation and helical polypeptide chain configuration in α-proteins. *J. Am. chem. Soc.*, **79**, 248. 1957.

Cohen, E. & Elvehjem, C. A. (1). The relation of iron and copper to the cytochrome and oxidase content of animal tissues. *J. biol. Chem.*, **107**, 97. 1934.

Cohn, F. (1). Uber den Einfluss der Muskelarbeit auf den Lactacidogengehalt in der roten und weissen Muskulatur des Kaninchens. *Z. physiol. Chem.*, **113**, 253. 1921.

Cohn, M. & Cori, G. T. (1). On the mechanism of action of muscle and potato phosphorylase. *J. biol. Chem.*, **175**, 89. 1948.

Cohnheim, O. & Uexkull, J. von. (1). Die Dauerkontraktion der glatten Muskeln. *Z. f. physiol. Chem.*, **76**, 314. 1911–12.

Coleman, D. L. & Ashworth, M. E. (1). Incorporation of glycine-1-C^{14} into nucleic acids and proteins of mice with hereditary muscular dystrophy. *Amer. J. Physiol.*, **197**, 839. 1959.

Colowick, S. P. & Kalckar, H. M. (1). The role of myokinase in transphorylations. 1. The enzymatic phosphorylation of hexoses by adenyl pyrophosphate. *J. biol. Chem.*, **148**, 117. 1943.

Colowick, S. P. & Sutherland, E. W. (1). Polysaccharide synthesis from glucose by means of purified enzymes. *J. biol. Chem.*, **144**, 423. 1942.

Colowick, S. P., Welch, M. S. & Cori, C. F. (1). Phosphorylation of glucose in kidney extracts. *J. biol. Chem.*, **133**, 359. 1940.

Connell, J. J. (1). The relative stabilities of the skeletal-muscle myosins of some animals. *Biochem. J.*, **80**, 503. 1961.

Connell, J. J. (2). Studies on the proteins of fish skeletal muscle. 2. Electrophoretic analysis of low ionic strength extracts of several species of fish. *Biochem. J.*, **55**, 378. 1953.

Connelly, C. M. & Chance, B. (1). Kinetics of reduced pyridine nucleotides in stimulating frog muscle and nerve. *Fed. Proc.*, **13**, 29. 1954.

Conover, T. E. & Bárány, M. (1). The absence of a myosin-like protein in liver mitochondria. *Biochim. biophys. Acta.*, **127**, 235. 1966.

Conti, G., Haenni, B., Laszt, L. & Rouiller, C. (1). Structure et ultrastructure de la cellule musculaire lisse de la paroi carotidienne à l'état de repos et à l'état de contraction. *Angiologica*, **1**, 119. 1964.

Corey, R. B. & Pauling, L. (1). Fundamental dimensions of polypeptide chains. *Proc. Roy. Soc.* B, **141**, 10. 1953.

Cori, C. F. (1). Phosphorylation of carbohydrates. In *A Symposium on respiratory enzymes*. pp. 175–189. Madison, Univ. of Wisconsin Press. 1941.

Cori, C. F., Colowick, S. P. & Cori, G. T. (1). The isolation and synthesis of glucose-1-phosphoric acid. *J. biol. Chem.*, **121**, 465. 1937.

Cori, C. F. & Cori, G. T. (1). Mechanism of formation of hexosemonophosphate in muscle and isolation of a new phosphate ester. *Proc. Soc. exper. Biol. & Med.*, **34**, 702. 1936.

Cori, C. F. & Cori, G. T. (2). Glycogen formation in the liver from d- and l- lactic acid. *J. biol. Chem.*, **81**, 389. 1929.

Cori, C. F., Cori, G. T. & Green, A. A. (1). Crystalline muscle phosphorylase. III. Kinetics. *J. biol. Chem.*, **151**, 39. 1943.

Cori, C. F. & Illingworth, B. (1). The prosthetic group of phosphorylase. *Proc. nat. Acad. Sci. U.S.A.*, **43**, 547. 1957.

Cori, G. T. (1). Glycogen structure and enzyme deficiencies in glycogen storage disease. *Harvey lectures*, 1952–3, p. 145.

Cori, G. T., Colowick, S. P. & Cori, C. F. (1). The action of nucleotides in the disruptive phosphorylation of glycogen. *J. biol. Chem.*, **123**, 381. 1938.

Cori, G. T. & Cori, C. F. (1). The formation of hexosephosphate esters in frog muscle. *J. biol. Chem.*, **116**, 119. 1936.

Cori, G. T. & Cori, C. F. (2). The kinetics of the enzymatic synthesis of glycogen from glucose-1-phosphate. *J. biol. Chem.*, **135**, 733. 1940.

Cori, G. T. & Cori, C. F. (3). The activating effect of glycogen on the enzymatic synthesis of glycogen from glucose-1-phosphate. *J. biol. Chem.*, **131**, 397. 1939.

Cori, G. T. & Cori, C. F. (4). Crystalline muscle phosphorylase. IV. Formation of glycogen. *J. biol. Chem.*, **151**, 57. 1943.

Cori, G. T. & Cori, C. F. (5). The enzymatic conversion of phosphorylase *a* to *b*. *J. biol. Chem.*, **158**, 321. 1945.

Cori, G. T. & Cori, C. F. (6). Glucose-6-phosphatase of the liver in glycogen storage disease. *J. biol. Chem.*, **199**, 661. 1952.

Cori, G. T. & Green, A. A. (1). Crystalline muscle phosphorylase. II. Prosthetic group. *J. biol. Chem.*, **151**, 31. 1943.

Cori, G. T. & Larner, J. (1). Action of amylo-1,6-glucosidase and phosphorylase on glycogen and amylopectin. *J. biol. Chem.*, **188**, 17. 1951.

Cori, O., Marcus, F. & Traverso-Cori, A. (1). Participation of adenine nucleotides in the proposed direct pathway of biosynthesis of *N*-phosphorylcreatine. *Nature*, **194**, 476. 1962.

Cori, O., Traverso-Cori, A., Lagarrigue, M. & Marcus, F. (1). Enzymic phosphorylation of creatine by 1:3 diphosphoglyceric acid. *Biochem. J.*, **70**, 633. 1958.

Cormier, M. J. & Totter, J. R. (1). Bioluminescence. *Ann. Rev. Biochem.*, **33**, 431. 1964.

Cornblath, M., Randle, P. J., Parmeggiani, A. & Morgan, H. E. (1). Regulation of glycogenolysis in muscle. Effects of glucagon and anoxia on lactate production, glycogen content, and phosphorylase activity in the perfused, isolated rat heart. *J. biol. Chem.*, **238**, 1592. 1963.

Corran, H. S., Green, D. E. & Straub, F. B. (1). On the catalytic function of heart flavoprotein. *J. Biochem.*, **33**, 793. 1939.

Corsi, A. (1). Some observations on the extra-protein of cross-striated muscle. *Biochim. biophys. Acta*, **25**, 640. 1957.

Corsi, A. & Perry, S. V. (1). Some observations on the localisation of myosin, actin and tropomyosin in the rabbit myofibril. *Biochem. J.*, **68**, 12. 1958.

Costantin, L. L., Franzini-Armstrong, C. & Podolsky, R. J. (1). Localisation of calcium-accumulating structures in striated muscle fibres. *Science*, **147**, 158. 1965.

Costantin, L. L., Podolsky, R. J. & Tice, L. W. (1). Calcium activation of frog slow muscle fibres. *J. Physiol.*, **188**, 261. 1967.

Craig, J. W. & Larner, J. (1). Influence of epinephrine and insulin on uridine diphosphate glucose-α-glucan transferase and phosphorylase in muscle. *Nature*, **202**, 971. 1964.

Crane, F. L. (1). Isolation and characterization of the coenzyme Q (ubiquinone) group and plastoquinone. In the CIBA Foundation Symposium on *Quinones in electron transport*, pp. 37–75. Ed. G. E. W. Wolstenholme & C. M. O'Connor. London, J. & A. Churchill. 1961.

Crane, F. L., Glenn, J. L. & Green, D. E. (1). Studies on the electron transfer system. IV. The electron transfer particle. *Biochim. biophys. Acta*, **22**, 475. 1956.

Crane, F. L., Hatefi, Y., Lester, R. L. & Widmer, C. (1). Isolation of a quinone from beef heart mitochondria. *Biochim. biophys. Acta*, **25**, 220. 1957.

Crane, F. L., Widmer, C., Lester, R. L. & Hatefi, Y. (1). Studies on the electron transport system. XV. Coenzyme Q (Q275) and the succinoxidase activity of the electron transport particle. *Biochim. biophys. Acta* **31**, 476. 1959.

Crane, R. K. & Lipmann, F. (1). The effect of arsenate on aerobic phosphorylation. *J. biol. Chem.*, **201**, 235. 1953.

Crane, R. K. & Sols, A. (1). The association of hexokinase with particular fractions of brain and other tissue homogenates. *J. biol. Chem.*, **203**, 273. 1953.

Crepax, P., Jacob, J. & Seldeslachts, J. (1). Contribution à l'étude des proteino-grammes electrophorétiques de'extraits de muscles contracturés. *Biochim. biophys. Acta*, **4**, 410. 1950.

Cretius, K. (1). Untersuchungen des Adenosintriphosphorsäure, Adenosin-diphosphorsäure und Adenosin-5-monophosphorsäure Gehalts menschlicher Skelett- und Uterusmuskulatur. *Z. Geburtsh. Gynaek.*, **149**, 131. 1957.

Cretius, K. (2). Der Kreatinphosphatgehalt menxchlicher Skelett- und Uterus muskulatur. *Z. Geburtsh. Gynaek.*, **149**, 114. 1957.

Crick, F. H. C. (1). Is α-keratin a coiled coil?. *Nature*, **170**, 882. 1952.

Criddle, R. S., Bock, R. M., Green, D. E. & Tisdale, H. (1). Physical characteristics of proteins of the electron transfer system and interpretation of the structure of the mitochondrion. *Biochemistry (Amer. Chem. Soc.)*, **1**, 827. 1962.

Croone, W. (1). *De ratione motus musculorum*. London. 1664.

Croone, W. (2). An Hypothesis of the Structure of a Muscle, and the reason of its Contraction; From Lectures to Barber Surgeons read in the Surgeon's Theatre Anno 1674, 1675. In Robert Hooke, *Philosophical Collections*, No. 2, Section 8, p. 22. 1675.

Cross, R. J., Taggart, J. V., Covo, G. A. & Green, D. E. (1). Studies on the cyclo-phorase system. VI. The coupling of oxidation and phosphorylation. *J. biol. Chem.*, **177**, 655. 1949.

Cruikshank, E. W. H. & McClure, G. S. (1). On the question of the utilisa-tion of amino-acids and fat by the mammalian heart. *J. Physiol.* **86**, 1. 1936.

Csapo, A. (1). The mechanism of myometrial function and its disorders. In *Modern trends in obstetrics and gynaecology*, 2nd series, Ed. K. Bowes, pp. 20–49. London and Washington D.C., Butterworth. 1955.

Csapo, A. (2). Actomyosin content of the uterus. *Nature*, **162**, 218. 1948.

Csapo, A. (3). Studies on adenosinetriphosphatase activity of uterine muscle. *Acta physiol. Scand.*, **19**, 100. 1950.

Csapo, A. (4). Actomyosin of uterus. *Amer. J. Physiol.*, **160**, 46. 1950.

Csapo, A. (5). Molecular structure and function of smooth muscle. In *Structure and Function of Muscle*, pp. 229–264, Ed. G. H. Bourne. New York and London, Academic Press. 1960.

Csapo, A., Erdös, T., Naeslund, J. & Snellman, O. (1). Preliminary note on acto-myosin from uterus studied on ultracentrifuge. *Biochim. biophys. Acta*, **5**, 53. 1950.

Csapso, A. & Gergely, J. (1). Energetics of uterine muscle contraction. *Nature*, **166**, 1078. 1950.

Csapo, A. & Herrmann, H. (1). Quantitative changes in contractile proteins of chick skeletal muscle during and after embryonic development. *Amer. J. Physiol.*, **165**, 701. 1951.

Currie, R. D. & Webster, H. L. (1). Preparation of 5'-adenylic acid deaminase based on phosphate-induced dissociation of rat actomyosin-deaminase com-plexes. *Biochim. biophys, Acta*, **64**, 30. 1962.

Cuvier, G. (1). *Leçons d'Anatomie comparée*. Vol. I. Paris, Baudouin. 1805.

Czok, R. & Bücher, T. (1). Crystalline enzymes from myogen of rabbit skeletal muscle. *Adv. in Protein Chem.*, **15**, 315. 1960.

Daemers-Lambert, C. (1). Action du chlorure de potassium sur le métabolisme des esters phosphorés et le tonus du muscle artériel (carotide de bovidé). *Angiologica*, **1**, 249. 1964.

Daemers-Lambert, C. (2). Action du fluorodinitrobenzene sur le métabolisme phosphore du muscle lisse arteriel pendant la stimulation electrique (carotide de bovidé). *Angiologica*, **6**, 1. 1969.

Daemers-Lambert, C. (3). Les composés phosphorés du muscle lisse vasculaire. Symp. int. *Biochimie de la paroi vasculaire*, Fribourg, 1968. Part II, p. 185. Karger, Basel. 1969.

Daemers-Lambert, C. (4). Dissociation par le fluorodinitrobenzine des effets ATP-asique, metabolique et contractile, liés à l'augmentation de la concentration en potassium extracellulaire dans le muscle lisse artériel (carotide de bovidé). *Angiologica*, **5**, 293. 1968.

Daemers-Lambert, C. & Roland, J. (1). Métabolisme des esters phosphorés pendant le développement et le maintien de la tension phasique du muscle lisse artériel (carotides de bovidé). *Angiologica*, **4**, 69. 1967.

Dainty, M., Kleinzeller, A., Lawrence, A. S. C., Miall, M., Needham, J., Needham, D. M. & Shen, S-C. Studies on the anomalous viscosity and flow-birefringence of protein solutions. III. Changes in these properties of myosin solutions in relation to adenosinetriphosphate and muscular contraction. *J. Gen. Physiol.*, **27**, 355. 1944.

Dakin, H. D. & Dudley, H. W. (1). An enzyme concerned with the formation of hydroxy acids from ketonic aldehydes. *J. biol. Chem.*, **14**, 155. 1913.

Dance, N. & Watts, D. C. (1). Comparison of creatine phosphotransferase from rabbit and brown-hare muscle. *Biochem. J.*, **84**, 114. 1962.

Danforth, W. H. (1) Glycogen synthetase activity in skeletal muscle. Interconversion of two forms and control of glycogen synthesis. *J. biol. Chem.*, **240**, 588. 1965.

Danforth, W. H. & Harvey, P. (1). Glycogen synthetase and control of glycogen synthesis in rat muscle. *Biochem. biophys. Res. Com.*, **16**, 466. 1964.

Danforth, W. H. & Helmreich, E. (1). Regulation of glycolysis in muscle. 1. The conversion of phosphorylase *b* to phosphorylase *a* in frog sartorius muscle. *J. biol Chem.*, **239**, 3133. 1964.

Danforth, W. H., Helmreich, E. & Cori, C. F. (1). The effect of contraction and of epinephrin on the phosphorylase activity of frog sartorious muscle. *Proc. Nat. Acad. Sci. U.S.A.*, **48**, 1191. 1962.

Danforth, W. H. & Lyon, J. B. (1). Glycogenolysis during tetanic contraction of isolated mouse muscles in the presence and absence of phosphorylase *a*. *J. biol. Chem.*, **239**, 4047. 1964.

Danilewsky, A. (1). Myosin, seine Darstellung, Eigenschaften, Umwandlung in Syntonin und Rückbildung aus demselben. *Z. physiol. Chem.*, **5**, 158. 1881.

Danilewsky, B. (1). Thermodynamische Untersuchungen der Muskel. *Arch. f. ges. Physiol.*, **21**, 109. 1880.

Daremberg, C. V. (1). (tr.) Oeuvres anatomiques, physiologiques et médicales de Galien. 2 vols. Paris. 1854–57.

Daremberg, C. V. & Ruelle, C. E. (1). *Oeuvres de Rufus d'Ephèse*. Paris, 1879.

Davenport, H. A. & Sacks, J. (1). Muscle phosphorus. II. The acid hydrolysis of lactacidogen. *J. biol. Chem.*, **81**, 469. 1929.

Davey, C. L. (1). Synthesis of adenylosuccinic acid in preparations of mammalian skeletal muscle. *Nature, Lond.* **183**, 995. 1959.

Davey, C. L. (2). The reamination of inosine monophosphate in skeletal muscle. *Arch. Biochem. Biophys.*, **95**, 296. 1961.

Davey, C. L. (3). An ion-exchange method of determining carnosine, anserine and their precursors in animal tissue. *Nature*, **179**, 209. 1957.

Davey, C. L. (4). The significance of carnosine and anserine in striated skeletal muscle. *Arch. Biochem. Biophys.*, **89**, 303. 1960.

Davies, R. E. (1). The role of ATP in muscle contraction. In *Muscle*, pp. 49–65. Ed. W. M. Paul, E. E. Daniel, C. M. Kay & G. Monckton. Oxford, Pergamon Press. 1965.

Davies, R. E. (2). On the mechanism of muscular contraction. In *Essays in Biochemistry*, pp. 29–55. Ed. P. N. Campbell & G. D. Greville. London and New York, Academic Press, publ. for the Biochem. Soc. 1965.

Davies, R. E. (3). In *Symposium on Muscle. Symposia biol. Hung.*, 8, 179. Budapest, Akadémiai Kiadó. 1968.

Davies, R. E. (4). A molecular theory of muscle contraction: calcium dependent contractions with hydrogen bond formation plus ATP-dependent extensions of part of the myosin-actin cross-bridges. *Nature*, 199, 1068. 1963.

Davies, R. E., Cain, D. & Delluva, A. M. (1). The energy supply for muscle contraction. *Ann. N.Y. Acad. Sci.*, 81, 468. 1959.

Davies, R. E., Cain, D. F., Infante, A. A., Klaupiks, D. & Eaton, W. A. (1). Changes in creatine, phosphocreatine, inorganic phosphate and adenosine triphosphate during single contractions of isolated muscles. In *Biochemistry of Muscle Contraction*, p. 463. Ed. J. Gergely. London, J. & A. Churchill. 1964.

Davies, R. E., Goldspink, G. & Larson, R. E. (1). ATP utilisation by fast and slow muscles during the development and maintenance of isometric tension. *J. Physiol.*, 206, 28 P. 1970.

Davies, R. E., Kushmerick, M. J. & Larson, R. E. (1). ATP, activation and the heat of shortening of muscle. *Nature*, 214, 148. 1967.

Davoren, P. R. & Sutherland, E. W. (1). The cellular location of adenyl cyclase in the pigeon erythrocyte. *J. biol. Chem.*, 238, 3016. 1963.

Dawson, D. M., Eppenberger, H. M & Kaplan, N. O. (1). The comparative enzymology of creatine kinase. II. Physical and chemical properties. *J. biol. Chem.*, 242, 210, 1968.

Dawson, D. M., Goodfriend, T. L. & Kaplan, N. O. (1). Lactic dehydrogenases: functions of the two types. *Science*, 143, 929. 1964.

Dawson, D. M. & Romanul, F. C. A. (1). Enzymes in muscle. II. Histochemical and quantitative studies. *Arch. Neurol.*, 11, 369. 1964.

Debus, A. G. (1). The Paracelsian aerial niter. *Isis*, 55, 179. 1964.

Debus, A. G. (2). *The English Paracelsians*. New York, Franklin Watts. 1965.

De Duve, C. (1). Lyosomes, a new group of cytoplasmic particles. In *Sub-cellular Particles*, Ed. T. Hayashi, p. 128. Ronald Press, New York, 1959.

De Duve, C., Pressman, B. C., Gianetto, R., Wattiaux, R. & Applemans, F. (1). Tissue fractionation studies. 6. Intracellular distribution patterns of enzymes in rat-liver tissue. *Biochem. J.*, 60, 604. 1955.

De Lange, R. J., Kemp, R. G., Riley, W. D., Cooper, R. A. & Krebs, E. G. (1). Activation of skeletal muscle phosphorylase kinase by adenosine triphosphate and adenosine 3′,5′-adenosine monophosphate. *J. biol. Chem.*, 243, 2200. 1968.

De Luca, H. F. & Engstrom, G. W. (1). Calcium uptake by rat kidney mitochondria. *Proc. Nat. Acad. Sci. U.S.A.*, 47, 1744. 1961.

Delbrück, A., Zebe, E. & Bücher, T. (1). Ueber Verteilungsmuster von Enzymen des Energie liefernden Stoffwechsels im Flugmuskel, Sprungmuskel und Fettkörper von *Locusta migratoria* und ihre cytologische Zuordnung. *Biochem. Z.*, 331, 273. 1959.

Dempsey, E. W., Wislocki, G. B. & Singer, M. (1). Some observations on the chemical cytology of striated muscle. *Anat. Record*, 96, 221. 1946.

Dempsey, M. E. & Boyer, P. D. (1). Catalysis of an inorganic phosphate-H_2O^{18} exchange by actomyosin and myosin. *J. biol. Chem.*, **236**, PC6. 1961.

Dempsey, M. E., Boyer, P. D. & Benson, E. S. (1). Characteristics of an orthophosphate oxygen exchange catalysed by myosin, actomyosin and muscle fibers. *J. biol. Chem.*, **238**, 2708. 1963.

Denny-Brown, D. E. (1). The histological features of striped muscle in relation to its functional activity. *Proc. Roy. Soc.* B, **104**, 371. 1929.

Descartes, René. (1) *Oeuvres*, ed. V. Cousin. Paris. 1824.

Despretz, C. (1). Recherches experimentales sur les causes de la chaleur animale. *Ann. de Chim. et de Phys.*, **26**, 337. 1824.

Deuticke, H.-J. (1). Über den Chemismus der Totenstarre. *Z. f. physiol. Chem.*, **149**, 259, 1925.

Deuticke, H.-J. (2). Kolloidzustandsänderungen der Muskelproteine beim Absterben und bei der Ermüdung. *Arch. f. ges. Physiol.*, **224**, 1. 1930.

Deuticke, H.-J. (3). Kolloidzustandsänderungen der Muskelproteine bei der Muskeltätigkeit. *Z. f. physiol. Chem.*, **210**, 91. 1930.

Dewan, J. G. & Green, D. E. (1). Co-enzyme factor – a new oxidation catalyst. *Biochem. J.*, **32**, 626. 1938.

Diamond, J. & Miledi, R. (1). A study of foetal and new-born rat muscle fibres. *J. Physiol.*, **162**, 393. 1962.

Dische, Z. (1). Zusammenhang zwischen der Synthese der Adenosinetriphosphorsäure und der oxydoreduktiven Umlagerung des Dioxyacetonphosphorsäure esters bei der Glykolyse. *Naturwissensch.*, **22**, 776. 1934.

Dische, Z. (2). Zwei Wege der Phosphorylierung der Glukose zu Hexosediphosphate in intakten Erythrozyten des Menschen. *Naturwissensch.*, **24**, 462. 1936.

Dische, Z. (3). Mit dem Hauptoxydoreduktion-prozess der Blut-glycolyse gekoppelte Synthese der Adenosinetriphosphorsäure. *Enzymologia*, **1**, 288. 1936–37.

Distèche, A. (1). Contribution à l'étude des échanges d'ions hydrogène au cours du cycle de la contraction musculaire. *Mem. Acad. roy. Belg.* Deuxième série, **32**, fasc. 1, 1. 1960–61.

Distèche, A. & Dubuisson, M. (1). Transient response of the glass electrode to pH step variations. *Rev. sci. Instruments*, **25**, 869. 1954.

Dixon, M. & Kenworthy, P. (1). D-Aspartate oxidase of kidney. *Biochim. biophys. Acta*, **146**, 54. 1967.

Dixon, M. & Kleppe, K. (1). D-Amino acid oxidase. I. Dissociation and recombination of the holoenzyme. *Biochim. biophys. Acta*, **96**, 357. 1965.

Dixon, M. & Kleppe, K. (2). D-Amino acid oxidase. II. Specificity, competitive inhibition and reaction sequence. *Biochim. biophys. Acta*, **96**, 368. 1965.

Dobie, W. M. (1). Observations on the minute structure and the mode of contraction of voluntary muscle fibres. *Ann. Mag. nat. Hist.* (series 2) **3**, 109. 1849.

Dobson, J. F. (1). Herophilus of Alexandria. *Proc. Roy. Soc. Med.*, **18**, 19. 1925.

Dobson, J. F. (2). Erasistratus. *Proc. Roy. Soc. Med.*, **20**, 21. 1927.

Doetsch, R. N. & Hageage, G. J. (1). Motility in procaryotic organisms: problems, points of view and perspectives. *Biol. Rev.*, **43**, 317. 1968.

Domonkos, J. & Latzkovits, L. (1). The metabolism of the tonic and tetanic muscles. III. Pyruvate metabolism of the tonic and tetanic muscles. *Arch. Biochem. Biophys.*, **95**, 147. 1961.

Dorogan, D. & Aratei, H. (1). Le contenu et le rôle physiologique de la myoglobin dans la musculature de l'intestin grêle chez le chien. *Analele Stiintifice ale. Univ. Cuza din Iaşi. Serie nouva. Sectiunea IIa*, **14**, 227. 1968.

Dorogan, D., Zerner-Faighelis, C. & Pieptu, M. (1). Le contenu et le rôle physiologique de l'aldolase dans la musculature de l'intestin grêle chez le chien. *Analele Stiintifice ale. Univ. Cuza din Iaşi. Serie nouva, Sectiunea IIa*, 14, 233. 1968.

Doty, P. & Gratzer, W. B. (1). In *Polyamino acids, Polypeptides and Proteins*, p. 111. Ed. M. A. Stahmann. University of Wisconsin Press, Madison. 1962.

Drabikowski, W. & Gergely, J. (1). The effect of the temperature of extraction on the tropomyosin content in actin. *J. biol. Chem.*, 237, 3412. 1962.

Drabikowski, W. & Gergely, J. (2). The role of sulfhydryl groups in the polymerisation and adenosine triphosphate binding of G-actin. *J. biol. Chem.*, 238, 640. 1963.

Drabikowski, W., Kuehl, W. M. & Gergely, J. (1). Inhibition of actin polymerization by mercurials, without removal of bound nucleotide. *Biochem. biophys. Res. Com.*, 5, 389. 1961.

Drabikowski, W., Maruyama, K., Kuehl, W. M. & Gergely, J. (1). Studies on the depolymerisation of F-actin. *J. gen. Physiol.*, 45, 595A. 1962.

Drabikowski, W. & Nonomura, Y. (1). The interaction of troponin with F-actin and its abolition by tropomyosin. *Biochim. biophys. Acta*, 160, 129. 1968.

Drabikowski, W., Nonomura, Y. & Maruyama, K. (1). Effect of tropomyosin on the interaction between F-actin and the 6S component of α-actinin. *J. Biochem.*, 63, 761. 1968.

Drabikowski, W. & Nowak, E. (1). Studies on sulphydryl groups of tropomyosin. *Acta biochim. Polonica*, 12, 61. 1965.

Drabikowski, W. & Nowak, E. (2). The interaction of α-actinin with F- actin and its abolition by tropomyosin. *European J. Biochem.* 5, 209. 1968.

Drahota, Z. & Gutmann, E. (1). Long-term regulatory influences of the nervous system on some metabolic differences in muscles of different function. In *The effect of Use and Disuse on neuro-muscular Function*. p. 143. Ed. E. Gutmann & P. Hnik. Prague, Czechoslovak Acad. Sci. 1963.

Draper, M. H. & Hodge, A. J. (1). Studies on muscle with the electron microscope. *Austral. Journ. Exp. Biol. and Med.*, 27, 465. 1949.

Draper, M. H. & Hodge, A. J. (2). Submicroscopic localisation of minerals in skeletal muscle by internal 'micro-incineration' within the electron microscope. *Nature*, 163, 576. 1949.

Draper, M. H. & Hodge, A. J. (3). Electron-induced micro-incineration with the electron microscope. I. Distribution of residual mineral content in vertebrate striated muscle. *Austral. J. exp. Biol. & Med.*, 28, 549. 1950.

Dreizen, P., Gersham, L. C., Trotta, P. P. & Stracher, A. (1). Myosin. Subunits and their interaction. *J. gen. Physiol.*, 50, No. 6, Part 2 in Proceedings of a symposium on *The contractile process*, p. 85. 1966–67.

Dreizen, P., Hartshorne, D. J. & Stracher, A. (1). The subunit structure of myosin. I. Polydispersity in 5M guanidine. *J. biol. Chem.*, 241, 443. 1966.

Dreyfus, J. C. (1). La permeabilité des muscles rouges et blancs aux acides aminés. Muscle normaux et atrophiés par section nerveuse. *Compt. rend. Soc. Biol.*, 159, 1317. 1965.

Dreyfus, J. C., Demos, J., Schapira, F. & Schapira, G. (1). La lacticodéshydrogenase musculaire chez le myopathe: persistance apparente du type foetal. *Compt. rend. Acad. Sci.*, 254, 4384. 1962.

Dreyfus, J. C., Schapira, G. & Schapira, F. (1). Biochemical study of muscle in progressive muscular dystrophy. *J. clin. Invest.*, 33, 794. 1954.

Dreyfus, J. C., Schapira, G., Schapira, F. & Demos, J. (1). Activités enzymatiques du muscle humain. *Clin. chim. Acta*, 1, 434. 1956.

Drummond, G. I., Harwood, J. P. & Powell, C. A. (1). Studies on the activation of phosphorylase in skeletal muscle by contraction and by epinephrine. *J. biol. Chem.* 244, 4235. 1969.

Drummond, G. I., Duncan, L. & Hertzman, E. (1). Effect of epinephrine on phosphorylase *b* kinase in perfused rat hearts. *J. biol. Chem.*, 241, 5899. 1966.

Drummond, G. I., Valadares, J. R. E. & Duncan, L. (1). Effect of epinephrine on contractile tension and phosphorylase activation in rat and dog hearts. *Proc. Soc. exp. Biol. Med.*, 117, 307. 1964.

Drury, A. N. (1). The physiological activity of nucleic acid and its derivatives. *Physiol. Rev.*, 16, 292. 1936.

Dubowitz, V. (1). Enzymatic maturation of skeletal muscle. *Nature*, 197, 1215. 1963.

Dubowitz, V. (2). Cross-innervated mammalian skeletal muscle: histo-chemical physiological and biochemical observations. *J. Physiol.*, 193, 481. 1967.

Dubowitz, V. & Newman, D. L. (1). Change in enzyme pattern after cross-innervation of fast and slow skeletal muscle. *Nature*, 214, 840. 1967.

Dubowitz, V. & Pearse, A. G. E. (1). Reciprocal relationship of phosphorylase and oxidative enzymes. *Nature*, 185, 701. 1960.

Dubrunfaut, M. (1). Note sur la saccharification des fécules par le malt. *Bull. des Sciences tech.* (5th section of the Bull. universal), 14, 326. 1830.

Dubuisson, M. (1). Differenciation électrophorétique de myosines dans les muscles au repos et fatigués, de Mammifères et de Mollusques. *Experientia*, 2, 1. 1946.

Dubuisson, M. (2). Separation, par voie chimique, des myosines α et β. *Experientia*, 2, 412. 1946.

Dubuisson, M. (3). Muscle activity and muscle proteins. *Biol. Rev.*, 25, 46. 1950.

Dubuisson, M. (4). Apparition d'une proteine nouvelle, la contractine, dans les extraits de muscles contractés. *Experientia*, 4, 437. 1948.

Dubuisson, M. (5). Modifications dans la structure physico-chimique de l'édifice contractile au cours du cycle de la contraction musculaire. *Biochim. biophys. Acta*, 4, 25. 1950.

Dubuisson, M. (6). Accessibilité, solubilité et association, *in situ*, de la myosine. *Experientia*, 3, 372. 1947.

Dubuisson, M. (7). Influence de la nature des ions sur l'extractabilité des protéines de muscles au repos ou contracturés. *Biochim. biophys. Acta*, 5, 489. 1950.

Dubuisson, M. (8). Contribution à l'étude de la transformation G-actine-F-actine. *Biochim. biophys. Acta*, 5, 426. 1950.

Dubuisson, M. (9). Untersuchungen über die Reaktionsänderung des Muskels im Verlauf der Tätigkeit. *Arch. f. ges. Physiol.*, 239, 314. 1937.

Dubuisson, M. (10). Studies on the chemical processes which occur in muscle before, during and after contraction. *J. Physiol.*, 94, 461. 1939.

Dubuisson, M. (11). Recherches sur la modifications de pH et sur les changements du point isoionique de la myosine pendant l'étirement et la contraction musculaire. *Arch. internat. Physiol.*, 50, 203. 1940.

Dubuisson, M. (12). Quelques considérations sur le muscle normal, au repos et en activité. In *The muscle, a study in biology and pathology*. Proceedings of the Symposium held at Royaumont. Ed. G. Schapira. 1950.

Dubuisson, M. (13). Les processus physico-chemiques de la contraction musculaire. *Ann. de Physiol.*, **15**, 448. 1939.

Dubuisson, M. & Fabry-Hamoir, C. (1). Sur les protéines de structure des muscles striés. *Experientia*, **3**, 102. 1950.

Dubuisson, M. & Jacob, J. (1). Electrophorése de protéines musculaires. *Experientia*, **1**, 272. 1945.

Dubuisson, M. & Pezeu, M. H. (1). Etude électrophorétique des myosines de muscles au repos et fatigués, de mollusques et de mammifères. *Comp. rend. Soc. Biol.*, **141**, 800. 1947.

Dubuisson, M. & Roubert, L. (1). Electrophorèse d'extraits protidiques de muscles de mollusques. *Comp. rend. Soc. Biol.*, **141**, 802. 1947.

Dubuisson, M. & Schulz, W. (1). Untersuchungen über die Reaktions-änderungen des Muskels im Verlauf der Tätigkeit im Zusammenhang mit dem chemischen Vorgängen. *Arch. f. ges. Physiol.*, **239**, 776. 1938.

Duckworth, W. L. H., Lyons, M. C. & Towers, B. (1). *Galen 'On Anatomical Procedures'; the later books.* Cambridge University Press. 1962.

Dudley, H. W. & Marrian, G. F. (1). The effect of insulin on the glycogen in the tissues of normal animals. *Biochem. J.*, **17**, 435. 1923.

Dulong, M. (1). Sur la chaleur animale. *Ann. de Chim. et de Phys.*, 3rd series, **1**, 440. 1841.

Durbin, R. P. & Jenkinson, D. H. (1). The calcium dependence of tension development in depolarized smooth muscle. *J. Physiol.*, **157**, 90. 1961.

Dworaczek, E. & Barrenscheen, H. K. (1). Über die chemische Umsetzung der glatten Muskulatur. *Biochem. Z.*, **292**, 388. 1937.

Dydynska, M. & Wilkie, D. R. (1). The chemical and energetic properties of muscle poisoned with fluorodinitrobenzene. *J. Physiol.*, **184**, 751. 1966.

Ebashi, F. & Yamanouchi, I. (1). Calcium accumulation and adenosinetriphosphatase of the relaxing factor. *J. Biochem.*, **55**, 504. 1964.

Ebashi, S. (1). A granule-bound relaxation factor in skeletal muscle. *Arch. Biochem. Biophys.*, **76**, 410. 1958.

Ebashi, S. (2). Kielley-Meyerhof granules and the relaxation of glycerinated muscle fibres. In *Conference on the chemistry of muscular contraction*, p. 89. Tokyo, Igaku-Shoin. 1958.

Ebashi, S. (3). Calcium binding and relaxation in the actomyosin system. *J. Biochem.*, **48**, 150. 1960.

Ebashi, S. (4). The role of the 'relaxing factor' in contraction-relaxation cycle of muscle. *Progr. theor. Phys.*, Suppl., **17**, 35. 1961.

Ebashi, S. (5). Third component participating in the superprecipitation of 'natural actomyosin'. *Nature*, **200**, 1010. 1963.

Ebashi, S. & Ebashi, F. (1). A new protein component participating in the superprecipitation of myosin B. *J. Biochem.*, **55**, 604. 1964.

Ebashi, S. & Ebashi, F. (2). A new protein factor promoting contraction of actomyosin. *Nature*, **203**, 645. 1964.

Ebashi, S. & Ebashi, F. (3). α-Actinin, a new structural protein from striated muscle. I. Preparation and action on actomyosin-ATP interaction. *J. Biochem.*, **58**, 7. 1965.

Ebashi, S. & Ebashi, F. (4). α-Actinin, a new structural protein from striated muscle. II. Action on actin. *J. Biochem.*, **58**, 13. 1965.

Ebashi, S., Ebashi, F. & Kodama, A. (1). Troponin as the Ca^{2+}-receptive protein in the contractile system. *J. Biochem.*, **62**, 137. 1967.

Ebashi, S. & Endo, M. (1). Calcium ion and muscle contraction. *Progr. in Biophys. mol. Biol.* **18**, 123. 1968.

Ebashi, S., Iwakura, H., Nakajima, H., Nakamura, R. & Ooi, Y. (1). New structural proteins from dog heart and chicken gizzard. *Biochem. Z.*, **345**, 201. 1966.

Ebashi, S. & Kodama, A. (1). A new factor promoting aggregation of tropomyosin. *J. Biochem.*, **58**, 107. 1965.

Ebashi, S. & Kodama, A. (2). A new factor promoting aggregation of tropomyosin. *J. Biochem.*, **59**, 425. 1966.

Ebashi, S. & Kodama, A. (3). Native tropomyosin-like action of troponin on trypsin-treated myosin B. *J. Biochem.*, **60**, 733. 1966.

Ebashi, S., Kodama, A. & Ebashi, F. (1). Troponin. 1. Preparation and physiological function. *J. Biochem.*, **64**, 465. 1968.

Ebashi, S. & Lipmann, F. (1). Adenosine triphosphate-linked concentration of calcium ions in a particulate fraction of rabbit muscle. *J. Cell Biol.*, **14**, 389. 1962.

Ebashi, S. & Maruyama, K. (1). Preparation and some properties of α-actinin-free actin. *J. Biochem.*, **58**, 20. 1965.

Ebashi, S., Otsuka, M. & Endo, M. (1). Calcium binding of the relaxing factor and the link between excitation and contraction. *Proc. 22nd Int. Physiol. Congr.*, *V.* II. Abstr. 899. 1962.

Ebashi, S., Toyokura, Y., Momoi, H. & Sugita, H. (1). High creatine phosphokinase activity of sera of progressive muscular dystrophy. *J. Biochem.*, **46**, 103. 1959.

Ebner, V. von. (1). *Untersuchungen über die Ursachen der Anisotropie organisierter Substanzen.* Leipzig, Engelmann. 1882.

Ebner, V. von. (2). Zur Frage der negativen Schwankung der Doppelbrechung bei der Muskelkontraktion. *Arch. f. ges. Physiol.*, **163**, 179, 1916.

Edman, K. A. P. (1). The relation between sarcomere length and active tension in isolated semitendinosus fibres of the frog. *J. Physiol.*, **183**, 407. 1966.

Edman, K. A. P. & Grieve, D. W. (1). The decremental propagation of the action potential and loss of mechanical response in frog sartorius muscle in the absence of calcium. *Experientia*, **19**, 40. 1963.

Edsall, J. T. (1). Studies in the physical chemistry of muscle globulin. II. On some physico-chemical properties of muscle globulin (myosin). *J. biol. Chem.*, **89**, 289. 1930.

Edsall, J. T. (2). In A discussion on muscular contraction and relaxation: their physical and chemical basis. *Proc. Roy. Soc.* B, **137**, 82. 1950.

Edsall, J. T., Greenstein, J. P. & Mehl, J. W. Denaturation of myosin. *J. Am. Chem. Soc.*, **61**, 1613. 1939.

Edwards, G. A., Ruska, H., Santos, P. de S. & Vallejo-Freire, A. (1). Comparative cytophysiology of striated muscle with special reference to the role of the endoplasmic reticulum. *J. biochem. biophys. Cyt.*, **2**, Suppl. 143. 1956.

Eggleton, M. G. [P.] (1). A physiological study of phosphagen in plain muscle. *J. Physiol.*, **82**, 79. 1934.

Eggleton, [M.] G. P. & Eggleton, P. (1). A method of estimating phosphagen and some other phosphorus compounds in muscle tissue. *J. Physiol.*, **68**, 193. 1929–30.

Eggleton, M. G. [P.] & Lovatt Evans, C. (1). The lactic acid content of the blood after muscular contraction under experimental conditions. *J. Physiol.*, **70**, 269. 1930.

Eggleton, P. & Eggleton, [M.] G.P. (1). The inorganic phosphate and a labile form of organic phosphate in the gastrocnemius of the frog. *Biochem. J.*, 21, 190. 1927.

Eggleton, P. & Eggleton, [M.] G. P. (2). The physiological significance of phosphagen. *J. Physiol.*, 63, 155. 1927.

Eggleton, P. & Eggleton, [M.] G. P. (3). Further observations on phosphagen. *J. Physiol.*, 65, 15. 1928.

Eichholz, D. (1). Aristotle's Theory of the formation of metals and minerals. *Classical Quarterly*, 43, 141. 1949.

Einbeck, H. (1). Über das Vorkommen der Fumärsaure in frischen Fleisch. *Z. physiol. Chem.*, 90, 301. 1914.

Eisenberg, E. & Moos, C. (1). The adenosine triphosphatase activity of acto-heavy meromyosin. A kinetic analysis of actin activation. *Biochemistry (Am. chem. Soc.)*, 7, 1486. 1968.

Eisenberg, E. & Moos, C. (2). The interaction of actin with myosin and heavy meromyosin in solution at low ionic strength. *J. biol. Chem.* 242, 2945. 1967.

Eisenberg, E., Zobel, C. R. & Moos, C. (1). Subfragment I of myosin: adenosine triphosphatase activation by actin. *Biochemistry (Amer. chem. Soc.)* 7, 3186. 1968.

Ellenbogen, E., Iyengar, R., Stern, H. & Olson, R. E. (1). Characterisation of myosin from normal dog heart. *J. biol. Chem.*, 235, 2642. 1960.

Elliott, G. F. (1). X-ray diffraction studies on striated and smooth muscle. *Proc. Roy. Soc. B.*, 160, 467. 1964.

Elliott, G. F. (2). Variations of the contractile apparatus in smooth and striated muscles. *J. gen. Physiol.*, 50, No. 6, p. 171. 1967.

Elliott, G. F., Hanson, J. & Lowy, J. (1). Paramyosin elements in lamellibrauch muscles. *Nature*, 180, 1291. 1957.

Elliott, G. F., Lowy, J. & Millman, B. M. (1). X-ray diffraction from living striated muscle during contraction. *Nature*, 206, 1357. 1965.

Elliott, G. F., Lowy, J. & Millman, B. M. (2). Low-angle X-ray diffraction studies of living striated muscle during contraction. *J. mol. Biol.*, 25, 31. 1967.

Elliott, G. F., Lowy, J. & Worthington, C. R. (1). An X-ray and light-diffraction study of the filament lattice of striated muscle in the living state and in rigor. *J. mol. Biol.*, 6, 295. 1963.

Elliott, G. F., Rome, E. M. & Spencer, M. (1). A type of contraction hypothesis applicable to all muscles. *Nature*, 226, 417. 1970.

Elliott, G. F. & Worthington, C. R. (1). Low-angle X-ray diffraction patterns of smooth and striated muscles. *J. Physiol.*, 149, 32P. 1959.

Elvehjem, C. A. (1). The role of iron and copper in the growth and metabolism of yeast. *J. biol. Chem.*, 90, 111. 1931.

Embden, G. (1). Säurebildung und Energielieferung bei der Muskelkontrakturen. *Ber. d. ges. Physiol.*, 32, 690. 1925.

Embden, G. (2). Untersuchungen über den Verlauf der Phosphorsäure- und Milchsäurebildung bei der Muskeltätigkeit. *Klin. Wochensch.*, 3, 1393. 1924.

Embden, G. (3). Neue Untersuchungen über die Tätigkeitssubstanzen der quergestreiften Muskulatur und den Chemismus der Muskelkontraktion. *Klin. Wochensch.*, 6, 628, 1927.

Embden, G. (4). Neue Untersuchungen über den Chemismus der Muskelkontraktion. *Klin. Wochensch.*, 9, 1337. 1930.

Embden, G. & Adler, E. (1). Über die Phosphorsäureverteilung in den weissen und roten Muskulatur des Kaninchens. *Z. f. physiol. Chem.*, 113, 201. 1921.

Embden, G., Carstensen, M. & Schumacher, H. (1). Über die Bedeutung der Adenylsäure für die Muskelfunktion. 4. Mitteilung: Spaltung und Wiederaufbau der Ammoniakbildenden Substanz bei der Mukseltatigkeit. *Z. f. physiol. Chem.*, 179, 186. 1928.

Embden, G. & Deuticke, H. J. (1). Über die Bedeutung der Phosphoglycerinsäure für die Glykolyse in der Muskulatur. *Z. f. physiol. Chem.*, 230, 29. 1934.

Embden, G. & Deuticke, H. J. (2). Über die Einwirkung von Fluorid und Bromessigsäure auf die Intermediärvorgänge bei der Glykolyse in der Muskulatur. *Z. f. physiol. Chem.* 230, 50. 1934.

Embden, G., Deuticke, H. J. & Kraft, G. (1). Über die intermediären Vorgänge bei der Glykolyse in der Muskulatur. *Klin. Wochensch.*, 12, 213. 1933.

Embden, G., Deuticke, H. J. & Kraft, G. (2). Über das Vorkommen einer optisch aktiven Phosphorglycerinsäure bei der Glykolyse in der Muskulatur. *Z. f. physiol. Chem.*, 230, 12. 1934.

Embden, G. & Grafe, E. (1). Über den Einfluss der Muskelarbeit auf die Phosphorsäureausscheidung. *Z. f. physiol. Chem.*, 113, 108. 1921.

Embden, G., Griesbach, W. & Schmitz, E. (1). Über Milchsäurebildung und Phosphorsäurebildung im Muskelpressaft. *Z. f. physiol. Chem.*, 93, 1. 1914.

Embden, G., Hirsch-Kauffmann, E., Lehnartz, E. & Deuticke, H. J. (1). Über den Verlauf der Milchsäurebildung beim Tetanus. *Z. f. physiol. Chem.*, 151, 207. 1926.

Embden, G. & Ickes, T. (1). Die Isolierung der Glycerinphosphorsäure aus fluoridvergiftetes Muskulatur. *Z. f. physiol. Chem.*, 230, 63. 1934.

Embden, G. & Jost, H. (1). Über die Spaltung des Lactacidogens bei der Muskelkontraktion. *Z. f. physiol. Chem.*, 179, 24. 1928.

Embden, G. & Jost, H. (2). Über die Spaltung des Lactacidogens bei der Muskelkontraktion. II. *Z. f. physiol. Chem.*, 203, 48. 1931.

Embden, G. & Jost, H. (3). Über die Zwischenstufen der Glykolyse in der quergestreiften Muskulatur. *Z. f. physiol. Chem.*, 230, 69. 1934.

Embden, G. & Jost, H. (4). Über kolloidchemisch Veränderungen bei der Muskelermüdung und ihre biologische Bedeutung. *Deutsch. med. Wochensch.*, 51, 636. 1925.

Embden, G., Kalberlah, F. & Engel, H. (1). Über Milchsäurebildung im Muskelpressaft. *Biochem. Z.*, 45, 45. 1912.

Embden, G. & Laquer, F. (1). Über die Chemie des Lactacidogen. I. Isolierungsversuche. *Z. f. physiol. Chem.*, 93, 94. 1914.

Embden, G. & Laquer, F. (2). Über die Chemie des Lactacidogens. *Z. f. physiol. Chem.*, 113, 1. 1921.

Embden, G. & Lawaczeck, H. (1). Über die Bildung anorganischer Phosphorsäure bei der Kontraktion des Froschmuskels. *Biochem. Z.*, 127, 181. 1922.

Embden, G. & Lehnartz, E. (1). Über die Bedeutung von Ionen für die Muskelfunktion. 1. Die Wirkung verschiedener Anionen auf den Lactacidogenwechsel im Froschmuskelbrei. *Z. f. physiol. Chem.*, 134, 243. 1924.

Embden, G. & Lehnartz, E. (2). Die zeitliche Verlauf der Milchsäurebildung bei der Muskelkontraktion. II. Mitteilung. *Z. f. physiol. Chem.*, 176, 231. 1928.

Embden, G., Lehnartz, E. & Hentschel, H. (1.) Die zeitliche Verlauf der Milchsäurebildung bei der Muskelkontraktion. *Z. f. physiol. Chem.*, 165, 255. 1927.

Embden, G. & Lehnartz, M. (1). Über die Ammoniakbildung bei längerem Aufenthalt von Froschmuskeln in Barkanscher Losung. *Z. f. physiol. Chem.*, 201, 273. 1931.

Embden, G. & Metz, E. (1). Über die Einwirkung der Halogenessigsäurevergiftung auf die Löslichkeit der Muskeleiweisskörper. *Z. f. physiol. Chem.*, **192**, 233. 1930.

Embden, G. & Norpath, L. (1). Über die Einwirkung der Bromessigsäurevergiftung auf die Ammoniakbildung in der Froschmuskulatur. *Z. f. physiol. Chem.*, **201**, 105. 1931.

Embden, G., Riebeling, C. & Selter, G. E. (1). Über die Bedeutung der Adenylsäure für die Muskelfunktion. 2. Mitteilung: Die Desaminierung der Adenylsäure durch Muskelbrei und die Ammoniakbildung bei der Muskelkontraktion. *Z. f. physiol. Chem.*, **179**, 149. 1928.

Embden, G. & Schmidt, G. (1). Über Muskeladenylsäure und Hefeadenylsäure. *Z. f. physiol. Chem.*, **181**, 130. 1929.

Embden, G. & Schmidt, G. (2). Berichtigung. *Z. f. physiol. Chem.*, **197**, 191. 1931.

Embden, G., Schmitz, E. & Meinke, P. (1). Über den Einfluss der Muskelarbeit auf den Lactacidogengehalt der quergestreiften Muskulatur. *Z. f. physiol. Chem.*, **113**, 10. 1921.

Embden, G. & Wassermeyer, H. (1). Über die Bedeutung der Adenylsäure für die Muskelfunktion. 3. Mitteilung. Das Verhalten der Ammoniakbildung bei der Muskelarbeit unter verschiedenen biologischen Bedingungen. *Z. f. physiol. Chem.*, **179**, 161. 1928.

Embden, G. & Wassermeyer, H. (2). Über die Bedeutung der Adenylsäure für die Muskelfunktion. 5. Mitteilung: Die Quelle des bei der Kontraktion gebildeten Ammoniaks. *Z. f. physiol. Chem.*, **179**, 226. 1928.

Embden, G. & Zimmermann, M. (1). Über die Chemie des Lactacidogens. IV. *Z. f. physiol. Chem.*, **141**, 225. 1924.

Embden, G. & Zimmermann, M. (2). Über die Chemie des Lactacidogens. V. *Z. f. physiol. Chem.*, **167**, 114. 1927.

Embden, G. & Zimmermann, M. (3). Beitrag zur Chemie der Muskulatur. *Ber. ü. d. gesamte Physiol. u. exper. Pharmakol.*, **38**, 157. 1927.

Embden, G. & Zimmermann, M. (4). Über die Bedeutung der Adenylsäure für die Muskelfunktion. 1. Mitteilung: Das Vorkommen von Adenylsäure in der Skelettmuskulatur. *Z. f. physiol. Chem.*, **167**, 137. 1927.

Endo, M. (1). Entry of a dye into the sarcotubular system of muscle. *Nature*, **202**, 1115. 1964.

Endo, M., Nonomura, Y., Masaki, T., Ohtsuki, I. & Ebashi, S. (1). Localization of native tropomyosin in relation to striation patterns. *J. Biochem.*, **60**, 605. 1966.

Engelhardt, V. A. (1). Enzymatic and mechanical properties of muscle proteins. *Yale J. Biol. Med.*, **15**, 21. 1942.

Engelhardt, V. A. (2). Adenosinetriphosphatase properties of myosin. *Adv. in Enzymology*, **6**, 147. 1946.

Engelhardt, V. A. (3). Mechanochemistry and enzymology. *Conference on the chemistry of muscular contraction*, p. 134. Tokyo, Igaku Shoin Ltd. 1958.

Engelhardt, V. A. (4). Ortho- und pyrophosphat im aeroben und anaeroben Stoffwechsel der Blutzellen. *Biochem. Z.*, **227**, 16. 1930.

Engelhardt, V. A. (5). Die Beziehungen zwischen Atmung und Pyrophosphatumsatz in Vogelerythrocyten. *Biochem. Z.*, **251**, 343. 1932.

Engelhardt, V. A. & Burnasheva, S. A. (1). On the localisation of the spermosin protein in sperm cells. *Biokhimiya*, **22**, 554, 1957.

Engelhardt, V. A. & Lyubimova, M. N. (1). Myosin and adenosinetriphosphatase. *Nature*, **144**, 668. 1939.

Engelhardt, V. A. & Lyubimova, M. N. (2). On the mechanochemistry of muscle. *Biokhimiya*, 7, 205. 1942.

Engelhardt, V. A., Lyubimova, M. N. & Meitina, R. A. (1). Chemistry and Mechanics of the muscle studied on myosin threads. *Compt. rend. de l'Acad. Sci. de l'URSS*, 30, 644. 1941.

Engelhardt, V. A. & Sakov, N. E. (1). On the mechanism of the Pasteur effect. *Biokhimiya*, 8, 9. 1943.

Engelmann, G. J. (1). Schwefelsäureausscheidung und Phosphorsäureausscheidung bei körperlicher Arbeit. *Arch. f. (Anat. u.) Physiol.*, p. 14. 1871.

Engelmann, T. W. (1). Contractilität und Doppelbrechung. *Arch. f. ges. Physiol.*, 11, 432. 1875.

Engelmann, T. W. (2). On the nature of muscular contraction. *Proc. Roy. Soc. B.*, 411. 1895.

Engelmann, T. W. (3). Mikroskopische Untersuchungen über die Quergestreifte Muskelsubstanz. II. *Arch. f. ges. Physiol.*, 7, 155. 1873.

Engelmann, T. W. (4). Mikrometrische Untersunchungen an contrahirten Muskelfasern. *Arch. f. ges. Physiol.*, 23, 571. 1880.

Engelmann, T. W. (5). Mikroscopische Untersuchungen über die quergestreifte Muskelsubstanz I. *Arch. f. ges. Physiol.*, 7, 33. 1873.

Engelmann, T. W. (6). Neue Untersuchungen über die mikroskopischen Vorgänge bei der Muskelkontraktion. *Arch. f. ges. Physiol.*, 18. 1. 1878.

Ennor, A. H. & Morrison, J. F. (1). Biochemistry of the phosphagens and related guanidines. *Physiol. Rev.*, 38, 631. 1958.

Ennor, A. H. & Rosenberg, H. (1). An investigation into the turnover rates of organophosphates. I. Exta-cellular space and intra-cellular inorganic phosphate in skeletal muscle. *Biochem. J.*, 56, 302. 1954.

Ennor, A. H. & Rosenberg, H. (2). An investigation into the turnover rates of organophosphates. 2. The rate of incorporation of ^{32}P into adenosine triphosphate and phosphocreatine in skeletal muscle. *Biochem. J.*, 56, 308. 1954.

Enomoto, M. (1). Genetic studies of paralysed mutants in *Salmonella*. I. Genetic fine structure of the MOT loci in *Salmonella typhimurium*. *Genetics*, 54, 715. 1966.

Eppenberger, H. M., Dawson, D. M. & Kaplan, N. O. (1). The comparative enzymology of creatine kinases. I. Isolation and characterisation from chicken and rabbit tissues. *J. biol. Chem.*, 242, 204. 1968.

Eppenberger, H. M., Eppenberger, M., Richterich, R. & Aebi, H. (1). The ontogeny of creatine kinase enzymes. *Dev. Biol.*, 10, 1. 1964.

Erb, W. (1). Über den 'juvenile Form' der progressive Muskelatrophie und ihre Beziehungen zur sogenannten Pseudohypertrophie der Muskeln. *Deutsch. Arch. klin. Med.*, 34, 467. 1884.

Erdös, T. (1). Rigor, contracture and ATP. *Stud. Inst. med. Chem. Univ. Szeged.*, III, 51. Ed. A. Szent Györgyi. Basel & New York, S. Karger, 1944.

Erdös, T. (2). On the relation of the activity and contraction of actomyosii threads. *Stud. Inst. med. Chem. Univ. Szeged.*, III, 57. 1944.

Erdös, T. & Snellman, O. (1). Electrophoretic investigations of crystallised myosin *Biochim. biophys. Acta*, 2, 642. 1948.

Erlander, S. R., Koffler, H. & Foster, F. F. (1). Physical properties of flagellin from *Proteus vulgaris*, a study involving the application of the Archibald sedimentation principle. *Arch. Biochem. Biophys.*, 90, 139. 1960.

Ermini, M. & Schaub, M. C. (1). Postnatal development of adenosine triphosphatases in red and white rat muscles. *Z. f. physiol. Chem.*, 349, 1266. 1968.

Ernst, E. (1). Untersuchungen über Muskelkontraktion. 1. Mitteilung. Volumänderung bei der Muskelkontraktion. *Arch. f. ges. Physiol.*, **209**, 613. 1925.

Ernst, E. (2). Untersuchungen über Muskelkontraktion. V. Mitteilung. Uber Zeitverhältnisse der Volumverminderung. *Arch. f. ges. Physiol.*, **214**, 240. 1926.

Ernst, E. (8). Untersuchungen über Muskelkontraktion. VI. Volumverminderung und Aktionsstrom. Beitrag zur Ionentheorie der Reizung. *Arch. f. ges. Physiol.*, **218**, 137. 1928.

Ernst, E. & Koczkás, J. (1). Die Volumverminderung des Muskels als Erregungserscheinung. *Arch. f. ges. Physiol.*, **235**, 389. 1935.

Ernst, E. & Koczkás, J. (2). Eigenfrequenz und Reversibilität der Volumverminderung des Muskels. *Arch. f. ges. Physiol.*, **239**, 691. 1938.

Ernst, E., Tigyi, J. & Örkényi, J. (1). Frequenz und Zeitverhältnisse der Volumerminderung des Muskels. *Acta physiol. Acad. Sci. Hung.*, **2**, 281. 1951.

Ernst, E., Tigyi, J. & Sebes, T. A. (1). Temperaturkoeffizient der Volumverminderung des Muskels. *Acta physiol. Acad. Sci. Hung.*, **6**, 181. 1954.

Ernster, L. (1). The phosphorylation occurring in the flavoprotein region of the respiratory chain. In *Proc. Fifth International Congress of Biochemistry*, Vol. 5, 115. Ed. E. C. Slater. Oxford, London, New York, Paris, Pergamon Press. 1961.

Essner, E., Novikoff, A. B. & Quintana, N. (1). Nucleoside phosphatase activities in rat cardiac muscle. *J. Cell Biol.*, **25**, 201. 1965.

Estabrook, R. W. & Maitra, P. K. (1). A fluorimetric method for the quantitative microanalysis of adenine mucleotides and pyridine nucleotides. *Anal. Biochem.*, **3**, 369. 1962.

Estabrook, R. W. & Sacktor, B. (1). α-Glycerophosphate oxidase of flight muscle mitochondria. *J. biol. Chem.*, **233**, 1014. 1958.

Este, J. E. & Moos, C. (1). A study of the exchangeability of actin-bound ADP during superprecipitation of IDP-actomyosin. *Fed. Proc.*, **24**, 400. 1965.

Euler, H. von & Adler, E. (1). Über die Komponente der Dehydrasesysteme. VI. Dehydrierung von Hexosen unter Mitwirkung von Adenosintriphosphorsäure. *Z. f. physiol. Chem.* **235**, 122. 1935.

Euler, H. von, Adler, E. & Günther, G. (1). Über die Komponenten der Dehydrasesysteme. XV. Zur Kenntnis der Dehydrierung von α-Glycerinphosphorsäure im Tierkörper. *Z. f. physiol. Chem.*, **249**, 1. 1937.

Euler, H. von, Adler, E., Günther, G. & Das, N. B. (1). Über den enzymatischen Abbau und Aufbau der Glutaminsäure. II. in tierischen Gewebe. *Zeitsch. physiol. Chem.*, **254**, 61. 1938.

Euler, H. von, Albers, H. & Schlenk, F. (1). Über die Co-Zymase. *Z. f. physiol. Chem.*, **237**, i. 1936.

Euler, H. von Albers, H. & Schlenk, F. (2). Chemische Untersuchungen an hochgereinigter Co-Zymase. *Z. f. physiol. Chem.*, **240**, 113. 1936.

Euler, H. von, & Myrbäck, K. (1). Der Anteil der Hexosemonophosphat am enzymatischen Zuckerabban. *Liebigs Ann.*, **464**, 56. 1928.

Euler, H. von & Myrbäck, K. (2). Gärungsprobleme. *Z. f. physiol. Chem.*, **181**, 1. 1929.

Euler, H. von & Myrbäck, K. (3). Co-zymase XVI. Weitere Isolierungsversuche. *Z. f. physiol. Chem.*, **184**, 163. 1929.

Euler, H. von & Myrbäck, K. (4). Gärungs-Co-Enzym (Co-Zymase) der Hefe. I. *Z. physiol. Chem.*, **131**, 179. 1923.

Euler, H. von & Nilsson, R. (1). Beiträge zur Kenntnis der Co-Zymase und der Co-Reduktase. *Z. f. physiol. Chem.*, **160**, 234. 1926.

Evans, D. H. L., Schild, H. O. & Thesleff, S. (1). Effects of drugs on depolarised plain muscle. *J. Physiol.*, **143**, 474. 1958.

Fabrenbach, W. R. (1). Sarcoplasmic reticulum: ultra-structure of the triadic junction. *Science*, **147**, 1308. 1965.

Fabricius Hieronymus (1). *Opera omnia anatomica et physiologica.* Leipzig. 1687.

Falloppio, Gabrielle. (1). *Opera genuina omnia.* Venice. 1606.

Fanburg, B. L., Drachman, D. B. & Moll, D. (1). Calcium transport in isolated sarcoplasmic reticulum during muscle maturation. *Nature*, **218**, 962. 1968.

Fanburg, B., Finkel, R. M. & Martonosi, A. (1). The role of calcium in the mechanism of relaxation of cardiac muscle. *J. biol. Chem.*, **239**, 2299. 1964.

Fanburg, B. & Gergely, J. (1). Studies on adenosine triophosphate-supported calcium accumulation by cardiac subcellular particles. *J. biol. Chem.*, **240**, 2721. 1965.

Farber, E., Sternberg, W. H. & Dunlop, C. E. (1). Biochemical localization of specific oxidative enzymes. I. Tetrazolium stains for diphosphopyridine nucleotide diaphorase and triphosphopyridine nucleotide diaphorase. *J. Histochem. Cytochem.*, **4**, 254. 1956.

Farquhar, M. & Palade, G. E. (1). Functional complexes in various epithelia. *J. Cell Biol.*, **17**, 375. 1962.

Farquharson, A. S. L. (1). (tr.). [Aristotle's] *De Motu Animalium.* In *The Works of Aristotle*, tr. ed. J. A. Smith & W. D. Ross. Vol. 5. Oxford, Clarendon Press. 1912.

Fawcett, D. W. & Revel, J. P. (1). The sarcoplasmic reticulum of a fast-acting fish muscle. *J. Cell Biol.*, **10**, (Suppl.) 89. 1961.

Fenn, W. O. (1). A quantiative comparison between the energy liberated and the work performed by the isolated sartorius muscle of the frog. *J. Physiol.*, **58**, 175. 1923.

Fenn, W. O. (2). The relation between the work performed and the energy liberated in muscular contraction. *J. Physiol.*, **58**, 373. 1923–24.

Fenn, W. O. (3). Electrolytes in muscle. *Physiol. Rev.*, **16**, 450. 1936.

Fenn, W. O. & Cobb, D. M. (1). Electrolyte changes in muscle during activity. *Amer. J. Physiol.*, **115**, 345. 1936.

Fenn, W. O. & Marsh, B. S. (1). Muscular force of different speeds of shortening. *J. Physiol.*, **85**, 277. 1935.

Fernández-Morán, H., Oda, T., Blair, P. V. & Green, D. E. (1). A macromolecular repeating unit of mitochondrial structure and function. Correlated electron microscopic and biochemical studies of isolated mitochondria and submitochondrial particles of beef heart muscle. *J. Cell Biol.*, **22**, 63. 1964.

Feuer, G., Molnár, F., Pettkó, E. & Straub, F. B. (1). Studies on the composition and polymerisation of actin. *Hung. Acta physiol.*, **1**, 150. 1948.

Feuer, G. & Wollemann, M. (1). Untersuchungen über die Mechanismus der Aktinpolymerisation I. *Acta physiol. Acad. Sci. Hung.*, **3**, 267. 1952.

Fick, A. (1). Einige Bemerkungen zu Engelmann's Abhandlung über den Ursprung der Muskelkraft. *Pfl. Arch. f. ges. Physiol.*, **53**, 606. 1893.

Fick, A. (2). *Mechanische Arbeit und Warmeentwicklung bei der Muskelthätigkeit.* Leipzig, F. A. Brockhaus. 1882.

Fick, A. (3). *Medizinische Physik.* Braunschweig, F. Viewee. 1885.

Fick, A. (4). Ueber das Wesen der Muskelarbeit. In Virchow & Holtzendorff's *Sammlung von wissenschaftliche Vorträge.* XII Serie, p. 328. 1877.

Fick, A. (5). Myothermische Fragen und Versuche. *Verhandl. d. Phys.-med. Gesellsch. zu Wurzburg*, **18**. Also *Gesam. Schriften, II*, p. 295. 1884.

Fick, A. (6). Neue Beiträge zur Kenntnis von der Wärme entwicklung im Muskel. *Arch. f. ges. Physiol.*, **51**, 541. 1892.

Fick, A. (7). Ueber der Aenderung der Elasticität des Muskels während der Zuckung. *Pfl. Arch. f. ges. Physiol.*, **4**, 301. 1871.

Fick, A. & Wislicenus, J. (1). Ueber die Entstehung der Muskelkraft. *Vierteljahrsschr. d. Züricher naturforsch. Gesellsch.*, **10**, 317. 1865.

Filliozat, J. (1). *La Doctrine classique de la Médicine Indienne; Ses Origines et ses Parallèles Grecs.* Paris, Imprimerie nationale. 1949.

Filo, R. S., Bohr, D. F. & Rüegg, J. C. (1). Glycermated skeletal and smooth muscle: calcium and magnesium dependence. *Science*, **147**, 1581. 1965.

Filo, R. S., Rüegg, J. C. & Bohr, D. F. (1). Actomyosin-like protein of arterial wall. *Amer. J. Physiol.*, **205**, 1247. 1963.

Finck, H. (1). Immunochemical studies on myosin III. Immunochemical comparisons of myosins from chicken skeletal, heart and smooth muscles. *Biochim. biophys. Acta*, **111**, 231. 1965.

Finck, H. (2). On the discovery of actin. *Science*, **160**, 332. 1968.

Finck, H. & Holtzer, H. (1). Attempts to detect myosin and actin in cilia and flagella. *Exp. Cell Res.*, **23**, 251. 1961.

Finck, H., Holtzer, H. & Marshall, J. M. (1). An immunochemical study of the distribution of myosin in glycerol-extracted muscle. *J. biophys. biochem. Cytol.*, Suppl. 2, 175. 1956.

Fine, I. H., Kaplan, N. O. & Kuftinec, D. (1). Developmental changes of the mammalian lactic dehydrogenases. *Biochemistry (Amer. chem. Soc.)*, **2**, 116. 1963.

Finkel, R. M. & Gergely, J. (1). Studies on cardiac myofibrillar adenosine triphosphatase. *J. biol. Chem.*, **236**, 1458. 1961.

Finlayson, B., Lymn, R. W. & Taylor, E. W. (1). Studies on the kinetics of formation and dissociation of the actomyosin complex. *Biochemistry (Am. chem. Soc.)*, **8**, 811. 1969.

Finlayson, B. & Taylor, E. W. (1). Hydrolysis of nucleoside triphosphates by myosin during the transient state. *Biochemistry (Am. chem. Soc.)*, **8**, 802. 1969.

Fischer, E. (1). The oxygen consumption of isolated muscles for isotonic and isometric twitches. *Amer. J. Physiol.*, **96**, 78. 1931.

Fischer, E. (2). Die Wärmebildung des Skeletsmuskels bei indirekter und direkter Reizung, sowie bei der Reflexzuckung. *Arch. f. ges. Physiol.*, **219**, 514. 1928.

Fischer, E. (3). The birefringence of striated and smooth mammalian muscles. *J. comp. cell. Physiol.*, **23**, 113. 1944.

Fischer, E. (4). Vertebrate smooth muscle. *Physiol. Rev.*, **24**, 467. 1944.

Fischer, E. (5). Zur Energetik der Muskelkontraktion. *Naturwissensch.*, **18**, 736. 1930.

Fischer, E. (6). Die Wärmebildung des Skeletmuskels bei aufgehobener Milchsäurebildung. *Arch. f. ges. Physiol.*, **226**, 500. 1931.

Fischer, E. & Ramsey, V. W. (1). Changes in protein content and in some physicochemical properties of the protein during muscular atrophies of various types. *Amer. J. Physiol.*, **145**, 571. 1945–46.

Fischer, E. & Ramsey, V. W. (2). The effect of daily electrical stimulation of normal and denervated muscles upon their protein content and upon some of the physicochemical properties of the protein. *Amer. J. Physiol.*, **145**, 583. 1945–46.

Fischer, E. H., Graves, D. J., Snyder, E. R., Crittenden & Krebs, E. G. (1). Structure of the site phosphorylated in the phosphorylase $b \to a$ reaction. *J. biol. Chem.*, **234**, 1698. 1959.

Fischer, E. H. & Krebs, E. G. (1). Conversion of phosphorylase b to phosphorylase a in muscle extracts. *J. biol. Chem.*, **121**. 1955.

Fischer, H. (1). Zur Physiologie der quergestreiften Muskeln der Saugetiere. *Arch., f. ges. Physiol.* **125**, 541. 1908.

Fischer, H. (2). Über Porphyrine und ihre Synthesen. *Ber. d. Deutsch. chem. Gesellsch.*, **60**, ii, 2611. 1927.

Fischer, H. O. L. & Baer, E. (1). Über die 3-Glyceraldehyd-phosphorsäure. *Ber. d. Deutsch. chem. Ges.*, **65**, 337. 1932.

Fiske, C. H. & Subbarow, Y. (1). The colorimetric determination of phosphorus. *J. biol. Chem.*, **66**, 375. 1925.

Fiske, C. H. & Subbarow, Y. (2). The nature of the inorganic phosphate in voluntary muscle. *Science*, **65**, 401. 1927.

Fiske, C. H. & Subbarow, Y. (3). The isolation and function of phosphocreatine. *Science*, **67**, 169. 1928.

Fiske, C. H. & Subbarow, Y. (4). Phosphocreatine. *J. biol. Chem.*, **81**, 629. 1929.

Fiske, C. H. & Subbarow, Y. (5). Phosphorus compounds of muscle and liver. *Science*, **70**, 381. 1929.

Fitzpatrick, K., Park, D. C., Pennington, R. J., Robinson, J. E. & Worsfold, M. (1). Further studies on muscle cathepsins. *Research on muscular dystrophy*. Proceedings of the fourth symposium on current research in muscular dystrophy. p. 374. London, Pitman, Medical Publishing Company. 1968.

Fleckenstein, A., Gerlach, E., Janke, J. & Marmier, P. (1). Die Inkorporation von markiertem Sauerstoff aus Wasser in die ATP-, Kreatinphosphat- und Orthophosphat-Fraktion intakter Muskeln bei Ruhe, tetanischer Reizung und Erholung. *Arch. f. ges. Physiol.*, **271**, 75. 1960.

Fleckenstein, A. & Janke, J. (1). Papierchromatographische Trennung von ATP, ADP und anderen Phosphor-Verbindungen im kontrahierten und erschlafften Froschmuskel. *Arch. f. ges. Physiol.*, **258**, 177 1953.

Fleckenstein, A., Janke, J. & Davies, R. E. (1). Der Austausch von radioaktivem Phosphat mit dem α-, β- und γ- Phosphor von ATP und mit Kreatinphosphor bei der Kontraktur des Froschrectus durch Acetylcholin, Nicotin und Succinyl-bischolin. *Arch. exp. Path. Pharm.*, **228**, 596. 1956.

Fleckenstein, A., Janke, J. & Davies, R. E. (2). Der Einbau von ^{32}P in die energierreichen Phosphorverbindungen bei ruhenden elektrisch gereizten und dauerkontrahierten Froschmuskeln. *Naturwiss.*, **43**, 185. 1956..

Fleckenstein, A., Janke, J., Davies, R. E. & Krebs, H. A. (1). Contraction of muscle without fission of adenosine triphosphate or creatine phosphate. *Nature*, **174**, 1081. 1954.

Fleckenstein, A., Janke, J., Lechner, G. & Bauer, G. (1). Zerfällt Adenosinetri-phosphat bei der Muskelkontraktion? *Arch. f. ges. Physiol.*, **259**, 246. 1954.

Fleisch, A. (1). Some oxidation processes of normal and cancerous tissue. *Biochem. J.*, **18**, 294. 1924.

Fleming, D. (1). Galen on the motions of the blood. *Isis*, **46**, 14. 1955.

Fleming, D. (2). William Harvey and the pulmonary circulation. *Isis*, **46**, 319. 1955.

Fletcher, W. M. (1). The survival respiration of muscle. *J. Physiol.*, **23**, 10. 1898–99.

Fletcher, W. M. (2). The influence of oxygen upon the survival respiration of muscle. *J. Physiol.*, **28**, 354. 1902.

Fletcher, W. M. (3). The relation of oxygen to the survival metabolism of muscle. *J. Physiol.*, **28**, 474. 1902.

Fletcher, W. M. (4). Lactic acid formation, survival respiration and rigor mortis in mammalian muscle. *J. Physiol.*, **47**, 361. 1913–14.

Fletcher, W. M. & Brown, J. M. (1). The carbon dioxide production of heat rigor in muscle and the theory of intra-molecular oxygen. *J. Physiol.*, **48**, 177. 1914.

Fletcher, W. M. & Hopkins, F. G. (1). Lactic acid in amphibian muscle. *J. Physiol.*, **35**, 247. 1907.

Fletcher, W. M. & Hopkins, F. G. (2). Croonian Lecture of 1915: The respiratory process in muscle and the nature of muscular motion. *Proc. Roy. Soc.* B, **89**, 444. 1917.

Flock, E. V. & Bollmann, J. L. (1). Adenosine triphosphate in muscles of rats studied with radioactive phosphorus. *J. biol. Chem.*, **152**, 371. 1944.

Flock, E. V. & Bollmann, J. L. (2). Resynthesis of muscle glycogen after exercise. *J. biol. Chem.*, **136**, 469. 1940.

Florkin, M. (1). *Théorie cellulaire dans l'Oeuvre de Théodore Schwann.* Paris, Hermann. 1960.

Florkin, M. (2). Personality and scientific research. Opening lecture, *XII Int. Congr. Hist. Sci.* Paris, 1968.

Fluharty, A. & Sanadi, D. R. (1). Evidence for a vicinal dithiol in oxidative phosphorylation. *Proc. Nat. Acad. Sci. U.S.A.*, **46**, 608. 1966.

Forsheit, A. B. & Hayashi, T. (1). The effects of colchicine on contractile proteins. *Biochim. biophys. Acta*, **147**, 546. 1967.

Foster, D. L. & Moyle, D. M. (1). A contribution to the study of the interconversion of carbohydrate and lactic acid in muscle. *Biochem. J.*, **15**, 672, 1921.

Foster, M. (1). *Lectures on the History of Physiology during the sixteenth, seventeenth and eighteenth Centuries.* Cambridge University Press. 1901.

Foster, M. (2). *Textbook of Physiology*, 3rd edition, p. 440. London, Macmillan & Co. 1879.

Foster, M. (3). *Claude Bernard.* London, Fisher Unwin. 1899.

Foulks, J. G. & Perry, F. A. (1). The time course of early changes in the rate of tension development in electrically-stimulated frog toe muscle: effects of muscle length, temperature and twitch-potentiators. *J. Physiol.*, **185**, 355. 1966.

Fourcroy, A. de (1). Sur l'existance de la matière albumineuse dans les végétaux. *Ann. de Chim.*, **3**, 252. 1789.

Fourcroy, A. de (2). Extrait d'un memoire ayant pour titre Recherches pour servir à l'histoire du gaz azote ou de la mofette, comme principe des matières animales. *Ann. de. Chim.*, **1**, 40. 1789.

Fowler, L. R. & Richardson, S. H. (1). Studies of the electron transfer system. L. On the mechanism of reconstitution of the mitochondrial electron transfer system. *J. biol. Chem.*, **238**, 456. 1963.

Fraenkel, G. (1). The distribution of vitamin B_T (carnitine) throughout the animal kingdom. *Arch. Biochem. Biophys.* **50**, 486. 1954.

Frank, G. B. (1). Inward movement of calcium as a link between electrical and mechanical events in contraction. *Nature*, **182**, 1800. 1958.

Frank, G. B. (2). Effects of changes in extracellular calcium concentration on the potassium-induced contracture of frog's skeletal muscle. *J. Physiol.*, **151**, 518, 1960.

Frank, G. B. (3). Role of extracellular calcium ions in excitation-contraction coupling in frog's skeletal muscle. In *Biophysics of physiological and pharmacological actions*. Ed. A. M. Shanes, p. 293. Publ. No. 69, Amer. Ass. Adv. Sci. Washington, D. C. 1961.

Frank, G. M. (1). Problems concerning the physical and physico-chemical basis of muscular contraction. *Izvestiya Akademii Nauk SSSR. Seriya Biologicheskaya*, No. 3, p. 335. (Trans. FASEB Translation Project) 1965.

Frank, G. M. (2). Some problems of the physical and physico-chemical bases of muscle contraction. *Proc. Roy. Soc.* B, **160**, 473. 1964.

Frank, O. (1). Thermodynamik des Muskels. *Erg. Physiol.*, **3**, 348. 1904.

Frankenhaeuser, B. (1). The effect of calcium on the myelinated nerve fibre. *J. Physiol.*, **137**, 245. 1957.

Frankenhaeuser, B. & Hodgkin, A. L. (1). The action of calcium on the electrical properties of squid axons. *J. Physiol.*, **137**, 218. 1957.

Franklin, K. J. (1) (tr.) [*William Harvey's*] *Exercitatio Anatomica de Motu Cordis et Sanguinis in Animalibus. Frankfurt, 1628*. Oxford, Blackwell. 1957.

Franzini-Armstrong, C. (1). Fine structure of sarcoplasmic reticulum and transverse tubular system in muscle fibers. *Fed. Proc.*, **23**, 887. 1964.

Franzini-Armstrong, C. & Porter, K. R. (1). The Z disc of skeletal muscle fibrils. *Zeits. f. Zellforsch. u. mikr. Anat.*, **61**, 661. 1964.

Franzini-Armstrong, C. & Porter, K. R. (2). Sarcolemmal invaginations and the T-system in fish skeletal muscle. *Nature*, **202**, 355. 1964.

Freeman, K. (1). *The Pre-Socratic Philosophers; a Companion to Diels' 'Fragmente der Vorsokratiker.'* Oxford, Blackwell, 1946.

Freeman, K. (2). *Ancilla to the Pre-Socratic Philosophers; a complete Translation of the Fragments in Diels' 'Fragmente der Vorsokratiker.'* Oxford, Blackwell, 1948.

Frieden, C. (1). L-Glutamate dehydrogenase. In *The Enzymes*. Vol. 7, pp. 1–24. Ed. P. D. Boyer, H. Lardy & K. Myrbäck. New York and London, Academic Press. 1963.

Friedkin, M. & Lehninger, A. L. (1). Esterification of inorganic phosphate coupled to electron transport between dihydrophosphopyridine nucleotide and oxygen. *J. biol. Chem.*, **178**, 611. 1949.

Friedman, D. L. & Larner, J. (1). Conversions of two activities of UDPG-glycogen transglucosylase. *Fed. Proc.*, **21**, 206. 1962.

Friedman, D. L. & Larner, J. (2). Studies on UDPG-α-glucan transglycosylase. III. Interconversion of two forms of muscle UDPG-α-glucan transglycosylase by a phosphorylation-dephosphorylation reaction sequence. *Biochemistry (Amer. Chem. Soc.)*, **2**, 669. 1963.

Friedman, D. L. & Larner, J. (3). Studies on uridine diphosphate glucose: α-1,4-glucan α-4-glucosyl transferase. VIII. Catalysis of the phosphorylation of muscle phosphorylase and transferase by separate enzymes. *Biochemistry (Am. chem. Soc.)*, **4**, 2261. 1965.

Friedman, S. & Fraenkel, G. (1). Reversible enzymatic acetylation of carnitine. *Arch. Biochem. Biophys.*, **59**, 491. 1955.

Friess, E. T. (1). The effect of a chelating agent on myosin ATPase. *Arch. Biochem. Biophys.*, **51**, 17. 1954.

Friess, E. T. & Morales, M. F. (1). Kinetic studies of the myosin-tripolyphosphate system. *Arch. Bioch. Biophys.*, **56**, 325. 1955.

Friess, E. T., Morales, M. F. & Bowen, W. J. (1). Some further observations on the interaction of EDTA with the myosin-ATP system. *Arch. Biochem. Biophys.*, **53**, 311. 1954.

Fritz, I. B. (1). The effects of muscle extracts on the oxidation of palmitic acid by liver slices and homogenates. *Acta physiol. Scand.*, **34**, 367. 1955.

Fritz, I. B. (2). Action of carnitine on long chain fatty acid oxidation by liver. *Amer. J. Physiol.*, **197**, 297. 1959.

Fritz, I. B. (3). Factors influencing the rates of long-chain fatty acid oxidation and synthesis in mammalian systems. *Physiol. Rev.*, **41**, 52. 1961.

Fritz, I. B. (4). An hypothesis concerning the role of carnitine in the control of interrelations between fatty acid and carbohydrate metabolism. *Perspectives in Biol. and Med.*, **10**, 643. 1967.

Fritz, I. B. & McEwen, B. (1). Effects of carnitine on fatty acid oxidation by muscle. *Science*, **129**, 334. 1959.

Fritz, I. B. & Marquis, N. R. (1). The role of acylcarnitine esters and carnitine palmityl transferase in the transport of fatty acyl groups across mitochondrial membranes. *Proc. Nat. Acad. Sci.*, **54**, 1226. 1965.

Fritz, I. B. & Yue, K. T. N. (1). Long-chain carnitine acyltransferase and the role of acylcarnitine derivatives in the catalytic increase of fatty acid oxidation induced by carnitine. *J. of Lipid Res.*, **4**, 279. 1963.

Froberg, S. O. (1). Determination of muscle lipids. *Biochim. biophys. Acta*, **144**, 83. 1967.

Fulton, J. F. & Cushing, H. (1). A bibliographical study of the Galvani and the Aldini writings on animal electricity. *Ann. Sci.*, **1**, 239. 1936.

Fürth, O. von. (1). Über die Eiweisskörper des Muskleplasmas. *Arch. f. exp. Path. u. Pharmak.*, **36**, 231. 1895.

Fürth, O. von. (2). Über die Einwirkung von Giften auf die Eiweisskörper des Muskelplasma und ihre Beziehung zur Muskelstarre. *Arch. f. expt. Path. und Pharmak.*, **37**, 389. 1896.

Fürth, O. von. (3). Die Kolloidchemie des Muskels und ihre Beziehungen zu den Problemen der Kontraktion und der Starre. *Ergeb. d. Physiol.*, **17**, 363. 1919.

Fürth, O. von. (4). Stoffwechsel des Herzens und des Muskels. *Oppenheimers Handbuch der Biochemie.* 2nd ed. p. 31. Jena, Gustav Fischer. 1925.

Fürth, O. von & Lenk, E. (1). Die Bedeutung von Quellungs- und Entquellungsvorgängen für den Eintritt und die Lösung der Totenstarre. *Arch. f. ges. Physiol.*, **23**, 341. 1911.

Furusawa, K. & Hartree, W. (1). The anaerobic delayed heat production in muscle. *J. Physiol.*, **62**, 203. 1926–27.

Furusawa, W. & Kerridge, P. M. T. (1). The hydrogen ion concentration of the muscles of the cat. *J. Physiol.*, **63**, 33. 1927.

Fusari, R. (1). Sur la structure des fibres musculaires striées. *Arch. Italienne Biol.*, **22**, 95. 1895.

Gabelova, N. A. (1). Evidence for conformational changes in actomyosin and myofibrils on ATP-contraction. *Inf. Exch. Grp.* **4**, No. 35. 1965.

Gad, J. (1). Zur Theorie der Erregungsvorgänge im Muskel. *Arch. f. Anat. u. Physiol.* (Müller's), p. 164. 1893.

Gaddie, R. & Stewart, C. P. (1). The role of glutathione in muscle glycolysis. *Biochem. J.*, **29**, 2101. 1935.

Gaetjens, E. & Bárány, M. (3). N-acetylaspartic acid in G-actin. *Biochim. biophys. Acta*, **117**, 176. 1966.

Gaetjens, E., Cheung, H. S. & Bárány, M. (1). The absence of free NH_2-terminal residues in L-myosin. *Biochim. biophys. Acta*, **93**, 188. 1964.

Galen, Claudius. See Kühn and Daremberg.

Galvani, A. (1). *Abhandlung über die Kräfte der Electricität bei der Muskelbewegung.* Leipzig, W. Engelmann. (Ostwald's Klassiker der exakten Wissenschaften. Nr. 52). 1894.

Ganotte, C. E. & Moses, H. L. (1). Light and dark cells as artifacts of liver fixation. *Laboratory Investigation*, **18**, 740. 1968.

Gansler, H. (1). Elektronmikroskopische Untersuchungen am Uterusmuskel der Ratte unter Follikelhormonwirkung. *Virchows Arch. Path. Anat.*, **329**, 235. 1956.

Gansler, H. (2). Phasenkontrast und elektron mikroskopische Untersuchungen zur Morphologie und Funktion der glatten Muskulatur. *Z. Zellforsch.*, **52**, 60. 1960.

Gansler, H. (3). Struktur und Funktion der glatten Muskulatur. II. Licht und elektronmikroskopische Befunde an Hohlorganen von Ratte, Meerschweinchen und Mensch. *Z. Zellforsch.*, **55**, 724. 1961.

Garamvölgyi, N. (1). The arrangement of the myofilaments in the insect flight muscle. I. *J. Ultrastr. Res.*, **13**, 409. 1965.

Garamvölgyi, N. (2). The arrangement of the myofilaments in the insect flight muscle. II. *J. Ultrastr. Res.*, **13**, 425. 1965.

Garamvölgyi, N. (3). In *Symposium on Muscle, Symposia biologica Hungarica.* Vol. 8, 248. Ed. E. Ernst & F. B. Straub. Akadémia Kiado, Budapest. 1968.

Garamvölgyi, N. & Belági, J. (1). Mechanical properties of the flight muscle of the bee, I. Resting elasticity and its ultrastructural interpretation. *Acta Biochim. Biophys. Acad. Sci. Hung.*, **3**, 195. 1968.

Garamvölgyi, N & Belági, J. (2). Mechanical properties of the flight muscle of the bee, III. The reversibility of extreme degrees of stretch. *Acta Biochim. Biophys. Acad. Sci. Hung.*, **3**, 299. 1968.

Garcia-Buñuel, L. & Garcia-Buñuel, V. M. (1). Connective tissue and the pentose phosphate pathway in normal and denervated muscle. *Nature*, **213**, 913. 1967.

Garland, P. B., Newsholme, E. A. & Randle, P. J. (1). Effect of fatty acids, ketone bodies, diabetes and starvation on pyruvate metabolism in rat heart and diaphragm muscle. *Nature*, **192**, 381. 1962.

Garland, P. B. & Randle, P. J. (1). Effects of alloxan diabetes and adrenaline on concentrations of free fatty acids in rat heart and diaphragm muscles. *Nature*, **199**, 381. 1963.

Garland, P. B. & Randle, P. J. (2). Regulation of glucose uptake by muscle. 10. Effects of alloxan-diabetes, starvation, hypophysectomy and adrenalectomy, and of fatty acids, ketone bodies and pyruvate, on the glycerol output and concentration of free fatty acids, long-chain fatty acyl-coenzyme A, glycerol phosphate and citrate-cycle intermediates in rat heart and diaphragm muscles. *Biochem. J.*, **93**, 678. 1964.

Garland, P. B., Randle, P. J. & Newsholme, E. A. (1). Citrate as an intermediary in the inhibition of phosphofructokinase in rat heart muscle by fatty acids, ketone bodies, pyruvate, diabetes and starvation. *Nature*, **200**, 169. 1963.

Garner, W. E. (1). The mechanism of muscular contraction. *Proc. Roy. Soc. B*, **99**, 40. 1925–26.

Gaspar-Godfroid, A. (1). L'activité adénosine triphosphatasique de la tonomyosine de carotides de bovidé. *Angiologica*, **1**, 12. 1964.

646 **Machina Carnis**

Gaspar-Godfroid, A. (2). Influence du β-mercaptoéthanol sur l'état d'agrégation et l'activité de la H-meromyosine de carotide de bovidé. *Biochim. biophys. Acta*, **167**, 622. 1968.

Gaspar-Godfroid, A., Hamoir, G. & Laszt, L. (1). Influence de la méthode de préparation sur les propriétés macromolécularies de l'actine de carotides de bovidés. *Angiologica*, **4**, 323. 1967.

Gaspar-Godfroid, A., Hamoir, G., & Laszt, L. (2). Isolement de la F-actine de carotides de bovidés par fractionnement de la tonoactomyosine au sulfate ammonique. *Angiologica*, **5**, 186. 1968.

Gasser, H. S. & Hill, A. V. (1). The dynamics of muscular contraction. *Proc. Roy. Soc.* B, **96**, 398. 1924.

Gautheron, D., Gaudemer, Y. & Zajdela, F. (1). Isolement de sarcosomes d'uterus de porc et leurs propriétés oxydophosphorylantes comparée à celle de sarcosomes de coeur. *Bull. Soc. chim. Biol.* (Paris), **43**, 193. 1961.

Gauthier, G. F. & Padykula, H. A. (1). Cytological studies of fiber types in skeletal muscle. *J. Cell Biol.*, **28**, 333. 1966.

Gehuchten, van A. (1). Etude sur la structure intime de la cellule musculaire striée. La cellule II. 1886.

Geiduschek, E. P. & Holtzer, A. (1). Application of light-scattering to biological systems: deoxyribonucleic acid and the muscle proteins. *Adv. in biol. and med. Physics*, **6**, 431. 1958.

Gellert, M. F., Hippel, P. H. von, Schachman, H. K. & Morales, M. F. (1). Studies on the contractile proteins of muscle. I. The ATP-myosin B interaction. *J. Am. chem. Soc.*, **81**, 1384. 1959.

Gemmill, C. L. (1). The effect of exercise on the acetone bodies in the blood of man on low carbohydrate diet. *Amer. J. Physiol.*, **108**, 55. 1934.

Gemmill, C. L. (2). The utilization of carbohydrate during aerobic activity in isolated frogs' muscles. *Amer. J. Physiol.*, **112**, 294. 1936.

Gemmill, C. L. (3). The fuel of muscular exercise. *Physiol. Rev.*, **22**, 32. 1942.

George, J. C. & Jyoti, D. (1). Histological features of the breast and leg muscle of bird and bat and their physiological and evolutionary significance. *J. Animal Morph. Physiol.* **2**, 1. 1955.

George, J. C. & Jyoti, D. (2). The lipid content and its reduction in the muscle and liver during long and sustained muscular activity. *J. Animal Morph., Physiol.* **2**, 10. 1955.

George, J. C. & Naik, J. M. (1). Relative distribution and chemical nature of the fuel store of the two types of fibres in the pectoralis major muscle of the pigeon. *Nature*, **181**, 709. 1958.

George, J. C. & Naik, J. M. (2). Relative distribution of mitochondria in the two types of fibres in the pectoralis major muscle of the pigeon. *Nature*, **181**, 709. 1958.

George, J. C. & Scaria, K. S. (1). On the occurrence of lipase in the skeletal muscles of vertebrates and its possible significance in sustained muscular activity. *J. Anim. Morph. Physiol.*, **3**, 91. 1956.

George, J. C. & Scaria, K. S. (2). Histochemical demonstration of lipase activity in the pectoralis major muscle of the pigeon. *Nature*, **181**. 783. 1958.

George, J. C. & Talesara, C. L. (1). A quantitative study of the distribution pattern of certain oxidizing enzymes and α-lipase in the red and white fibres of the pigeon breast muscle. *J. cell. comp. Physiol.*, **58**, 253. 1961.

Gerendás, M. (1). Technisches über Myosinfäden nebst einigen Beobachtungen über ihre Kontraktion. *Studies from the Inst. Med. Chem. Univ. Szeged.*, **1**, 47. 1943.

Gerendás, M. & Matoltsy, A. G. (1). Double refraction of the N-protein. (1). *Hung. Acta Physiol.*, **1**, 124. 1947.

Gerendás, M., Szarvas, P. & Matoltsy, A. G. (1). Spectroscopic investigation of the N-protein in muscle. *Hung. Acta Physiol.*, **1**, 121. 1947.

Gergely, J. (1). The interaction between actomyosin and adenosinetriphosphate. Light scattering studies. *J. biol. Chem.*, **220**, 917. 1956.

Gergely, J. (2). Muscle proteins and energy utilisation. *Ann. New York Acad. Sci.*, **72**, 538. 1959.

Gergely, J. (3). A note on the thermodynamics of muscle contraction. *Enzymologia*, **14**, 220. 1950.

Gergely, J. (4). Relation of ATPase and myosin. *Fed. Proc.*, **9**, 176. 1950.

Gergely, J. (5). Studies on myosin-adenosine triphosphatase. *J. biol. Chem.*, **200**, 543. 1953.

Gergely, J., Gouvea, M. A. & Karibian, D. (1). Fragmentation of myosin by chymotrypsin. *J. biol. Chem.*, **212**, 165. 1954.

Gergely, J., Gouvea, M. A. & Martonosi, A. (1). Studies on actin. II. Partially polymerised actin solution. *J. biol. Chem.*, **235**, 1704. 1960.

Gergely, J., Kaldor, G. & Briggs, F. N. (1). Participation of a dialysable cofactor in the relaxing factor system of muscle. II. Studies with myofibrillar ATPase. *Biochim. biophys. Acta*, **34**, 218. 1959.

Gergely, J. & Kohler, H. (1). Light scattering studies on the stepwise formation and dissociation of myosin B. In *Conference on the chemistry of muscular contraction*, p. 14. Tokyo, Igaku Shoin Ltd. 1958.

Gergely, J. & Kohler, H. (2). Molecular parameters of cardiac myosin. *Fed. Proc.*, **16**, 185. 1957.

Gergely, J. & Laki, K. (1). A thermodynamic discussion of the length-tension diagram of muscle. *Enzymologia*, **14**, 272, 1950.

Gergely, J. & Maruyama, K. (1). The binding of inorganic phosphate to myosin in the presence of adenosine triphosphate. *J. biol. Chem.*, **235**, 3174. 1960.

Gergely, J., Pragay, D., Scholz, A. F., Seidel, J., Sreter, F. A. & Thompson, M. M. (1). Comparative studies on white and red muscle. In *Molecular Biology of muscular Contraction*. Pp. 145–159. Ed. S. Ebashi, F. Oosawa, T. Sekine & Y. Tonomura. Tokyo, Ikagu Shoin, Ltd. 1965.

Gersham, L. C., Dreizen, P. & Stracher, A. (1). Subunit structure of myosin II. Heavy and light alkali components. *Proc. nat. Acad. Sci. U.S.A.*, **56**, 966. 1966.

Geske, G., Ulbrecht, M. & Weber, H. H. (1). Der Einfluss der Mg-Konzentration auf die Substrathemmung der ATP-Spaltung als angebliche Ursache der Erschlaffung des Muskels. *Arch. exper. Path. u. Pharm.*, **230**, 301. 1957.

Ghosh, B. N. & Mihalyi, E. (1). Binding of ions by myosin. II. The effect of neutral salts on the acid-base equilibria of myosin. *Arch. Biochem. Biophys.*, **41**, 107. 1952.

Gibbons, I. R. (1). Studies on the protein components of cilia from Tetrahymena pyriformis. *Proc. nat. Acad. Sci. U.S.A.*, **50**, 1002. 1963.

Gibbons, I. R. (2). Studies on the adenosine triphosphatase activity of 14S and 30S dynein from cilia of *tetrahymena*. *J. biol. Chem.*, **241**, 5590. 1966.

Gibbons, I. R. & Rowe, A. J. (1). Dynein: A protein with adenosine triphosphatase activity from cilia. *Science*, **149**, 424. 1965.

Gierke, E. von. (1). Hepato-Nephromegalia glycogenica (Glykogenspeicher-krankheit der Leber und Nieren). *Beiträg. path. Anat.*, **82**, 497. 1929.

Gilmour, D. (1). Activity in relation to pH of myosin-5-nucleotidase. *Nature*, **186**, 295. 1960.

Gilmour, D. & Calaby, J. H. (1). Physical and enzymic properties of actomyosins from the femoral and thoracic muscles of an insect. *Enzymologia*, **16**, 23. 1953.

Gilmour, D. & Gellert, M. (1). The binding of p-chlormercuribenzoate by myosin. *Arch. Biochem. Biophys.*, **93**, 605. 1961.

Girardier, L., Reuben, J. P., Brandt, P. W. & Grundfest, H. (1). Evidence for anion-permselective membrane in crayfish muscle fibres and its possible role in excitation-contraction coupling. *J. gen. Physiol.*, **47**, 189. 1963–4.

Gladner, J. A. & Falk, J. E. (1). Carboxypeptidase B. II. Mode of action of protein substrates and its application to carboxyl terminal group analysis. *J. biol. Chem.*, **231**, 393. 1958.

Glaister, G. & Kerly, M. (1). The oxygen consumption and carbohydrate metabolism of the retractor muscle of the foot of *Mytilus edulis*. *J. Physiol.*, **87**, 56. 1936.

Gleiss, W. (1). Ein betrag zur Muskelchemie. *Arch. f. ges. Physiol.*, **41**, 69. 1887.

Glisson, F. (1). *Tractatus de Ventriculo et Intestini*.... Brome, London. 1677.

Glynn, I. M. (1). Transport adenosinetriphosphatase in electric organ. The relation between ion transport and oxidative phosphorylation. *J. Physiol.*, **169**, 452. 1963.

Godeaux, J. Action des toxiques de guerre et des vésicants sur les fils de myosine. *Bull. Soc. Roy. Sci. Liège*, **13**, 216. 1944.

Godlewski, E. (1). Die Entwicklung des Skelet- und Herzmuskelgewebes der Säugetiere. *Arch. mikr. Anat.*, **60**, 11. 1902.

Golarz, M. N. & Bourne, G. H. (1). Histochemical evidence for a possible primary biochemical lesion in muscular dystrophy. *J. Histochem. Cytochem.*, **11**, 286. 1963.

Golarz, M. N. & Bourne, G. H. (2). The histochemistry of muscular dystrophy. In *Muscular dystrophy in man and animals*, ed. G. H. Bourne & M. N. Golarz, pp. 90–156. Basel and New York, S. Karger. 1963.

Goldacre, R. J. & Lorch, I. J. (1). Folding and unfolding of protein molecules in relation to cytoplasmic streaming, amoeboid movement and osmotic work. *Nature*, **166**, 497. 1950.

Goldman, D. S. (1). Studies in the fatty acid oxidising system of animal tissues. VII. The β-ketoacyl coenzyme A cleavage enzyme. *J. biol. Chem.*, **208**, 345. 1954.

Goldspink, G. (1). Biochemical and physiological changes associated with the postnatal development of the biceps brachi. *Comp. Biochem. Physiol.*, **7**, 157. 1962.

Goldspink, G. (2). Studies on postembryonic growth and development of skeletal muscle. *Proc. roy. Irish Acad.*, **62** B, 135. 1962.

Goll, D. E., Mommaerts, W. F. H. M., Reedy, M. K. & Seraydarian, K. (1). Studies on α-actinin-like proteins liberated during trypsin digestion of α-actinin and of myofibrils. *Biochim. biophys. Acta*, **175**. 174. 1969.

Gollnick, P. D. & Hearn, G. R. (1). Lactic dehydrogenase activities of heart and skeletal muscle of exercised rats. *Amer. J. Physiol.*, **201**, 694. 1961.

Goodall, F. R. (1). Degradative enzymes in the uterine myometrium of rabbits under different hormonal conditions. *Arch. Biochem. Bioph.*, **112**, 403. 1965.

Goodall, M. C. (1). Carnosine phosphate as phosphate donor in muscular contraction. *Nature*, **178**, 539. 1956.

Goodall, M. C. & Szent-Györgyi, A. G. (1). Relaxing factors in muscle. *Nature*, **172**, 84. 1953.

Goodford, P. J. (1). The calcium content of the smooth muscle of the guinea-pig taenia coli. *J. Physiol.*, **192**, 145. 1967.

Gordon, A. M., Huxley, A. F. & Julian, F. J. (1). The length-tension diagram of single vertebrate striated muscle fibres. *J. Physiol.*, **171**, 28P. 1964.

Gordon, A. M., Huxley, A. F. & Julian, F. J. (2). Tension development in highly stretched vertebrate muscle fibres. *J. Physiol.*, **184**, 143. 1966.

Gordon, A. M., Huxley, A. F. & Julian, F. J. (3). The variation in isometric tension with sarcomere length in vertebrate muscle fibres. *J. Physiol.*, **184**, 170. 1966.

Gorodissky, H. (1). Über die anaerobe Resynthese von Phosphokreatin nach der Reizung isolierter Froschmuskeln. *Z. f. physiol. Chem.*, **175**, 261. 1928.

Gould, M. K. & Rawlinson, W. A. (1). Biochemical adaptation as a response to exercise. 1. Effect of swimming on the levels of lactic dehydrogenase, malic dehydrogenase and phosphorylase in muscles of 8-, 11- and 15-week-old rats. *Biochem. J.*, **73**, 41. 1959.

Gözsy, B. & Szent-Györgyi, A. (1). Über den Mechanismus der Hauptatmung des Taubenbrustmuskels. *Z. physiol. Chem.*, **224**, 1. 1934.

Graaf, G. L. A., Hudson, A. J. & Strickland, K. P. (1). Aldolase and 'aminoacyl arylamidase' in the denervated rat gastrocnemius muscle. *Can. J. Biochem.*, **43**, 699. 1965.

Graham, G. N. (1). Direct extraction of actin in the fibrous form from fish muscle. *Nature*, **203**, 405. 1964.

Graham, G. N. & Wilson, R. (1). A partial characterisation of Fn-actin. *Biochem. J.*, **95**, 7P. 1965.

Grant, R. J., Cohen, L. B., Clark, E. E. & Hayashi, T. (1). A low viscosity G-actin preparation. *Biochem. biophys. Res. Com.*, **16**, 314. 1964.

Gratzer, W. B. & Lowey, S. (1). Effect of substrate on conformation of myosin. *J. biol. Chem.*, **244**, 22. 1969.

Gray, J. (1). The movements of sea-urchin spermatozoa. *J. exp. Biol.*, **32**, 775. 1955.

Green, D. E. (1). The mitochondrial electron-transfer system. In *Comprehensive Biochemistry*, Vol. 14, pp. 309–326. Ed. M. Florkin & E. H. Stotz. Amsterdam, London & New York, Elsevier Pub. Co. 1966.

Green, D. E. (2). Co-enzyme Q and electron transport. In CIBA Foundation symposium on *Quinones in electron transport*, pp. 130–159. London, J. & A. Churchill. 1961.

Green, D. E. (3). Fatty acid oxidation in soluble systems of animal tissues. *Biol. Rev.*, **29**, 330. 1954.

Green, D. E. (4). α-Glycerophosphate dehydrogenase. *Biochem. J.*, **30**, 629. 1936.

Green, D. E., Asai, J., Harris, R. A. & Penniston, J. T. (1). Conformational basis of energy transformations in membrane systems. III. Configurational changes in the mitochondrial inner membrane induced by changes in functional states. *Arch. Biochem. Biophys.*, **125**, 684. 1968.

Green, D. E., Loomis, W. F. & Auerbach, V. H. (1). Studies on the cyclophorase system. 1. The complete oxidation of pyruvic acid to carbon dioxide and water. *J. biol. Chem.*, **172**, 389. 1948.

Green, D. E., Mii, S., Mahler, H. R. & Bock, R. M. (1). Studies on the fatty acid oxidising system of animal tissues. III. Butyryl coenzyme A dehydrogenase. *J. biol. Chem.*, **206**, 1. 1954.

Green, D. E., Needham, D. M. & Dewan, J. G. (1). Dismutations and oxido-reductions. *Biochem. J.*, **31**, 2327. 1937.

Greenstein, J. P. & Edsall, J. T. (1). The effect of denaturing agents on myosin. 1. Sulfhydryl groups as estimated by porphyrindin titration. *J. biol. Chem.*, **133**, 397. 1940.

Greven, K. (1). Über den Mechanismus der Regulierung der Kontraktionsstärke beim glatten Muskel durch tetanische und quantitative (räumliche) Summation. *Z. Biol.*, **106**, 377. 1953–4.

Greville, G. D. & Needham, D. M. (1). Effect of 2:4-dinitropheonl and phenyl-mercuric acetate on enzymic activity of myosin. *Biochim. biophys. Acta*, **16**, 284. 1955.

Greville, G. D. & Reich, E. (1). Effects of 2:4-dinitrophenol and other agents on the nucleotide triphosphatase activities of L-myosin. *Biochim. biophys. Acta*, **20**, 440. 1956.

Griffiths, D. E. (1). Oxidative phosphorylation. In *Essays in Biochemistry*, pp. 91–120. Ed. for the Biochem. Soc. by P. N. Campbell & G. D. Greville. London & New York, Academic Press. 1965.

Griffiths, D. E., Morrison, J. F. & Ennor, A. H. (1). The distribution of guanidines, phosphagens and N-amidino phosphokinases in Echinoids. *Biochem. J.*, **65**, 612. 1957.

Griffiths, D. E. & Wharton, D. C. (1). Studies of the electron transport system. XXXV. Purification and properties of cytochrome oxidase. *J. biol. Chem.*, **236**, 1850. 1961.

Griffiths, D. E. & Wharton, D. C. (2). Studies of the electron transport system. XXXVI. Properties of copper in cytochrome oxidase. *J. biol. Chem.*, **236**, 1857. 1961.

Grillo, T. A. I. (1). A histochemical study of phosphorylase in the tissues of the chick embryo. *J. Histoch. Cytoch.*, **9**, 386. 1961.

Grillo, T. A. I. & Ozone, K. (1). Uridine diphosphate glucose-glycogen synthetase activity in the chick embryo. *Nature*, **195**, 902. 1962.

Grimstone, A. V. & Klug, A. (1). Observations on the substructure of flagellar fibres. *J. Cell Science*, **1**, 351. 1966.

Gröschel-Stewart, U. (1). Vergleichende Untersuchungen an kontraktilen Proteinen aus glatter und quergestreifter Muskulatur. Habilitationsschrift, Universität, Würzburg. 1969.

Gröschel-Stewart, U. & Doniach, D. (1). Immunological evidence for human myosin iso-enzymes. *Immunology*, **17**, 991. 1969.

Gröschel-Stewart, U., Rüdiger, H. & Turba, F. (1). Zur selektive Hemmung und Lokalisierung der Calcium-sensitiven Adnosintriphosphatase im H-meromyosin. *Biochem. Z.*, **339**, 539. 1964.

Gröschel-Stewart, U. & Turba, F. (1). Zuordnung der Adenosintriphosphatase und des Actinbindungsvermögens des H-Meromyosins zu bestimmten SH-Bezirken. *Biochem. Z.*, **337**, 109. 1963.

Grubhofer, N. & Weber, H. H. (1). Über Actin-Nucleotide und die Funktion und Bindung der Nucleotidphosphate in G- und F-Actin. *Z. f. Naturforsch.*, **16B**, 435. 1961.

Gruda, J., Kakol, I. & Niemierko, W. (1). Direct transfer of orthophosphate from adenosine triphosphate to myosin and H-meromyosin. *Acta. biochim. Pol.*, **9**, 215. 1962.

Gruda, J., Kakol, I. & Rzysko, C. (1). Phosphomyosin and phospho-H-meromyosin formation during splitting of ATP. *Bull. Acad. Pol. Sci.*, **8**, 129. 1960.

Grützner, P. (1). Über die Reizwirkung der Stöhrerschen Maschine auf Nerv und Muskel. *Arch. f. ges. Physiol.*, **41**, 256. 1887.

Grützner, P. (2). Die glatten Muskeln. *Erg. Physiol.*, **3**, II, 12. 1904.

Guba, F. (1). Effect of halogen ions on F-actin. *Nature*, **165**, 439. 1950.

Guba, F. (2). In *Symposium on Muscle*, pp. 48, 104, 234, 250. *Symposia biologica Hungarica*, Vol. 8. 1968.

Günther, H. (1). Über den Muskelfarbstaff. *Virchows Arch. f. path. Anat. u. Physiol.*, **230**, 146. 1921.

Guth, L. & Watson, P. K. (1). The influence of innervation on the soluble proteins of slow and fast muscles of the rat. *Exp. Neurol.*, **17**, 107. 1967.

Gutmann, E. (1). Evidence for the trophic function of the nerve cell in neuromuscular relations. In *The effect of use and disuse on neuromuscular functions*. Proceedings of a symposium, Liblice. Ed. E. Gutmann & P. Hnik. Prague, Publishing House of the Czechoslovak Acad. Sci. 1963.

Gutmann, E., Bass, A., Vodička, Z. & Vrbová, G. (1). Nervous control of 'trophic' processes in striated muscle. *Physiol. Bohemoslov.*, **5**, 14. 1956.

Gutmann, E., Hanikova, M., Hajek, I., Klicpera, M. & Syrovy, I. (1). The postdenervation hypertrophy of the diaphragm. *Physiol. Bohemoslov.*, **15**, 508. 1966.

Gutmann, E. & Hanzlikova, V. (1). Contracture responses of fast and slow mammalian muscles. *Physiol. Bohemoslov.*, **15**, 404. 1966.

Gutmann, E. & Sandow, A. (1). Caffeine-induced contracture and potentiation of contraction in normal and denervated rat muscle. *Life Sciences*, **4**, 1149. 1965.

Gutmann, E. & Syrový, I. (1). Metabolic differentiation of the anterior and posterior latissimus dorsi of the chicken during development. *Physiol. Bohemoslov.*, **16**, 232. 1967.

Gutmann, E. & Zak, R. (1). Nervous regulation of nucleic acid level in crossstriated muscle. Changes in denervated muscle. *Physiol. Bohemoslov.*, **10**, 493. 1961.

Gutmann, E. & Zak, R. (2). Nervous regulation of nucleic acid levels in crossstriated muscle. Resynthesis of nucleic acids and proteins in normal and denervated muscle. *Physiol. Bohemoslov.*, **10**, 501. 1961.

Hahn, A. (1). Zur Thermodynamik des Erholungsvorgänges im Muskel. *Zeitsch. f. Biol.*, **91**, 444. 1931.

Hahn, A. (2). Über Dehydrierungsvorgänge im Muskel. *Z. f. Biol.*, **92**, 355. 1932.

Hahn, A., Fischbach, E. & Haarmann, W. (1). Über die Dehydrierung der Milchsäure. *Z. f. Biol.*, **88**, 516. 1929.

Hahn, L. A., Hevesy, G. C. & Rebbe, O. H. (1). Do the potassium ions inside the muscle cells and blood corpuscles exchange with those present in the plasma? *Biochem. J.*, **33**, 1549. 1939.

Hajdu, S. (1). Behaviour of frog and rat muscle at higher temperatures. *Enymologia*, **14**, 187. 1950.

Hajdu, S. (2). The action of chloroform on the $\Delta F'$ curve of the frog muscle. *Enzymologia*, **14**, 194. 1950–51.

Hajdu, S. & O'Sullivan, R. B. (1). The $\Delta F'$ slope of frog muscle. *Enzymologia*, **14**, 182. 1950.

Hájek, I., Gutmann, E., Klicpera, M. & Syrový, I. (1). The incorporation of S³⁵ methionine into proteins of denervated and reinnervated muscle. *Physiologia Bohemoslov*, 15, 148. 1966.

Hájek, I., Gutmann, E. & Syrový, I. (1). Changes of proteolytic activity of proteins following denervation in the anterior and posterior latissimus dorsi of the chicken. *Physiol. Bohemoslov.*, 15, 1. 1966.

Hales, S. (1). *Statical essays*. Vol. II. London, W. Innys & R. Manby. 1733.

Hall, C. E., Jakus, M. A. & Schmitt, F. O. (1). The structure of certain muscle fibrils as revealed by the use of electron stains. *J. applied Phys.*, 16, 459. 1945.

Hall, C. E., Jakus, M. A. & Schmitt, F. O. (2). An investigation of cross-striations and myosin filaments in muscle. *Biol. Bull.*, 90, 32. 1946.

Haller, A. von (1). A dissertation on the sensible and irritable parts of animals. J. Nourse, London. Reprinted in the *Bull. Hist. of Med.*, 4, 651, 1936.

Haller, A. von (1 a). *Memoires sur la Nature sensible et irritable des Parties du Corps animal*. Lausanne, Bousquet et Ce. 1761.

Haller, A. von (2). *Elementa Physiologiae*, vol. 4, Lib. x, *Cerebrum et Nervei*. Lausanne, 1766.

Halliburton, W. D. (1). On muscle-plasma. *J. Physiol.*, 8, 133. 1887.

Ham, A. W. (1). *Histology*. 5th edition. Philadelphia, Pitman Medical Publishing Co. Ltd. London, J. P. Lippincott Co. 1965.

Hama, K. & Porter, R. K. (1). An application of high voltage electron microscopy to the study of biological materials. *J. de Micr*, 8, 149. 1969.

Hamoir, G. (1). Electrophoretic study of the muscle structural proteins. *Disc. of the Faraday Soc.*, 13, 116. 1953.

Hamoir, G. (2). Fish tropomyosin and fish nucleotropomyosin. *Biochem. J.*, 48, 146. 1951.

Hamoir, G. (3). Further investigations on fish tropomyosin and fish nucleotropomyosin. *Biochem. J.*, 50, 140. 1951.

Hamoir, G. (4). The muscle proteins of the vascular wall. *Angiologica*, 6, 190. 1969.

Hamoir, G. (5). The comparative biochemistry of fish sarcoplasmic proteins. *Acta zool. et path. Antverpiensa*, 46, 69. 1968.

Hamoir, G. & Gaspar-Godfroid, A. (1). Comparaison de la thermostabilité de la tonoactomyosine et de la tonomyosine de carotides de bovidé et de leur groupement atépasique avec celle des protéines correspondantes du muscle strié. *Angiologica*, 1, 317. 1964.

Hamoir, G., Gaspar-Godfroid, B. & Laszt, L. (1). Changements d'état d'agrégation et de dissociation de la tonoactomyosine de carotides de bovidés sous l'influence de la force ionique et de l'ATP. *Angiologica*, 2, 44. 1965.

Hamoir, G. & Konosu, S. (1). Carp myogens of white and red muscles. General composition and isolation of low-molecular-weight components of abnormal amino acid composition. *Biochem. J.*, 96, 85. 1965.

Hamoir, G. & Laszt, L. (1). Tonomyosin of arterial muscle. *Nature*, 193, 682. 1962.

Hamoir, G. & Laszt, L. (2). La tropomyosine B de carotides de bovidés. *Biochim. biophys. Acta*, 59, 365. 1962.

Hansford, R. G. & Chappell, J. B. (1). The effect of Ca²⁺ on the oxidation of glycerolphosphate by blowfly flight-muscle mitochondria. *Biochem. biophys. Res. Com.*, 27, 686. 1967.

Hanson, J. (1). Axial period of actin filaments. *Nature*, 213, 353. 1967.

Hanson, J. (2). Changes in the cross-striation of myofibrils during contraction induced by adenosine triphosphate. *Nature*, 169, 530. 1952.

Hanson, J. (3). X-ray diffraction of muscle. *Quart. Rev. Biophys.*, 1, 177. 1968.

Hanson, J. (4). Elongation of cross-striated myofibrils. *Biochim. biophys. Acta*, 20, 289. 1956.

Hanson, J. & Huxley, H. E. (1). Structural basis of the cross-striations in muscle. *Nature*, 172, 530. 1953.

Hanson, J. & Huxley, H. E. (2). The structural basis of contraction in striated muscle. *Symp. Soc. Exp. Biol.*, 9, 228. 1955.

Hanson, J. & Huxley, H. E. (3). Quantitative studies in the structure of cross-striated myofibrils: II. Investigations by biochemical techniques. *Biochim. biophys. Acta*, 23, 250. 1957.

Hanson, J. & Lowy, J. (1). The structure of F-actin and of actin filaments isolated from muscle. *J. mol. Biol.*, 6, 46. 1963.

Hanson, J. & Lowy, J. (2). The structure of actin filaments and the origin of the axial periodicity in the I-substance of vertebrate striated muscle. *Proc. Roy. Soc. B*, 160, 449. 1964.

Hanson, J. & Lowy, J. (3). Comparative studies on the structure of contractile systems. *Circulation Research*, Suppl. II. to vols 14 & 15, pt. II, p. 4. 1964.

Hanson, J. & Lowy, J. (4). The presence of a double array of myofilaments in certain invertebrate smooth muscles. *J. Physiol.*, 137, 42P. 1957.

Hanson, J. & Lowy, J. (5). Structure of smooth muscles. *Nature*, 180, 906. 1957.

Hanson, J. & Lowy, J. (6). Evidence for a sliding filament contractile mechanism in tonic smooth muscles of lamellibrauch molluscs. *Nature*, 184, 286. 1959.

Hanson, J. & Lowy, J. (7). The structure of the muscle fibres in the translucent part of the adductor of the oyster *Crassostrea angulata*. *Proc. Roy. Soc. B*, 154, 173. 1961.

Hanson, J. & Lowy, J. (8). The problem of location of myosin in vertebrate smooth muscle. *Proc. Roy. Soc. B*, 160, 523. 1964.

Hanson, J. & Lowy, J. (9). Structure and function of the contractile apparatus in the muscles of invertebrate animals. In *Structure and function of muscle*. Vol. I, pp. 265–335. Ed. G. H. Bourne. New York and London, Acad. Press. 1960.

Hanson, J., Lowy, J., Huxley, H. E., Bailey, K., Kay, C. M. & Rüegg, J. C. (1). Structure of molluscan tropomyosin. *Nature*, 180, 1134. 1957.

Harary, I. & Farley, B. (1). *In vitro* studies of single isolated beating heart cells. *Science*, 131, 1674. 1960.

Harary, I. & Slater, E. C. (1). Studies on single beating heart cells. VIII. The effect of oligomycin, dinitrophenol and ouabain on the beating rate. *Biochim. biophys. Acta*, 99, 227. 1965.

Harden, A. (1). *Alcoholic Fermentation*. 4th Edition. London, Longmans Green & Co. 1932.

Harden, A. (2). The function of phosphate in alcoholic fermentation. *Nature*, 125, 313. 1930.

Harden, A. & Henley, F. R. (1). The equation of alcoholic fermentation. II. *Biochem. J.* 23, 230. 1929.

Harden, A. & Young, W. J. (1). The alcoholic ferment of yeast-juice. *Proc. Roy. Soc. B*, 77, 405. 1906.

Harden, A. & Young, W. J. (2). The alcoholic ferment of yeast juice. Part III. The function of phosphates in the fermentation of glucose by yeast-juice. *Proc. Roy. Soc. B*, 80, 299. 1908.

Harden, A. & Young, W. J. (3). The alcoholic ferment of yeast juice. Part VI. The effect of arsenates and arsenites on the fermentation of the sugars by yeast-juice. *Proc. Roy. Soc. B*, 83, 451. 1911.

Harden, A. & Young, W. J. (4). The alcoholic ferment of yeast-juice. Part II. The co-ferment of yeast-juice. *Proc. Roy. Soc. B.*, **78**, 369. 1906.

Harris, E. J, (1). The output of ⁴⁵Ca from frog muscle. *Biochim. biophys. Acta*, **23**, 80. 1957.

Harris, M. & Suelter, C. H. (1). A simple chromatographic procedure for the preparation of rabbit-muscle myosin A free from AMP deaminase. *Biochim. biophys. Acta*, **133**, 393. 1967.

Harris, R. A., Penniston, J. T., Asai, J. & Green, D. E. (1). The conformational basis of energy conservation in membrane sytems. II. Correlation between conformational change and functional states. *Proc. nat. Acad. Sci. U.S.A.*, **59**, 830, 1968.

Harris, R. J. C. (1). Ed. *Protein Biosynthesis*. London & New York, Academic Press. 1961.

Harting, J. (1). Oxidation of acetaldehyde by glyceraldehyde-3-phosphate dehydrogenase of rabbit muscle. *Fed. Proc.*, **10**, 195. 1951.

Harting, J. & Velick, S. F. (1). Transfer reactions of acetyl phosphate catalysed by glyceraldehyde-3-phosphate dehydrogenase. *J. biol. Chem.*, **207**, 867. 1954.

Hartmann, H. (1). Die Änderungen des Muskelvolumens bei der tetanischen Kontraktion als Ausdruck der chemischen Vorgänge im Muskel. *Biochem. Zeitsch.*, **270**, 164. 1934.

Hartree, W. (1). A revised analysis of the initial heat production of muscle. *J. Physiol.*, **79**, 492. 1933.

Hartree, W. (2). A negative phase in the heat production of muscle. *J. Physiol.*, **77**, 104. 1932.

Hartree, W. & Hill, A. V. (1). The recovery heat-production in muscle. *J. Physiol.*, **56**, 367. 1922.

Hartree, W. & Hill, A. V. (2). The energy liberated by an isolated muscle during the performance of work. *Proc. Roy. Soc. B*, **104**, 1. 1929.

Hartree, W. & Hill, A. V. (3). The anaerobic delayed heat production after a tetanus. *Proc. Roy. Soc. B*, **103**, 207. 1928.

Hartree, W. & Hill, A. V. (3a). The factors determining the maximum work and the mechanical efficiency of muscle. *Proc. Roy. Soc. B*, **103**, 234. 1928.

Hartree, W. & Hill, A. V. (4). The regulation of the supply of energy in muscular contraction. *J. Physiol.*, **55**, 133. 1921.

Hartshorne, D. J. & Mueller, H. (1). Separation and recombination of the ethylene glycol bis (β-aminoethyl ether)-N,N′tetraacetic acid-sensitizing factor obtained from a low ionic strength extract of natural actomyosin. *J. biol. Chem.*, **242**, 3089. 1967.

Hartshorne, D. J. & Mueller, H. (2). The preparation of tropomyosin and troponin from natural actomyosin. *Biochim. biophys. Acta*, **175**, 301. 1969.

Hartshorne, D. J. & Perry, S. V. (1). A chromatographic and electrophoretic study of sarcoplasm from adult and foetal-rabbit muscles. *Biochem. J.*, **85**, 171. 1962.

Hartshorne, D. J., Perry, S. V. & Davies, V. (1). A factor inhibiting the ATPase activity and superprecipitation of actomyosin. *Nature*, **209**, 1352. 1966.

Hartshorne, D. J., Perry, S. V. & Schaub, M. C. (1). A protein factor inhibiting the magnesium-activated adenosine triphosphatase of desensitized actomyosin. *Biochem. J.*, **104**, 907. 1967.

Hartshorne, D. J., Theiner, M. & Mueller, H. (1). Studies on troponin. *Biochim. biophys. Acta*, **175**, 320. 1969.

Harvey, E. N. (1). *A history of luminescence*. Philadelphia, The American Philosophical Society. 1957.

Harvey, E. N. (2). Bioluminescence. In *Comparative Biochemistry*, Vol. II, pp. 545–591. Ed. M. Florkin & H. S. Mason. New York and London, Academic Press. 1960.

Harvey, W. (1). *De motu locali animalium*. 1627. See Whitteridge (1).

Harvey, W. (2). *Second essay to Jean Riolan*. 1649. See Franklin.

Harvey, W. (3). *The Anatomical Lectures of William Harvey*. See Whitteridge (2).

Hasselbach, W. (1). Die Diffusionskonstante des Adenosintriphosphats im Inneren der Muskelfaser. *Z. f. Naturforsch.*, 7b, 334. 1952.

Hasselbach, W. (2). Die Umwandlung von Aktomyosin-ATPase in L-Myosin-ATPase durch Aktivatoren und die resultierenden Aktivierungseffekte. *Z. f. Naturforsch.*, 7b, 163. 1952.

Hasselbach, W. (3). Die Wechselwirkung verschiedener nukleosidtriphosphat mit aktomyosin im Gelzustand. *Biochim. biophys. Acta*, 20, 355. 1956.

Hasselbach, W. (4). Die Nucleosidtriphosphatase-aktivität von L-Myosin und Aktomyosin in Abhängigkeit von den ionalen Bedingungen. *Biochim. biophys. Acta*, 25, 365. 1957.

Hasselbach, W. (5). Die Bindung von Adenosinediphosphat, von anorganischen Phosphat und von Erdalkalien an die Strukturproteine des Muskels. *Biochim. biophys. Acta*, 25, 562. 1957.

Hasselbach, W. (6). Relaxing factor and the relaxation of muscle. *Progr. in Biophys. and mol. Biol.*, 14, 169. 1964.

Hasselbach, W. (7). Relaxation and the sarcolemmal calcium pump. *Fed. Proc.*, 23, 909. 1964.

Hasselbach, W. (8). Elektronmikroskopische Untersuchungen an Muskelfibrillen bei totaler und partieller Extraktion des L-Myosins. *Z. f. Naturforsch.*, 8b, 449. 1953.

Hasselbach, W. & Elfvin, L.-G. (1). Structural and chemical asymmetry of the calcium-transporting membranes of the sarcotubular system as revealed by electron microscopy. *J. Ultrastr. Res.* 17, 598. 1967.

Hasselbach, W. & Ledermair, O. (1). Der Kontraktionscyclus der isolierten kontractilen Strukturen der Uterusmuskulatur und seine Besonderheiten. *Pfl. Arch. ges. Physiol.*, 267, 532. 1958.

Hasselbach, W. & Makinose, M. (1). Die Calciumpumpe der 'Erschlaffungsgrana' des Muskels und ihre Abhängigkeit von der ATP-Spaltung. *Biochem. Z.*, 333, 518. 1961.

Hasselbach, W. & Makinose, M. (2). ATP and active transport. *Biochem. biophys. Res. Com.*, 7, 132. 1962.

Hasselbach, W. & Makinose, M. (3). Über den Mechanismus des Calciumtransportes durch die Membranen des sarkoplasmatischen Reticulums. *Biochem. Z.*, 339, 94. 1963.

Hasselbach, W. & Schneider, G. (1). Der L-Myosin- und Aktingehalt des Kaninchenmuskels. *Biochem. Z.*, 321, 461. 1951.

Hasselbach, W. & Seraydarian, K. (1). The role of sulfhydryl groups in calcium transport through the sarcoplasmic membranes of skeletal muscle. *Biochem. Zeitsch.*, 345, 159. 1966.

Hasselbach, W. & Weber, H. H. (1). Der Einfluss des M-B-Faktors auf die Kontraktion des Fasermodells. *Biochim. biophys. Acta*, 11, 160. 1953.

Hatano, S. & Oosawa F. (1). Isolation and characterization of plasmodium actin. *Biochim. biophys. Acta*, **127**, 488. 1966.

Hatano, S. & Tazawa, M. (1). Isolation, purification and characterization of myosin B from myxomycete plasmodium. *Biochim. biophys. Acta*, **154**, 507. 1968.

Hatano, S., Totsuka, T. & Oosawa, F. (1). Polymerization of plasmodium actin. *Biochim. biophys. Acta*, **140**, 109. 1967.

Hatefi, Y., Haavik, A. G., Fowler, L. R. & Griffiths, D. E. (1). Studies of the electron transfer system. XLII. Reconstitution of the electron transfer system. *J. biol. Chem.*, **237**, 2661. 1962.

Hatefi, Y., Haavik, A. G. & Griffiths, D. E. (1). Studies of the electron transport system. XL. Preparation and properties of mitochondrial DPNH-coenzyme Q reductase. *J. biol. Chem.*, **237**, 1676. 1962.

Hatefi, Y., Haavik, A. G. & Griffiths, D. E. (2). Studies of the electron transport system. XLI. Reduced coenzyme Q-cytochrome C reductase. *J. biol. Chem.*, **237**, 1681. 1962.

Hatefi, Y., Haavik, A. G. & Jurtshuk, P. (1). Studies of the electron transport system. XXX. DPNH-cytochrome C reductase. *Biochim. biophys. Acta*, **52**, 106. 1961.

Havel, R. J., Naimark, A. & Borchgrevink, C. F. (1). Turnover rate and oxidation of free fatty acids of blood plasma in man during exercise: Studies during continuous infusion of palmitate-1-C^{14}. *J. clin. Invest*, **42**, 1054. 1963.

Hayashi, T. (1). Contractile properties of compressed monolayers of actomyosin. *J. gen. Physiol.*, **36**, 139. 1952–53.

Hayashi, T. & Rosenbluth, R. (1). Studies on actin. II. Polymerisation and the bound nucleotide. *Biol. Bull.*, **119**, 294. 1960.

Hayashi, T. & Rosenbluth, R. (2). Actin polymerisation by direct transphosphorylation. *Bioch. Bioph. Res. Com.*, **8**, 20. 1962.

Hayashi, T., Rosenbluth, R. & Lamont, H. C. (1). Studies of fibres of acto- and paramyosin from lamellibrauch muscle. *Biol. Bull.*, **117**, 396. 1959.

Hayashi, T., Rosenbluth, R., Satir, P. & Vozick, M. (1). Actin participation in actomyosin contraction. *Biochim. biophys. Acta*, **28**, 1. 1958.

Hayashi, Y. & Tonomura, Y. (1). Dependence of activity of myofibrillar ATPase on sarcomere length and calcium ion concentration. *J. Biochem.*, **63**, 101. 1968.

Hearn, G. R. & Wainio, W. W. (1). Succinic dehydrogenase activity of the heart and skeletal muscle of exercised rats. *Amer. J. Physiol.*, **185**, 348. 1956.

Hearn, G. R. & Wainio, W. W. (2). Aldolase activity of the heart and skeletal muscle of exercised rats. *Amer. J. Physiol.*, **190**, 206. 1957.

Heidenhain, R. (1). *Mechanische Leistung, Wärmeentwicklung und Stoffumsatz bei der Muskeltatigkeit.* Leipzig, Breitkopf & Härtel. 1864.

Heilbrun, L. V. & Wiercinski, F. J. (1). The action of various cations on muscle protoplasm. *J. cell. comp. Physiol.*, **29**, 15. 1947.

Heinz, E. & Holton, F. (1). Die Abhangigheit der ATP-Spaltung von der ATP-Konzentration und die Spannungsentwicklung des Fasermodells. *Zeitsch. f. Naturforsch.*, **7b**, 386. 1952.

Helmholtz, H. (1). Physiologische Wärmeentscheinungen. *Fortschr. d. Physik*, p. 232. 1847.

Helmholtz, H. (2). Über die Erhaltung der Kraft. A lecture delivered in Berlin. Published in 1889 in Ostwald's *Klassiker der exakten Wissenschaften*. Leipzig, Wilhelm Engelmann. 1847.

Helmreich, E. & Cori, C. F. (1). The role of adenylic acid in the activation of phosphorylase. *Proc. Nat. Acad. Sci. U.S.A.*, **51**, 131. 1964.

Helmreich, E. & Cori, C. F. (2). Regulation of glycolysis in muscle. *Adv. in Enz. Regulation.*, **3**, 91. 1965.

Hensay, J. (1). Der Einfluss verschiedener Puffergemische auf die Löslichkeit von Eiweisskörpern des lebensfrischen, ermudeten, absterbenden und starren Muskels. *Arch. f. ges. Physiol.*, **224**, 44. 1930.

Hermann, D. L. (1). *Physiologie des Menschen.* Berlin, Hirschfeld. 1874.

Hermann, [D.] L. (2). *Untersuchungen über den Stoffwechsel der Muskeln, ausgehend von Gaswechsel derselben.* Berlin, Hirschfeld. 1867.

Hermann, [D.] L. (3). *Weitere Untersuchungen zur Physiologie der Muskeln und Nerven.* Berlin, Hirschfeld, 1867.

Hermann, D. [L.] (4). *Handbuch der Physiologie. Part I. Der Bewegungsapparate.* Leipzig, F.C.W. Vogel. 1879.

Hermansen, L., Hultman, E. & Saltin, B. (1). Muscle glycogen during prolonged severe exercise. *Acta physiol. Scand.*, **71**, 129. 1967.

Herriott, R. M., Anson, M. L. & Northrop, J. H. (1). Reaction of enzymes and proteins with mustard gas (bis(β-chlorethyl)) sulfide. *J. gen. Physiol.*, **30**, 185. 1945–47.

Herrmann, H. & Cox, W. W. (1). Content of inorganic phosphate and phosphate esters in muscle tissue of chick embryo. *Amer. J. Physiol.*, **165**, 711. 1951.

Herrmann, H. & Nicholas, J. S. (1). Quantitative changes in muscle protein fractions during rat development. *J. expt. Zool.*, **107**, 165. 1948.

Herrmann, H. & Nicholas, J. S. (2). Enzymatic liberation of inorganic phosphate from adenosine-triphosphate in developing rat muscle. *J. exp. Zool.*, **107**, 177. 1948.

Herrmann, H., Nicholas, J. S. & Vosgian, M. E. (1). Liberation of inorganic phosphate from adenosinetriphosphate by fractions derived from developing rat muscle. *Proc. Soc. exp. Biol. & Med.*, **72**, 454. 1949.

Hers, H. G. (1). α-Glucosidase deficiency in generalised glycogen-storage disease (Pompe's disease). *Biochem. J.*, **86**, 11. 1963.

Hers, H. G. (2). Glycogen storage disease; Type II. In the Ciba Foundation Symposium *Control of glycogen metabolism*, p. 354. Ed. by W. J. Whelan & M. P. Cameron. London, J. & A. Churchill Ltd. 1964.

Hers, H. G., Verhue, W. & Mathieu, M. (1). The mechanism of action of amylo-1, 6-glucosidase. In Ciba Foundation Symposium *Control of glycogen metabolism*, p. 151. Ed. by W. J. Whelan & M. P. Cameron. London, J. & A. Churchill Ltd. 1964.

Hertzog, R. O. & Jancke, W. (1). Röntgenographische Untersuchungen am Muskel. *Naturwissensch.*, **14**, 1223. 1926.

Herz, R. & Weber, A. (1). Caffeine inhibition of Ca uptake by muscle reticulum. *Fed. Proc.*, **24**, 208. 1965.

Hess, A. (1). The structure of slow and fast extrafusal muscle fibres in the extraocular muscles and their nerve endings in guinea pigs. *J. cell. comp. Physiol.*, **58**, 63. 1961.

Hess, A. & Pilar, G. (1). Slow fibres in the extraocular muscles of the cat. *J. Physiol.*, **169**, 780. 1963.

Hess, R. & Pearse, A. G. E. (1). Dissociation of uridine diphosphate glucoseglycogen transglucosylase from phosphorylase activity in individual muscle fibres. *Proc. Soc. exp. Biol. Med.*, **107**, 569. 1961.

Heumann, H.-G. & Zebe, E. (1). Über die Funktionsweise glatter Muskelfasern. Elektron mikroskopische Untersuchungen am Byssusretraktor (ABRM) von *Mytilus edulis. Zeitsch. f. Zellforsch.*, **85**, 534. 1968.

Heuson-Stiennon, J.-A. (1). Intervention de polysomes dans la synthèse des myofilaments du muscle embryonnaire du rat. *J. Micros.*, **3**, 229. 1964.

Hevesy, G. & Rebbe, O. (1). Molecular rejuvenation of tissue. *Nature*, **141**, 1097. 1938.

Hevesy, G. & Rebbe, O. (2). Rate of penetration of phosphate into muscle cells. *Acta Physiol. Scand.*, **1**, 171. 1940–41.

Heywood, S. M., Dowben, R. M. & Rich, A. (1). The identification of polyribosomes synthesizing myosin. *Proc. nat. Acad. Sci. U.S.A.*, **57**, 1002. 1967.

Heywood, S. M. & Rich, A. (1). In vitro synthesis of native myosin, actin and tropomyosin from embryonic chick polyribosomes. *Proc. Nat. Acad. Sci. U.S.A.*, **59**, 590. 1968.

Hicks, R. M. & Kerly, M. (1). Transaminase activity in the perfused rat heart. *J. Physiol.*, **150**, 621. 1960.

Hierons, R. & Meyer, A. (1). Willis's place in the history of muscle physiology. *Proc. roy. Soc. Med.*, **57**, 687. 1964.

Higashi, S. & Oosawa, F. (1). Conformational changes associated with polymerisation and nucleotide binding in actin molecules. *J. mol. Biol.*, **12**, 843. 1965.

Hikada, T., Osa, T. & Twarog, B. M. (1). The action of 5-hydroxytryptamine on *Mytilus* smooth muscle. *J. Physiol.*, **192**, 869.

Hill, A. V. (1). The heat production of surviving amphibian muscles during rest, activity and rigor. *J. Physiol.*, **44**, 466. 1912.

Hill, A. V. (2). The energy degraded in the recovery processes of stimulated muscles. *J. Physiol.*, **46**, 28. 1913.

Hill, A. V. (3). The oxidative removal of lactic acid. *J. Physiol.*, **48**, x, 1914.

Hill, A. V. (4). The recovery heat-production in oxygen after a series of twitches. *Proc. Roy. Soc.* B, **103**, 183. 1928.

Hill, A. V. (5). Die Beziehungen zwischen der Wärmebildung und den im Muskel stattfindenden chemischen Prozessen. *Erg. d. Physiol.*, **15**, 340. 1916.

Hill, A. V. (6). The energetics of relaxation in a muscle twitch. *Proc. Roy. Soc.* B, **136**, 211. 1949–50.

Hill, A. V. (7). The absolute mechanical efficiency of the contraction of an isolated muscle. *J. Physiol.*, **46**, 435. 1913.

Hill, A. V. (8). The heat-production in prolonged contractions of an isolated frog's muscle. *J. Physiol.*, **47**, 305. 1913–14.

Hill, A. V. (9). The heat-production in isometric and isotonic twitches. *Proc. Roy. Soc.* B, **107**, 115. 1930–31.

Hill, A. V. (10). Length of muscle, and the heat and tension developed in an isometric contraction. *J. Physiol.*, **60**, 237. 1925.

Hill, A. V. (11). *Trails and Trials in Physiology.* London, Edward Arnold, Ltd. 1965.

Hill, A. V. (12). The absolute value of the isometric heat coefficient Tl/H in a muscle twitch, and the effect of stimulation and fatigue. *Proc. Roy. Soc.* B, **103**, 163. 1928.

Hill, A. V. (12a). *Adventures in Biophysics.* Oxford University Press. 1931.

Hill, A. V. (13). Muscular exercise. *Nature*, **112**, 77. 1923.

Hill, A. V. (14). The surface-tension theory of muscle contraction. *Proc. Roy. Soc.* B, **98**, 506. 1925.

Hill, A. V. (15). The heat of shortening and the dynamic constants of muscle. 126, 136. 1938–39.

Hill, A. V. (16). Is relaxation an active process? *Proc. Roy. Soc.* B, **136**, 420. 1949–50.

Hill, A. V. (17). The series elastic component of muscle. *Proc. Roy. Soc.* B, **137**, 273. 1950.

Hill, A. V. (18). The heat of activation and the heat of shortening in a muscle twitch. *Proc. Roy. Soc.* B, **136**, 195. 1949–50.

Hill, A. V. (19). The onset of contraction. *Proc. Roy. Soc.* B, **136**, 242. 1949–50.

Hill, A. V. (20). Does heat production precede mechanical response in muscular contraction? *Proc. Roy. Soc.* B, **137**, 269. 1950.

Hill, A. V. (21). The development of the active state of muscle during the latent period. *Proc. Roy. Soc.* B, **137**, 320. 1950.

Hill, A. V. (22). The earliest manifestation of the mechanical response of striated muscle. *Proc. Roy. Soc.* B, **138**, 339. 1951.

Hill, A. V. (23). Mechanics of the contractile element in muscle. *Nature*, **166**, 415. 1950.

Hill, A. V. (24). The instantaneous elasticity of active muscle. *Proc. Roy. Soc.* B, **141**, 161. 1953.

Hill, A. V. (25). The effect of load on the heat of shortening of muscle. *Proc. Roy. Soc.* B, **159**, 297. 1964.

Hill, A. V. (26). Adenosine triphosphate and muscular contraction. *Nature*, **163**, 320. 1949.

Hill, A. V. (27). A discussion on muscular contraction and relaxation: their physical and chemical basis. Introduction. *Proc. Roy. Soc.* B, **137**, 40. 1950.

Hill, A. V. (28). On the time required for diffusion and its relation to processes in muscle. *Proc. Roy. Soc.* B, **135**, 446. 1948.

Hill, A. V. (29). The effect of tension in prolonging the active state in a twitch. *Proc. Roy. Soc.* B, **159**, 589. 1964.

Hill, A. V. (30). The variation of total heat production in a twitch with velocity of shortening. *Proc. Roy. Soc.* B, **159**, 596. 1964.

Hill, A. V. (31). Production and absorption of work by muscle. *Science*, **131**, 897. 1960.

Hill, A. V. (32). A further challenge to biochemists. *Biochem. Z.* **345**, 1. 1965.

Hill, A. V. (33). *First and last Experiments.* Cambridge University Press, London. 1970.

Hill, A. V. (34). The maximum work and mechanical efficiency of human muscles and their most economical speed. *J. Physiol.*, **56**, 19. 1922.

Hill, A. V. (35). The mechanical efficiency of frog's muscle. *Proc. Roy. Soc.* B, **127**, 434. 1939.

Hill, A. V. (36). The negative delayed heat production in stimulated muscle. *J. Physiol.*, **158**, 178. 1961.

Hill, A. V. (37). Is muscular relaxation an active process? *Nature, Lond.*, **166**, 646. 1950.

Hill, A. V. (38). The transformations of energy and the mechanical work of muscle. *Proc. phys. Soc. of London*, **51**, 1. 1939.

Hill, A. V. & Hartree, W. (1). The four phases of heat-production of muscle. *J. Physiol.*, **54**, 84. 1920–21.

Hill, A. V. & Howarth, J. V. (1). The reversal of chemical reactions in contracting muscle during an applied stretch. *Proc. Roy. Soc.* B, **151**, 169. 1959.

Hill, A. V. & Kupalov, P. S. (1). The vapour pressure of muscle. *Proc. Roy. Soc.* B, 106, 445. 1930.

Hill, A. V. & Lupton, H. (1). Muscular exercise, lactic acid, and the supply and utilization of oxygen. *Quart. J. Med.*, 16, 135. 1922-23.

Hill, A. V. & Woledge, R. C. (1). An examination of absolute values in myothermic measurements. *J. Physiol.*, 162, 311. 1962.

Hill, D. K. (1). The time course of the oxygen consumption of stimulated muscle. *J. Physiol.*, 98, 207. 1940.

Hill, D. K. (2). The time-course of evolution of oxidative recovery heat of frog's muscle. *J. Physiol.*, 98, 454. 1940.

Hill, D. K. (3). The anaerobic recovery heat production of frog's muscle at $0°$ C. *J. Physiol.*, 98, 460. 1940.

Hill, D. K. (4). Changes in transparency of a muscle during a twitch. *J. Physiol.*, 108, 292. 1949.

Hill, D. K. (5). The space accessible to albumin within the striated muscle fibre of the toad. *J. Physiol.*, 175, 275. 1964.

Hill, D. K. (6). The location of adenine nucleotide in the striated muscle of the toad. *J. Cell Biol.*, 20, 435. 1964.

Hill, R. (1). Oxygen affinity of muscle haemoglobin. *Nature*, 132, 897. 1933.

Hill, R. (2). Oxygen dissociation curves of muscle haemoglobin. *Proc. Roy. Soc.* B, 120, 472. 1936.

Hill, T. L. & Morales, M. F. (1). On 'high energy phosphate bonds' of biochemical interest. *J. Amer. chem. Soc.*, 73, 1656. 1951.

Himwich, H. E., Koskoff, Y. D. & Nahum, L. H. (1). Studies in carbohydrate metabolism. I. A glucose-lactic acid cycle involving muscle and liver. *J. biol. Chem.*, 85, 571. 1929-30.

Hippel, P. H. von, Gellert, M. F. & Morales, M. F. (1). Studies on the contractile proteins of muscle. II. Polymerisation reactions in the myosin B system. *J. Am. chem. Soc.*, 81, 1393. 1959.

Hippocrates [Hippocratic Corpus]. On the Nature of Bones; On the Sacred Disease; Epidemics; On Maladies. See Littré.

Hobson, G. E. & Rees, K. R. (1). The annelid phosphagens. *Biochem. J.*, 61, 549. 1955.

Hobson, G. E. & Rees, K. R. (2). The annelid phosphokinases. *Biochem. J.*, 65, 305. 1957.

Hodge, A. J. (1). The fine structure of striated muscle. A comparison of insect flight muscle with vertebrate and invertebrate skeletal muscle. *J. biophys. biochem. Cytol.*, 2 Suppl., 131. 1956.

Hodge, A. J., Huxley, H. E. & Spiro, D. (1). Electron microscope studies on ultrathin sections of muscle. *J. exp. Med.*, 99, 201.

Hodgkin, A. L. (1). The ionic basis of electrical activity in nerve and muscle. *Biol. Rev.*, 26, 339. 1951.

Hodgkin, A. L. (2). Ionic movements and electrical activity in giant nerve fibres. *Proc. Roy. Soc.* B, 148, 1. 1958.

Hodgkin, A. L. (3). *The Conduction of the Nervous Impulse.* Liverpool, University Press. 1965.

Hodgkin, A. L. & Horowicz, P. (1). Movements of Na and K in single muscle fibres. *J. Physiol.*, 145, 405. 1959.

Hodgkin, A. L., Huxley, A. F. & Katz, B. (1). Ionic currents underlying activity in the giant axon of the squid. *Arch. Sci. physiol.*, 3, 129. 1949.

Hoet, J. P. & Kerridge, P. M. T. (1). Observations on the muscles of normal and moulting crustacea. *Proc. Roy. Soc. B*, **100**, 116. 1926.

Hoet, J. P. & Marks, H. P. (1). Observations on the onset of rigor mortis. *Proc. Roy. Soc. B*, **100**, 72. 1926.

Hoffmann-Berling, H. (1). Adenosintriphosphat als Betriebsstoff von Zellbewegungen. *Biochim. biophys. Acta*, **14**, 182. 1954.

Hoffmann-Berling, H. (2). Other mechanisms producing movement. In *Comparative Biochemistry*, 2, 341. Ed. M. Florkin & H. S. Mason. Academic Press, London and New York. 1960.

Hoffmann-Berling, H. (3). Geisselmodelle und Adenosinetriphosphat (ATP). *Acta biochim. biophys.*, **16**, 146. 1955.

Hoffmann-Berling, H. (4). Die Bedeutung des Adenosintriphosphat für die Zell- und Kernteilungsbewegungen in der Anaphase. *Biochim. biophys. Acta*, **15**, 226. 1954.

Hoffmann-Berling, H. (5). Die Glycerin wasserextrahierte Telophasezelle als Model der Zytokinese. *Biochim. biophys. Acta*, **15**, 332. 1954.

Hoffmann-Berling, H. (6). Der Mechanismus eines neuen, von der Muskelkontraktion verschiedenen, Kontraktionszyclus. *Biochim. biophys. Acta*, **27**, 247. 1958.

Hoffmann-Berling, H. (7). Das kontraktile Eiweiss undifferentzierter Zellen. *Biochim. biophys. Acta*, **19**, 453. 1956.

Hofmeister, F. (1). Zur Lehre von Wirkung der Salze. Fünfte Mittheilung. Untersuchungen über den Quellungsvorgang. *Arch. exp. Pathol. und Pharmak.*, **27**, 395. 1890.

Hogenhuis, L. A. H. & Engel, W. K. (1). Histochemistry and cytochemistry of experimentally denervated guinea pig muscle. I. Histochemistry. *Acta anat.*, **60**, 39. 1965.

Hohorst, H. J., Reim, M. & Bartele, H. (1). Creatine kinase equilibrium in muscle and the significance of ATP and ADP levels. *Biochem. biophys. Res. Com.*, **7**, 142. 1962.

Holland, D. L. (1). Some aspects of skeletal muscle development. Ph.D. Thesis, Birmingham Univ. 1968.

Holland, D. L. & Perry, S. V. (1). The adenosine triphosphatase and calcium ion-transporting activities of the sarcoplasmic reticulum of developing muscle. *Biochem. J.*, **114**, 161. 1969.

Hollmann, S. (1). Über die anaerobe Glycolyse in der Uterusmuskulatur. *Z. physiol. Chem.*, **284**, 89. 1949.

Holloszy, J. O. (1). Biochemical adaptations in muscle. *J. biol. Chem.* **242**, 2278. 1967.

Hollwede, W. & Weber, H. H. (1). Alkalibindung und isoelectrischer Punkt des Myosins. *Biochem. Z.*, **295**, 205. 1937–38.

Holmes, F. L. (1). Elementary analysis and the origins of physiological chemistry. *Isis*, **54**, 175. 1963.

Holmes, F. L. (2). Introduction. In *Animal Chemistry by Justus Liebig: facsimile of the Cambridge edition of 1842*. The *Sources of Science* No. 4. New York and London, Johnson Reprint Corporation. 1964.

Holmgren, E. (1). Untersuchungen über die morphologisch nachweisbaren stofflichen Umsetzungen der quergestreiften Muskelfasern. *Arch. f. mikros. Anat.*, **75**, 240. 1910.

Holmyard, E. J. (1). *Alchemy*. London, Penguin Books Ltd. 1957.

Holtzer, A. (1). On the spontaneous aggregation of myosin. *Arch. Biochem. Biophys.*, 64, 507. 1956.

Holtzer, A., Clark, R. & Lowey, S. (1). The conformation of native and denatured tropomyosin B. *Biochemistry*, 4, 2401. 1965.

Holtzer, A. & Lowey, S. (1). On the molecular weight, size and shape of the myosin molecule. *J. Am. chem. Soc.*, 78, 5954. 1956.

Holtzer, A. & Lowey, S. (2). The molecular weight, size and shape of the myosin molecule. *J. Am. chem. Soc.*, 81, 1370. 1959.

Holtzer, A., Wang, T.-Y. & Noelken, M. E. (1). The effect of various monvalent anions on myosin B solutions. The identification of actin as a product of ATP action. *Biochim. biophys. Acta*, 42, 452. 1960.

Holtzer, H. (1). Aspects of chondogenesis and myogenesis. *Synthesis of molecular and cellular structure.* Symposium of the Society for the Study of Development and Growth, p. 35. Ed. D. Rudnick. New York, Ronald Press Co. 1961.

Holtzer, H. & Abbott, J. (1). Contraction of glycerinated embryonic myoblasts. *Anat. Rec.*, 131, 417. 1958.

Honig, C. R. (1). Control of smooth-muscle actomyosin by phosphate and 5'-AMP: possible role in metabolic autoregulation. *Microvasc. Res.*, 1, 133. 1968.

Honig, C. R. & Stamm, A. C. (1). Relaxing systems of cardiac muscle. *Fed. Proc.*, 23, 927. 1964.

Hoogeboom, G. H., Schneider, W. C. & Pallade, G. E. (1). Cytochemical studies of mammalian tissues. I. Isolation of intact mitochondria from rat liver; some biochemical properties of mitochondria and submicroscopic particulate material. *J. biol. Chem.*, 172, 619. 1948.

Hooke, R. (1). *Micrographia.* London, Martin & Allestry. 1665.

Hoole, S. (1). (tr.). *Select works of A. van Leeuwenhoek.* London. 1798.

Hooton, B. T. & Watts, D. C. (1). Adenosine 5'-triphosphate-creatine phosphotransferase from dystrophic mouse skeletal muscle. *Biochem. J.*, 100, 637. 1966.

Hooton, B. T. & Watts, D. C. (2). Levels of protein and non-protein sulphydryl groups in the skeletal muscle of normal and dystrophic Bar Harbour mice. *Clin. chim. Acta*, 16, 173. 1967.

Hopkins, F. G. (1). The dynamic side of Biochemistry. *Ann. Rep. Brit. Ass.*, p. 652. 1913.

Hopkins, F. G. (2). Herter Lecture: The chemical dynamics of muscle. *Johns Hopkins Hosp. Bull.*, 32. 1921.

Hopkins, F. G. (3). Glutathione. Its influence in the oxidation of fats and proteins. *Biochem. J.*, 19, 787. 1925.

Hopkins, F. G. (4). Some chemical aspects of life. *Rep. British Ass.*, p. 1, 1933.

Hoppe-Seyler, F. (1). *Handbuch der physiologisch- und pathologisch-chemischen Analyse für Ärzte und Studirende.* Berlin, August Hirschwald. 1865.

Hoyle, G. (1). In *Symposium on Muscle. Symposia biologica Hungarica*, Vol. 8, 34, 53. Ed. E. Ernst & F. B. Straub. Akadémia Kiadó, Budapest. 1968.

Huennekens, F. M. & Whiteley, H. R. (1). Phosphoric acid anhydrides and other energy-rich compounds. In *Comparative Biochemistry*, Vol. I, p. 107. Ed. M. Florkin & H. S. Mason. New York and London, Academic Press, 1960.

Huijing, F. & Slater, E. C. (1). The use of oligomycin as an inhibitor of oxidative phosphorylation. *J. Biochem.* (Japan), 49, 493. 1961.

Hultman, E. (1). Studies on muscle metabolism of glycogen and active phosphate in man with special reference to exercise and diet. *Scand. J. clin. & Labor. Invest.*, **19**, Suppl. No. 94. 1967.

Hultman, E. & Bergström, J. (1). Muscle glycogen synthesis in relation to diet studied in normal subjects. *Acta med. Scand.*, **182**, 109. 1967.

Hunter, A. (1). Creatine and creatinine. Longmans, Green & Co. London. 1928.

Hunter, F. E. (1). Anaerobic phosphorylation due to a coupled oxidation-reduction between α-ketoglutaric acid and oxalacetic acid. *J. biol. Chem.*, **177**, 361. 1949.

Hunter, F. E. (2). Oxidative phosphorylation during electron transport. In *Phosphorus Metabolism*. p. 297. Ed. by W. D. McElroy & B. Glass. Baltimore, The Johns Hopkins Press. 1951.

Hunter, F. E. & Spector, S. (1). Effect of dinitrophenol on phosphorylations coupled with oxidation of α-ketoglutarate. *Fed. Proc.*, **10**, 201. 1951.

Huriaux, F. (1). Digestion de la tonomyosine de carotides de bovidé par la trypsine et propriétés des L- and H-méromyosines correspondantes. *Angiologica*, **2**, 153. 1965.

Huriaux, F., Pechère, J.-F. & Hamoir, G. (1). Propriétés et composition de la tonomyosine de carotides de bovidé. *Angiologica*, **2**, 15. 1965.

Hürthle, K. (1). Über die Struktur der quergestreiften Muskelfasern von Hydrophilus im ruhenden und tätigen Zustand. *Arch. f. ges. Physiol.*, **126**, 1. 1909.

Hürthle, K. (2). Über die Struktur des quergestreiften Muskels im ruhenden und tätigen Zustand und über seine Aggregatzustand. *Biol. Centralblatt*, **27**, 112. 1907.

Huxley, A. F. (1). Muscle structure and theories of contraction. *Progr. in Bioph.*, **7**, 255. 1957.

Huxley, A. F. (2). Local activation of striated muscle from frog and crab. *J. Physiol.*, **135**, 17. 1956.

Huxley, A. F. (3). Local activation of muscle. *Ann. N.Y. Acad. Sci.*, **81**, 446. 1959.

Huxley, A. F. (4). Muscle. *Ann. Rev. Physiol.*, **26**, 131. 1964.

Huxley, A. F. & Niedergerke, R. (1). Structural changes in muscle during contraction. Interference microscopy of living muscle fibres. *Nature*, **173**, 971. 1954.

Huxley, A. F. & Niedergerke, R. (2). Measurement of the striations of isolated muscle fibres with the interference microscope. *J. Physiol.* **144**, 403. 1958.

Huxley, A. F. & Peachey, L. D. (1). The maximum length for contraction in vertebrate striated muscle. *J. Physiol.*, **156**, 150. 1961.

Huxley, A. F. & Straub, R. W. (1). Local activation and interfibrillar structures in striated muscle. *J. Physiol.*, **143**, 40. 1958.

Huxley, A. F. & Taylor, R. E. (1). Function of Krause's membrane. *Nature*, **176**, 1068. 1955.

Huxley, A. F. & Taylor, R. E. (2). Local activation of striated muscle fibres. *J. Physiol.*, **144**, 426. 1958.

Huxley, H. E. (1). Electron microscope studies on the structure of natural and synthetic protein filaments from striated muscle. *J. mol. Biol.*, **7**, 281. 1963.

Huxley, H. E. (2). In discussion on 'Size and shape factors in colloidal systems.' *Disc. Faraday Soc.*, **11**, 148. 1951.

Huxley, H. E. (3). X-ray analysis and the problem of muscle. *Proc. Roy. Soc. B*, **141**, 59. 1953.

Huxley, H. E. (4). Electron miscroscope studies of the organisation of the filaments in striated muscle. *Biochim. biophys. Acta*, **12**, 387. 1953.

Huxley, H. E. (5). The double array of filaments in cross-striated muscle. *J. of biophys. biochem. Cyt.*, **3**, 631. 1957.

Huxley, H. E. (6). Introduction to *The structure of striated muscle*. In *Biochemistry of muscle contraction*. p. 303. Ed. J. Gergely. London, J. & A. Churchill. 1964.

Huxley, H. E. (7). Structural evidence concerning the mechanism of contraction in striated muscle. In *Muscle*. p. 3. Ed. W. M. Paul, E. E. Daniel, C. M. Kay & G. Monckton. Oxford, Pergamon Press. 1965.

Huxley, H. E. (8). Structural arrangements and the contraction mechanism in striated muscle. *Proc. Roy. Soc.* B, **160**, 442. 1964.

Huxley, H. E. (9). Muscle cells. In *The Cell*, Vol. IV, p. 365. Ed. J. Brachet & A.D. Mirsky. New York & London, Academic Press. 1960.

Huxley, H. E. (10). In *Symposium on Muscle. Symp. biol. Hung.* 8, p. 249. Ed. E. Ernst & F. B. Straub. Akadémia Kiadó, Budapest. 1968.

Huxley, H. E. (11). Structural difference between resting and rigor muscle; evidence from intensity changes in the low-angle equatorial X-ray diagram. *J. mol. Biol.*, **37**, 507. 1968.

Huxley, H. E. (12). Evidence for continuity between the central elements of the triads and extracellular space in frog sartorius muscle. *Nature*, **202**, 1067. 1964.

Huxley, H. E. (13). The mechanism of muscular contraction. *Science*, **164**, 1356. 1969.

Huxley, H. E. (14). Ph.D. thesis. University of Cambridge. 1952.

Huxley, H. E. (15). The contractile structure of cardiac and skeletal muscle. *Circulation*, **24**, 328. 1961.

Huxley, H. E. & Brown, W. (1). The low-angle X-ray diagram of vertebrate striated muscle and its behaviour during contraction and rigor. *J. mol. Biol.*, **30**, 383. 1967.

Huxley, H. E., Brown W. & Holmes, K. C. (1). Constancy of axial spacings in frog sartorius muscle during contraction. *Nature*, **206**, 1358. 1965.

Huxley, H. E. & Hanson, J. (1). Changes in the cross-striations of muscle during contraction and stretch and their structural interpretation. *Nature*, **173**, 973. 1954.

Huxley, H. E. & Hanson, J. (2). Quantitative studies of the structure of cross-striated myofibrils. I. Investigations by interference microscopy. *Biochim. biophys. Acta*, **23**, 229. 1957.

Huxley, H. E. & Perutz, M. F. (1). Polypeptide chains in frog sartorius muscle. *Nature*, **167**, 1054. 1951.

Huxley, T. H. (1). *The Crayfish*. London, Kegan Paul. 1880.

Huys, J. (1). Isolement et propriétés de l'actomyosine d'utérus de vache. *Arch. internat. Physiol. Biochim.*, **68**, 445. 1960.

Huys, J. (2). Isolement et propriétés de la tonoactomyosine d'uterus de vache. *Arch. internat. Physiol. Biochim.*, **69**, 677. 1961.

Huys, J. (3). Données nouvelles sur l'actomyosine d'utérus humain gravide. *Bull. Soc. Roy. Belge de Gynécol. et d'Obstétr.*, **33**, 429. 1963.

Ikehara, M., Ohtsuka, E., Kitagawa, S. & Tonomura, Y. (1). Interaction between synthetic ATP analogues and actomyosin systems. II. *Biochim. biophys. Acta*, **82**, 74. 1964.

Ikehara, M., Ohtsuka, E., Kitagawa, S., Yagi, K. & Tonomura, Y. (1). Interaction between synthetic ATP analogs and actomyosin systems. *J. Amer. chem. Soc.*, **83**, 2679. 1961.

Illingworth, B. & Brown, D. H. (1). Action of amylo-1,6-glucosidase on low molecular weight substrates and the assay of this enzyme in glycogen storage disease. *Proc. nat. Acad. Sci. U.S.A.*, **48**, 1619. 1962.

Illingworth, B. & Brown, D. H. (2). Glycogen storage diseases, types III, IV and VI. In the Ciba Foundation Symposium *Control of glycogen metabolism*, p. 336. Ed. W. J. Whelan & M. P. Cameron. London, J. & A. Churchill Ltd. 1964.

Illingworth, B. & Cori, G. T. (1). Structure of glycogens and amylopectins. III. Normal and abnormal human glycogen. *J. biol. Chem.*, **199**, 653. 1952.

Illingworth, B., Cori, G. T. & Cori, C. F. (1). Amylo-1,6-glucosidase in muscle tissue in generalised glycogen storage disease. *J. biol. Chem.*, **218**, 123. 1956.

Illingworth, B., Jansz, H. S., Brown, D. H. & Cori, C. F. (1). Observations on the function of pyridoxal-5-phosphate in phosphorylase. *Proc. nat. Acad. Sci. U.S.A.*, **44**, 1180. 1958.

Illingworth, B., Larner, J. & Cori, G. T. (1). Structure of glycogens and amylopectins. I. Enzymatic determination of chain length. *J. biol. Chem.*, **199**, 631. 1952.

Imamura, K., Kanazawa, T., Tada, M. & Tonomura, Y. (1). The pre-steady state of the myosin-adenosine triphosphate system. III. Properties of the intermediate. *J. Biochem.*, **57**, 627. 1965.

Imamura, K., Tada, M. & Tonomura, Y. (1). The pre-steady state of the myosin-adenosine triphosphate system. IV. Liberation of ADP from the myosin-ATP system and effects of modifiers on the phosphorylation of myosin. *J. Biochem.* **59**, 280. 1966.

Infante, A. A. & Davies, R. E. (1). ATP breakdown during a single isotonic twitch of frog sartorius muscle. *Biochem. biophys. Res. Comm.*, **9**, 410. 1962.

Infante, A. A. & Davies, R. E. (2). The effect of 2,4 dinitro-fluorobenzene on the activity of striated muscle. *J. biol. Chem.*, **240**, 3996. 1965.

Infante, A. A., Klaupiks, D. & Davies, R. E. (1). Relation between length of muscle and breakdown of phosphocreatine in isometric tetanic contractions. *Nature*, **201**, 620. 1964.

Infante, A. A., Klaupiks, D. & Davies, R. E. (2). Length, tension and metabolism during short isometric contractions of frog sartorius muscles. *Biochim. biophys. Acta*, **88**, 215. 1964.

Infante, A. A., Klaupiks, D. & Davies, R. E. (3). Phosphorylcreatine consumption during single-working contraction of isolated muscle. *Biochim. biophys. Acta*, **94**, 504. 1965.

Infante, A. A., Klaupiks, D. & Davies, R. E. (4). Adenosine triphosphate change in muscles doing negative work. *Science*, **144**, 1577. 1964.

Iodice, A. A., Leong, V. & Weinstock, I. M. (1). Separation of cathepsins A and D of skeletal muscle. *Arch. Biochem. Biophys.*, **117**, 477. 1966.

Iodice, A. A. & Weinstock, I. M. (1). Purification and properties of a proteolytic enzyme from normal and dystrophic muscle. *Fed. Proc.*, **23**, 544. 1964.

Iodice, A. A. & Weinstock, I. M. (2). Cathepsin A in nutritional dystrophy and hereditary muscular dystrophy. *Nature*, **207**, 1102. 1965.

Iskikawa, H., Bischoff, R. & Holtzer, H. (1). Formation of arrowhead complexes with heavy meromyosin in a variety of cell-types. *J. Cell Biol.*, **43**, 312. 1969.

Ivanov, I. I., Mirovich, N. I., Jakhova, Z. N. & Takachinsky, S. E. (1). Fractions of myofibrillar proteins of various types of muscle. *Biokhimiya* (Eng. vers.), **27**, 94. 1962.

Ivanov, I. I., Mikrovich, N. I., Moissejeva, V. P., Parshina, E. A., Tukachinsky, S.E., Yuriev, V. A., Zhakhova, Z. N. & Zinovieva, I. P. (1). *Acta physiol. Acad. Sci. Hungaricae*, **16**, 1. 1959.

Iyengar, M. R., Glauser, S. C. & Davies, R. E. (1). An ATP-induced conformational change in the acto-H-meromyosin system. *Biochem. biophys. Res. Com.*, **16**, 379. 1964.

Iyengar, M. R. & Olson, R. E. (1). The amino acid composition of dog-heart myosin. *Biochim. biophys. Acta*, **94**, 371. 1965.

Iyengar, M. R. & Weber, H. H. (1). The relative affinities of nucleotides to G-actin and their effects. *Biochim. biophys, Acta*, **86**, 543. 1964.

Jacob, J. (1). IV. Différenciation électrophorétique des protéines musculaires de la grenouille et du lapin. *Bull. Soc. roy. des Sci. de Liège*, p. 242. 1945.

Jacob, J. (2). Electrophorèse de protéines musculaires de lapin. *Experientia*, **2**, 110, 1946.

Jacob, J. (3). The electrophoretic analysis of protein extracts from striated rabbit muscle. *Biochem. J.*, **41**, 83. 1947.

Jacob, J. (4). Etude électrophoretique des variations de composition d'extraits musculaires de lapin sous l'influence de la fatigue et de la contracture par le monobromacétate de soude. *Experientia*, **3**, 241. 1947.

Jacobsohn, K. P. (1). Zur Thermodynamik des Systems der Fumarase. *Biochem. Z.*, **274**, 167. 1934.

Jakus, M. A. & Hall, C. E. (1). Studies of actin and myosin. *J. biol. Chem.*, **167**, 705. 1947.

Jannsen, S. & Jost, H. (1). Über den Wiederaufbau des Kohlehydrats im Warmblütermuskel. *Z. physiol. Chem.*, **148**, 41. 1925.

Jen, M.-H., Hsü, T.-C. & Tsao, T.-C. (1). The sulphydryl groups of rabbit tropomyosin and the amino-acid sequence near these groups. *Scientia Sinica*, **14**, 81. 1965.

Jen, M.-H. & Tsao, T.-C. (1). A comparative chemical study of tropomyosins from different sources. 1. Amino-acid composition and N-terminal structure. *Scientia Sinica*, **6**, 317. 1957.

Jen, M.-H., Wen, H.-Y. & Niu, C.-I. (1). A comparative chemical study of tropomyosins from different sources. II. C-terminal structure. *Acta biochim. Sinica*, **1**, 167. 1958.

Jevons, F. R. (1). Boerhaave's Biochemistry. *Med. Hist*, **6**, 343. 1962.

Jewell, B. R. (1). The nature of the phasic and the tonic responses of the anterior byssal retractor muscle of *Mytilus*. *J. Physiol.*, **149**, 154. 1959.

Jewell, B. R., Pringle, J. W. S. & Rüegg, J. C. (1). Oscillatory contraction of insect fibrillar muscle after glycerol extraction. *J. Physiol.*, **173**, 6. 1964.

Jewell, B. R. & Rüegg, J. C. (1). Oscillatory contraction of insect fibrillar muscle after glycerol extraction. *Proc. Roy. Soc.* B, **164**, 428. 1966.

Jewell, B. R. & Wilkie, D. R. (1). An analysis of the mechanical components in frog's striated muscle. *J. Physiol.*, **143**, 515. 1958.

Jinnai, D. (1). Functional differentiation of skeletal muscle. *Acta Med. Okayama*, **14**, 159. 1960.

Jöbsis, F. F. (1). Mechanical activity of striated muscle. *Symposia biologica Hungarica*, 8. *Symposium on Muscle*, p. 151. Ed. E. Ernst & F. B. Straub. Budapest, Akad. Kiadó. 1968.

Jöbsis, F. F. (2). Early kinetics of the cytochrome B response to muscular contraction. *Ann. N. Y. Acad. Sci.*, **81**, 505. 1959.

Jöbsis, F. F. (3). Spectrophotometric studies on intact muscle. II. Recovery from contractile activity. *J. gen. Physiol.*, **46**, 929. 1963.

Jöbsis, F. F. & O'Connor, M. J. (1). Calcium release and re-absorption in the sartorius muscle of the toad. *Biochem. biophys. Res. Comm.*, **25**, 246. 1966.

Johnels, A. F. (1). On the origin of the electric organ in *Malapterurus electricus*. *Quart. J. micr. Sci.* **97**, 455. 1956.

Johnson, P., Harris, C. I. & Perry, S. V. (1). 3-Methylhistidine in actin and other muscle proteins. *J. Biochem.*, **105**, 361. 1967.

Johnson, P. & Landolt, H. R. (1). Myosin, actin and their interaction. *Faraday Soc. Disc.*, **11**, 179. 1951.

Johnson, P., Napper, D. H. & Rowe, A. J. (1). Sedimentation studies on polymerised actin solutions. *Biochim. biophys. Acta*, **74**, 365. 1963.

Johnson, P. & Rowe, A. J. (1). The sedimentation of myosin. *Biochem. J.*, **74**, 432. 1960.

Johnson, P. & Rowe, A. J. (2). The spontaneous transformation reactions of myosin. *Biochim. biophys. Acta*, **53**, 343. 1961.

Johnson, P. & Rowe, A. J. (3). The intrinsic viscosity of myosin and the interpretation of its hydrodynamic properties. *Biochem. J.*, **79**, 524. 1961.

Johnson, P. & Rowe, A. J. (4). An ultracentrifuge study of the actin–myosin interaction. In *Biochemistry of muscle contraction*. p. 279. Ed. J. Gergely. London, J. & A. Churchill. 1964.

Johnson, W. H., Kahn, J. S. & Szent-Györgyi, A. G. (1). Paramyosin and contraction of 'catch muscles'. *Science*, **130**, 160. 1959.

Johnson, W. H. & Szent-Györgyi, A. G. (1). The molecular basis for the 'catch' mechanism in molluscan muscle. *Biol. Bull.*, **117**, 382. 1959.

Johnson, W. H. & Twarog, B. M. (1). The basis for prolonged contractions in molluscan muscles. *J. gen. Physiol.*, **43**, 941. 1960.

Jones, J. M. & Perry, S. V. (1). The biological activity of sub-fragment I prepared from heavy meromyosin. *Biochem. J.*, **100**, 120. 1966.

Jones, W. H. S. (1). *Hippocrates* tr. and ed. The Loeb Classics, Heinemann, London. 1948.

Jordan, H. E. (1). The structural changes in striped muscle during contraction. *Physiol. Rev.*, **13**, 301. 1933.

Jordan, W. K. & Oster, G. (1). On the nature of the interaction between actomyosin and ATP. *Science*, **108**, 188. 1948.

Joseph, M., Cohn, W. E. & Greenberg, D. M. (1). Studies in mineral metabolism with the aid of artificial radio-active isotopes. II. Absorption, distribution and excretion of potassium. *J. biol. Chem.*, **128**, 673. 1939.

Judah, J. D. (1). The action of 2:4-dinitrophenol on oxidative phosphorylation. *Biochem. J.*, **49**, 271. 1951.

Kafiani, K. A. & Engelhardt, V. A. (1). Contractile properties of film threads of pure myosin. *Dokl. Akad. Nauk. S.S.S.R.*, **92**, 385. 1953. (See *Abstr. Amer. Chem. Soc.*, **48**, 4014. 1954.)

Kafiani, K. A. & Poglazov, B. F. (1). Contractile properties of film filaments of myosin. *Dokl. Akad. Nauk. S.S.S.R.*, **126**, 414. 1959. (See *Abstr. Amer. Chem. Soc.*, **54**, 3548. 1960.)

Kahn, J. S. & Johnson, W. H. (1). The localization of myosin and paramyosin in the myofilaments of the byssus retractor muscle of *Mytilus edulis*. *Biochem. biophys. Arch.*, **86**, 138. 1960.

Kakol, I., Gruda, J. & Rzysko, Cz. (1). Chromatographic fractionation of H-meromyosin ATPase. *Abstr. 5th Int. Congr. Biochem.*, p. 256. 1961.

Kalckar, H. M. (1). The role of myokinase in transphosphorylations. II. The enzymatic action of myokinase on adenine nucleotides. *J. biol. Chem.*, 148, 127. 1943.

Kalckar, H. M. (2). Phosphorylation in kidney tissue. *Enzymologia*, 2, 47. 1937–8.

Kalckar, H. M. (3). The nature of the phosphoric esters formed in kidney extracts. *Biochem. J.*, 33, 631. 1939.

Kalckar, H. M. (4). The nature of energetic coupling in biological syntheses. *Chem. Rev.*, 28, 71. 1941.

Kalckar, H. M. (5). Adenylpyrophosphatase and myokinase. *J. biol. Chem.*, 153, 355. 1944.

Kalckar, H. M., Dehlinger, J. & Mehler, A. (1). Rejuventation of phosphate in adenine nucleotides. II. The rate of rejuvenation of labile phosphate compounds in muscle and liver. *J. biol. Chem.*, 154, 275. 1944.

Kalckar, H. M. & Rittenberg, D. (1). Rejuvenation of muscle adenylic nitrogen in vivo studied with isotopic nitrogen. *J. biol. Chem.*, 170, 455. 1947.

Kaminer, B. (1). Synthetic myosin filaments from vertebrate smooth muscle. *J. mol. Biol.*, 39, 257. 1969.

Kamiya, N. (1). The mechanism of cytoplasmic movement in a myxomycete plasmodium. *Symp. Soc. Exp. Biol.*, 22, 199. 1968.

Kamp, F. (1). Muskelkontraktion und Löslichkeit der Muskeleiweisskörper. *Biochem. Z.*, 307, 228. 1940–41.

Kanazawa, T. & Tonomura, Y. (1). The pre-steady state of the myosin-adenosine triphosphatase system. I. Initial rapid liberation of inorganic phosphate. *J. Biochem.*, 57, 604. 1965.

Kaplan, N. O. (1). Multiple forms of enzymes. *Bact. Rev.*, 27, 155. 1963.

Kaplan, N. O. & Cahn, R. D. (1). Lactic dehydrogenases and muscular dystrophy in the chicken. *Proc. nat. Acad. Sci. U.S.A.*, 48, 2123. 1962.

Kaplan, N. O. & Goodfriend, T. L. (1). Role of the two types of lactic dehydrogenase. *Adv. in Enz. Reg.*, 2, 203. 1964.

Kar, N. C. & Pearson, C. M. (1). Creatine phosphokinase isoenzymes in muscle in human myopathies. *Amer. J. clin. Med.*, 43, 207. 1965.

Karpatkin, S., Helmreich, E. & Cori, C. F. (1). Regulation of glycolysis in muscle. II. Effect of stimulation and epinephrin in isolated frog sartorius muscle. *J. biol. Chem.*, 239, 3139. 1964.

Karpatkin, S., Helmreich, E. & Cori, C. F. (2). The effect of anaerobiosis, insulin and stimulation on hexokinase activity of frog sartorius muscle. *Fed. Proc.*, 24, 423. 1965.

Kasai, M., Asakura, S. & Oosawa, F. (1). The cooperative nature of G-F transformation of actin. *Biochim. biophys. Acta*, 57, 22. 1962.

Kasai, M., Asakura, S. & Oosawa, F. (2). The G-F equilibrium in actin solutions under various conditions. *Biochim. biophys. Acta*, 57, 13. 1962.

Kasai, M., Nakano, E. & Oosawa, F. (1). Polymerisation of actin free from nucleotides and divalent kations. *Biochim. biophys. Acta*, 94, 494. 1965.

Kasai, M. & Oosawa, F. (1). Removal of nucleotides from F-actin. *Biochim. biophys. Acta*, 75, 223. 1963.

Kasavina, B. S. & Torchinsky, M. (1). A microelectrophoretic study of the protein composition of muscle tissue in ontogenesis. *Biochimia*, 21, 510. 1956.

Katz, A. M. (1). Influence of tropomyosin upon the reactions of actomyosin at low ionic strength. *J. biol. Chem.*, 239, 3304. 1964.

Katz, A. M. (2). Purification and properties of a tropomyosin-containing protein fraction that sensitizes reconstituted actomyosin to calcium-binding agents. *J. biol. Chem.*, **241**, 1522. 1966.

Katz, A. M. (3). Regulation of cardiac muscle contractility. *J. gen. Physiol.*, **50**, No. 6, pt. 2, p. 185. 1967. Symposium on *The Contractile Process*.

Katz, A. M. & Mommaerts, W. F. H. M. (1). The sulfhydryl groups of actin. *Biochim. biophys. Acta*, **65**, 82. 1962.

Katz, A. M., Repke, D. I. & Cohen, B. R. (1). Control of the activity of highly purified cardiac actomyosin by Ca^{++}, Na^+ and K^+. *Circ. Res.*, **19**, 1062. 1966.

Katz, B. (1). The transmission of impulses from nerve to muscle, and the subcellular units of synoptic action. *Proc. Roy. Soc.* B, **155**, 455. 1962.

Katz, J. (1). Die mineralischen Bestandteile des Muskelfleisches. *Pfl. Arch. ges. Physiol.*, **63**, 1. 1896.

Katz, J. R. (1). Die Gesetze der Quellung. Eine biochemische und kolloidchemische Studie. *Kolloidchem. Beihefte*, **9**, 1. 1917.

Katz, J. R. & De Rooy, A. (1). Kristallinität des Fibrins. *Naturwiss.*, **21**, 559. 1933.

Kaufman, S. (1). Studies on the mechanism of the reaction catalysed by the phosphorylating enzyme. *J. biol. Chem.*, **216**, 153. 1955.

Kaufman, S., Gilvarg, C., Cori, O. & Ochoa, S. (1). Enzymatic oxidation of α-ketoglutarate and coupled phosphorylation. *J. biol. Chem.*, **203**, 869. 1953.

Kay, C. M. (1). A re-examination of the molecular characteristics of G-actin. *Biochim. biophys. Acta*, **43**, 259. 1960.

Kay, C. M. (2). Some physico-chemical properties of *Pinna nobilis* tropomyosin. *Biochim. biophys. Acta*, **27**, 469. 1958.

Kay, C. M. (3). Physico-chemical studies on cardiac and skeletal myosin A. In *Muscle*, ed. by W. M. Paul, E. E. Daniel, C. M. Kay & G. Monckton. Oxford, Pergamon Press. 1965.

Kay, C. M. & Bailey, K. (1). Light scattering in solutions of native and guanidinated rabbit tropomyosin. *Biochim. biophys. Acta*, **40**, 149. 1960.

Kay, C. M. & Brahms, J. (1). The influence of ethylene glycol on the enzymatic adenosine triphosphatase activity and molecular conformation of fibrous proteins. *J. biol. Chem.*, **238**, 2945. 1963.

Kay, C. M., Green, W. A. & Oikawa, K. (1). Influence of solvent composition on cardiac and skeletal myosin as determined by optical rotatory dispersion measurements. *Arch. Biochem. Biophys.*, **108**, 189. 1964.

Kay, C. M. & Pabst, H. F. (1). Physicochemical properties of γ-myosin from dystrophic muscle. *J. biol. Chem.*, **237**, 727. 1962.

Kay, H. D. (1). The phosphatases of mammalian tissue. *Biochem. J.*, **22**, 855. 1928.

Keatinge, W. R. (1). Ionic requirements for arterial action potential. *J. Physiol.*, **194**, 169. 1968.

Keatinge, W. R. (2). Sodium flux and electrical activity of arterial smooth muscle. *J. Physiol.*, **194**, 183. 1968.

Keech, D. B. & Utter, M. F. (1). Pyruvate carboxylase. I. Nature of reaction. *J. biol. Chem.*, **238**, 2609. 1963.

Keech, D. B. & Utter (2). Pyruvate carboxylase II. Properties. *J. biol. Chem.*, **238**, 2609. 1963.

Keilin, D. (1). *The History of Cell Respiration and Cytochrome.* Cambridge University Press. 1966.

Keilin, D. (2). On cytochrome, a respiratory pigment, common to animals, yeast and higher plants. *Proc. Roy. Soc.* B, **98**, 312. 1925.

Keilin, D. (3). Cytochrome and the respiratory enzymes. *Proc. Roy. Soc.* B, **104**, 206. 1929.

Keilin, D. (4). Cytochrome and intracellular oxidation. *Proc. Roy. Soc.* B, **106**, 418. 1930.

Keilin, D. & Hartree, E. F. (1). Cytochrome *a* and cytochrome oxidase. *Nature*, **141**, 870. 1938.

Keilin, D. & Hartree, E. F. (2). Cytochrome and cytochrome oxidase. *Proc. Roy. Soc. B*, **127**, 167. 1939.

Keilin, D. & Hartree, E. F. (3). Relationships between certain components of the cytochrome system. *Nature*, **176**, 200. 1955.

Keilin, D. & Hartree, E. F. (4). Succinic dehydrogenase-cytochrome system of cells. Intracellular respiratory system catalysing aerobic oxidation of succinic acid. *Proc. Roy. Soc.* B, **129**, 277. 1940.

Keilin, D. & Hartree, E. F. (5). Activity of the succinic dehydrogenase-cytochrome system in different tissue preparations. *Biochem. J.*, **44**, 205. 1949.

Keilin, D. & Hartree, E. F. (6). Relationship between certain components of the cytochrome system. *Nature*, **176**, 200. 1955.

Keller, P. J. & Cori, G. T. (1). Enzymic conversion of phosphorylase *a* to phosphorylase *b*. *Biochem. biophys. Acta*, **12**, 235. 1953.

Kelly, R. E. & Rice, R. V. (1). Localization of myosin filaments in smooth muscle. *J. Cell Biol.*, **37**, 105. 1968.

Kelly, R. E. & Rice, R. V. (2). Ultrastructural studies on the contractile mechanism of smooth muscle. *J. Cell Biol.*, **42**, 683. 1969.

Kemp, R. G. & Krebs, E. G. (1). Binding of metabolites of phosphofructokinase. *Biochemistry* (Am. chem. Soc.), **6**, 423. 1967.

Kendrick-Jones, J. & Perry, S. V. (1). The enzymes of adenine nucleotide metabolism in developing skeletal muscle. *Biochem. J.*, **103**, 207. 1967.

Kendrick-Jones, J. & Perry, S. V. (2). Enzymatic adaptation to contractile activity in skeletal muscle. *Nature*, **208**, 1068. 1965.

Kendrick-Jones, J. & Perry, S. V. (3). Protein synthesis and enzyme response to contractile activity in skeletal muscle. *Nature*, **213**, 406. 1967.

Kennedy, E. P. & Lehninger, A. L. (1). Intracellular structures and the fatty acid oxidase system of rat liver. *J. biol. Chem.*, **172**, 847. 1948.

Kennedy, E. P. & Lehninger, A. L. (2). Oxidation of fatty acids and tricarboxylic acid cycle intermediates by isolated rat liver mitochondria. *J. biol. Chem.*, **279**, 957. 1949.

Kerkut, G. A. (1). The transfer of ^{14}C-labelled material from CNS⇌muscle along a nerve trunk. *Comp. Biochem. and Physiol.*, **23**, 729. 1967.

Keynes, R. D. (1). Electric organs. In *The Physiology of Fishes*, p. 323. Ed. M. E. Brown. New York, Academic Press. 1957.

Keynes, R. D. & Aubert, X. (1). Energetics of the electric organ. *Nature*, **203**, 261. 1964.

Keynes, R. D. & Martins-Ferreira, H. (1). Membrane potentials in the electroplates of the electric eel. *J. Physiol.*, **119**, 315. 1953.

Kielley, W. W. (1). Studies on the structure of myosin. In *Molecular biology of muscular contraction*, p. 24. Ed. S. Ebashi, F. Oosawa, T. Sekine & Y. Tonomura. Tokyo, Igaku Shoin Ltd. and Amsterdam, Elsevier Publ. Co. 1965.

Kielley, W. W. & Barnett, L. M. (1). The identity of the myosin sub-units. *Biochem. biophys. Acta*, **51**, 589. 1961.

Kielley, W. W. & Bradley, L. (1). The relationship between sulfhydryl groups and the activation of myosin adenosinetriphosphatase. *J. biol. Chem.*, **218**, 653. 1956.

Kielley, W. W. & Harrington, W. F. (1). A model for the myosin molecule. *Biochim. biophys. Acta*, **41**, 401. 1960.

Kielley, W. W., Kalckar, H. M. & Bradley, L. B. (1). The hydrolysis of purine and pyrimidine nucleoside triphosphates by myosin. *J. biol. Chem.*, **219**, 95. 1956.

Kielley, W. W., Kimura, M. & Cooke, J. P. (1). The active site of myosin. *Abs. 6th Internat. Congress Biochem.*, p. 634. New York. 1964.

Kielley, W. W. & Meyerhof, O. (1). Studies on adenosinetriphosphatase of muscle. II. A new magnesium-activated adenosinetriphosphatase. *J. biol. Chem.*, **176**, 591. 1948.

Kinoshita, N., Kanazawa, T., Onishi, H. & Tonomura, Y. (1). The pre-steady state of the myosin-adenosine triphosphate system. IX. Effect of F-actin on the myosin-ATP system. *J. Biochem.*, **65**, 567. 1969.

Kinoshita, S., Andoh, B. & Hoffmann-Berling, H. (1). Das Erschlaffungs-system von Fibroblastenzellen. *Biochim. biophys. Acta*, **79**, 88. 1964.

Kinoshita, S. & Hoffmann-Berling, H. (1). Lokale Kontraktion als Ursache der Plasmateilung von Fibroblasten. *Biochim. biophys. Acta.*, **79**, 98. 1964.

Kinoshita, S. & Yakaki, I. (1). The behaviour and localization of intracellular relaxing system during cleavage in the sea-urchin egg. *Exp. Cell Res.*, **47**, 449. 1967.

Kipnis, D. M. & Cori, C. F. (1). Studies of tissue permeability. VI. The penetration and phosphorylation of 2-deoxy-glucose in the diaphragm of diabetic rats. *J. biol. Chem.*, **235**, 3070. 1960.

Kipnis, D. M., Helmreich, E. & Cori, C. F. (1). Studies of tissue permeability. IV. The distribution of glucose between plasma and muscle. *J. biol. Chem.*, **234**, 165. 1959.

Kisch, B. (1). Nachweis von Phosphagen im elektrischen Organ von *Torpedo*. *Biochem. Z.*, **225**, 183. 1930.

Kitagawa, S., Drabikowski, W. & Gergely, J. (1). Exchange and release of the bound nucleotide of F-actin. *Arch. Biochem. Biophys.*, **125**, 706. 1968.

Kitagawa, S., Martonosi, A. & Gergely, J. (1). Release of actin-bound ^{14}C-ADP and superprecipitation of actomyosin. *Fed. Proc.*, **24**, 598. 1965.

Kleinzeller, A. (1). Adenosine- and inosine-nucleotides in the phosphorus metabolism of muscle. *Biochem. J.*, **36**, 729. 1942.

Klimek, R. & Parnas, J. K. (1). Adenylsäure und Adeninnucleotid. *Biochem. Z.*, **252**, 392. 1932.

Klingenberg, M. & Bücher, T. (1). Flugmuskelmitochondrien aus *Locusta migratoria* mit Atmungskontrolle. Aufbau und Zusammensetzung der Atmungskette. *Biochem. Z.*, **331**, 312. 1959.

Klingenberg, M. & Slenczka, W. (1). Atmungsaktivität von Mitochondrien verschiedener Organe mit Glycerin-1-P im Vergleich zu Substraten des Tricarbonsäurecyclus. *Biochem. Z.*, **331**, 334. 1959.

Klingenberg, M. & Slenczka, W. (2). Pyridinnucleotide in Lebermitochondrien. Eine Analyse ihre Redox-Beziehunger. *Biochem. Z.*, **331**, 486. 1959.

Klingenberg, M., Slenczka, W. & Ritt, E. (1). Vergleichende Biochemie der Pyridinnucleotid-Syteme in Mitochondrien verschiedener Organe. *Biochem. Z.*, **332**, 47. 1959.

Klug, A. (1). The design of self-assembling systems of equal units. *Symp. Int. Soc. Cell. Biol.*, **6**, 1. Ed. K. B. Warren. New York and London, Academic Press. 1967.

Knappeis, G. G. & Carlsen, F. (1). The ultrastructure of the Z disc in skeletal muscle. *J. Cell Biol.*, **13**, 323. 1962.

Knappeis, G. G. & Carlsen, F. (2). The ultrastructure of the M line in skeletal muscle. *J. Cell Biol.*, **38**, 202. 1968.

Knoll, P. (1). Über protoplasmaarme und protoplasmareiche Muskulatur. *Denksch. d. Österreichischen Akad, d. Wissensch.*, **58**, 633. 1891.

Knoll, P. & Hauer, A. (1). Über das Verhalten der protoplasmaarmen und protoplasmareichen, quergestreiften Muskelfasern unter pathologischen Verhältnissen. *Sitzungsber. d. k. Akad. Wien*, **101**. Abt. III, 315. 1892.

Knoop, F. (1). Der Abbau aromatischer Fettsäuren im Tierkörper. *Beitr. chem. Physiol. Path.*, **6**, 150. 1904.

Koeppe, O. J., Boyer, P. D. & Stulberg, M. P. (1). On the occurrence, equilibrium and site of acyl-enzyme formation of glyceraldehyde-3-phosphate dehydrogenase. *J. biol. Chem.*, **219**, 569. 1956.

Kohn, R. R. (1). Mechanism of protein loss in denervation muscle atrophy. *Amer. J. Path.*, **45**, 435. 1964.

Koike, M., Reed, L. J. & Carroll, W. R. (1). α-Keto acid dehydrogenation complexes. 1. Purification and properties of pyruvate and α-keto-glutarate dehydrogenation complexes of *Escherichia coli*. *J. biol. Chem.*, **235**, 1924. 1960.

Kölliker, A. (1). *Manual of Human Histology.* See Busk & Huxley.

Kölliker, A. (2). *Mikroscopische Anatomie.* Vol. 2., part 1. p. 248. Leipzig, Wilhelm Engelman. 1850.

Kölliker, A. (3). Ueber den Bau und die Verbreitung der glatten Muskeln. *Z. f. Zool.*, **1**, 48. 1849.

Kominz, D. R. (1). Contribution in the *Symposium on Muscle*, p. 105. Budapest, ed. E. Ernst & F. B. Straub. Akadémia Kiadó. 1968.

Kominz, D. R., Carroll, W. R., Smith, E. N. & Mitchell, E. R. (1). A subunit of myosin. *Arch. Biochem. Biophys.*, **79**, 191. 1959.

Kominz, D. R., Hough, A., Symonds, P. & Laki, K. (1). Amino acid composition of actin myosin, tropomyosin and the meromyosins. *Arch. Biochem. Biophys.*, **50**, 148. 1954.

Kominz, D. R., Mitchell, E. R., Nihei, T. & Kay, C. M. (1). The papain digestion of skeletal myosin A. *Biochemistry* (Amer. Chem. Soc.), **4**, 2373. 1965.

Kominz, D. R. & Saad, F. (1). Uterine myosin. *Fed. Proc.*, **15**, 112. 1956.

Kominz, D. R., Saad, F., Gladner, J. A. & Laki, K. (1). Mammalian tropomyosins. *Arch. Biochem. Biophys.*, **70**, 16. 1957.

Kominz, D. R., Saad, F. & Laki, K. (1). Vertebrate and invertebrate tropomyosins. *Nature*, **179**, 206. 1957.

Kondo, K. (1). Ueber Milchsäurebildung im Muskelpresssaft. *Biochem. Z.*, **45**, 63. 1912.

Konosu, S., Hamoir, G. & Pechère, J.-F. (1). Carp myogens of white and red muscles. Properties and amino acid composition of the main low-molecular-weight proteins. *Biochem. J.*, **96**, 98. 1965.

Korey, S. (1). Some factors influencing the contractility of a non-conducting fibre preparation. *Biochim. biophys. Acta*, **4**, 58. 1950.

Korkes, S., del Campillo, A., Gunsalus, I. C. & Ochoa, S. (1). Enzymatic synthesis of citric acid. IV. Pyruvate as acetyle donor. *J. biol. Chem.*, **193**, 721. 1951.

Kornberg, A. & Pricer, W. E. (1). Enzymatic synthesis of the coenzyme A derivatives of long chain fatty acids. *J. biol. Chem.*, **204**, 329. 1953.

Korr, I. M., Wilkinson, P. N. & Chornock, F. W. (1). Axonic delivery of neuroplasmic components to muscle cells. *Science*, 155, 342. 1967.

Korzybski, T. & Parnas, J. K. (1). Observations sur les échanges des atomes du phosphore renfermé dans l'acide adénosine triphosphorique, dans l'animal vivant, à l'aide du phosphor marqué par du radiophosphor [32]P. *Bull. Soc. Chim. biol.*, 21, 713. 1939.

Koschtojanz, C. & Rjabinowskaja, A. (1). Beitrag zur Physiologie des Skeletmuskels der Säugetiere auf verschiedenen Stadien ihren individuellen Entwicklung. *Arch. f. ges. Physiol.*, 235, 416. 1935.

Koshland, D. E., Budenstein, Z. & Kowalsky, A. (1). Mechanism of hydrolysis on adenosine triphosphate catalysed by purified muscle proteins. *J. biol. Chem.*, 211, 279. 1954.

Koshland, D. E. & Clarke, E. (1). Mechanism of hydrolysis of adenosinetriphosphate catalysed by lobster muscle. *J. biol. Chem.*, 205, 917. 1953.

Koshland, D. E. & Levy, H. M. (1). Evidence for an intermediate in ATP hydrolysis by myosin. In *Biochemistry of muscle contraction*. p. 87. Ed. J. Gergely. London, J. & A. Churchill Ltd. 1964.

Kozloff, L. M. & Lute, M. (1). A contractile protein in the tail of bacteriophage T2. *J. biol. Chem.*, 234, 539. 1959.

Krahl, M. (1). *The Action of Insulin on Cells.* New York & London, Academic 1961.

Kraus, H. M., Eijk, H. G. van & Westenbrink, H. G. K. (1). A study of G-actin. *Biochim. biophys. Acta*, 100, 193. 1965.

Krause, W. (1). Mikroskopische Untersuchungen über die quergestreifte Muskelsubstanz. *Nachrichten d. kön. Gesellsch. der Univ. Göttingen. Mitt. d. path. Inst.*, 17, 357. 1868.

Krause, W. (2). *Allgemeine und milkroskopische Anatomic.* Hannover. 1876.

Krebs, E. G. & Fischer, E. H. (1). Phosphorylase activity of skeletal muscle extracts. *J. biol. Chem.*, 216, 113. 1955.

Krebs, E. G., Graves, D. J. & Fischer, E. H. (1). Factors affecting the activity of muscle phosphorylase *b* kinase. *J. biol. Chem.*, 234, 2867. 1959.

Krebs, E. G., Kent, A. B. & Fischer, E. H. (1). The muscle phosphorylase *b* kinase reaction. *J. biol. Chem.*, 231, 73. 1958.

Krebs, E. G., Love, D. S., Bratvold, G. E., Trayser, K. A., Meyer, W. L. & Fischer, E. H. (1). Purification and properties of rabbit skeletal muscle phosphorylase *b* kinase. *Biochemistry* (Am. Chem. Soc.), 3, 1022. 1964.

Krebs, H. A. (1). The intermediary stages in the biological oxidation of carbohydrate. *Adv. in Enz.*, 3, 191. 1943.

Krebs, H. A. (2). Metabolism of amino acids. III. Deamination of amino acids. *Biochem. J.*, 29, 1620. 1935.

Krebs, H. A. (3). Cyclic processes in living matter. *Enzymologia*, 12, 88. 1946.

Krebs, H. A. (4). Gluconeogenesis. *Proc. Roy. Soc.* B, 159, 545. 1964.

Krebs, H. A. & Eggleston, L. V. (1). The oxidation of pyruvate in pigeon breast muscle. *Biochem. J.*, 34, 442. 1940.

Krebs, H. A. & Hems, R. (1). Some reactions of adenosine and inosine phosphates in animal tissues. *Biochim. biophys. Acta.*, 12, 172. 1953.

Krebs, H. A. & Johnson, W. A. (1). The role of citric acid in intermediate metabolism in animal tissues. *Enzymologia*, 4, 148. 1937.

Krebs, H. A. & Woodford, M. (1). Fructose 1,6-diphosphatase in striated muscle. *Biochem. J.*, 94, 436. 1965.

Kritzmann, M. G. (1). The enzyme transferring the amino group of glutamic acid. *Biochimia*, 3, 603. 1938.

Kritzmann, M. G. (2). The enzyme system transferring the amino group of aspartic acid. *Nature*, 143, 603. 1939.

Kritzmann, M. & Samarina, O. (1). Reversible splitting of glutamic aminopherase. *Nature*, 158, 104. 1946.

Kronecker, H. & Stirling, W. (1). The genesis of tetanus. *J. Physiol.*, 1, 384. 1878–9.

Krüger, P. (1). *Tetanus und Tonus der quergestreiften Skelettmuskeln der Wirbeltiere und des Menschen.* Leipzig, Akad. Verlagsges. 1952.

Kruh, J., Dreyfus, J. C., Schapira, G. & Gey, G. O. (1). Abnormalities of muscle protein metabolism in mice with muscular dystrophy. *J. clin. Invest.*, 39, 1180. 1960.

Kubo, S., Tokura, S. & Tonomura, Y. (1). On the active site of myosin A-adenosine triphosphatase. 1. Reaction of the enzyme with trinitrobenzenesulfonate. *J. biol. Chem.*, 235, 2835. 1960.

Kubo, S., Tokuyama, H. & Tonomura, Y. (1). On the active site of myosin A-adenosine triphosphatase. V. Partial solution of the chemical structure around the binding site of trinitrobenzenesulfonate. *Biochim. biophys. Acta.*, 100, 459. 1965.

Kuby, S. A. & Mahowald, T. A. (1). Studies on the ATP-transphosphorylases. *Fed. Proc.*, 18, 267. 1959.

Kuffler, S. W. & Vaughan Williams, E. M. (1). Small-nerve junctional potentials. The distribution of small motor nerves to frog skeletal muscle, and the membrane characteristics of the fibres they innervate. *J. Physiol.*, 121, 289. 1953.

Kuffler, S. W. & Vaughan Williams, E. M. (2). Properties of the 'slow' skeletal muscle fibres of the frog. *J. Physiol.*, 121, 318. 1953.

Kühn, K. G. (1). (tr.) [*Galen's*] *Opera Omnia.* Latin with orig. Gk. text. 20 vols. Leipzig. 1821–33.

Kuhn, R. & Baur, H. (1). Der Milchsäuregehalt des Muskels im Insulin- und Hungertod. *Muenchener med. Wochenschr.*, 71, 541. 1924.

Kühne, W. (1). Untersuchungen über Bewegungen und Veränderungen der contractilen Substanzen. *Arch. f. Anat., Physiol. u. wissensch. Med.*, p. 748. 1859.

Kühne, W. (2). *Untersuchungen über das Protoplasma und die Contractilität.* Leipzig, W. Engelmann. 1864.

Kühne, W. (3). On the origin and causation of vital movement. *Proc. Roy. Soc.* 44, 220. 1888.

Kühne, W. (4). Farbstoff der Muskeln. *Arch. f. path. Anat. u. Physiol.*, 33, 79. 1865.

Kühne, W. (5). Erfahrungen und Bemerkungen über Enzyme und Fermente. *Untersuch. a. d. physiol. Inst. der Univ. Heidelberg*, 1, 291. 1878.

Kulbertus, H. (1). Etude des échanges d'ions H^+ au cours de la contraction des parois carotidienne soumises à différents agents vasomoteurs. *Angiologica*, 1, 275. 1964.

Kulbertus, H. (2). Contribution à l'étude de l'elasticité des parois carotidiennes de bovidés soumises à différents agents vasomoteurs. *Arch. internat. Physiol. Biochim.*, 71, 540. 1963.

Kumagai, H., Ebashi, S. & Takeda, F. (1). Essential relaxing factor in muscle other than myokinase and creatine phosphokinase. *Nature*, 176, 166. 1955.

Kundig, W., Ghosh, S. & Roseman, S. (1). Phosphate bound to histidine in a protein as an intermediate in a novel phosphotransferase system. *Proc. Nat. Acad. Sci. U.S.A.*, 52, 1067. 1964.

Kundig, W., Kundig, F. D., Anderson, B. & Roseman, S. (1). Restoration of active transport of glycosides in *Escherichia coli* by a component of a phosphotransferase system. *J. biol. Chem.*, **241**, 3243. 1966.

Kurtz, A. L. & Luck, J. M. (1). Studies on annelid muscle. II. Observations on annelid phosphagen. *Proc. Soc. exp. Biol. Med.*, **37**, 299. 1937–38.

Kuschinsky, G. & Turba, F. (1). Über den Chemismus von Zustandsänderungen des Aktomyosins. *Experientia*, **6**, 103. 1950.

Kuschinsky, G. & Turba, F. (2). Über die Rolle der SH Gruppen bei Vorgängen am Aktomyosin, Myosin und Aktin. *Biochim. biophys. Acta.*, **6**, 426. 1951.

Kuschinsky, G. & Turba, F. (3). Über die Bedeutung von Sulfhydrylgruppen für Prozesse am Aktomyosin. *Naturwissensch.*, **37**, 425. 1950.

Kushmerick, M. J. & Davies, R. E. (1). The chemical energetics of muscle contraction. II. The chemistry, efficiency and power of maximally working sartorius muscle. *Proc. Roy. Soc. B.*, **174**, 315. 1969.

Kushmerick, M. J., Larson, R. E. & Davies, R. E. (1). The chemical energetics of muscle. I. Activation heat, heat of shortening and ATP utilization for activation-relaxation processes. *Proc. Roy. Soc. B.*, **174**, 293. 1969.

Kutscher, F. (1). Über einige Extraktstoffe des Flusskrebses. Zugleich ein Betrag zur Kenntnis der Kreatinbildung im Tier. *Z. f. Biol.*, **64**, 240. 1914.

Kutscher, F. & Ackermann, D. (1). Vergleichend-physiologische Untersuchungen von Extrakten verscheidener Tierklassen auf tierische Alkaloide, eine Zusammenfassung. *Z. Biol.*, **84**, 181. 1926.

Laki, K. (1). Adenosinetriphosphatase of muscle. *Studies from the Institute of Medical Chemistry, Univ. of Szeged.*, III, 16. 1944.

Laki, K. (2). Nature of meromyosins. *Science*, **128**, 653. 1958.

Laki, K. (3). The composition of contractile muscle proteins. *J. comp. cell. Physiol.*, **49**, Suppl. 1, p. 249. 1957.

Laki, K. (4). A tropomyosin like fragment of denatured myosin. *Conf. on the Chemistry of Muscular Contraction.* Tokyo, Igaku Shoin Ltd., p. 77. 1957.

Laki, K., Bowen, W. J. & Clark, A. (1). The polymerisation of proteins. Adenosine-triphosphate and the polymerisation of actin. *J. gen. Physiol.*, **33**, 437. 1949–50.

Laki, K. & Carroll, W. R. (1). Size of the myosin molecule. *Nature*, **175**, 389. 1955.

Laki, K., Maruyama, E. & Kominz, D. R. (1). Evidence for the interaction between tropomyosin and actin. *Arch. Biochem. Biophys.*, **98**, 323. 1962.

Laki, K., Spicer, S. S. & Carroll, W. R. (1). Evidence for a reversible equilibrium between actin, myosin and actomyosin. *Nature*, **169**, 328. 1952.

Laki, K. & Standaert, J. (1). The minimal molecular weight of actin estimated with the use of carboxypeptidase A. *Arch. Biochem. Biophys.* 86, 16. 1960.

Lamb, D. R., Peter, J. B., Jeffress, R. N. & Wallace, A. H. (1). Glycogen, hexokinase and glycogen synthetase adaptations to exercise. *Amer. J. Physiol.*, **217**, 1628. 1969.

Landon, M. & Perry, S. V. (1). Tryptic peptides from L-myosin and the meromyosins. *Biochem. J.*, **88**, 9P. 1963.

Lane, B. P. (1). Alterations in the cytologic detail of intestinal smooth muscle in various stages of contraction. *J. Cell Biol.*, **27**, 199. 1965.

Lane, B. P. & Rhodin, J. A. G. (1). Cellular interrelationships and electrical activity in two types of smooth muscle. *J. Ultrastr. Res.*, **10**, 470. 1964.

Lange, G. (1). Über die Dephosphoryherung von Adenosinetriphosphat zu Adenosinediphosphat während der Kontraktionsphase von Froschrectusmuskel. *Biochem. Z.*, **326**, 172. 1955.

Langer, G. A. (1). Calcium exchange in dog ventricular muscle. Relation of frequency of contraction and maintenance of contractility. *Circ. Res.*, **17**, 78. 1965.

Langley, J. N. (1). Observations on denervated muscle. *J. Physiol.*, **50**, 335. 1915–16.

Langley, J. N. (2). Observations on denervated and regenerating muscle. *J. Physiol.*, **51**, 377. 1917.

Langley, J. N. & Hashimoto, M. (1). Observations on the atrophy of denervated muscle. *J. Physiol.*, **52**, 15. 1918–19.

Lankester, E. R. (1). Über das Vorkommen von Haemoglobin in den Muskeln der Mollusken und die Verbreitung desselben in den lebendigen Organismen. *Arch. f. ges. Physiol.*, **4**, 315. 1871.

Lapique, L. (1). *L'excitabilité en Fonction du Temps*. Paris. 1928.

Laquer, F. (1). Über die Bildung von Milchsäure und Phosphorsäure im Froschmuskel. *Z. f. physiol. Chem.*, **93**, 60. 1914.

Laquer, F. (2). Über den Abbau der Kohlehydrate im quergestreiften Muskel. I. *Z. f. physiol. Chem.*, **116**, 169. 1921.

Laquer, F. (3). Über den Abbau der Kohlehydrat im quergestreiften Muskel. II. *Z. f. physiol. Chem.*, **122**, 26. 1922.

Laquer, F. & Meyer, P. (1). Über den Abbau der Kohlehydrate im quergestreiften Muskel. III. Mitteilung. *Z. f. physiol. Chem.*, **124**, 211. 1923.

Lardy, H. A. (1). Regulation of energy yielding processes in muscle. In *The Physiology and Biochemistry of Muscle as a Food.* p. 31. Ed. by E. J. Briskey, R. G. Casseus & J C. Trautman. Wisconsin. University of Wisconsin Press. 1966.

Lardy, H. A. & Elvehjem, C. A. (1). Biological oxidations and reductions. *Ann. Rev. Biochem.*, **14**, 1. 1945.

Lardy, H. A. Hansen, R. G. & Phillips, P. H. (1). The metabolism of bovine epidymal spermatozoa *Arch. Biochem.*, **6**, 41. 1945.

Lardy, H. A., Johnson, D. & McMurray, W. C. (1). Antibiotics as tools for metabolic studies. I. A survey of toxic antibiotics in respiratory phosphorylative and glycolytic systems. *Arch. Biochem. Biophys.*, **78**, 587. 1958.

Lardy, H. A. & McMurray, W. C. (1). The mode of action of oligomycin. *Fed. Proc.*, **18**, 209. 1959.

Lardy, H. A. & Parks, R. E. (1). Influence of ATP concentration on rates of some phosphorylation reactions. In *Enzymes: Units of biological Structure and Function.* p. 584. Ed. O. H. Gaebler. New York, Academic Press. 1956.

Lardy, H. A. & Wellman, H. (1). Oxidative phosphorylations: role of inorganic phosphate and acceptor systems in control of metabolic rates. *J. biol. Chem.*, **195**, 215. 1952.

Lardy, H. A. & Wellman, H. (2). The catalytic effect of 2,4-dinitrophenol on adenosinetriphosphate hydrolysis by cell particles and soluble enzymes. *J. biol. Chem.*, **201**, 357. 1953.

Larner, J. (1). The action of branching enzymes on outer chains of glycogen. *J. biol. Chem.*, **202**, 491. 1953.

Larner, J., Illingworth, B., Cori, G. T. & Cori, C. F. (1). Structure of glycogens and amylopectens. II. Analysis of enzymatic degradation. *J. biol. Chem.*, **199**, 641. 1952.

Larner, J. & Villar-Palasi, C. (1). Enzymes in a glycogen storage myopathy. *Proc. nat. Acad. Sci. U.S.A.*, **45**, 1234. 1959.

Larner, J., Villar-Palasi, C., Goldberg, N. D., Bishop, J. S., Huijing, F., Wenger, J. I., Sasko, H. & Brown, N. B. (1). Hormonal and non-hormonal control of glycogen synthesis – control of transferase and transferase *I* kinase. *Adv. in Eng. Regulation*, **6**, 409. 1968.

Larner, J., Villar-Palasi, C. & Richman, D. J. (1). Insulin-stimulated glycogen formation in rat diaphragm. *Ann. New York Acad. Sci.*, **82**, 345. 1959.

Larner, J., Villar-Palasi, C. & Richman, D. J. (2). Insulin-stimulated glycogen formation in rat diaphragm. Levels of tissue intermediates in short-time experiments. *Arch. Biochem. Biophys.*, **86**, 56. 1960.

Laszt, L. (1). Properties of vessel muscle proteins extracted with water or salt solutions of low ionic strength. *Nature*, **189**, 230. 1961.

Laszt, L. (2). Was ist Gefässtonus? *Angiologica*, **1**, 346. 1964.

Laszt, L. & Hamoir, G. (1). Etude par électrophorèse et ultracentrifugation de la composition protéinique de la couche musculaire des carotides de bovidé. *Biochim. biophys. Acta.*, **50**, 430. 1961.

Laudahn, G. & Heyck, H. (1). Muskelenzym-Befunde bei der progressive Muskeldystrophie. In *Myopathien*. p. 165. Ed. R. Beckmann. Stuttgart, Georg Thieme. 1965.

Lavoisier, A. L. & de Laplace, P. S. (1). Two Memoirs on Heat. In *Ostwald's Klassiker der exakten Wissenschaften*, No. 40. 1780, 1784.

Lawrence, A. S. C., Needham, J. & Shen, S.-C. (1). Studies on the anomalous viscosity and flow-birefringence of protein solutions. I. General behaviour of proteins subjected to shear. *J. gen. Physiol.*, **27**, 201. 1944.

Lawrie, R. A. (1). Biochemical differences between red and white muscle. *Nature*, **170**, 122. 1952.

Lawrie, R. A. (2). The activity of the cytochrome system in muscle and its relation to myoglobin. *Biochem. J.*, **55**, 298. 1953.

Lawrie, R. A. (3). The relation of energy-rich phosphate in muscle to myoglobin and to cytochrome-oxidase activity. *Biochem. J.*, **55**, 305. 1953.

Lawrie, R. A. (4). Effect of enforced exercise on myoglobin content in muscle. *Nature*, **171**, 1069. 1953.

Layzer, R. B., Rowland, L. P. & Ranney, H. M. (1). Muscle phosphofructokinase deficiency. *Arch. Neurol.*, **17**, 512. 1967.

Leadbeater, L. & Perry, S. V. (1). The effect of actin on the magnesium-activated adenosine triphosphatase of heavy meromyosin. *Biochem. J.*, **87**, 233. 1963.

Ledermair, O. (1). Der Uterusmuskel in der Schwangerschaft. *Arch. Gynaek.* **192**, 109. 1959.

Lee, C.-P., Azzone, G. F. & Ernster, L. (1). Evidence for energy-coupling in non-phosphorylating electron transport particles from beef-heart mitochondria. *Nature*, **201**, 152. 1964.

Lee, H. D. P. (1). *Aristotle's Meteorologica*, Loeb Classics Series. Heinemann, London 1952.

Lee, K. S. (1). Present status of cardiac relaxing factor. *Fed. Proc.*, **24**, 1432. 1965.

Lee, K. S., Ladinsky, H., Choi, S. J. & Kasuya, Y. (1). Studies on the in vitro interaction of electrical stimulation and Ca^{++} movement in sarcoplasmic reticulum. *J. gen. Physiol.*, **49**, 689. 1965–66.

Leeuwenhoek, A. van. (1). The collected letters of Antoni van Leeuwenhoek. Ed. by a Committee of Dutch scientists. Amsterdam, Swets & Zeitlinger. 1939.

Leeuwenhoek, A. van. (2). See Hoole.

Leeuwenhock, A. van (3). More observations. *Phil. Trans. Roy. Soc.*, **9**, 186. 1674.

Lehmann, C. G. (1). *Physiological Chemistry*. Tr. by G. E. Day. London, Harrison & Son. 1851.

Lehmann, H. (1). Über die enzymatische Synthese der Kreatinphosphorsäure durch Umesterung der Phosphobrenztraubensäure. *Biochem. Z.*, **281**, 271. 1935.

Lehmann, H. (2). Über die Umesterung des Adenylsäuresystems mit Phosphagenen. *Biochem. Z.*, **286**, 336. 1936.

Lehmann, H. & Griffiths, (1). Creatinephosphokinase activity in semen. *Lancet*, Vol. II, p. 498. 1963.

Lehnartz, E. (1). Über die nachträgliche Milchsäurebildung beim Tetanus und bei der Zuckung isolierter Froschmuskeln. *Z. f. physiol. Chem.*, **197**, 55. 1931.

Lehnartz, E. (2). Über die Verknüpfung des Aufbaus und Abbaus der Tätigkeitssubstanzen des Muskels. *Klin. Wochensch.*, **7**, 1645. 1928.

Lehnartz, E. (3). Über die Einwirkung von Fluorid auf die intermediären Vorgänge bei der Glykolyse in der Hefe. *Z. f. physiol. Chem.*, **230**, 90. 1934.

Lehnartz, M. (1). Besteht ein bestimmtes Verhältnis zwischen Milchsäure- und Ammoniakbildung bei der Muskelkontraktion? *Z. f. physiol. Chem.*, **184**, 183. 1929.

Lehninger, A. L. (1). The relationships of the adenosine polyphosphates to fatty acid oxidation in homogenised liver preparations. *J. biol. Chem.*, **157**, 363. 1945.

Lehninger, A. L. (2). On the activation of fatty acid oxidation. *J. biol. Chem.*, **161**, 437. 1945.

Lehninger, A. L. (3). The oxidation of higher fatty acids in heart muscle suspensions. *J. biol. Chem.*, **165**, 131. 1946.

Lehninger, A. L. (4). Esterification of inorganic phosphate coupled to electron transport between diphosphopyridine nucleotide and oxygen. *J. biol. Chem.*, **178**, 625. 1949.

Lehninger, A. L. (5). Phosphorylation coupled to oxidation of dihydrodiphosphopyridine nucleotide. *J. biol. Chem.*, **190**, 345. 1951.

Lehninger, A. L. (6). Oxidative phosphorylation. In *The Harvey Lectures, 1953–1954*, **49**, 176. 1955.

Lehninger, A. L. (7). *The Mitochondrion. Molecular basis of structure and function.* W. A. Benjamin. New York, 1964.

Leicester, H. M. (1). Biochemical Concepts among the ancient Greeks. *Chymia* (Philad.), **7**, 9. 1961.

Leicester, H. M. (2). *The historical background of chemistry*. New York, John Wiley & Sons. 1956.

Lejeune, N., Thinès-Sempoux & Hers, H. G. (1). Tissue fractionation studies. 16. Intracellular distribution and properties of α-glucosidases in rat liver. *Biochem. J.*, **86**, 16. 1963.

Leloir, L. F. (1). The biosynthesis of glycogen, starch and other polysaccharides. *Harvey Lectures*, Series 56, p. 23. 1961.

Leloir, L. F. & Cardini, C. E. (1). Biosynthesis of glycogen from uridine diphosphate glucose. *J. Am. chem. Soc.*, **79**, 6340. 1957.

Leloir, L. F., Olavarría, J. M., Goldenberg, S. H. & Carminatti, H. (1). Biosynthesis of glycogen from uridine diphosphate glucose. *Arch. Biochem. Biophys.*, **81**, 508. 1959.

Levene, P. A. (1). Properties of the nucleotides obtained from yeast nucleic acid. *J. Biol. Chem.*, **41**, 483. 1920.

Levene, P. A. & Jacobs, W. A. (1). Über die Inosinsäure. *Ber. d. Deutsch chem. Gesellsch.*, **44**, 746. 1911.

Levi, G. & Chèvremont, M. (1). Transformation structurale des éléments des muscle squelettiques pendant leur croissauce in vitro. Relations entre mito-chondries et myofibrilles. *Arch. de Biol.*, **52**, 523. 1941.

Levin, A. & Wyman, J. (1). The viscous elastic properties of muscle. *Proc. Roy. Soc.* B, **101**, 218. 1927.

Levine, R. & Goldstein, M. S. (1). On the mechanism of action of insulin. *Progr. in Hormone Res.*, **11**, 343. 1955.

Levintow, L. & Meister, A. (1). Reversibility of the enzymatic synthesis of glutamine. *J. biol. Chem.*, **209**, 265. 1954.

Levy, H. M. & Koshland, D. E. (1). Evidence for an intermediate in the hydrolysis of ATP by muscle proteins. *J. Am. chem. Soc.*, **80**, 3614. 1958.

Levy, H. M. & Koshland, D. E. (2). Mechanism of hydrolysis of adenosine-triphosphate by muscle proteins and its relation to muscular contraction. *J. biol. Chem.*, **234**, 1102. 1959.

Levy, H. M. & Ryan, E. M. (1). Comparative effects of 2,4-dinitrophenol and p-chlormercuribenzoate on the magnesium-moderated nucleotide triphosphatase activity of L-myosin. *Biochim. biophys. Acta*, **46**, 193. 1961.

Levy, H. M. & Ryan, E. M. (2). Evidence that calcium activates the contraction of actomyosin by overcoming substrate inhibition. *Nature*, **205**, 703. 1965.

Levy, H. M. & Ryan, E. M. (3). Heat inactivation of the relaxing site of acto-myosin: prevention and reversal by dithiothreitol. *Science*, **156**, 73. 1967.

Levy, H. M., Sharon, N. & Koshland, D. E. (1). Purified muscle proteins and the walking rate of ants. *Proc. Nat. Acad. Sci. U.S.A.*, **45**, 785. 1959.

Levy, H. M., Sharon, N., Lindemann, E. & Koshland, D. E. (1). Properties of the active site in myosin hydrolysis of adenosine triphosphate as indicated by the ^{18}O-exchange reaction. *J. biol. Chem.*, **235**, 2628. 1960.

Levy, H. M., Sharon, N., Ryan, E. M. & Koshland, D. E. (1). Effect of temperature on the rate of hydrolysis of ATP and ITP by myosin with and without modifiers. Evidence for a change in protein conformation. *Biochim. biophys. Acta*, **56**, 118. 1962.

Lewis, G. M., Stewart, K. M. & Spencer-Peet, J. (1). Absence of the liver enzyme, UDPG-glycogen 1,4-transglucosylase, as a cause of infantile hypoglycaemia. *Biochem. J.*, **84**, 115P. 1962.

Lewis, M. S., Maruyama, K., Carroll, W. R., Kominz, D. R. & Laki, K. (1). Physical properties and polymerisation reactions of native and inactivated G-actin. *Biochemistry* (Am. chem. Soc.), **2**, 34. 1963.

Lewis, M. S. & Saroff, H. A. (1). The binding of ions to muscle proteins. Measure-ments of the binding of potassium and sodium ions to myosin A, myosin B and actin. *J. Am. Chem. Soc.*, **79**, 2112. 1957.

Lewis, S. E. & Fowler, K. S. (1). In vitro synthesis of phosphoarginine by blowfly muscle. *Nature*, **194**, 1178. 1962.

Leydig, F. (1). Zur Anatomie der Insecten. *Arch. f. Anat., Physiol. wiss. Med.*, **149**, 1859.

Liddell, E. G. T. (1). *The Discovery of Reflexes.* Oxford, Clarendon Press. p. 31. 1960.

Lieben, F. (1). *Geschichte der physiologischen Chemie.* Leipzig and Wien, Franz Duticke. 1935.

Lieberman, I. (1). Involvement of guanosine triphosphate in the synthesis of adenylosuccinate from inosine-5-phosphate. *J. Amer. Chem. Soc.*, 78, 251. 1956.

Liebig, G. v. (1). Ueber die Respiration der Muskeln. *Arch. f. Anat. Physiol. u. Wissensch. Med.* (Müller's). p. 393. 1850.

Liebig, J. v. (1). Ueber die Erscheinungen der Gärung, Fäulniss und Verwesung und ihre Ursachen. *Liebigs Ann. d. Pharm.*, 30, 250. 1839.

Liebig, J. v. (2). Ueber die Gärung und die Quelle der Muskelkraft. I. Die Alkoholgärung. *Ann. d. Chem. u. Pharm.*, 153, 1. 1870.

Liebig, J. v. (3). *Animal Chemistry or Organic Chemistry in its applications to Physiology and Pathology.* tr. W. Gregory. London, Taylor and Walton. 1842.

Liebig, J. v. (4). Ueber die Gärung und die Quelle der Muskelkraft. III. Die Quelle der Muskelkraft. *Ann. d. Chem. u. Pharm.*, 153, 137. 1870.

Liebig, J. v. (5). Über die Bestandtheile der Flüssigkeit des Fleisches. *Ann. Chem. Pharmac.*, 62, 257. 1847.

Lillie, R. S. (1). *Protoplasmic Action and Nervous Action.* Chicago, Univ. Chicago Press. 1923.

Linnane, A. W. (1). A soluble component required for oxidative phosphorylation by a sub-mitochondrial particle from beef heart muscle. *Biochim. biophys. Acta.*, 30, 221. 1958.

Lipmann, F. (1). Über den Tätigkeitsstoffwechsel des fluoridvergifteten Muskels. *Biochem. Z.*, 227, 110. 1930.

Lipmann, F. (2). Metabolic generation and utilisation of phosphate bond energy. *Adv. in Enz.*, 1, 99. 1941.

Lipmann, F. (3). Coupling between pyruvic acid dehydrogenation and adenylic acid phosphorylation. *Nature*, 143, 281. 1939.

Lipmann, F. (4). Acetylation of sulfanilamide by liver homogenates and extracts. *J. biol. Chem.*, 160, 173. 1945.

Lipmann, F. (5). Biological oxidations and reductions. *Ann. Rev. Biochem.*, 12, 1. 1943.

Lipmann, F. (6). Metabolic process patterns. In *Currents in Biochemical Research*, p. 137. Ed. D. E. Green. New York, Interscience Publishers. 1946.

Lipmann, F. (7). The chemistry and thermodynamics of phosphate bonds. Introductory remarks. In *Phosphorus Metabolism*, p. 521. Ed. W. D. McElroy & B. Glass. Johns Hopkins Press, Baltimore. 1951.

Lipmann, F., Kaplan, N. O., Novelli, G. D., Tuttle, L. C. & Guirard, B. M. (1). Coenzyme for acetylation, a pantothenic acid derivative. *J. biol. Chem.*, 167, 869. 1947.

Lipmann, F., Kaplan, N. O., Novelli, G. D., Tuttle, L. C. & Guirard, B. M. (2). Isolation of coenzyme A. *J. biol. Chem.*, 186, 235. 1950.

Lipmann, F. & Lohmann, K. (1). Über die Umwandlung der Harden-Youngschen Hexosediphosphorsäure und die Bildung von Kohlenhydratphosphorsäureestern in Froschmuskelextrakt. *Biochem. Z.*, 222, 389. 1930.

Lipmann, F. & Meyerhof, O. (1). Über die Reaktionsänderung des tätigen Muskels. *Biochem. Z.*, 227, 84. 1930.

Lissmann, H. W. (1). Continuous electric signals from the tail of a fish, *Gymnarchus niloticus* Cuv. *Nature*, 167, 201. 1951.

Littré, E. (1). (tr.) *Oeuvres complètes d'Hippocrate.* 10 vols. Paris. 1839–61.

Lochner, A. & Brink, A. J. (1). Oxidative phosphorylation and glycolysis in the hereditary muscular dystrophy of the Syrian hamster. *Clin. Sci.*, 33, 409. 1967.

Locker, R. H. (1). C-terminal groups in myosin, tropomyosin and actin. *Biochim. biophys. Acta*, **14**, 533. 1954.

Locker, R. H. & Hagyard, C. J. (1). Small subunits of myosin. *Arch. Biochem. Biophys.*, **120**, 454. 1967.

Loewy, A. G. (1). A theory of covalent bonding in muscle contraction. *J. theoret. Biol.*, **20**, 164. 1968.

Loewy, A. G. (2). An actomyosin-like substance from the plasmodium of a myxomycete. *J. cell comp. Physiol.*, **40**, 127. 1952.

Lohmann, K. (1). Über die Isolierung verschiedener natürlicher Phosphorsäure-verbindungen und die Frage ihre Einheitlichkeit. *Biochem. Z.*, **194**, 306. 1928.

Lohmann, K. (2). Über das Vorkommen und den Umsatz von Pyrophosphat im Muskel. *Naturwissensch.*, **16**, 298. 1928.

Lohmann, K. (3). Über das Vorkommen und den Umsatz von Pyrophosphat in Zellen. *Biochem. Z.*, **202**, 466. 1928.

Lohmann, K. (4). Über das Vorkommen und den Umsatz von Pyrophosphat in Zellen. II. Die Menge der leicht hydrolysierbaren P-Verbindung in tierischen und pflanzlichen Zellen. *Biochem. Z.*, **203**, 164. 1928.

Lohmann, K. (5). Über das Vorkommen und den Umsatz von Pyrophosphat in Zellen. III. Das physiologische Verhalten des Pyrophosphats. *Biochem. Z.*, **203**, 172. 1928.

Lohmann, K. (6). Über die Pyrophosphatfraktion im Muskel. *Naturwissensch.*, **17**, 624. 1929.

Lohmann, K. (7). Darstellung der Adenylpyrophosphorsäure aus Muskulatur. *Biochem. Z.*, **233**, 460. 1931.

Lohmann, K. (8). Der Einfluss des Coferments der Milchsäurebildung auf die Aufspaltung von Kohlenhydratphosphorsäureestern im Muskelextrakt. *Biochem. Z.*, **241**, 50. 1931.

Lohmann, K. (9). Über die enzymatische Aufspaltung der Kreatinphosphorsäure; zugleich ein Beitrag zum Chemismus der Muskelkontraktion. *Biochem. Z.*, **271**, 264. 1934.

Lohmann, K. (10). Über die Aufspaltung der Adenylpyrophosphorsäure und Argininphosphorsäure in Krebsmuskulatur. *Biochem. Z.*, **282**, 109. 1935.

Lohmann, K. (11). Untersuchungen zur Konstitution der Adenylpyrophosphorsäure. *Biochem. Z.*, **254**, 381. 1932.

Lohmann, K. (12). Konstitution der Adenylpyrophosphorsäure and Adeninediphosphorsäure. *Biochem. Z.*, **282**, 120. 1935.

Lohmann, K. (13). Über die Bildung und Aufspaltung von Phosphorsäureestern in der Muskulatur in Gegenwart von Fluorid, Oxalat, Citrat und Arseniat. *Biochem. Z.*, **222**, 324. 1930.

Lohmann, K. (14). Beitrag zur enzymatischen Umwandlung von synthetischem Methylglyoxal in Milchsäure. *Biochem. Z.*, **254**, 332. 1932.

Lohmann, K. (15). Über Phosphorylierung und Dephosphorylierung. Bildung der natürliche Hexosemonophosphorsäure aus ihren Komponenten. *Biochem. Z.*, **262**, 137. 1933.

Lohmann, K. (16). Zuckerabbau in der tierische Zelle. *Oppenheimers Handbuch der Biochemie*, I, p. 915. Jena, Gustav Fischer. 1933.

Lohmann, K. (17). Untersuchungen an Oktopusmuskulatur. Isolierung und enzymatisches Verhalten von Adenylpyrophosphorsäure and Argininphosphorsäure. *Biochem. Z.*, **286**, 28. 1936.

Lohmann, K. (18). Vergleichende Untersuchungen über das Coferment der Milchsäurebildung und die alkoholischen Gärung. *Biochem. Z.*, **241**, 67. 1931.

Lohmann, K. & Meyerhof, O. (1). Über die enzymatische Unwandlung von Phosphoglycerinsäure in Brentzraubensäure und Phosphorsäure. *Biochem. Z.*, **273**, 60. 1934.

Lohmann, K. & Schuster, P. (1). Untersuchungen über die Cocarboxylase. *Biochem. Z.*, **294**, 188. 1937.

Loo, E. van. (1). (tr.) Rooseboom's *Microscopium*. Communication No. 95 from the National Museum for the History of Science, Leiden. Leiden. 1956.

Loomis, W. F. & Lipmann, F. (1). Reversible inhibition of the coupling between phosphorylation and oxidation. *J. biol. Chem.*, **173**, 807. 1948.

Lorand, L. (1). Adenosinetriphosphate-creatine transphosphorylase as relaxing factor of muscle. *Nature*, **172**, 1181. 1953.

Lovatt Evans, C. L. (1). The physiology of plain muscle. *Physiol. Rev.*, 6, 358. 1926.

Lovatt Evans, C. L. (2). Studies on the physiology of plain muscle. II. The oxygen usage of plain muscle, and its relation to tonus. *J. Physiol.*, **58**. 22, 1923.

Lovatt Evans, C. L. (3). Studies on the physiology of plain muscle. IV. The lactic acid content of plain muscle under various conditions. *Biochem. J.*, **19**, 1115. 1925.

Lowey, S. (1). Myosin substructure: Isolation of a helical subunit from heavy meromyosin. *Science*, **145**, 597. 1964.

Lowey, S. (2). Comparative study of the α-helical muscle proteins. Tyrosyl titration and effect of pH on conformation. *J. biol. Chem.*, **140**, 2421. 1965.

Lowey, S. & Cohen, C. (1). Studies on the structure of myosin. *J. mol. Biol.*, **4**, 293. 1962.

Lowey, S., Goldstein, L., Cohen, C. & Luck, S. M. (1). Proteolytic degradation of myosin and the meromyosins by a water-insolbule polyanionic derivative of trypsin. *J. mol. Biol.*, **23**, 287. 1967.

Lowey, S. & Holtzer, A. (1). The aggregation of myosin. *J. Am. chem. Soc.*, **81**, 1378. 1959.

Lowey, S. & Holtzer, A. (2). The homogeneity and molecular weights of the meromyosins and their relative proportions in myosin. *Biochim. biophys. Acta.*, **34**, 470. 1959.

Lowey, S., Kucera, J. & Holtzer, A. (1). On the structure of the paramyosin molecule. *J. mol. Biol.*, **7**, 234. 1963.

Lowey, S., Slayter, H. S., Weeds, A. G. & Baker, H. (1). The substructure of the myosin molecule. I. Subfragments of myosin by enzymic degradation. *J. mol. Biol.*, **42**, 1. 1969.

Lowry, O. H. & Passonneau, J. V. (1). Kinetic evidence for multiple binding sites on phosphofructokinase. *J. biol. Chem.*, **241**, 2268. 1966.

Lowry, T. M. (1). *Historical Introduction to Chemistry*. Macmillan, London. 1936.

Lowy, J. & Hanson, J. (1). Ultrastructure of invertebrate smooth muscle. *Physiol. Rev.*, **42**, Suppl. 5, 34. 1962.

Lowy, J. & Hanson, J. (2). Electron microscope studies of bacterial flagella. *J. mol. Biol.*, **11**, 293. 1965.

Lowy, J. & Millman, B. M. (1). Contraction and relaxation in smooth muscles of lamellibrauch molluscs. *Nature*, **183**, 1730. 1959.

• Lowy, J. & Millman, B. M. (2). The contractile mechanism of the anterior byssus retractor muscle of *Mytilus edulis*. *Phil. Trans. Roy. Soc.* B, **246**, 105. 1963.

Lowy, J., Millman, B. M. & Hanson, J. (1). Structure and function in smooth tonic muscles of lamellibrauch molluscs. *Proc. Roy. Soc.* B, **160**, 525. 1964.

Lowy, J., Poulson, F. R. & Vibert, P. J. (1). Myosin filaments in vertebrate smooth muscle. *Nature*, **225**, 1053.

Lowy, J. & Small, J. V. (1). The organization of myosin and actin in vertebrate smooth muscle. *Nature*, **227**, 46. 1970.

Lowy, J. & Spencer, M. (1). Structure and function of bacterial flagella. *Symp. Soc. exp. Biol.*, **22**, 215. 1968.

Luchi, R. J., Kritcher, E. M. & Conn, H. L. (1). Molecular characteristics of canine cardiac myosin. *Circ. Res.*, **16**, 74. 1965.

Luck, S. M. & Lowey, S. (1). Equilibrium binding of ADP to deaminase-free myosin. *Fed. Proc.*, **27**, 519. 1968.

Luh, W. & Henkel, E. (1). Gehalt und Verteilung von Phosphotransferasen in menschlicher Skelett- und Uterusmuskulatur. *Z. Geburtsh. Gynaek.*, **163**, 279. 1965.

Lundholm, L. & Mohme-Lundholm, E. (1). Studies on the effects of drugs upon the lactic acid metabolism and contraction of vascular smooth muscle. *Acta physiol. Scand.*, **55**, 45. 1962.

Lundholm, L. & Mohme-Lundholm, E. (2). Dissociation of contraction and stimulation of lactic acid production in experiments on smooth muscle under anaerobic conditions. *Acta physiol. Scand.*, **57**, 111. 1963.

Lundholm, L. & Mohme-Lundholm, E. (3). Contraction and glycogenolysis of smooth muscle. *Acta physiol. Scand.*, **57**, 125. 1963.

Lundholm, L. & Mohme-Lundholm, E. (4). Energetics of isometric and isotonic contraction in isolated vascular smooth muscle under anaerobic conditions. *Acta physiol. Scand.*, **64**, 275. 1965.

Lundsgaard, E. (1). Untersuchungen über Muskelkontraktion ohne Milchsäure. *Biochem. Z.*, **217**, 162. 1930.

Lundsgaard, E. (2). Weitere Untersuchungen über Muskelkontraktionen ohne Milchsäurebildung. *Biochem. Z.*, **227**, 51. 1930.

Lundsgaard, E. (3). Über die Einwirkung der Monoiodoessigsäure auf den Spaltungs- und Oxydationsstoffwechsel. *Biochem. Z.*, **220**, 8. 1930.

Lundsgaard, E. (4). Über die Bedeutung der Argininphosphorsäure für den Tätigkeitsstoffwechsel der Crustaceenmuskeln. *Biochem. Z.*, **230**, 10. 1931.

Lundsgaard, E. (5). Über die Energetik der anaeroben Muskelkontraktion. *Biochem. Z.*, **233**, 322. 1931.

Lundsgaard, E. (6). Phosphagen- und Pyrophosphatumsatz in jodessigsäurevergifteten Muskeln. *Biochem. Z.*, **269**, 308. 1934.

Lundsgaard, E. (7). Hemmung von Esterifizierungs Vorgängen als Ursache der Phlorrizinwirkung. *Biochem. Z.*, **264**, 209. 1933.

Lundsgaard, E. (8). The biochemistry of muscle. *Ann. Rev. Biochem.*, **7**, 377. 1938.

Lundsgaard, E. (9). The ATP content of resting and active muscle. *Proc. Roy. Soc.* B, **137**, 73. 1950.

Lundsgaard, E. (10). Observations on a factor determining the metabolic rate of the liver. *Biochim. biophys. Acta.*, **4**, 322. 1950.

Lundsgaard, E. (11). On the mode of action of insulin. *Uppsala Läkareforenigens Förhandlinger*, **45**, 143. 1939.

Lüttgau, H. C. & Niedergerke, R. (1). The antagonism between Ca and Na ions on the frog's heart. *J. Physiol.*, **143**, 486. 1958.

Lyding, G. (1). Untersuchungen über den Lactacidogenphosphorsäure- und Restphosphorsäuregehalt von Hühne- und Taubenmuskeln. *Z. f. physiol. Chem.*, **113**, 223. 1921.

Lymn, R. W. & Taylor, E. W. (1). Transient state phosphate production in the hydrolysis of nucleoside triphosphates by myosin. *Biochemistry*, **9**, 2975. 1970.

Lynen, F. & Ochoa, S. (1). Enzymes of fatty acid metabolism. *Biochim. biophys. Acta*, **12**, 299. 1953.

Lynen, F. & Reichert, F. (1). Zur chemische Struktur der aktivierten Essigsäure. *Angewandte Chem.*, **63**, 47. 1951.

Lynen, F., Reichert, E. & Rueff, L. (1). Zum biologischen Abbau der Essigsäure. VI. 'Aktivierte essigsäure', ihre Isolierung aus Hefe und ihre chemische Natur. *Liebigs Ann. d. Chem.*, **574**, 1. 1951.

Lynn, W. S. (1). Effects of cations, polyanions and sulfhydryl reagents on muscle proteins. *Arch. Biochem. Biophys.*, **110**, 262. 1965.

Lythgoe, B. & Todd, A. R. (1). Structure of adenosine di- und triphosphate. *Nature*, **155**, 695. 1945.

Lyubimova, M. N. (1). Characteristics of plant motion systems. In *Molecular Biology. Problems and Perspectives* (in Russian). Ed. A. E. Braunstein, A. N. Nesmeyanov & I. E. Tamm. Moscow, Nauka. 1964.

Lyubimova, M. N., Demyanovskaya, I. B., Pedarovich, I. B. & Itomlenskite, I. V. (1). Participation of ATP in the motor function of the *Mimosa pudica* leaf. *Biokhimiya* (Eng. vers.), **29**, 663. 1964.

Lyubimova, M. N. & Engelhardt, V. A. (1). Adenosinetriphosphatase and myosin. *Biokhimiya*, **4**, 716. 1939.

Maar, W. (1). ed. [Stensen's] *Opera philosophica*. Copenhagen. 1910.

Macallum, A. B. (1). On the distribution of potassium in animal and vegetable cells. *J. Physiol.*, **32**, 95. 1905.

Macallum, A. B. (2). The origin of muscular energy: thermodynamic or chemodynamic? *J. biol. Chem.*, **14**, 7. 1913.

McArdle, B. (1). Myopathy due to a defect in muscle glycogen breakdown. *Clin. Sci.*, **10**, 3. 1951.

MacArthur, I. (1). Structure of α-keratin. *Nature*, **152**, 38. 1943.

McCaman, M. W. (1). Enzyme studies in skeletal muscle in mice with hereditary muscular dystrophy. *Amer. J. Physiol.*, **205**, 897. 1963.

McCaman, M. W. & McCaman, R. E. (1). Effects of denervation on normal and dystrophic muscle: DNA and nucleotide enzymes. *Amer. J. Physiol.*, **209**, 495. 1965.

McDougall, W. (1). On the structure of cross-striated muscle, and a suggestion as to the nature of its contraction. *J. Anat. and Physiol.*, **31**, 410 and p. 539. 1897.

McElroy, W. D. (1). The energy source for bioluminescence in an isolated system. *Proc. nat. Acad. Sci. U.S.A.*, **33**, 342. 1947.

McElroy, W. D. & Seliger, H. H. (1). The chemistry of light emission. *Adv. in Enz.*, **25**, 119. 1963.

Macfarlane, M. G. (1). The action of arsenate on hexosediphosphatase. *Biochem. J.*, **24**, 1051. 1930.

McKie, D. (1). Fire and the *flamma vitalis*: Boyle, Hooke & Mayow. In *Science, Medicine and History, Essays in honour of Charles Singer*. p. 469. Ed. E. A. Underwood. Oxford. 1953.

MacMunn, C. A. (1). On myohaematin, an intrinsic muscle-pigment of vertebrates and invertebrates, on histohaematin, and on the spectrum of the suprarenal bodies. *J. Physiol.*, **5**, xxiv. 1884–5.

MacMunn, C. A. (2). Researches on myohaematin and the histohaematins. *Phil. Trans. Roy. Soc.*, **177**, 267. 1886.

MacMunn, C. A. (3). Further observations on myohaematin and the histohaematins. *J. Physiol.*, **8**, 51. 1887.

MacMunn, C. A. (4). Über das Myohaematin. *Z. physiol. Chem.*, **13**, 497. 1889.

McSwiney, B. A. & Robson, J. M. (1). The response of smooth muscle to stimulation of the vagus nerve. *J. Physiol.*, **68**, 124. 1929.

Machin, K. E. (1). Wave propagation along flagella. *J. exp. Biol.*, **35**, 796. 1958.

Machin, K. E. & Pringle, J. W. S. (1). The physiology of insect flight muscle. II. Mechanical properties of a beetle flight muscle. *Proc. Roy. Soc.* B, **151**, 204. 1959.

Machin, K. E. & Pringle, J. W. S. (2). The physiology of insect flight muscle. III. The effect of sinusoidal changes of length on a beetle flight muscle. *Proc. Roy. Soc.* B., **152**, 311. 1960.

Maddox, C. E. R. & Perry, S. V. (1). Differences in the myosins of the red and white muscles of the pigeon. *Biochem. J.*, **99**, 8P. 1966.

Madsen, N. B. (1). Allosteric properties of phosphorylase *b*. *Biochem. biophys. Res. Com.*, **15**, 390. 1964.

Madsen, N. B. & Cori, C. F. (1). The binding of adenylic acid by muscle phosphorylase. *J. biol. Chem.*, **224**, 899. 1957.

Madsen, N. B. & Cori, C. F. (2). The interaction of phosphorylase with protamine. *Biochim. biophys. Acta.*, **15**, 516. 1954.

Magendie, M. F. (1). Sur la propriétés nutritive des substances qui ne contiennent pas d'azote. *Ann. de Chim. et de Phys.*, **3**, 66. 1816.

Mahler, H. R. (1). Studies in the fatty acid oxidising system of animal tissues. IV. The prosthetic group of butyryl coenzyme A dehydrogenase. *J. biol. Chem.*, **206**, 13. 1954.

Mahler, H. R., Wakil, S. J. & Bock, R. M. (1). Studies on fatty acid oxidation. 1. Enzymatic activation of fatty acids. *J. biol. Chem.*, **204**, 453. 1953.

Mahler, R. (1). Glycogen synthesis from pyruvate in muscle. *Nature*, **209**, 616. 1966.

Makinose, M. (1). Die Nucleosidtriphosphat-nucleosiddiphosphat-transphosphorylase-Aktivität der Vesikel des sarkoplasmatischen Reticulums. *Biochem. Z.*, **345**, 80. 1966.

Makinose, M. & Hasselbach, W. (1). Die Abhängigkeit der Granawirkung von der Art des Aktomyosinsystems und Gergely's co-Factor. *Biochim. biophys. Acta.*, **43**, 239. 1960.

Makinose, M. & Hasselbach, W. (2). Der Einfluss von Oxalat auf den Calcium-Transport isolierter Vesikel des sarkoplasmatischen Reticulum. *Biochem. Z.*, **343**, 360. 1965.

Makinose, M. & The, R. (1). Calcium-Akkumulation und Nucleosidtriphosphat-Spaltung durch die Vesikel des sarkoplasmatischen Reticulum. *Biochem. Z.*, **343**, 383. 1965.

Maleknia, N., Ebersolt, E., Schapira, G. & Dreyfus, J. C. (1). Modifications en sens inverse de la teneur en myoglobine dans les muscles rouge et blanc après section nerveuse. *Bull. Soc. Chim. Biol.*, **48**, 905. 1966.

Mandel, P. & Kempf, E. (1). Les nucleotides libres du tissu aortique. *Biochim. biophys. Acta.*, **51**, 184. 1961.

Mann, T. (1). Studies on the metabolism of semen. I. General aspects, occurrence and distribution of cytochrome, certain enzymes and coenzymes. *Biochem. J.*, **39**, 451. 1945.

Mann, T. (2). Über die Verkettung der chemische Vorgänge im Muskel. VII. Die Phosphatabspaltung aus Phosphorglycerinsäure in Fluoridvergifteten Muskelbrei. *Biochem. Z.*, **279**, 82. 1935.

Manners, D. J. & Wright, A. (1). A case of limit dextrinosis. *Biochem. J.*, **79**, 18P. 1961.

Mansour, T. E. (1). Studies on heart phosphofructokinase: purification, inhibition and activation. *J. biol. Chem.*, **238**, 2285. 1963.

Mansour, T. E., Clague, M. E. & Beernink, K. D. (1). Effect of cyclic 3,5-AMP on heart phosphofructokinase. *Fed. Proc.* **21**, 238. 1962.

Mansour, T. E., Wakid, N. W. & Sprouse, H. M. (1). Purification, crystallisation and properties of activated sheep heart phosphofructokinase. *Biochem. biophys. Res. Com.*, **19**, 721. 1961.

Mansour, T. E., Wakid, N. W. & Sprouse, H. M. (2). Studies on heart phosphofructokinase. Purification, crystallization and properties of sheep heart phosphofructokinase. *J. biol. Chem.*, **241**, 1512. 1965.

Marceau, F. (1). Recherches sur la morphologie, l'histologie et la physiologie comparée des muscles adductors des mollusques acéphalés. *Arch. Zool. exp. gen.*, **2** (5th series), 295. 1909.

Maréchal, G. (1). *La metabolisme de la phosphorylcreatine et de l'adenosine triphosphate durant la contraction musculaire*. Bruxelles, Editions Arscia. 1964.

Maréchal, G. & Beckers-Bleukx, G. (1). Adenosine triphosphate and phosphorylcreatine breakdown in resting and stimulated muscles after treatment with 1-fluoro-2,4-dinitrobenzene. *Biochem. Z.*, **345**, 286. 1966.

Maréchal, G. & Mommaerts, W. F. H. M. (1). The metabolism of phosphocreatine during an isometric tetanus in the frog sartorius muscle. *Biophys. biochim. Acta.*, **70**, 53. 1963.

Margaria, R. (1). An apparent change of pH on stretching a muscle. *J. Physiol.*, **82**, 496. 1934.

Margaria, R. & Edwards, H. T. (1). The removal of lactic acid from the body during recovery from muscular exercise. *Amer. J. Physiol.*, **107**, 681. 1934.

Margaria, R., Edwards, H. T. & Dill, E. B. (1). The possible mechanisms of contracting and paying the oxygen debt and the role of lactic acid in muscular contraction. *Amer. J. Physiol.*, **106**, 689. 1933.

Margreth, A., Muscatello, U. & Andersson-Cedergren, E. (1). A morphological and biochemical study on the regulation of carbohydrate metabolism in the muscle cell. *Exp. Cell Res.*, **32**, 484. 1962.

Mark, J. S. T. (1). Uterine smooth muscle. *Anat. Rec.*, **125**, 473. 1956.

Marsh, B. B. (1). The effects of adenosine triphosphate on the fibre volume of a muscle homogenate. *Nature*, **167**, 1065. 1951.

Marsh, B. B. (2). The effects of adenosine triphosphate on the fibre volume of a muscle homogenate. *Biochim. biophys. Acta.*, **9**, 247. 1952.

Marsh, B. B. (3). Rigor mortis in beef. *J. Science of Food and Agric.*, **5**, 70. 1954.

Marshall, J. M., Holtzer, H., Finck, H. & Pepe, F. (1). The distribution of protein antigens in striated myofibrils. *Exp. Cell Res.* Suppl. **7**, 219. 1959.

Martin, D S. (1). The relation between work performed and heat liberated by the isolated gastrocnemius, semi-tendinosus and tibialis anticus muscles of the frog. *Amer. J. Physiol.*, **83**, 543. 1928.

Martin, E. G., Field, J. & Hall, V. E. (1). Metabolism following anoxemia. I. Oxygen consumption and blood lactates after experimentally induced exercise. *Amer. J. Physiol.*, **88**, 407. 1929.

Martinez, R. J., Brown, D. M. & Glazer, A. N. (1). The formation of bacterial flagella. III. Characterization of the subunits of the flagella of *Bacillus subtilis* and *Spirillum serpeus*. *J. mol. Biol.*, **28**, 45. 1967.

Martius, C. (1). Über den Abbau der Citronensäure. *Z. physiol. Chem.*, **247**, 104. 1937.

Martius, C. & Knoop, F. (1). Der physiologische Abbau der Citronensäure. *Z. physiol. Chem.*, **246**, 1. 1937.

Martonosi, A. (1). The specificity of the interaction of ATP with G-actin. *Biochim. biophys. Acta*, **47**, 163. 1962.

Martonosi, A. (2). Studies on actin. VII. Ultracentrifugal analysis of partially polymerised actin solutions. *J. biol. Chem.*, **237**, 279. 1962.

Martonosi, A. & Feretos, R. (1). Sarcoplasmic reticulum. I. The uptake of Ca^{2+} by sarcoplasmic reticulum fragments. *J. biol. Chem.*, **239**, 648. 1964.

Martonosi, A. & Feretos, R. (2). Correlation between adenosine triphosphatase activity and Ca^{2+} uptake. *J. biol. Chem.*, **239**, 659. 1964.

Martonosi, A. & Gouvea, M. A. (1). Studies on actin. VI. The interaction of nucleoside triphosphates with actin. *J. biol. Chem.*, **236**, 1345. 1961.

Martonosi, A. & Gouvea, M. A. (2). Studies on actin. V. Chemical modification of actin. *J. biol. Chem.*, **236**, 1338. 1961.

Martonosi, A., Gouvea, M. A. & Gergely, J. (1). Studies on actin. I. The interaction of C^{14} labelled adenine nucleotides with actin. *J. biol. Chem.*, **235**, 1700. 1960.

Martonosi, A., Gouvea, M. A. & Gergely, J. (2). Studies on actin. III. G–F transformation of actin and muscular contraction (experiments in vivo). *J. biol. Chem.*, **235**, 1707. 1960.

Maruyama, K. (1). A new protein-factor hindering network formation of F-actin in solution. *Biochim. biophys. Acta*, **94**, 209. 1965.

Maruyama, K. (2). Interaction of tropomyosin with actin. A flow birefringence study. *Arch. Biochem. Biophys.*, **105**, 142. 1964.

Maruyama, K. (3). Adenosinetriphosphatase activity of the contractile protein from the body wall muscle of the echinoid *Urechis unicinctus*. *Enzymologia*, **17**, 90. 1954.

Maruyama, K. (4). Interaction of the contractile protein from a sea-anemone with adenosine nucleotides. *Sci. Papers Coll. gen. Educ. Univ. Tokyo*, **6**, 95. 1956.

Maruyama, K. (5). A further study of insect actomyosin. *Sci. Papers Coll. gen. Educ. Univ. Tokyo*, **7**, 213. 1957.

Maruyama, K. & Ebashi, S. (1). α-Actinin, a new structural protein from striated muscle. II. Action on actin. *J. Biochem.*, **58**, 13. 1965.

Maruyama, K. & Gergely, J. (1). Removal of the bound calcium of G-actin by ethylenediamine-tetracetate (EDTA). *Biochem. biophys. Res. Com.*, **6**, 245. 1961.

Maruyama, K., Hama, H. & Ishikawa, Y. (1). Some physico-chemical properties of KI extracted F-actin. *Biochim. biophys. Acta*, **94**, 200. 1965.

Maruyama, K. & Ishikawa, Y. (1). Effect of temperature and potassium chloride concentration on the onset of superprecipitation of actomyosin. Adenosinetriphosphatase studies. *J. Biochem.*, **55**, 110. 1964.

Maruyama, K. & Pringle, J. W. S. (1). The effect of ADP on the ATPase activity of insect actomyosin of low ionic strength. *Arch. Biochem. Biophys.*, **120**, 225. 1967.

Maruyama, K., Pringle, J. W. S. & Tregear, R. T. (1). The calcium sensitivity of ATPase activity of myofibrils and actomyosins from insect flight and leg muscles. *Proc. Roy. Soc.* B., **169**, 229. 1968.

Masaki, T., Endo, M. & Ebashi, S. (1). Localization of the 6S component of α-actinin at the Z-band. *J. Biochem.*, **62**, 630. 1967.

Masaki, T., Takaiti, T. & Ebashi, S. (1). M-substance, a new protein constituting the M-line of myofibrils., *J. Biochem.*, **64**, 909. 1968.

Masoro, E. J., Rowell, L. B. & McDonald, R. M. (1). Skeletal muscle lipids. 1. Analytical method and composition of monkey gastrocnemius and soleus. *Biochim. biophys. Acta*, **84**. 493. 1964.

Massey, V. (1). The identity of diaphorase and lipoyl dehydrogenase. *Biochim. biophys. Acta*, **37**, 314. 1960.

Massey, V. (2). The composition of the ketoglutarate dehydrogenase complex. *Biochim. biophys. Acta*, **38**, 447. 1960.

Matačić, S. & Loewy, A. G. (1). The identification of isopeptide cross links in insoluble fibrin. *Biochem. biophys. Res. Comm.*, **30**, 356. 1968.

Matoltsy, A. G. & Gerendás, M. (1). Isotropy in the I striation of striated muscle. *Nature*, **159**, 502. 1947.

Matoltsy, A. G. & Gerendás, M. (2). On the nature of cross-striation. *Hung. Acta. Physiol.*, **1**, 116. 1947.

Matteucci, C. (1). Sur le courant électrique ou propre de la grenouille. *Ann. Chim. Phys.*, **68**, 93. 1838.

Matteucci, C. (2). Sur le courant électrique des muscles des animaux vivants ou recemment tués. *Compt. rend de l'Acad. Sci.*, **16**, 197. 1843.

Matteucci, C. (3). Recherches sur les phénomènes physique et chimique de la contraction musculaires. *Ann. de Chim. et de Phys.*, **47**, 129. 1856.

Mayer, J. R. (1). Bemerkungen über die Kräfte der unbelebten Natur. *Ann. d. Chem. und Pharm.*, **42**, 233. 1842.

Mayer, J. R. (2). *Die organische Bewegung in ihrem Zusammenhang mit dem Stoffwechsel.* 1845. (Ostwalds Klassiker der exakten Wissenschaften 180). Ed. A. v. Oettingen. Leipzig. 1911.

Mayer, S. (1). Von der functionellen Verschiedenheit der peripherischen Nerven. In *Handbuch der Physiologie*, ed. L. Hermann. p. 197. Leipzig, F. C. W. Vogel. 1879.

Mayer, S. E. & Moran, N. C. (1). Relation between pharmacologic augmentation of cardiac contractile force and the activation of myocardial glycogen phosphorylase. *J. Pharm. & exp. Therapeutics*, **129**, 271. 1960.

Mayow, J. (1). On muscular motion and animal spirits. Fourth treatise in *Medical-physical works*. Alembic Club Reprint No. 17. Published by the Alembic Club. Tr. A.C.B. & L.D. 1907.

Mazia, D., Chaffee, R. R. & Iverson, R. M. (1). Adenosine triphosphatase in the mitotic apparatus. *Proc. nat. Acad. Sci. U.S.A.*, **47**, 788. 1961.

Mazia, D. & Ruby, A. (1). Dissolution of erythrocyte membranes in water and comparison of the membrane protein with other structural proteins. *Proc. nat. Acad. Sci. U.S.A.*, **61**, 1005. 1968.

Mehl, J. W. (1). The oxidation and reduction of muscle ATPase. *Science*, **31**, 292. 1944.

Mehl, J. W. (2). Studies on the proteins of smooth muscle. I. *J. biol. Chem.*, **123**, lxxxiii. 1938.

Meigs, E. B. (1). The structure of the element of cross-striated muscle, and the changes of form which it undergoes during contraction. *Z. f. allgemeine Physiol.*, **8**, 81. 1908.

Meinrenken, W. (1). Besonderheiten der Erschlaffung des isolierten contractilen Apparates von fibrillären (oscillierenden) Insektenflugmuskeln. *Arch. f. ges. Physiol.*, **294**, R45. 1967.

Meissner, G. (1). Bericht über die Fortschritte der Physiologie im Jahre 1864. *Ber. ü. d. Fortschritte d. Anat. u. Physiol.*, p. 427. 1864.

Menkes, J. H. & Csapo, A. (1). Changes in the adenosine triphosphate and creatine phosphate content of the rabbit uterus throughout sexual maturation and after ovulation. *Endocrinology*, **50**, 37. 1952.

Meshkina, M. P. & Karyavkina, O. E. (1). Participation of carnosine and anserine in the glycolytic and oxidative processes in muscle tissue. *Biokhimiya* (Eng. vers.), **30**, 74. 1967.

Metzzer, B., Helmreich, E. & Glaser, L. (1). The mechanism of activation of skeletal muscle phosphorylase *a* by glycogen. *Proc. Nat. Acad. Sci. U.S.A.*, **57**, 994. 1967.

Meyer, H. K. & Mark, H. (1). Über den Aufbau des Seiden-Fibroin. *Ber. d. Deutsch. chem. Ges.*, **61**, 1932. 1928.

Meyer, K. (1). Über die Reinigung des milchsäurebildenden Ferments. *Biochem. Z.*, **193**, 139. 1928.

Meyer, K. & Weber, H. H. (1). Das kolloidale Verhalten der Muskeleiweisskörper. V. Das Mengenverhältnis der Muskeleiweisskörper in seiner Bedeutung für die Struktur des quergestreiften Kaninchenmuskels. *Biochem. Z.*, **266**, 137. 1933.

Meyer, K. H. (1). Über Feinbau, Festigkeit und Kontraktilität tierische Gewebe. *Biochem. Z.*, **214**, 253. 1929.

Meyer, K. H. (2). Die Muskelkontraktion. *Biochem. Z.*, **217**, 433. 1930.

Meyer, K. H. (3). Über Feinbau, Festigkeit und Kontraktilität tierischer Gewebe. *Experientia*, **7**, 361. 1951.

Meyer, K. H. (4). Thermoelastic properties of several biological systems. *Proc. Roy. Soc. B*, **139**, 498. 1951–52.

Meyer, W. L., Fischer, E. H. & Krebs, E. G. (1). Activation of skeletal muscle phosphorylase *b* kinase by Ca^{2+}. *Biochemistry* (Am. chem. Soc.), **3**, 1033. 1964.

Meyerhof, O. (1). Zur Verbrennung der Milchsäure in der Erholungsperiode des Muskels. *Arch. f. ges. Physiol.*, **175**, 88. 1919.

Meyerhof, O. (2). Das Schicksal der Milchsäure in der Erholungsperiode des Muskels. *Arch. f. ges. Physiol.*, **182**, 284. 1920.

Meyerhof, O. (3). Kohlehydrat- und Milchsäure-umsatz im Froschmuskel. *Arch. f. ges. Physiol.*, **185**, 11. 1920.

Meyerhof, O. (4). *Die chemischen Vorgänge im Muskel*. Berlin, Julius Springer. 1930.

Meyerhof, O. (5). Ueber die Beziehung der Milchsäure zur Wärmebildung und Arbeitsleistung des Muskels in der Anaerobiose. *Arch. f. ges. Physiol.*, **182**, 232. 1920.

Meyerhof, O. (6). Milchsäurebildung und mechanische Arbeit. *Arch. f. ges. Physiol.*, **191**, 128. 1921.

Meyerhof, O. (7). Ueber die Milchsäurebildung in der zerschnittenen Muskulatur. *Arch. f. ges. Physiol.*, **188**, 114. 1921.

Meyerhof, O. (8). Ueber den Ursprung der Kontraktionswärme. *Arch. f. ges. Physiol.*, **195**, 22. 1922.

Meyerhof, O. (9). Die Energieumwandlungen im Muskel. VII. Weitere Untersuchungen über den Ursprung der Kontraktionswärme. *Arch. f. ges. Physiol.*, **204**, 295. 1924.

Meyerhof, O. (10). Über die enzymatische Milchsäurebildung im Muskelextrakt. *Biochem. Z.*, **178**, 395. 1926.

Meyerhof, O. (11). Über die enzymatische Milchsäurebildung im Muskelextrakt. II. Die Spaltung der Polysaccharide und der Hexosediphosphorsäure. *Biochem. Z.*, **178**, 462. 1926.

Meyerhof, O. (12). Über die enzymatische Milchsäurebildung im Muskelextrakt. III. Mitteilung: Die Milchsäure aus den gärfähige Hexosen. *Biochem. Z.*, **183**, 176. 1927.

Meyerhof, O. (13). Die zeitliche Verlauf der Milchsäurebildung bei der Muskelkontraktion. *Klin. Wochenschr.*, **10**, 214. 1931.

Meyerhof, O. (14). Neue Versuche zur Energetik der Muskelkontraktion. *Naturwissensch.*, **19**, 923. 1931.

Meyerhof, O. (15). Über die Intermediärvorgänge der enzymatischen Kohlehydratspaltung. *Erg. d. Physiol.*, **39**, 10. 1937.

Meyerhof, O. (16). Über die Wirkungsweise der Hexokinase. *Naturwissensch.*, **23**, 850. 1935.

Meyerhof, O. (17). Über die Synthese der Kreatinphosphorsäure im Muskel und die 'Reaktionsform' des Zuckers. *Naturwissensch.*, **25**, 443. 1937.

Meyerhof, O. (18). The chemistry of the anaerobic recovery of muscle. *New England J. of Med.*, **220**, 49. 1939.

Meyerhof, O. (19). Sur l'isolement de l'acide 3-glycéraldéhyde phosphorique biologique au cours de la dégradation enzymatique de l'acide hexose diphosphorique. *Bull. Soc. Chim. biol.*, **20**, 1033. 1938.

Meyerhof, O. (20). Energy relationships in glycolysis and phosphorylation. *Ann. New York Acad. Sci.*, **45**, 377. 1945.

Meyerhof, O. (21). Über die Verbreitung der Argininphosphorsäure in der Muskulatur der Wirbellosen. *Arch. Sci. biol. Napoli*, **12**, 536. 1928.

Meyerhof, O. (22). Über die Volumenschwankung bei der Muskelkontraktion. *Naturwiss.*, **20**, 977. 1932.

Meyerhof, O. (23). Bemerkungen zu der Arbeit von E. Ernst and J. Koczkás: Eigenfrequenz und Reversibilität der Volumverminderung des Muskels. *Arch. f. ges. Physiol.*, **240**, 386. 1938.

Meyerhof, O. (24). The significance of oxidations for muscular contraction. *Biol. Symp.*, **3**, 239. Ed. J. Cattell. Lancaster, Pa., Jaques Cattell Press. 1941.

Meyerhof, O. (25). Über die enzymatische Spaltung des Traubenzuckers und anderer Hexosen im Muskelextrakt. *Die Naturwissensch.*, **14**, 756. 1926.

Meyerhof, O. (26). Über die Änderung des osmotischen Drucks des Muskels bei Ermüdung und Starre. *Biochem. Z.*, **226**, 1. 1930.

Meyerhof, O. (27). Aldolase and isomerase. In *The Enzymes*, Vol. II, part I. Ed. Sumner, J. B. & Myrbäck, K. 1st edition. New York, Academic Press. 1951.

Meyerhof, O. (28). Über einige Probleme der Muskelphysiologie. *Naturwissensch.*, **12**, 1137. 1924.

Meyerhof, O. & Grollman, A. (1). Weitere Versuche über den Zusammenhang zwischen chemischem Umsatz und osmotischer Druckzunahme im Muskel. *Biochem. Z.*, **241**, 21. 1931.

Meyerhof, O. & Hartmann, H. (1). Uber die Volumenschwankung bei der Muskelkontraktion. *Biochem. Z.*, **234**, 722. 1934.

Meyerhof, O. & Himwich, H. E. (1). Beiträge zum Kohlenhydratstoffwechsel des Warmblütermuskels, insbesondere nach einseitiger Fetternährung. *Arch. f. ges. Physiol.*, **205**, 414. 1924.

Meyerhof, O. & Junowicz-Kocholaty, R. (1). The equilibria of isomerase and aldolase, and the problem of the phosphorylation of glyceraldehyde phosphate. *J. biol. Chem.*, **149**, 71. 1943.

Meyerhof, O. & Kiessling, W. (1). Über das Auftreten und den Umsatz der α-Glycerinphosphorsäure bei der enzymatischen Kohlenhydratspaltung. *Biochem. Z.*, **264**, 40. 1933.

Meyerhof, O. & Kiessling, W. (2). Über die Isolierung der isomeren Phosphoglycerinsäuren (Gylcerinsäure-2-phosphorsäure und Glycerinsäure-3-phosphorsäure) aus Gärensätzen und ihr enzymatisches Gleichgewicht. *Biochem. Z.*, **276**, 239. 1935.

Meyerhof, O. & Kiessling, W. (3). Über den Hauptweg der Milchsäurebildung in der Muskulatur. *Biochem. Z.*, **283**, 83. 1935.

Meyerhof, O. & Kiessling, W. (4). Über die enzymatische Umwandlung von Glycerinaldehydphosphorsäure in Dioxyacetonphosphorsäure. *Biochem. Z.*, **279**, 40. 1935.

Meyerhof, O. & Kiessling, W. (5). Die Umesterungsreaktion der Phosphobrenztraubensäure bei der alkoholischen Zuckergärung. *Biochem. Z.*, **281**, 249. 1935.

Meyerhof, O., Kiessling, W. & Schulz, W. (1). Über die Reaktionsgleichungen der alkoholischen Gärung. *Biochem. Z.*, **292**, 25. 1937.

Meyerhof, O. & Lehmann, H. (1). Über die Synthese der Kreatinphosphorsäure durch Umesterung der Phosphobrenztraubensäure. *Naturwissensch.*, **23**, 337. 1935.

Meyerhof, O. & Lohmann, K. (1). Über die Vorgänge bei der Muskelermüdung. *Biochem. Z.*, **168**, 128. 1926.

Meyerhof, O. & Lohmann, K. (2). Über die enzymatische Milchsäurebildung im Muskelextrakt. IV. Mitteilung: Die Spaltung der Hexosemonophosphorsäuren. *Biochem. Z.*, **185**, 113. 1927.

Meyerhof, O. & Lohmann, K. (3). Über die zeitliche Zusammensetzung von Kontraktion und Milchsäurebildung im Muskel. *Arch. f. ges. Physiol.*, **210**, 791. 1925.

Meyerhof, O. & Lohmann, K. (4). Über die Charakterisierung der Hexosemonophosphorsäure und ihr Verhalten bei der zellfreien Gärung. *Naturwissensch.*, **14**, 1277. 1926.

Meyerhof, O. & Lohmann, K. (5). Über den Ursprung der Kontraktionswärme. *Naturwissensch.*, **15**, 670. 1927.

Meyerhof, O. & Lohmann, K. (6). Über eine neue Aminophosphorsäure. *Naturwissensch.*, **16**, 47. 1928.

Meyerhof, O. & Lohmann, K. (7). Über die natürlichen Guanidinophosphorsäuren (Phosphagene) in der quergestreiften Muskulatur. I. Das physiologische Verhalten der Phosphagene. *Biochem. Z.*, **196**, 22. 1928.

Meyerhof, O. & Lohmann, K. (8). Über die natürlichen Guanidinophosphorsäure (Phosphagene) in der quergestreiften Muskulatur. II. Die physikalisch-chemischen Eigenschaften der Guanidinophosphorsäure. *Biochem. Z.*, **196**, 49. 1928.

Meyerhof, O. & Lohmann, K. (9). Über energetische Wechselbeziehungen zwischen dem Umsatz der Phosphorsäureester im Muskelextrakt. *Biochem. Z.*, **253**, 431. 1932.

Meyerhof, O. & Lohmann, K. (10). Über die enzymatische Gleichgewichtsreaktion zwischen Hexosediphosphorsäure und Dioxyacetonphosphorsäure. *Biochem. Z.*, 271, 89. 1934.

Meyerhof, O. & Lohmann, K. (11). Über die enzymatische Gleichgewichtsreaktion zwischen Hexosediphosphorsäure und Dioxyacetonphosphorsäure. II. *Biochem. Z.*, 273, 73. 1934.

Meyerhof, O. & Lohmann, K. (12). Über die enzymatische Gleichgewichtsreaktion zwischen Hexosediphosphorsäure und Dioxyacetonphosphorsäure. III. *Biochem. Z.*, 275, 430. 1935.

Meyerhof, O. & Lohmann, K. (13). Über Atmung und Kohlehydratumsatz tierische Gewebe. III. Uber den Unterschied von d- und l- Milchsäure für Atmung und Kohlehydratsynthese im Organismus. *Biochem. Z.*, 171, 421. 1926.

Meyerhof, O. & Lohmann, K. (14). Über die Energetik der anaeroben Phosphagensynthese (Kreatinphosphorsäure) im Muskelextract. *Naturwissensch.*, 19, 575. 1931.

Meyerhof, O., Lohmann, K. & Meier, R. (1). Über die Synthese des Kohlehydrats im Muskel. *Biochem. Z.*, 157, 459. 1925.

Meyerhof, O., Lohmann, K. & Meyer, K. (1). Über das Koferment der Milchsäurebildung im Muskel. *Biochem. Z.*, 237, 437. 1931.

Meyerhof, O., Lohmann, K. & Schuster, P. (1). Über die Aldolase, ein Kohlenstoffverknüpfendes Ferment. I. *Biochem. Z.*, 286, 301. 1936.

Meyerhof, O., Lohmann, K. & Schuster, P. (2). Über die Aldolase, ein Kohlenstoffverknupfendes Ferment. II. *Biochem. Z.*, 286, 319. 1936.

Meyerhof, O., Lundsgaard, E. & Blaschko, H. (1). Über die Energetik der Muskelkontraktion bei aufgehobener Milchsäurebildung. *Naturwissensch.*, 18, 787. 1930.

Meyerhof, O., Lundsgaard, E. & Blaschko, H. (2). Über die Energetik der Muskelkontraktion bei aufgehobener Milchsäurebildung. *Biochem. Z.*, 236, 326. 1931.

Meyerhof, O., McCullagh, R. D. & Schulz, W. (1). Neue Versuche über den kalorischen Quotienten der Milchsäure. *Arch. f. ges. Physiol.*, 224, 230. 1930.

Meyerhof, O. & McEachern, D. (1). Über anaerobe Bildung und Schwund von Brenztraubensäure in der Muskulatur. *Biochem. Z.*, 260, 417. 1933.

Meyerhof, O. & Meier, R. (1). Die Verbrennungswärme des Glykogens. *Biochem. Zeitsch.*, 150, 233. 1924.

Meyerhof, O. & Möhle, W. (1). Über die Volumenschwankung des Muskels in Zusammenhang mit dem Chemismus der Kontraktion. III. Die Volumenschwankung bei verschiedenen Kontraktionsformen. *Biochem. Z.*, 260, 469. 1933.

Meyerhof, O. & Möhle, W. (2). Über die Volumenschwankung des Muskels in Zusammenhang mit dem Chemismus der Kontraktion. III. Uber die Volumenänderung bei chemischen Vorgänge im Muskel. *Biochem. Z.*, 261, 252. 1933.

Meyerhof, O. & Möhle, W. (3). Über die Volumenschwankung des Muskels als Ausdruck der chemischen Vorgänge. *Biochem. Z.*, 284, 1. 1936.

Meyerhof, O. & Möhle, W. (4). Über den reversibilen Anteil der Volumenkonstriktion des Muskels. *Arch. f. ges. Physiol.*, 236, 533. 1935.

Meyerhof, O., Möhle, W. & Schulz, W. (1). Über die Reaktionsänderung des Muskels im Zusammenhang mit Spannungsentwicklung und chemischem Umsatz. *Biochem. Z.*, 246, 284. 1932.

Meyerhof, O. & Nachmansohn, D. (1). Neue Beobachtungen über den Umsatz des 'Phosphagens' im Muskel. *Naturwissensch.*, **16**, 726. 1928.

Meyerhof, O. & Nachmansohn, D. (2). Über die Synthese der Kreatinphosphorsäure in lebenden Muskel. *Biochem. Z.*, **222**, 1. 1930.

Meyerhof, O. & Oesper, P. (1). The mechanism of the oxidative reaction in fermentation. *J. biol. Chem.*, **170**, 1. 1947.

Meyerhof, O. & Ohlmeyer, P. (1). Über die Rolle der Co-Zymase bei der Milchsäurebildung im Muskelextrakt. *Biochem. Z.*, **290**, 334. 1937.

Meyerhof, O., Ohlmeyer, P. & Möhle, W. (1). Über die Koppelung zwischen Oxydoreduktion und Phosphatversesterung bei der anaeroben Kohlenhydratspaltung. *Biochem. Z.*, **297**, 90. 1938.

Meyerhof, O., Ohlmeyer, P. & Möhle, W. (2). Über die Koppelung zwischen Oxydoreduktion und Phosphatveresterung bei der anaeroben Kohlenhydratspaltung. II. *Biochem. Z.*, **297**, 113. 1938.

Meyerhof, O. & Schulz, W. (1). Ueber das Verhältnis von Milchsäurebildung und Sauerstoffverbrauch bei der Muskelkontraktion. *Arch. f. ges. Physiol.*, **217**, 547. 1927.

Meyerhof, O. & Schulz, W. (2). Über das Verhältnis von Milchsäurebildung und Kreatinphosphorsäurespaltung bei der anaeroben Tätigkeit des Muskels. *Biochem. Z.*, **236**, 54. 1931.

Meyerhof, O. & Schulz, W. (3). Über die Energieverhältnisse bei der enzymatischen Milchsäurebildung und der Synthese der Phosphagene. *Biochem. Z.*, **281**, 292. 1935.

Meyerhof, O. & Schulz, W. (4). Über die Warmetönung der Aldolkondensation der Hexose-1-6-phosphorsäure. *Biochem. Z.*, **289**, 87. 1936.

Meyerhof, O., Schulz, W. & Schuster, P. (1). Über die enzymatische Synthese der Kreatinphosphorsäure und die biologische 'Reaktionsform' des Zuckers. *Biochem. Z.*, **293**, 309. 1937.

Meyerhof, O. & Suranyi, J. (1). Über die Wärmetönungen der chemischen Reaktionsphasen im Muskel. *Biochem. Z.*, **191**, 106. 1927.

Michelson, A., Russell, E. S. & Harman, P. J. (1). Dystrophia muscularis: a hereditary primary myopathy in the house mouse. *Science*, **41**, 1079. 1955.

Middlebrook, W. R. (1). Individuality of the meromyosins. *Science*, **130**, 621. 1959.

Middlebrook, W. R. (2). The action of trypsin on acetylmyosin. *Fed. Proc.*, **20**, 359. 1961.

Mihalyi, E. (1). The dissociation curves of crystalline myosin. *Enzymologia*, **14**, 224. 1950.

Mihalyi, E. (2). Trypsin digestion of myosin proteins. II. The kinetics of the digestion. *J. biol. Chem.*, **201**, 197. 1953.

Mihalyi, E. & Harrington, W. F. (1). Studies on the tryptic digestion of myosin. *Biochim. biophys. Acta.*, **36**, 447. 1959.

Mihalyi, E. & Szent-Györgyi, A. G. (1). Trypsin digestion of muscle proteins. I. Ultracentrifugal analysis of the process. *J. biol. Chem.*, **201**, 189. 1953.

Mihalyi, E. & Szent-Györgyi, A. G. (2). Trypsin digestion of muscle proteins. III. Adenosinetriphosphatase activity and actin-binding capacity of the digested myosin. *J. biol. Chem.*, **201**, 211. 1953.

Miki-Noumura, T. & Oosawa, F. (1). An actin-like protein of the sea-urchin eggs. 1. Its interaction with myosin from rabbit striated muscle. *Exp. Cell Res.*, **56**, 274. 1969.

Miller, A. (1). An axial period in tropomyosin. *J. mol. Biol.*, **12**, 280. 1965.

Millikan, G. A. (1). The kinetics of muscle haemoglobin. *Proc. Roy. Soc.* B, **120**, 366. 1936.

Millikan, G. A. (2). Experiments on muscle haemoglobin *in vivo;* the instantaneous measurement of muscle metabolism. *Proc. Roy. Soc.* B, **123**, 218. 1937.

Millikan, G. A. (3). Muscle haemoglobin. *Physiol. Rev.*, **19**, 503. 1939.

Millman, B. M. & Elliott, G. F. (1). X-ray diffraction from contracting molluscan muscle. *Nature*, **206**, 824. 1965.

Mines, G. R. (1). On functional analysis by the action of electrolytes. *J. Physiol.*, **46**, 188. 1913.

Minihan Nauss, K. & Davies, R. E. (1). Changes in inorganic phosphate and arginine during the development, maintenance and loss of tension in the anterior byssus retractor muscle of *Mytilus edulis*. *Biochem. Z.*, **345**, 173. 1966.

Mirsky, A. E. (1). Sulfhydryl and disulphide groups of proteins. IV. Sulfhydryl groups of the proteins of muscle. *J. gen. Physiol.*, **19**, 559. 1936.

Mitchell, P. (1). Coupling of phosphorylation to electron and hydrogen transfer by a chemiosmotic type of mechanism. *Nature*, **191**, 144. 1961.

Mitchell, P. (2). Chemiosmotic coupling in oxidative and photosynthetic phosphorylations. *Biol. Rev.*, **41**, 445. 1966.

Mitchell, P. (3). Conduction of protons through the membrane of mitochondria and bacteria by uncouplers of oxidative phosphorylation. *Biochem. J.*, **81**, 24P. 1961.

Mitchell, P. (4). Translocations through natural membranes. *Adv. in Enz.*, **29**, 33. 1967.

Mitchell, P. & Moyle, J. (1). Stoichiometry of proton translocation through the respiratory chain and adenosine triphosphatase systems of rat liver mitochondria. *Nature*, **208**, 147. 1965.

Mitchell, P. & Moyle, J. (2). Proton-transport phosphorylation: some experimental tests. In *Biochemistry of Mitochondria*, p. 53. Ed. E. C. Slater, Z. Kaniuga & L. Wojtczak. Warsaw, Polish Scientific Publishers. London & New York, Academic Press.

Mitchell, P. & Moyle, J. (3). Acid-base titration across the membrane system of rat-liver mitochondria. *Biochem. J.*, **104**, 588. 1967.

Mitchell, P. & Moyle, J. (4). Respiration-driven proton translocation in rat liver mitochondria. *Biochem. J.*, **105**, 1147. 1967.

Mitscherlich, E. (1). Ueber die Aetherbildung. *Ann. d. Phys. und Chem.*, **31**, 273. 1834.

Mitscherlich, E. (2). Über die chemische Zersetzung und Verbindung mittelst Contactsubstanzen. *Ann. d. Phys.*, **55**, 209. 1842.

Miyasaki, K., Toyoda, R., Tomino, H., Yoshimatsu, M., Saijo, K., Katsunuma, N. & Fujino, A. (1). Electrophoretic patterns of creatine kinase isozyme during development or in the case of skeletal muscular dystrophy. *Chem. Abst.*, **64**, 1152c. 1966.

Mommaerts, W. F. H. M. (1). Quantitative studies on some effects of adenyltriphosphate on myosin B. *Studies from Inst. Med. Chem. Univ. Szeged*, **1**, 37. 1943.

Mommaerts, W. F. H. M. (2). On the nature of the forces acting between the components of the muscle fibrillum. I. The disaggregation of actomyosin by adenosinetriphosphate and some other reagents. *Arkiv Kemi, Mineral. Geol. A.*, **19**. No. 18. 1945.

Mommaerts, W. F. H. M. (3). Does adenosine triphosphate cause a contraction or a disaggregation of dissolved actomyosin? *Exptl. Cell Res.*, **2**, 133. 1951.

Mommaerts, W. F. H. M. (4). The reaction between actomyosin and adenosine triphosphate. *J. gen. Physiol.*, **31**, 361. 1947.

Mommaerts, W. F. H. M. (5). A consideration of experimental facts pertaining to the primary reaction in muscular activity. *Biochim. biophys. Acta*, **4**, 50. 1950.

Mommaerts, W. F. H. M. (6). Reversible polymerisation and ultra-centrifugal purification of actin. *J. biol. Chem.*, **188**, 559. 1951.

Mommaerts, W. F. H. M. (7). The molecular transformation of actin. I. Globular actin. *J. biol. Chem.*, **198**, 445. 1952.

Mommaerts, W. F. H. M. (8). The molecular transformation of actin. III. The participation of nucleotides. *J. biol. Chem.*, **198**, 469. 1952.

Mommaerts, W. F. H. M. (9). Stoichiometric and dynamic implications of the participation of actin and ATP in the contraction process. *Biochim. biophys. Acta*, **7**, 477. 1951.

Mommaerts, W. F. H. M. (10). Molecular alterations in myofibrillar proteins. In *The physiology and biochemistry of muscle as a food*, p. 277. Madison, Milwaukee & London, Univ. of Wisconsin Press. 1966.

Mommaerts, W. F. H. M. (11). Is adenosine triphosphate broken down during a single muscle twitch? *Nature*, **174**, 1083. 1954.

Mommaerts, W. F. H. M. (12). Investigation of the presumed breakdown of adenosine triphosphate and phosphocreatine during a single muscle twitch. *Am. J. Physiol.*, **182**, 585. 1955.

Mommaerts, W. F. H. M. (13). Discussion on The energetics of muscle contraction. In *Biochemistry of Muscle Contraction*, p. 456. Ed. J. Gergely. J. & A. Churchill, London. 1964.

Mommaerts, W. F. H. M. (14). Phosphate metabolism in the activity of skeletal and cardiac muscle. In *Phosphorus Metabolism*, Vol. I, p. 551. Ed. W. D. McElroy & B. Glass. Baltimore, Johns Hopkins Press. 1951.

Mommaerts, W. F. H. M. (15). Ultraviolet circular dichroism of myosin. *J. mol. Biol.*, **15**, 377. 1966.

Mommaerts, W. F. H. M. (16). The molecular transformations of actin. II. The polymerisation process. *J. biol. Chem.*, **198**, 459. 1952.

Mommaerts, W. F. H. M. & Aldrich, B. B. (1). Determination of the molecular weight of myosin. *Biochim. biophys. Acta*, **28**, 627. 1958.

Mommaerts, W. H. F. M., Buller, A. J. & Seraydarian, K. The modification of some biochemical properties of muscle by cross-innervation. *Proc. Nat. Acad. Sci. U.S.A.*, **64**, 128. 1969.

Mommaerts, W. F. H. M. & Green, I. (1). Adenosinetriphosphatase systems of muscle. III. A survey of the adenosinetriphosphatase activity of myosin. *J. biol. Chem.*, **208**, 833. 1954.

Mommaerts, W. F. H. M., Illingworth, B., Pearson, C. M., Guillory, R. J. & Seraydarian, K. (1). A functional disorder associated with the absence of phosphorylase. *Proc. Nat. Acad. Sci. U.S.A.*, **45**, 791. 1959.

Mommaerts, W. F. H. M., Olmsted, M., Seraydarian, K. & Wallner, A. (1). Contraction with and without demonstrable splitting of energy-rich phosphate in turtle muscle. *Biochim. biophys. Acta*, **63**, 82. 1962.

Mommaerts, W. F. H. M. & Rupp, J. C. (1). Dephosphorylation of adenosine-triphosphate in muscular contraction. *Nature*, **158**, 957. 1951.

Mommaerts, W. F. H. M. & Schilling, M. O. (1). Interruption of muscular contraction by rapid cooling. *Am. J. Physiol.*, **182**, 579. 1955.

Mommaerts, W. F. H. M. & Seraydarian, K. (1). A study of the adenosine triphosphatase activity of myosin and actomyosin. *J. gen. Physiol.*, **30**, 201. 1947.

Mommaerts, W. F. H. M. & Seraydarian, K. (2). The quantity of chemical change in muscle associated with shortening and work. *Fed. Proc.*, **19**, 254. 1960.

Mommaerts, W. F. H. M., Seraydarian, K. & Maréchal, G. (1). Work and chemical change in isotonic muscular contractions. *Biochim. biophys. Acta.*, **57**, 1. 1962.

Mommaerts, W. F. H. M., Seraydarian, K. & Wallner, A. (1). Demonstration of phosphocreatine splitting as an early reaction in contraction of frog sartorius muscle. *Biochim. biophys. Acta.*, **63**, 75. 1962.

Mommaerts, W. F. H. M. & Wallner, A. (1). The breakdown of adenosine triphosphate in the contraction cycle of the frog sartorius muscle. *J. Physiol.*, **193**, 343. 1967.

Monckton, G. & Nihei, T. (1). Some biochemical changes in muscular dystrophy and their possible genetical significance. *Ann. N.Y. Acad. Sci.*, **138**, 329. 1966.

Monod, J., Changeux, J. P. & Jacob, F. (1). Allosteric proteins and cellular control systems. *J. mol. Biol.*, **6**, 306. 1963.

Monod, J., Wyman, J. & Changeux, J. P. (1). On the nature of allosteric transitions: a plausible model. *J. mol. Biol.*, **12**, 88. 1965.

Monro, A. (1). *Experiments on the Nervous System.* Edinburgh, Bell & Bradford. 1793.

Moog, F. (1). Adenylpyrophosphatase in brain, liver, heart and muscle of chick embryos and hatched chicks. *J. expt. Zool.*, **105**, 209. 1947.

Moore, D. H. & Ruska, H. (1). Electron microscopic study of mammalian cardiac muscle. *J. biophys. biochem. Cyt.*, **3**, 261. 1957.

Moos, C. (1). Can creatine kinase phosphorylate the myofibril-bound nucleotide of muscle? *Biochim. biophys. Acta.*, **93**, 85. 1964.

Moos, C., Estes, J. & Eisenberg, E. (1). Exchange of F-actin-bound nucleotide in the presence and absence of myosin. *Biochem. biophys. Res. Com.*, **23**, 347. 1966.

Morales, M. F. (1). Mechanisms of muscle contraction. *Reviews Mod. Phys.*, **31**, 426. 1959.

Morales, M. & Botts, J. (1). A model for the elementary process in muscle action. *Arch. Biochem. Biophys.*, **37**, 283. 1952.

Morales, M. & Botts, J. (2). Energetics and molecular mechanisms in muscle action. I. Outline of a theory of muscle action, and some of its experimental basis. *Discussions of the Faraday Soc.*, **13**, 125. 1952.

Morales, M. F., Botts, J., Blum, J. J. & Hill, T. L. (1). Elementary processes in muscle action: an examination of current concepts. *Physiol. Rev.*, **35**, 475. 1955.

Moreland, B., Watts, D. C. & Virden, R. (1). Phosphagen kinases and evolution in the echinodermata. *Nature*, **214**, 458. 1967.

Morey, K. S., Tarczy-Hornoch, K., Richards, E. G. & Duane Brown, W. (1). Myosin from dystrophic and control chicken muscle. *Arch. Biochem. Biophys.*, **119**, 491. 1967.

Morgan, H. E., Henderson, M. J., Regen, D. M. & Park, C. R. (1). Regulation of glucose uptake in muscle. 1. The effects of insulin and anoxia on glucose transport and phosphorylation in the isolated perfused heart of normal rats. *J. biol. Chem.*, **236**, 253. 1961.

Morgan, H. E. & Parmeggiani, A. (1). Regulation of glycolysis in muscle. II. Control of glycogen phosphorylase reaction in isolated perfused heart. *J. biol. Chem.*, **239**, 2435. 1964.

Morgan, H. E. & Parmeggiani, A. (2). Regulation of glycolysis in muscle. III. Control of muscle glycogen phosphorylase activity. *J. biol. Chem.*, **239**, 2440. 1964.

Morgan, H. E. & Parmeggiani, A. (3). Regulation of glycogenolysis in muscle. In Ciba Symposium: *Control of glycogen metabolism*, p. 254. Ed. W. J. Whelan & M. P. Cameron. J. A. Churchill, London. 1964.

Morgan, H. E., Randle, P. J. & Regen, D. M. (1). Regulation of glucose uptake by muscle. 3. The effects of insulin, anoxia, salicylate and 2:4-dinitrophenol on membrane transport and intracellular phosphorylation of glucose in the isolated rat heart. *Biochem. J.*, **73**, 573. 1959.

Morita, F. (1). Interaction of heavy meromyosin with substrate. II. Rate of formation of ATP-induced ultraviolet difference spectrum of heavy meromyosin measured by stopped-flow method. *Biochim. biophys. Acta.*, **172**, 319. 1969.

Morita, F. & Yagi, K. (1). Spectral shift in heavy meromyosin induced by substrate. *Bioch. biophys. Res. Com.*, **22**, 297. 1966.

Mörner, K. A. H. (1). Beobachtungen über den Muskelfarbstoff. *Nord. med. Ark.*, **30**, 1. 1896.

Morrison, J. F. & Doherty, M. D. (1). The biosynthesis of N-phosphorylcreatine: an investigation of the postulated alternative pathway. *Biochem. J.*, **79**, 433. 1961.

Morrison, J. F., Griffiths, D. E. & Ennor, A. H. (1). Biochemical evolution: position of the tunicates. *Nature*, **178**, 359. 1956.

Morton, R. A. (1). Ubiquinone. *Nature*, **182**, 1764. 1958.

Moyle, D. M. (1). A quantiative study of succinic acid in muscle. I. *Biochem. J.*, **18**, 351. 1924.

Moyle, J. & Dixon, M. (1). Purification of the *isocitric* enzyme (triphosphopyridine nucleotide-linked isocitric dehydrogenase-oxalosuccinic carboxylase. *Biochem. J.*, **63**, 549. 1956.

Mozołowski, W., Mann, T. & Lutwak, C. (1). Über den Ammoniakgehalt und die Ammoniakbildung im Muskel und deren Zusammenhang mit Funktions- und Zustandsänderung. *Biochem. Z.*, **231**, 290. 1931.

Mozołowski, W. & Sobczuk, B. (1). Ammoniakbildung und Pyrophosphatzerfall im Muskel. II. *Biochem. Z.*, **265**, 41. 1933.

Mueller, H. (1). Molecular weight of myosin and meromyosins by Archibald experiments performed with increasing speed of rotations. *J. biol. Chem.*, **239**, 797. 1964.

Mueller, H. (2). Characterization of the molecular region containing the active sites of myosin. *J. biol. Chem.*, **240**, 3816. 1965.

Mueller, H. (3). EGTA-sensitizing activity and molecular properties of tropomyosin prepared in presence of a sulfhydryl protecting agent. *Biochem. Z.*, **345**, 300. 1966.

Mueller, H., Franzen, J., Rice, R. V. & Olson, R. E. (1). Characterization of cardiac myosin from the dog. *J. biol. Chem.*, **239**, 1447. 1964.

Mueller, H. & Perry, S. V. (1). The chromatographic behaviour and adenosine triphosphatase activities of the meromyosins. *Biochim. biophys. Acta.*, **40**, 187. 1960.

Mueller, H. & Perry, S. V. (2). Chromatography of the meromyosins on diethylaminoethylcellulose. *Biochem. J.*, **80**, 217. 1961.

Mueller, H. & Perry, S. V. (3). The degradation of heavy meromyosin by trypsin. *Biochem. J.*, 85, 431. 1962.

Mueller, H., Theiner, M. & Olson, R. E. (1). Macromolecular fragments of canine cardiac muscle obtained by tryptic digestion. *J. biol. Chem.*, 239, 2153. 1964.

Mugikura, H., Miyazaki, E. & Nagai, T. (1). The influence of oxarsan As III on ATPase and superprecipitation of actomyosin, and the significance of ATPase in the syneresis of actomyosin. *Enzymologia*, 17, 321. 1954–56.

Mühlrad, A., Fábián, F. & Biro, N. A. (1). On the activation of myosin ATPase by EDTA. *Biochim. biophys. Acta.*, 89, 186. 1964.

Mulder, G. J. (1). Zusammensetzung von Fibrin, Albumin, Leimzucker, Leucin u.s.w. *Ann. d. Pharm.*, 28, 72. 1838.

Müller, J. (1). *Handbuch der Physiologie des Menschen.* Coblenz, J. Hölscher. 1837–40.

Munch-Petersen, A. (1). Dephosphorylation of adenosine triphosphate during the rising phase of a muscle twitch. *Acta physiol. Scand.*, 29, 202. 1953.

Munch-Petersen, A. (2). Investigation of the properties and mechanism of the uridine diphosphate glucose pyrophosphorylase reaction. *Acta. chem. Scand.*, 9, 1523. 1955.

Munch-Petersen, A., Kalckar, H. M., Cutolo, E. & Smith, E. E. B. (1). Uridyl transferases and the formation of uridine triphosphate. Enzymic production of uridine triphosphate: uridine diphosphoglucose pyrophosphorolysis. *Nature*, 172, 1036. 1953.

Muñoz, J. M. & Leloir, L. F. (1). Fatty acid oxidation by liver enzymes. *J. biol. Chem.*, 147, 355. 1943.

Murad, F., Chi, Y.-M., Rall, T. W. & Sutherland, E. W. (1). Adenylcyclase. III. The effect of catechol amines and choline esters on the formation of adenosine 3′,5′-phosphate by preparations from cardiac muscle and liver. *J. biol. Chem.*, 237, 1233. 1962.

Muralt, A. von (1). Über das Verhalten der Doppelbrechung des quergestreiften Muskels während der Kontraktion. *Arch. f. ges. Physiol.*, 230, 299. 1932.

Muralt, A. von. (2) Kontraktionshypothesen und Feinstruktur des Muskels. *Koll. Z.*, 63, 228. 1933.

Muralt, A. von (3). Zusammenhänge zwischen physikalischen und chemischen Vorgängen bei der Muskelkontraktion. *Ergeb. d. Physiol.*, 37, 407. 1935.

Muralt, A. von (4). Lichtdurchlässigkeit und Tätigkeitsstoffwechsel des Muskels. II. Mitteilung. *Arch. ges. Physiol.*, 234, 653. 1934.

Muralt, A. von & Edsall, J. T. (1). Studies in the physical chemistry of muscle globulin. III. The anisotropicity of myosin and the angle of isocline. *J. biol. Chem.*, 89, 315. 1930.

Muralt, A. von & Edsall, J. T. (2). Studies in the physical chemistry of muscle globulin. IV. The anisotropy of myosin and double refraction of flow. *J. biol. Chem.*, 89, 351. 1930.

Myers, D. K. & Slater, E. C. (1). The enzymic hydrolysis of adenosine triphosphate by liver mitochondria. I. Activities at different pH values. *Biochem. J.*, 67, 558. 1957.

Myrbäck, K. (1). Zur Kenntnis der Co-Zymase. III. *Z. f. physiol. Chem.*, 225, 199. 1934.

Nachmansohn, D. (1). Über den Zerfall der Kreatinphosphorsäure in Zusammenhang mit der Tätigkeit des Muskels. *Biochem. Z.*, 196, 73. 1928.

Nachmansohn, D. (2). Über den Zerfall der Kreatinphosphorsäure im Zusammenhang mit der Tätigkeit des Muskels. II. *Biochem. Z.*, 208, 237. 1929.

Nachmansohn, D. (3). Über den Zerfall der Kreatinphosphorsäure im Zusammenhang mit der Tätigkeit des Muskels. III. Umsatzgrösse und Erregungsgeschwindigketi. *Biochem. Z.*, 213, 262. 1929.

Nachmansohn, D. (4). *Chemical and molecular basis of nerve activity*. p. 83. New York and London, Academic Press. 1959.

Nachmansohn, D., Cox, R. T., Coates, C. W. & Machado, A. L. (1). Action potential and enzyme activity in the electrical organ of *Electrophorus electricus*. *J. Neurophysiol.*, 6, 383. 1943.

Nachmansohn, D. & Machado, A. L. (1). The formation of acetylcholine. A new enzyme: 'choline acetylase'. *J. Neurophysiol.*, 6, 397. 1943.

Nachmansohn, D. & Rothenberg, M. A. (1). Studies on cholinesterase. I. On the specificity of the enzyme in nervous tissue. *J. biol. Chem.*, 158, 653. 1945.

Nachmias, V. T. & Padykula, H. A. (1). A histochemical study of normal and denervated red and white muscles of the rat. *J. biophys. biochem. Cyt.*, 4, 47. 1958.

Naeslund, J. & Snellman, O. (1). Untersuchungen über die Kontractilität im Corpus, Isthmus und in der Cervix uteri unter normalen Verhältnissen und während der Schwangerschaft und der Entbindung. *Arch. Gynaek.*, 180, 137. 1951.

Nagai, T., Makinose, M. & Hasselbach, W. (1). Der physiologische Erschlaffungsfaktor und die Muskelgrana. *Biochim. biophys. Acta*, 43, 223. 1960.

Nagai, T., Uchida, K. & Yasuda, M. (1). Some further properties of the muscle relaxing-factor system and the separation of the effective substance. *Biochim. biophys. Acta*, 56, 205. 1962.

Nagy, B. (1). Optical rotatory dispersion of G-actin-adenosine-5-diphosphate. *Biochim. biophys. Acta*, 115, 498. 1966.

Nagy, B. & Jencks, W. P. (1). Optical rotatory dispersion of G-actin. *Biochemistry* (Am. chem. Soc.) 1, 987. 1962.

Nakajima, A. & Horn, L. (1). Electrical activity of vascular smooth muscle. *Amer. J. Physiol.*, 213, 25. 1967.

Nakajima, H. (1). Some properties of a contractile protein in a myxomycete plasmodium. *Protoplasma*, 52, 413. 1966.

Namm, D. H. & Mayer, S. E. (1). Effects of epinephrine on cardiac cyclic 3',5'-AMP, phosphorylase kinase, and phosphorylase. *Mol. Pharm.*, 4, 61. 1968.

Nanninga, L. B. (1). The binding of magnesium, calcium and chlorine ions to heavy and light meromyosins. *Arch. Biochem. Biophys.*, 70, 346. 1957.

Nanninga, L. B. (2). On the interaction between myosin A and F-actin. *Biochim. biophys. Acta*, 82, 507. 1964.

Nanninga, L. B. (3). Investigation of the effect of Ca ions on the splitting of ATP by myosin. *Biochim. biophys. Acta*, 36, 191. 1959.

Nanninga, L. B. & Mommaerts, W. F. H. M. (1). Studies on the formation of an enzyme-substrate complex between myosin and adenosine triphosphate. *Proc. Nat. Acad. Sci. U.S.A.*, 46, 1155. 1960.

Nanninga, L. B. & Mommaerts, W. F. H. M. (2). Kinetic constants of the interaction between myosin and adenosinetriphosphate. *Proc. Nat. Acad. Sci. U.S.A.*, 46, 1166. 1960.

Nass, M. M. K. (1). Developmental changes in frog actomyosin characteristics. *Dev. Biol.*, 4, 289. 1962.

Nasse, O. (1). Chemie und Stoffwechsel der Muskeln. In [Hermann's] *Handbuch der Physiologie*. Part 1. *Der Bewegungsapparate*. p. 263. Leipzig, F. C. W. Vogel. 1879.

Nasse, O. (2). Beiträge zur Physiologie der contractilen Substanz. *Arch. f. ges. Physiol.*, **2**, 97. 1869.

Nasse, O. (3). Bemerkungen zur Physiologie der Kohlehydrate. *Arch. f. ges. Physiol.*, **14**, 473. 1877.

Nastuk, W. L. & Hodgkin, A. L. (1). The electrical activity of single muscle fibres. *J. cell. comp. Physiol.*, **35**, 39. 1950.

Needham, D. M. (1). Energy-yielding reactions in muscle contraction. *Enzymologia*, **5**, 158. 1938.

Needham, D. M. (2). Chemical cycles in muscle contraction. In *Perspectives in Biochemistry*, p. 201. Ed. J. Needham & D. E. Green. Cambridge University Press. 1937.

Needham, D. M. (3). The adenosinetriphosphatase activity of myosin preparations. *Biochem. Journ.*, **36**, 113. 1942.

Needham, D. M. (4). Myosin and adenosinetriphosphate in relation to muscle contraction. *Biochim. biophys. Acta.*, **4**, 42. 1950.

Needham, D. M. (5). A quantitative study of succinic acid in muscle. II. The metabolic relationships of succinic, malic and fumaric acids. *Biochem. J.*, **21**, 739. 1927.

Needham, D. M. (6). A quantitative study of succinic acid in muscle. III. Glutamic and aspartic acids as precursors. *Biochem. J.*, **24**, 208. 1930.

Needham, D. M. (7). Red and white muscle. *Physiol. Rev.*, **6**, 1. 1926.

Needham, D. M. & van Heyningen, W. E. (1). Linkage of chemical changes in muscle. *Nature*, **135**, 585. 1935.

Needham, D. M. & van Heyningen, W. E. (2). The linkage of chemical changes in muscle extract. *Biochem. J.*, **29**, 2040. 1935.

Needham, D. M. & Lu, G. D. (1). The specificity of coupled esterification of phosphate in muscle. *Biochem. J.*, **32**, 2040. 1938.

Needham, D. M., Needham, J., Baldwin, E. & Yudkin, J. (1). A comparative study of the phosphagens, with some remarks on the origin of vertebrates. *Proc. Roy. Soc.*, B, **110**, 260. 1932.

Needham, D. M. & Pillai, R. K. (1). Coupling of dismutations with esterification of phosphate in muscle. *Nature*, **140**, 64. 1937.

Needham, D. M. & Pillai, R. K. (2). The coupling of oxidoreductions and dismutations with esterification of phosphate in muscle. *Biochem. J.*, **31**, 1837. 1937.

Needham, D. M. & Shoenberg, C. F. (1). Biochemistry of the myometrium. In *Cellular Biology of the Uterus*. p. 291. Ed. R. M. Wynn. New York, Appleton-Century-Crofts. 1967.

Needham, D. M. & Shoenberg, C. F. (2). Proteins of the contractile mechanism of mammalian smooth muscle and their possible location in the cell. *Proc. Roy. Soc.* B, **160**, 433. 1964.

Needham, D. M. & Williams, J. M. (1). Some properties of uterus actomyosin and myofilaments. *Biochem. J.*, **73**, 171. 1959.

Needham, D. M. & Williams, J. M. (2). Proteins of the uterine contractile mechanism. *Biochem. J.*, **89**, 552. 1963.

Needham, D. M. & Williams, J. M. (3). The proteins of the dilution precipitate obtained from salt extracts of pregnant and non-pregnant uterus. *Biochem. J.*, **89**, 534. 1963.

Needham, J. (1). *Science and Civilisation in China*. 7 vols. in 10 parts. Cambridge University Press. 1954– .

Needham, J. (2). *History of Embryology*. Cambridge University Press. 2nd ed. 1959.

Needham, J. (3). Frederick Gowland Hopkins. *Perspectives in Biology and Medicine*, 6, 1. 1962.

Needham, J. (4). *Chemical Embryology*. Vol. 3. Cambridge University Press. 1931.

Needham, J. (5). *Order and Life*. The Terry Lectures. Cambridge University Press. 1936.

Needham, J., Kleinzeller, A., Miall, M., Dainty, M., Needham, D. M. & Lawrence, A. S. C. (1). Is muscle contraction essentially an enzyme-substrate combination? *Nature*, 150, 46. 1942.

Needham, J. & Lu, Gwei-Djen. (1). Medicine and culture in China. In *Medicine and Culture*, Symposium of the Wellcome Historical Medical Museum and Library and the Wenner-Gren Foundation, London. 1969.

Needham, J., Shen, S.-C., Needham, D. M. & Lawrence, A. S. C. (1). Myosin birefringence and adenylpyrophosphate. *Nature*, 147. 766.

Negelein, E. & Brömel, H. (1). R-Diphosphoglycerinsäure, ihre Isolierung und Eigenschaften. *Biochem. Z.*, 303, 132. 1939.

Nelson, D. G. & Benson, E. S. (1). On the structural continuities of the transverse tubular system of rabbit and human myocardial cells. *J. Cell Biol.*, 16, 297. 1963.

Nelson, L. (1). Enzyme distribution in fragmented bull spermatazoa. *Biochim. biophys. Acta*, 14, 312. 1954.

Nelson, L. (2). Chemical morphology of the contractile system in spermatozoa. *Ann. Histochim. Suppl.* 2, 283. 1962.

Nelson, L. (3). Cytochemical aspects of spermatozoan motility. (1). In *Spermatozoan Motility*. p. 171. Washington, Amer. Ass. Advancement Sci., 1962.

Nemetchek-Gansler, H. (1). Ultrastructure of the myometrium, In *Cellular Biology of the Uterus*, p. 353. Ed. by R. M. Wynn. New York, Appleton-Century-Crofts. 1967.

Neubauer, O. & Fromherz, K. (1). Über den Abbau bei der Hefegärung. *Z. f. physiol. Chem.*, 70, 326. 1910–11.

Neuberg, C. (1). Über die Zerstörung von Milchsäurealdehyd und Methylglyoxal durch tierische Organe. *Biochem. Z.*, 49, 502. 1913.

Neuberg, C. (2). Weitere Untersuchungen über die biochemische Umwandlung von Methylglyoxal in Milchsäure nebst Bemerkungen über die Entstehung der verschiedenen Milchsäure in der Natur. *Biochem. Z.*, 51, 484. 1913.

Neuberg, C. & Kerb, J. (1). Über zuckerfrei Hefegärungen. XII. Über die Vorgänge bei der Hefegärung. *Biochem. Z.*, 53, 406. 1913.

Neuberg, C. & Kerb, J. (2). Über zuckerfrei Hefegärungen. XIII. Zur Frage der Aldehydbildung bei der Gärung von Hexosen sowie bei der sog. Selbstgärung. *Biochem. Z.*, 58, 158. 1914.

Neuberg, C. & Kobel, M. (1). Die desmolytische Bildung von Methylglyoxal durch Hefenenzym. *Biochem. Z.*, 203, 463. 1928.

Neuberg, C. & Reinfurth, E. (1). Natürlich und erzwungene Glycerinbildung bei der alkoholischen Gärung. *Biochem. Z.*, 92, 234. 1918.

Neurath, H., Greenstein, J. P., Putnam, F. W. & Erickson, J. O. (1). The chemistry of protein denaturation. *Chem. Rev.*, 34, 157. 1944.

Newsholme, E. A. & Randle, P. J. (1). Regulation of glucose uptake by muscle. 5. Effects of anoxia, insulin, adrenaline and prolonged starving on concentrations of hexose phosphates in isolated rat diaphragm and perfused isolated rat heart. *Biochem. J.*, **80**, 665. 1961.

Newsholme, E. A., Randle, P. J. & Manchester, K. L. (1). Inhibition of the phosphofructokinase reaction in perfused rat heart by respiration of ketone bodies, fatty acids and pyruvate. *Nature*, **193**, 270. 1962

Newton, A. A. & Perry, S. V. (1). Incorporation of nitrogen[15] in the 6-NH_2 group of adenosine triphosphate by muscle extracts. *Nature, Lond.*, **179**, 49. 1957.

Newton, A. A. & Perry, S. V. (2). The incorporation of ^{15}N into adenine nucleotides and their formation from inosine monophosphate by skeletal-muscle preparations. *Biochem. J.*, **74**, 127. 1960.

Newton, B. A. & Kerridge, D. (1). Flagellar and ciliary movement in microorganisms. *Symp. Soc. gen. Microbiol.*, **15**, 220. 1965.

Newton, I. (1). *Opticks: or, a Treatise of the Reflections, Refractions, Inflections and Colours of Light.* London, Innys. 1718.

Nichol, C. J. (1). Sulfhydryl and disulfide concentrations in dystrophic muscle. *Can. J. Biochem.*, **42**, 1643. 1964.

Niedergerke, R. (1). Movements of Ca in beating ventricles of the frog heart. *J. Physiol.*, **167**, 551. 1963.

Niedergerke, R. (2). Movements of Ca in frog heart ventricles at rest and during contractures. *J. Physiol.*, **167**, 515. 1963.

Nielsen, S. O. & Lehninger, A. L. (1). Oxidative phosphorylation in the cytochrome system of mitochondria. *J. Amer. chem. Soc.*, **76**, 3861. 1954.

Nihei, T. & Kay, C. M. (1). Isolation and properties of an enzymatically active fragment from papain-digested myosin. *Biochim. biophys. Acta.*, **160**, 46. 1968.

Nilsson, R. (1). Einige Betrachungen über den glykolytischen Kohlenhydratabbau. *Biochem. Z.*, **258**, 198. 1933.

Noda, H. & Maruyama, K. (1). An attempt to demonstrate the separation of F-actin from myosin B by the action of ATP. *J. Biochem.*, **48**, 723. 1960.

Noda, L. (1). Adenosine triphosphate-adenosine monophosphate transphosphorylase. III. Kinetic studies. *J. biol. Chem.*, **232**, 237. 1958.

Noda, L. & Bono, V. (1). Nucleotides of myofibril. *Fed. Proc.*, **23**, 529. 1964.

Noll, D. & Weber, H. H. (1). Polarisationsoptik und molekularer Feinbau der Q-Abschnitte des Froschmuskels. *Arch. f. ges. Physiol.*, **235**, 234. 1935.

Nonomura, Y. (1). A study on the physico-chemical properties of α-actinin. *J. Biochem.*, **61**, 796. 1967.

Nonomura, Y. (2). Myofilaments in smooth muscles of guinea pig's taenia coli. *J. Cell. Biol.*, **39**, 741. 1968.

Nonomura, Y., Drabikowski, W. & Ebashi, S. (1). The localization of troponin in tropomyosin paracrystals. *J. Biochem.*, **64**, 419. 1968.

Noorden, v. & Embden, G. (1). Einige Probleme des intermediären Kohlenhydratstoffwechsels. *Zentralbl. f. d. ges. Physiol. u. Path. d. Stoffwechsels*, p. 1. 1906.

Ochoa, S. (1). Efficiency of aerobic phosphorylation in cell-free heart extracts. *J. biol. Chem.*, **151**, 492. 1943.

Ochoa, S. (2). Coupling of phosphorylation with oxidation of pyruvate in brain. *J. biol. Chem.*, **138**, 751. 1941.

Ochoa, S., Mehler, A. H. & Kornberg, A. (1). Biosynthesis of dicarboxylic acids by carbon dioxide fixation. 1. Isolation and properties of an enzyme from pigeon liver catalysing the reversible oxidative decarboxylation of l-malic acid. *J. biol. Chem.*, **174**, 979. 1948.

Ochoa, S., Stern, J. R. & Schneider, M. C. (1). Enzymatic synthesis of citric acid. II. Crystalline condensing enzyme. *J. biol. Chem.*, **193**, 691. 1951.

Oertmann, E. (1). Ueber den Stoffwechsel entbluteter Frösche. *Arch. f. ges. Physiol.*, **15**, 381. 1877.

Oesper, P. (1). Sources of the high-energy content in energy-rich phosphate bonds. *Arch. Biochem. Biophys.*, **27**, 255. 1950.

Offer, G. W. (1). Myosin: An N-acetylated protein. *Biochim. biophys. Acta.*, **90**, 193. 1964.

Offer, G. W. (2). The N-terminus of myosin. I. Studies on N-acetylpeptides from a pronase digest of myosin. *Biochim. biophys. Acta.*, **111**, 191. 1965.

Offer, G. W. (3). *Structure and enzymatic properties of myosin.* Ph.D. Dissertation. University of Cambridge. 1964.

Offer, G. W. (4). The antagonistic action of magnesium ions and ethylenediaminetetraacetate on myosin A ATPase (potassium activated). *Biochim. biophys. Acta.*, **89**, 566. 1964.

Offer, G. W. & Starr, R. L. (1). A parallel two-chain structure for the myosin rod. *Symposium on Interactions between subunits and biological macromolecules.* International Union for pure and applied biophysics. p. 23. 1968.

Ogata, T. (1). A histochemical study of the red and white muscle fibers. Part I. Activity of the succinoxidase system in muscle fibers. *Acta Med. Okayama*, **12**, 216. 1958.

Ogata, T. (2). A histochemical study of the red and white muscle fibers. Part II. Activity of the cytochrome oxidase in muscle fibers. *Acta Med. Okayama*, **12**, 228. 1958.

Ogata, T. (3). A histochemical study of the red and white muscle fibers. Part III. Activity of the diphosphopyridine nucleotide diaphorase and triphosphopyridine nucleotide diaphorase in muscle fibers. *Acta Med. Okayama*, **12**, 233. 1958.

Ogata, T. (4). The differences in some labile constituents and some enzymatic activities between the red and the white muscles. *J. Biochem.* (Tokyo), **47**, 726. 1960.

Ogawa, Y. (1). Synthesis of skeletal muscle proteins in early embryos and regenerating tissue of chick and *Triturus. Exp. Cell Res.*, **26**, 269. 1962.

Ogle, W. (1). (tr.) [Aristotle's] *De Partibus Animalium* in *The Works of Aristotle,* tr. ed. J. A. Smith & W. D. Ross. Vol. 5, Oxford, Clarendon Press. 1912.

Ohnishi, T. (1). Role of contractile proteins in biological membranes. *Abstracts, 6th Int. Cong. Biochem.* Vol. VIII, p. 662. 1964.

Ohnishi, T. & Ebashi, S. (1). The velocity of calcium binding of isolated sarcoplasmic reticulum. *J. Biochem.*, **55**, 598. 1964.

Ohnishi, T. & Ohnishi, T. (1). Extraction of actin- and myosin-like proteins from liver mitochondria. *J. Biochem.*, **52**, 230. 1962.

Ohshima, Y., Maruyama, K. & Noda, H. (1). Developmental changes in chick muscle contractile proteins. In *Molecular biology of muscular contraction*, p. 132. Ed. S. Ebashi, F. Oosawa, T. Sekine & Y. Tonomura. Tokyo and Osaka, Igaku Shoin Ltd.; Amsterdam, Elsevier Publishing Co. 1965.

Ohtsuki, I., Masaki, T., Nonomura, Y. & Ebashi, S. (1). Periodic distribution of troponin along the thin filament. *J. Biochem.*, **61**, 817. 1967.

O'Kane, D. J. & Gunsalus, I. C. (1). Pyruvic acid metabolism. A factor required for oxidation by *Streptococcus faecalis*. *J. Bact.*, **56**, 499. 1948.

Okuno, G., Hizukuri, S. & Nishikawa, M. (1). Activities of glycogen synthetase and UDPG-pyrophosphorylase in muscle of a patient with a new type of glycogenosis caused by phosphofructokinase deficiency. *Nature*, **212**, 1490. 1966.

Olson, E., Vignos, P. J., Woodlock, J. & Perry, T. (1). Oxidative phosphorylation of skeletal muscle in human muscular dystrophy. *J. Lab. clin. Med.*, **71**, 220. 1968.

Olson, R. E. & Piatnek, D. A. (1). Conservation of energy in cardiac muscle. *Ann. N.Y. Acad. Sci.*, **72**, 466. 1958–9.

Ooi, T., Mihashi, K. & Kobayashi, H. (1). On the polymerisation of tropomyosin. *Arch. Biochem. Biophys.*, **98**, 1. 1962.

Oosawa, F. (1). Fibrous and globular aggregations of charged macromolecules. *J. Polymer Sci.*, **26**, 29. 1957.

Oosawa, F. (2). *Mechanochemistry of actin. Sixth Internat. Congr. of Biochemistry*, New York. 1964.

Oosawa, F., Asakura, S., Asai, H., Kasai, M., Kobayashi, S., Mihashi, K., Ooi, T., Taniguchi, M. & Nakano, E. (1). Structure and function of actin polymers. In *Biochemistry of muscle contraction*. p. 158. Ed. J. Gergely. London, J. & A. Churchill Ltd. 1964.

Oosawa, F., Asakura, S., Higashi, S., Kasai, M., Kobayashi, S., Nakano, E., Ohnishi, T. & Taniguchi, M. (1). Morphogenesis and motility of actin polymers. In *Molecular biology of muscular contraction*, p. 56. Ed. S. Ebashi, F. Oosawa, T. Sekine & Y. Tonomura. Tokyo, I. Shoin Ltd.; Amsterdam, Elsevier Publ. Co. 1965.

Oosawa, F., Asakura, S., Hotta, K., Imai, N. & Ooi, T. (1). G–F transformation of actin as a fibrous condensation. *J. Polymer Sci.*, **37**, 323. 1959.

Oosawa, F., Asakura, S. & Ooi, T. (1). Physical chemistry of muscle protein 'actin'. *Progr. Theor. Phys.*, Suppl. **17**, 14. 1961.

Oosawa, F. & Kasai, M. (1). A theory of linear and helical aggregations of macromolecules. *J. mol. Biol.*, **4**, 10. 1961.

Opie, L. H. & Newsholme, E. A. (1). The inhibition of skeletal-muscle fructose, 1,6-diphosphatase by adenosine monophosphate. *Biochem. J.*, **104**, 353. 1967.

Opie, L. H. & Newsholme, E. A. (2). The activities of fructose 1,6-diphosphatase, phosphofructokinase and phosphoenolpyruvate carboxykinase in white and red muscle. *Biochem. J.*, **103**, 391. 1967.

Oppenheimer, H., Bárány, K., Hamoir, G. & Fenton, J. (1). Succinylation of myosin. *Arch. Biochem. Biophys.*, **120**, 108. 1967.

Oppenheimer, H., Bárány, K. & Milhorat, A. T. (1). Myosin from mice with hereditary muscular dystrophy. *Proc. Soc. exp. Biol. & Med.*, **116**, 877. 1964.

Ostern, P., Baranowski, T. & Reis, J. (1). Sur la formation de l'acide adenosine-triphosphorique et sur le rôle des phosphagènes. *Compt. rend. Soc. Biol.*, **118**, 1414. 1935.

Ostern, P., Baranowski, T. & Reis, J. (2). Über die Verkettung der chemischen Vorgänge im Muskel. VIII. *Biochem. Z.*, **279**, 85. 1935.

Ostern, P. & Guthke, J. A. (1). Les transformations initiales de la glycogenolyse. La fonction de l'ester hexosemonophosphorique. *Compt. rend. Soc. Biol.*, **121**, 282. 1936.

Ostern, P., Guthke, J. A. & Terszakowec, J. (1). Les transformations initiales de la glycogenolyse. La fonction de l'ester hexose monophosphorique. *Compt. rend. Soc. Biol.*, **121**, 1133. 1936.

Ostwald, W. (1). Elektrische Eigenschaften halbdurchlassige Scheidewände. *Z. f. phys. Chem.*, **6**, 71. 1890.

Ostwald, W. (2). Uber Oxydationen mittels freien Sauerstoffs. *Z. f. phys. Chem.*, **34**, 248, 1900.

Ouellet, L., Laidler, K. J. & Morales, M. F. (1). Molecular kinetics of muscle adenosinetriphosphatase. *Arch. Biochem.*, **39**, 37. 1952.

Overton, E. (1). Beitrage zur allgemeinen Muskel- und Nervenphysiologie II. Über die Unentbehrlichkeit von Natrium – (oder Lithium-) Ionen für den Contractionsact des Muskels. *Arch. f. ges. Physiol.*, **92**, 346. 1902.

Özand, P. & Narahara, H. T. (1). Regulation of glycolysis in muscle. III. Influence of insulin, epinephrine and contraction on phosphofructokinase activity in frog skeletal muscle. *J. biol. Chem.*, **239**, 3146. 1964.

Ozawa, E. & Ebashi, S. (1). Requirement of Ca ion for the stimulating effect of cyclic 3′,5′-AMP on muscle phosphorylase *b* kinase. *J. Biochem.*, **62**, 285. 1967.

Ozawa, E., Hosoi, K. & Ebashi, S. (1). Reversible stimulation of muscle phosphorylase *b* kinase by low concentrations of calcium ions. *J. Biochem.*, **61**, 531. 1967.

Page, S. G. (1). Filament lengths in resting and excited muscle. *Proc Roy. Soc.* B, **160**, 460. 1964.

Page, S. G. (2). The organisation of the sarcoplasmic reticulum in frog muscle. *J. Physiol.*, **175**, 10P. 1964.

Page, S. G. (3). A comparison of the fine structure of frog slow and twitch muscles. *J. Cell Biol.*, **26**, 477. 1965.

Page, S. G. & Huxley, H. E. (1). Filament lengths in striated muscle. *J. Cell Biol.*, **19**, 369. 1963.

Pagel, W. (1). *Paracelsus. An introduction to philosophical medicine in the era of the Renaissance.* Basel and New York, S. Karger. 1958.

Pagel, W. (2). Harvey and Glisson on irritability with a note on van Helmont. *Bull. Hist. of Med.*, **41**, 497. 1967.

Palay, S. L., McGee-Russell, S. M., Gordon, S. & Grillo, M. A. (1). Fixation of neural tissues for electron microscopy by perfusion with osmium tetroxide. *J. Cell Biol.*, **12**, 385. 1962.

Panner, B. J. & Honig, C. R. (1). Filament ultrastructure and organisation in vertebrate smooth muscle. *J. Cell Biol.*, **35**, 303. 1967.

Pápai, M. B., Székessy-Hermann, V. & Szöke, K. (1). A natural inhibitor of actin polymerisation. *Abs. of 1st meeting of Fed. of Eur. Bioch. Societies*, p. 51. 1964.

Park, C. R., Post, R. L., Kalman, C. F., Wright, J. H., Johnson, L. H. & Morgan, H. E. (1). The transport of glucose and other sugars across cell membranes and the effect of insulin. *Ciba Foundation Colloquia*, **9**, 240. 1956.

Parker, C. J. & Gergely, J. (1). Soluble relaxing factor from muscle. *J. biol. Chem.*, **235**, 3449. 1960.

Parker, K. D. & Rudall, K. M. (1). Structure of the silk of *Chrysopa* egg-stalks. *Nature*, **179**, 905. 1957.

Parkes, E. A. (1). Further experiments on the effect of diet and exercise on the elimination of nitrogen. *Proc. Roy. Soc.* B, **19**, 349. 1870–71.

Parmeggiani, A. & Bowman, R. H. (1). Regulation of phosphofructokinase activity by citrate in normal and diabetic muscle. *Biochem. biophys. Res. Com.*, **12**, 268. 1963.

Parmeggiani, A. & Morgan, H. E. (1). Effect of adenine nucleotides and inorganic phosphate on muscle phosphorylase activity. *Biochem. biophys. Res. Com.*, **9**, 252. 1962.

Parnas, J. K. (1). Ueber das Wesen der Muskelerholung. *Zentralbl. f. Physiol.*, 30, 1. 1915.

Parnas, J. K. (2). Über die Ammoniakbildung im Muskel und ihren Zusammensetzung mit Funktion und Zustandsänderung. VI. Mitteilung: Der Zusammensetzung der Ammoniakbildung mit der Umwandlung des Adeninnucleotids zu Inosinsäure. *Biochem. Z.*, 206, 16. 1929.

Parnas, J. K. (3). L'enchaînement des processus enzymatiques dans le tissu musculaire. *Bull. Soc. Chim. biol.*, 18, 53. 1936.

Parnas, J. K. (4). Über die Verkettung der chemischen Vorgänge im Muskel. *Klin. Wochensch.*, 14, 1017. 1935.

Parnas, J. K. (5). Der Mechanismus der Glykogenolyse im Muskel. *Erg. d. Enzymforsch.*, 6, 57. 1937.

Parnas, J. K. (6). Energetik glatter Muskeln. *Arch. f. ges. Physiol.*, 134, 441. 1910.

Parnas, J. K. (7). Le métabolisme du muscle en activité. Societé de Biologie, Reunion plénière. 1929.

Parnas, J. K. & Baranowski, T. (1). Sur les phosphorylations initiales de glycogène. *Compt. rend. Soc. Biol.*, 120, 307. 1935.

Parnas, J. K. & Lewínski, W. (1). Über den Ammoniakgehalt und die Ammoniakbildung im Muskel. XXII. Über den Zusammenhang zwischen Ammoniakbildung und Muskeltätigkeit unter aerobic Bedingungen. *Biochem. Z.*, 276, 399. 1935.

Parnas, J. K., Lewínski, W., Jaworska, J. & Umschweif, B. (1). Über den Ammoniakgehalt und die Ammoniakbildung im Froschmuskel. VII. *Biochem. Z.*, 228, 366. 1930.

Parnas, J. K. & Mozołowski, W. (1). Über die Ammoniakgehalt und die Ammoniakbildung im Muskel und deren Zusammenhang mit Funktion und Zustandsänderung. *Biochem. Z.*, 184, 399. 1927.

Parnas, J. K., Mozołowski, W. & Lewínski, W. (1). Über die Ammoniakgehalt und die Ammoniakbildung im Blute. IX Miteilung. Der Zusammenhang des Blutammoniak mit der Muskelarbeit. *Biochem. Z.*, 188, 15. 1927.

Parnas, J. K., Mozołowski, W. & Lewínski, W. (2). Über die Ammoniakbildung im isolierten Muskel und ihren Zusammenhang mit der Muskelarbeit. *Klin. Wochensch.*, 6, 1710. 1927.

Parnas, J. K. & Ostern, P. La Mecanisme de la glycogenolyse. *Bull. Soc. Chim. biol.*, 18, 1471. 1936.

Parnas, J. K., Ostern, P. & Mann, T. (1). Über die Verkettung der chemischen Vorgänge im Muskel. *Biochem. Z.*, 272, 64. 1934.

Parnas, J. K., Ostern, P. & Mann, T. (2). Über die Verkettung der chemischen Vorgänge im Muskel. II. *Biochem. Z.*, 275, 74. 1934.

Parnas, J. K., Ostern, P. & Mann, T. (3). Über die Verkettung der chemischen Vorgänge im Muskel. III. Die Phosphat Ubertragung durch Brenztraubensaure. *Biochem. Z.*, 275, 163. 1935.

Parnas, J. K., Sobczuk, B. & Mejbaum, W. (1). Le mechanisme de la suppression de l'ammoniogenèse dans le muscle par l'acide pyruvique. *Compt. rend. Soc. Biol.*, 121, 701. 1936.

Parnas, J. K. & Wagner, R. (1). Ueber den Kohlehydratumsatz isolierter Amphibienmuskeln und über die Beziehungen zwischen Kohlehydratschwund und Milchsäurebildung im Muskel. *Biochem. Z.*, 61, 387. 1914.

Parnas, J. K. & Wagner, R. (2). Beobachtungen über Zuckerneubildung. 1. Nach Versuchen, die an einem Falle besonderer Kohlenhydratstoffwechselstörung angestellt wurden. *Biochem. Z.*, 127, 55. 1922.

Parrish, R. G. & Mommaerts, W. F. H. M. (1). Studies on myosin II. Some molecular-kinetic data. *J. biol. Chem.*, **209**, 901. 1954.

Parrish, R. G. & Mommaerts, W. F. H. M. (2). Instantaneous reversible depolymerisation of actin. *Arch. Biochem. Biophys.*, **31**, 459. 1951.

Parsons, C. & Porter, R. K. (1). Muscle relaxation: Evidence for an intrafibrillar restoring force in vertebrate striated muscle. *Science*, **153**, 426. 1966.

Partington, J. R. (1). *A Short History of Chemistry.* London, Macmillan. 1957.

Partington, J. R. (2). *A History of Chemistry.* Vol. IV. London, Macmillan. 1962.

Passonneau, J. V. & Lowry, O. H. (1). Phosphofructokinase and the Pasteur effect. *Biochem. biophys, Res. Com.*, **7**, 10. 1962.

Passonneau, J. V. & Lowry, O. H. (2). P-fructokinase and the control of the citric acid cycle. *Biochem. biophys. Res. Com.*, **13**, 372. 1963.

Passonneau, J. V. & Lowry, O. H. (3). The role of phosphofructokinase in metabolic regulation. *Adv. in Enz. Regulation*, **2**, 265. 1964.

Pasteur, M. L. (1). Expériences et vues nouvelles sur la nature des fermentations. *Compt. rend. heb. Acad. Sci. Paris*, **52**, 1260. 1861.

Pattengale, P. K. & Holloszy, J. O. (1). Augmentation of skeletal muscle myoglobin by a program of treadmill running. *Amer. J. Physiol.*, **213**, 783. 1967.

Paukul, E. (1). Die Zuckungsformen von Kaninchenmuskeln verschiedener Farbe und Struktur. *Arch. f. Anat. u. Physiol.*, p. 100. 1904.

Paul, M. H. & Sperling, E. (1). Cyclophorase system XXIII. Correlation of cyclophorase activity and mitochondrial density in striated muscle. *Proc. Soc. exp. Biol. Med.*, **79**, 352. 1952.

Pauli, W. (1). Ueber den Zusammenhang von elektrischen, mechanischen und chemischen Vorgänge im Muskel. *Kolloidchem. Beihefte*, **3**, 361. 1912.

Pauli, W. & Valkó, E. (1). *Kolloidchemie der Eiweisskörper.* 2nd edition, pp. 166, 207. Dresden and Leipzig, T. Steinkopff. 1933.

Pauling, L. & Corey, R. B. (1). Two hydrogen-bonded spiral configurations of the polypeptide chain. *J. Am. Chem. Soc.*, **72**, 5349. 1950.

Pauling, L. & Corey, R. B. (2). The structure of hair, muscle and related proteins. *Proc. Nat. Acad. Sci. U.S.A*, **37**, 261. 1951.

Pauling, L. & Corey, R. B. (3). Compound helical configurations of polypeptide chains: structure of proteins of the α-keratin type. *Nature*, **171**, 59. 1953.

Pauling, L., Corey, R. B. & Branson, H. R. (1). The structure of proteins: two hydrogen bonded helical configurations of the polypeptide chain. *Proc. Nat. Acad. Sci. U.S.A.*, **37**, 205. 1951.

Pautard, F. G. E. (1). The fundamental molecular event in muscular contraction. *Nature*, **182**, 788. 1958.

Payen, A. & Persoz, J. F. (1). Memoire sur la diastase, les principaux produits de ses réactions, et leur application aux arts industriels. *Ann. de. Chim. et de Phys.*, **53**, 73. 1833.

Peachey, L. D. (1). The sarcoplasmic reticulum and transverse tubules of the frog's sartorius. *J. Cell Biol.*, **25**, No. 3, Pt. 2, 209. 1965.

Peachey, L. D. & Huxley, A. F. (1). Transverse tubules in crab muscle. *J. Cell Biol.*, **23**, 70. 1964.

Peachey, L. D. & Huxley, A. F. (2). Structural identification of twitch and slow striated muscle fibers of the frog. *J. Cell Biol.*, **13**, 177. 1962.

Peachey, L. D. & Porter, K. R. (1). Intracellular impulse conduction in muscle cells. *Science*, **129**, 721. 1959.

Pearce, G. W. (1). Tissue culture and electron microscopy in muscle disease. In *Disorders of voluntary muscle*, p. 230. Ed. J. N. Walton. Boston, Little, Brown & Co. 1964.

Pearce, G. W. (2). Electron microscopy in the study of muscular dystrophy. In *Muscular Dystrophy in Man and Animals*. p. 159. Ed. G. H. Bourne & M. N. Golarz. Basel and New York, S. Karger. 1963.

Pearson, C. M. (1). Muscular dystrophy. *Amer. J. Med.*, 35, 632. 1963.

Pearson, C. M. & Rimer, D. G. (1). Evidence for direct utilisation of fructose in working muscle in man. *Proc. Soc. exp. Biol. Med.*, 100, 671. 1959.

Pearson, D. J. & Tubbs, P. K. (1). Carnitine and derivatives in rat tissues. *Biochem. J.*, 105, 953. 1967.

Pechère, J.-F. & Focant, B. (1). Carp myogens of white and red muscles. Gross isolation on sephadex columns of the low-molecular-weight components and examination of their participation in anaerobic glycogenolysis. *Biochem. J.*, 96, 113. 1965.

Pellegrino, C. & Bibbiani, C. (1). Increase of muscle permeability to aldolase in several experimental atrophies. *Nature*, 204, 483. 1964.

Pellegrino, C. & Franzini, C. (1). An electron miscroscope study of denervation atrophy in red and white muscles. *J. Cell Biol.*, 17, 327. 1963.

Penefsky, H. S., Pullman, M. E., Datta, A. & Racker, E. (1). Partial resolution of the enzymes catalysing oxidative phosphorylation. II. Participation of a soluble adenosine triphosphate in oxidative phosphorylation. *J. biol. Chem.*, 235, 3330. 1960.

Peng, C.-M., Kung, T.-H., Hsiung, L.-M. & Tsao, T.-C. (1). Electron microscopical studies of tropomyosin and paramyosin. II *Scientia Sinica*, 14, 219. 1965.

Pennington, R. J. (1). Biochemistry of dystrophic muscle. Mitochondrial succinate-tetrazolium reductase and adenosine triphosphatase. *Biochem. J.*, 80, 649. 1961.

Pennington, R. J. (2). Biochemistry of dystrophic muscle. 2. Some enzyme changes in dystrophic mouse muscle. *Biochem. J.*, 88, 64. 1963.

Pennington, R. J. (3). 5' Adenylic acid deaminase in dystrophic mouse muscle. *Nature*, 192, 884. 1961.

Pennington, R. J. (4). Biochemical aspects of muscle disease. In *Disorders of Voluntary Muscle*, p. 255. Ed. J. N. Walton. London, J. & A. Churchill, Ltd. 1964.

Penniston, J. T., Harris, R. A., Asai, J. & Green, D. E. (1). The conformational basis of energy transformations in membrane systems I. Conformational changes in mitochondria. *Proc. nat. Acad. Sci. U.S.A.*, 59, 624. 1968.

Pepe, F. A. (1). Some aspects of the structural organisation of the myofibril as revealed by antibody-staining methods. *J. Cell Biol.*, 28, 505. 1966.

Pepe, F. A. (2). The myosin filament I. Structural organisation from antibody staining observed in electron microscopy. *J. mol. Biol.*, 27, 303. 1967.

Pepe, F. A. (3). The myosin filament II. Interaction between myosin and actin filaments observed using antibody staining in fluorescent and electron microscopy. *J. mol. Biol.*, 27, 227. 1967.

Perry, S. V. (1). The chromatography of L-myosin on diethylaminoethylcellulose. *Biochem. J.*, 74, 94. 1960.

Perry, S. V. (2). Relation between chemical and contractile function and structure of the skeletal muscle cell. *Physiol. Rev.*, 36, 1. 1956.

Perry, S. V. (3). The adenosinetriphosphatase activity of myofibrils isolated from skeletal muscle. *Biochem. J.*, 48, 257. 1951.

Perry, S. V. (4). The ATP-ase activity of isolated myofibrils. *Biochem. J.*, **47**, xxxviii. 1950.

Perry, S. V. (5). Muscle proteins in contraction. In *Muscle*, p. 29. Ed. by W. M. Paul, E. E. Daniel, C. M. Kay & G. Monckton. London and New York, Pergamon Press. 1965.

Perry, S. V. (6). The protein components of the isolated myofibril. *Biochem. J.*, **55**, 114. 1953.

Perry, S. V. (7). The structure and interactions of myosin. *Progr. in Biophys. & mol. Biol.*, **17**, 325. 1967.

Perry, S. V. (8). In the discussion of the paper by Heyck & Laudahn in *Excerpta Medica Foundation, International Congress Series*, No. 147, p. 242. 1967.

Perry, S. V. (9). Biochemical adaptation during development and growth in skeletal muscle. In the *Symposium on Physiology and Biochemistry of Muscle as a Food*. Univ. of Wisconsin Press; Madison, 1966.

Perry, S. V. (10). The adenosinetriphosphatase activity of lipoprotein granules isolated from skeletal muscle. *Biochim. biophys. Acta.*, **8**, 499. 1952.

Perry, S. V. (11). The bound nucleotide of the isolated myofibril. *Biochem. J.*, **51**, 495. 1952.

Perry, S. V. & Corsi, A. (1). Extraction of proteins other than myosin from the isolated rabbit myofibril. *Biochem. J.*, **68**, 5. 1958.

Perry, S. V. & Cotterill, J. (1). The action of thiol reagents on the adenosine-triphosphatase activities of heavy meromyosin and L-myosin. *Biochem. J.*, **96**, 224. 1965.

Perry, S. V. & Cotterill, J. (2). The action of thiol inhibitors on the interaction of F-actin and heavy meromyosin. *Biochem. J.*, **92**, 603. 1964.

Perry, S. V. & Cotterill, J. (3). The interaction of actin and heavy meromyosin. *Biochem. J.*, **88**, 9P. 1963.

Perry, S. V. & Cotterill, J. (4). Interaction of actin and myosin. *Nature*, **206**. 1965.

Perry, S. V., Davies, V. & Hayter, D. (1). 'Natural' tropomyosin and the factor sensitizing actomyosin adenosine triphosphatase to ethylenedioxybis(ethyleneamino)-tetra-acetic acid. *Biochem. J.*, **99**, 1c. 1966.

Perry, S. V. & Grey, T. C. (1). A study of the effects of substrate concentration and certain relaxing factors on the magnesium-activated myofibrillar adenosine triphosphatase. *Biochem. J.*, **64**, 184. 1956.

Perry, S. V. & Grey, T. C. (2). Ethylenediaminetetra-acetate and the adenosinetriphosphatase activity of actomyosin systems. *Biochem. J.*, **64**, 5P. 1956.

Perry, S. V. & Hartshorne, D. J. (1). The effect of use and disuse on neuromuscular functions. *Proc. of Sym. at Liblice*. p. 491. Ed. Gutmann & Hník. Czechoslovak Acad. of Sciences. 1963.

Perry, S. V. & Landon, M. (1). Electrophoretic studies on the products of proteolytic digestion of myosin. In *Biochemistry of muscle contraction*. p. 36. Ed. J. Gergely. London, J. & A. Churchill. 1964.

Perry, S. V. & Zydowo, M. (1). The nature of the extra protein fraction from myofibrils of striated muscle. *Biochem. J.*, **71**, 220. 1959.

Perry, S. V. & Zydowo, M. (2). A ribonucleoprotein of skeletal muscle and its relation to the myofibril. *Biochem. J.*, **72**, 682. 1959.

Perutz, M. F. (1). New X-ray evidence on the configuration of polypeptide chains. *Nature*, **167**, 1053. 1951.

Peters, R. A. (1). The heat production of fatigue and its relation to the production of lactic acid in amphibian muscle. *J. Physiol.*, **47**, 243. 1913.

Peters, R. A. (2). The study of enzymes in relation to selective toxicity in animal tissues. *Symp. Soc. exp. Biol.*, **3**, 36. 1949.

Peters, R. A. (3). Co-ordinative biochemistry of the cell and tissues. *The Harben Lectures*, reprinted from the *Journal of State Medicine*. London, T. G. Scott & Son, Ltd. 1929.

Peters, R. A., Shorthouse, M. & Murray, L. R. (1). Enolase and fluorophosphate. *Nature*, **202**, 1331. 1964.

Pette, D. & Bücher, T. (1). Proportionskonstante Gruppen in Beziehung zur Differenzierung der Enzymaktivitätsmuster von Skelett-Muskeln des Kaninchens. *Z. physiol. Chem.*, **331**, 180. 1963.

Pettenkofer, M. & Voit, C. (1). Untersuchungen über den Stoffverbrauch des normalen Menschen. *Zeitsch. f. Biol.*, **2**, 459. 1866.

Pettko, E. & Straub, F. B. (1). The active principle of muscle extracts increasing the performance of the hypodynamic frog's heart: adenosinetriphosphate. *Hung. Acta. Physiol.*, **2**, 114. 1949.

Pflüger, E. (1). Beiträge zur Lehre von der Respiration. I. Ueber die physiologische Verbrennung in den lebendigen Organismen. *Arch. f. ges. Physiol.*, **10**, 251. 1875.

Pflüger, E. (2). Über die physiologische Verbrennung in den lebendigen Organismen. *Arch. f. ges. Physiol.*, **10**, 641. 1875.

Pillai, R. K. (1). Dephosphorylation in muscle extracts. *Biochem. J.*, **32**, 1087. 1938.

Pinaev, G. P. (1). Change in the shape and size of actomyosin particles of the striated muscles in ontogenesis. *Biokhimiya* (Eng. vers.), **30**, 15. 1965.

Piras, R., Rothman, L. B. & Cabib, E. (1). Regulation of muscle glyogen synthetase by metabolites. Differential effects on I and D forms. *Biochemistry U.S.A.*, **7**, 56. 1968.

Piras, R. & Staneloni, R. (1). In vivo regulation of rat muscle glycogen synthetase activity. Biochemistry, **8**, 2153. 1969.

Podolsky, R. J. (1). The maximum sarcomere length for contraction of isolated myofibrils. *J. Physiol.*, **170**, 110. 1964.

Podolsky, R. J. (2). The chemical thermodynamics and molecular mechanism of muscular contraction. *Ann. New York Acad. Sci.*, **72**, 522. 1959.

Podolsky, R. J. (3). Mechanochemical basis of muscular contraction. *Fed. Proc.*, **21**, 965. 1962.

Podolsky, R. J. (4). Introduction to 'Theories of muscle contraction.' In *Biochemistry of muscle contraction*. p. 477. Ed. J .Gergely. London, J. & A. Churchill, Ltd. 1964.

Podolsky, R. J. & Contantin, L. L. (1). Regulation by calcium of the contraction and relaxation of the muscle fibers. *Fed. Proc.*, **23**, 933. 1964.

Podolsky, R. J. & Morales, M. F. (1). The enthalpy change of adenosine triphosphate hydrolysis. *J. biol. Chem.*, **218**, 945. 1956.

Pogson, C. I. & Randle, P. J. (1). The control of rat-heart phosphofructokinase by citrate and other regulators. *Biochem. J.*, **100**, 683. 1966.

Pohl, J. (1). Zur Lehre von der Wirkung substituirter Fettsäuren. *Arch. f. exper. Path. u. Pharmac.*, **24**, 142. 1888.

Polis, B. D. & Meyerhof, O. (1). Studies on adenosine triphosphatase in muscle. I. Concentration of the enzyme on myosin. *J. biol. Chem.*, **169**, 389. 1947.

Pompe, (1). Hypertrophie idiopathique du coeur. *Ann. Anat. path.* **10**, 23. 1933.

Pordage, S. (1). (tr.) [*T. Willis'*] *De Motu Musculorum.* London. 1684.

Pordage, S. (2). (tr.) [*T. Willis'*] *Of Feavers, contained in The Remaining Medical Works of that Famous and Renowned Physician Dr Thomas Willis.* London. 1684.

Portal, A. (1). *Histoire de l'Anatomie et de Chirugie.* Vol. III. Paris. 1770.

Porter, K. R. (1). Sarcoplasmic reticulum in muscle cells of *Amblystoma* larvae. *J. biophys. biochem. Cyt.*, 2, Suppl. 163. 1956.

Porter, K. R. (2). The sarcoplasmic reticulum. Its recent history and present status. *J. biophys. biochem. Cyt.*, 10, Suppl. 219. 1961.

Porter, K. R. & Palade, G. E. (1). Studies on the endoplasmic reticulum. III. Its form and distribution in striated muscle cells. *J. biophys. biochem. Cyt.*, 3, 269. 1957.

Portzehl, H. (1a). Masse und Masze des L-Myosins. *Z. Naturforsch.*, 5b, 75. 1950.

Portzehl, H. (1). Muskelkontraktion und Modellkontraktion. II. *Zeitsch. f. Naturforsch.*, 6b, 355. 1951.

Portzehl, H. (2). Der Arbeitszyclus geordneter Aktomyosinsysteme (Muskel und Muskelmodelle). *Zeitsch. f. Naturforsch.*, 7b, 1. 1952.

Portzehl, H. (3). Gemeinsame Eigenschaften von Zell- und Muskelkontraktilität. *Biochim. biophys. Acta.*, 14, 195. 1954.

Portzehl, H. (4). Bewirkt das System Phosphokreatin-phosphokinase die Erschlaffung des lebenden Muskels? *Biochim. biophys. Acta.*, 24, 474. 1957.

Portzehl, H. (5). Die Bindung des Erschlaffungsfaktor von Marsh an die Muskelgrana. *Biochim. biophys. Acta.*, 26, 373. 1957.

Portzehl, H., Caldwell, P. C. & Rüegg, J. C. (1). The dependence of contraction and relaxation of muscle fibres from the crab *Maia squinado* on the internal concentration of free calcium ions. *Biochim. biophys. Acta.*, 79, 581. 1964.

Portzehl, H., Schramm, G. & Weber, H. H. (1). Aktomyosin und seine Komponenten I. *Zeitsch. f. Naturforsch.*, 5b, 61. 1950.

Portzehl, H. & Weber, H. H. (1). Zur Thermodynamik der ATP-Kontraktion des Aktomyosinfadens. *Zeitsch. f. Naturforsch.*, 5b, 123. 1950.

Posner, J. B., Stern, R. & Krebs, E. G. (1). *In vivo* response of skeletal muscle glycogen phosphorylase, phosphorylase *b* kinase and cyclic AMP to epinephrine administration. *Biochem. biophys. Res. Com.*, 9, 293. 1962.

Posner, J. B., Stern, R. & Krebs, E. G. (2). Effects of electrical stimulation and epinephrine on muscle phosphorylase, phosphorylase *b* kinase, and adenosine 3',5'-phosphate. *J. biol. Chem.*, 240, 982. 1965.

Post, R. L., Morgan, H. E. & Park, C. R. (1). Regulation of glucose uptake in muscle. III. The interaction of membrane transport and phosphorylation in the control of glucose uptake. *J. biol. Chem.*, 236, 269. 1961.

Potter, V. R. & Reif, A. E. (1). Inhibition of an electron transport component by antimycin A. *J. biol. Chem.*, 194, 287. 1952.

Prágay, D. (1). Aktin-Fraktionen, Aktinpolymerisation. *Naturwissensch.*, 44, 397. 1957.

Prewitt, M. A. & Salafsky, B. (1). Effect of cross-innervation on biochemical characteristics of skeletal muscles. *Amer. J. Physiol.*, 213, 295. 1967.

Pringle, J. W. S. (1). The excitation and contraction of the flight muscles of insects. *J. Physiol.*, 108, 226. 1949.

Pringle, J. W. S. (2). The mechanism of the myogenic rhythm of certain insect striated muscle. *J. Physiol.*, 124, 269. 1954.

Pringle, J. W. S. (3). *Insect Flight.* Cambridge University Press. 1957.

Pringle, J. W. S. (4). Myogenic rhythms. In *Recent Advances in Insect Physiology.* p. 99. Ed. B. T. Scheer. Eugene, Oregon. University of Oregon Publications. 1957.

Pringle, J. W. S. (5). The contractile mechanism of insect fibrillar muscle. *Progr. in Biophys. and Mol. Biol.,* **17,** 3. 1967.

Pringle, J. W. S. (6). Models of muscle. *Symp. Soc. exp. Biol.,* **14,** 42. 1960.

Prout, W. (1). *Chemistry, meteorology and the function of digestion considered with reference to natural theology.* London, William Pickering. 1834.

Pryor, M. G. M. (1). Mechanical properties of fibres and muscles. *Progr. Biophys.,* **1,** 216. 1950.

Pryor, M. G. M. (2). Heat exchange in a muscle model. *Nature,* **171,** 213. 1953.

Pullman, M. E. & Schatz, G. (1). Mitochondrial oxidations and energy coupling. *Ann. Rev. Biochem.,* **36,** 539. 1967.

Purkinje, J. E. (1). Beobachtungen, betreffend die innerste Struktur der Nerven. In *Opera Omnia,* II, p. 88. Prague, Purkynova Spolecriost, 1937. 1838.

Raaflaub, J. (1). Komplexbildner als Cofaktoren isolierter Zellgranula. *Helv. chim. Acta.,* **38,** 27. 1955.

Racker, E. & Krimsky, I. (1). The mechanism of oxidation of aldehydes by glyceraldehyde-3-phosphate dehydrogenase. *J. biol. Chem.,* **198,** 731. 1952.

Rainford, P., Noguchi, H. & Morales, M. (1). Hydrostatic pressure and the contractile system. *Biochemistry* (Am. chem. Soc.) **4,** 1958. 1965.

Rall, T. W. & Sutherland, E. W. (1). Formation of a cyclic adenine ribonucleotide by tissue particles. *J. biol. Chem.,* **232,** 1065. 1957.

Rall, T. W. & Sutherland, E. W. (2). Adenylcyclase. II. The enzymatically catalyzed formation of adenosine 3′,5′-phosphate and inorganic pyrophosphate from adenosine triphosphate. *J. biol. Chem.,* **237,** 1228. 1962.

Rall, T. W., Sutherland, E. W. & Berthet, J. (1). The relationship of epinephrine and glucagon to liver phosphorylase. IV. Effect of epinephrine and glucagon on the reactivation of phosphorylase in liver homogenates. *J. biol. Chem.,* **224,** 463. 1957.

Ramsey, R. W. & Street, S. F. (1). The isometric length-tension diagram of isolated skeletal muscle fibers of the frog. *J. cell. comp. Physiol.,* **15,** 11. 1940.

Randle, P. J. (1). Monosaccharide transport in muscle and its regulation. In *Membrane transport and metabolism,* p. 431. Symposium at Prague, ed. A. Kleinzeller & A. Kotyk. Academic Press, London. 1960.

Randle, P. J. & Smith, G. H. (1). Regulation of glucose uptake by muscle. I. The effects of insulin, anaerobiosis and cell poisons on the uptake of glucose and release of potassium by isolated rat diaphragm. *Biochem. J.,* **70,** 490. 1958.

Randle, P. J. & Smith, G. H. (2). Regulation of glucose uptake by muscle. 2. The effects of insulin, anaerobiosis and cell poisons on the penetration of isolated rat diaphragm by sugars. *Biochem. J.,* **70,** 501. 1958.

Ranke, J. (1). *Tetanus.* Leipzig, Engelmann. 1865.

Ranvier, M. L. (1). Des muscles rouges et des muscles blancs chez les rongeurs. *Compt. rend. Acad. Sci.,* **77,** 1030. 1873.

Ranvier, M. L. (2). Traité technique d'histologie. Librairie F. Savoy. Paris, 1875.

Rapkine, L. (1). Sulphydryl groups and enzymic oxidoreduction. *Biochem. J.,* **32,** 1729. 1938.

Rauh, F. (1). Die Latenzzeit des Muskelelements. *Z. f. Biol.,* **76,** 25. 1922–23.

Rawlinson, W. A. & Gould, M. K. (1). Biochemical adaptation as a response to exercise. 2. Adenosinetriphosphatase and creatine phosphokinase activity in muscles of exercised animals. *Biochem. J.,* **73,** 44. 1959.

Read, J. (1). *Prelude to Chemistry.* London, G. Bell & Co. 1936.

Read, W. O. & Johnson, D. C. (1). Creatine phosphokinase activity in heart and skeletal muscle of foetal rabbits. *Proc. Soc. exp. Biol. Med.,* **102,** 740. 1959.

Redfearn, E. R. (1). The possible role of ubiquinone (coenzyme Q) in the respiratory chain. In the CIBA Symposium on *Quinones in electron transport.* p. 346. London, J. & A. Churchill. 1961.

Reed, L. J. (1). Chemistry and function of lipoic acid. In *Comprehensive Biochemistry,* Vol. **14,** p. 99. Ed. M Florkin & E. H. Stotz. Amsterdam, London, New York, Elsevier Publishing Co. 1966.

Reed, L. J., De Busk, B. G., Gunsalus, I. C. & Hornberger, C. S. (1). Crystalline α-lipoic acid: a catalytic agent associated with pyruvate dehydrogenase. *Science,* **114,** 93. 1951.

Reedy, M. K. (1). In Discussion on the physical and chemical basis of muscular contraction. *Proc. Roy. Soc.* B, **160,** 458. 1964.

Reedy, M. K., Holmes, K. C. & Tregear, R. T. (1). Induced changes in orietation of the cross-bridges of glycerinated insect flight muscle. *Nature,* **207,** 1276. 1965.

Rees, M. K. & Young, M. (1). Studies on the isolation and molecular properties of homogeneous globular actin. *J. biol. Chem.,* **242,** 4449. 1967.

Reid, E. (1). Membrane systems. In *Enzyme Cytology,* p. 321. Ed. D. B. Roodyn. London and New York, Academic Press. 1967.

Reid, R. A., Moyle, J. & Mitchell, P. (1). Synthesis of adenosine triphosphate by a protonmotive force in rat liver mitochondria. *Nature,* **212,** 257. 1966.

Reil, J. C. (1). Von der Lebenskraft. *Arch. f. die Physiol.,* **1,** 8. 1796.

Renaud, F. L., Rowe, A. J. & Gibbons, I. R. (1). Some properties of the protein forming the outer fibers of cilia. *J. Cell Biol.,* **36,** 79. 1968.

Reporter, M. C., Konigsberg, I. R. & Strehler, B. L. (1). Kinetics of accumulation of creatine phosphokinase activity in developing embryonic skeletal muscle *in vivo* and in monolayer culture. *Exp. Cell Res.,* **30,** 410. 1963.

Retzius, G. (1). Muskelfibrille und Sarkoplasma. *Biol. Untersuchungen. Neue Folge,* **1,** 51. 1890.

Retzius, G. (2). Zur Kenntniss der quergestreiften Muskelfaser. *Biol. Untersuchungen,* p. 1. 1881.

Revel, J. P. (1). The sarcoplasmic reticulum of the bat cricothyroid muscle. *J. biophys. biochem. Cyt.,* **12,** 571. 1962.

Rey, C. (1). Les esters phosphorés des muscles du lombric. *Biochim. biophys. Acta.,* **19,** 300. 1956.

Rhodes, W. C. & McElroy, W. D. (1). The synthesis and function of luciferyladenylate and oxyluciferyl-adenylate. *J. biol. Chem.,* **233,** 1528. 1958.

Rhodin, J. A. G. (1). Fine structure of vascular walls in mammals. With special reference to a smooth muscle component. *Physiol. Rev.,* **42** Suppl. 5, pt. II, 48. 1962.

Rice, R. V. (1). Conformation of individual macromolecular particles from myosin solutions. *Biochim. biophys. Acta.,* **52,** 602. 1961.

Rice, R. V. (2). Electron microscopy of macro-molecules from myosin solutions. In *Biochemistry of Muscle Contraction.* p. 41. Ed. J. Gergely. London, J. & A. Churchill Ltd. 1964.

Rice, R., Brady, A. C., Depue, R. H. & Kelly, R. E. (1). Morphology of individual macro-molecules and their ordered aggregates by electron microscopy. *Biochem. Z.,* **345,** 270. 1966.

Richterich, R. (1). Zur Biochemie der progressive Muskeldystrophie. In *Myopathien*, p. 187. Ed. R. Beckmann, Stuttgart, Georg Thieme. 1965.

Riesser, O. (1). Beiträge zur Physiologie des Kreatins. *Z. physiol. Chem.*, **120**, 189. 1922.

Rinaldi, R. A. & Baker, W. R. (1). A sliding filament model of amoeboid motion. *J. theor. Biol.*, **23**, 463. 1969.

Rinaudo, M. T. & Bruno, R. (1). α-Glucan-phosphorylase in the striated muscle of the developing embryo of *Gallus domesticus*. *Enzymologia*, **31**, 45. 1968.

Ringer, S. (1). A further contribution regarding the influence of the different constituents of the blood on the contraction of the heart. *J. Physiol.*, **4**, 29. 1883.

Ringler, R. L., Minekami, S. & Singer, T. P. (3). Studies on the respiratory chain-linked reduced nicotinamide adenine dinucleotide dehydrogenase. II. Isolation and molecular properties of the enzyme from beef heart. *J. biol. Chem.*, **238**, 801. 1963.

Riseman, J. & Kirkwood, J. G. (1). Remarks on the physico-chemical mechanism of muscular contraction and relaxation. *J. Amer. Chem Soc.*, **70**, 2820. 1948.

Ritchie, A. D. (1). *The Comparative Physiology of Muscular Tissue*. Cambridge, Cambridge University Press. 1928.

Robbins, E. A. & Boyer, P. D. (1). Determination of the equilibrium of the hexokinase reaction and the free energy of hydrolysis of adenosine triphosphate. *J. biol. Chem.*, **224**, 121. 1957.

Robbins, P. W., Traut, R. T. & Lipmann, F. (1). Glycogen synthesis from glucose, glucose-6-phosphate, and uridine diphosphate glucose in muscle preparations. *Proc. Nat. Acad. Sci. U.S.*, **45**, 6. 1959.

Roberts, Ff. (1). Degeneration of muscle following nerve injury. *Brain*, **39**, 296. 1916.

Roberts, H. S. (1). Mechanisms of cytokinesis: a critical review. *Quart. Rev. Biol.*, **36**, 155. 1961.

Robertson, J. D. (1). Some features of the ultrastructure of reptilian skeletal muscle. *J. biophys. biochem. Cyt.*, **2**, 369. 1956.

Robin, Y. & Thoai, N.-V. (1). Sur une nouvelle guanidine monosubstituée biologique, l'hypotaurocyamine (acide 2-guanido-éthanesulfinique) et le phosphagène correspondant. *Biochim. biophys. Acta*, **63**, 481. 1962.

Robinson, B. (1). *Dissertation on the Aether of Sir Isaac Newton*. Dublin, Ewing & Smith. 1743.

Robinson, D. S. (1). Changes in the protein composition of chick muscle during development. *Biochem. J.*, **52**, 621. 1952.

Robinson, D. S. (2). Changes in the nucleoprotein content of chick muscle during development. *Biochem. J.*, **52**, 628. 1952.

Robinson, D. S. (3). A study of the adenosinetriphosphatase activity of developing chick muscle. *Biochem. J.*, **52**, 633. 1952.

Robison, G. A., Butcher, R. W., Øye, I., Morgan, H. E. & Sutherland, E. W. (1). The effect of epinephrine on adenosine 3',5'-phosphate levels in the isolated perfused rat heart. *Mol. Pharm.*, **1**, 168. 1965.

Robison, R. (1). A new phosphoric ester produced by the action of yeast juice on hexoses. *Biochem. J.*, **16**, 809. 1922.

Roche, J. & Robin, Y. (1). Sur les phosphagènes des éponges. *Comp. rend. Soc. Biol.*, **148**, 1541. 1954.

Roche, J., Robin, Y., di Jeso, F. & Thoai, N.-V. (1). Sur la présence de phosphohypotaurocyamine chez l'Arenicole, *Arenicola marina*. *Comp. rend. Soc. Biol.*, **156**, 830. 1962.

Roche, M., Benedict, J. D., Yü, T. F., Bien, E. J. & Stetten, D. (1). Origin of urinary creatine in progressive muscular dystrophy. *Metabolism*, 1, 13. 1952.

Roeder, K. D. (1). Movements of the thorax and potential changes in the thoracic muscles of insects during flight. *Biol. Bull.*, 100, 95. 1951.

Rogozkin, V. A. (1). Phosphorylation of creatine. *Biochimia*, 28, 347. 1963.

Rollett, A. (1). Untersuchungen über den Bau der quergestreiften Muskelfasern. I. *Denksch. d. Kais. Akad. Wissensch. Wien. Math. Naturwiss. Klasse*, 49, 81. 1885.

Rollett, A. (2). Untersuchungen über den Bau der quergestreiften Muskelfasern. II. *Denksch. d. Kais. Akad. Wissensch. Wien. Math. Naturwiss. Klasse*, 51, 23. 1886.

Rollett, A. (3). Untersuchungen über Contraction und Doppelbrechung der quergestreiften Muskelfasern. *Denksch. Kais. Acad. d. Wissensch. Wien. Math. Naturwiss. Klasse*, 58, 41. 1891.

Rollett, A. (4). Anatomische und physiologische Bemerkungen über die Muskeln der Fledermaus. *Sitzungsber. d. K. Akad. Wien*, 98, Abt. III, 169. 1890.

Romanul, C. F. A. (1). Enzymes in muscle. I. Histochemical studies of enzymes in individual muscle fibres. *Arch. Neurol.*, 11, 355. 1964.

Romanul, C. F. A. & Van der Meulen, J. P. (1). Reversal of the enzyme profiles of muscle fibres in fast and slow muscles by cross-innervation. *Nature*, 212, 1369. 1966.

Rome, E. (1). Light and X-ray diffraction studies of the filament lattice of glycerol-extracted rabbit psoas muscle. *J. mol. Biol.*, 27, 591. 1967.

Rome, E. (2). X-ray diffraction studies of the filament lattice of striated muscle in various bathing media. *J. mol. Biol.*, 37, 331. 1968.

Ronzoni, E. & Ehrenfest, E. (1). The effect of dinitrophenol on the metabolism of frog muscle. *J. biol. Chem.*, 115, 749. 1936.

Rooseboom, M. (1). *Microscopium*. See E. van Loo.

Rosell-Perez, M. & Larner, J. (1). Studies on UDPG-α-glucan transglucosylase. V. Two forms of the enzyme in dog skeletal muscle and their interconversion. *Biochemistry* (Am. chem. Soc.), 3, 81. 1964.

Rosell-Perez, M., Villar-Palasi, C. & Larner, J. (1). Studies on UDPG transglucosylase. I. Preparation and differentiation of two activities of UDPG-glycogen transglucosylase from rat skeletal muscle. *Biochemistry* (Am. chem. Soc.), 1, 763. 1962.

Rosell-Perez, M., Villar-Palasi, C. & Larner, J. (2). Studies on UDPG transglucosylase. IV. Purification and characteristics of two forms from rabbit skeletal muscle. *Biochemistry* (Am. chem. Soc.), 3, 75. 1964.

Rosenbeck, K. & Doty, P. (1). The far ultraviolet absorption spectra of polypeptide and protein solutions and their dependence on conformation. *Proc. Nat. Acad. Sci. U.S.A.*, 47, 1775. 1961.

Rossum, G. D. V. van (1). The effect of oligomycin on cation transport in slices of rat liver. *Biochem. J.*, 84, 35. 1962.

Rostgaard, J. & Behnke, O. (1). Fine structural localization of adenine nucleoside phosphatase activity in the sarcoplasmic reticulum and the T system of rat myocardium. *J. Ultrastr. Res.*, 12, 579. 1965.

Rothschild, P. (1). Gilt das Alles-oder-nichts-Gesetz für den Tätigkeitsstoffwechsel von Einzelzuckungen des Muskels? *Biochem. Z.*, 222, 31. 1930.

Rouget, C. (1). Mémoire sur les tissus contractile et la contractilité. *J. de Physiol. de l'Homme et des Animaux*, 6, 647. 1863.

Rowe, A. J. (1). The contractile proteins of skeletal muscle. *Proc. Roy. Soc. B*, 160, 437. 1964.

Rowsell, E. V. (1). Transaminations with L-glutamate and α-oxoglutarate in fresh extracts of animal issues. *Biochem. J.*, **64**, 235. 1956.

Rowsell, E. V. (2). Transaminations with pyruvate and other α-keto acids. *Biochem. J.*, **64**, 246. 1956.

Rozsa, G., Szent-Györgyi, A. & Wyckoff, R. W. G. (1). The electron microscopy of F-actin. *Biochim. biophys. Acta*, **3**, 561. 1949.

Rozsa, G., Szent-Györgyi, A. & Wyckoff, R. W. G. (2). The fine structure of myofibrils. *Exp. Cell Res.*, **1**, 194. 1950.

Rubner, M. (1). Die Quelle der tierischen Wärme. *Zeitsch. f. Biol.*, **30**, 73. 1894.

Rubulis, A., Jacobs, R. D. & Hughes, E. C. (1). Glycogen synthetase in mammalian uterus. *Biochim. biophys. Acta*, **99**, 584. 1965.

Rudall, K. M. (1). The structure of epidermal protein. *Symp. on fibrous proteins*, p. 15. Soc. Dyers and Colourists. 1946.

Rüegg, J. C. (1). Die Reinigung der Myosin-ATP-ase eines glatten Muskels. *Helv. Physiol. Acta*, **15**, C33. 1957.

Rüegg, J. C. (2). The proteins associated with contraction in lamellibranch 'catch' muscle. *Proc. Roy. Soc.* B, **154**, 209. 1961.

Rüegg, J. C. (3). The possible function of invertebrate tropomyosin. *Biochem. J.*, **69**, 46P. 1958.

Rüegg, J. C. (4). On the tropomyosin-paramyosin system in relation to the viscous tone of lamellibrauch 'catch' muscle. *Proc. Roy. Soc.* B, **154**, 224. 1961.

Rüegg, J. C. (5). Actomyosin inactivation by thiourea and the nature of viscous tone in a molluscan smooth muscle. *Proc. Roy. Soc.* B, **158**, 177. 1963.

Rüegg, J. C. (6). Tropomyosin-paramyosin system and 'prolonged contraction' in a molluscan smooth muscle. *Proc. Roy. Soc.* B, **160**, 536. 1964.

Rüegg, J. C. (7). Smooth muscle tone. *Physiol. Rev.*, **51**, 201. 1971.

Rüegg, J. C. (8). Contractile mechanisms of smooth muscle. In *Aspects of motility*. Symp. Soc. Exp. Biol., **22**, p. 43. 1968.

Rüegg, J. C. (9). Oscillatory mechanism in fibrillar insect flight muscle. *Experientia*, **24**, 529. 1968.

Rüegg, J. C. (10). Tropomyosin and tonus in lamellibranch adductor muscles. *Biochim. biophys. Acta*, **35**, 278. 1959.

Rüegg, J. C. & Strassner, E. (1). Fraktionierung der Globuline aus den glatten Muskeln der Arterienwand. *Helv. physiol. Acta*, **21**, C57. 1963.

Rüegg, J. C., Strassner, E. & Schirmer, E. H. (1). Extraktion und Reinigung von Arterienactomyosin, Actin und Extraglobulin. *Biochem. Z.*, **343**, 70. 1965.

Rüegg, J. C., Straub, R. W. & Twarog, B. M. (1). Inhibition of contraction in a molluscan smooth muscle by thiourea, an inhibitor of the actomyosin contractile mechanism. *Proc. Roy. Soc.* B, **158**, 156. 1963.

Rüegg, J. C. & Stumpf, H. (1). Activation of the myofibrillar ATPase activity by extension of glycerol extracted insect fibrillar muscle. *Arch. f. ges. Physiol*, **305**, 34. 1969.

Rüegg, J. C. & Tregear, R. T. (1). Mechanical factors affecting the ATPase activity of glycerol-extracted insect fibrillar muscle. *Proc. Roy. Soc.* B, **165**, 497. 1966.

Rüegg, J. C. & Weber, H. H. (1). Kontraktionszyclus und Sperrtonus. In *Perspectives in Biology*, p. 301. Ed. L. F. Leloir & S. Ochoa. 1963.

Rufus of Ephesus (1). See Daremberg & Ruelle (1).

Rupp, J. C. & Mommaerts, W. F. H. M. (1). The scattering of light in myosin solutions. II. A determination of the molecular weight. *J. biol. Chem.*, **224**, 277. 1957.

Saad, F. M. & Kominz, D. R. (1). The N-terminal group of rabbit tropomyosin. *Arch. Biochem. Biophys.*, **92**, 541. 1961.

Sacks, J. (1). Radioactive phosphorus as a tracer in anaerobic contraction. *Am. J. Physiol.*, **129**, 227. 1940.

Sacks, J. (2). The absence of phosphate transfer in oxidative muscular contraction. *Am. J. Physiol.*, **140**, 316. 1943–44.

Sacks, J. & Altshuler, C. H. (1). Radioactive phosphorus studies on striated and cardiac muscle metabolism. *Am. J. Physiol.*, **137**, 750. 1942.

Sacks, J. & Sacks, W. C. (1). Carbohydrate changes during recovery from muscular contraction. *Amer. J. Physiol.*, **122**, 565. 1935.

Sacks, J. & Sacks, W. C. (2). Fundamental chemical changes in contracting muscle. II. Changes in lactic acid, phosphocreatine and hexosemonophosphate in mammalian muscle. *Amer. J. Physiol.*, **105**, 687. 1933.

Sacks, J. & Sacks, W. C. (3). The resynthesis of phosphocreatine after muscular contraction. *Amer. J. Physiol.*, **112**, 116. 1935.

Sacks, J. & Sacks, W. C. (4). Recovery from muscular activity and its bearing on the chemistry of contraction. *Amer. J. Physiol.*, **122**, 215. 1938.

Sacktor, B. (1). Investigations on the mitochondria of the housefly *Musca domestica*. II. Oxidative enzymes with special reference to malic oxidase. *Arch. Biochem. Biophys.*, **45**, 349. 1953.

Sacktor, B. & Cochran, D. G. (1). The respiratory metabolism of insect flight muscle. I. Manometric studies of oxidation and concomitant phosphorylation with sarcosomes. *Arch. Biochem. Biophys.*, **74**, 266. 1958.

Sacktor, B. & Hurlbut, E. C. (1). Regulation of metabolism in working muscle *in vivo*. II. Concentration of adenine nucleotides, arginine phosphate, and inorganic phosphate in insect flight muscle during flight. *J. biol. Chem.*, **241**, 632. 1966.

Sacktor, B. & Wormser-Shavit, E. (1). Regulation of metabolism in working muscle *in vivo*. I. Concentrations of some glycolytic, tricarboxylic acid cycle, and amino acid intermediates in insect flight muscle during flight. *J. biol. Chem.*, **241**, 624. 1966.

Sakov, N. E. (1). Transesterification with fructose monophosphate and mineralisation of ATP in muscle extract. *Biokhimiya*, **6**, 163. 1941.

Salter, W. T. (1). Certain physico-chemical characteristics of muscle globulin. *Proc. Soc. expt. Biol. Med.*, **24**, 116. 1926–27.

Samaha, E. J. & Gergely, J. (1). Ca++ uptake and ATPase activity of human sarcoplasmic reticulum. *J. clin. Investigation*, **44**, 1423. 1965.

Samuels, A. (1). The immuno-enzymology of muscle proteins. I. General features of myosin and 5′-adenylic acid deaminase. *Arch. Biochem. Biophys.*, **92**, 497. 1961.

Sanadi, D. R., Gibson, D. M. & Ayengar, P. (1). Guanosine triphosphate, the primary product of phosphorylation coupled to the breakdown of succinyl coenzyme A. *Biochim. biophys. Acta*, **14**, 434. 1954.

Sanadi, D. R., Gibson, D. M., Ayengar, P. & Ouellet, L. (1). Evidence for a new intermediate in the phosphorylation coupled to α-ketoglutarate oxidation. *Biochim. biophys. Acta*, **13**, 146. 1954.

Sanadi, D. R., Langley, M. & Searls, R. S. (1). α-Ketoglutarate dehydrogenase. VI. Reversible oxidation of dihydrothioctamide by diphosphopyridine nucleotide. *J. biol. Chem.*, **234**, 178. 1959.

Sanadi, D. R., Langley, M. & White, F. (1). α-Ketoglutaric dehydrogenase. VII. The role of thioctic acid. *J. biol. Chem.*, **234**, 183. 1959

Sanadi, D. R. & Littlefield, J. W. (1). Studies on α-ketoglutaric oxidase. 1. Formation of active succinate. *J. biol. Chem.*, **193**, 683. 1951.

Sanadi, D. R., Littlefield, J. W. & Bock, R. M. (1). Studies on α-ketoglutaric oxidase. II. Purification and properties. *J. biol. Chem.*, **197**, 851. 1952.

Sandow, A. (1). Studies on the latent period of muscular contraction. Method. General properties of latency relaxation. *J. cell. comp. Physiol.*, **24**, 221. 1944.

Sandow, A. (2). Latency relaxation and a theory of muscular mechanochemical coupling. *Ann. N. Y. Acad. Sci.*, **47**, 895. 1946–47.

Sandow, A. (3). Excitation-contraction coupling in muscular response. *Yale J. Biol. Med.*, **25**, 176. 1952.

Sandow, A. (4). Excitation-contraction coupling in skeletal muscle. *Pharm. Rev.*, **17**, 265. 1965.

Sandow, A. (5). Energetics of muscular contraction. In *Biophysics of Physiological and Pharmacological Actions*. Ed. A. M. Shanes. Am. Assn. for the Adv. of Sci. 1961.

Sandow, A. (6). Latency relaxation: a brief analytical review. *Med. Coll. Virginia Quarterly*, **2**, 82. 1966.

Sarkar, M., Sarkar, S. & Kozloff, L. M. (1). Tail components of T2 bacteriophage. I. Properties of the isolated contractile sheath. *Biochemistry*, **3**, 511. 1964.

Sarkar, N. K. (1). The effect of ions and adenosinetriphosphate on myosin and actomyosin. *Enzymologia*, **14**, 237. 1950–51.

Sarkar, N. K., Szent-Györgyi, A. G. & Varga, L. (1). Adenosine triphosphatase activity of the glycerol extracted muscle fibres. *Enzymologia*, **14**, 267. 1950–51.

Sarno, J., Tarendash, A. & Stracher, A. (1). Carboxyl terminal residues of myosin and heavy meromyosin. *Arch. Biochem. Biophys.*, **112**, 378. 1965.

Sarton, G. (1). Introduction to the History of Science. 3 volumes in 5 parts. Williams & Wilkins, Philadelphia (for the Carnegie Institution of Washington). 1927–47.

Sartorelli, L., Fromm, H. J., Benson, R. W. & Boyer, P. D. (1). Direct and ^{18}O-exchange measurements relevant to possible activated or phosphorylated states of myosin. *Biochemistry* (Am. chem. Soc.), **5**, 2877. 1966.

Satoh, T. (1). Über die Hydrolyse der Adenosinetriphosphorsäure durch Phosphomonoesterase und Pyrophosphatase. *J. Biochem.*, **21**, 19. 1935.

Schaechtelin, G. (1). Der Einfluss von Calcium und Natrium auf die Kontraktur des M. rectus abdominis. *Arch. f. ges. Physiol.*, **274**, 295. 1961.

Schäfer, E. A. (1). *The Essentials of Histology*. 8th edition. London, Longmans, Green & Co. 1910.

Schäffner, A. & Berl, H. (1). Über die Phosphorylierungssysteme der alkoholischen Gärung. *Z. f. physiol. Chem.*, **238**, 111. 1936.

Schapira, F. (1). Variations pathologique en rapport avec l'ontogenèse. Des formes moleculaires multiples de la lactico-dehydrogenase, de la créatinekinase et de l'aldolase. *Bull. Soc. Chim. biol.*, **49**, 1647. 1967.

Schapira, F., Schapira, G. & Dreyfus, J.-C. (1). Hypoaldolasémie chez la souris myopathique. *Comp. rend. Acad. Sci.*, **245**, 753. 1957.

Schapira, G. & Dreyfus, J.-C. (1). Biochemistry of progressive muscular dystrophy. In *Muscular Dystrophy in Man and Animals*, p. 47. Ed. by G. H. Bourne & M. N. Golarz. 1963.

Schapira, G., Dreyfus, J.-C., Schapira, F. & Kruh, J. (1). Glycogenolytic enzymes in human progressive muscular dystrophy. *Amer. J. phys. Med.*, **34**, 313. 1955.

Schatz, G. (1). Mitochondrial oxidative phosphorylation. *Angew. Chem. internat. Edit.*, **6**, 1035. 1967.

Schaub, M. C., Hartshorne, D. J. & Perry, S. V. (1). The adenosinetriphosphatase activity of desensitized actomyosin. *Biochem. J.*, **104**, 263. 1967.

Schaub, M. C., Hartshorne, D. J. & Perry, S. V. (2). Effect of tropomyosin on the calcium-activated adenosine triphosphatase of actomyosin. *Nature*, **215**, 635. 1967.

Schaub, M. C. & Perry, S. V. (1). The relaxing protein system of striated muscle. Resolution of the troponin complex into inhibitory and calcium-ion sensitising factors, and their relationship to tropomyosin. *Biochem. J.*, **115**, 993. 1969.

Schick, A. F. & Hass, G. M. (1). The properties of mammalian striated myofibrils isolated by an enzymic method. *J. exp. Med.*, **91**, 655. 1950.

Schipiloff, C. (1). Ueber die Entstehungsweise der Muskelstarre. *Centralblatt f. d. med. Wissensch.*, **20**, 291. 1882.

Schipiloff, C. & Danilewsky, A. (1). Über die Natur der anisotropen Substanzen des quergestreiften Muskeln und ihre räumliche Verteilung im Muskelbündel. *Z. f. physiol. Chem.*, **5**, 349. 1881.

Schirmer, R. H. (1). Die Besonderheiten des contractilen Proteins der Artieren. *Biochem. Z.*, **343**, 269. 1965.

Schlenk, F. & Fisher, A. (1). Studies on glutamic-aspartic transaminase. *Arch. Biochem. Biophys.*, **12**, 69. 1947.

Schlenk, F. & Snell, E. E. (1). Vitamin B_6 and transamination. *J. biol. Chem.*, **157**, 425. 1945.

Schliselfeld, L. & Bárány, M. (1). The binding of ATP to myosin. *Biochemistry* (Am. Chem. Soc.) **7**, 3206. 1968.

Schliselfeld, L. H., Conover, T. E. & Bárány, M. (1). Preparation and properties of a complex of adenosine triphosphate with myosin, actomyosin and subfragment 1. *Biochemistry* (Amer. chem. Soc.), **9**, 1133. 1970.

Schlossmann, O. (1). Über das Verhalten des Kreatingehaltes des Froschmuskels bei der Arbeit. *Z. physiol. Chem.*, **139**, 87. 1924.

Schmid, R. & Mahler, R. (1). Chronic progressive myopathy with myoglobinuria: demonstration of a glycogenolytic defect in the muscle. *J. clin. Invest.*, **38**, 2044. 1959.

Schmid, R., Robbins, P. W. & Traut, R. R. (1). Glycogen synthesis in muscle lacking phosphorylase. *Proc. nat. Acad. Sci. U.S.A.*, **45**, 1236. 1959.

Schmidt, G. (1). Über fermentative Desaminierung im Muskel. *Z. f. physiol. Chem.*, **179**, 243. 1928.

Schmidt, W. J. (1). Über die Doppelbrechung der I-Glieder der quergestreiften Myofibrillen und das Wesen der Querstreifung überhaupt. *Zeitsch. Zellforsch.*, **21**, 224. 1934.

Schmitt, F. O. (1). Tissue ultrastructure analysis: Polarised light method. In *Medical Physics*, p. 1586. Ed. O. Glasser. Chicago, Year Book Publishers. 1944.

Schönheimer, R. (1). Über eine einenartige Störung des Kohlehydrat-Stoffwechsels. *Z. physiol. Chem.*, **182**, 148. 1929.

Schramm, G. & Weber, H. H. (1). Über monodisperse Myosinlösungen. *Koll. Zeitsch.*, **100**, 242. 1942.

Schultz, P. (1). Die glatte Musculatur der Wirbeltiere. I. Ihr Bau. *Arch. f. Anat. u. Physiol.*, *Physiol. Abt*, 517. 1895.

Schultz, P. (2). Quergestreifte und längsgestreifte Muskeln. *Arch. f. Anat. u. Physiol.*, *Physiol. Abt*, 329. 1897.

Schultz, P. (3). Zur Physiologie der längsgestreiften (glatten) Musklen der Wirbelthiere. *Arch. Anat. Physiol., Physiol. Suppl.*, **1**. 1903.

Schultz, P. (4). Ueber den Einfluss der Temperaturen auf die Leistungsfähigkeit der längsgestreiften Muskeln der Wirbelthiere. *Arch. Anat. Physiol.*, p. 1. 1897.

Schwalm, H. & Cretius, K. (1). Über den Gehalt der menschlichen Uterus muskulatur an contractilen Proteinen, an wasserlöslichen Proteinen und Stromaeiweiss. *Arch. Gynaek.*, **191**, 271. 1958.

Schwann, T. (1). Ueber das Wesen des Verdauungsprocesses. *Ann. d. Physik u. Chemie*, **38**, 358. 1836.

Schwann, T. (2). Vorläufige Mittheilung, betreffend Versuche über die Weingärung und Faulniss. *Ann. Phys. Lpz.*, **41**, 184. 1837.

Schwann, T. (3). *Mikroskopische Untersuchungen über die Übereinstimmung in der Struktur und dem Wachstum der Tiere und Pflanzen*. Berlin, Sander'shen Buchhandlung, 1839.

Schwartz, A. & Oschmann, A. (1). Contribution au problème du mechanisme des contractures musculaires. Le taux de l'acide phosphorique musculaire libre dans les contractures des animaux empoisonnées par l'acide monobromacétique. *Compt. rend. Soc. Biol.*, **91**, 275. 1924.

Schwartz, A. & Oschmann, A. (2). Contribution au problème du mechanisme des contractures musculaires. Le taux de l'acide lactique musculaire dans les contractures des animaux empoisonnés par l'acide. *Compt. rend. Soc. Biol.*, **92**, 169. 1925.

Schwarzenbach, G. & Ackermann, H. (1). Komplexone V. Die Äthylendiamintetraessigsäure. *Helv. chim. Acta.*, **30**, 1798. 1947.

Scott, G. H. (1). Distribution of mineral ash in striated muscle cells. *Proc. Soc. exp. Biol. & Med.*, **29**, 349. 1931–32.

Scott, G. H. & Packer, D. M. (1). An electron microscope study of magnesium and calcium in striated muscle. *Anat. Record*, **74**, 31. 1939.

Seery, V. L., Fischer, E. H. & Teller, D. C. (1). A reinvestigation of the molecular weights of glycogen phosphorylase. *Biochemistry*, (Amer. chem. Soc.), **6**, 3315. 1967.

Segal, H. L. & Boyer, P. D. (1). The role of sulfhydryl groups in the activity of D-glyceraldehyde 3-phosphate dehydrogenase. *J. biol. Chem.*, **204**, 265. 1953.

Séguin, A. & Lavoisier, A. L. (1). *Oeuvres de Lavoisier*. Vol. II, p. 688. (Mémoires de l'Acad. des Sciences 1789, p. 185). Paris. 1862.

Seidel, D. & Weber, H. H. (1). Die Aktomyosinstruktur im glatten Muskel der Vertebraten. *Pfl. Arch. ges. Physiol.*, **297**, 1. 1967.

Seidel, J. C. (1). Comments on the existence of a soluble relaxing substance and on co-factors of the relaxation of muscle. *Fed. Proc.*, **23**, 901. 1964.

Seidel, J. C. & Gergely, J. (1). Studies on the myofibrillar adenosine triphosphatase with calcium-free adenosine triphosphate. I. The effect of ethylene-diaminetetraacetate, calcium, magnesium and adenosine triphosphate. *J. biol. Chem.*, **238**, 3648. 1963.

Seidel, J. C. & Gergely, J. (2). Studies on myofibrillar adenosine triphosphatase with calcium-free adenosine triphosphate. II. Concerning the mechanism of inhibition by the fragmented sarcoplasmic reticulum. *J. biol. Chem.*, **239**, 3331. 1964.

Seidel, J. C., Sréter, F. A. & Gergely, J. (1). Comparative studies of myofibrils, myosin and actomyosin from red and white rabbit skeletal muscle. *Biochem. biophys. Res. Com.*, **17**, 662. 1964.

Seidel, J. C., Sréter, F. A. & Gergely, J. (2). ATPase activities of myosin from red and white muscles. *Fed. Proc.*, **24**, 400. 1965.

Sekine, T. (1). The active site of myosin adenosine triphosphatase. In *Molecular biology of muscular contraction.* p. 33. Ed. S. Ebashi, F. Oosawa, T. Sekine & Y. Tonomura. Tokyo, Igaku Shoin. 1965.

Sekine, T., Barnett, L. M. & Kielley, W. W. (1). The active site of myosin adenosine triphosphatase. *J. biol. Chem.*, **237**, 2769. 1962.

Sekine, T. & Kielley, W. W. (1). The enzymic properties of N-ethylmaleimide modified myosin. *Biochim. biophys. Acta*, **81**, 336. 1964.

Sekine, T. & Yamaguchi, M. (1). Effect of ATP on the binding of N-ethylmaleimide to SH groups in the active centre of myosin ATPase. *J. Biochem.*, **63**, 196.

Sekiya, K., Mii, S., Takeuchi, K. & Tonomura, Y. (1). The optical rotatory dispersion of myosin A. V. Re-examination of changes in the optical rotatory power of myosin caused by adenosine triphosphate, inorganic pyrophosphate and p-chlorimercuribenzoate. *J. Biochem.*, **59**, 584. 1966.

Sekiya, K. & Tonomura, Y. (1). Change in ultraviolet absorption spectrum of H-meromyosin induced by its binding with substrate and competitive inhibitor. *J. Biochem.*, **61**, 787. 1967.

Sekuzu, I., Orii, Y. & Okunuki, K. (1). Studies on cytochrome C_1. I. Isolation, purification and properties of cytochrome C_1 from heart muscle. *J. Biochem.*, **48**, 214. 1960.

Selby, C. C. & Bear, R. S. (1). The structure of the actin-rich filaments of muscles according to X-ray diffraction. *J. biophys. biochem. Cyt.*, **2**, 71. 1956.

Seliger, H. H. & McElroy, W. D. (1). Spectral emission and quantum yield of firefly bioluminescence. *Arch. Biochem. Biophys.*, **88**, 136. 1960.

Selwyn, M. J. (1). *Studies on mitochondrial ATPase.* Ph.D. Thesis. University of Cambridge. p. 160. 1963.

Seravin, L. N. (1). New techniques for the preparation of contracting models from the stalks of *Carchesium polypinium. Biokhimiya*, **28**, 493. 1963.

Seraydarian, K., Briskey, E. J. & Mommaerts, W. F. H. M. (1). The modification of actomyosin by α-actinin. I. A survey of experimental conditions. *Biochim. biophys. Acta*, **133**, 399. 1967.

Seraydarian, K., Mommaerts, W. F. H. M. & Wallner, A. (1). The amount and compartmentalisation of adenosine diphosphate in muscle. *Biochim. biophys. Acta*, **65**, 443. 1962.

Seraydarian, K., Mommaerts, W. F. H. M., Wallner, A. & Guillory, R. J. (1). An estimation of the true inorganic phosphate content of frog sartorius muscle. *J. biol. Chem.*, **236**, 2071. 1961.

Severin, S. E. (1). Proc. 6th Inter. Congr. Biochem. Vol. 33. p. 45. New York. 1964.

Severin, S. E., Georgievskaya & Ivanow, V. I. (1). Synthesis and some properties of phosphocarnosine. *Biokhimiya*, **12**, 35. 1947.

Sexton, A. W. & Gersten, J. W. (1). Isometric tension differences in fibers of red and white muscle. *Science*, **157**, 199. 1967.

Shanes, A. M. & Bianchi, C. P. (1). Radiocalcium release by stimulated and potassium-treated sartorius muscles of the frog. *J. gen. Physiol.*, **43**, 481. 1960.

Shchesno, T. Y. (1). Effect of strenuous exercise on content of nucleic acids and other compounds in functionally distinct rabbit muscles. *Fed. Proc.* **24**, Trans. Suppl., T90. 1965.

Shelanski, M. L. & Taylor, E. W. (1). Properties of the protein sub-unit of central-pair and outer doublet microtubules of sea-urchin flagella. *J. Cell Biol.*, 38, 304. 1968.

Shelley, H. J. (1). Glycogen reserves and their changes at birth and in anoxia. *Brit. med. Bull.*, 17, 137. 1961.

Shen, S.-C. (1). The interaction of myosin and adenosinetriphosphatase. *Ann. N.Y. Acad. Sci.*, 47, 875. 1946–47.

Sheng, P.-K. & Tsao, T.-C. (1). A comparative study of nucleotropomyosins from different sources. *Scientia Sinica*, 4, 157. 1955.

Sheng, P.-K., Tsao, T.-C. & Peng, C.-M. (1). The electrophoretic behaviour of nucleotropomyosins from different sources and the nuclear base composition of their pentose nucleic acid components. *Scientia Sinica*, 5, 675. 1956.

Shifrin, S. & Kaplan, N. O. (1). Co-enzyme binding. *Adv. Enzymol.*, 22, 337. 1960.

Shipp, J. C., Opie, L. H. & Challoner, D. (1). Fatty acid and glucose metabolism in the perfused heart. *Nature*, 189, 1018. 1961.

Shoenberg, C. F. (1). An electron microscope study of smooth muscle in pregnant uterus of the rabbit. *J. biophys. biochem. Cyt.*, 4, 609. 1958.

Shoenberg, C. F. (2). Some electron microscope observations on the contraction mechanism in vertebrate smooth muscle. In *Intern. Congr. Electron Microscopy*, p. M.8. Ed. by S. S. Breeze. New York & London, Academic Press. 1962.

Shoenberg, C. F. (3). Contractile proteins of vertebrate smooth muscle. *Nature*, 206, 526. 1965.

Shoenberg, C. F. (4). An electron microscope study of the influence of divalent ions on myosin filament formation in chicken gizzard extracts and homogenates. *Tissue and Cell*, 1, 83. 1969.

Shoenberg, C. F. (5). A study of myosin filaments in extracts and homogenates of vertebrate smooth muscle. Int. Symp. Biochemistry of the vascular wall, Fribourg, 1968. *Angiologica*, 6, 233. 1969.

Shoenberg, C. F., Rüegg, J. C., Needham, D. M., Schirmer, R. H. & Nemetchek-Gansler, H. (1). A biochemical and electron microscope study of the contractile proteins of vertebrate smooth muscle. *Biochem. Z.* 345, 255. 1966.

Sibley, J. A. & Lehninger, A. L. (1). Aldolase in the serum and tissues of tumour-bearing animals. *J. nat. Cancer Inst.*, 9, 303. 1948–49.

Siebold, C. T. von (1). *Lehrbuch der vergleichenden Anatomie der wirbellos Tiere.* Berlin, Veit & Comp. 1848.

Simard-Duquesne, N. & Couillard, P. (1). Amoeboid movement II. Research on contractile proteins in *Amoeba proteus*. *Exp. Cell Res.*, 28, 92. 1962.

Simard-Duquesne, N. & Couillard, P. (2). Amoeboid movement. I. Reactivation of glycerinated models of *Amoeba proteus* with adenosinetriphosphate. *Exp. Cell Res.*, 28, 85. 1962.

Simmons, N. S., Cohen, C., Szent-Györgyi, A. G., Wetlaufer, D. B. & Blout, E. R. (1). A conformation-dependent Cotton effect in α-helical polypeptides and proteins. *J. Am. Chem. Soc.*, 83, 4767. 1961.

Simon, E. J., Gross, C. S. & Lessell, I. M. (1). Turnover of muscle and liver proteins in mice with muscular dystrophy. *Arch. Biochem. Biophys.*, 96, 41. 1962.

Simpson, F. O. & Oertelis, S. J. (1). The fine structure of sheep myocardial cells; sarcolemmal invaginations and the transverse tubular system. *J. Cell Biol.*, 12, 91. 1962.

Singer, C. (1). *A Short History of Biology.* Oxford. 1931.

Singer, C. (2). *Galen 'On anatomical procedures'*. For the Wellcome Historical Medical Museum. Oxford, Oxford University Press. 1956.

Singer, C. & Underwood, E. A. (1). *A Short History of Medicine*. Second edition. Oxford, Clarendon Press. 1962.

Singer, T. P. (1). Flavoprotein dehydrogenases of the respiratory chain. In *Comprehensive Biochemistry*, Vol. 14, 127. Ed. M. Florkin & E. H. Stotz. Amsterdam, London & New York, Elsevier Publishing Co. 1966.

Singer, T. P. & Barron, E. S. (1). Effect of sulfhydryl reagents on adenosinetriphosphatase activity of myosin. *Proc. Soc. exp. Biol. Med.*, 56, 120. 1944.

Singer, T. P., Kearney, E. B. & Bernath, P. Studies on succinic dehydrogenase. II. Isolation and properties of the dehydrogenase from beef heart. *J. biol. Chem.*, 223, 599. 1956.

Siperstein, M. D. (1). Glycolytic pathways. Their relation to the synthesis of cholesterol and fatty acids. *Diabetes*, 7, 181. 1958.

Sjöstrand, F. S. (1). The connections between A and I band filaments in striated frog muscle. *J. ultrastr. Res.*, 7, 225. 1962.

Sjöstrand, F. S. & Andersson, E. (1). The ultrastructure of skeletal myofilaments at various conditions of shortening. *Exp. Cell Res.*, 11, 493. 1956.

Slater, E. C. (1). A respiratory catalyst required for the reduction of cytochrome *c* by cytochrome *b*. *Biochem. J.*, 45, 14. 1949.

Slater, E. C. (2). Oxidative phosphorylation. In *Comprehensive Biochemistry*, Vol. 14, 327. Ed. M. Florkin & E. H. Stotz. Amsterdam, London & New York, Elsevier Publishing Co. 1966.

Slater, E. C. (3). Phosphorylation coupled with the reduction of cytochrome *c* by α-ketoglutarate in heart muscle granules. *Nature*, 166, 982. 1950.

Slater, E. C. (4). Phosphorylation coupled with the oxidation of α-ketoglutarate by heart-muscle sarcosomes. 3. Experiments with ferricytochrome *c* as hydrogen acceptor. *Biochem. J.*, 59, 392. 1955.

Slater, E. C. (5). Mechanism of phosphorylation in the respiratory chain. *Nature*, 172, 975. 1953.

Slater, E. C. (6). Mechanism of oxidative phosphorylation. *Rev. pure & applied Chem.*, 8, 221. 1958.

Slater, E. C. (7). The respiratory chain and oxidative phosphorylation: some of the unsolved problems. In *Biochemistry of Mitochondria*, p. 1. Ed. E. C. Slater, Z. Kaninga & L. Wojtczak. London & New York, Academic Press. 1967.

Slater, E. C. (8). An evaluation of the Mitchell hypothesis of chemiosmotic coupling in oxidative and photo-synthetic phosphorylation. *Eur. J. Biochem.*, 1, 317. 1967.

Slater, E. C. & Hülsmann, W. C. (1). Control of rate of intracellular respiration. In Ciba Foundation Symposium on the *Regulation of Cell Metabolism*, p. 58. Ed. G. E. W. Wolstenholme & C. M. O'Connor. 1959.

Slayter, H. S. & Lowey, S. (1). Substructure of the myosin molecule as visualized by electron microscopy. *Proc. nat. Acad. Sci. U.S.A.*, 58, 1611. 1967.

Small, P. A., Harrington, W. F. & Kielley, W. W. (1). The electrophoretic homogeneity of the myosin sub-units. *Biochim. biophys. Acta.*, 49, 1961.

Smith, B. (1). Changes in the enzymes chemistry of skeletal muscle during experimental denervation and reinervation. *J. Neurol., Neurosurg. & Psychiat.*, 28, 99. 1965.

Smith, D. S. (1). Reticular organizations within the striated muscle cell. An historical survey of light microscope studies. *J. biophys. biochem. Cyt.*, 10, Suppl. 61. 1961.

Smith, D. S. (2). The structure of insect fibrillar flight muscle. A study made with special reference to the membrane systems of the fibre. *J. biophys. biochem. Cyt.*, **10**, Suppl. 123. 1961.

Smith, D. S. (3). The organisation and function of the sarcoplasmic reticulum and T system of muscle cells. *Progr. Biophys. & mol. Biol.*, **16**, 107. 1966.

Smith, E. C. See Bate Smith, E. C.

Smith, E. E., Taylor, P. M. & Whelan, W. J. (1). Enzymic processes in glycogen metabolism. In *Carbohydrate metabolism and its disorders*, ed. F. Dickens, P. J. Randle & W. J. Whelan, W. J. London & New York. 1968.

Smoller, M. & Fineberg, R. A. (1). Studies of myosin in hereditary muscular dystrophy in mice. *J. clin. Invest.*, **44**, 615. 1965.

Smythe, C. V. & Gerischer, W. (1). Über die Vergärung von Hexosemonophosphorsäure. *Biochem. Z.*, **260**, 414. 1933.

Snell, E. E. (1). The vitamin B_6 group. V. The reversible interconversion of pyridoxal and pyridoxamine by transamination reactions. *J. Amer. chem. Soc.*, **67**, 194. 1945.

Snell, E. E., Brown, G. M., Peters, V. J., Craig, J. A., Wittle, E. L., Moore, J. A., McGlohan, V. M. & Bird, O. D. (1). Chemical nature and synthesis of the *Lactobacillus bulgaricus* factor. *J. Amer. chem. Soc.*, **72**, 5349. 1950.

Snellman, O. & Erdös, T. (1). Ultracentrifugal analysis of crystallised myosin. *Biochim. biophys. Acta.*, **2**, 650. 1948.

Snellman, O. & Erdös, T. (2). Ultracentrifugal studies of F-actomyosin. *Biochim. biophys. Acta.*, **3**, 523. 1949.

Snoswell, A. M. (1). The mechanism of the reduction of mitochondrial diphosphorpyridine nucleotide by succinate in rabbit heart sarcosomes. *Biochim. biophys. Acta.*, **52**, 216. 1961.

Sonnenblick, E. H., Spiro, D. & Cottrell, T. S. (1). Fine structural changes in heart muscle in relation to the length-tension curve. *Proc. Nat. Acad. Sci. U.S.A.*, **49**, 193. 1963.

Sotavalta, O. (1). Recordings of the high wing-stroke and thoracic vibration frequency in some midges. *Biol. Bull.*, **104**, 439. 1953.

Speidel, C. C. (1). Studies of living muscles. II. Histological changes in single fibres of striated muscle during contraction and clotting. *Amer. Journ. Anat.*, **65**, 471. 1939.

Speyer, J. F. & Dickman, S. R. (1). On the mechanism of action of aconitase. *J. biol. Chem.*, **220**, 193. 1956.

Spicer, S. S. (1). Gel formation caused by adenosinetriphosphate in actomyosin solutions. *J. biol. Chem.*, **190**, 257. 1951.

Spicer, S. S. & Bowen, W. J. (1). Reactions of inosine- and adenosine-triphosphates with actomyosin and myosin. *J. biol. Chem.*, **188**, 741. 1951.

Spicer, S. S. & Gergely, J. (1). Studies on the combination of myosin with actin. *J. biol. Chem.*, **188**, 179. 1951.

Spiro, D. (1). The filamentous fine structure of striated muscle at different stages of shortening. *Exp. Cell Res.*, **10**, 562. 1956.

Spiro, D. (2). Ultrastructure of heart muscle. *Trans. N. Y. Acad. Sci.* (2) **24**, 879. 1961–2.

Spitzer, W. (1). Die Zuckerzerstörende Kraft des Blutes und der Gewebe. *Arch. f. ges. Physiol.*, **60**, 303. 1895.

Spronck, A. C. (1). Evolution temporelle de l'hydrolyse de la phosphocreatine et de la synthèse d'hexosediphosphate pendant et après cinq secousses simple à

0° C, chez le sartorius de *Rana temporaria* intoxiqué par l'acide m onoiodo-acétique. *Arch. internat. Physiol. Biochim.*, **73**, 241. 1965.

Sréter, F. A. (1). Comparative studies on white and red muscle fr actions. *Fed. Proc.*, **23**, 930. 1964.

Sréter, F. A. & Gergely, J. (1). Comparative studies of the Mg acti vated ATPase activity and Ca uptake of fractions of white and red muscle homogenates. *Biochem. biophys. Res. Com.*, **16**, 438. 1964.

Sréter, F. A., Ikemoto, N. & Gergely, J. (1). Studies on the fragmented sarco-plasmic reticulum of normal and dystrophic mouse muscle. *Excerpta Medica International Congress Series* No. **147**, p. 289. 1966.

Sréter, F. A., Seidel, J. C. & Gergely, J. (1). Studies on the myosin from red and white skeletal muscles of the rabbit. *J. biol. Chem.*, **241**, 5772. 1966.

Stagni, N. & de Bernard, B. (1). Lysosomal enzyme activity in rat and beef skeletal muscle. *Biochim. biophys. Acta.*, **170**, 129. 1968.

Stare, F. J. & Baumann, C. A. (1). The effect of fumarate on respiration. *Proc. Roy. Soc. B*, **121**, 338. 1936–37.

Stasny, J. T. & Crane, F. L. (1). The effect of sonic oscillation on the structure and function of beef heart mitoc hondria. *J. Cell Biol.*, **22**, 49. 1964.

Stave, U. (1). Age-dependent changes of metabolism. I. Studies of enzyme patterns of rabbit organs. *Biologia Neonatorum*, **6**, 128. 1964.

Steiger, G. J. & Rüegg, J. C. (1). Energetics and 'efficiency' in the isolated contractile machinery of an insect fibrillar muscle at various frequencies of oscillation. *Arch. f. ges. Physiol.*, **307**, 1. 1969.

Steiner, R. F., Laki, K. & Spicer, S. (1). Light scattering studies on some muscle proteins. *J. Polymer Sci.*, **8**, 23. 1952.

Stella, G. (1). The combination of carbon dioxide with muscle: its heat of neutrali-sation and its dissociation curve. *J. Physiol.*, **68**, 49. 1929–30.

Stensen, N. (1). *De musculis et glandulis observationem specimen.* Amsterdam, Petrum le Grand. 1664. See also Maar (1).

Stensen, N. (2). *Elementorum myologiae specimen, sen musculi descripto geometrica.* Florence 1667. See also Maar (1).

Stephens, R. E. (1). Fluorescent antimyosin labelling in locally contracted chick myofibrils. *Biol. Bull.*, **127**, 390. 1964.

Stephens, R. E. (2). Analysis of muscle contraction by ultraviolet micro-beam disruption of sarcomere structure. *J. Cell Biol.*, **25**, No. 2, Pt. 2, 129. 1965.

Stephens, R. E. (3). On the structural protein of flagellar outer fibres. *J. mol. Biol.*, **32**, 277. 1968.

Stephens, R. E. (4). Reassociation of micro-tubule protein. *J. mol. Biol.*, **33**, 517. 1968.

Stephens, R. E., Renaud, F. L. & Gibbons, I. R. (1). Guanine nucleotide asso-ciated with the protein of the outer fibers of flagella and cilia. *Science*, **156**, 1606. 1966.

Stern, J. R., Ochoa, S. & Lynen, F. (1). Enzymatic synthesis of citric acid. V. Reaction of acetyl CoA. *J. biol. Chem.*, **198**, 313. 1952.

Stern, J. R., Shapiro, B., Stadtman, E. R. & Ochoa, S. (1). Enzymatic synthesis of citric acid. III. Reversibility and mechanism. *J. biol. Chem.*, **193**, 703. 1951.

Stintzing, R. (1). Untersuchungen über die Mechanik der physiologischen Kohlensäurebildung. *Arch. f. ges. Physiol.*, **18**, 388. 1878.

Stracher, A. (1). Disulfide-sulfhydryl interchange studies on myosin A. *J. biol. Chem.*, **239**, 1118. 1964.

Stracher, A. (2). Characterisation of the sulfhydryl residues involved in the binding of ATP and actin to myosin A. In *Muscle*, p. 85. Ed. W. M. Paul, E. E. Daniel, C. M. Kay & G. Monckton. Oxford, Pergamon Press. 1965.

Stracher, A. (3). Evidence for histidine at the active site of myosin A. *J. biol. Chem.*, **240**, PC958. 1965.

Straub, F. B. (1). Actin. *Stud. Inst. med. Chem. Univ. Szeged*, II, 3. 1943.

Straub, F. B. (2). Actin II. *Stud. Inst. med. Chem. Univ. Szeged*, III, 23. 1943.

Straub, F. B. (3). On the specificity of the ATP effect. *Stud. Inst. med. Chem. Univ. Szeged*, III, 38. 1944.

Straub, F. B. (4). Isolation and properties of a flavo-protein from heart muscle tissue. *Biochem. J.*, **33**, 787. 1939.

Straub, F. B. & Feuer, G. (1). Adenosinetriphosphate the functional group of actin. *Biochim. biophys. Acta.*, **4**, 455. 1950.

Ströbel, G. (1). Doppelbrechungsänderungen bei der aktiven Kontraktion des Fasermodells (aus Kaninchen-Psoas). *Z. f. Naturforsch.*, 7b, 102. 1952.

Strohman, R. C. & Samorodin, A. J. (1). The requirements for adenosinetriphosphate binding to globular actin. *J. biol. Chem.*, **237**, 363. 1962.

Stubbs, S. St. G. & Blanchaer, M. C. (1). Glycogen phosphorylase and glycogen synthetase activity in red and white skeletal muscle of the guinea pig. *Can. J. Biochem.*, **43**, 463. 1965.

Stübel, H. (1). Die Ursache der Doppelbrechung der quergestreiften Muskelfaser. *Pfl. Arch. f. ges. Physiol.*, **201**, 629. 1923.

Stübel, H. & Liang, T.-Y. (1). Untersuchungen über die Veränderungen der Doppelbrechung bei verschiedenen Starrezustanden des Muskels. *Ch. J. Physiol.*, 2, 139. 1928.

Sumner, J. B. & Myrbäck, K. (1). *The Enzymes*. First edition. New York, Academic Press. 1951.

Sundermeyer, J. F., Gudbjarnason, S., Wendt, V. E., den Bakker, P. B. & Bing, R. J. (1). Myocardial metabolism in progressive muscular dystrophy. *Circulation*, **24**, 1348. 1961.

Suranyi, J. (1). Über die Zusammenhang von Spannung und Milchsäurebildung bei der tetanischen Kontraktion des Musckels. *Arch. f. ges. Physiol.*, **214**, 228. 1926.

Sutherland, E. W., Cohn, M., Posternak, T. & Cori, C. F. (1). The mechanism of the phosphoglucomutase reaction. *J. biol. Chem.*, **180**, 1285. 1949.

Sutherland, E. W., Colowick, S. P. & Cori, C. F. (1). The enzymatic conversion of glucose-6-phosphate to glycogen. *J. biol. Chem.*, **140**, 309. 1941.

Sutherland, E. W. & Cori, C. F. (1). Effect of hyperglycemic-glycogenolytic factor and epinephrine on liver phosphorylase. *J. biol. Chem.*, **188**, 531. 1951.

Sutherland, E. W., Posternak, T. & Cori, C. F. (1). Mechanism of the phosphoglyceric mutase reaction. *J. biol. Chem.*, **181**, 153. 1949.

Sutherland, E. W. & Robison, G. A. (1). The role of cyclic 3′,5′ AMP in response to catecholamines and other hormones. *Pharmac. Rev.*, **18**, 145. 1966.

Sutherland, E. W. & Wosilait, W. D. (1). Inactivation and activation of liver phosphorylase. *Nature*, **175**, 169. 1955.

Swammerdam, J. (1). The History of Insects, Pt. II, in *Biblia Naturae*. Trans. T. Floyd. Seyffert, London. 1758.

Swan, R. C. (1). Univalent ion fluxes in resting skeletal muscle fibres. In *Biophysics of Physiological and Pharmacological Actions*. Ed. A. M. Shanes. Washington, Am. Ass. Adv. Sci. 1961.

Swanson, J. R. & Yount, R. G. (1). A manganese stimulated, nucleotide dependent ¹⁸O-inorganic phosphate exchange reaction catalyzed by heavy meromyosin. *Biochem. biophys. Res. Com.*, **19**, 765. 1965.

Syrový, I. & Gutmann, E. (1). Metabolic differentiation of the anterior and posterior latissimus dorsi of the chick during development. *Nature*, **213**, 937. 1967.

Syrový, I., Hájek, I. & Gutmann, E. (1). Factors affecting the proteolytic activity in denervated muscle. *Physiol. Bohemoslov.*, **15**, 7. 1966.

Syrový, I., Hájek, I. & Gutmann, E. (2). Degradation of proteins of M. latissimus dorsi anterior and posterior of the chicken. *Physiol. Bohemoslov.*, **14**, 17. 1965.

Szent-Györgyi, A. (1). The contraction of myosin threads. *Stud. Inst. med. Chem. Univ. Szeged*, **1**, 17. 1943.

Szent-Györgyi, A. (2). Studies on muscle. *Acta physiol. Scand.*, **9**, Suppl. 25, p. 25. 1945.

Szent-Györgyi, A. (3). Free energy relations and contraction of actomyosin. *Biol. Bull.*, **96**, 140. 1949.

Szent-Györgyi, A. (4). Observations on actomyosin. *Stud. Inst. med. Chem. Univ. Szeged*, III, 86. 1944.

Szent-Györgyi, A. (5). *Chemistry of muscular contraction*. New York, Acad. Press. Inc. 1951.

Szent-Györgyi, A. (6). Thermodynamics of muscle. *Enzymologia*, **14**, 177. 1950.

Szent-Györgyi, A. (7). Contraction and the chemical structure of the muscle fibril. *J. coll. Sci.*, **1**, 1. 1946.

Szent-Györgyi, A. (8). Über den Mechanismus des Succin- und Paraphenylendiaminoxydation. Ein Betrag zur Theorie der Zellatmung. *Biochem. Z.*, **150**, 195. 1924.

Szent-Györgyi, A. (9). Über die Bedeutung der Fumarsäure für die tierische Gewebsatmung Einleitung. Übersicht, Methoden. *Z. f. physiol. Chem.*, **236**, 1. 1935.

Szent-Györgyi, A. (10). Actomyosin. *Studies on Muscle*, from the *Inst. of med. Chem., Univ. of Szeged*. p. 39. 1944.

Szent-Györgyi, A. (11). *Nature of Life*. New York, Academic Press. 1948.

Szent-Györgyi, A. & Banga, I. (1). Adenosinetriphosphatase. *Science*, **93**, 158. 1941.

Szent-Györgyi, A. G. (1). A new method for the preparation of actin. *J. biol. Chem.*, **192**, 361. 1951.

Szent-Györgyi, A. G. (2). Meromyosins, the subunits of myosin. *Arch. Biochem. Biophys.*, **42**, 305. 1953.

Szent-Györgyi, A. G. (3). The reversible depolymerisation of actin by potassium iodide. *Arch. Biochem. Biophys.*, **31**, 97. 1951.

Szent-Györgyi, A. G. (4). In 'Symposium on Muscle', p. 244, (discussion.) Akadémiai Kiadó, Budapest. 1968.

Szent-Györgyi, A. G., Benesch, R. E. & Benesch, R. (1). Cysteine and cystine content of muscle protein fractions. In *Sulfur in Proteins*, p. 291. Ed. R. Benesch, R. E. Benesch, P. D. Boyer, I. M. Klotz, W. R. Middlebrook & D. R. Schwartz. New York & London, Academic Press. 1959.

Szent-Györgyi, A. G. & Borbiro, M. (1). Depolymerisation of light meromyosin by urea. *Arch. Biochem. Biophys.*, **60**, 186. 1956.

Szent-Györgyi, A. G. & Cohen, C. (1). Role of proline in polypeptide chain configuration of proteins. *Science*, **126**, 697. 1957.

Szent-Györgyi, A. G., Cohen, C. & Philpott, D. E. (1). Light meromyosin Fraction I: a helical molecule from myosin. *J. mol. Biol.*, **2**, 133. 1960.

Szent-Györgyi, A. G., Holtzer, H. & Johnson, W. H. (1). Localization of myosin in chick myofibrils determined at various sarcomere lengths with the aid of antibody. In *Biochemistry of muscle contraction*, p. 354. Ed. J. Gergely. London, J. & A. Churchill Ltd. 1964.

Szent-Györgyi, A. G. & Johnson, W. H. (1). An alternative theory for contraction of striated muscles. In *Biochemistry of muscle contraction*, p. 485. Ed. J. Gergely. London, J. & A. Churchill Ltd. 1964.

Szent-Györgyi, A. G., Mazia, D. & Szent-Györgyi, A. (1). On the nature of the cross-striation of body muscle. *Biochim. biophys. Acta*, **16**, 339. 1955.

Szent-Györgyi, A. G. & Prior, G. (1). Exchange of nucleotide bound to actin in superprecipitated actomyosin. *Fed. Proc.*, **24**, 598. 1965.

Szent-Györgyi, A. G. & Prior, G. (2). Exchange of adenosine diphosphate bound to actin in superprecipitated actomyosin and contracting myofibrils. *J. mol. Biol.*, **15**, 515. 1966.

Takane, R. (1). Über Atmung und Kohlehydratumsatz tierische Gewebe. II. Atmung und Kohlehydratumsatz in Leber und Muskel des Warmblüters. *Biochem. Z.*, **171**, 403. 1926.

Takeuchi, T. (1). Histochemical demonstration of phosphorylase. *J. Histochem. Cytochem.*, **4**, 84. 1956.

Takeuchi, T. (2). Histochemical demonstration of branching enzyme (amylo-1,4 → 1,6-transglucosidase. *J. Histochem. Cytochem.*, **6**, 208. 1958.

Takeuchi, T. & Glenner, G. (1). Histochemical demonstration of a pathway for polysaccharide synthesis from uridine diphosphoglucose. *J. Histochem. Cytochem.*, **8**, 227. 1960.

Talke, H., Arese, R. & Hohorst, H. J. (1). Etudes comparées de l'équilibre entre phosphagène et ATP dans le différents tissus. *Comp. rend. Soc. Biol.*, **46**, 219. 1964.

Tan, P.-H., Sun, C. J. & Lin, N. C. (1). Properties of some chemically and enzymically modified tropomyosins. 2. The role of sulphydryl groups in the polymerisation of tropomyosin. *Acta biochim. Sinica*, **1**, 287. 1958.

Tan, P.-H. & Tsao, T.-C. (1). Properties of some chemically and enzymically modified tropomyosins. *Scientia Sinica*, **6**, 1049. 1958.

Tappel, A. L., Zalkin, H., Caldwell, K. A., Desai, I. D. & Shibko, S. (1). Increased lysosomal enzymes in genetic muscular dystrophy. *Arch. Biochem. Biophys.*, **96**, 340. 1962.

Tarui, S., Okuno, G., Ikura, Y., Tanaka, T., Suda, M. & Nishikawa, M. (1). Phosphofructokinase deficiency in skeletal muscle. *Biochem. biophys. Res. Com.*, **19**, 517. 1965.

Taxi, J. (1). Etude au microscope électronique de l'innervation du muscle lissé intestinal, comparée à celle de quelques autres muscles lisses de mammifères. *Arch. Biol.*, **75**, 301. 1964.

Taylor, F. Sherwood (1). *The Alchemists*. London, Heinemann. 1951.

Taylor, E. W., Lymn, R. W. & Moll, G. (1). Myosin-product complex and its effect on the steady-state rate of nucleotide triphosphate hydrolysis. *Biochemistry* (Am. chem. Soc.), **9**, 2984. 1970.

Teich, M. (1). On the historical foundations of modern biochemistry. *Clio Medica*, **1**, 41. 1965.

Temkin, O. (1). The classical roots of Glisson's doctrine of irritation. *Bull. Hist. of Med.*, **38**, 297. 1964.

Theorell, H. (1). Krystallinisches Myoglobin. 1. Kristallisieren und Reinigung des Myoglobin sowie vorläufige Mitteliung über sein Molekulargewicht. *Biochem. Z.*, **252**, 1.

Theorell, H. (2). Reindarstellung der Wirkungsgruppe des gelben Ferments. *Biochem. Z.*, **275**, 344. 1935.

Thin, G. (1). On the minute anatomy of muscle and tendon, and some notes regarding the structure of the cornea. *Edinburgh med. J.*, **20**, Pt. 1, p. 238. 1874.

Thoai, N.-V. (1). Sur la taurocyamine et la glycocyamine phosphokinases. *Bull. Soc. Chim. biol.*, **39**, 197. 1957.

Thoai, N.-V., di Jeso, F. & Robin, Y. (1). Sur l'isolement et la synthèse d'une nouvelle guanidine monosubstitutueé biologique, l'acide guanidinoéthyl-methyl-phosphorique, et sur le phosphagène correspondant, l'acide *N'*-phosphoryl-guanidinoethyl-methyl-phosphorique. *Comp. rend. Acad. Sci.*, **256**, 4525. 1963.

Thoai, N.-V. & Pradel, L.-A. (1). Taurocyamine phosphokinase. Purification et caractères généraux. *Bull. Soc. Chim. biol.*, **44**, 641. 1962.

Thoai, N.-V. & Robin, Y. (1). Metabolisme des dérivés guanidylés. IV. Sur une nouvelle guanidine monosubstituée biologique: l'ester guanidoéthylserylphosphorique (lombricine) et le phosphagène correspondant. *Biochim. biophys. Acta*, **14**, 76. 1954.

Thoai, N.-V., Robin, Y. & Pradel, L.-A. (1). Hypotaurocyamine phosphokinase. Compariason avec la taurocyamine phosphokinase. *Biochim. biophys. Acta*, **73**, 437. 1963.

Thoai, N.-V., & Roche, J. (1). Sur la biochimie comparée des phosphagènes et leur répartition chez les animaux. *Biol. Rev.*, **39**, 214. 1964.

Thoai, N.-V., Roche, J., Robin, Y. & Thiem, N.-V. (1). Sur la présence de la glycocyamine (acide guanidylacétique), de la taurocyamine (guanidyltaurine) et des phosphagène correspondants dans les muscles de vers marins. *Biochim. biophys. Acta*, **11**, 593. 1953.

Thoai, N.-V., Roche, J., Robin, Y. & Thiem, N.-V. (2). Sur le phosphagène de *Lumbricus terrestris. Comp. rend. Soc. Biol.*, **147**, 1670. 1953.

Thompson, C. M. & Wolpert, L. (1). The isolation of motile protoplasm from *Amoeba proteus. Exp. Cell Res.*, **32**, 156. 1963.

Thomson, W. H. S. & Guest, K. E. (1). A trial of therapy by nucleosides and nucleotides in muscular dystrophy. *J. Neurol. Neurosurg. Psychiat.*, **26**, 111. 1963.

Thunberg, T. (1). Zur Kenntnis der Einwirkung tierische Gewebe auf Methylenblau. *Skand. Arch. Physiol.*, **35**, 163. 1917.

Thunberg, T. (2). Zur Kenntnis des intermediären Stoffwechsels und der dabei wirksamen Enzyme. *Skand. Arch. Physiol.*, **40**, 1. 1920.

Thunberg, T. (3). The hydrogen-activating enzymes of the cells. *Quart. Rev. Biol.*, **5**, 318. 1930.

Tiegs, O. W. (1). Studies on plain muscle. *Austr. J. exp. Biol. med. Sci.*, **1**, 131. 1925.

Tiegs, O. W. (2). The nerve net of plain muscle, and its relation to automatic rhythmic movements. *Austr. J. exp. Biol. med. Sci.*, **2**, 157. 1926.

Tiegs, O. W. (3). The function of creatine in muscular contraction. *Austr. J. exp. Biol. med. Sci.*, **2**, 1. 1925–26.

Tiegs, O. W. (4). On the mechanism of muscular action. *Austr. J. exp. Biol. med. Sci.*, **1**, 11. 1924.

Tisdale, H. D., Wharton, D. C. & Green, D. E. (1). Studies of the electron transport system. LIII. The isolation and composition of succinic-coenzyme Q reductase and succinic-cytochrome c reductase. *Arch. Biochem. Biophys.*, 102, 114. 1963.

Tobin, R. B. & Coleman, W. A. (1). A family study of phosphorylase deficiency in muscle. *Ann. int. Med.*, 62, 313. 1965.

Todrick, A. & Walker, E. (1). Sulphydryl groups in proteins. *Biochem. Journ.*, 31, 292. 1937.

Toenniessen, E. & Brinkmann, E. (1). Über den oxydativen Abbau der Kohlehydrate im Säugetiermuskel, ins besondere über die Bildung von Bernsteinsäure aus Brenztraubensäure. *Z. f. physiol. Chem.*, 187, 137. 1930.

Tokiwa, T. & Tonomura, Y. (1). The pre-steady state of the myosin-adenosine triphosphate system. II. Initial rapid absorption and liberation of hydrogen followed by a stopped-flow method. *J. Biochem.*, 57, 616. 1965.

Tokuyama, H., Kubo, S. & Tonomura, Y. (1). Molecular properties of fraction S-1 from a trypsin digest of myosin. *Biochem. Z.*, 345, 57. 1966.

Tomita, T. (1). Current spread in the smooth muscle of the guinea-pig vas deferens. *J. Physiol.*, 189, 163. 1967.

Tonomura, Y., Appel, P. & Morales, M. (1). On the molecular weight of myosin. II. *Biochemistry* (Amer. chem. Soc.), 5, 515. 1966.

Tonomura, Y. & Kanazawa, T. (1). Formation of a reactive myosin-phosphate complex as a key reaction in muscle contraction. *J. biol. Chem.*, 240, PC4110. 1965.

Tonomura, Y., Kanazawa, T. & Sekiya, K. (1). Phosphorylation and transconformation of myosin by adenosine triphosphate. *Ann. Rep. Sci. Works. Fac. Sci. Osaka Univ.*, 12, 1. 1964.

Tonomura, Y. & Kitagawa, S. (1). The intial phase of actomyosin-triphosphatase. *Biochim. biophys. Acta.*, 26, 15. 1957.

Tonomura, Y. & Kitagawa, S. (2). The initial phase of actomyosin-adenosinetriphosphatase. *Biochim. biophys. Acta.*, 40, 135. 1960.

Tonomura, Y., Kitagawa, S. & Yoshimura, J. (1). The initial phase of myosin A-adenosinetriphosphatase and the possible phosphorylation of myosin. *J. biol. Chem.*, 237, 3660. 1962.

Tonomura, Y., Kubo, S. & Imamura, K. (1). A molecular model for the interaction of myosin with adenosine triphosphate. In *Molecular biology of muscular contraction*. Ed. S. Ebashi, F. Oosawa, T. Sekine & Y. Tonomura. Tokyo & Osaka, Igaku Shoin Ltd. 1965.

Tonomura, Y., Nakamura, H., Kinoshita, N., Onishi, H. & Shigekawa, M. (1). The pre-steady state of the myosin-adenosine triphosphate system. X. The reaction mechanism of the myosin-ATP system and molecular mechanism of muscle contraction. *J. Biochem.*, 66, 599. 1969.

Tonomura, Y., Sekiya, K. & Imamura, K. (1). The optical rotatory dispersion of myosin A. I. Effect of inorganic salts. *J. biol. Chem.*, 237, 3110. 1962.

Tonomura, Y., Sekiya, K. & Imamura, K. (2). The optical rotatory dispersion of myosin A. IV. Conformational changes in meromyosins. *Biochim. biophys. Acta.*, 78, 690. 1963.

Tonomura, Y., Sekiya, K. & Imamura, K. (3). The optical-rotatory dispersion of myosin A. II. The effect of dioxane and p-chlormercuribenzoate. *Biochim. biophys. Acta.*, 69, 296. 1963.

Tonomura, Y., Sekiya, K., Imamura, K. & Tokiwa, T. (1). The optical rotatory dispersion of myosin A. III. Effect of adenosine triphosphate and inorganic pyrophosphate. *Biochim. biophys. Acta.*, 69, 305. 1963.

Tonomura, Y., Tokiwa, T. & Shimada, T. (1). Role of ADP bound to F-actin in superprecipitation and enzymatic activity of actomyosin. *J. Biochem.*, **59**, 322. 1966.

Tonomura, Y., Tokura, S. & Sekiya, K. (1). Binding of myosin A to F-actin. *J. biol. Chem.*, **237**, 1074. 1962.

Tonomura, Y., Yagi, K., Kubo, S. & Kitagawa, S. (1). A molecular mechanism of muscle contraction. *J. Res. Inst. for Catalysis. Hokkaido Univ.*, **9**, 256. 1961.

Tonomura, Y., Yagi, K. & Matsumiya, H. (1). Contractile proteins from adductors of Pecten. I. Some enzymic and physicochemical properties. *Arch. Biochem. Biophys.*, **59**, 76. 1955.

Tonomura, Y. & Yoshimura, J. (1). Binding of *p*-chlormercuribenzoate to actin. *J. Biochem.*, **51**, 259. 1962.

Tonomura, Y. & Yoshimura, J. (2). Removal of bound nucleotide and calcium of G-actin by treatment with ethylene diamine tetraacetic acid. *J. Biochem.*, **50**, 79. 1961.

Tower, S. S. (1). Trophic control of non-nervous tissues by the nervous system: a study of muscle and bone innervated from an isolated and quiescent region of spinal cord. *J. comp. Neurol.*, **67**, 241. 1937.

Townes, M. M. & Brown, D. E. S. (1). The involvement of pH, adenosine triphosphate, calcium and magnesium in the contraction of the glycerinated stalks of *Vorticella*. *J. cell. comp. Physiol.*, **65**, 261. 1965.

Traube, M. (1). Die chemische Theorie der Fermentwirkungen und der Chemismus der Respiration. *Ber. d. Deutsch. Chem. Gesellsch.*, **10**, 1984. 1877.

Traube, M. (2). Ueber die Beziehung der Respiration zur Muskeltätigkeit und die Bedeutung der Respiration überhaupt. *Arch. f. Path. Anat. u. Physiol.*, **21**, 386. 1861.

Traut, R. R. & Lipmann, F. (1). Activation of glycgoen synthetase by glucose-6-phosphate. *J. biol. Chem.*, **238**, 1213. 1963.

Trayer, I. P., Harris, C. I. & Perry, S. V. (1). 3-Methyl histidine and adult and foetal forms of skeletal muscle myosin. *Nature*, **217**, 452. 1968.

Trayer, I. P. & Perry, S. V. (1). The myosin of developing skeletal muscle. *Biochem. Z.*, **345**, 87. 1966.

Trayer, I. P., Perry, S. V. & Teale, F. W. J. (1). Structural differences between myosins of adult and foetal rabbit skeletal muscle. Abstr. 7th Int. Congr. Biochem; Tokyo, p. 943. 1967.

Trotta, P. P., Dreizen, P. & Stracher, A. (1). Studies on subfragment-I, a biologically active fragment of myosin. *Proc. nat. Acad. Sci. U.S.A.*, **61**, 659. 1968.

Trucco, R. E. (1). Enzymatic synthesis of uridine diphosphate glucose. *Arch. Biochem. Biophys.*, **34**, 482. 1951.

Tsao, T.-C. (1). Fragmentation of the myosin molecule. *Biochim. biophys. Acta.*, **11**, 368. 1953.

Tsao, T.-C. (2). The molecular dimensions and the monomer-dimer transformation of actin. *Biochim. biophys. Acta.*, **11**, 227. 1953.

Tsao, T.-C. (3). The interaction of actin, myosin and adenosine triphosphate. *Biochim. biophys. Acta.*, **11**, 236. 1953.

Tsao, T.-C. & Bailey, K. (1). The extraction, purification and some chemical properties of actin. *Biochim. biophys. Acta.*, **11**, 102. 1953.

Tsao, T.-C., Bailey, K. & Adair, G. S. (1). The size, shape and aggregation of tropomyosin particles. *Biochem. J.*, **49**, 27. 1951.

Tsao, T.-C., Hsü, K., Jen, M. H., Pan, C. H., Tan, P. N., Tao, T. C., Wen, H. Y. & Niu, C. I. (1). Isolation and properties of a new structural protein of muscle. *Scientia Sinica*, **7**, 637. 1958.

Tsao, T.-C., Kung, T.-H., Peng, C.-M., Chang, Y.-S. & Tsou, Y.-S. (1). Electron microscopical studies of tropomyosin and paramyosin. I. *Scientia Sinica*, Peking, **14**, 91. 1965.

Tsao, T.-C., Tan, P.-H. & Peng, C.-M. (1). A comparative physico-chemical study of tropomyosins from different sources. *Scientia Sinica*, Peking, **5**, 91. 1956.

Tsao, T.-C., Tsou, Y.-S., Lu, Z.-X., Kung, T.-H., Pan, C.-H., Li, Q.-L., Ku, H.-J. & Tsao, H.-T. (1). Demonstration of the existence of tropomyosin and actin in the thin filaments of striated muscle by direct isolation. *Scientia Sinica*, **14**, 1707. 1965.

Ts'o, P. O. P., Bonner, J., Eggman, L. & Vinograd, J. (1). Extraction of plasmodium of myxomycete *Physarum polycephalum*. *J. gen. Physiol.*, **39**, 325. 1955–56.

Ts'o, P. O. P., Eggman, L. & Vinograd, J. (1). The isolation of myxomyosin, an ATP-sensitive protein from the plasmodium of a myxomycete. *J. gen. Physiol.*, **39**, 801. 1955–56.

Ts'o, P. O. P., Eggman, L. & Vinograd, J. (2). Physical and chemical studies of myxomyosin, an ATP-sensitive protein in cytoplasm. *Biochim. biophys. Acta.*, **25**, 532. 1957.

Tsou, C. L. (1). On the cyanide inactivation of succinic dehydrogenase and the relation of succinic dehydrogenase to cytochrome *b*. *Biochem. J.*, **49**, 512. 1951.

Tsou, Y.-S., Lu, Z.-X. & Tsao, T.-C. (1). Rabbit SH-tropomyosin. *Acta biochim. et biophys. Sin.*, **4**, 622. 1964.

Tubbs, P. K., Pearson, D. J. & Chase, J. F. A. (1). Assay of carnitine and coenzyme *A* and their derivatives in tissues. In *Recent Research on Carnitine*, p. 117. Ed. G. Wolf. Cambridge, Mass., M.I.T. Press. 1965.

Tunik, B. & Holtzer, H. (1). The distribution of muscle antigens in contracted myofibrils determined by fluorescin-labelled antibodies. *J. biophys. biochem. Cytol.*, **11**, 67. 1961.

Turba, F. & Kuschinsky, G. (1). Über die Differenzierung der Sulfhydrylgruppen von Aktomyosin, Myosin und Aktin. *Biochim. biophys. Acta.*, **8**, 76. 1952.

Twarog, B. M. (1). Responses of a molluscan smooth muscle to acetylcholine and 5-hydroxytryptamine. *J. cell. comp. Physiol.*, **44**, 141. 1954.

Twarog, B. M. (2). Effects of acetylcholine and 5-hydroxytryptamine on the contraction of a molluscan smooth muscle. *J. Physiol.*, **152**, 236. 1960.

Twarog, B. M. (3). The regulation of catch in molluscan muscle. *J. gen. Physiol.*, **50**, 157. 1967.

Twarog, B. M. (4). Factors influencing contraction and catch in *Mytilus* smooth muscle. *J. Physiol.*, **192**, 847. 1967.

Twarog, B. M. (5). Excitation of *Mytilus* smooth muscle. *J. Physiol.*, **192**, 857. 1967.

Tyler, F. H. (1). Muscular dystrophies. In *The Metabolic Basis of Inheritable Disease*, pp. 939, 962. Ed. J. B. Stanbury, J. B. Wyngarden & D. S. Fredickson, McGraw Hill, New York. 1966.

Uchida, K., Mommaerts, W. F. H. M. & Meretsky, D. (1). Myosin in association with preparations of sarcotubular vesicles from muscle. *Biochim. biophys. Acta.*, **104**, 287. 1965.

Uexkull, J. von (1). Studien über den Tonus. VI. Z. Biol., 58, 305. 1912.

Ui, M. (1). A role of phosphofructokinase in pH-dependent regulation of glycolysis. Biochim. biophys. Acta., 124, 310. 1966.

Ulbrecht, G. & Ulbrecht, M. (1). Der isolierte Arbeitscyclus glatter Muskulatur. Zeitsch. f. Naturforsch., 7b, 434. 1952.

Ulbrecht, G. & Ulbrecht, M. (2). Die Verkurzungsgeschwindigkeit und der Nutzeffekt der ATP-Spaltung während der Kontraktion des Fasermodells. Biochim. biophys. Acta., 11, 138. 1953.

Ulbrecht, G. & Ulbrecht, M. (3). Phosphat-Austausch zwischen ATP und AD^{32}P durch hochgereinigte Aktomyosin-Preparate und gewaschene Muskelfibrillen. Biochim. biophys. Acta., 25, 100. 1957.

Ulbrecht, M. (1). Beruht der Phosphat-Austausch zwischen Adenosintriphosphat und Adenosine–[^{32}P] diphosphat in gereinigten Fibrillen und Actomyosin-prepäraten auf einer Vereinigung durch Muskel-grana? Biochim. biophys. Acta., 57, 438. 1962.

Ulbrecht, M., Grubhofer, N., Jaisel, F. & Walter, S. (1). Die erschöpfende Reinigung von Aktin-Präparaten. Zahl und Art der phosphathaltigen prosthetischen Gruppen von G- und F-Aktin. Biochim. biophys. Acta., 45, 443. 1960.

Utewski, A. (1). Zur Frage nach dem Schicksal der Brenztraubensäure bei der Autolyse des Muskelgewebes. Biochem. Z., 215, 406. 1929.

Utter, M. F. & Keech, D. B. (1). Formation of oxaloacetate from pyruvate and CO_2. J. biol. Chem., 235, PC17. 1960.

Utter, M. F. & Keech, D. B. (2). Pyruvate carboxylase. 1. Nature of the reaction. J. biol. Chem., 238, 2603. 1963.

Utter, M. F., Keech, D. B. & Scrutton, M. C. (1). A possible role for acetyl CoA in the control of gluconeogenesis. Adv. in Enz. Regulation, 2, 49. 1964.

Utter, M. F. & Kurahashi, K. (1). Phosphopyruvate carboxykinase. J. biol. Chem., 207, 821. 1959.

Varga, E., Köver, A., Kovács, T., Jókay, I. & Szilagyi, T. (1). Differentiation of myosins extracted from tonic and tetanic muscles on the basis of their antigenic properties. Acta physiol. Acad. Sci. Hung., 22, 21. 1962.

Varga, L. (1). The relation of temperature and muscular contraction. Hung. Acta Physiol., 1, 1. 1946.

Varga, L. (2). Observations on the glycerol-extracted musculus psoas of the rabbit. Enzymologia, 14, 196. 1950.

Varga, L. (3). Observations on the glycerol-extracted musculus psoas of the rabbit at higher temperatures. Enzymologia, 14, 212. 1950.

Vaughan, H. & Newsholme, E. A. (1). The effect of calcium ions on the activities of hexokinase, phosphofructokinase and fructose-1,6-diphosphatase from vertebrate and insect muscles. Biochem. J., 114, 81. 1969.

Velick, S. F. (1). The metabolism of myosin, the meromyosins, actin and tropomyosin in the rabbit. Biochim. biophys. Acta., 20, 228. 1956.

Veratti, E. (1). Investigations on the fine structure of striated muscle fiber. J. biophys. biochim. Cyt., 10, Suppl. 1. 1961. Tr. by C. Bruni, H. S. Bennett & D. de Koven from Mem. reale Ist Lombardo, 19, 87. 1902.

Verbinskaia, N. A., Borsuk, V. N. & Kreps, E. N. (1). Biochemistry of muscle contraction in Cucumaria frondosa. Arch. Sci. Biol. U.S.S.R., 38, 369. 1935.

Verpoorte, J. A. & Kay, C. M. (1). Optical rotatory dispersion and enzymic studies on the tryptic digestion of rabbit skeletal and cardiac myosins and their macro-molecular fragments. Arch. Biochem. Biophys., 113, 53. 1966.

Verzar, F. (1). The gaseous metabolism of striated muscle in warm-blooded animals. *J. Physiol.*, **44**, 243. 1912.

Vesalius, A. (1). *De Humano Corporis Fabrica*. Basel, 1543.

Vignais, P. M., Vignais, P. V. & Lehinger, A. L. (1). Restoration of ATP-induced contraction of 'aged' mitochondria by phosphatidyl inositol. *Biochem. biophys. Res. Com.*, **11**, 313. 1963.

Vignais, P. V., Vignais, P. M., Rossi, C. S. & Lehninger, A. L. (1). Restoration of ATP-induced contraction of mitochondria by 'contractile protein'. *Biochem. biophys. Res. Com.*, **11**, 307. 1963.

Vignos, P. J. & Lefkowitz, M. (1). A biochemical study of certain skeletal muscle constituents in human progressive muscular dystrophy. *J. clin. Invest.*, **38**, 873. 1959.

Vignos, P. J. & Warner, J. L. (1). Glycogen, creatine and high energy phosphate in human muscle disease. *J. Lab. clin. Med.*, **62**, 579. 1963.

Villafranca, G. W. de (1). A study on the nature of the *A* band of cross-striated muscle. *Arch. Biochem. Biophys.*, **61**, 378. 1956.

Villafranca, G. W. de (2). Adenosinetriphosphatase activity in developing rat muscle. *J. exp. Zool.*, **127**, 367. 1954.

Villar-Palasi, C. & Larner, J. (1). A uridine coenzyme-linked pathway of glycogen synthesis in muscle. *Biochim. biophys. Acta.*, **30**, 449. 1958.

Villar-Palasi, C. & Larner, J. (2). Levels of activity of the enzymes of the glycogen cycle in rat tissues. *Arch. Biochem. Biophys.*, **86**, 270. 1960.

Villar-Palasi, C. & Larner, J. (3). Insulin treatment and increased UDPG-glycogen transglucosylase activity in muscle. *Arch. Biochem. Biophys.*, **94**, 436. 1961.

Villar-Palasi, C. & Larner, J. (4). Feedback control of glycogen metabolism in muscle. *Fed. Proc.*, **25**, 583. 1966.

Vincenzi, D. L. De & Hedrick, J. L. (1). Reevaluation of the molecular weights of glycogen phosphorylases *a* and *b* using Sephadex gel filtration. *Biochemistry* (Am. chem. Soc.), **6**, 3489. 1967.

Virden, R. & Watts, D. C. (1). The distribution of guanosine–adenosine triphosphate phosphotranferases, and adenosine triphosphatase in animals from several phyla. *Comp. Biochem. Physiol.*, **13**, 161. 1964.

Vogt, M. (1). Die Isolierung von Methylglyoxal als Zwischensubstanz bei der Glykolyse. *Klin. Wochensch.*, **8**, 793. 1929.

Voit, C. (1). Ueber die Entwicklung der Lehre von der Quelle der Muskelkraft und einiger Theile der Ernährung seit 25 Jahren. *Z. f. Biol.*, **6**, 305. 1870.

Volfin, P., Clauser, H. & Gautheron, D. (1). Influence of oestradiol and progesterone injections on the acid-soluble phosphate fractions of the rat uterus. *Biochim. biophys. Acta.*, **24**, 137. 1957.

Volfin, P., Clauser, C., Gautheron, D. & Eboué, D. (1). Influence de l'oestradiol et de la progestérone sur le taux des nucléotides et de la phosphocréatine dans l'utérus de rat. *Bull. Soc. Chim. biol.*, **43**, 107. 1961.

Volta, A. (1). On the electricity excited by the mere contact of conducting substances of different kinds. *Phil. Trans.*, **90**, 403. 1800.

Vrbova, G. & Gutmann, E. (1). Sur l'importance de l'état fonctionnel des centre nerveux pour la vitesse de la resynthèse de glycogène dans le muscle strié. *J. de Physiol.*, **48**, 751. 1956.

Wahl, R. & Kozloff, L. M. (1). The nucleotide triphosphate content of various bacteriophages. *J. biol. Chem.*, **237**, 1953. 1962.

Wainio, W. W., van der Welde, C. & Shimp, N. F. (1). Copper in cytochrome c oxidase. *J. biol. Chem.*, **234**, 2433. 1959.

Wajzer, J., Nekhorocheff, J. & Dondon, J. (1). Désamination des nucléotides adényliques pendant la contraction musculaire. *Comp. rend. Acad. Sci.*, **246**, 3694. 1958.

Wajzer, J., Weber, R., Lerique, J. & Nekhorocheff, J. (1). Reversible degradation of adenosine triphosphate to inosinic acid during a single muscle twitch. *Nature*, **178**, 1287. 1956

Wakabayashi, T. & Ebashi, S. (1). Reversible change in physical state of troponin induced by calcium ion. *J. Biochem.*, **64**, 731. 1968.

Wakid, N. W. (1). Cytoplasmic fractions of the rat myometrium. I. General description and some enzymic properties. *Biochem. J.*, **76**, 88. 1960.

Wakid, N. W. & Mansour, T. E. (1). Factors influencing the stability of heart phosphofructokinase. *Mol. Pharm.*, **1**, 53. 1965.

Wakil, S. J., Green, D. E. & Mahler, H. R. (1). Studies in the fatty acid oxidising system of animal tissues. VI. β-Hydroxyacyl coenzyme *A* dehydrogenase. *J. biol. Chem.*, **207**, 631. 1954.

Wakil, S. J. & Mahler, H. R. (1). Studies in the fatty acid oxidising system of animal tissues. V. Unsaturated fatty acid acyl coenzyme *A* hydrase. *J. biol. Chem.*, **207**, 125. 1954.

Walaas, O. & Walaas, E. (1). The content of adenosinetriphosphate in uterine muscle of rats and rabbits. *Acta physiol. Scand.*, **21**, 1. 1950.

Wald, G. (1). Life in the second and third periods; or, why phosphorus and sulfur for high-energy bonds? From *Horizons in Biochemistry*, p. 127. Ed. M. Kasha & B. Pullman. New York, Academic Press. 1962.

Walker, G. J. & Whelan, W. J. (1). The mechanism of carbohydrase action. 8. Structures of the muscle phosphorylase limit dextrins of glycogen and amylopectin. *Biochem. J.*, **76**, 264. 1960.

Walsh, J. (1). Of the electric property of the Torpedo. *Phil. Trans.*, **63**, 461. 1773.

Walton, J. N. (1). Clinical aspects of human muscular dystrophy. In *Muscular Dystrophy in Man and Animals*, p. 263. Ed. G. H. Bourne. Basel & New York, S. Karger. 1963.

Walton, J. N. (2). Some clinical and genetic aspects of the muscular dystrophies. In *Muscle*, p. 501. Ed. by W. M. Paul, E. E. Daniel, C. M. Kay & G. Monckton. Oxford, Pergamon Press. 1965.

Wang, J. H. & Graves, D. J. (1). The relationship of the dissociation to the catalytic activity of glycogen phosphorylase *a*. *Biochemistry* (Amer. chem. Soc.), **3**, 1437. 1964.

Wang, T.-Y., Tsou, C.-L. & Wang, Y.-L. (1). Studies on succinic dehydrogenase. I. Isolation, purification and properties. *Scientia Sinica*, **5**, 73. 1956.

Warburg, O. (1). Über Eisen, den Sauerstoff übertragenden Bestandteil des Atmungsferment. *Ber. deutsch. chem. Gesellsch.*, **58**, (1), 1001. 1925.

Warburg, O. (2). Über die Wirkung des Kohlenoxyds auf den Stoffwechsel der Hefe. *Biochem. Z.*, **177**, 470. 1926.

Warburg, O. (3). Über die chemische Konstitution des Atmungsferment. *Naturwiss.*, **16**, 345. 1928.

Warburg, O. (4). Über die Grundlage der Wielandschen Atmungstheorie. *Biochem. Z.*, **142**, 518. 1923.

Warburg, O. & Christian, W. (1). Isolierung und Kristallisation des Gärungsferment Enolase. *Biochem. Z.*, **310**, 384. 1941–42.

Warburg, O. & Christian, W. (2). Über Aktivierung der Robisonschen Hexose-Mono-Phosphorsäure in roten Blutzellen und die Gewinnung aktivierender Fermentlösungen. *Biochem. Z.*, **242**, 206. 1931.

Warburg, O. & Christian, W. (3). Pyridin, der wasserstoffübertragende Bestandteil von Gärungsfermenten. (Pyridin-Nucleotide.) *Biochem. Z.*, **287**, 291. 1936.

Warburg, O. & Christian, W. (4). Isolierung und Kristallisation der Proteins des oxydierenden Gärungsferments. *Biochem. Z.*, **303**, 40. 1939–40.

Warburg, O. & Christian, W. (5). Über ein neues Oxydationsferment und sein Absorptionsspektrum. *Biochem. Z.*, **254**, 438. 1932.

Warburg, O. & Christian, W. (6). Ein zweites sauerstoffübertragendes Ferment und sein Absorptionsspektrum. *Naturwiss.*, **20**, 688. 1932.

Warburg, O. & Christian, W. (7). Über das gelbe Ferment und seine Wirkungen. *Biochem. Z.*, **266**, 377. 1933.

Warburg, O., Christian, W. & Griese, A. (1). Wasserstoff übertragendes Co-Ferment, seine Zusammensetzung und Wirkungsweise. *Biochem. Z.*, **282**, 157. 1935.

Warburg, O. & Negelein, E. (1). Über das Absorptionsspektrum des Atmungsferment. *Biochem. Z.*, **214**, 64. 1929.

Ward, P. C. J., Edwards, C. & Benson, E. S. (1). Relation between adenosine triphosphatase activity and sarcomere length in stretched glycerol-extracted frog skeletal muscle. *Proc. Nat. Acad. Sci. U.S.A.*, **53**, 1377. 1965.

Warner, J. R., Rich, A. & Hall, C. E. (1). Electron microscope studies of ribosomal clusters synthesizing haemoglobin. *Science*, **138**, 1399. 1962.

Watanabe, M. I. & Williams, C. M. (1). Mitochondria in the flight muscles of insects. I. Chemical composition and enzymatic content. *J. gen. Physiol.*, **34**, 675. 1951.

Watanabe, S. & Sleator, W. (1). EDTA relaxation of glycerol-treated muscle fibres and the effects of magnesium, calcium and manganese ions. *Arch. Biochem. Biophys.*, **68**, 80. 1957.

Watts, D. C. & Rabin, B. R. (1). A study of the 'reactive' sulphydryl groups of adenosine 5′-triphosphate-creatine phosphotransferase. *Biochem. J.*, **85**, 507. 1962.

Weber, A. (1). The ultracentrifugal separation of L-myosin and actin in an actomyosin sol under the influence of ATP. *Biochim. biophys. Acta*, **19**, 345. 1956.

Weber, A. (2). Muskelkontraktion und Modellkontraktion. *Biochim. biophys. Acta*, **7**, 214. 1951.

Weber, A. (3). On the role of calcium in the activity of adenosine 5′-triphosphate hydrolysis by actomyosin. *J. biol. Chem.*, **234**, 2764. 1959.

Weber, A. (4). Energized calcium transport and relaxing factors. *Current topics in bioenergetics.*, **1**, 203. 1966.

Weber, A. & Hasselbach, W. (1). Die Erhöhung der Rate der ATP-Spaltung durch Myosin- und Actomyosin-Gele bei Beginn der Spaltung. *Biochim. biophys. Acta*, **15**, 237. 1954.

Weber, A. & Herz, R. (1). The binding of calcium to actomyosin systems in relation to their biological activity. *J. biol. Chem.*, **238**, 599. 1963.

Weber, A., Herz, R. & Reiss, I. (1). On the mechanism of the relaxing effect of fragmented sarcoplasmic reticulum. *J. gen. Physiol.*, **46**, 679. 1962–63.

Weber, A., Herz, R. & Reiss, I. (2). Study of the kinetics of calcium transport by isolated fragmented sarcoplasmic reticulum. *Biochem. Z.*, **345**, 329. 1966.

Weber, A., Herz, R. & Reiss, I. (3). Role of calcium in contraction and relaxation. *Fed. Proc.*, **23**, 896. 1964.

Weber, A., Herz, R. & Reiss, I. (4). The mechanism of action of cardiac relaxing factor. *Fed. Proc.*, **22**, 228. 1963.

Weber, A., Herz, R. & Reiss, I. (5). The nature of the cardiac relaxing factor. *Biochim. biophys. Acta.*, **131**, 188. 1967.

Weber, A., Herz, R. & Reiss, I. (6). The regulation of myofibrillar activity by calcium. *Proc. Roy. Soc.* B, **160**, 489. 1964.

Weber, A. & Weber, H. H. (1). Zur Thermodynamik der Kontraktion des Fasermodells. *Biochim. biophys. Acta.*, **7**, 339. 1951.

Weber, A. & Weber, H. H. (2). Zur Thermodynamik der ATP-Kontraktion am Fasermodell. *Z. f. Naturforsch.*, **5b**, 124. 1950.

Weber, A. & Winicur, S. (1). The role of calcium in the superprecipitation of actomyosin. *J. biol. Chem.*, **236**, 3198. 1961.

Weber, E. (1). Muskelbewegung. In *Handwörterbuch der Physiologie* p. 1. Ed. R. Wagner. Braunschweig, F. Bieweg und Sohn. 1846.

Weber, H. H. (1). Das kolloidale Verhaltung der Muskeleiweisskörper. I. Isoelektrischer Punkt und Stabilitätsbedingungen des Myogens. *Biochem. Z.*, **158**, 443. 1925.

Weber, H. H. (2). Das kolloidale Verhalten der Muskeleiweisskörper. II. Isoelektrischer Punkt und Löslichkeit des Myosins. *Biochem. Z.*, **158**, 473. 1925.

Weber, H. H. (3). Der Feinbau und die mechanischen Eigenschaften des Myosinfadens. *Arch. f. Physiol.*, **235**, 205. 1935.

Weber, H. H. (4). Das kolloidale Verhalten der Muskeleiweisskörper. III. Physikochemische Konstanten des Myogens. *Biochem. Z.*, **189**, 407. 1927.

Weber, H. H. (5). Die Muskelkontraktion. *Biochem. Z.*, **217**, 430. 1930.

Weber, H. H. (6). Die Wirkung von Adenosintriphosphat auf die kontraktilen Proteine und die Kontraktion von Muskeln und Zellen. *Fortschritte d. Zool.*, **10**, 304. 1956.

Weber, H. H. (7). Is the contracting muscle in a new elastic equilibrium? *Proc. Roy. Soc.* B, **139**, 512. 1951–52.

Weber, H. H. (8). Das kontraktile System von Muskel und Zellen. *Rappts. 3rd. Congr. Intern. Biochim. Bruxelles*, p. 81. 1955.

Weber, H. H. (9). *The motility of muscle and cells.* Cambridge, Mass., Harvard Univ. Press. 1958.

Weber, H. H. (10). Muscle contraction and muscle proteins. *Proc. Roy. Soc.* B, **137**, 50. 1950.

Weber, H. H. (11). Interaction of myosin and actin. In *Biochemistry of Muscle Contraction*, p. 193. Ed. J. Gergely. J. & A. Churchill, London. 1964.

Weber, H. H. (12). The relaxation of the contracted actomyosin system. *Ann. N.Y. Acad. Sci.*, **81**, 409. 1959.

Weber, H. H. (13). Muskeleiweisskörper und Eigenschaften des Muskels. *Naturwissensch.*, **27**, 33. 1939.

Weber, H. H. (14). Die Bjerrumsche Zwitterionen-theorie und die Hydratation der Eiweisskörper. *Biochem. Z.*, **218**, 1. 1930.

Weber, H. H. & Nachmansohn, D. (1). Die Unabhängigkeit der Eiweisshydratation von der Eiweissionisation. *Biochem. Z.*, **204**, 215. 1929.

Weber, H. H. & Portzehl, H. (1). The transference of the muscle energy in the contraction cycle. *Progress in Biophys. and biophys. Chem.*, **4**, 60. 1954.

Weber, H. H. & Portzehl, H. (2). Muscle contraction and fibrous muscle proteins. *Adv. in Protein Chem.*, 7, 161. 1952.

Weber, H. H. & Stöver, R. (1). Das Kolloidale Verhalten der Muskeleiweisskörper. *Biochem. Z.*, 259, 269. 1933.

Webster, H. L. (1). Direct deamination of adenine diphosphate by washed myofibrils. *Nature*, 172, 453. 1953.

Weeds, A. G. (1). The thiol sequences and subunits of light meromyosin fraction I. *Biochem. J.*, 104, 44P. 1967.

Weeds, A. G. (2). Small sub-units of myosin. *Biochem. J.*, 105, 25c. 1967.

Weeds, A. G. (3). Light chains of myosin. *Nature*, 223, 1362. 1969.

Weeds, A. G. & Hartley, B. S. (1). A chemical approach to the sub-structure of myosin. *J. mol. Biol.*, 24, 307. 1967.

Weeds, A. G. & Hartley, B. S. (2). Selective purification of the thiol peptides of myosin. *Biochem. J.*, 107, 531. 1968.

Weibull, C. (1). Some chemical and physical-chemical properties of the flagella of *Proteus vulgaris*. *Biochim. biophys. Acta.*, 2, 351. 1948.

Weinstock, I. M. (1). Comparative biochemistry of myopathies. *Ann. N.Y. Acad. Sci.*, 138, 199. 1966.

Weinstock, I. M., Epstein, S. & Milhorat, A. T. (1). Enzyme studies in muscular dystrophy. III. In hereditary muscular dystrophy in mice. *Proc. Soc. exp. Biol. Med.*, 99, 272. 1958.

Weise, E. (1). Untersuchungen zur Frage der Verteilung und der Bindungsart des Calcium im Muskel. *Arch. exp. Path. Pharm.*, 176, 367. 1934.

Weis-Fogh, T. (1). A rubber-like protein in insect cuticle. *J. exp. Biol.*, 37, 889. 1960.

Weis-Fogh, T. (2). Fat combustion and metabolic rate of flying locusts (*Schistocerca gregaria* Forskål). *Phil. Trans. Roy. Soc.* B, 237, 1. 1952.

Weizäcker, V. (1). Myothermic experiments in salt-solutions in relation to the various stages of a muscular contraction. *J. Physiol.*, 48, 396. 1914.

West, J. J., Nagy, B. & Gergely, J. (1). Free adenosine diphosphate as an intermediary in the phosphorylation by creatine phosphate of adenosine diphosphate bound to actin. *J. biol. Chem.*, 242, 1140. 1967.

White, I. G. & Griffiths, D. E. (1). Guanidines and phosphagens of semen. *Austr. J. exp. Biol. med. Sci.*, 26, 97. 1958.

Whitehouse, M., Mocksi, H. & Gurin, S. (1). The synthesis and biological properties of fatty acyl adenylates. *J. biol. Chem.*, 226, 813. 1957.

Whitteridge, G. (1). (ed. and trans.) [W. Harvey's] *De motu locali animalium*. Cambridge University Press, London. 1959.

Whitteridge, G. (2). (ed. and trans.) *The anatomical lectures of William Harvey*. Edinburgh and London, E. & S. Livingstone Ltd. 1964.

Wieland, H. (1). Über Hydrierung und Dehydrierung. *Ber. deutsch. chem. Gesellsch.* 45, 484. 1912.

Wieland, H. (2). Über den Mechanismus der Oxydationsvorgänge. *Ber. deutsch. chem. Gesellsch.*, 46, 3327. 1913.

Wieland, H. (3). Über den Mechanismus der Oxydationsvorgänge. *Erg. d. Physiol.*, 20, 477. 1922.

Wieme, R. J. & Herpol, J. E. (1). Origin of the lactate dehydrogenase isoenzyme pattern found in the serum of patients having primary muscular dystrophy. *Nature*, 194, 287. 1962.

Wieme, R. J. & Lauryssens, M. J. (1). Lactate dehydrogenase multiplicity in normal and diseased human muscle. *Lancet*, **1**, 433. 1962.

Wiener, O. (1). Die Theorie des Mischkörpers für das Feld der stationären Strömung. I. Die Mittelwertsätze für Kraft, Polarisation und Energie. *Abh. d. kön. Sachs. Gesellsch. d. Wissensch. Math.-phys. Kl.*, **32**, 507. 1912.

Wilbrandt, W. & Koller, H. (1). Die Calcium-Wirkung am Froschherzen als Funktion des Ionengleichgewichts zwischen Zellmembran und Umgebung. *Helv. physiol. Acta.*, **6**, 208. 1948.

Wilhelmi, D. (1). Über das Verhalten des Lactacidogens bei ermüdender Reizung von isolierten Froschgastrocnemim. *Z. f. physiol. Chem.*, **203**, 34. 1931.

Wilkie, D. R. (1). Heat, work and chemical change in muscle. *Proc. Roy. Soc.* B, **160**, 476. 1964.

Wilkie, D. R. (2). Energetic aspects of muscular contraction. In *Symposium on Muscle, Symp. Biol. Hung.*, **8**, 207. Akadémiai Kiadó, Budapest. 1968.

Wilkie, D. R. (3). Muscle. *Ann. Rev. Physiol.*, **28**, 17. 1966.

Wilkie, D. R. (4). Heat, work and phosphorylcreatine break-down in muscle. *J. Physiol.*, **195**, 157. 1968.

Williams, R. J., Lyman, C. M., Goodyear, G. H., Truesdail, J. H. & Holaday, D. (1). 'Pantothenic acid', a growth determinant of universal biological occurrence. *J. Amer. chem. Soc.*, **55**, 2912. 1933.

Williams, R. J. & Major, R. T. (1). The structure of pantothenic acid. *Science*, **91**, 246. 1940.

Williamson, J. R. (1). Kinetic studies of epinephrine effects in the perfused rat heart. *Pharmac. Rev.*, **18**, 205. 1966.

Williamson, J. R. & Jamieson, D. (1). Metabolic effects of epinephrine in the perfused rat heart. I. Comparison of redox states, tissue pO_2, and force of contraction. *Mol. Pharm.*, **2**, 191. 1966.

Williamson, J. R. & Krebs, H. A. (1). Acetoacetate as fuel of respiration in the perfused rat heart. *Biochem. J.*, **80**, 540. 1961.

Willis, T. (1). *Of Feavers*. See S. Pordage.

Willis, T. (2). *De Motu Musculorum*. See Pordage.

Wilson, A. C., Cahn, R. D. & Kaplan, N. O. (1). Functions of the two forms of lactic dehydrogenase in the breast muscle of birds. *Nature*, **197**, 331. 1963.

Wilson, L. G. (1). Erasistratus, Galen and the Pneuma. *Bull. Hist. Med.*, **33**, 310. 1959.

Wilson, L. G. (2). William Croone's Theory of muscular contraction. *Notes and Records of the Royal Society of London*, **16**, 158. 1961.

Wilson, L. G. (3). The transformation of ancient concepts of respiration in the seventeenth century. *Isis*, **51**, 161. 1960.

Winegard, S. (1). Autoradiographic studies of intracellular calcium in frog skeletal muscle. *J. gen. Physiol.*, **48**, 455. 1965.

Winegard, S. (2). The possible role of calcium in excitation-contraction coupling of heart muscle. *Circulation*, **24**, No. 2, pt. 2, 523. 1961.

Winnick, E. R. & Winnick, T. (1). Protein synthesis in skeletal muscle with emphasis on myofibrils. *J. biol. Chem.*, **235**, 2657. 1960.

Wohlfarth-Bottermann, K. E. (1). Protoplasmadifferenzierungen und ihre Bedeutung für die Protoplasmaströmung. *Protoplasma*, **54**, 514. 1962.

Wöhlisch, E. (1). Eine kolloidosmotische Hypothese der Muskelkontraktion. *Verhandl. d. physikal-med. Gesellsch. Würzburg*, Neue Folge **50**, 163. 1925.

Wöhlisch, E. (2). Morphologie und Mechanik der Muskelfaser. *Kolloidzeitsch.*, **96**, 261. 1941.

Wolfson, R., Yakulis, V., Coleman, R. D. & Heller, P. (1). Studies on fetal myoglobin. *J. Lab. clin. Med.*, **69**, 728. 1967.

Wolpert, L. (1). Cytoplasmic streaming and ameoboid movement. In *Function and Structure of Micro-organisms. Fifteenth. Sym. Soc. gen. Microbiol.*, p. 270. 1965.

Woods, E. F. (1). Peptide chains of tropomyosin. *Nature*, **207**, 82. 1965.

Woods, E. F. (2). The dissociation of tropomyosin by urea. *J. mol. Biol.*, **16**, 581. 1966.

Woods, E. F., Himmelfarb, S. & Harrington, W. F. (1). Studies on the structure of myosin in solution. *J. biol. Chem.*, **238**, 2375. 1963.

Woods, H. J. (1). The contribution of entropy to the elastic properties of keratin, myosin and some other high polymers. *J. Coll. Sci.*, **1**, 407. 1946.

Woodward, A. A. (1). The release of radioactive ^{45}Ca from muscle during stimulation. *Biol. Bull.*, **97**, 264. 1949.

Worthington, C. R. (1). Large axial spacings in striated muscle. *J. mol. Biol.*, **1**, 398. 1959.

Worthington, C. R. (2). X-ray diffraction studies on the large-scale molecular structure of insect muscle. *J. mol. Biol.*, **3**, 618. 1961.

Wosilait, W. D. (1). Studies on the organic phosphate moiety of liver phosphorylase. *J. biol. Chem.*, **233**, 597. 1958.

Wróblewski, A. (1). Über den Buchner'schen Hefepresssaft. *J. f. prakt. Chem.*, **64**, 1, 1901.

Wu, R. (1). The effect of azide and oligomycin on inorganic phosphate transport in slices of rat kidney. *Biochem. biophys. Acta.*, **82**, 212. 1964.

Yagi, K. & Noda, L. (1). Phosphate transfer to myofibrils by ATP-creatine transphosphorylase. *Biochim. biophys. Acta.*, **43**, 249. 1960.

Yagi, K. & Yazawa, Y. (1). Isolation and characterisation of a subfragment of myosin A. *J. Biochem.*, **60**, 450. 1966.

Yakishiji, E. & Okunuki, K. (1). Uber eine neue Cytochromkomponente und ihre Funktion. *Proc. imp. Acad. Japan*, **16**, 299. 1940.

Yamaguchi, M. & Sekine, T. Sulfhydryl groups involved in the active site of myosin A adenosine triphosphate. I. Specific blocking of the SH group responsible for the inhibitory phase in the 'biphasic response' of the catalytic activity. *J. Biochem.*, **59**, 24. 1966.

Yamashita, T., Soma, Y., Kobayashi, S., Sekine, T., Titani, K. & Narita, K. (1). The amino acid sequence at the active site of myosin A adenosine triphosphatase activated by EDTA. *J. Biochem.*, **55**, 576. 1964.

Yamashita, Y., Soma, Y., Kobayashi, S. & Sekine, T. (1). The amino acid sequence at the active site of myosin A adenosine triphosphatase activated by Ca^{2+}. *J. Biochem.*, **57**, 460. 1964.

Yanagisawa, T. (1). Studies on guanidine phosphoryl-transferases of Echinoderms. III. Comparative biochemistry of some enzymes contained in muscular tissue. *J. Fac. Sci. Univ. Tokyo.* Section IV. 9, 9. 1960.

Yanagisawa, T. (2). Studies on Echinoderm phosphagens. II. Change in the content of creatine phosphate after stimulation of sperm motility. *Exp. Cell Res.*, **46**, 348. 1967.

Yasui, B., Fuchs, F. & Briggs, F. N. (1). The role of the sulfhydryl groups of tropomyosin and troponin in the calcium control of actomyosin contractility. *J. biol. Chem.*, **243**, 735. 1968.

Yen, L.-F., Han, Y.-S. & Shih, T.-C. (1). Purification of the plant contractile protein and its ATPase properties. *Kexue Tong Bao*, (foreign language edition), 17, 138. 1966.

Yen, L.-F. & Shih, T.-C. (1). The presence of a contractile protein in higher plants. *Scientia Sinica*, 14, 601. 1965.

Yoshimura, J., Matsumiya, H. & Tonomura, Y. (1). Studies on the G–F transformation of actin. *Ann. Rep. Sci. Works, Fac. Sci. Osaka Univ.*, 11, 51. 1963.

Young, D. M., Himmelfarb, S. & Harrington, W. F. (1). On the structural assembly of the polypeptide chains of heavy meromyosin. *J. biol. Chem.*, 240, 2428. 1965.

Young, D. M., Himmelfarb, S. & Harrington, W. F. (2). The relationship of the meromyosins to the molecular structure of myosin. *J. biol. Chem.*, 239, 2822. 1964.

Young, H. L. & Pace, N. (1). Some physical and chemical properties of crystalline α-glycerophosphate dehydrogenase. *Arch. Biochem. Biophys.*, 75, 125. 1958.

Young, H. L., Young, W. & Edelman, I. S. (1). Electrolyte and lipid composition of skeletal and cardiac muscle in mice with hereditary muscular dystrophy. *Amer. J. Physiol.*, 197, 487. 1959.

Young, L. G. & Nelson, L. (1). Viscometric analysis of the contractile proteins of mammalain spermatozoa. *Exp. Cell Res.*, 51, 34. 1968.

Young, M. (1). Studies on the structural basis of the interaction of myosin and actin. *Proc. nat. Acad. Sci. U.S.A.*, 58, 2393. 1967.

Young, W. J. (1). Über die Zusammensetzung der durch Hefepresssaft gebildeten Hexosephosphorsäure. II. *Biochem. Z.*, 32, 177. 1911.

Yount, R. G. & Koshland, D. E. (1). Properties of the O^{18} exchange reaction catalysed by heavy meromyosin. *J. biol. Chem.*, 238, 1708. 1963.

Yphantis, D. (1). Equilibrium ultracentrifugation of dilute solutions. *Biochemistry* (Amer. chem. Soc.), 3, 297. 1964.

Yudkin, W. H. (1). Transphosphorylation in echinoderms. *J. cell. comp. Physiol.*, 44, 507. 1954.

Zaalishvili, M. M. & Mikadze, G. V. (1). The role of actin in muscle tissue and some questions of the theory of muscle contraction. *Biochimia*, 24, 566. (trans. Consultants' Bureau, New York). 1959.

Žak, R. & Gutmann, E. (1). Lack of correlation between synthesis of nucleic acids and proteins in denervated muscle. *Nature*, 185, 766. 1960.

Žak, R., Gutmann, E. & Vrbovà, G. (1). Quantitative changes of muscle proteins after stimulation of the muscle. *Experientia*, 13, 80. 1957.

Zalkin, H. & Racker, E. (1). Partial resolution of the enzymes catalysing oxidative phosphorylation. V. Properties of coupling factor 4. *J. biol. Chem.*, 240, 4017. 1965.

Zalkin, H., Tappel, A. L., Desai, I., Caldwell, K. A. & Peterson, D. W. (1). Increased lysosomal enzymes in muscular dystrophy. *Fed. Proc.*, 20, 303. 1961.

Zebe, E. (1). Die α-Glycerophosphatoxidation des Heuschreckenbrustmuskels (*Locusta migratoria*). *Experientia*, 12, 68. 1956.

Zebe, E. C. & McShan, W. H. (1). Lactic and α-glycerophosphate dehydrogenases in insects. *J. gen. Physiol.*, 40, 779. 1957.

Zebe, H., Delbrück, A. & Bücher, T. (1). Glycerophosphat-dehydrogenase unter Zellphysiologischen Aspekten. *Angew. Chem.*, 69, 65. 1957.

Zebe, H., Delbrück, A. & Bücher, T. (2). Über den Glycerin-1-P-Cyclus im Flugmuskel von *Locusta migratoria*. *Biochem. Z.*, 331, 254. 1959.

Ziegler, D. M. & Doeg, K. A. (1). The isolation and properties of a soluble succinic-coenzyme Q reductase from beef heart mitochondria. *Biochem. biophys. Res. Com.*, **1**, 344. 1959.

Ziegler, D. M. & Doeg, K. A. (2). Studies on the electron transport system. XLIII. The isolation of a succinic-coenzyme Q reductase from beef heart mitochondria. *Arch. Biochem. Biophys.*, **97**, 41. 1962.

Zierler, K. H. (1). Aldolase leak from muscle of mice with hereditary muscular dystrophy. *Bull. Johns Hopkins Hosp.*, **102**, 17. 1958.

Zierler, K. H. (2). Potassium flux and further observations on aldolase flux in dystrophic mouse muscle. *Bull. Johns Hopkins Hosp.*, **108**, 208. 1961.

Ziff, M. (1). Reversible inactivation of adenosinetriphosphatase. *J. biol. Chem.*, **153**, 24. 1944.

Ziff, M. & Moore, D. H. (1). Electrophoresis, sedimentation and ATPase activity of myosin. *J. biol. Chem.*, **153**, 653. 1944.

Zobel, C. R. & Carlson, F. D. (1). An electron microscopic investigation of myosin and some of its aggregates. *J. mol. Biol.*, **7**, 78. 1963.

AUTHOR INDEX

As this index does not cover footnotes, certain papers of particular authors which happen to have been referred to only in them, will not be found here, hence occasional gaps in the sequence of their reference numbers. But all are listed in the Bibliography.

Tonomura, Y., Kubo, S. & Imamura, K., (1) 187
Tonomura, Y., Sekiya, K., Imamura, K. & Tokiwa, T., (1) 270
Tonomura, Y., Tokura, S. & Sekiya, K., (1) 225, 269, 286
Tonomura, Y., Yagi, K., Kubo, S. & Kitagawa, S., (1) 280
Tonomura, Y., Yagi, K. & Matsumiya, H., (1) 515
Tonomura, Y. & Yoshimura, J., (1) 216, 224; (2) 224
Tower, S. S., (1) 484
Townes, M. M. & Brown, D. E. S., (1) 586
Traube, M., (*fl.* 1877), 37; (1) 32; (2) 34, 36
Traut, R. R. & Lipmann, F., (1) 444
Trayer, I. P., Harris, C. I. & Perry, S. V., (1) 474
Trayer, I. P. & Perry, S. V., (1) 472, 474, 479
Trotta, P. P., Dreizen, P. & Stracher, A., (1) 211, 288
Trucco, R. E., (1) 443
Tsao, T.-C., 191, 205, 206, 228, 517; (1) 155, 194, 199; (2) 215
Tsao, T.-C. & Bailey, K., (1) 215, 221, 223
Tsao, T.-C., Bailey, K. & Adair, G. S., 231; (1) 230, 233
Tsao, T.-C., Tan, P.-H. & Peng, C.-M., (1) 557
Ts'o, P. O. P., 579, 580
Tsou, C. L., (1) 387
Tubbs, P. K., 402
Tunik, B. & Holtzer, H., (1) 258
Turba, F. & Kuschinsky, G., (1) 164, 286
Twarog, B. M., 532; (1) 528
Tyler, F. H., (1) 506

Uchida, K., Mommaerts, W. F. H. M. & Meretsky, D., (1) 587
Uexkull, J. von, (1) 524
Ulbrecht, G. & Ulbrecht, M., (1, 2) 165; (3) 264
Ulbrecht, M., (1) 265
Utewski, A., (1) 107

Varga, E., 462
Varga, L., (1) 178; (2) 179
Velick, S. F., (1) 234
Veratti, E., 600; (1) 331-2
Verbinskaia, N. A., Borsuk, V. N. & Kreps, E. N., (1) 519
Verpoorte, J. A. & Kay, C. M., (1) 271
Verzar, F., (1) 45
Versalius, (*fl.* 1543), 12-13, 26

Vignais, P. M., 587
Vignos, P. J. & Lefkowitz, M., (1) 503
Vignos, P. J. & Warner, J. L., (1) 503
Villafranca, G. W. de, (1) 243; (2) 472
Villar-Palasi, C. & Larner, J., (1) 442, 443; (2, 3) 443; (4) 447
Vogt, M., (1) 107
Voit, C., (*fl.* 1865), (1) 36
Volta, Alessandro, (1) 311, 595

Wahl, R. & Kozloff, L. M., (1) 585
Wainio, W. W., van der Welde, C. & Shimp, N. F., (1) 385
Wajzer, J., 351
Wakabayashi, T. & Ebashi, S., (1) 296
Wakid, N. W., (1) 560
Wakil, S. J., Green, D. E. & Mahler, H. R., (1) 401
Wakil, S. J. & Mahler, H. R., (1) 400
Wald, G., (1) 126
Walker, G. J. & Whelan, W. J., (1) 428, 510, 511
Walsh, J., (*fl.* 1773), (1) 595
Walton, J. N., (1, 2) 499
Wang, T.-Y., Tsou, C.-L. & Wang, Y.-L., (1) 387
Warburg, O., 381-2, 384-5
Warburg, O. & Christian, W., 118, 412; (1) 112; (2, 3) 117; (4) 121; (5) 386; (6) 386, 403; (7) 386
Ward, P. C., Edwards, C. & Benson, E. S., (1) 247
Warner, J. R., Rich, A. & Hall, C. E., (1) 475
Watanabe, M. I. & Williams, C. M., (1) 536
Watts, D. C. & Rabin, B. R., (1) 482
Weber, A., (1) 157; (2) 159-60; (3) 321, 323
Weber, A. & Hasselbach, W., (1) 280
Weber, A. & Herz, R., (1) 324
Weber, A., Herz, R. & Reiss, I., (1) 324; (2) 328, 329; (3) 324, 336; (4) 468; (5) 469
Weber, A. & Weber, H. H., (1) 165, 185-6; (2) 183
Weber, A. & Winicur, S., (1) 324, 326
Weber, E., (*fl. c.* 1840), 55; (1) 53
Weber, H. H., 159, 165, 166, 172, 179, 266; (1) 50, 144; (2) 135, 144; (3) 137; (4) 142; (5) 145; (6) 169, 171, 186, 187; (8) 263; (10) 155, 285; (13) 139; (14) 142
Weber, H. H. & Portzehl, H., (1) 160, 166, 186, 187; (2) 197
Weber, H. H. & Stöver, R., (1) 197, 199
Webster, H. L., (1) 272, 372
Weeds, A. G., (1, 2, 3) 211
Weeds, A. G. & Hartley, B. S., (1) 211

SUBJECT INDEX

A bands, *see* anisotropic bands
A filaments, constancy of length of, 245, 255
A substance, 130, 170, 171–2, 243
ADP, *see* adenosinediphosphate
ATP, *see* adenosinetriphosphate
ATPase, *see* adenosinetriphosphatase
acescence, 12
acetaldehyde in glycolysis, 106, 118
acetyl-CoA, formation of, 395, 400
acetylcholine, 591
acetylmyosin, 203
acid–alkali concept of Sylvius, 12
acid maltase, and glycogen storage disease, 511
aconitase, in the oxidation of citric to α-ketoglutaric acid, 397, 406
actin, 133, 150ff, 163ff, 169, 172, 182, 200, 209, 212ff
  amino-acid composition of, 216
  ATP in, 213–14
  binding of Ca to, 213, 224–5
  electron-microscope examination of, 191, 226–7
  estimation of amount in muscle, 190–1
  factors concerned in controlling the size of particles of, 217
  histidine residues in, 225
  interaction with myosin, *see* interaction
  3-methylhistidine in hydrolysates of, 482
  from non-muscular motile sources, 580–1
  part played in contraction theories, 187–9, 291–4
  polymerisation, 213, 214, 219, 221ff
  prepared from dystrophic rabbits, 218
  purification methods, 214, 215
  in red and white muscle, 461
  SH groups in, 223ff, 289
  sliding mechanism and, 238ff
  susceptibility to oxidation, 213
  tyrosine residues in, 222, 225
  in vertebrate and smooth muscle, 552–3
  from wide range of animals compared, 216
  X-ray diffraction and, 226
F-actin, 150, 156ff, 164, 171, 182, 188, 191, 193, 213–14, 216ff

binding of ADP to, 218ff
binding of Ca to, 225
depolymerisation of, 220, 225
polymerisation of G- to F-, 150, 217, 220ff
three components found, 216–17
ultraviolet absorption spectra of, 222
Fn-actin, 218
G-actin, 150, 158, 188, 189, 213ff, 218ff
  bound ATP of, 213–14, 224
  effect of loss of Ca from, 224
  exchangeability of ATP on, 218
  molecular weight of, 214, 215–16
  optical rotatary dispersion of, 222–3
  SH groups in, 216
  treatment with trinitrobenzenesulphonate, 225
  ultraviolet absorption spectra of, 222
G–ADP actin, 215
actin-like protein in non-muscular function
  from *Physarum*, 579
  from sea-urchin eggs, 589
  from sea-urchin sperm flagella, 583–4
  in spindle fibres, 589
  from *Tetrahymena* cilia, 583
α-actinin, 299–300
  location in the Z line, 300
β-actinin, 217
action potential in stimulated nerve and muscle, 314–15, 336
'activation' of actin, 150
activation heat, 176, 357, 361, 362
active sites on myosin, 271ff, 286–9, 290–1
  in alkali light-chain component of subfragment 1, 211–12
  in subfragment 1, 209
active state of muscle, 176, 267, 360–61, 538ff.
  and Ca release, 334
active transport, 327, 593
actomyosin, 133, 146ff, 164, 187, 213
  action of ATP on, 156–8, 160, 183, 290
  action of trypsin on, 200ff
  ATPase activity of, 151–4, 167, 188–9, 298
  correlation between ATPase activity and speed of contraction, 483
  in developing muscle, 473–4

Printed in the United States
By Bookmasters